Studies in Systems, Decision and Control

Volume 375

The series "Studies in Systems, Decision and Control" (SSDC) covers both new developments and advances, as well as the state of the art, in the various areas of broadly perceived systems, decision making and control–quickly, up to date and with a high quality. The intent is to cover the theory, applications, and perspectives on the state of the art and future developments relevant to systems, decision making, control, complex processes and related areas, as embedded in the fields of engineering, computer science, physics, economics, social and life sciences, as well as the paradigms and methodologies behind them. The series contains monographs, textbooks, lecture notes and edited volumes in systems, decision making and control spanning the areas of Cyber-Physical Systems, Autonomous Systems, Sensor Networks, Control Systems, Energy Systems, Automotive Systems, Biological Systems, Vehicular Networking and Connected Vehicles, Aerospace Systems, Automation, Manufacturing, Smart Grids, Nonlinear Systems, Power Systems, Robotics, Social Systems, Economic Systems and other. Of particular value to both the contributors and the readership are the short publication timeframe and the worldwide distribution and exposure which enable both a wide and rapid dissemination of research output.

Indexed by SCOPUS, DBLP, WTI Frankfurt eG, zbMATH, SCImago.

All books published in the series are submitted for consideration in Web of Science.

More information about this series at https://link.springer.com/bookseries/13304

Alexandr Alexandrovich Kuznetsov ·
Oleksandr Volodymyrovych Potii ·
Nikolay Alexandrovich Poluyanenko ·
Yurii Ivanovich Gorbenko · Natalia Kryvinska

Stream Ciphers in Modern Real-time IT Systems

Analysis, Design and Comparative Studies

Springer

Alexandr Alexandrovich Kuznetsov
Department of Information
and Communication Systems Security
Faculty of Computer Science
V. N. Karazin Kharkiv National University
Kharkiv, Ukraine

Nikolay Alexandrovich Poluyanenko
Department of Information
and Communication Systems Security
Faculty of Computer Science
V. N. Karazin Kharkiv National University
Kharkiv, Ukraine

Natalia Kryvinska
Department of Information Systems
Faculty of Management
Comenius University
Bratislava, Slovakia

Oleksandr Volodymyrovych Potii
JSC Institute of Information Technology
Kharkiv, Ukraine

Yurii Ivanovich Gorbenko
Department of Information
and Communication Systems Security
Faculty of Computer Science
V. N. Karazin Kharkiv National University
Kharkiv, Ukraine

ISSN 2198-4182 ISSN 2198-4190 (electronic)
Studies in Systems, Decision and Control
ISBN 978-3-030-79769-0 ISBN 978-3-030-79770-6 (eBook)
https://doi.org/10.1007/978-3-030-79770-6

This Springer imprint is published by the registered company Springer Nature Switzerland AG
The registered company address is: Gewerbestrasse 11, 6330 Cham, Switzerland

Contents

Symbols, Signs, Units, Abbreviations and Terms

AS	Automated system
ANF	Algebraic normal form
BSC	Block symmetric cipher
MT	Mersenne twister
GOST 28147	Standard encryption using BSC
PRSG	Pseudorandom sequence generator
DRBG	Deterministic random bit generator
IS	Information system
IT	Information technology
ITS	Information telecommunication system
«Kalyna»	Symbol for BSC, determined by State Standard of Ukraine 7624:2014
CIP	Cryptographic information protection
CI	Correlation immunity
CP	Cryptographic protocol
DC	Distribution criterion
M-sequence	Pseudorandom sequence, which is formed by LFSR or NLFSR, which has the maximum possible period at a given value of the register size and corresponds to the properties De Bruijn sequence
M-NLFSR	Nonlinear-feedback shift register, generating M-sequence
M-LFSR	Linear-feedback shift register, generating M-sequence
PRS	Pseudorandom sequence
PLIC	Programmable logic integrated circuit
De Bruijn sequence	De Bruijn sequence of element order from binary field $GF(2)$ is a sequence of the period, in which different L-processions appear only once
SSC	Symmetric stream cipher
SE	Stream encryption
SFMT	Simple fast Mersenne twister

LFSR	Linear-feedback shift register
NLFSR	Nonlinear-feedback shift register
NLFSR of the second order	NLFSR, in function of which's feedback the nonlinearity of second order only is used, i.e., the product amounts to not more as two tiles of the given register
NLFSR *r*-order	NLFSR, in function of which's feedback the nonlinearity of *r*-order is used, i.e., the product from *r* tiles of the given register
FA	Finite-state automaton
SBC	Symmetric block conversion
SLAE	System of linear algebraic equation
SAC	Strict avalanche criterion
SCC	Symmetric cryptographic conversion
SISC	System of information stream conversion
«Strumok»	Identification code of the stream cipher fixed by DSTU 8845:2019
LAT	Linear approximation table
DDT	Difference distribution table
AES	Advanced Encryption Standard
ASIC	Application-specific integrated circuit
CBC	Cipher block chaining (BSC application mode: cipher block chaining)
CCM	Counter with cipher block chaining message authentication code (BSC application mode: counter with cipher block chaining message authentication code)
CFB	Cipher feedback (BSC application mode: prohibition with feedback after the cipher text)
CMAC	Symmetric key block cipher-based message authentication code (BSC application mode: elaboration of a message authentication code)
CPU	Central processing unit (central processor)
CryptMT	Software-oriented stream cipher
CTR	Counter mode (BSC application mode: prohibition)
Decim	Bit stream cipher
DES	Data Encryption Standard
DIEHARD	Testing methodology for research of statistical security of contemporary cryptography algorithms
ECB	Electronic codebook (BSC application mode: simple substitution)
Enocoro	Hardware-oriented bit stream cipher
eSTREAM	International project for detection of new bit stream ciphers, usable for wide application, organized by European Union
FIPS-197	Deciphering standard with BSC usage
FPGA	Field-programmable gate array

FSM	Finite-state machine
GCM and GMAC	Galois/Counter Mode and Galois Message Authentication Code (BSC application mode: elaboration of a message authentication code with prohibition and without prohibition)
GPU	Graphics Processing Unit (graphic processor)
Grain	Hardware-oriented bit stream cipher
HC-128	Software-oriented bit stream cipher
KCipher-2	SSC, stream symmetric cipher
KW	Key wrapping (BSC application mode: key ciphering)
LEs	Logic elements (logic elements in FPLD (field-programmable logic device))
MICKEY	Hardware-oriented bit stream cipher
MUGI	Stream cipher
NESSIE	New European Schemes for Signatures, Integrity, and Encryption
NIST	National Institute of Standards and Technology
NIST STS	NIST Statistical Test Suite (modality of testing for research of statistical security of contemporary cryptoalgorithms)
OFB	Output feedback (BSC application mode: inhibition with feedback after enciphering keystream)
Permutation	Transposition
Rabbit	Bit stream cipher
RC4	Software-oriented bit stream cipher
RFID	Radio frequency identification
Salsa 20	Software-oriented bit stream cipher
ShiftBytes	Line shift modification
SNOW 2.0	Stream cipher
SOSEMANUK	Software-oriented bit stream cipher
SSL	Secure socket layer (cryptographic protocol, which ensures an establishing of a secure connection between client and server)
SubBytes	Modification of a substitution
Substitution	Substitution
TLS	Transport layer security (cryptographic protocol, which enables a secure data sending into Internet)
Trivium	Hardware-oriented bit stream cipher
WEP	Wired Equivalent Privacy (algorithm for assuring of Wi-Fi network security)
WPA	Wi-Fi Protected Access (security protocol for protection of wireless networks)
XTS	XOR encrypt XOR (XEX) tweakable block cipher (BSC application mode: index linked replacing)
$\oplus$	Operation of the logic exclusive OR (XOR)

&	Operation of the logic AND (AND)
¬	Operation of the logic negation (NOT)
>>	Right shift of indicated number of bits operator
<<	Left shift of indicated number of bits operator
>>>	Cyclic right shift of indicated number of bits operator
<<<	Cyclic left shift of indicated number of bits operator

Chapter 1
Introduction

This research was supported by the Ukrainian National Research Foundation, under the Grant 2020.01/0351.

The monograph is an updated and improved version of the book published in Ukrainian:

Потокові шифри: монографія / О. О. Кузнецов, О. В. Потій, М. О. Полуяненко, Ю. І. Горбенко; за загальною редакцією І. Д. Горбенка.—Харків: Видавництво «Форт», 2019.—544 с., ISBN 978–617-630–059-5. The monography is translated and edited by Yevhen Hodun and Larysa Ovsiannikova.

The most used cryptographic mechanism, which is applied to the contemporary information and telecommunication systems (ICTs) for protection of important informations, is ciphering [1, 2]. This is an inverse cryptographic modification, which is used for opened data (Plaintext), aiming to hide its substantial content from unauthorized user (abuser). Reciprocally unambiguous encryption-decryption processes of the open text and the encrypted text (Ciphertext) are parametrized with key data, which for symmetric cryptoconversion coincide, and namely they define security level of the ciphering algorithm applied.

Contemporary symmetric cryptoconversions (SCC) are used for protecting information practically in all cryptographic appliances, in particular: guarantee of confidentiality and integrity of information and messages throughout all phases of their life cycle; encryption in ITS in various working modes, depending on actual requirements; generating of pseudorandom sequences (PRS); cryptographical authentication protocols, detection of stealthy and keys, secret distribution etc., when high claims to complexity (fast operation) are put; cryptographic protocols of electronic digital signature etc. The contemporary SCC are mostly applied for protecting of information and telecommunication systems and technologies, in particular state-owned informations, privacy data, hidden and commercial information, and other communications, which are liable to be protected according to laws, orders and decisions, and other regulations.

A. A. Kuznetsov et al., *Stream Ciphers in Modern Real-time IT Systems*,
Studies in Systems, Decision and Control 375,
https://doi.org/10.1007/978-3-030-79770-6_1

Among the symmetric cryptoconversions, hold a specific place the bit stream algorithms, in which the information is supplied and processed endless stream, i. e. as a sequence, which hypothetically can have an infinite length. The main advantage of this conversion is establishing of a determined dependence between single symbols of a bit stream, what enables to ensure an additional protection from imposing of misleading information or false working modes of the protective equipment or of the finite devices of telecommunication systems and networks. Consequently, the stream cryptoconversion usually enjoys more confidence during protection of important informational resources in various appliances, in particular for protecting privacy, important secret and non-public information, military and State secret, non-public and other classified facts, protecting level of which is defined by different regulations.

Among the definitive advantages of stream cryptoalgorithms should be pointed out the increased velocity indicators, and makes their utilization the most efficient for protecting control and communication channels in military purposes, in state administration, in banking sector etc. There exists, as well, a possibility of parallelizing of defined processes of bitstream cryptoconversion. In particular, the bitstream enciphering consists usually of addition to the information data the so called enciphering keystream—PRS (Pseudo-Random Sequence) with defined cryptographic properties. Deciphering at the receiving side consists in subtraction from the received data of the same keystream, i. e., if the PRS (keystream) is timely formed, then the encryption/decryption can be performed simultaneously by simultaneous initiation of several computing systems. Correspondingly, grows considerable the velocity of such a cryptoconversion, in comparison with the usual (block) methods.

In contemporary ITS, for cryptographic information protection (CIP), besides of bitstream enciphering, the block symmetric ciphering (BSC) modes are widely used, the efficiency of which is defined not only by properties of base conversion (electronic codebook or simple substitution), but also by distinctive features of selected mode of BSC-usage. This is conditioned by distinctive features of stream cryptographic conversion, when the processed data are supplied as a stream of interrelated symbols, so that imposition of a misleading information or of false working modes of finitestate ITS equipment is complicated substantially.

The today's challenges of elaboration and standardization of bitstream cryptoconversions are being resolved in many technologically developed countries of the World. However, the greatest confidence enjoy and the most used are the Worldwide known cryptoalgorithms, defined as standards of the International organisation for standardization, of the National Institute for Standards and Technology (NIST) USA, of the counterpart organizations in Ukraine and Russia [2–14]:

ISO/IEC 180,334 «Information technology—Security techniques—Encryption algorithms—Part 4: Stream ciphers» [2], in which are defined: common PRC-models; stream modes of BSC-usage for building synchronous (OFB- and CTR-modes) and asynchronous (CFB-mode) stream cryptoconversions; dedicated keystream generators—being immediately developed PRC (algorithms MUGI, SNOW 2.0, Rabbit, DEcim, KCipher-2);

ISO/IEC 10,116 «Information technology—Security techniques—Modes of operation for an n-bit block cipher» [11], in which the stream modes of BSC are defined: cipher block chaining (Cipher Block Chaining—CBC), inhibition with feedback after the cipher-text (Cipher Feedback—CFB), inhibition with feedback after the enciphering keystream (Output Feedback—OFB), inhibition (Counter Mode——CTR);

NIST Special Publication 800–38 «Block Cipher Modes» [13] and DSTU 7624:2014 «Information technologies. Cryptographic information protection. Symmetric block conversion algorithm» [3], in which ten modes are defined: the modes of simple substitution (Electronic Codebook——ECB); inhibition (Counter Mode—CTR); inhibition with feedback after the cipher-text (Cipher Feedback); elaboration of message authentication code (Symmetric Key Block Cipher-Based Message Authentication Code); chaining of cipher blocks (Cipher Block Chaining—CBC); inhibition with feedback after the enciphering keystream (Output Feedback); elaborating of message authentication code with inhibition and without inhibition (Galois/Counter Mode and Galois Message Authentication Code); inhibition and message authentication codes Counter with Cipher Block Chaining-Message Authentication Code; indexed substitution (XOR Encrypt XOR (XEX) Tweakable Block Cipher); key enciphering (Key Wrapping) [15, 16], among them bitstream modes CBC, CFB, OFB, CTR;

NIST Special Publication 800-90A «Recommendation for Random Number Generation Using Deterministic Random Bit Generators» [14], in which the functional models of determined generators of random bits [DRBG] are defined; conceptual mechanisms and baseline requirements to DRBG, and namely: DRBG, based on hash-functions (Hash_DRBG and HMAC_DRBG); DRBG, based on the BSC-usage (CTR_DRBG) and DRBG, based on appliance of number-theoretical problems (Dual Elliptic Curve Deterministic RBG—Dual_EC_DRBG);

DSTU GOST 28,147:2009 «Information processing systems. Cryptographic protection. Algorithm of cryptographic conversion» (Information processing systems. Cryptographic protection. Algorithm of cryptographic conversion) [17], in which in the function of the second and third mode the bitstream cryptoconversions of inhibition (CTR) and inhibition with feedback after the cipher-text (CFB) are used;

GOST R ISO/IEC 10,116–93 «Information technology. Working modes for the algorithm of n-digit block enciphering» [18], in which three bitstream modes are defined: CBC, CFB, OFB.

To the specified list of items ought to be added also the PSC-algorithms, which are named winners eSTREAM—an international project for detecting new widely usable bitstream ciphers, organized by the European Union. To the winners eSTREAM are also attributed the algorithms, which satisfy imposed requirements [4–10]:

- software-oriented PSC: HC-128; Rabbit; Salsa20/12; SOSEMANUK;
- hardware-oriented PSC: Grain; MICKEY; Trivium.

It should be noted that also the international standard ISO/IEC 29,192 «Information technology—Security techniques—Lightweight cryptography» [12], dedicated to enunciating algorithms of so-called lightweight (short-resource) cryptography

(Engl. Lightweight cryptography). These are cryptoalgorithms, realization of which is oriented on low-price, unpretentious devices concerning operating velocity and memory spaces, i. e. the respective algorithms of cryptographic conversion do not necessitate sophisticated calculations or large-sized memory stores. The third part of the standard is dedicated to bitstream ciphers, and namely to the algorithms Enocoro [19] and Trivium [10].

In the monograph there is given an analysis and comparative researches of the bitstream BSC-modes, especially: CTR, CFB, CBC, OFB. The specifications of mentioned modes are known, they are presented in numerous national and international standards, so as in regulations [11, 15–17, 20]. Therefore, the national standards of the USA and of Ukraine define the biggest list of BSC-usage modes, i. e. in these regulations are anticipated the most spacious using area of the cryptographic protecting of information by the aid of BSC. Thus, the analysis and comparative researches of stream BSC-modes we shall perform ulteriorly through the use of the terminology and of the main base statements, defined in the standards NIST Special Publication 800–38 «Block Cipher Modes» and DSTU 7624:2014 «Information technologies. Cryptographic information protection. Algorithm of symmetric block conversion».

Developing, research and validation of the using conditions of contemporary bitstream cryptoalgorithms is extremely complicated and work-consuming task. Being standardized on the national and/or international levels, the PSC owes to ensure a high level of stability (inclusive in circumstances of possible application of quantum cryptoanalysis), to have a sufficient operating velocity level and to work effectively on various computing platforms. These positive qualities are usually ensured by the efficiency of different PSC-components, for instance:

- by secret key unwrapping scheme (on key schedule);
- by selected base algorithm structure;
- by using logic operations and functional conversions;
- by linear conversions and non-linear complications nodes etc.

To the base for building majority of cryptographic systems is put a multiple using, relatively to simple conversions, of so-called cryptographic typicals. The founder of the mathematical theory of secret systems, the noted mathematician and electrician Claude Shannon proposed to use substitutions [Engl. Substitutions] and permutations [Engl. Permutations] as extremely simple and computationally effective cryptographic typicals [21]. Afterwards, the technique for building of cryptographic typicals has been developed in numerous scientific publications, in particular at construction of Boole functions, of non-linear complications nodes, of mixing block, of logical functions, of modular operations, of different functional conversions etc. Therefore, at this time, the models, the methods, and the numerical algorithms for constructing cryptographic typicals has been separated into an individual scientific domain, which serves as an important and extremely powerful section of the contemporary information protection theory [1].

The developing trend of contemporary SPC demonstrates that the most perspective at this time are the PRS-generators, based on registers with non-linear feedback. Such generators as these possess a high operating velocity, a simple and low-cost hardware or software realization. The registers with non-linear feedback raise substantially the complexity of sequences formed, and in that way the cryptographic stability [22]. Into the profile of perspective bitstream ciphers for systems with limited resources of the international contest eSTREAM were included several cryptographic typicals, to which belong the bitstream ciphers Trivium and Grain. The base element of mentioned algorithms are the registers with non-linear feedbacks, which, on the one part, ensure the efficiency of hardware realization, and on the other part—enable the counteraction to algebraic attacks [23].

A special topicality the shift registers with non-linear feedbacks (NLFSR) have gained in environments with limited limited calculating resources. The contemporary development of pocket computer devices, the usage of radio-frequency tags (RFID), of industrial controllers, of sensor units and of smart cards has become a quite widespread phenomenon. The transition from desktop/server environments to little instruments (built-in systems, RFID, sensor networks, health care systems, distributed control systems, Internet of things, the cyberphysical systems and the smart electric grid) provoke a set of questions concerning security and privacy. The usual cryptographic standards had been optimized regarding to large dimension calculating systems, in this connexion the existing cryptographic technologies were hardly, but sometimes impossibly, to use for devices with limited resources. In connection with this the NIST raised the question regarding standardization of algorithms in lightweight cryptography, which is usable for satisfaction of systems with limited environment, for creation of ultracheap service applications and for working with limited energy supply [24].

The essential synthesis task required of the PRS based on the shift register with linear and non-linear feedbacks is to define the generator structure with preprogrammed requirements to statistical and structural properties of the sequence formed by it. In general case, the non-linear function due to be in accord with the maximal linear sophistication, to meet the balance of probabilities [25], to ensure the minimal deviation of the ideal auto-correlation; the combinational feedback circuit must have a minimum allowed computational complexity [26] etc. The realization of a synthesis algorithm of an optimal non-linear generator, even with a relatively not large number of shift-register digits, becomes so labour-intensive that the possibility of its application is practically excluded [26].

The continual improvement of calculating technologies, the appearance of new mathematical methods of cryptoanalysis—all these advance new claims to cryptographic safety of PRS generating systems. For this reason, during elaborating of new enciphering algorithms for small resource cryptography the two main items have been pointed up: the criteria objectivation for a PRS-generator structure and the questions, directly linked with the synthesis of generators themselves, which would comply with the criteria selected.

As regarding the first item, there are numerous fundamental works, the existent proceedings however do not cover the entirety of possible cryptographic damages that are growing continuously. The item of registers synthesis with targeted constructional characteristics, which would meet the proposed criteria of cryptographic stability, as for today is nearly not resolved. The existing methods are based as few as on building-up of a de Brujn sequence, with following renewal of the feedbacks structure, which shape the built-up sequence.

No existing methods allow to shape NLFSR with preprogrammed constructional characteristics. In addition to it, the existing methods are very demanding of time expenditure, what doesn't enable looking for registers of sufficiently great size. Thus, the found by this time maximum non-linear polynomial, by aid of which the de Brujn sequence is shaped, has all in all the 33-d power.

The main objectives of the present work are:

- analysis and comparative researches of contemporary algorithms of stream cryptoconversion;
- analysis and comparative researches of contemporary methods of constructing PSC and BSC stream modes;
- researches of standardized algorithms of bitstream cryptoconversion;
- researches of cryptographic elements (cryptographic typicals) PSC and of BSC stream modes;
- researches of PSC key systems and of BSC stream modes;
- developing of RNFB synthesis at the cost of validation and of practical application of necessary conditions and claims for shaping the maximum period PRS with preprogrammed constructional characteristics;
- analytical and empirical estimation of stability index of the synthesized NLFSR, their experimental verification and implementation while in elaborating of schemes of bitstreamd symmetric ciphering.

In the first chapter there are given the materials after results of analysis and comparative researches of the contemporary cryptoconversion algorithms. Especially the criteria and indices of efficiency of stream cryptoconversion are validated, the criteria and indices of stability (security) are analysed, that characterize the system property to resist various attacks; software and hardware realization indices and criteria, that characterize the efficiency of respective realization; the constructional-technological indices and criteria, that characterize the construction transparency, the perspectivity of respective solutions, their stability margin etc.

In the second and the third chapters there are researched the primary approaches to the building-up and analysis of PRC. Their classification is performed, the contemporary methods of stream cryptoconversions constructing are analysed, and their properties are researched.

In the fourth and the fifth chapters there are adduced the results of analysis and comparative researches of BSC-stream modes, in particular the modes of inhibition, of inhibition with feedback after the cipher text, of cipher-blocks chaining, of inhibition with feedback after the ciphering keystream.

For research of statistical safety of contemporary cryptographic algorithms the most frequently are used the test methods NIST STS, DIEHARD and others [27–29].

Their essence lies in the verification of hypothesis about a random character of the initial sequence of researched cryptoalgorithm, i. e. in the verification of assumption that the data being formed do not differ in statistical sense from any hypothetical «random» sequence. In the fifth chapter there are adduced the results of statistical testing of ciphers GOST 28,147–89, TDEA, FIPS-197, Camellia and Kalyna, which have shown that the initial sequences of these BSC do not differ in statistical sense from any hypothetical «random» sequence.

In the chapters six—nine there are adduced the results of researches of standardized stream cryptoconversion, expounded, in particular, in national and international regulations and searching projects [4–10, 30]: ISO/IEC 18,033–4, NIST SP Special Publication 800-90A, eSTREAM, ISO/IEC 29,192:2012. In these chapters there are presented the results of analysis of various stream cryptoconversion algorithms and of comparative researches of statistical safety (by various testing procedures), and of computing complexity for stream conversion realization.

In the sixth chapter there are researched the stream ciphers MUGI, SNOW 2.0, Rabbit, DEcim, KCipher-2, constructed after the scheme of key streams generators, specification of which is lied down in the international standard ISO/IEC 18,033–4. The specified ciphers are being built-up both with using of linear shift register technique, and of cryptographic elements of block ciphers. In different generators there are used the complicating non-linear functions, which are being constructed by using non-linear substitution blocks, in particular, the most frequently used are the complicating units BSC AES (FIPS-197). This allows, on the one part, to simplify substantially the analysis of cipher stability by using a verified mathematical tool and some known statements concerning certain cryptographic properties; on other part, allows using of still existing cryptotypicals and, possibly, their software realization, what fact accelerates substantially the implementation of algorithms on various calculating systems.

Besides with specialized key-streams generators MUGI, SNOW 2.0, Rabbit, DEcim, KCipher-2 the international standard ISO/IEC 18,033–4 includes also general PRC-models and stream-modes of BSC-usage for building-up of synchronous (modes OFB and CTR) and asynchronous (mode CFB) bitstream based cryptoconversions. In the standard are specified the protection levels of bitstream based cryptoconversions as well, and indicated a possible compromise with the safety level and software or hardware realization. Even this chapter is the most interesting from the point of view of a possible realization of bitstream based cryptoconversions in various computing systems, inasmuch it allows to adopt certain protection mechanisms taking into consideration the special aspects of environmental building-up for using and functional destination of an ITS.

The USA standard NIST Special Publication 800-90A defines some functional models of determined random bits generators (DRBG), the conceptual tools and base requirements to such generators, the mechanisms of separate functions and the specifications of certain algorithms, in special of generators, which are based on the use of a hashcoding, BSC and of theoretical-numerical problems. The DRBG researches, defined by this standard, are presented in the Chap. 7.

As the winners of the international project eSTREAM are presented 7 PRC-algorithms, which are being devided in two functional groups. The first group are the software-oriented stream ciphers, to which belong: HC-128; Rabbit; Salsa20/12; SOSEMANUK. To the second group—the hardware-oriented ciphers—belong the algorithms Grain; MICKEY and Trivium. In the eighth chapter is presented the analysis of structure and of main used functions for conversion of software-oriented stream ciphers (HC-128, Rabbit, Salsa20/12, SOSEMANUK, CryptMT) and of hardware-oriented ciphers, to which belong the algorithms Grain, MICKEY and Trivium.

The international standard ISO/IEC 29,192–3 «Information technology—Security techniques—Lightweight cryptography» is dedicated to the presentation of stream algorithms for lightweight (small-resource) cryptography, the realization of which is oriented on cheap and non-pretentious as demanding operating velocity and storage capacity devices. Among the specified algorithms the stream cipher Trivium is also taken in consideration as a winner of the international project eSTREAM, of which properties are detail studied, and the algorithm advantages and lacks are mentioned. An other stream cipher Enocoro (being researched in the ninth chapter) is very similar with MUGI, which has been specified in the international standard ISO/IEC 18,033–4 and is, in reality, its simplified version for simple systems with small calculating resources. The application in Enocoro of a non-linear complicacy function, that is constructed taking into consideration an S-Block BSC AES (FIPS-197), is also directed to ensuring of necessary stability indices together with high requirements for forming velocity of an initial PRS.

In addition to that, in the ninth chapter there is being researched the ciphering algorithm RC4, which is used in some widespread standards and ciphering protocols (for instance WEP, WPA, SSL and TLS).

In the tenth chapter there are presented some comparative researches of statistical safety indices of various PRC and of their computing efficiency. The researches results after various methodologies of statistical testing and of research of the velocity of PRS-forming are presented.

The eleventh chapter is dedicated to researching of main methods of NLFSR-constructing, to the analysis of existent methods of NLFSR-synthesis, which are forming a maximum period sequence.

In the twelfth chapter there are formulated and validated the necessary conditions for building of a second order NFLSR, which shape a maximum period sequence. The indicated conditions consist in defining of Requirements for feedback coefficients, observance of which is a necessary condition for shaping by the register of a maximum period sequence. The quantitative estimation of each Requirement in selection of NFLSR, which do not generate any M-sequence.

In the thirteenth chapter a method of synthesis of an NFLSR, which generates an M-sequence, is described, with targeted constructive characteristics and a hardware and software complex, which realizes this method. Researches of any possibility of hardware realization of an NFLSR are performed, based on a programmable logic integrated circuit (PLIC) and on a graphic processing unit (GPU). The obtained non-linear polynomials of 32-nd power, by which using is formed an M-sequence are presented.

The using of the proposed method allows to designate, at the designing stage, a wanted quantity of feedback, which will be used in the register, and a maximum quantity and disposing of the possible feedback coefficients as well, hereby to carry out the looking for NFLSR, which shapes an M-sequence according to preset technical criteria. For instance, in all found NFLSRs with dimensionality of 32 tiles the feedbacks are absent from the first 12 tiles, thanks to which a realization of simultaneous formation of 13 input bits is possible, at the cost of multiple realization of the feedback function, what will enable to to increase the PRS-generating velocity by up to 13 times.

In the fourteenth chapter there is formulated and experimentally verified the model of estimation of the cryptographic stability of the NFLSR, as a base element of the stream symmetric ciphering, which lies in elaborating of a system of criteria and stability indices of a pseudo-random sequence, formed by a NFLSR.

In the fifteenth chapter there are presented the results of substantiation of the structure of the stream symmetric cipher, of the key system and of the main cryptographic elements of a new stream cipher «Strumok». This algorithm, as for today, is already adopted as the national standard of Ukraine DSTU 8845 with stream symmetric ciphering [31].

References

1. Wembo Mao. Modern cryptography. Theory and practice. Per. with English—M.: Izd. Williams House, p. 768 (2005)
2. ISO/IEC 180334:2011. Information technology—Security techniques—Encryption algorithms—Part 4: Stream ciphers. [Electronic resource]. Access mode: http://www.iso.org/iso/home/store/catalogue_ics/catalogue_detail_ics.htm?csnumber=54532
3. Development of a new block symmetric cipher: report for the first stage of R&D "Algorithm" (intermediate)/JSC "IIT"; ker. ID. Gorbenko—Kharkiv, **4**, 304 (2014)
4. The eSTREAM Project eSTREAM Phase 3. HC (Portfolio Profile 1). [Electronic resource]. Access mode: http://www.ecrypt.eu.org/stream/hcpf.html
5. The eSTREAM Project eSTREAM Phase 3. Rabbit (Portfolio Profile 1). [Electronic resource]. Access mode: http://www.ecrypt.eu.org/stream/ rabbitpf.html
6. The eSTREAM Project—eSTREAM Phase 3. Salsa20 (Portfolio Profile 1). [Electronic resource]. Access mode: http://www.ecrypt.eu.org/stream/salsa20pf.html
7. The eSTREAM Project—eSTREAM Phase 3. SOSEMANUK (Portfolio Profile 1). [Electronic resource]. Access mode: http://www.ecrypt.eu.org/stream/sosemanukpf.html
8. The eSTREAM Project—eSTREAM Phase 3. Grain (Portfolio Profile 2). [Electronic resource]. Access mode: http://www.ecrypt.eu.org/stream/grainpf.html
9. The eSTREAM Project—eSTREAM Phase 3. MICKEY (Portfolio Profile 2). [Electronic resource]. Access mode: http://www.ecrypt.eu.org/stream/mickeypf.html
10. The eSTREAM Project—eSTREAM Phase 3. Trivium (Portfolio Profile 2). [Electronic resource]. Access mode: http://www.ecrypt.eu.org/stream/triviumpf.html

11. ISO/IEC 10116:2004. Information technology – Security techniques – Modes of operation for an nbit block cipher. [Electronic resource]. – Access mode: http://www.iso.org/iso/home/store/catalogue_tc/catalogue_detail.htm?csnumber=38761
12. ISO/IEC 291923:2012. Information technology – Security techniques – Lightweight cryptography – Part 3: Stream ciphers. [Electronic resource]. – Access mode: http://www.iso.org/iso/home/store/catalogue_tc/catalogue_detail.htm?csnumber=56426
13. NIST Special Publication 80038A. Recommendation for Block Cipher Modes of Operation. [Electronic resourse]. – Access mode: http://csrc.nist.gov/publications/nistpubs/80038a/sp80038a.pdf.
14. NIST Special Publication 80090A. Recommendation for Random Number Generation Using Deterministic Random Bit Generators. [Electronic resourse]. – Access mode: http://csrc.nist.gov/publications/nistpubs/80090A/SP80090A.pdf.
15. DSTU 7624: 2014 Information technologies. Cryptographic information protection. Symmetric block transformation algorithm. - Kyiv: Ministry of Economic Development of Ukraine. - 227 p.
16. A New Encryption Standard of Ukraine: The Kalina Block Cipher. [Electronic resource]. - Access mode: https://eprint.iacr.org/2015/650.pdf
17. DSTU GOST 28147: 2009. Information processing systems. Cryptographic protection. Cryptographic transformation algorithm [Electronic resource]. Access mode: http://document.ua/docs/tdoc16773.php
18. GOST R ISO / IEC 1011693. Information technology. Operating modes for n-bit block encryption algorithm [Electronic resource]. Access mode: http://gostrf.com/normadata/1/4294818/4294818194.pdf
19. Pseudorandom Number Generator Enocoro. [Electronic resourse]. – Access mode: http://www.cryptrec.go.jp/english/cryptrec_13_spec_cypherlist_files/PDF/23_00espec.pdf.
20. NIST Special Publication 80038. Recommendation for Block Cipher Modes of Operation. [Electronic resource]. - Access mode: http://csrc.nist.gov
21. Shannon, K.: Works on information theory and cybernetics. - M .: ИЛ, 1963. - 830 c. (Section—Theory of Communication in Secret Systems).
22. Domashev, A.V., Gruntovich, M.M., Popov, V.O., Pravikov, D.I., Prokofiev, I.V., Scherbakov, A.Y.: Programming information security algorithms. M .: Publishing house “Nolidzh”, p. 416 (2002)
23. Knellwolf, S., Meier, W., NayaPlasencia, M.: Conditional Differential Cryptanalysis of NLFSRBased Cryptosystems. [Electronic resourse]. 2010. URL: https://link.springer.com/chapter/https://doi.org/10.1007/9783642173738_8 (Date: 01.08.2017)
24. McKay, K.A., Bassham, L., Turan, M.S., Mouha, N.: Report on lightweight cryptography. National Institute of Standards and Technology Internal Report 8114. [Electronic resourse]. (2017). URL: https://doi.org/10.6028/NIST.IR.8114 (Date: 01.08.2017)
25. Lansky, V.G.: One of the criteria for the synthesis of the optimal structure of a nonlinear GPSDP. Bulletin of the Vyatka Scientific Center of the Verkhne-Volzhsk Branch of the Russian Academy of Science. Series: Information Processing Problems, **1**(4) (2003). Kirov: VNC.—S.124–126
26. Lansky, V.G., Lansky, A.M., Shmakova, N.A.: Synthesis of nonlinear pseudorandom binary sequence generators. Development of information systems. [Electronic resource]. Vestnik ITARK No. 1. 2015. URL: http://itarctica.ru/system/attachments/uploads/000/000/073/original/SYNTHESISOFNON-LINEARGENERATORS.pdf (Date of beast: 01.08.2017)
27. NIST Special Publication 80022. A Statistical Test Suite for Random and Pseudorandom Number Generators for Cryptographic Applications. [Electronic resourse]. – Access mode: http://csrc.nist.gov/publications/nistpubs/80022rev1a/SP80022rev1a.pdf.
28. Dieharder: A Random Number Test Suite - free implementation of statistical tests (including diehard and NIST tests) in C. [Electronic resource]. - Access mode: http://www.phy.duke.edu/~rgb/General/dieharder.php
29. The Marsaglia Random Number CDROM including the Diehard Battery of Tests of Randomness. [Electronic resource]. - Access mode: http://stat.fsu.edu/pub/diehard.

30. Stanica P. Some Results Concerning Generalized Bent Functions [Electronic resourse]. – Access mode: http://eprint.iacr.org/2011/290.pdf.
31. DSTU 8845: 2019. Information technology. Cryptrographic information protection. Algorithm of symmetric streaming re-creation. Kiev; Subsidiary "UkrNDNTS", p. 50 (2019)

Chapter 2
Criteria and Indices Substantiation of the Stream Cryptoconversion Efficiency

The efficiency of the functioning of cryptographic data conversion schemes must be estimated from the point of view of their immunity to various methods of information disclosure, as well from the point of view of imposed restrictions on the cryptoconversion velocity and on the using particularity. Therefore, the known efficiency indices and criteria of can be divided into two basic groups:

- the unconditional criteria and indices, which characterize critical important system properties, non-fulfilment or reduction of which will result into failure of system main functions. To cryptographic means of information protection should be ascribed the indices and criteria of stability (safety), characterizing the system resistivity against the influence of various attacks;
- the conditional efficiency criteria and indices, which characterize the wanted properties, the fulfilment of which will enable to improve certain efficiency indices, but non-fulfilment or reducing below the admissible limits won't cause any violation of system functioning. To conditional criteria and indices should be ascribed the software and hardware-realization ones, which characterize the efficiency of software and hardware realization, i. e. the utility of BSC-schemes and of stream BSC-modes. To conditionals should be ascribed also the constructional-technological ones, which characterize the construction transparency, the prospectivity of respective decisions, the margin of their stability etc. (see Fig. 2.1).

Stability indices of stream cryptoconversions characterize the stability of schemes against various analysis methods and are unconditional indices, inasmuch as the usage expediency of every information protection schemes is based above all on the estimation of stability against at present known methods (attacks) of analysis, which are applied to the class of researched schemes. The non-fulfilment of these unconditional indices may cause a violation of the preset safety level of protected information, i. e. a failure of the system main functional destination.

A. A. Kuznetsov et al., *Stream Ciphers in Modern Real-time IT Systems*,
Studies in Systems, Decision and Control 375,
https://doi.org/10.1007/978-3-030-79770-6_2

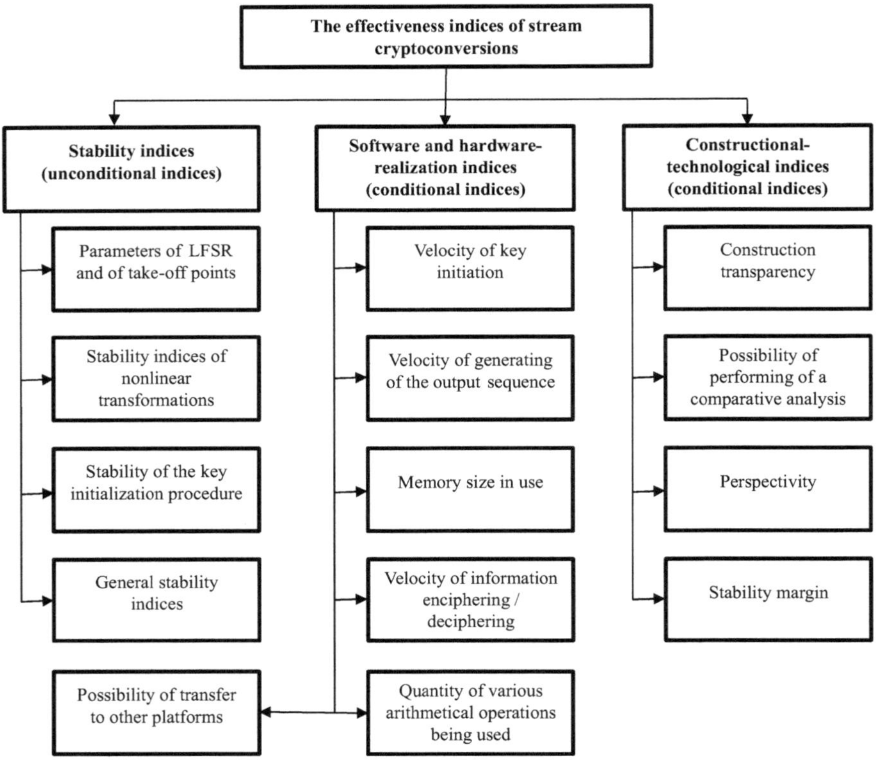

Fig. 2.1 The effectiveness indices of stream cryptoconversions

Software and hardware-realization indices characterize the efficiency of software and hardware realization of stream cryptoconversions and are indices, which characterize the utility of PRC and of stream BSC-modes.

The cryptoconversions in use must have, undoubtedly, a necessary stability against various analysis methods, it is wanted, however, that they wouldn't have any high hardware realization cost, wouldn't have any low operating velocity etc. One of the most important advantages of stream conversion of information, thanks to which they have got a wide circulation, is a high velocity of information processing. Consequently, the schemes of stream ciphering (BSC), having high indices of realization effectiveness, will be selected, highly likely, for practical programmes.

The SC-schemes utilization is considered frequently from the position of unification (universality): for example, one and the same scheme can be used equal successfully both in software, in hardware, and in smart-cards. Some domains of software impose very high requirements to the link speed (gigabit traffics), for other using domains of software the minimum hardware platform is acceptable and/or pocket-sized hardwares (mobile phones, smart-cards etc.). Ideally, the concerned schemes have to be versatile, i. e. to be realized on more than one platform. Therefore, while defining the realization effectiveness, it is expediently to select indices, which reflect the effectiveness of functioning in specified domains of software, inasmuch as one and the same scheme unlikely can be effectively realized on all platforms and in all programmes.

Constructive-technological indices characterize the construction transparency, the schemes prospectivity, the margins of their stability, the ability of the known analysis methods for estimating of characteristics of stream cryptoconversions, the possibility of carrying out of a comparative PSC-analysis and of stream BSC-modes, and, consequently, the confidence level of users.

Let us to examine in detail each group of parameters, to substantiate the criteria and indices of stream cryptoconversions efficiency.

2.1 Criteria and Indices of the Stream Cryptoconversions Stability

The analysis of the known proceedings [1, 2–6] enabled us to separate a multitude of criteria and of indices of stream cryptoconversions stability. Their general classification is presented in Fig. 2.2. Let us to examine in detail each group of indices and criteria of stability estimation.

2.1.1 Indices Demonstrating the Parameters of Shift-Registers and of Take-Off Points for Non-Linear Functions and Feedbacks

For building-up of stream cryptoconversions the technique of shift-registers with linear feedbacks (LFSR). The LFSR-parameters and the take-off points for non-linear functions and feedbacks are immediately influencing on efficiency indices of stream cryptoconversions.

Primitivity of generating polynomial. The primitive polynomial of n power is an irreducible polynomial, which is dividing x, although isn't dividing $x^{2^{n-1}}+1$ for any such d, which divides 2^n-1. The power of the polynomial is the length of the shift register. Only the registers, which use in quality of a generating polynomial the primitive polynomial, are maximum period registers (pass through all the possible 2^n-1 states) and generate maximum period sequences, so-called m-sequences.

The selection criterion is a logical swapping Yes/No.

Reciprocally simple powers of generating polinomials (for combining genetators). During selection of generating polynomials, which have reciprocally simple powers, the linear complexity (see index linear complexity) and the generated sequence period is maximum one. So, $\Lambda(z) = f(\Lambda_1, \ldots, \Lambda_N)$, where Λ_j is the linear complexity of j-th generator at $j = 1, \ldots, N$.

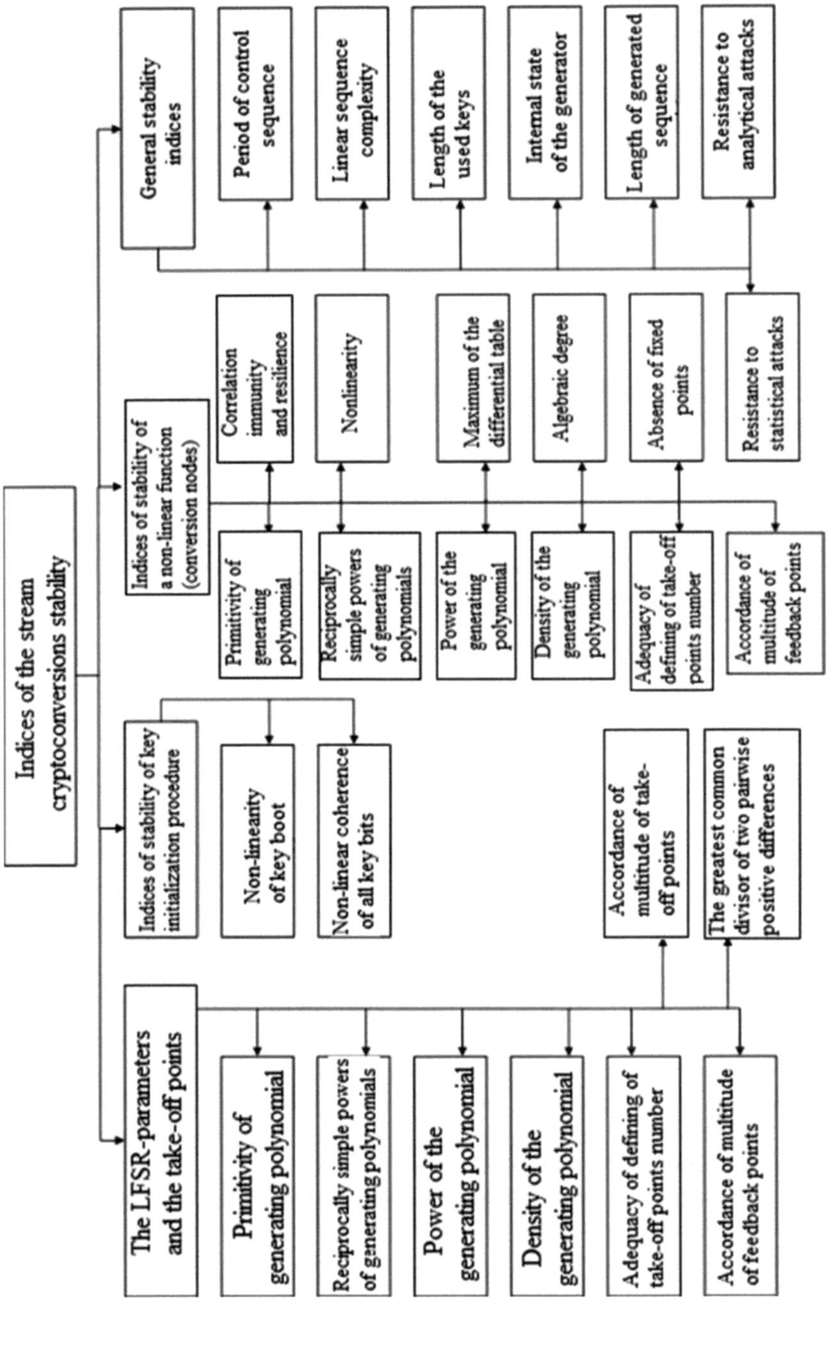

Fig. 2.2 Indices of stability of stream cryptoconversions

The selection criterion is a logical swapping Yes/No.

Power of the generating polynomial. If the generating polynomial is selected correctly, the polynomial power n determines the generated sequence period. As for today, for resisting the analytical attacks the minimum register length have to be not less than 128 bits.

As a selection criterion we shall consider the bit-length of the register.

Density of the generating polynomial. For resisting to correlation attacks, based on lightweight parity verification, the quantity of non-zero coefficients k in the generating polynomial has to be about $n/2$.

As a selection criterion we shall consider $k \to n/2$.

Adequacy of defining of take-off points number for a non-linear function (for filter-generators). For resistance against analytical attacks r must be $r' \leq \sqrt{2n}$.

The selection criterion is $r \to r'$.

The accordance of multitude of feedback points with a new multitude of positive differences ΔB. The multitude is named a full multitude of positive differences, if all positive pairwise differences between its elements are different. The necessary accordance ensures the resistance against analytical attacks.

The selection criterion is a logical swapping Yes/No.

The accordance of multitude of take-off points for a non-linear function with a new multitude of positive differences $\Delta\Gamma$ (for filter-generators). The necessary accordance ensures resistance against analytical attacks.

The selection criterion is a logical swapping Yes/No.

The greatest common divisor of two pairwise (neighboring) positive differences must be equal to 1.

The selection criterion is a logical swapping Yes/No.

2.1.2 Indices of Stability of Key Initialization Procedure

Non-linearity of key boot. The operations non-linearity increases their stability against algorithms, based on using of the key boot.

The selection criterion is a logical swapping Yes/No.

Non-linear coherence of all key bits. Each bit of the initialized register is a result of non-linear conversions of all key bits. Wanted is a realization, at which each bit of the initial register is a function of a non-linear conversions of all key bits, in such a case:

$$I\left(K_{out}^{n}; K_{in}^{n-k}\right) \to 0$$

where n—key length; k—number of bits of the initialized key, which are introduced through non-linear conversions, $k \to n$; K_{out}^{n} and K_{in}^{n-k}—the initialized key, which we are introducing properly. Consequently, the knowing of the key, which we are introducing, excludes the knowing of the initialized key (at guaranteed stability of non-linear conversions).

The selection criterion is a logical swapping Yes/No.

2.1.3 Indices of Stability of a Non-Linear Function (of Non-Linear Conversion Nodes)

The basic cryptographic indices of a non-linear function (of non-linear conversion nodes are:

- balanced state;
- correlation immunity and resilience (for streame ciphers) / circulation criterion and strict avalanche criterion (for block ciphers);
- nonlinearity;
- maximum of the linear approximation table;
- maximum of the differential table;
- avalanche: autocorrelation / index of sum of squares (sum-of-squareindicator);
- algebraic degree;
- linear redundance;
- absence of fixed points;
- absence of linear structures.

It exists a number of other indices, for example such as:

(1) Independence between input and output data.
(2) Independence between output and input data.
(3) Independence between output and output data.
(4) Dynamic independence between input and output data.
(5) Dynamic independence between output and input data.
(6) Dynamic independence between output and output data.

These indices won't be taken into consideration in this work.

We'll present definitions and analytic expressions for the main indices of cryptographic stability of a non-linear function (swapping nodes). For this purpose we'll introduce some main signs and definitions of the Boole algebra mathematical apparatus.

Let be $V_2^n = \mathrm{GF}(2)^n$, where n—the dimensional vector space The Boolean function $f(x)$ from n variables—that is a reflexion $f(x) : V_2^n \to V_2$, where $x = (x_1, \ldots, x_n)$.

It is significant that during the examination of cryptographic properties of an S-block it is not enough to examine only the properties of separate components of its initial Boolean functions (co-ordinate Boolean functions). It is necessary also to examine the cryptographic properties of all linear combinations of co-ordinate functions.

The Boolean functions can be presented in some different forms. The two most general presentation forms are a truth table and a polar truth table.

Truth table of a Boolean function from n variables—that is a binary output vector of the function, which incorporates 2^n elements, each element belongs to {0; 1}.

Polar truth table of a Boolean function from n variables is designated $\hat{f}(x)$—that is a vector, which incorporates 2^n elements, each element belongs to {−1; 1}. It is built from $f(x)$ on the $\hat{f}(x) = 1 - 2f(x)$ or way $\hat{f}(x) = (-1)^{f(x)}$, what is equivalently.

Let us denote $h_w(f)$ with the Hamming weight of Boolean function *f(x)* from n variables. Then $h_w(f) = \sum_{x=0}^{2^n-1} f(x)$ is the number of unities in of the truth table *f(x).*

Let us denote $d_H(f, g)$ with the Hamming distance between two Boolean functions *f(x)* and *g(x)* from n variables. Then

$$d_H(f, g) = \sum_{x=0}^{2^n-1} (f(x) \oplus g(x))$$

where ⊕ denotes the addition after modulus 2. Therefore, $d_H(f, g)$—is the number of different elements in corresponding positions of the two truth tables.

Algebraic normal form (ANF) of a Boolean function from n variables is written as:

$$f(x) = a_0 \oplus a_1x_1 \oplus a_2x_2 \oplus \ldots \oplus a_nx_n \oplus a_{12}x_1x_2 \oplus a_{13}x_1x_3 \oplus \ldots$$
$$\oplus\, a_{(n-1)n}x_{n-1}x_n \oplus \ldots \oplus a_{123\ldots n}x_1x_2x_3 \ldots x_n,$$

where the coefficients compose the elements of the ANF *f(x)*. Each representation of the ANF matches up with the unique table of a Boolean function.

Let us denote as *Deg(f)—the algebraic degree of the Boolean function f(x).* The algebraic degree is—the number of variables in the greatest term of an ANF function, which has a non-zero coefficient.

Let be $S = (f_1, f_2, \ldots, f_m)$—*S*-block (non-linear swapping node) with the size $n \times m$, where $f_i (i = 1, \ldots, m)$—the co-ordinate Boolean functions from n variables. Let be g_i g_i—the multitude of all linear combinations $f_i (i = 1, \ldots, m)$ (which includes the function f_i).

Then the *algebraic degree of a swapping block S,* $Deg(S_{n,m})$ is being defined in this way:

$$Deg(S_{n,m}) = \min_{g}\{Deg(g_j)\}\ (j = 1, \ldots, 2^m - 1).$$

Selection criterion is the maximization of the algebraic degree. For a regular *S*-block $n \times m$ the algebraic degree *Deg* can't exceed $n-1$.

Let be w and x—vectors belonging to V_2^n. The Boolean function presented as

$$l_w(x) = w \cdot x = w_1x_1 \oplus w_2x_2 \oplus \ldots \oplus w_nx_n$$

is named affine function from n variables. Obvious, if $w_0 = 0$, so the function is linear.

The multitude of affine functions from n variables consists of the multitude of all linear functions from n variables and their complementation. The affine functions have the algebraic degree 1.

WHT, Walsh-Hadamard transform of the polar truth table of a Boolean function $\hat{f}(x)$, which we'll assign as $\widehat{F}(x)$,—this is the correlation dimension of a function and of the entire multitude of all linear functions. Let us define it in such a way:

$$\widehat{F}(x) = \sum\nolimits_x (-1)^{f(x)}(-1)^{l_w(x)} = \sum\nolimits_x \hat{f}(x)\widehat{l_w}(x)$$

where $\widehat{l_w}(x) \in \{1, -1\}$—is a polar presentation of a linear function.

Thus, $\widehat{F}(x)$—is a real number in the interval $[(-2)^n, 2^n]$. The vector, representing the Walsh-Hadamard transform of a function, is named *Walsh-Hadamard spectrum.*

The Boolean function $f(x)$ from n variables is named *balanced*, if $d_H(f) = 2^{n-1}$ or

$$\#\{x | f(x) = 0\} = \#\{x | f(x) = 1\},$$

i. e. the numbers zero amount in the truth table of a function is equal with the numbers one amount.

In the terms of the Walsh-Hadamard transform the Boolean function is balanced then and only then, when $\widehat{F}(x) = 0$. We'll mention as well that the algebraic degree of a balanced function cannot exceed $n-1$.

An *S-block is balanced* (or *bijective*), if all its component Boolean functions and their linear combinations are balanced.

The *nonlinearity of a Boolean function* $f(x)$ from n variables, which we shall denote as *NL*(f),—that is the minimum Hamming distance from the function $f(x)$ to the multitude of all affine functions from n variables:

$$N_f = \min\{d(f, \varphi)\},$$

where φ—is the multitude of affine functions.

The mathematical correlation between the non-linearity of a Boolean function $f(x)$ from n variables and the Walsh-Hadamard transform of this function is being written so:

$$NL(f) = \frac{1}{2}\left(2^n - WHT_{\max}\right)$$

where WHT_{max}—is the maximum value in the Walsh-Hadamard spectrum.

For a balanced function f above $V_n (n \geq 3)$ the non-linearity N_f can be achieved:

$$Nf \leq \begin{cases} 2^{n-1} - 2^{\frac{n}{2}-1} - 2, & n = 2k \\ \left|\left|2^{n-1} - 2^{\frac{n}{2}-1}\right|\right|, & n = 2k+1 \end{cases}$$

where $|x|$—is the maximum even integer, which is less or equal to x.

The upper limit of non-linearity for an arbitrary function f above V_n can achieve:

$$N_f \leq 2^{n-1} - 2^{\frac{n}{2}-1}$$

The non-linearity of a function is an important index, inasmuch as the nonobservance of this pointer enables the realization of a correlating attack, which uses the correlation of this function with the multitude of weak functions. During the building of stable Boolean functions it is necessary to ensure their minimal correlation with the multitude of all weak functions, i.e. it is necessary that the non-linearity of the given function approaches to the upper limit of non-linearity.

The selection criterion is $N_f = \max_{N_j}\{N_1, \ldots, N_r\}$, N_j—the nonlinearity of a j-th function; $j = 1, \ldots, r$.

The non-linearity of an S-block $S_{n \times m}$, $\mathrm{NL}(S_{n,m})$ is being defined as a minimal non-linearity on the non-linearity multitude of co-ordinate Boolean functions and of all their linear combinations:

$$\mathrm{NL}(S_{n,m}) = \min_{g}\{NL(g_j)\} \quad (j = 1, \ldots, 2^m - 1).$$

The selection criterion is a non-linearity NL maximization of an S-block.

Let us examine an S-block of dimensionality $n \times n$. In case of an odd n the S-block is named maximum-nonlinear, if the Walsh-Hadamard spectrum of all its co-ordinate functions and of all their linear combinations has only three values $\left\{0, \pm 2^{\frac{n+1}{2}}\right\}$. If n is even, so the maximum attainable nonlinearity for S is equal $2^{n-1} - 2^{\frac{n}{2}}$.

The function f possesses a correlating immunity of the order k, if the initial sequence of the function $y \in Y$ does not depend statistically of any sub-multitude from k input co-ordinates:

$$\forall\{x_1, \ldots, x_k\} P(y \in Y/\{x_1, \ldots, x_k\} \in \mathrm{X}) = P(y \in Y).$$

The equivalent definition of the correlative immunity in the terms of the Walsh transform: the function f over the field $GF(_2{}^n)$ possesses the correlative immunity (CI) of the order k, CI(k), if its Walsh transform complies with the equation $F(\omega) = 0$ for every $\omega \in V_n$ such that $1 \leq W(\omega) \leq k$.

$$\forall \omega \in Vn,\ F(\omega) = 0,\ CI(f) = k.$$

The correlative-immune function of the k-th order—is a function which possesses a correlative immunity of the order k. The balanced correlative-immune functions are being named elastic functions.

The function f over the field $GF(2^n)$ complies with:

- The distribution criterion with relation to the vector α, DC(α), if the function $f(\mathrm{x}) \oplus f(\mathrm{x} \oplus \alpha)$ is balanced, $x \in V_n$, where $x = (x_1, x_2, \ldots, x_n)$:

$$P(f(\mathrm{x}) = f(\mathrm{x} \oplus \alpha)) = \frac{1}{2};$$

- The distribution criterion of the degree k DC(k), if the distribution criterion related to all vectors $\alpha \in V_n$ at $1 \le W(\omega) \le k$ is complied.

$$P(f(\mathrm{x}) = f(\mathrm{x} \oplus \alpha)) = \frac{1}{2} \text{ and } \forall\alpha : 1 \le W(\alpha) \le k;$$

- The strict avalanche criterion (SAC), if f complies with the distribution criterion of the degree $_1$:

$$P(f(\mathrm{x}) = f(\mathrm{x} \oplus \alpha)) = \frac{1}{2}\forall\alpha : W(\alpha) = 1$$

The function f above the field $GF(2^n)$ is called a bent-function, if

$$2^{-n/2} \sum_{x \in V_n} (-1)^{f(x) \oplus \langle \beta, x \rangle} = \pm 1$$

for every $\beta \in V_n$ where $\langle \beta, x \rangle$—being a scalar product of the vectors $\beta = (\beta_1, \beta_2, \ldots, \beta_n)$ and $x = (x_1, x_2, \ldots, x_n)$, i. e. $\langle \beta, x \rangle = \beta_1 x_1 \oplus \ldots \oplus \beta_n x_n$.

The sequence of a bent-function is being called *bent-sequence.* For the bent-functions are valid such affirmations:

- $\langle \varepsilon, l \rangle = \pm 2^{\frac{n}{2}}$ for any affine sequence of l length $_2{}^n$;
- $f(x) \oplus f(x \oplus \alpha)$ balanced $\forall\alpha \in Vn$, $W(\alpha) \neq 0$;
- $f(x) \oplus \langle \alpha, x \rangle$ takes on the value of one $2^{n-1} \pm 2^{\frac{n}{2}-1}$ times $\forall\alpha \in Vn$;
- $f(x) \oplus h(x)$, where $h(x)$—the affine function, is also a bent-function.

The autocorrelative function $\hat{r}(s)$ for $s \in 0 \ldots 2^n - 1$ is being defined as

$$\hat{r}(s) = \sum_{\mathrm{x}=0}^{2^n-1} \hat{f}(x)\hat{f}(x \oplus s)$$

It is spoken that the function f complies with the circulation characteristic m, if

$$(1 \le |s| \le m) \Rightarrow |\hat{r}(s)| = 0.$$

Analogically, the autocorrelation $AC\ (f)$ of the function f is being defined as the modulus of the greatest value of $\hat{r}(s)$:

$$\mathrm{AC}(f) = \max_{s \neq 0} \left| \sum_x \hat{f}(x) \hat{f}(x \oplus s) \right| = \max_{s \neq 0} \left| \hat{r}(s) \right|$$

The selection criterion is an autocorrelation AC (f) minimization.

The bent-function achieves the zero-autocorrelation (and more than that, the maximum non-linearity).

The autocorrelation $\hat{r}(s)$ ensures a leakage of information stream from input to output of the function. In some cases, the very strong interconnection can be presented as a linear structure (for which are valid the equations $f(x) \oplus f(x \oplus s) = 1$ або $f(x) \oplus f(x \oplus s) = 0$). The cryptoanalytical significance of this fact (for a block cipher) is demonstrated in. It is usually avoided. In the fundamental work by Meyer and Staffelbach [7] the minimum distance of the function for any function with a linear structure is defined as the nonlinearity dimension. The autocorrelation AC(f) has been proposed as global avalanche characteristic, inasmuch as it includes all autocorrelations, not only corresponding with s with all partial (low) Hamming weights.

The concept of avalanche effect (avalanche) is being expressed as a certain dependence of the output on the input. The autocorrelation, the index of sum of squares, and also the distribution criterion are dimensions of the function avalanche characteristics.

Let us assign as $d_\alpha f(x)$ the derivative of the Boolean function $f(x)$ in direction $\alpha, \alpha \in V_2^n$. Then

$$d_\alpha f(x) = f(x) \oplus f(x \oplus \alpha) = \hat{f}(x) \hat{f}(x \oplus \alpha)$$

The derivative of the Boolean function gives a dimension of effect (influence) on the function output in case of the function input change in direction α. Otherwise speaking, the derivative of the Boolean function is the output function difference at the preset input difference.

The avalanche state of the Boolean functions can be expressed in the same terms as the global characteristic, through the sum of squares index σ, which is being calculated as follows:

$$\sigma = \sum\nolimits_\alpha \hat{r}^2(\alpha),$$

where $\alpha \in \{1, \ldots, 2^n - 1\}$.

During the estimation of the S-blocks avalanche properties the autocorrelation AC is being defined.

The autocorrelation of an S-block, $\mathrm{AC}(S_{n,m})$, is defined as follows:

$$\mathrm{AC}\left(S_{n,m}\right) = \underset{g}{max}\left|\widehat{r_{g_j}}(\alpha)\right|$$

where $\alpha \in \{1, \ldots, 2^n - 1\}$ and $j = 1, \ldots, 2^m - 1$.

The selection criterion is an autocorrelation AC minimization.

It is spoken that the Boolean function $f(x)$ from n variables complies with a strict avalanche criterion for each s, such as $\mathrm{d_H(s)} = 1$, valid is the expression $\sum_x f(x) \oplus \mathrm{f(x \oplus s)} = 2^{n-1}$. It is affirmed, analogically, that the Boolean function $f(x)$ complies with the strict avalanche criterion (SAC), then and only then when the autocorrelation function $\hat{r}_f(s)$ includes zero-values in all positions s, at which $\mathrm{d_H(s)} = 1$.

If the Boolean function complies with SAC, then its derivative function for all inputs and on all values of the input shift s with the Hamming weight 1 is balanced.

The algebraic degree of each variable x_i, $\deg(f, x_i)$, is being defined by the degree of the longest term of the function, which contains x_i

$$deg(f, x_i) = \underset{0 \le i \le n}{max}\, deg\left(term_{x_i}\right)$$

The function terms number, $term(f)$, is being defined by the terms number, of which the function consists, being presented in an algebraic normal form:

$$term(f) = \underset{0 \le o < 2^n}{min}\ term_j$$

The function terms number, which contain the variable x_i, $term_{x_i}(f)$ is being defined as

$$term_{x_i}(f) = \underset{j}{min}\, term_j(x_i)$$

According to [8–10], the introduced coefficient of uniform minimization of the cross-correlation:

$$k_{um} = \frac{k_{cc}}{r * c_{cc}}$$

where

k_{cc}— the limit value of cross-correlation;
r— the specific weight of non-zero values of cross-correlation coefficients;
c_{cc}— the average value of cross-correlation coefficients.

The presented values look like as follows:

$$r = \left|\frac{B}{2^n} - 2^{n-1}\right|$$

$$c_{cc} = \frac{\sum_{i=1}^{2^n} c_i}{2^n}$$

$$k_{cc=}\frac{|1-2^n| * (c_n + c_{n+2})}{2} \text{ for } V_{n+1}$$

$$\text{and } k_{cc=}\left|1-2^n\right| * c_n \text{ for } V_n$$

where

B— the general amount of non-zero values of cross-correlation coefficients;
c_i— the value of cross-correlation coefficient of the function with the multitude of all affine functions L_w;
c_n— the correlation coefficient of the bent-function over V_n with the multitude of all affine function over V_n;
c_{n+2}— the correlation coefficient of the bent-function over V_{n+2} with the multitude of all affine function over V_{n+2}.

The absolute value of the correlation function, C_f, is being defined as the maximum value of all correlation coefficients

$$Cf = \max|c(f, l_i)|$$

and determines the maximum value of correlation coefficients of the function f over V_n, $l_i \in l$, where l_i, L—being a linear function and the multitude of all linear functions respectively.

The amount of vectors num_1 and num_2, at which the function doesn't comply/complies with the distribution criterion, is being determined as

$$num_1 = nPC(k),$$

$$num_2 = PC(k).$$

The amount of vectors num_3 and num_4, at which the function doesn't comply/complies with the correlative immunity is being defined as

$$num_3 = nCI(k),$$

$$num_4 = CI(k).$$

The distribution criterion is a generalization of the strict avalanche criterion.

It is spoken that the Boolean function from n variables complies with the *distribution criterion (DC, PC, PropagationCriteria)* of the degree *k*, DC(k), related to the non-zero vector $\alpha \in V_2^n$, if

$$\sum_{x} f(x) \oplus f(x \oplus \alpha) = 2^{N-1}$$

$\forall\alpha$ such that $1 \leq d_H(\alpha) \leq k$.

Consequently, it is spoken that the Boolean function complies with DC(k), if the complementation t of input bits ($1 \leq t \leq k$) causes the change of each input with the probability ½. The distribution criterion of the degree 1, DC($_1$), is equivalent with the strict avalanche criterion.

It is spoken that S complies with the SAC, if all $g_j (j = 1, \ldots, 2^m - 1)$ comply with SAC. Analogically, an S-block complies with the distribution criterion of the degree k, DC(k), if all $g_j (j = 1, \ldots, 2^m - 1)$ comply with DC(k).

The selection criterion is a maximization of the degree of the distributing of DC.

If exists an affine transformation, which connects two Boolean functions, is spoken that these functions are affine equivalent and stay in one equivalent class.

Two Boolean functions $f(x)$ and $g(x)$ from n variables are considered equivalent, if exists a non-degenerate binary matrix D, two binary vectors a and b, which contain n elements, and a binary constant c, such as:

$$g(x) = f(Dx \oplus a) \oplus b * x \oplus c,$$

where

$$b * x = b_1 x_1 \oplus b_2 x_2 \oplus \ldots \oplus b_{1n} x_n$$

is a linear function from x.

If the vectors b and c—are null ones, so the functions are being called *affine equivalent*.

It is spoken that the S-block doesn't have any linear redundancy, if all the functions g_j aren't affine equivalent, otherwise it is spoken that S is linear redundant.

Let be f—a function at V_n. The vector $\alpha \in V_2$ being called the linear structure f, if $f(x) \oplus f(x \oplus \alpha)$—is constant. It is spoken that an S-block doesn't contain any *linear structures*, if all its component functions and all their linear combinations doesn't contain any linear structures.

Let be F—an S-block $n \times s$, where $n \geq s$. Let be Δ—the maximum value in the difference table (differential table) of an S-block (without consideration of the first element of the first line), and namely:

$$\Delta = \max_{\alpha \in V_n} \max_{\beta \in V_s} |x| F(x) \oplus F(x \oplus \alpha)|$$

Then is spoken that F is Δ uniform.

The selection criterion is a minimization of the Δ– uniformity. The minimum possible value of the Δ–uniformity is equal 2.

For an S-block $n \times m$ the correlation between its autocorrelation AC and the Δ–uniformity is being written as follows:

$$\Delta \geq 2^{n-m} + 2^{-m} AC.$$

It is spoken that an S-block contains some fixed points or some opposite fixed points, if $S(X) = X$ or $S(X) = \overline{X}$, respectively, where the vector $X \in V_n$.

For an S-block $n \times m$ the values of the *linear approximations table (LAT)* is being defined in such a way:

$$LAT(\alpha, \beta) = \#\{X \in Z_2^n | \alpha * X = \beta * S(X)\} - 2^{n-1}$$

where $\in Z_2^n$, $\beta \in Z_2^m \backslash \{0\}$.

For an S-block $n \times m$ the correlation between its nonlinearity *NL* and the maximum of the linear approximations table is being written as follows:

$$NL = 2^{n-1} \max_{\alpha,\beta \neq 0} |LAT|$$

The selection criterion is a minimization of *NL*.

2.1.4 General Stability Indices

Period of control sequence. As period of sequence T is being named the amount of bits until that moment, when the sequence will be repeated. The period—is one of the most important characteristics for studying properties of the control sequence. If the period of the control sequence will occur too much short, so different parts of the clear text will appear transformed identically, what presents a redoutable weakness of the scheme. When knowing a fragment of the clear text, an analyst will be able to reconstruct the respective fragment of the control sequence. This fact, that the discovered interval of the control sequence appears in the multitude of other places of the control sequence, enables the analyst to discover successfully other closed texts. Besides, if there are only transformed texts, closed by the same control sequence, so they can be added pairwise for getting a sequence, which is tantamount with the sum of open messages and doesn't depend on the control sequence. Having the necessary statistics of the open text, the open text of messa‘ges can be reconstructed, and the control sequence itself as well.

In as much as the binary LFSR with a length of n bits can stay in one of the $2^n - 1$ internal states, so theoretically it can have a period $2^n - 1$.

The question, how large must be the period of the control sequence, is debatable in the open literature, but the resolution depends always on the specialized using area. For the systems of stream information conversion (SISC), which are working with a velocity of 1 MB/s, the sequence with the period of 2^{32} will repeat itself after

2^9 seconds or 8,5 min, for this reason the means of information closing with such a period can't be considered as adequate protection.

Traditionally, for stream cryptoconversions the two stability levels are distinguished:

- «high», when the scheme must have a key length $l_{key} \geq 256$ bits and an internal memory $M_h \geq 256$ bits;
- «normal», when the scheme must have a key length $l_{key} \geq 128$ bits and an internal memory $M_n \geq 128$ bits;

Consequently, the registers length for ensuring the necessary stability must consist minimally of 256 and 128 bits respectively. Thus, we can write down that the periods length T_h for schemes with a «high» stability level, and T_n for schemes with a «normal» stability level must be

$$T_h \geq 2^{256} \text{ bits;}$$

$$T_n \geq 2^{128} \text{ bits.}$$

A scrupulous estimation of the control sequence, formed by the scheme, is indispensable when elaborating of any SISC. From a practical point of view, the control sequence must be long enough in order that in the further majority of cases will be hardly probable a repeated using of the same fragment during closing of information.

The selection criterion is $T_f \max_{T_j}\{T_1, \ldots, T_r\}$, where T_j—the period of the j-th scheme; $j = 1, \ldots,$ r.

Linear sequence complexity. The linear complexity $\Lambda\left(s^l\right)$ of the sequence $s^l = s_0, s_1, \ldots, s_l$—the length L of the shortest shift register, which generates the preset periodical sequence s^l, when the first L digits of the sequence s^l is an initial completing of the register.

The given criterion is a basic criterion. Any sequence, which can be generated by a finite automaton (linear or non-linear) over an finite field, has a finite linear complexity. Consequently, a building of algorithms is possible, which determine the linear complexity of any sequence, independent of the generation method. So, there exists an effective Berlekamp—Massey algorithm [11, 12], which finds quickly the shortest register after studying of only $2L$ bits of the control sequence, generated by an LFSR.

The linear complexity in its essence is a complexity dimension of the generated sequence. When examining a certain interval of the sequence, the analyst tries on the base of existing sequence to build his own sequence. Determination of the sequence linear complexity enables on the base of the found linear equivalent to construct a scheme that would generate a sequence analogically to examined one, herewith the knowledge about the scheme structure, which generates the given sequence, is superfluously. As a result, the greatest linear complexity of the control sequence is a necessary, but not sufficient condition for a practical stability of SISC.

The selection criterion is $\Lambda_f \max_{\Lambda_j}\{\Lambda_1, \ldots, \Lambda_r\}$, where Λ_j—is the linear complexity of the j-th scheme; $j = 1, \ldots, r$.

Length of the used keys. The length of the keys is determined by the resistance against forceful attacks and gives $k_h \geq 256$ bits for a scheme with a «high» stability level and $k_n \geq 128$ bits for schemes with a «normal» stability level.

The selection criterion is $k_f \max_{k_j}\{k_1, \ldots, k_r\}$, where k_j being the key length of j-th scheme with a preset stability level; $j = 1, \ldots, r$.

Internal state of the generator (internal memory) M has to be not less as 2 k bits at the key length of k bits. This index reflects the stability against attacks «time–memory».

The selection criterion is a logical swapping Yes/No.

Length of generated sequence $l_n$may not exceed a certain threshold value $N_{\max}$, $l_n < N_{\max}$. This index reflects the stability against correlative attacks, which enables after a known fragment of the control sequence to reconstruct the initial completing of the LFSR. The aim of introducing of $N_{\max}$ is non-admission of generation of any sequence, the length of which would guarantee using correlation of a sequence with the LFSR-sequence.

The selection criterion is a logical swapping Yes/No.

Index, which characterizes the scheme stability against the analytical deciphering methods. Generally, namely the stability of the scheme against the analysis methods determines the efficiency of the SISC-functioning.

The selection criterion is a logical swapping Yes/No.

Index, which characterizes the scheme stability against the statistical methods of analysis. The statistical property of a control sequence is one of the components, which define the stability of a conversion scheme. The scheme stability depends on how much close it approximates the generator of random numbers, i. e. on how much the control sequence will be calculative unpredictable and won't differ from a really random sequence. A universally recognized estimation methodology of statistical properties of cryptoconversions as for today is the methodology proposed by *NIST* [13]. In this methodology after a choice is built-up a statistical portrait of the source, which contains a probability value $P_{ij} \in [0, 1]$—the probability values obtained as a testing result of an i-th sequence with a j-th test, $i = 1, \ldots, m$ and $j = 1, \ldots, q$. The recommended values: $m = 100$ sequences of 10^6 bits (the general body makes up 10^8 bits $q = 188$. Further away, when analyzing the statistical portrait, the following criteria are used. As the first selection criterion is considered the coefficient of passing through of tests by the sequences:

$$r_j = \frac{\#\{P_{ij} \geq \alpha | i = 1, 2, \ldots, m|\}}{m}$$

where the probabilities are taken, which get into the interval of credibility $[r_{\max}, r_{\min}]$,

$$r_{max(min)} = \hat{p} \pm \sqrt{\frac{\hat{p}(1-p)}{m}}$$

where $\hat{p} = 1 - \alpha$, α—the significance level, which is equal to 0,01.

The second criterion is based on the calculative value χ^2:

$$\chi_j^2 = \sum_{k=0}^{10} \frac{(F_k - m/10)^2}{m/10}.$$

Commonly is considered that the scheme has passed the proof, if the coefficients values r_j for all $j = 1, \ldots, q$ stay within the interval of credibility $[r_{max}, r_{min}]$ and the condition $\chi_j^2 > 0,0001$ for all $j = 1, \ldots, q$ is fulfilled.

2.2 Software and Hardware-Realization Indices

On the base of performed researches for (SISC) let us to determine the following multitude realization effectiveness.

Velocity of key initiation, cycles/bytes, —reflects the velocity of scheme initiation execution—amount of downloaded bytes n_{bd} of the key during one full shifting cycle of the register.

The selection criterion is $n_{bd} = \max_{n_j}\{n_1, \ldots, n_r\}$, where n_j—the amount of downloaded bytes n_{63} of the key during one full shifting cycle of the register of the *j*-th scheme; $j = 1, \ldots, r$.

Velocity of generating of the output sequence, cycles/bytes—reflects the velocity of scheme initiation executing—amount of downloaded bytes m_{bd} of the key during one full shifting cycle of the register.

The selection criterion is $m_{bd} = \max_{m_j}\{m_1, \ldots, m_r\}$, where m_j—amount of generated bytes of an initial sequence in a full cycle of the shift register of the *j*-th scheme; $j = 1, \ldots, r$.

Memory size in use, Kb. Reflects the memory size which is used RAM and ROM. (RAM—Random Access Memory; ROM—read-only memory).

The selection criterion is $\nu_{mem} = \max_{\nu_j}\{\nu_1, \ldots, \nu_r\}$, where ν_j—the memory size used in the *j*-th scheme; $j = 1, \ldots, r$.

Velocity of information enciphering/deciphering, Mbytes/s.

The selection criterion is $s = \max_{s_j}\{s_1, \ldots, s_r\}$, where s_j—the velocity of information *enciphering/deciphering* of the *j*-th scheme; $j = 1, \ldots, r$.

Quantity of various arithmetical operations being used.

The selection criterion is $op = \max_{op_j}\{op_1, \ldots, op_r\}$, where the quantity of various arithmetical operations being used in the *j*-th scheme; $j = 1, \ldots, r$.

Index of translation into other platforms.

The selection criterion is a logical swapping Yes/No.

For hardware realization can be considered as additional indices: the product dimensions, the product costs, the connectivity with other apparatus-types, the realization possibility with microcircuits of different types.

2.3 Constructive-Technological Indices

Let us define the following multitude of constructive-technological indices of the realization effectiveness:

Construction transparency. The selection criterion is a logical swapping Yes/No.

Possibility of performing of a comparative analysis. The selection criterion is a logical swapping Yes/No.

Perspectivity. The selection criterion is a logical swapping Yes/No.

Stability margin. The selection criterion is a logical swapping Yes/No.

As for today, for estimation of constructive-technological indices the methods of expert opinions can be used.

2.4 Generalization of Indices and Criteria of Stream Cryptoconversion

The examined indices and criteria of SSC and of stream BSC-modes criteria enable, taking into consideration concrete conditions of functioning of information protection systems and information-telecommunication systems to proceed on the whole an integral estimation of the stream cryptoconversions efficiency.

The presented indices characterize sufficiently complete a concrete SSC-scheme and/or a tstream BSC-mode. The introduced efficiency criteria as a whole with estimation methodologies enable to carry out a comparison of various schemes of stream cipher. The examined multitude of indices and criteria, as a whole, will enable to build up an effective instrumentarium for estimating schemes of stream cryptoconversions. In the Table 2.1 is given a generalized system of indices and criteria for effectiveness estimation.

As the table shows, the vector of efficiency indices contains rather many particular indices. As for today, the questions of mutual influence of some indices remains not researched up to the end. The research of complicated dependences and elaborating of the building methods of optimal schemes for stream cryptoconversion are, at present time, actual tasks.

An other interesting direction of the research is elaborating of complex indices, which integrates several particular indices and characterizes at once several different properties of schemes. In this case, in event of ensuring of the target level of quality

Table 2.1 Indices and criteria for effectiveness estimation of stream cryptoconversion

Effectiveness indices of stream cryptoconversion functioning	Criteria for effectiveness estimation of stream cryptoconversion functioning
Stability indices	
Indices characterizing the parameters of register and of take-off points for a non-linear function and for feedbacks	
Primitiveness of generating polynomial	Yes/No
Reciprocally simple degrees of the generating polynomial	Yes/No
Degree of the generating polynomial, n	$n \geq 128$ bits
Density of the generating polynomial, k	$k \to n/2$
Rectitude of detecting of take-off points for a non-linear funcrtion	Yes/No
Accordance of the multitude of feedbacks B with the full multitude of positive differences ΔB	Yes/No
Accordance of the multitude of take-off points Γ with the full multitude of positive differences ΔΓ	Yes/No
The greatest common divisor of two even (neighboring) positive differences has to be equal 1	Yes/No
Indices of non-linear functions stability	
Balancedness	Yes/No
Non-linearity, N_f	$N_f = \max_{N_j}\{N_1, \ldots, N_r\}$
Algebraic degree, $deg(f)$	$deg(f) = \max_{deg(f_j)}\{deg(f_1), \ldots, deg(f_r)\}$
Coefficient of the uniform minimization of correlation, kum	$K_{um} = \min_{k_{um\ j}}\{k_{um\ 1}, \ldots, k_{um\ r}\}$
Absolute correlation value of function, C_f	$C_f = \min_{l_i}\left\{C_{f\ 1}, \ldots, C_{f\ r}\right\}$
Quantity of vectors, at which the function doesn't comply with the circulation criterion, num_1	$num_1 = \min_{num_{1\ j}}\{num_{1\ 1}, \ldots, num_{1\ r}\}$
Quantity of vectors, at which the function has a linear structure, num_2	$num_2 = \min_{num_{2\ j}}\{num_{2\ 1}, \ldots, num_{2\ r}\}$
Indices of stability of the key initiation procedure	
Non-linearity of key boot operations	Yes/No
Each bit of the initialized register is a result of non-linear conversions of all key bits	Yes/No
General indices of stability	
Period of current key, T_f	$T_f = \max_{T_j}\{T_1, \ldots, T_r\}$
Sequence complexity, Λ_f	$\Lambda_f = \max_{\Lambda_j}\{\Lambda_1, \ldots, \Lambda_r\}$
Length of used keys, k_f	$k_f = \max_{k_j}\{k_1, \ldots, k_r\}$

(continued)

Table 2.1 (continued)

Effectiveness indices of stream cryptoconversion functioning	Criteria for effectiveness estimation of stream cryptoconversion functioning
Internal generator state (internal memory) M	
has to be not less than 2 k bits at the key length of k bits	Yes/No
The length of the generated sequence, l_n, may not exceed a certain threshold value $N_{\max}$	Yes/No
Index characterizing the scheme stability against the analytical deciphering methods	Yes/No
Indices characterizing the scheme stability against the statistical analysis methods, r_j and χ_j^2ara>	$[r_{\max} \leq rj \leq r_{\min}]$ $\chi_j^2 > 0,0001$
Software and hardware-realization indices	
Key initiation,n_{bd}	$n_{bd} = \max_{n_j}\{n_1, \ldots, n_r\}$
Generation of output sequence,m_{bd}	$m_{bd} = \max_{m_j}\{m_1, \ldots, m_r\}$
Size of used memory, v_{mem}	$\nu_{mem} = \max_{\nu_j}\{\nu_1, \ldots, \nu_r\}$
Velocity of information enciphering/deciphering, s	$s = \max_{s_j}\{s_1, \ldots, s_r\}$
Quantity of various used arithmetical operations, op	$\mathrm{op} = \max_{\mathrm{op}_j}\{\mathrm{op}_1, \ldots, \mathrm{op}_r\}$
Transfer on other platforms	Yes/No
Constructional-technological indices	
Construction transparency	Yes/No
Possibility of executing of comparative analysis	Yes/No
Perspectivity	Yes/No
Stability margin	Yes/No

estimation (accuracy, credibility and so on), the indices amount is being reduced and the methodology estimation of schemes efficiency is being simplified.

Finally, actual remains the task of elaborating of methodology of the effectiveness estimation based on using of system analysis methods. In the area of estimation of cryptographic schemes was dominating quite a long time the parametrical approach to the effectiveness estimation on the base of stability indices. Evidently, such an approach actually has exhausted itself already. A more perspective is using of methods of defining of systemic indices, a common usage of calculation methods of quantitative indices and defining of qualitative (for instance, using of decomposing methods with following expert evaluations). The researches in this direction will enable to develop effective methods of estimation of stream cryptoconversion schemes.

2.5 Conclusions and Recommendations

1. Among symmetric cryptographic transformations, a special place is occupied by SSCs and stream modes of application of BSCs, in which information is submitted and processed in the form of an infinite stream, that is, a sequence that can hypothetically be of infinite length. The main advantage of this conversion is the dependency between individual symbols of the data stream, which provides additional protection against the imposition of false information, or false modes of protection equipment or terminal equipment of telecommunication systems and networks. Consequently, cryptographic streaming is typically more trusted with users because protected data streams cannot be distorted in any way due to the intentional or unintentional action of users and attackers, or any accidental natural factors or factors.
2. The performance of cryptographic data conversion schemes should be evaluated from the point of view of their resistance to different methods of disclosure, as well as taking into account the imposed restrictions on the speed of cryptoconversions and application features. Among the unconditional advantages of streaming crypto algorithms are the increased rates of cryptographic stability and speed with the ability to parallelize certain processes of streaming cryptoconversions. In particular, information flow encryption typically consists of adding to the information data a range of encryption—PRS with certain cryptographic properties. The decryption on the receiving side is to subtract the same keystream from the obtained data, that is, if the pseudo-random sequence (keystream) is formed in advance, then the encryption / decryption can be performed in parallel, using multiple computing systems. Accordingly, the speed of such stream cryptoconversions is significantly increasing compared to traditional (block) methods.
3. During the research, the indicators and criteria for the effectiveness of SSC and stream modes of BSC were substantiated, which are defined by two main groups: unconditional criteria and indicators characterizing the critical properties of the system, failure of which or lowering below the permissible limits will lead to failure to perform the basic functions of the system and stability (security) criteria that characterize vulnerability to attack); conditional indicators and criteria characterizing the desirable properties, the performance of which will improve individual performance indicators, and failure to comply or lower below the admissible limits will not lead to disruption of the functioning of the system (software and hardware and performance criteria and indicators, structural technology and others).
4. The considered indicators and criteria for the efficiency of SE and stream modes of BSC allow taking into account the specific conditions of functioning of the information security system and information and telecommunication system as a whole to carry out a comprehensive assessment of the efficiency of stream cryptoconversions. The above indicators characterize the specific circuit of the SE and / or the flow mode of the BSC quite fully. The above performance

criteria, together with the evaluation techniques, make it possible to compare different stream encryption schemes. In general, the considered set of indicators and performance criteria allows you to build an effective toolkit for evaluating stream cryptoconversion.

References

1. Mao, W.: Modern cryptography. Theory and Practice / Per. with English - M .: Izd. Williams House, p. 768 (2005)
2. Schneier, B.: Applied cryptography. Protocols, algorithms, source texts in SI language. M.: Triumf, p. 797 (2002)
3. Menezes, A.J., van Oorschot, P.C., Scott, A.: Vanstone. Handbook of Applied Cryptography – CRC Press, p. 794 (1997)
4. AlShehri, K.N.: Encryption Primitives and their Application to Stream Ciphers Design: Master's Degree Thesis / K.N. AlShehri. King Saud University, p. 121 (2007)
5. Encyclopedia of cryptography and security. [Editorinchief Henk C.A. van Tilborg]. Springer, p. 697 (2005)
6. Feistel, H.: Cryptography and computer privacy. Sci. Am. **228**(5), 15–23 (1973)
7. Maier, W.: Nonlinearity criteria for cryptographic functions. In: Maier, W., Staffelbach, O. (eds.) Advances in Cryptology, EUROCRYPT'89, Lecture Notes in Computer Science, Vol. 434, pp. 549–562, Springer, 1990
8. Orlova, S.Y.: Methods for assessing the effectiveness of current ciphers. Legal, Regulatory and Metrological Support of Information Security in Ukraine: scientific and technical collection. Vip. 9, pp. 141–152 (2004)
9. Potiy, A.V., Izbenko, Y.A.: System of indicators for assessing the effectiveness of the operation of streaming encryption schemes. Radio Engineering: All-Ukrainian interdepartmental scientific and technical collection. 123, pp. 146–158 (2003)
10. Orlova, S.Y.: Scientific and methodological bases of evaluation of streaming encryption systems with enhanced requirements for durability and speed: Dis ... Cand. tech. Sciences: 05.13.21. Kharkiv National Institute of Radio Electronics. H., p. 152 (2005)
11. Berlekamp, E.R.: Algebraic Coding Theory. McGraw Hill, NY, p. 474 (1968)
12. McWilliams, F.J.: The Theory of ErrorCorrecting Codes, McWilliams, F.J., Sloane, N.J. (eds.), p. 762, NorthHolland, 1978
13. NIST Special Publication 80022. A Statistical Test Suite for Random and Pseudorandom Number Generators for Cryptographic Applications. [Electronic resourse]. Access mode: http://csrc.nist.gov/publications/nistpubs/80022rev1a/SP80022rev1a.pdf

Chapter 3
Analysis of Stream Cryptoconversion Principles

3.1 Basic Approaches to Construction and Analysis of SSC and BSC Stream Modes

A detailed analysis of modern methods of SSC construction was carried out in one of the most authoritative scientific works of the Swiss cryptographer R. Ruppel [1]. It outlines four approaches to constructing streaming cryptoconversion schemes, namely:

- Theoretical and informational approach, which uses general theoretical propositions, mathematical models and assumptions on information theory and secret systems theory, developed by world-renowned American scientists K. Shannon;
- Theoretical system approach, applying the methods of systems theory and system analysis, in particular to determine the criteria and indicators of SSC efficiency;
- Theoretical complexity approach based on the use of complexity theory methods, in particular the problem of deciphering an attacker, is associated with the solution of some extremely complex mathematical problem;
- Theoretical and stochastic approach aimed at the construction of so-called randomized streaming cryptoconversions.

In the case of the theoretical and informational approach, certain assumptions are made from K. Shannon's theory of secret systems, in particular, that the cryptanalyst (the attacker) has unlimited time and computing resources. Its purpose is to conduct cryptanalysis, which is to determine the message (or specific key) in the presence of only cryptograms and a priori probabilities for different keys and messages. System stability is considered reduced (and the system is considered compromised, respectively) if there is a "unique" solution for cryptograms: one message with probability is almost one (this event is practically reliable), for all other messages the probability is almost zero (these events are almost impossible). The assumption of an infinite computational and temporal capacity of the enemy means that the position of stability in the Shannon model does not depend on the complexity of the encryption

A. A. Kuznetsov et al., *Stream Ciphers in Modern Real-time IT Systems*,
Studies in Systems, Decision and Control 375,
https://doi.org/10.1007/978-3-030-79770-6_3

or decryption method. The encryption system is considered to be absolutely stable if the open and closed message texts are statistically independent. In other words, a cryptanalyst is not approaching success in cryptanalysis even after he has intercepted a cryptogram. It is considered that the encryption system is ideally stable if the cryptanalyst cannot find a single solution for plain text, no matter how much he can get for analyzing the ciphertext.

The purpose of the **theoretical system approach** is to create confidence that each new cryptosystem creates a complex and previously unknown problem for a cryptanalyst. The dissection of such a system is a more complicated problem than cryptanalysis of systems based on any known problem such as factorization or discrete logarithm. The purpose of the developer is to show that none of the fundamental cryptoanalytic principles (eg, "divide and reveal", statistical analysis, etc.) are unsuitable for the new system. In order to hinder cryptanalysis based on these fundamental principles, a set of common design criteria for encryption keystream generators has been developed. Examples of such criteria are period, linear complexity, statistical criteria, nonlinearity criteria for Boolean functions. Since these criteria are inherently theoretical systems, it is accordingly considered to be an approach to systems theory. The streaming cryptoconversion system is designed to meet directly the set of design criteria chosen. To date, this approach is most common for the practical development of SSC.

In the theoretical system approach, the main goals of SSC developers are:

1. Development of such methods and crypto-primitives (building blocks) that have demonstrable properties with respect to certain theoretical systems indicators, such as, for example, sequence period, linear complexity, etc.
2. The study of cryptanalytic principles and the development of constructive solutions that would prevent attacks based on these principles. Such fundamental cryptanalytic principles are:
   - replacement and approximation, preferably linear components;
   - split-and-reveal correlation attacks;
   - the use of statistical weaknesses.

To prevent cryptanalytic attacks based on these fundamental principles, a set of common design criteria for encryption keystream generators has been developed and updated:

- long period;
- sequence complexity criteria: linear complexity, quadratic span, maximum order complexity, etc.;
- satisfactory statistical properties;
- mixing and scattering;
- criteria for nonlinearity of transformations.

Each developed SSC scheme must satisfy a specific set of design criteria. In the case of a theoretical system approach, ciphers are developed to directly meet these criteria. This approach is by far the most common approach to practical sustainability.

However, it should be noted that the main problem is that compliance with these criteria does not guarantee the cryptographic stability of SSC—their application is necessary, but is not a sufficient condition for the stability of the schemes to cryptanalysis methods.

In the case of **a theoretical complexity approach**, all calculations are parameterized with a stability parameter (eg key length) and an asymptotic analysis is performed. Only those algorithms whose operating time can be expressed as a polynomial of the input size are considered computationally acceptable. Cryptanalysis in this context is:

– or the process of predicting a single digit of the encryption range;
– or the process of distinguishing a sequence of encryption keystream from a truly random sequence.

The purpose of the developer is to build their encryption system based on some problem, the solution of which is computationally unattainable (or equivalent). The encryption generator is defined as perfect if it is either unpredictable or indistinguishable by all polynomial time statistical tests. However, a perfect generator is a hypothetical device, so far, nothing is known about the existence of such a device. All the proposed generators are built on the expected complexity of one of several known problems such as discrete logarithm, quadratic deduction, inversion of crypto-algorithm Rivest–Shamir–Adleman (RSA) and the like.

The theoretical and stochastic approach is related to the construction of so-called randomized SSC. Instead of ensuring that the cryptanalytic process will require unreachable labor costs, the developer may try to prove that the cryptanalytic task is of unimaginable size. The essence of this approach can be reduced to the number of bits that cryptanalyst needs to analyze during cryptanalysis, while the secret key is small. This can be done with the help of large random bits used for encryption and decryption for all plaintext. The key here specifies which parts of the large randomizer to use, while the opponent, who does not know the secret key, is forced to comb through the entire array of random data. The robustness of a randomized streaming encryption system can be evaluated (below) by the average number of bits that a cryptanalyst should analyze before its chances of identifying a key or plaintext are improved over a purely random assumption. Different interpretations of this result are possible. The expected number of bit tests is the lower limit for the number of steps that anyone who performs an algorithm cryptanalysis has to perform, and this leads to the idea of computational stability. But the expected number of bit tests is also the lower limit for the number of bits an opponent must see before his a posteriori probabilities for different keys improve, and thus leads to the idea of information-theoretic stability.

3.2 Formalization of Stream Cryptoconversions by Their General Classification

The theory of automatons is usually used to describe the operation of SSC or streaming modes of BSC.

Let X denote the plaintext alphabet, Y denote the ciphertext alphabet, Z denote the ciphertext alphabet, S denote the space of states (internal states) of the stream cipher, K denote the key space. Let x_i, y_i, z_i and s_i denote the characters of plaintext, ciphertext, keystream, and internal state at time and respectively. The key $k \in K$ is chosen according to the probability distribution P_k. Typically, the key is selected according to even distribution, but in some cases it may not be possible to select the key at random.

In general, SSC can be described by the equations:

$$s_{i+1} = F(k, s_i, x_i),$$

$$y_i = f(k, s_i, x_i),$$

where F is the function of the next state and f is the output function.

Usually

$$y_i = x_i + F(k_i, s_i),$$

this condition is necessary and sufficient for the decryption to function without delay. In this case, encryption is usually provided according to the scheme shown in Fig. 3.1.

Sequence

$$\{Z_i = F(k, s_i) : i > 1\}$$

is called the encryption keystream (or keystream encryption). To ensure stable encryption, the keystream should be as random as possible. In fact, keystream encryption is an extended key in a pseudorandom sequence (PRS) of a hypothetically infinite length. It is the randomness of the keystream, its irreversibility and unpredictability that builds secure encryption.

Known methods of stream cryptoconversion can be divided into two general groups:

- self-synchronizing ciphers;
- synchronous stream cryptoconversion.

Self-synchronizing ciphers or *asynchronous stream ciphers*, unlike synchronous ones, have the ability to continue decrypting correctly when the encryption sequences generated by the receiving encoder (decoder) fall out of sync with the keystream

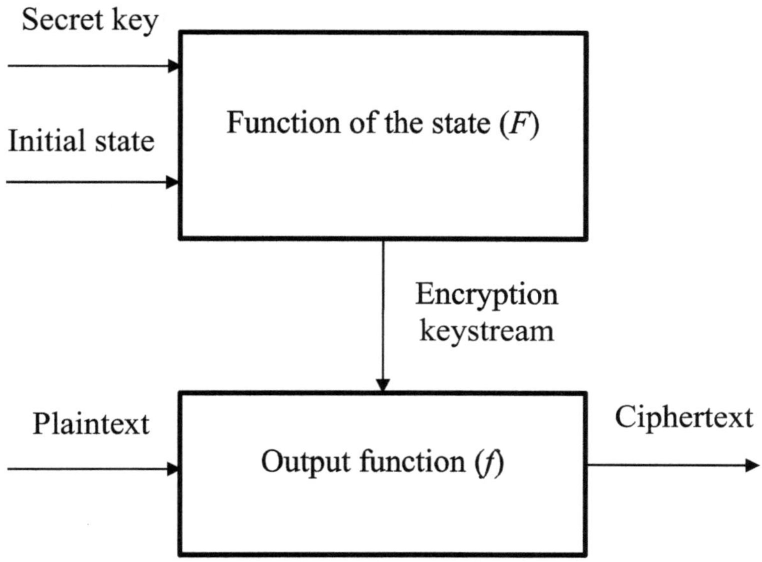

Fig. 3.1 General scheme of stream encryption

transmitter. For such stream ciphers, the function that determines the next state of the cryptosystem takes as input some of the ciphertext that was generated before.

The most common mode of self-synchronizing streaming encryption is ciphertext feedback mode. For example, this might be Cipher Feedback (CFB) standardization mode [2–5]. The following state and output function are given as follows:

$$s_{i+1} = F(y_i, y_{i-1}, \ldots, y_{i-N}),$$
$$z_i = f(k, s_i).$$

Therefore, the status of such stream cipher is given by the preceding N ciphertext characters. Cryptographic stability is concentrated in the output function f.

The cipher is self-synchronized, that is, the receiving side during decryption is automatically synchronized without knowing the clock frequency of the encoder. To do this, it is sufficient to correctly accept from the channel a state consisting of N consecutive digits.

Self-synchronizing ciphers have the property of limited error propagation. The channel distorted character remains in the internal state of the stream decoder to decrypt N consecutive plaintext characters (as long as the distorted character moves through the internal decoder state). After receiving N consecutive undistorted characters from the communication channel, the stream decoder is again able to decrypt correctly.

The most common type of stream ciphers is synchronous stream ciphers; in open cryptography, the vast majority of works on the design of stream encryption schemes are devoted to the study and design of this particular type of cipher. The use of self-synchronizing stream ciphers is more characteristic of military information security systems; there is a small amount of work on this type of scheme.

Let us analyze the methods of stream cryptoconversion belonging to the class of synchronous, examine the features of their construction and compare them by their basic properties.

In synchronous stream cryptoconversion, cryptographic keystream is formed regardless of the plaintext stream and the ciphertext stream. Two rules apply when operating a keystream generator:

$$s_{i+1} = F(k, s_i),$$
$$z_i = f(k, s_i).$$

The initial state of s_0 may be a function of the key k, and possibly of some randomized variable. The purpose of the keystream generator is to deploy a short random key k into a long pseudorandom sequence $z_\ell = z_1, z_2, \ldots, z_\ell$.

In a concise form, this can be written as follows:

$$G : K \to Z^\ell,$$
$$z^\ell = G(k),$$

which emphasizes the functional relationship between the key k and the keystream flux z^ℓ .

For a binary key k^n of length n and a binary keystream stream of length ℓ, the sequence generator (n, ℓ)—is a function from $\{0, 1\}^n$ in $\{0, 1\}^\ell$ such that $z^\ell = G(k^n)$.

Synchronous stream ciphers can be further classified according to the mode in which they operate:

- counter mode;
- output feedback mode.

Counter mode. The next counter mode function is independent of the key bits, but is guaranteed to pass through the entire state space (or much of it).

$$s_{i+1} = F(s_i);$$
$$z_i = f(k, s_i).$$

Examples of such a function F are conventional meters and LFSR with a maximum period length. The level of cryptographic stability is substantiated by the properties of the original function f.

Exit Feedback Mode (Internal Feedback). The output function in this mode is independent of the key.

$$s_{i+1} = F(k, s_i);$$
$$z_i = f(s_i).$$

Very often the function f simply consists of a 1 bit predicate of the current state (for example, the youngest bit or the parity bit). Sometimes consider the variant of such mode, in which the key k sets only the initial state:

$$s_0 = k;$$
$$s_{i+1} = F(s_i);$$
$$z_i = f(s_i).$$

The synchronous stream cipher has the property of not propagating errors. Deciphering a distorted ciphertext character only affects the corresponding plaintext digit. While this property may seem desirable, it has other sides. First, in this case, the ability to detect the decryption error is limited, but more importantly, the adversary is able to make controlled changes to the ciphertext, knowing exactly what changes in the corresponding plaintext will cause.

From a practical point of view, what is even more important is that the encryption and decryption devices must work in a consistent manner, since the encryption cannot be successful until the encryption and decryption sequences are synchronized. If a symbol is distorted or lost during transmission, the recipient will only detect meaningless data from the point where the synchronization failed. Synchronization recovery usually requires finding possible shifts between the clock of the recipient and the sender. Typically, synchronization is achieved by inserting special "tokens" into the transmitted message, resulting in a bit of ciphertext missed during transmission that results in incorrect decryption only until one of the tokens is received. Another solution is to reinitialize the states of both the encoder and the decoder under some previously agreed condition.

Synchronous stream cryptoconversions have become most widespread, since keystream encryption is formed regardless of the plaintext stream and the ciphertext stream. This enables the use of a particular PRS generator to pre-generate keystream encryption and to perform plaintext cryptoconversion very quickly by means of a simple conversion function (for example, modulo two). It is also possible to apply parallelization, which will further increase the speed of stream cryptoconversion. Thus, the development of efficient synchronous stream cryptoconversion is reduced to the construction of PRS generator with the specified cryptographic properties of the original sequence (key stream).

3.3 Conclusions and Recommendations

1. The analysis of the known approaches to the design of stream cryptoconversion schemes showed that the greatest applications were obtained: a theoretical and informational approach using general theoretical positions, mathematical models and assumptions from information theory and secret systems theory; a system-theoretic approach that uses methods of systems theory and systems

analysis, in particular to determine the criteria and performance indicators of SSC; a theoretically complex approach based on the use of complexity theory methods, in particular the problem of deciphering an attacker is associated with the solution of some extremely complex mathematical problem; theoretical and stochastic approach aimed at the construction of so-called randomized stream cryptoconversions.

2. The most common method for describing the operation of SSC or streaming modes BSC is the use of automation theory. In general, streaming cryptoconversion is represented by equations that use a formalized representation of the next state function and the output function. The output function is usually implemented by adding (for example, Module 2) the result of the next state function (gamut of encryption) to the information sequence. This condition is necessary and sufficient for the stream to function without delay.
3. To ensure stable encryption, the keystream should be as random as possible. In fact, keystream encryption is an extended key to a pseudorandom sequence (PRS) of a hypothetically infinite length. It is the randomness of the keystream, its irreversibility and unpredictability, and is based on ensuring secure encryption.
4. The conducted analysis showed that known methods of stream cryptoconversion can be divided into two general groups: self-synchronized ciphers (asynchronous ciphers); synchronous stream cryptoconversions. Asynchronous ciphers have the ability to continue decrypting correctly when the encryption sequences generated by the receiving encoder (decoder) fall out of sync with the gamut of the transmitting encoder. For such stream ciphers, the function that determines the next state of the cryptosystem takes as input some of the ciphertext that was generated before. In synchronous streaming, cryptographic keystream is formed regardless of the plaintext stream and the ciphertext stream. The synchronous streaming cipher has the property of not propagating errors. These advantages indicate the prospect of practical application of synchronous stream cryptoconvernsions; their analysis and study is an important and priority area for further research.

References

1. Rueppel, R.A.: Analysis and design of stream ciphers. SpringerVerlag, Berlin (1986), 244 p.
2. GOST R ISO/IEC 1011693.: Information technology. Operating modes for n-bit block encryption algorithm [Electronic resource]. Access mode: http://gostrf.com/normadata/1/4294818/4294818194.pdf
3. Dolmatov, V.: GOST 28147-89: Encryption, Decryption, and Message Authentication Code (MAC) Algorithms. RFC Editor (2010). https://doi.org/10.17487/rfc5830

4. ISO/IEC 10116:2004.: Information technology—security techniques—modes of operation for an nbit block cipher [Electronic resource]. Access mode: http://www.iso.org/iso/home/store/catalogue_tc/catalogue_detail.htm?csnumber=38761
5. NIST Special Publication 80038A.: Recommendation for block cipher modes of operation [Electronic resourse]. Access mode: http://csrc.nist.gov/publications/nistpubs/80038a/sp80038a.pdf

Chapter 4
Analysis of Synchronous Stream Cryptoconversions

4.1 General Classification of Synchronous Stream Cryptoconversions

Synchronous SSC is the largest and most widespread class of modern stream cryptoconversions that are used to build cryptographic protections in modern information and telecommunication systems and networks. The general classification of known methods of construction of synchronous SSC is shown in Fig. 4.1.

Different approaches to the construction of synchronous stream cryptoconversions will be analyzed, the features of their structure will be studied, the advantages and disadvantages will be evaluated.

4.2 Stream Encryption Methods Established on Use of Registers with Steady Motion

Methods based on the use of schemes with steady motion of registers have become among the methods of streaming the information of widespread dissemination. The advantages of such methods are simplicity of both software and hardware implementations, high stability of transformations and high speed.

As the analysis showed [1–5], most of the methods of stream cryptoconversion based on the use of schemes with steady motion of registers, is divided into two classes (see Fig. 4.1):

- methods built using filter generators;
- methods built using combination generators.

In the first case, the construction of PRS generators is based on the use of a single LFSR, usually very long, and nonlinear cryptoconversion is applied to the contents

A. A. Kuznetsov et al., *Stream Ciphers in Modern Real-time IT Systems*,
Studies in Systems, Decision and Control 375,
https://doi.org/10.1007/978-3-030-79770-6_4

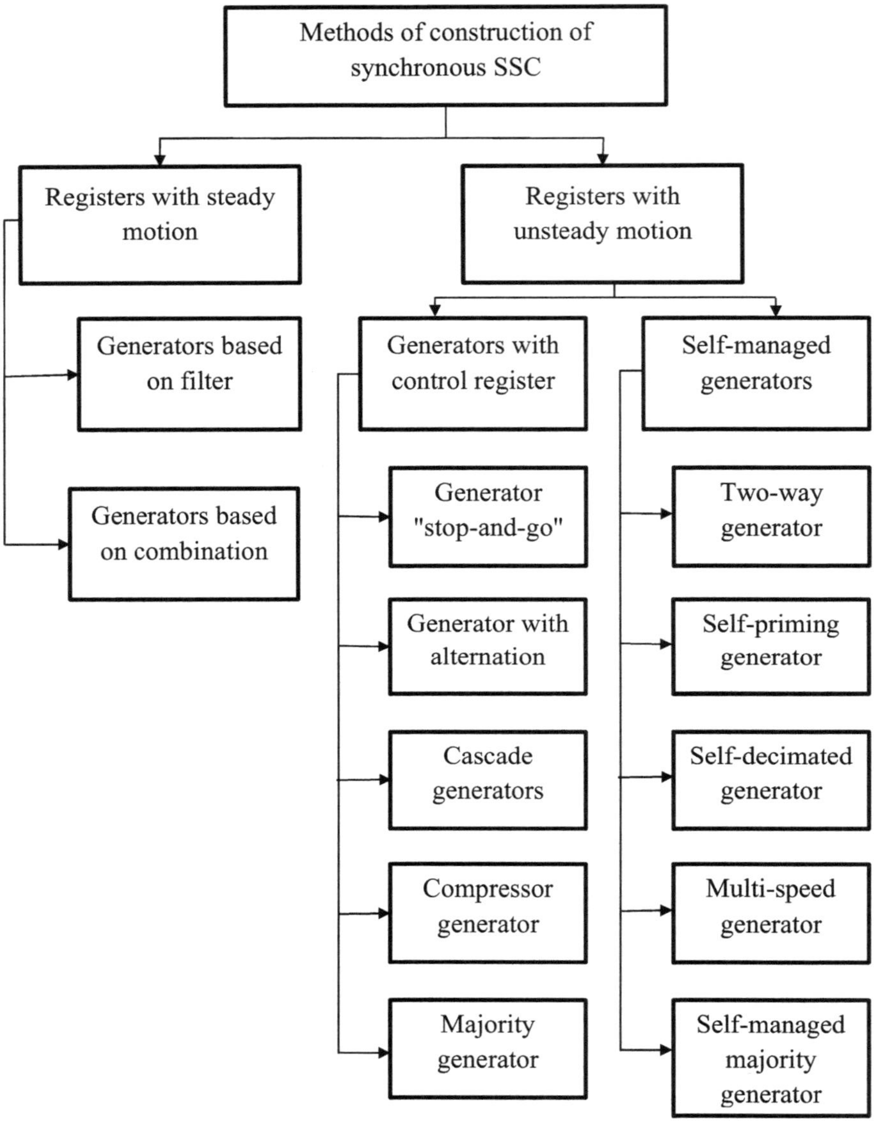

Fig. 4.1 General classification of known methods of construction of synchronous SSC

of this register. Multiple registers are used in combination generators, usually of much smaller length, but non-linear cryptoconversion is applied to the output of each register.

4.2.1 *Synchronous Stream Cryptoconversions with Steady Motion According to the Scheme of the Filter Generator*

Methods for stream information conversion based on the use of filter generators, in the general case, accept the use of LFSR, the output sequence of which enters the input of a nonlinear complication node which directly performs cryptoconversions. At each conversion cycle, data is removed from certain cells of the LFSR, after which the obtained data are transformed by cryptographic nonlinear transformation (nonlinear complication node, block S). The result of the operation is a key stream that is summarized by some arithmetic (eg, Module 2, i.e, XOR operation) with the open test, and the sequence thus obtained is the desired closed test (cipher).

A generalized diagram of the filter generator is shown in Fig. 4.2.

The figure schematically shows long-range radar LFSR. The nonlinear transformation in the generator is applied to the contents of this register, that is, the so-called nonlinear filtering of data that is taken from individual cells of the register. If the length of the LFSR is equal to m and the filtering function is of the order n, then the linear complexity of the formed PRS (key stream) cannot exceed.

$$S = \sum_{i=1}^{n} C_m^i.$$

For the sake of requirements of the length of the period of the formed PRS, it is necessary to use an LFSR with a bit length m of at least 128 bits.

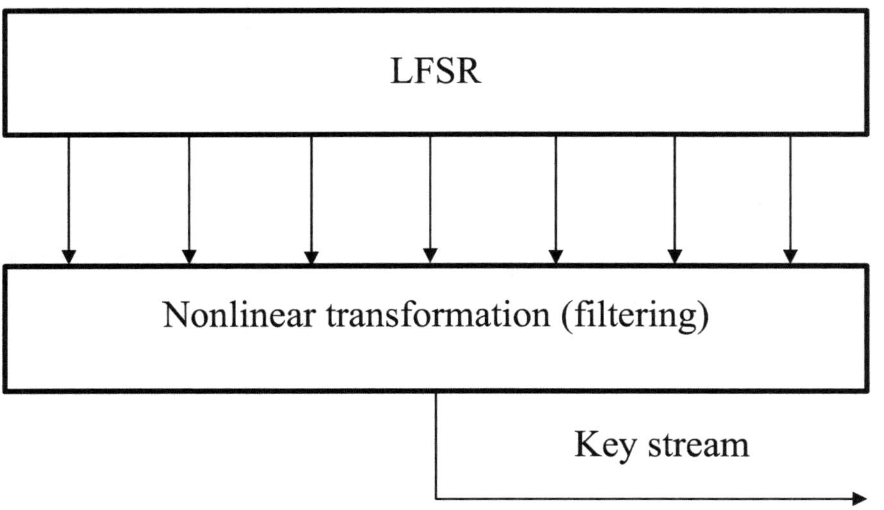

Fig. 4.2 General scheme of the generator on nonlinear filtration

The advantage of the filter generator lies in the fact that the use of a single shift register makes simple hardware implementation, which in turn allows to achieve a high rate of cryptoconversions. It is this class of methods that is used in applications that are subject to stringent conversion/data transfer speed requirements.

4.2.2 Synchronous Stream Cryptoconversions with Steady Motion According to the Scheme of the Combination Generator

The difference between the methods of stream conversion of the information based on the use of combinatorial generators from the methods based on the use of filter generators is that they involve the use of several parallel LFSR, the output sequences of which are similarly fed to the input of nonlinear transformation (complication node). At each conversion cycle, data is extracted from the outputs of the LFSR, after which the selected data is converted by cryptographic nonlinear transformation (nonlinear S block). The result of the operation is a key stream, which is also summed up by some open-test arithmetic (eg, XOR operation), and the sequence thus obtained is the required closed test. The classic scheme of the combination generator is shown in Fig. 4.3.

The simplest example of a combinator is the so-called Geff generator, which uses three LFSR [6, 7], one of which controls the combination of the other two, that is, a

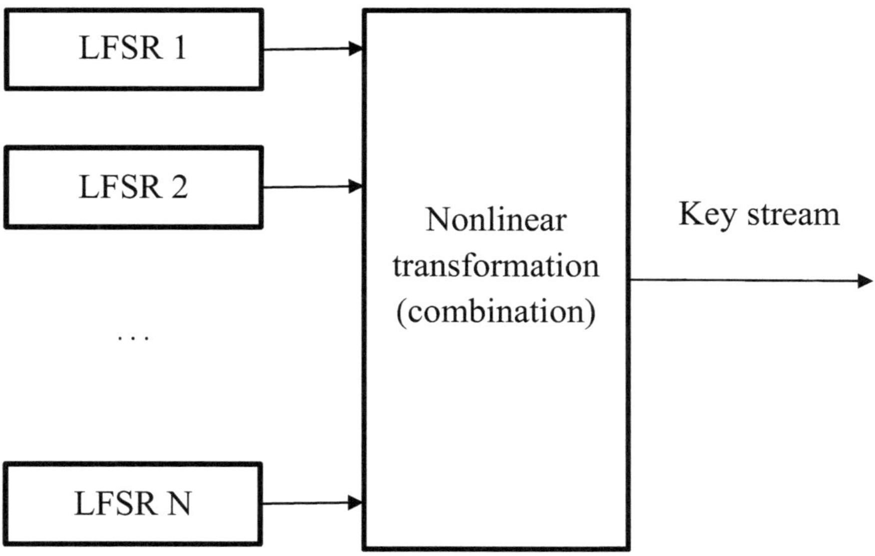

Fig. 4.3 General scheme of the combination generator

nonlinear transform is a common multiplexer (Fig. 4.4a). Let us denote the outputs of the three LFSR as x_1, x_2, and x_3 respectively. Then the rule of combining the outputs x_1 and x_2 (LFSR 1 and LFSR 2) with the help of the multiplexer controlled by the output x_3 LFSR 3 can be written as a function (Fig. 4.4b):

$$f(x_1, x_2, x_3) = x_1x_3 \oplus x_2(1 \oplus x_3) = x_1x_3 \oplus x_2x_3 \oplus x_2$$

If binary LFSR with lengths m_1, m_2 and m_3, respectively, are used in the Geff generator (Fig. 4.4), then the length of the output PRS (key stream) will be equal to:

$$L = (2^{m_1} - 1)(2^{m_2} - 1)(2^{m_2} - 1)$$

The linear complexity of PRS is accordingly equal

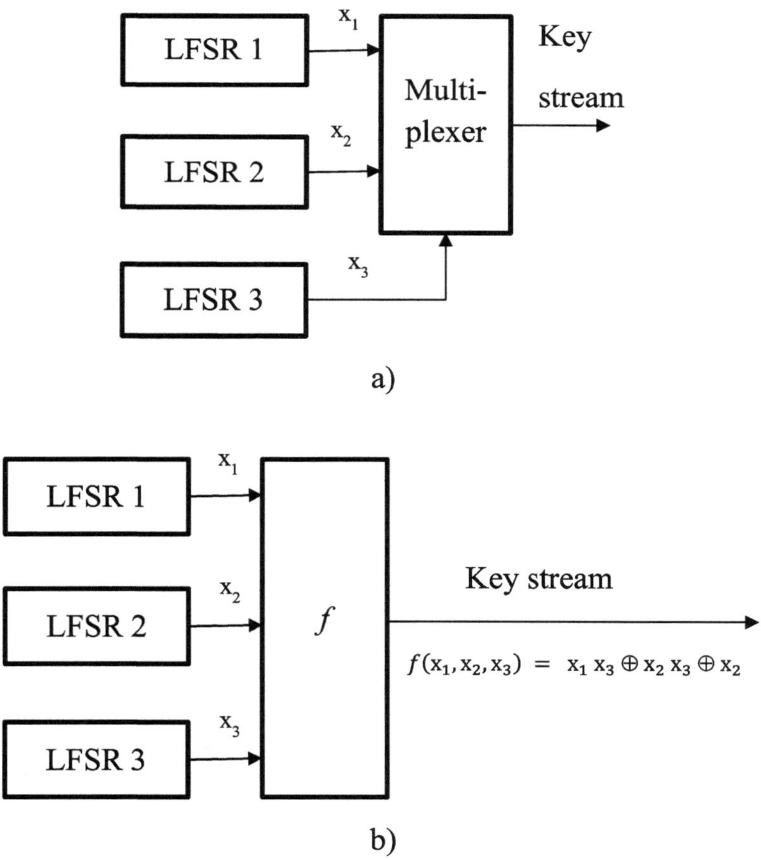

Fig. 4.4 General scheme of the Geff generator

$$L = m_1 m_3 + m_2 m_3 + m_2$$

It should be noted the high period length and linear complexity of PRS for the Geff generator. However, its cryptographic stability is low, as the formed PRS is highly correlated with the output of one of the LFSR. Indeed, let's denote as $P(f(x_1, x_2, x_3) = x_1)$ the probability of such an event when the symbol of the original PRS coincides with the original symbol of LFSR. Based on the formula $f(x_1, x_2, x_3) = x_1 x_3 \oplus x_2(1 \oplus x_3) = x_1 x_3 \oplus x_2 x_3 \oplus x_2$ we see that this event is bound to happen when $x_3 = 1$ or when $x_3 = 0$ and at the same time $x_2 = x_1$. That is, we have:

$$P(f(x_1, x_2, x_3) = x_1) = P(x_3 = 1) + P(x_3 = 0) * P(x_2 = x_1).$$

But each LFSR works independently and the characters on their outputs are equally probable. That is, $P(x_3 = 1) = 0.5$, $P(x_3 = 0) = 0.5$, $P(x_2 = x_1) = 0.5$, so we have:

$$P(f(x_1, x_2, x_3) = x_1) = 0.5 + 0.5 * 0.5 = 0.75,$$

that is, in 75% of cases, the original PRS will coincide with the release of LFSR 1, which greatly simplifies cryptanalysis.

This example clearly demonstrates the principles of construction of a combination generator, as well as the difficulties that arise when choosing a nonlinear conversion.

4.3 Streaming Encryption Methods Established on the Use of Registers with Unsteady Motion

4.3.1 Generator with Control Register "stop-and-go"

The StopandGo, BothPiper generator uses the LFSR 1 output to control the LFSR 2. The LFSR 2 changes its state at some point in time ti only if the LFSR 1 output at the time t_{i-1} is equal to one. The scheme of the simplest variant of the generator "stop-and-go" is shown in Fig. 4.5.

The generator contains three shift registers, the control register LFSR 1 generates a control sequence of clock symbols that control the shift LFSR 2 by changing its clock frequency. LFSR 3 as well as LFSR 1 is controlled by the input sequence of clock symbols. The outputs of LFSR 2 and LFSR 3 are functionally transformable (not necessarily nonlinear; conventional modular two may be used). Functional conversion results in a key output stream.

The stop-and-go generator is very easy to implement, however, such a key stream generation scheme has vulnerabilities to correlative disclosure.

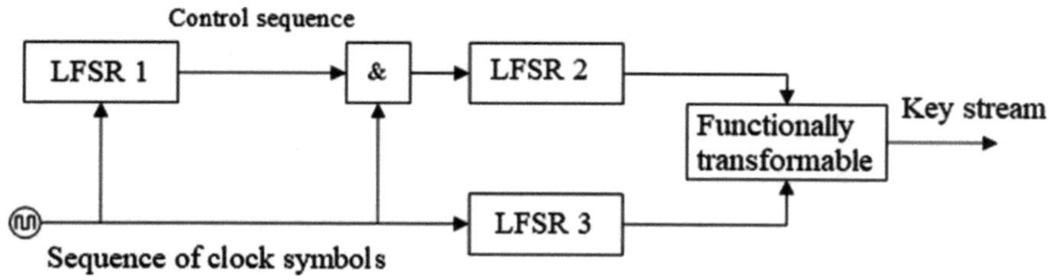

Fig. 4.5. General scheme of the generator "stop-and-go"

4.3.2 Generator with Control Register "stop-and-go" with Alternation

In order to increase the cryptoconversion of the considered generator, a "stop-and-go" generator was proposed, in which there is an alternation of the displacements of different LFSR. In particular, in the simplest version, three shift registers of different lengths are used (Fig. 4.6). The first register controls the clock frequency of the 2nd and 3rd registers, that is, LFSR 2 changes its state when the output of LFSR 1 is equal to one, and LFSR 3—when the output of LFSR 1 is zero. The output of the generator is a functional operation (for example, adding modulo two) of the outputs LFSR 2 and LFSR 3.

The alternator with a control register "stop-and-go" with alternation has shifted a long period and great linear complexity. There are ways of correlating LFSR 1 analysis, but this does not weaken the cryptographic properties in a critical way.

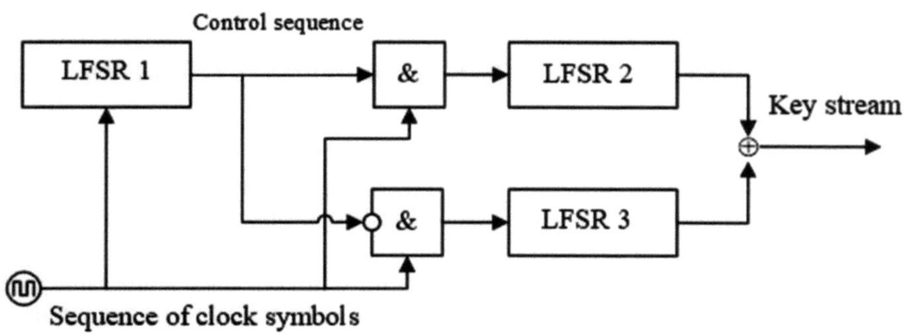

Fig. 4.6 General scheme of the generator "stop-and-go" with alternating displacements

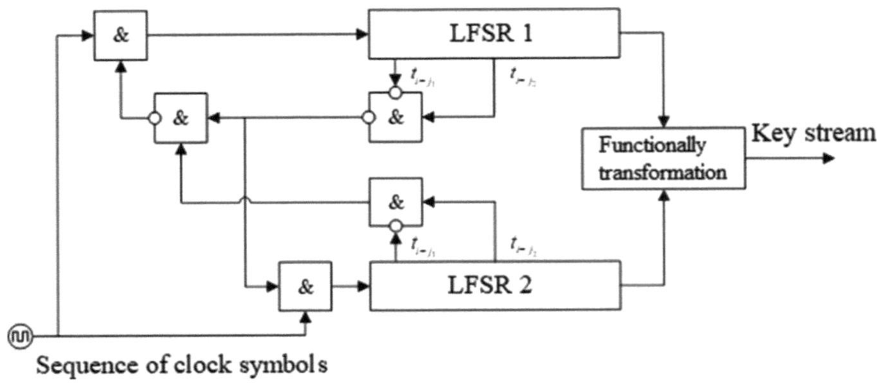

Fig. 4.7 General scheme of two-way generator "stop-and-go"

4.3.3 *Two-Way stop-and-go Generator*

A complicated offset scheme is used in a two-way stop-and-go generator that uses 2 shift registers of the same length. If the output of LFSR 1 at some point in time t_{i-j1} is zero, and at time t_{i-j2}—units, then LFSR 2 does not shift at time ti. If the output of LFSR 2 at time t_{i-j1} is zero, and at time t_{i-j2}—units, and if this register is shifted at time t_i, then at the same time LFSR 1 is not shifted. The scheme of such two-way generator "stop-and-go" is shown in Fig. 4.7.

The linear complexity of the output PRS of the two-way generator "stop-and-go" is approximately equal to the period of the formed sequence.

4.3.4 *Cascade Generators with Control Registers (Holmann Cascades)*

The scheme of the cascade generator (Fig. 4.8) is an improved version of the generator "stop-and-go". It consists of a sequence of LFSR, the offset of each of which is governed by the output of the previous LFSR. If the output of LFSR 1 at some point in time is a unit, then it is displaced LFSR 2. If the output LFSR 2 at the same time is a unit, then shifts LFSR 3, and so on. The output of the last LFSR is the output of the generator.

Shown in Fig. 4.8, the scheme demonstrates the use of a cascade of registers which can be used to generate sequences with long periods, high linear complexity and good statistical properties. For example, if the length of all LFSR is equal to and equal to m, then the period of the system with n LFSR is equal to $L = (2^m - 1)^n$, and the linear complexity $S = m(2^m - 1)^{n-1}$.

It should be noted that cascade generators are sensitive to the method of autopsy, to the so-called locking), when the cryptanalyst, initially restoring the input sequence

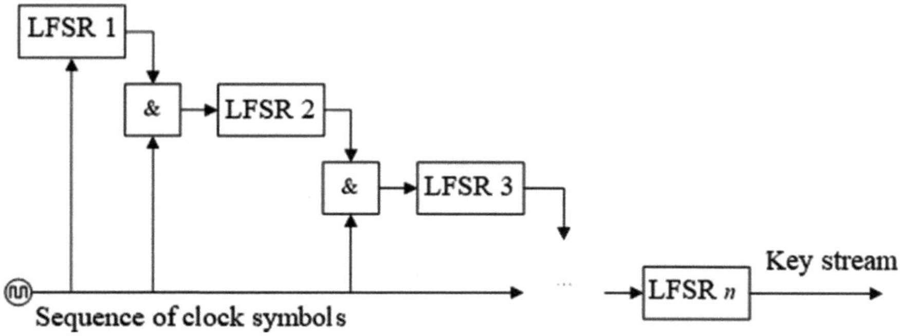

Fig. 4.8 General scheme of cascade stop-and-go generator

of the last register in the cascade, breaks the whole cascade, register by register. With increasing number of n registers, a sequence is generated approaching casual, that is, it is better to use more short length LFSR than less long LFSR.

4.3.5 Compressor Generator with Control Register

There are also compression and self-compressing generators.

Shrinking contains two shift registers LFSR 1 and LFSR 2. Both registers are synchronized equally, however LFSR 1 is the control shift register that controls the output of the LFSR register 2. If the generator is clocked, LFSR is 1, then the output of the generator will be the output of the LFSR 2. Otherwise, when the clock pulse is applied, both bits are reset, the operation starts again. The scheme of the compression generator is shown in Fig. 4.9.

The compression generator can be broken if the feedback polynomials are sparse. In addition, the speed of its generation is not constant. The self-compressing generator requires about half the memory space than the first, but it runs twice as fast.

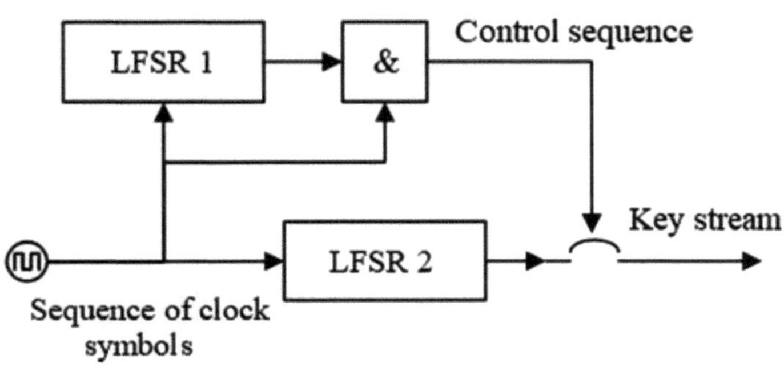

Fig. 4.9 General scheme of compression generator

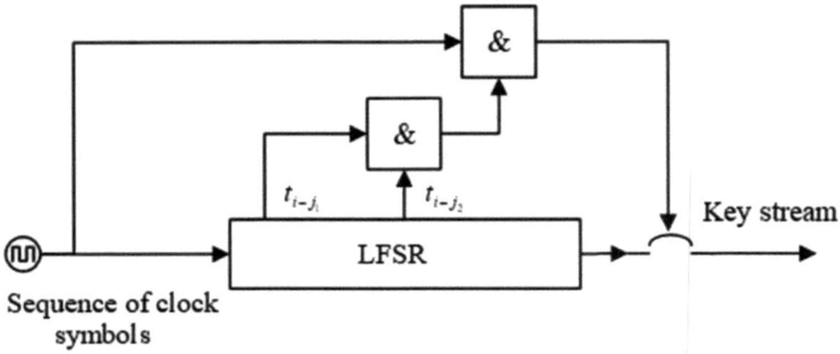

Fig. 4.10 General scheme of self-compressing generator

The compression generator is simple, well-scaled and has good protective properties. To improve its security, the following conditions apply:

- the lengths of the LFSR 1 and LFSR 2 registers should be relatively simple numbers;
- it is advisable to use a hidden connection between the LFSR 1 and LFSR 2 registers.

4.3.6 *Self-Priming Generator*

In self-shrinking generator compared to shrinking (with control register) instead of checking the outputs of two registers for 0 or 1, 2 bits of one shift register are checked. One scheme of such a self-compressing generator is shown in Fig. 4.10.

If at some points in time t_{i-j1} and t_{i-j2} are equal to one, then the output of the generator is the output of the radar. If the output of LFSR at time t_{i-j1} or at time t_{i-j2} is zero, then the output of LFSR is reset and nothing is output to the generator output.

4.3.7 *Self-Decimated Generator*

Self-decimated generators control their own clock speed. The scheme of the self-generating generator in the general form is shown in Fig. 4.11.

There are two types of generators: the Ruppel generator and the Chambers-Colman generator.

The first type of generator was proposed by Raynor Ruppel [8, 9]. It consists of a shift register and a linear feedback circuit that shifts this register depending on what the output values of the shift register are. If the output of the LFSR is equal to

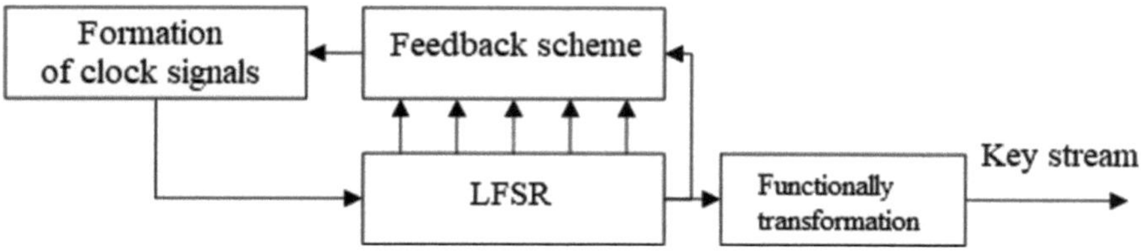

Fig. 4.11 General diagram of a self-decimated generator

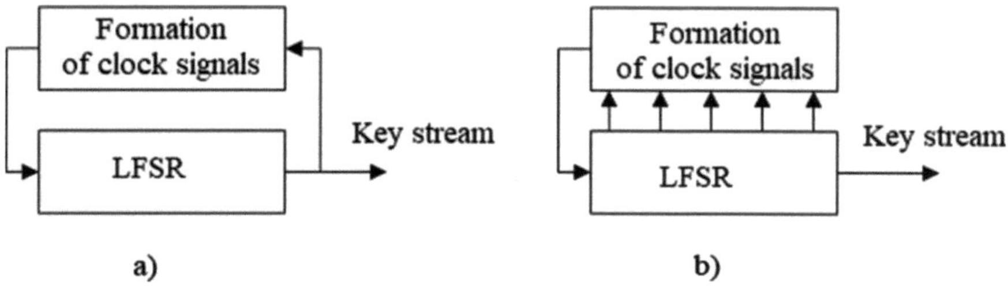

Fig. 4.12 Scheme of self-decimated generator: **a** Ruppel; **b** Chambers-Colman

one, then the register shifts with one frequency, and if the output is zero, then with another. The scheme of the self-destructive Ruppel generator is shown in Fig. 4.12a.

The second type of self-generating generators was proposed by Bill Chambers and Dieter Colman (Fig. 4.12b). The feedback circuit receives not the output itself, but the result of the XOR operation of certain bits of the LFSR, or, for example, the result of their non-linear filtering. In Fig. 4.12 The clock rate rule is generally referred to as some conversion of the offset or contents of the shift register to generate clock signals.

4.3.8 Multi-speed Generator with Internal Product (or with Internal Nonlinear Conversion)

The multi-speed internal product generator uses two shift registers LFSR 1 and LFSR 2. The LFSR 2 clock frequency is d times higher than in LFSR 1. Some bits of these registers are multiplied by the AND operation. The multiplication results are added by the XOR operation, and the resulting sequence is the key stream. The scheme of the multi-speed generator with internal product is shown in Fig. 4.13.

The multi-speed generator with internal product has high linear complexity and good statistical properties. However, its state can be determined from the output sequence of length $m_1 + m_2 + log_2 d$, where m_1 and m_2 are the lengths of the LFSR 1 and LFSR 2, respectively, and d is the ratio of the clock frequencies of the first and second clock signals (see Fig. 4.13).

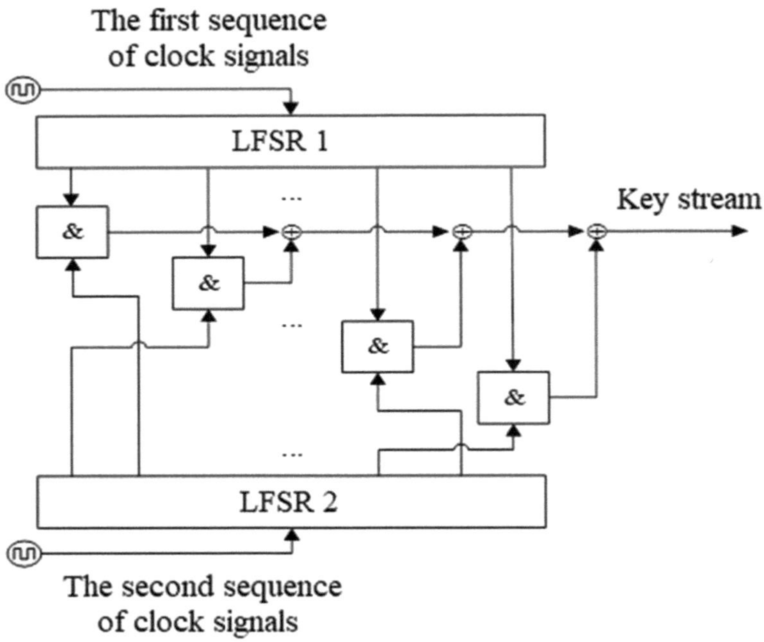

Fig. 4.13 General scheme of the multi-speed generator with inner work

The scheme of the multi-speed generator with internal product can be summarized in the case of application of some nonlinear function of conversion of contents and outputs of shift registers. The scheme of such a generator is shown in Fig. 4.14.

The generator contains a nonlinear transformation, which can be considered as a case of nonlinear filtering of read data from the contents and outputs of LFSR 1 and LFSR 2 respectively. In this sense, such a generator is similar to the one discussed above, a filter generator with a uniform movement of registers, but here two different LFSR are used and each of them shifts with a different clock frequency. This obviously complicates the process of cryptanalysis of streaming transformation and increases the linear complexity of the formed PRS.

4.3.9 Majority Generator

The majority (or threshold) generator consists of a large number of shift registers, the outputs of which are supplied to the device that implements the function of majority output. The peculiarity of such a generator is that each of the shift registers gives its output a_i, $i = 1, 2, \ldots, n$ as variables of the majority function, where n is the number of LFSR of the majority generator. For example, for the three shift registers with outputs a_1, a_2, a_3, respectively, the majority function looks like:

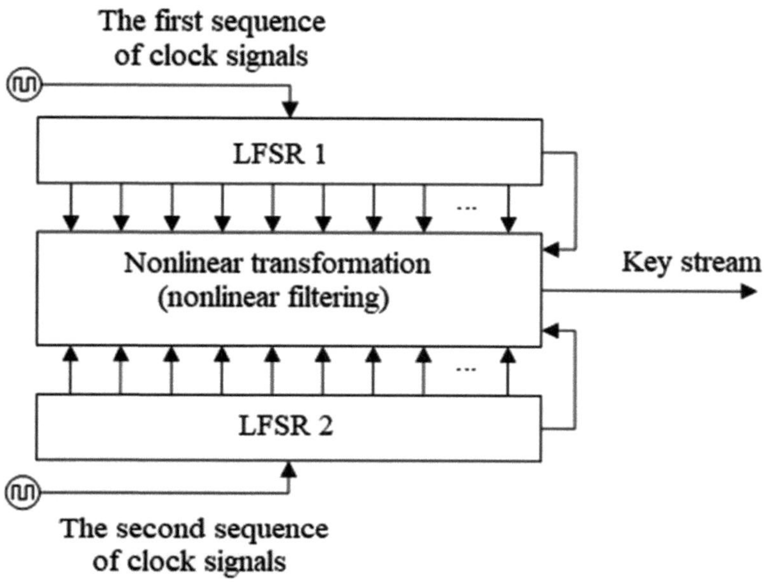

Fig. 4.14 General scheme of the multi-speed generator with internal nonlinear transformation

$$f(\mathrm{a}_1, \mathrm{a}_2, \mathrm{a}_3) = (\mathrm{a}_1 \wedge \mathrm{a}_2) \vee (\mathrm{a}_1 \wedge \mathrm{a}_3) \vee (\mathrm{a}_2 \wedge \mathrm{a}_3)$$

and only those registers are shifted, the clock bits of which are equal to the value of $b = f(\mathrm{a}_1, \mathrm{a}_2, \mathrm{a}_3)$, that is, the shift of registers occurs for most values of such bits: 0 or 1.

The generalized scheme of the majority generator with registers LFSR 1, LFSR 2, …, LFSR n and the majority function $f(\mathrm{a}_1, \mathrm{a}_2, \mathrm{a}_3)$ is shown in Fig. 4.15.

The use of the majority function allows to obtain a generator with a variable number of RPMs that shift at a certain point in time. The linear complexity of the formed PRS is equal to.

$$S = \sum_{i,j=1}^{n} m_i m_j$$

where m_i is the length of *i* register.

The disadvantages of the majority generator are the ability to directly search and correlate analysis (each output bit gives some information about the status of the shift register).

The input of the majority transformation can be submitted data that is not read from the outputs of LFSR, as from some cells of these LFSR. The result of the majority conversion b is compared with the value of the synchronization point of

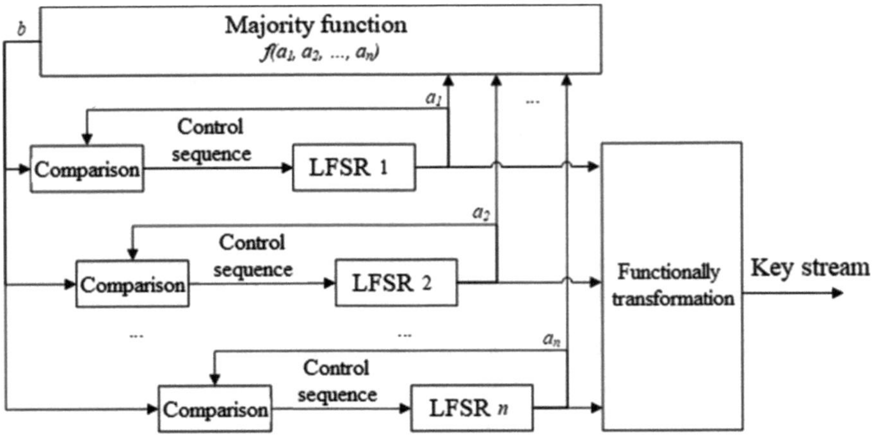

Fig. 4.15 General scheme of the majority generator

each of the LFSR, ie only the registers whose synchronization bits are equal to the value of b are shifted. Thus, the movement of registers occurs in most values of bits: *0* or *1*. The general scheme of such a majority generator with synchronization points $a_1, a_2, \ldots, a_n$ is shown in Fig. 4.16.

Similar majority generators with synchronization points are used in *A5* streaming encryption algorithms [9–11]. The scheme of PRS formation according to the algorithm in A5 is shown in Fig. 4.17.

A further development of the majority approach is the additional introduction of the control register, the values of the synchronization points of which are the output

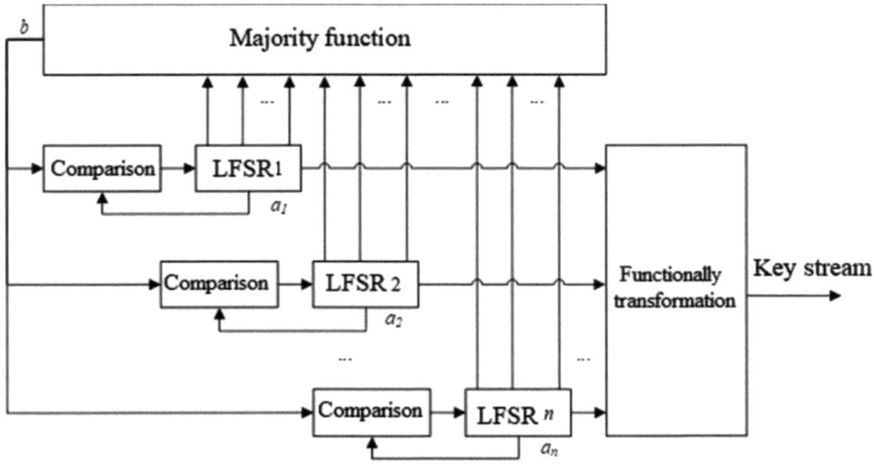

Fig. 4.16 General scheme of the majority generator with points of synchronization

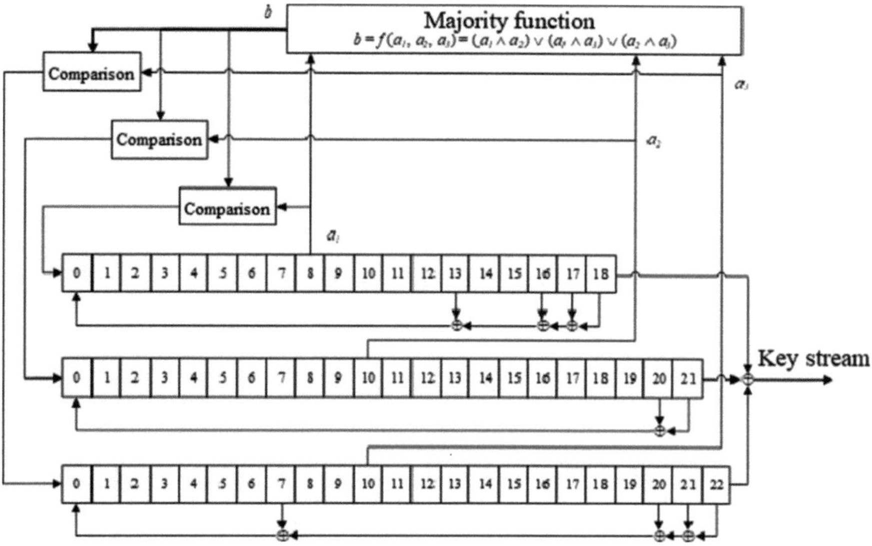

Fig. 4.17 Streaming A5 encryption scheme

for the majority transformation, and which determine the control sequence for the movement of other registers. The general scheme of the majority generator with the control register is shown in Fig. 4.18.

On this principle the algorithm of SSC A5/2 is constructed, the scheme of which is shown in Fig. 4.19. To increase the complexity of the formed PRS, this algorithm also applies additional majority functions for each of the shift registers. The outputs

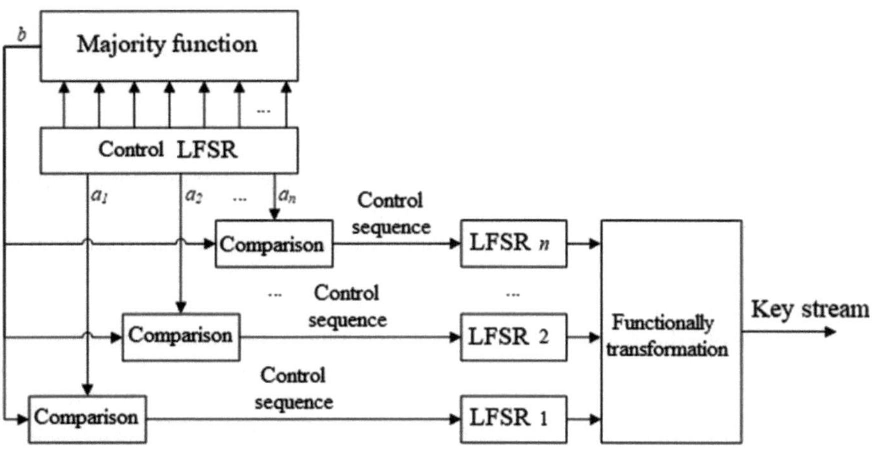

Fig. 4.18 General scheme of majority generator with control register

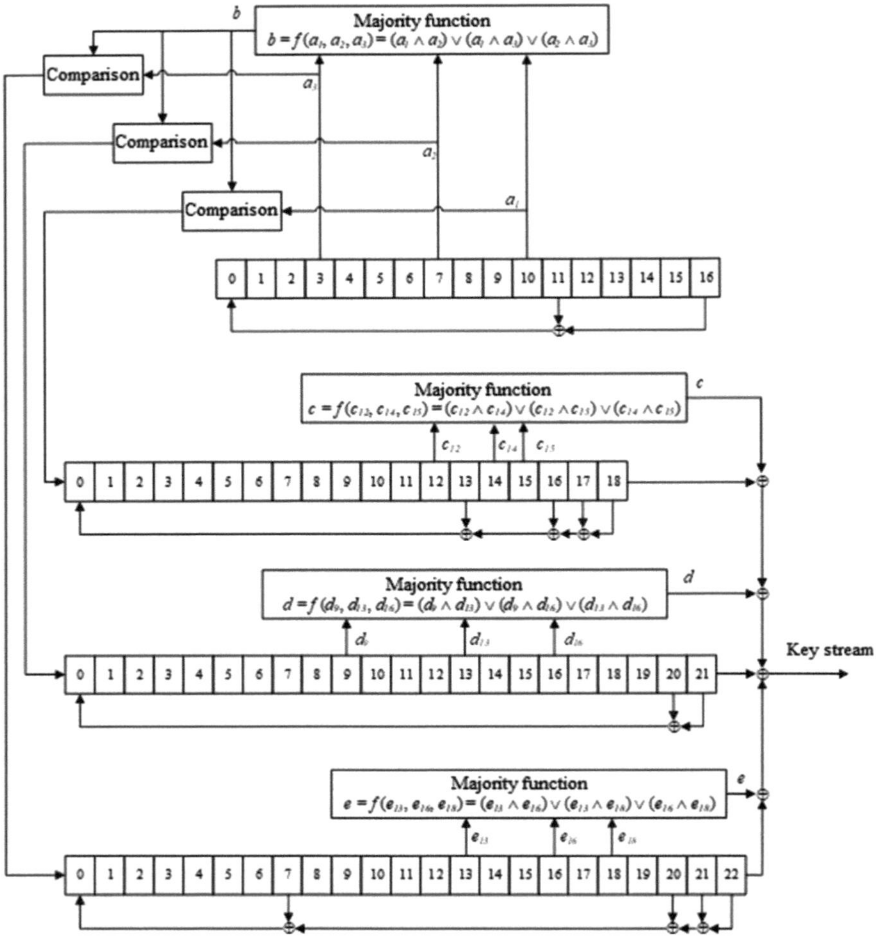

Fig. 4.19 Streaming A5/2 encryption scheme

of these major transformations are added to the outputs of the shift registers, resulting in the formation of PRS—the output key stream.

It should be noted that despite the complicated scheme of transformation of the algorithm of SSC A5/2 its stability is called into question. In particular, this is attributed to the very small length of the shift register (17 bits). At least in the previous version, the majority function for generating the control sequence was calculated immediately by three registers 19, 22 and 23 bits respectively. That is, it is believed that the developers of A5/2 aimed at reducing the stability of the SSC algorithm by simplifying the rule of formation of the control sequence which manages shifts of LFSR generator.

4.4 Conclusions and Recommendations

1. Synchronous streaming cryptoconversions have become most widespread in PSS, since keystream encryption is formed regardless of the plaintext stream and the ciphertext stream. This makes it possible, using a particular PRS generator, to pre-generate keystream encryption and to perform plaintext cryptoconversion very quickly by a simple conversion function (for example, adding two). It is also possible to apply parallelization, which will further increase the speed of streaming cryptoconversion. That is, the development of an efficient synchronous streaming cryptoconversion is reduced to the construction of a PRS generator with the specified cryptographic properties of the source sequence (key stream).
2. The analysis showed that the known methods of construction of synchronous PSS are divided into two large groups: with uniform movement of shift registers and PRS generators with uneven movement (shift) of registers. In the first case, all LFSR used in the PRS generator are shifted by the same clock signals, ie the movement of registers is uniform (uniform). This greatly simplifies the implementation of generators and usually increases their performance. Compared to this approach, in non-uniform motion generators, different LFSR can be shifted by different clock signals. This is done either through the use of the so-called control register (one of the used in the scheme of the LFSR generator performs the function of controlling the clock speed of other LFSR), or through the construction of generators in which the clock frequency of individual LFSR is determined by the logic of feedbacks or certain functional transformations between outputs the contents of the LFSR.
3. Among the methods with uniform movement of registers are the so-called filter generators and combinatorial generators. In the first case, the formation of PRS is based on the use of a single LFSR, usually very long, and nonlinear cryptoconversion is applied to the contents of this register. Multiple registers are used in combination generators, usually of much smaller length, but non-linear cryptoconversion is applied to the output of each register. Advantages of methods with uniform movement of registers are simplicity both the software, and hardware implementations, high stability of transformations and high speed.
4. Methods of constructing synchronous PSS with uneven movement of registers, as a rule, have a more complex structure, which reduces the performance of the respective generators. However, cryptographic analysis of such algorithms is usually much more complicated, in particular, the linear complexity of the formed PRS is usually higher than in the schemes with uniform register movement.

References

1. Menicocci R.: Short Gollmann Cascade generators may be insecure. In: Proceedings of the 4th IMA Conference on Cryptography and Coding, Cirencester (1993)
2. Håstad, J., Mats Näslund. BMGL: Synchronous Keystream Generator with Provable Security. http://www.cryptonessie.org
3. Coppersmith, D., Krawczyk, H., Mansour, Y.: The shrinking generator. Advances in Cryptology: CRYPTO'93, Springer Verlag (1994)
4. Reeds, J.A., Sloane, N.J.A.: Shiftregister synthesis [Electronic resourse]. Access mode: http://web.archive.org/web/20120829171715/http://www2.research.att.com/~njas/doc/1218shift.pdf
5. Rueppel, R.A.: Analysis and design of stream ciphers. Berlin, SpringerVerlag. 244 p. (1986)
6. Anin, B.Y.: Protection of computer information. SPb .: BKhVSt. Petersburg, 384 c (2000)
7. Alferov, A.P, Zubov, A.Y., Kuzmin, A.S., Cheremushkin, A.V.: Fundamentals of cryptography, Helios ARV, 480 p (2005)
8. Gorbenko, I.D, Gorbenko, Y.I.: Applied cryptology: Monograph. Kharkiv: KNURE, Fort, 1st and 2nd editions, 868 p (2012)
9. Gorbenko, I.D., Gorbenko, Y.I.: Applied cryptology: Textbook. Kharkiv, KNURE, Fort, 1st and 2nd editions, 878 p (2012)
10. Gorbenko, Y.I., Potiy, A.V., Izbenko, Y.A., Orlova, S.Y.: Analysis of current encryption schemes submitted to the European competition NESSIE//Legal, regulatory and metrological support of the information protection system in Ukraine: scientific and technical collection, Vip. 5, pp. 92–110 (2002)
11. Ross Anderson: A5—The GSM encryption algorithm [Electronic resourse]. Access mode: http://www.ussrback.com/crypto/source/algorithms/A5GSMAlgorithm.txt

Chapter 5
Analysis of Stream Modes for Block Symmetric Ciphers

Generalities about various modes of BSC utilization are presented in Table 5.1. As a base have being selected the specified in contemporary national and international GOST 28147-89, ISO/IEC 10116, DSTU 7624 «Kalyna» and NIST SP 800-38 modes [1–5]. Besides names of the mode and the ensured security service, in Table 5.1 are given the English language name of the mode and the respective standard, where the respective mode is specified.

In the following subunits the main conversions of the basic mode (electronic code book) are being researched accordingly with the specification of the national standard DSTU 7624 «Kalyna» [3, 4] and of each of the stream modes of utilization of the block symmetric ciphering, which are being presented in Table 5.1.

5.1 Electronic Codebook Mode—ECB

The simple substitution mode, which is defined in NIST Special Publication 800-38 «Block Cipher Modes», complies with the first utilization mode of the DSTU GOST 28147:2009 «Information processing systems. Cryptographic protection. Algorithm of cryptographic modification» DSTU 7624 «Kalyna» and *ISO/IEC* 10116 «*Information technology—Security techniques—Modes of operation for an n-bit block cipher*». A general scheme for ciphering and deciphering in the simple substitution mode is presented in Fig. 5.1. The simple substitution mode (electronic codebook) is being used as the main building element (basic conversion).

This is the simplest, from the point of view of realization, mode of the BSC-using, which lies in ensuring of confidentiality of separate blocks of clear text by its ciphering after the introduced secret key K. This means that each block of clear text is being correlated with a block of cipher-text, what represents a certain analogy with attribution of a code-word in the ciphering book.

Formally, the simple substitution mode is defined as follows.

A. A. Kuznetsov et al., *Stream Ciphers in Modern Real-time IT Systems*,
Studies in Systems, Decision and Control 375,
https://doi.org/10.1007/978-3-030-79770-6_5

Table 5.1 General data about known modes of BSC utilization

No.	Mode name		Security service	Normative document, which contains the specification of the mode
1	Electronic codebook	ECB	Confidentiality	GOST 28147-89, *NIST SP* 800-38A, *ISO/IEC* 10116
2	Counter	CTR	Confidentiality	GOST 28147-89, *NIST SP* 800-38A, *ISO/IEC* 10116
3	*Cipher feedback*	*CFB*	Confidentiality	GOST 28147-89, *NIST SP* 800-38A, *ISO/IEC* 10116
4	*Symmetric key block cipher-based message authentication code*	*CMAC*	Integrity	GOST 28147-89, *NIST SP* 800-38B
5	Cipher block chaining	*CBC*	Confidentiality	*NIST SP* 800-38B, *ISO/IEC* 10116
6	*Output feedback*	*OFB*	Confidentiality	*NIST SP* 800-38B, *ISO/IEC* 10116
7	*Galois/counter mode and Galois message authentication code*	*GCM and GMAC*	Confidentiality and integrity	*NIST SP* 800-38D
8	*Counter with cipher block chaining-message authentication code*	*CCM*	Confidentiality and integrity	*NIST SP* 800-38C
9	*XOR encrypt XOR (XEX) Tweakable block cipher*	*XTS*	Confidentiality	*NIST SP* 800-38E
10	*Key wrapping*	*KW*	Confidentiality	*NIST SP* 800-38F

Ciphering in simple substitution mode:

$$\begin{aligned} &\text{for } j = -1 \ldots n \\ &\quad C_j = CIPH_K(P_j). \end{aligned}$$

Deciphering in simple substitution mode:

$$\begin{aligned} &\text{for } j = -1 \ldots n \\ &\quad P_j = CIPH^{-1}K(C_j) \end{aligned}$$

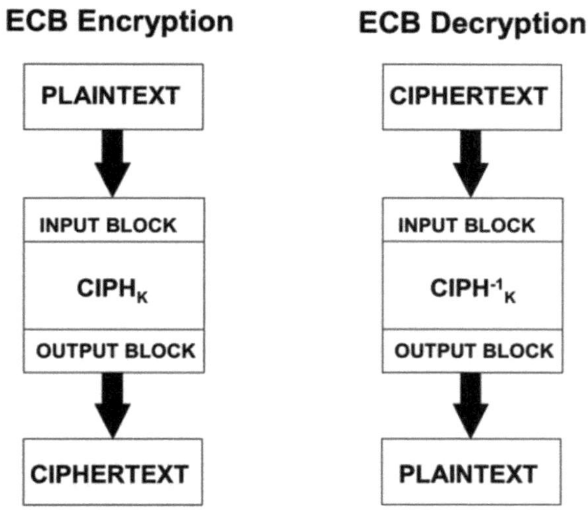

Fig. 5.1 Scheme for ciphering and deciphering in electronic codebook mode

While using the simple substitution mode each activation of the ciphering function $CIPHK/(Pj)$ is utilized immediately and independently for every block of the clear text Pj. The resulting sequence of output blocks Cj is the ciphered text.

In the case of deciphering in the mode of simple substitution the inverse cipher function is being applied immediately and independently to each block of the ciphered text C_j. The resulting sequence of output blocks Pj is the open text.

When applying the simple substitution mode, the multiple activations of the functions $CIPH_K(P_j)$ and $CIPH_K^{-1}(C_j)$ can be realized in parallel. That is the main advantage of this mode.

In the simple substitution mode, for the introduced secret key any block of the clear text during enciphering is being transformed always into one and the same block of the ciphered text. If this property is not wanted in a specific application, then is the simple substitution mode not usable. The simple substitution mode is an obligatory constituent of all other application modes of the block symmetric ciphering.

Let us examine the simple substitution mode, which is determined in the national standard DSTU 7624:2014 «Information technologies. Cryptographic information protection. Algorithm of symmetric block modification», analyze some main cryptoconversions, which determine this basic ciphering mode.

5.1.1 General Parameters

The basic enciphering modification $T_{l,k}^{(K)}$ is a parametrized ciphering key K reflection $T_{l,k}^{(K)} : V_l \times V_k \rightarrow V_l$ $l, k \in (128, 256, 512)$, herewith $k = l$ or $k = 2l$, that was being realized in the form of iterative application of a row of functions, which handle the input argument $x \in V_l$ as a matrix of internal state of the size $8 \times c$ bytes, which

Table 5.2 Parameters of basic cryptoconversion

No.	Block size (l)	Key length (k)	Number of transformation iterations (t)	Number of columns in the matrix (c)
1	128	128	10	2
2		256	14	
3	256	256	14	4
4		512	18	
5	512	512	18	8

contains the elements of the field $GF\left(2^8\right)$. The basic deciphering modification $U_{l,k}^{(K)}$ is a parametrized ciphering key K reflection, inverse to $T_{l,k}^{(K)}$, also realized in the form of an iterative modification. The dependence on the iterations number (t) during realization of modifications $T_{lk}^{(K)}$ and $U_{lk}^{(K)}$, number of columns in the internal state matrix (c) on the block size (l) and on the length of the ciphering key (k) is being presented in Table 5.2

The working mode of the cryptographic algorithm is designated as follows «Kalyna-*l*/*k*-mode sign—mode parameters» (parameters absent for some modes), where l—size of the basic modification block, k—the key length. For instance, Kalyna-256/512-CCM-32,128 determines using of basic modification with the block size 256 bits, the key length 512 bits, using in the mode of elaborating of a message authentication code and inhibition, the length of confidential (and open) part of the message is always less than 232 bites, the length of the message authentication code is equal with 128 bits. The simple substitution mode coincides with the basic modification, therefore besides the sign «Kalyna-*l*/*k*-ECB» can be used the sign «Kalyna-*l*/*k*» [6].

5.1.2 Introducing of Input and Output Data, Internal Modification State

The basic modification executes the processing of input block with the length of l bits (clear text in the case of enciphering or cipher-text in the case of deciphering).

The internal state matrix is denoted as $G = \left(g_{i,j}\right)$, $g_{i,j} \in GF\left(2^8\right)$, where $i = \overline{0, 7}$, $j = \overline{0, c-1}$. The recording of the bytes $B_1, B_2, \ldots B_{l8}$ $T_{l,k}^{(K)}$ and $U_{l,k}^{(K)}$ into the matrix and reading from it is being executed columnwise. An example of bytes recording for the internal state for $l = 512 (k = 512, c = 8)$ see in Table 5.3.

Table 5.3 Internal state structure

Input sequence							
B_1	B_9	B_{17}	B_{25}	B_{33}	B_{41}	B_{49}	B_{57}
B_2	B_{10}	B_{18}	B_{26}	B_{34}	B_{42}	B_{50}	B_{58}
B_3	B_{11}	B_{19}	B_{27}	B_{35}	B_{43}	B_{51}	B_{59}
B_4	B_{12}	B_{20}	B_{28}	B_{36}	B_{44}	B_{52}	B_{60}
B_5	B_{13}	B_{21}	B_{29}	B_{37}	B_{45}	B_{53}	B_{61}
B_6	B_{14}	B_{22}	B_{30}	B_{38}	B_{46}	B_{54}	B_{62}
B_7	B_{15}	B_{23}	B_{31}	B_{39}	B_{47}	B_{55}	B_{63}
B_8	B_{16}	B_{24}	B_{32}	B_{40}	B_{48}	B_{56}	B_{64}
Internal state of the basic transformation							
g0,0	g0,1	g0,2	g0,3	g0,4	g0,5	g0,6	g0,7
g1,0	g1,1	g1,2	g1,3	g1,4	g1,5	g1,6	g1,7
g2,0	g2,1	g2,2	g2,3	g2,4	g2,5	g2,6	g2,7
g3,0	g3,1	g3,2	g3,3	g3,4	g3,5	g3,6	g3,7
g4,0	g4,1	g4,2	g4,3	g4,4	g4,5	g4,6	g4,7
g5,0	g5,1	g5,2	g5,3	g5,4	g5,5	g5,6	g5,7
g6,0	g6,1	g6,2	g6,3	g6,4	g6,5	g6,6	g6,7
g7,0	g7,1	g7,2	g7,3	g7,4	g7,5	g7,6	g7,7

5.1.3 Enciphering

5.1.3.1 Modification Structure

The basic enciphering modification $T_{i,k}^{(K)}$ is defined as follows:

$$T_{l,k}^{(K)} : \eta_l^{(K_t)} \psi_l \, \tau_l \, \pi_l' \left(\prod_{v=1}^{t-1} \left(K_l^{(K_v)} \psi_l \, \tau_l \, \pi_l' \right) \right) \eta_l^{(K_0)}$$

where

- l size of the internal state of the block cipher (in bits);
- K ciphering key;
- k length of the ciphering key (in bits);
- $\eta_l^{(K_v)}$ function of addition of the cycle key $K_v (v \in \{0, t\})$ after the modulus $_2{}^{64}$;
- π_l' layer of non-linear bijective reflecting, which executes the processing of vectors, prescribed over V_8 (a bytes substitution);
- τ_l permutation of elements $g_{i,j} \in GF(2^8)$ of the internal state (a cyclic shift of lines to the right at the matrix presentation;
- ψ_l linear modification (multiplying of the linear modification matrix by the internal state matrix over a finite field;

ψ_l function of the addition of the cycle key $K_v (v \in \{1, 2, \ldots, t-1\})$ after the modulus 2 (involutive modification).

In the functions π_l',τ_1 and ψ_1 the input argument $x \in V_l$ and the output value, $\chi(x) \in V_l$, $\chi \in \{\pi_l', \tau_l, \psi_l\}$ are being considered as some matrices $8 \times c$ bytes sized (see Table 5.2). The functions $\eta_l^{(K_v)}$ and $\kappa_l^{(K_v)}$ have two input arguments $x \in V_l$ (internal state of the cipher and $K_v \in V_l$ (cycle key of the i-th iteration) and the output value $\chi(x, K_v) \in V_l$, $\chi \in \left\{\eta_l^{(K_v)}, \kappa_l^{(K_v)}\right\}$, herewith the input arguments and the output value are being considered as the $8 \times c$ bytes sized matrices.

5.1.3.2 Function of Cycle Key Addition K_v After the Modulus 2^{64}

$\eta_l^{(K_v)}$ executes the addition after the modulus 2^{64} of the columns of the internal state matrix $G = (g_{i,j})$ and of the columns of cycle key $K_v = \left(k_{i,j}^v\right)$, herewith the result is also a $8 \times c$ bytes sized matrix (by internal state after addition).

While executing addition, the least significant bytes possess a minor index, i.e. the format little endian is used.

5.1.3.3 Layer of Non-linear Bijective Reflection

The function π_l' executes changing of each element $g_{i,j}$ of the internal state matrix $G = (g_{i,j})$ to $\pi_{i \bmod 4}(g_{i,j})$, where $\pi_s \colon V_8 \to V_8$, $s \in \{0, 1, 2, 3\}$—the substitutions, presented in Table 5.4.

For instance, let be $g_{0,0} = 0x23$, then $\pi_0(0x23) = 0x4F$.

For executing of the modification can be used another substitutions set, different from the presented in Table 5.4. In this case the substitutions set must be delivered and applied in due order.

5.1.3.4 Permutation of Elements

The function τ_1 executes a cyclic right shift of the lines of the state matrix $G = (g_{i,j})$. The amount of the shift elements depends on the line number $i \in \{0, 1, \ldots, 7\}$, on the block size $l \in \{128, 256, 512\}$ and is calculated by the formula $\delta_i = \left\lfloor \frac{i \cdot l}{512} \right\rfloor$.

For example, the 5-th line of the state matrix of the cipher with a 256-bits block is right-shifted by 2 elements.

5.1.3.5 Linear Modification

During the calculation of the function ψ_1 result, each element g_{ij} of the internal state matrix $G = (g_{i,j})$ is considered as an element of the finite field $GF(2^8)$, created by

Table 5.4 Tables for changing for the layer of non-linear bijective reflection (hexadecimal representation)

Substitution π_0

A8	43	5F	06	6B	75	6C	59	71	DF	87	95	17	F0	D8	09
6D	F3	1D	CB	C9	4D	2C	AF	79	E0	97	FD	6F	4B	45	39
3E	DD	A3	4F	B4	B6	9A	0E	1F	BF	15	E1	49	D2	93	C6
92	72	9E	61	D1	63	FA	EE	F4	19	D5	AD	58	A4	BB	A1
DC	F2	83	37	42	E4	7A	32	9C	CC	AB	4A	8F	6E	04	27
2E	E7	E2	5A	96	16	23	2B	C2	65	66	0F	BC	A9	47	41
34	48	FC	B7	6A	88	A5	53	86	F9	5B	DB	38	7B	C3	1E
22	33	24	28	36	C7	B2	3B	8E	77	BA	F5	14	9F	08	55
9B	4C	FE	60	5C	DA	18	46	CD	7D	21	B0	3F	1B	89	FF
EB	84	69	3A	9D	D7	D3	70	67	40	B5	DE	5D	30	91	B1
78	11	01	E5	00	68	98	A0	C5	02	A6	74	2D	0B	A2	76
B3	BE	CE	BD	AE	E9	8A	31	1C	EC	F1	99	94	AA	F6	26
2F	EF	E8	8C	35	03	D4	7F	FB	05	C1	5E	90	20	3D	82
F7	EA	0A	0D	7E	F8	50	1A	C4	07	57	B8	3C	62	E3	C8
AC	52	64	10	D0	D9	13	0C	12	29	51	B9	CF	D6	73	8D
81	54	C0	ED	4E	44	A7	2A	85	25	E6	CA	7C	8B	56	80

(continued)

Table 5.4 (continued)

Substitution π_1

CE	BB	EB	92	EA	CB	13	C1	E9	3A	D6	B2	D2	90	17	F8
42	15	56	B4	65	1C	88	43	C5	5C	36	BA	F5	57	67	8D
31	F6	64	58	9E	F4	22	AA	75	0F	02	B1	DF	6D	73	4D
7C	26	2E	F7	08	5D	44	3E	9F	14	C8	AE	54	10	D8	BC
1A	6B	69	F3	BD	33	AB	FA	D1	9B	68	4E	16	95	91	EE
4C	63	8E	5B	CC	3C	19	A1	81	49	7B	D9	6F	37	60	CA
E7	2B	48	FD	96	45	FC	41	12	0D	79	E5	89	8C	E3	20
30	DC	B7	6C	4A	B5	3F	97	D4	62	2D	06	A4	A5	83	5F
2A	DA	C9	00	7E	A2	55	BF	11	D5	9C	CF	0E	0A	3D	51
7D	93	1B	FE	C4	47	09	86	0B	8F	9D	6A	07	B9	B0	98
18	32	71	4B	EF	3B	70	A0	E4	40	FF	C3	A9	E6	78	F9
8B	46	80	1E	38	E1	B8	A8	E0	0C	23	76	1D	25	24	05
F1	6E	94	28	9A	84	E8	A3	4F	77	D3	85	E2	52	F2	82
50	7A	2F	74	53	B3	61	AF	39	35	DE	CD	1F	99	AC	AD
72	2C	DD	D0	87	BE	5E	A6	EC	04	C6	03	34	FB	DB	59
B6	C2	01	F0	5A	ED	A7	66	21	7F	8A	27	C7	C0	29	D7

(continued)

Table 5.4 (continued)

Substitution π_2

93	D9	9A	B5	98	22	45	FC	BA	6A	DF	02	9F	DC	51	59
4A	17	2B	C2	94	F4	BB	A3	62	E4	71	D4	CD	70	16	E1
49	3C	C0	D8	5C	9B	AD	85	53	A1	7A	C8	2D	E0	D1	72
A6	2C	C4	E3	76	78	B7	B4	09	3B	0E	41	4C	DE	B2	90
25	A5	D7	03	11	00	C3	2E	92	EF	4E	12	9D	7D	CB	35
10	D5	4F	9E	4D	A9	55	C6	D0	7B	18	97	D3	36	E6	48
56	81	8F	77	CC	9C	B9	E2	AC	B8	2F	15	A4	7C	DA	38
1E	0B	05	D6	14	6E	6C	7E	66	FD	B1	E5	60	AF	5E	33
87	C9	F0	5D	6D	3F	88	8D	C7	F7	1D	E9	EC	ED	80	29
27	CF	99	A8	50	0F	37	24	28	30	95	D2	3E	5B	40	83
B3	69	57	1F	07	1C	8A	BC	20	EB	CE	8E	AB	EE	31	A2
73	F9	CA	3A	1A	FB	0D	C1	FE	FA	F2	6F	BD	96	DD	43
52	B6	08	F3	AE	BE	19	89	32	26	B0	EA	4B	64	84	82
6B	F5	79	BF	01	5F	75	63	1B	23	3D	68	2A	65	E8	91
F6	FF	13	58	F1	47	0A	7F	C5	A7	E7	61	5A	06	46	44
42	04	A0	DB	39	86	54	AA	8C	34	21	8B	F8	0C	74	67

(continued)

Table 5.4 (continued)

Substitution π_3

68	8D	CA	4D	73	4B	4E	2A	D4	52	26	B3	54	1E	19	1F
22	03	46	3D	2D	4A	53	83	13	8A	B7	D5	25	79	F5	BD
58	2F	0D	02	ED	51	9E	11	F2	3E	55	5E	D1	16	3C	66
70	5D	F3	45	40	CC	E8	94	56	08	CE	1A	3A	D2	E1	DF
B5	38	6E	0E	E5	F4	F9	86	E9	4F	D6	85	23	CF	32	99
31	14	AE	EE	C8	48	D3	30	A1	92	41	B1	18	C4	2C	71
72	44	15	FD	37	BE	5F	AA	9B	88	D8	AB	89	9C	FA	60
EA	BC	62	0C	24	A6	A8	EC	67	20	DB	7C	28	DD	AC	5B
34	7E	10	F1	7B	8F	63	A0	05	9A	43	77	21	BF	27	09
C3	9F	B6	D7	29	C2	EB	C0	A4	8B	8C	1D	FB	FF	C1	B2
97	2E	F8	65	F6	75	07	04	49	33	E4	D9	B9	D0	42	C7
6C	90	00	8E	6F	50	01	C5	DA	47	3F	CD	69	A2	E2	7A
A7	C6	93	0F	0A	06	E6	2B	96	A3	1C	AF	6A	12	84	39
E7	B0	82	F7	FE	9D	87	5C	81	35	DE	B4	A5	FC	80	EF
CB	BB	6B	76	BA	5A	7D	78	0B	95	E3	AD	74	98	3B	36
64	6D	DC	F0	59	A9	4C	17	7F	91	B8	C9	57	1B	E0	61

an irreducible polynomial $J(x) = x8 + x4 + x3 + x2 + 1$, or *0x11d* in hexadecimal notation.

Each element of the resulting state matrix $W = \left(w_{i,j}\right)$ is being obtained as a result of the length vectors 8 multiplication over the finite field $GF(2^8)$ by the formula

$$w_{i,j} = (v >>> i) \otimes G_j,$$

where $v = (0x01, 0x01, 0x05, 0x01, 0x08, 0x06, 0x07, 0x04)$—the vector, which creates a circulant matrix of the MVD-code and consists of a sequence of byte constants in hexadecimal notation, which are being treated as elements of the field $GF(2^8)$, herewith the cyclic shift is being executed relating to the vector elements over the finite field;

G_j—the j-th column of the state matrix $G = \left(g_{i,j}\right)$.

5.1.3.6 Function of Addition of the Cyclic Key K_v by the Modulus 2

The function $\kappa_l^{(K_1)}$ has two input arguments: $x \in V_l$ (internal state cipher) and $K_v \in V_l$ (cyclic key of the ν-th iteration), of which each is being presented as a $8 \times c$ bytes sized matrix.

$\kappa_l^{(K_v)}$ executes the bitwise addition (after the modulus 2) of the columns of the internal state matrix $G = \left(g_{ij}\right)$ and of the columns of the cyclic key matrix $K_v = \left(k_{i,j}^v\right)$, herewith the result is also a $8 \times c$ bytes sized matrix (internal state after addition).

5.1.4 Deciphering

5.1.4.1 Conversion Structure

The base conversion $U_{l,k}^{(K)}$ has been defined as follows:

$$U_{l,k}^{(K)} =_{-1} \eta_l^{(K_0)} \left(\prod_{v=t-1}^{1} \left({}_{-1}\pi'_{l-1} \tau_{l-1} \psi_l \, \kappa_l^{(K_v)} \right) \right)_{-1} \pi'_{l-1} \tau_{l-1} \psi_{l-1} \eta_l^{(K_t)},$$

where

l	size of internal state block cipher (in bits);
K	ciphering key;
k	length of the ciphering key (in bits);

- ${}_{-1}\eta_l^{(K_v)}$ function of subtraction of the cyclic key $K_v(v \in \{0, t\})$ after the modulus 264 (inverse to $\eta_l^{(K_v)}$);
- ${}_{-1}\psi_l$ inverse linear conversion (multiplication of the inverse linear conversion matrix by the internal state matrix over the finite field);
- ${}_{-1}\tau_l$ inverse permutation of the elements $g_{i,j} \in GF(2^8)$ of the internal state (cyclic left shift of lines at matrix representation);
- ${}_{-1}\pi_l'$ layer of inverse non-linear bijective reflection, which executes the processing of vectors, given over V_8 (inverse byte stuffing);
- $\kappa_l^{(K_v)}$ involutive function of the cyclic key addition $K_v(v \in \{1, 2, \ldots, t-1\})$ after the modulus 2 (identical for enciphering and deciphering).

As in the case of enciphering, in the functions ${}_{-1}\pi', {}_{-1}\tau_l$ and ${}_{-1}\psi_l$ the input argument and the output value $\chi(x) \in V_l$, $\chi \in \{{}_{-1}\pi', {}_{-1}\tau_l, {}_{-1}\psi_l\}$ are being considered as some $8 \times c$ bytes sized matrices.

The function ${}_{-1}\eta_l^{(K_v)}$ has two input arguments $x \in V_l$ (cipher internal state) and $K_v \in V_l$ (cyclic key of i-th iteration) and the output value $(x, K_v) \in V_b$, herewith the input arguments and the output value are being considered as some $8 \times c$ bytes sized matrices.

5.1.4.2 Function of Subtraction of the Cyclic Key K_v After the Modulus 2^{64}

${}_{-1}\eta_l^{(K_v)}$ is an inverse one related to $\eta_l^{(K_v)}$ and executes a subtraction after the modulus 2^{64} columns of the cyclic key matrix $K_v = \left(k_{i,j}^v\right)$ from the columns of the internal state matrix $G = \left(g_{ij}\right)$, herewith the result is also a $8 \times c$ bytes sized matrix (an internal state after subtraction).

While subtraction, the minor significant bytes have a smaller index, i.e. the format little endian is used.

5.1.4.3 Layer of the Inverse Bijective Reflection

The function ${}_{-1}\pi_l'$ executes the substitution of each element g_{ij} of the internal state matrix $G = \left(g_{i,j}\right)$ into ${}_{-1}\pi_{i \bmod 4}\left(g_{i,j}\right)$, where ${}_{-1}\pi_s : V_8 \to V_8, s \in \{0, 1, 2, 3\}$—substitutions, presented in Table 5.5.

For example, let be $g_{0,0} = 0xA3$, then ${}_{-1}\pi_0(0xA3) = 0x22$.

In the case of the using of stuffings, different from those presented in Table 5.5, are used the corresponding inverse.

Table 5.5 Changing tables for the layer of inverse non-linear bijective reflection (hexadecimal representation)

Substitution $-1\pi_0$

A4	A2	A9	C5	4E	C9	03	D9	7E	0F	D2	AD	E7	D3	27	5B
E3	A1	E8	E6	7C	2A	55	0C	86	39	D7	8D	B8	12	6F	28
CD	8A	70	56	72	F9	BF	4F	73	E9	F7	57	16	AC	50	C0
9D	B7	47	71	60	C4	74	43	6C	1F	93	77	DC	CE	20	8C
99	5F	44	01	F5	1E	87	5E	61	2C	4B	1D	81	15	F4	23
D6	EA	E1	67	F1	7F	FE	DA	3C	07	53	6A	84	9C	CB	02
83	33	DD	35	E2	59	5A	98	A5	92	64	04	06	10	4D	1C
97	08	31	EE	AB	05	AF	79	A0	18	46	6D	FC	89	D4	C7
FF	F0	CF	42	91	F8	68	0A	65	8E	B6	FD	C3	EF	78	4C
CC	9E	30	2E	BC	0B	54	1A	A6	BB	26	80	48	94	32	7D
A7	3F	AE	22	3D	66	AA	F6	00	5D	BD	4A	E0	3B	B4	17
8B	9F	76	B0	24	9A	25	63	DB	EB	7A	3E	5C	B3	B1	29
F2	CA	58	6E	D8	A8	2F	75	DF	14	FB	13	49	88	B2	EC
E4	34	2D	96	C6	3A	ED	95	0E	E5	85	6B	40	21	9B	09
19	2B	52	DE	45	A3	FA	51	C2	B5	D1	90	B9	F3	37	C1
0D	BA	41	11	38	7B	BE	D0	D5	69	36	C8	62	1B	82	8F

(continued)

Table 5.5 (continued)

Substitution $-1\pi_0$															
A4	A2	A9	C5	4E	C9	03	D9	7E	0F	D2	AD	E7	D3	27	5B
Substitution $_{-1}\pi_1$															
83	F2	2A	EB	E9	BF	7B	9C	34	96	8D	98	B9	69	8C	29
3D	88	68	06	39	11	4C	0E	A0	56	40	92	15	BC	B3	DC
6F	F8	26	BA	BE	BD	31	FB	C3	FE	80	61	E1	7A	32	D2
70	20	A1	45	EC	D9	1A	5D	B4	D8	09	A5	55	8E	37	76
A9	67	10	17	36	65	B1	95	62	59	74	A3	50	2F	4B	C8
D0	8F	CD	D4	3C	86	12	1D	23	EF	F4	53	19	35	E6	7F
5E	D6	79	51	22	14	F7	1E	4A	42	9B	41	73	2D	C1	5C
A6	A2	E0	2E	D3	28	BB	C9	AE	6A	D1	5A	30	90	84	F9
B2	58	CF	7E	C5	CB	97	E4	16	6C	FA	B0	6D	1F	52	99
0D	4E	03	91	C2	4D	64	77	9F	DD	C4	49	8A	9A	24	38
A7	57	85	C7	7C	7D	E7	F6	B7	AC	27	46	DE	DF	3B	D7
9E	2B	0B	D5	13	75	F0	72	B6	9D	1B	01	3F	44	E5	87
FD	07	F1	AB	94	18	EA	FC	3A	82	5F	05	54	DB	00	8B
E3	48	0C	CA	78	89	0A	FF	3E	5B	81	EE	71	E2	DA	2C
B8	B5	CC	6E	A8	6B	AD	60	C6	08	04	02	E8	F5	4F	A4
F3	C0	CE	43	25	1C	21	33	0F	AF	47	ED	66	63	93	AA

(continued)

Table 5.5 (continued)

Substitution $_{-1}\pi_2$

45	D4	0B	43	F1	72	ED	A4	C2	38	E6	71	FD	B6	3A	95
50	44	4B	E2	74	6B	1E	11	5A	C6	B4	D8	A5	8A	70	A3
A8	FA	05	D9	97	40	C9	90	98	8F	DC	12	31	2C	47	6A
99	AE	C8	7F	F9	4F	5D	96	6F	F4	B3	39	21	DA	9C	85
9E	3B	F0	BF	EF	06	EE	E5	5F	20	10	CC	3C	54	4A	52
94	0E	C0	28	F6	56	60	A2	E3	0F	EC	9D	24	83	7E	D5
7C	EB	18	D7	CD	DD	78	FF	DB	A1	09	D0	76	84	75	BB
1D	1A	2F	B0	FE	D6	34	63	35	D2	2A	59	6D	4D	77	E7
8E	61	CF	9F	CE	27	F5	80	86	C7	A6	FB	F8	87	AB	62
3F	DF	48	00	14	9A	BD	5B	04	92	02	25	65	4C	53	0C
F2	29	AF	17	6C	41	30	E9	93	55	F7	AC	68	26	C4	7D
CA	7A	3E	A0	37	03	C1	36	69	66	08	16	A7	BC	C5	D3
22	B7	13	46	32	E8	57	88	2B	81	B2	4E	64	1C	AA	91
58	2E	9B	5C	1B	51	73	42	23	01	6E	F3	0D	BE	3D	0A
2D	1F	67	33	19	7B	5E	EA	DE	8B	CB	A9	8C	8D	AD	49

(continued)

Table 5.5 (continued)

Substitution $_{-1}\pi_2$

45	D4	0B	43	F1	72	ED	A4	C2	38	E6	71	FD	B6	3A	95
82	E4	BA	C3	15	D1	E0	89	FC	B1	B9	B5	07	79	B8	E1

Substitution $_{-1}\pi_3$

B2	B6	23	11	A7	88	C5	A6	39	8F	C4	E8	73	22	43	C3
82	27	CD	18	51	62	2D	F7	5C	0E	3B	FD	CA	9B	0D	0F
79	8C	10	4C	74	1C	0A	8E	7C	94	07	C7	5E	14	A1	21
57	50	4E	A9	80	D9	EF	64	41	CF	3C	EE	2E	13	29	BA
34	5A	AE	8A	61	33	12	B9	55	A8	15	05	F6	03	06	49
B5	25	09	16	0C	2A	38	FC	20	F4	E5	7F	D7	31	2B	66
6F	FF	72	86	F0	A3	2F	78	00	BC	CC	E2	B0	F1	42	B4
30	5F	60	04	EC	A5	E3	8B	E7	1D	BF	84	7B	E6	81	F8
DE	D8	D2	17	CE	4B	47	D6	69	6C	19	99	9A	01	B3	85
B1	F9	59	C2	37	E9	C8	A0	ED	4F	89	68	6D	D5	26	91
87	58	BD	C9	98	DC	75	C0	76	F5	67	6B	7E	EB	52	CB
D1	5B	9F	0B	DB	40	92	1A	FA	AC	E4	E1	71	1F	65	8D
97	9E	95	90	5D	B7	C1	AF	54	FB	02	E0	35	BB	3A	4D
AD	2C	3D	56	08	1B	4A	93	6A	AB	B8	7A	F2	7D	DA	3F
FE	3E	BE	EA	AA	44	C6	D0	36	48	70	96	77	24	53	DF
F3	83	28	32	45	1E	A4	D3	A2	46	6E	9C	DD	63	D4	9D

5.1.4.4 Inverse Permutation of Elements

The function ${}_{-1}\tau_l$ executes a cyclic right shift of the lines of the state matrix $G = \left(g_{i,j}\right)$. The amount of shifting elements depends on the line number $i \in \{0, 1, .., 7\}$, on the block size $l \in \{128, 256, 512\}$ and is being calculated by the formula

$$\delta_i = \frac{i \cdot l}{512}$$

For example, the 4-th line of the cipher state matrix with a 128-bits block is being shifted to the left by 1 element.

5.1.4.5 Inverse Linear Conversion

While calculating of the result of the function ${}_{-1}\psi_i$ each element $g_{i,j}$ of the internal state matrix $G = \left(g_{i,j}\right)$ is considered as an element of the finite field $GF(2^8)$, that has been created by the irreducible polynomial $\mathrm{J}(x) = x^8 + x^4 + x^3 + x^2 + 1$, or $0x11d$ in hexadecimal representation.

Each element of the resulting state matrix ${}_{-1}W = \left({}_{-1}w_{i,j}\right)$ is obtained as a result of the multiplication of the length vectors 8 over the finite field $GF(2^8)$ by the formula

$${}_{-1}w_{i,j} = ({}_{-1}v >>> i) \otimes G_j$$

where ${}_{-1}v = (0xAD, 0x95, 0x76, 0xA8, 0x2F, 0x49, 0xD7, 0xCA)$—the vector, which creates the circulant matrix of the MVD-code and consists from a sequence of byte constants in hexadecimal representation, which are being treated as some elements of the field $GF(2^8)$, herewith the cyclic shifting is being executed relating to the vector elements over the finite field;

G_j—the j-th column of the state matrix $G = \left(g_{ij}\right)$.

5.1.5 Forming of Cyclic Keys

5.1.5.1 Auxiliary Key $\mathbf{K}_\sigma$

The auxiliary key K_σ has the size of the internal state cipher (l bits), is represented as a $8 \times c$ bytes sized matrix (see Table 5.2) and is being formed on the base of the ciphering key K and by utilizing the conversion $\Theta^{(K)}$:

$$\Theta^{(K)} = \psi_l\ \tau_l\ \pi_l'\ \eta_l^{(K_\alpha)}\ \psi_l\ \tau_l\ \pi_l'\ \kappa_l^{(K_\omega)}\ \psi_l\ \tau_l\ \pi_l'\ \eta_l^{(K_\alpha)}$$

where $\eta_l^{(\cdot)}$, π_l', τ_l, ψ_l, $\kappa_l^{(\cdot)}$—conversions, described in the paragraphs 5.1.3.2–5.1.3.6.

When the key length and the block size of the basic conversion coincide ($k = l$), $K_\alpha = K_\omega = K$ (the second argument of the functions $\eta_l^{(\cdot)}$ and $kappa_l^{(\cdot)}$ is the ciphering key).

If the key length and the block size of the basic conversion does not coincide ($k = 2 \cdot l$) then $K_\alpha \| K_\omega = K$, i.e $K_\alpha = L_{l,l/2}(K)$ and $K_w = R_{l,l/2}(K)$.

For calculating of the value of the auxiliary key K_σ to the input of the conversion $\Theta^{(K)}$ is supplied the l-bit value $\left(\frac{l+k}{64} + 1\right)$, represented in the format *little endian.*

5.1.5.2 Auxiliary Keys with the Even Indices

Each of the cyclic keys $K_0, K_1, \ldots, K_t$ has the size of the cipher internal state (l bits), is supplied as a $8 \times c$ bytes sized matrix and is formed on the base of the ciphering key K, of the auxiliary key K_σ and of the own index i.

The cyclic keys K_i with the even indices (i $\in \{0, 2, \ldots, t\}$) are being formed with the use of the conversion $\Xi^{(K,K_\sigma,i)}$:

$$\Xi^{(K,K_\sigma,i)} = \eta_l^{(\varphi,(K_\sigma)} \ \psi_l \ \tau_l \ \pi_l' \ \kappa_l^{(\varphi_i(K_\sigma)} \ \psi_l \ \tau_l \ \pi_l' \ \eta_l^{(\varphi_i(K_\sigma))}$$

where $\eta_l^{(\cdot)}$, π_l', τ_l, ψ_l, $\kappa_l^{(\cdot)}$—conversions, described in the paragraphs 5.1.3.2–5.1.3.6.

$\varphi_i(K_\sigma)$ the parametrized by the index of an even cyclic key (i) conversion of the auxiliary key K_σ (argument of conversion), which is defined as $\varphi_i = \eta_l^{(\theta \lll (i/2)}$ for the value $\vartheta = \mu_l^{(0x00010001...0001)}$, that has a cipher internal state length.

When the key length and the size of the basic conversion block coincide ($k = l$) for the forming of the cyclic key K_i on the input of the $\Xi^{(K,K_\sigma,i)}$ conversion is being supplied the value ($K >>> 32 \cdot i$), where K—the ciphering key.

If the key length and the size of the basic conversion block do not coincide ($k = 2 \bullet 1$), then for the forming of the cyclic key K_i on the input of the $\Xi^{(K,K_\sigma,i)}$ conversion is being supplied the value $L_{k,l}(K >>> 16 \cdot i)$ for the forming of the cyclic keys K_i with the indices factors of $4 (i = 0, 4, 8, \ldots)$; $R_{k,l}\left(K >>> 64 \cdot \left\lfloor \frac{i}{4} \right\rfloor\right)$ for the forming of the cyclic keys K_i with the indices non-factors of $4 (i = 2, 4, 8, \ldots)$, where K—the ciphering key.

5.1.5.3 Cyclic Keys with Odd Indices

Each of the cyclic keys with odd indices from the precedent key with an even index in accordance with the relationships:

$$K_i = \left(K_{i-1} <<< \left(\frac{l}{4} + 24\right)\right),$$

where l—size of the block cipher internal state (in bits), $i \in \{1, 3, .., t - 1\}$.

5.2 Counter Mode—CTR

The inhibition mode, which is defined in NIST Special Publication 800-38 «Block Cipher Modes» and DSTU 7624 «Kalyna», matches with the general ciphering scheme in accordance with the second using mode of the DSTU GOST 28147:2009 «Information processing systems. Cryptographic protection. Algorithm of cryptographic conversion» and the fifth mode *CTR* from *ISO/IEC* 10116 «Information technology—Security techniques—Modes of operation for an n-bit block cipher». The general enciphering and deciphering scheme in the inhibition mode is designed in Fig. 5.2.

The inhibition mode is meant for confidentiality ensuring by the way of the quick enciphering of the clear text blocks with the possibility of the using of parallel calculations. This way lies in the ciphering of the set of input blocks, which are a counter output, and in the addition (with the aid of the operation of the EXCLUSIVE OR ⊕) of the resulting blocks (keystream blocks) to the blocks of the clear text. Each block on the counter output must be different from another block. This property is not

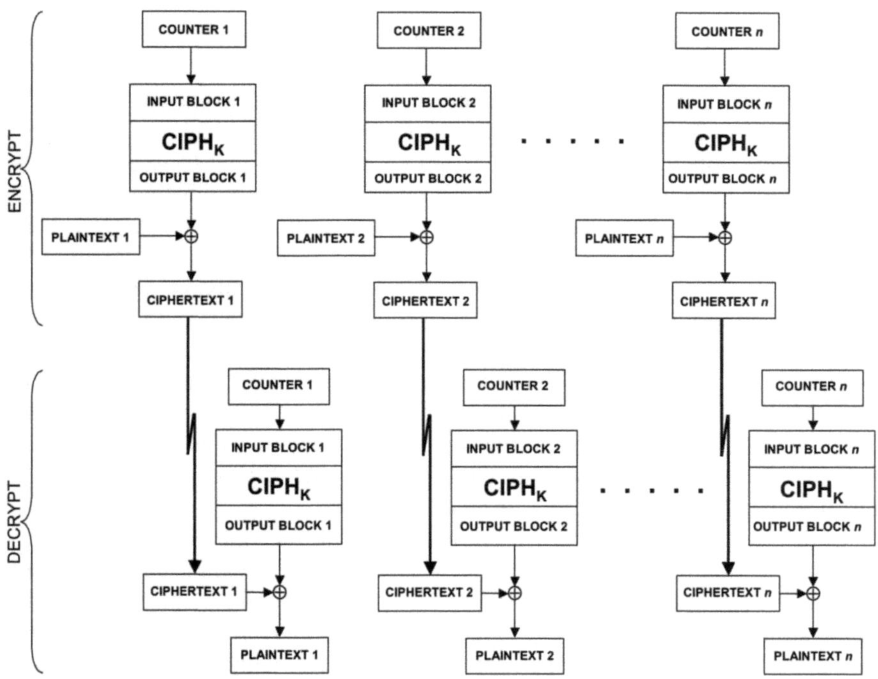

Fig. 5.2 Ciphering and deciphering scheme in the counter mode

restricted to a single informational message, i.e. for all messages liable for ciphering by using of the same key, the outputs of the counter must differ from each other.

Let us denote the blocks from the counter output as $T_1, T_2, \ldots, T_n$. The obtained blocks sequence $T_1, T_2, \ldots, T_n$ is being processed by the inhibition mode thuswise.

Ciphering in inhibition mode:

$$\begin{aligned} &\text{for } j = 1, 2 \ldots n \\ &\quad O_j = CIPH_K(T_j); \\ &\text{for } j = 1, 2 \ldots n - 1 \\ &\quad C_j = P_j \oplus O_j; \\ &C_n^* = P_n^* \oplus MSB_u(O_n). \end{aligned}$$

Deciphering in inhibition mode:

$$\begin{aligned} &\text{for } j = 1, 2 \ldots n \\ &\quad O_j = CIPH_K(T_j); \\ &\text{for } j = 1, 2 \ldots n - 1 \\ &\quad P_j = C_j \oplus O_j; \\ &p_n^* = C_n^* \oplus MSB_u(O_n). \end{aligned}$$

In the case of ciphering in the inhibition mode each activation of the ciphering function $CIPH_K(T_j)$ is utilized immediately and independently to each input block of the counter T_j. The resulting sequence of output blocks O_j is added (through the operation $\oplus$) to the blocks of the clear text P_j, the result of which is a sequence of blocks of the ciphered text C_j. To the last block of the clear text P_n which, possibly, is filled by u bits, are added (through the operation $\oplus$) the u most significant bits (u the most left) of the block O_n. The rest from b–u bits of the block O_n is being truncated.

The deciphering in inhibition mode is executed analogically, i.e. each initiation of the ciphering function $CIPH_K(T_j)$ is utilized immediately and independently to each output block of the counter T_j. Afterwards the resulting sequence of output blocks O_j is added (through the operation $\oplus$) to the blocks of the cipher-text C_j, of which result is the sequence of the open text blocks P_j. To the last block of the cipher-text C_j, which, possibly, is the partial block filled by u bits is added (through the operation $\oplus$) u most significant bits (the u most left bits) of the block O_n. The rest from b–u bits of the block O_n is being truncated.

In the case of using of the inhibition mode both for ciphering, also for deciphering, the multiple initiations of respective functions $CIPH_K(T_j)$ can be realized in parallel. Analogically, the block of the open text, which corresponds to any specific block of the enciphered text, can be processed independently on another block of the open text, if a corresponding counter block has been defined. Besides, the initiations of the cipher functions $CIPH_k(Tj)$ can be applied for the preliminary-formed counter blocks for the further quick enciphering or deciphering of data blocks.

Let us examine the specification of the inhibition mode, which is defined in the national standard DSTU 7624:2014 «Information technologies. Cryptographic protection of information. Algorithm of symmetric block conversion».

5.2.1 General Parameters

The mode ensures a message confidentiality by means of ciphering. The ciphering executes a direct imaging of the message $M\,(|M| \geq 1)$ into a cipher-text C, $|C| = |M|$ and an inverse imaging of the cipher-text C into the message M. The requirements for multiplicity of the message length to the size of the basic conversion block are not imposed. The mode parameters are a ciphering key K, $|K| = k$, and a synchro-reference S, $|S| = l$. Complementary requirements concerning synchro-reference are not being imposed.

5.2.2 Enciphering

The message $M(M \geq 1)$ is being presented as a sequence of blocks:

$$M = m_1 \| m_2 \| \ldots \| m_n\,,\ |m_i| = l \text{ for } i = 1, 2, \ldots, n-1,\ 1 \leq |m_n| \leq l.$$

The initial value of the counter $s_0(|s_0| = l)$ is being calculated as $s_0 = T_{l,k}^{(K)}(S)$.

Each of the blocks of the cipher-text is being calculated in accordance with the relationship: $c_i = m_i \oplus L_{l,|m_i|}\left(T_{l,k}^{(K)}\left(L_{l,l/2}(s_0 + i) \| R_{l,l/2}(s_0)\right)\right)$ for $i = 1, 2, \ldots, n$ $|c_i| = |m_i|$.

The result of the message enciphering is the cipher-text $C = c_1 \| c_2 \| c_n$.

5.2.3 Deciphering

The cipher-text $C(|C| \geq 1)$ is presented as a sequence of blocks:

$$C = c_1 \| c_2 \| \ldots \| c_n\,,\ |c_i| = l \text{ for } i = 1, 2, .., n-1,\ 1 \leq |c_n| \leq l.$$

The initial counter value s_0 is calculated as $s_0 = T_{l,k}^{(K)}(S)$.

Each of the message blocks is being calculated in accordance with the relationship: $m_i = c_i \oplus L_{l,|c_i|}\left(T_{l,k}^{(K)}\left(L_{l,l/2}(s_0 + i) \| R_{l,l/2}(s_0)\right)\right)$ for $i = 1, 2, \ldots, n$. The result of deciphering is the message $M = m_1 \| m_2 \| \ldots \| m_n$.

5.3 Cipher Feedback Mode—CFB

The inhibition mode with the feedback after the cipher-text, defined in the *NIST Special Publication 800-38 «Block Cipher Modes»* and in the DSTU 7624 «Kalyna», matches with the general ciphering scheme in accordance with the third mode of DSTU GOST 28147:2009 «Information processing systems. Cryptographic protection. Algorithm of cryptographic conversion» and with the third *CFB*-mode from *ISO/IEC* 10116 *«Information technology—Security techniques—Modes of operation for an n-bit block cipher»*.

The general scheme of ciphering and deciphering in the *CFB*-mode is presented in the Fig. 5.3. This mode is destined for ensuring of the confidentiality by ciphering of the data stream with multiplying of errors and disabling of manipulations with separate blocks of the open text.

In the base of the mode lies the feedback utilization, which is realized after the cipher-text, i.e. to each block of the open text is added (through the operation $\oplus$) the ciphering result of the anterior cipher-text. In fact, this mode repeats the anterior inhibition mode and differs from it by the forming rule of the keystream—it is formed not as a result of ciphering of the output sequences of the counter, but as a result of ciphering of the anterior block of the cipher-text.

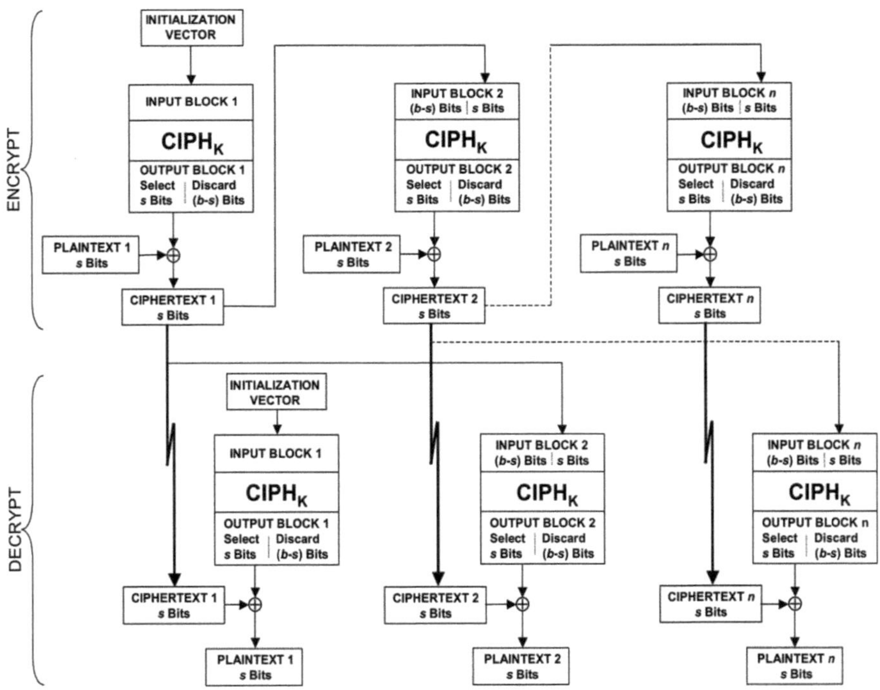

Fig. 5.3 Scheme of ciphering and deciphering in the Cipher feedback mode after the cipher-text

The inhibition mode with the feedback after the cipher-text requires an initial value (initiation vector) *IV* as an initial value while forming of the keystream. This vector can not be secret one, but it must be unpredictable.

The inhibition mode with the feedback after the cipher-text requires also an additional parameter—an integer number s, such as $1 \le s \le b$. After the specification of this mode, each segment of the open text $P_j^{\#}$ values of and each segment of the cipher-text consists of s bits. The values of s are being indicated sometimes in the mode name, for instance, the 1-digit *CFB*-mode, 8-digit *CFB*-mode, 64-digit *CFB*-mode or 128-digit *CFB*-mode.

Formally, the inhibition mode with the feedback after the cipher-text is determined in such a way.

Ciphering in the inhibition mode with the feedback after the cipher-text:

$$\begin{aligned}
&I_1 = IV;\\
&\text{for } j = 2 \ldots n\\
&\quad I_j = LSB_{b-s}(I_{j-1})C_{j-1}^{\#};\\
&\text{for } j = 1, 2 \ldots n\\
&\quad O_j = CIPH_K(I_j);\\
&\text{for } j = 1, 2 \ldots n\\
&\quad C_j^{\#} = P_j^{\#} \oplus MSB_s(O_j).
\end{aligned}$$

The deciphering in the inhibition mode with the feedback after the cipher-text:

$$\begin{aligned}
&I_1 = IV;\\
&\text{for } j = 2 \ldots n\\
&\quad I_j = LSB_{b-s}(I_{j-1})\, C_{j-1}^{\#}\\
&\text{for } j = 1, 2 \ldots n\\
&\quad O_j = CIPH_K(I_j);\\
&\text{for } j = 1, 2 \ldots n\\
&\quad P_j^{\#} = C_j^{\#} \oplus MSB_s(O_j).
\end{aligned}$$

In the case of the ciphering by the first input block for the forming of the keystream is an initiation vector, i.e. $l_1 = IV$. The application over it of the cipher function gives a block of the keystream O_j, of which s major bits are being added (through the operation $\oplus$) to the first block of the open text, i.e. $C\#j = P\#j \oplus MSBs(Oj)$. b–s of the minor digits of the *IV* are being concatenated with the s-bits of the first segment after the cipher text in order to form the second input block $l_j = LSB_{b-s}(l_j - 1)\ C_{j-1}^{\#}$ for generating of the second block of the keystream $O_j = CIPH_K(l_j)$. An alternative way of creating of the second input block l_j lies in those that the bits of the first input blocks are shifted cyclically by s positions to the left, and then the segment of the enciphered text changes s minor digits of the result.

The described process is being repeated with the successive input blocks until all the segments of the ciphered text will be formed. Each input block l_j, which is destined for the forming of the keystream-block, is being ciphered, but *s* major bits as a keystream are being added (through the operation $\oplus$) to the open text block. The result is a segment of the cipher-text which is used for the forming of the next keystream-block. In such a way, while forming of the keystream the feedback with the anterior segment of the cipher-text is being realized. This feedback can be described in the terms of individual segments, if:

$$I_j = (i_1, i_2, \ldots, i_b)$$

is the *j*-th input block for forming of the *j*-th block of the keystream, and

$$C_j = (c_1, c_2, \ldots, c_s)$$

is the *j*-th segment of the ciphered text, then the $(j + 1)$-th input block is

$$I_j = (i_{s+1}, i_{s+2}, \ldots, i_b, c_1, c_2, \ldots, c_s).$$

In the case of deciphering after the cipher-text, the first input block for forming of the keystream is the initiation vector, i.e. $l_1 = IV$.

Each following input block for forming of the keystream, as in the ciphering case, is being linked with the *b–s* minor digits of the anterior input block $LSB_{b-s}(I_{j-1})$ and with the *s* major bits of the anterior ciphered segment $C\#j - 1$, i. e. $I_j = LSB_{b-s}(l_{j-1})/C^{\#}_{j-1}$.

The ciphering function is applied to each input block *lj*, that gives a block of the keystream $O_j = CIPH_K(I_j)$. *s* major bits of the keystream are being added (through the operation $\oplus$) to the respective segments of the cipher-text $C\#\, j$, what enables to get a segment of the open text $P\#j = C\#j \oplus MSBs(Oj)$.

In the case of ciphering in the *CFB*-mode, the input block l_j (besides the first $l_1 = IV$) depends on the ciphering result of the anterior block, consequently the ciphering of separate blocks cannot be realized in parallel. In the case of deciphering and if the respective blocks of the cipher-text are known, the initiations of the ciphering function can be realized in parallel.

Let us examine the specification of the inhibition mode with the feedback after the cipher-text, which is defined in the national standard DSTU 7624:2014 «Information technologies. Cryptographic information protection. Algorithm of symmetric block conversion».

5.3.1 General Parameters

The mode ensures the message confidentiality by the method of ciphering. The ciphering executes a direct imaging of the message $M(|M| \geq 1)$ into a cipher-text, $C\ |C| = |M|$, and an inverse reflecting of the cipher-text C into the message M. The requirements concerning the multiplicity of the message length relating to the size of the basic conversion block are not being imposed.

The mode parameters are the ciphering key $K, |K| = k$, the synchro-reference S, $|S| = l$ and the completive value q, which determines the bits quantity of the message that are being processed by an application of the basic conversion, $q \in \{1, 8, 64, 128, 256, 512\ q \leq l\}$. The recommended value of the parameter is $q = l$. A completive requirement for the synchro-reference in this mode is the occasionality, inclusive the unpredictability of the value, which will be applied to any message, up to its forming.

5.3.2 Enciphering

The message $M(M \geq 1)$ is supplied as a blocks sequence:

$$M = m_1 \| m_2 \| \ldots \| m_n\,,\ |m_i| = q \text{ for } i = 1, 2, \ldots, n-1,\ 1 \leq |m_n| \leq q.$$

The value is being established $c_0^{\#} = T_{l,k}^{(K)}(S)$, $\left|c_0^{\#}\right| = l$.

Each of the blocks of the cipher-text $ci(|ci| = q)$ is being calculated in accordance with the relationship $c_i = m_i \oplus R_{l,|m_i|}\left(C_{i-1}^{\#}\right)$ for $i = 1, 2, \ldots, n$ and $c_i^{\#} = T_{l,k}^{(K)}\left(L_{l,l-q}\left(C_{i-1}^{\#}\right) \| c_i\right)$ for $i = 1, 2, \ldots, n-1$, $\left|c_i^{\#}\right| = l$.

A result of the message enciphering is the cipher-text $C = c_1 \| c_2 \| \ldots \| c_n$.

5.3.3 Deciphering

The cipher-text $C(|C| \geq 1)$ is being supplied as a sequence of blocks: $C = c_1 \| c_2 \| \ldots \| c_n, \cdot |c_i| = q$ for $i = 1, 2, \ldots, n-1$, $1 \leq |c_n| \leq q$.

The value is being established $c_0^{\#} = T_{l,k}^{(K)}(S)$.

Each of the message blocks is being calculated in accordance with the relationship $m_i = c_i \oplus R_{l,|c_i|}\left(c_{i-1}^{\#}\right)$ for $i = 1, 2, \ldots, n$ and $c_i^{\#} = T_{l,k}^{(K)}\left(L_{l,l-q}\left(c_{i-1}^{\#}\right) \| c_i\right)$ for $i = 1, 2, \ldots, n-1$.

The result of the deciphering is the message $M = m_1 \| m_2 \| \ldots . \| m_n$.

5.4 Cipher Block Chaining—CBC

The mode of the cipher-blocks chaining is defined in the NIST Special Publication 800-38 «Block Cipher Modes» and in the DSTU 7624 «Kalyna», destined for ensuring of confidentiality, in which the ciphering process is built through integrating («chain forming») of the open text blocks with anterior blocks of the ciphered text. It complies with the second *CBC*-mode from the *ISO/IEC* 10116 «*Information technology—Security techniques—Modes of operation for an n-bit block cipher*».

The mode of cipher-blocks chaining requires, that initiation vector *IV* would be integrated with the first block of the open text. Herewith the vector *IV* must not be secret, but it has to be unpredictable.

The mode of cipher-blocks chaining is being defined in such a way.

Ciphering in the mode of cipher-blocks chaining:

$$
\begin{aligned}
&C_1 = CIPH_K(P_1 \oplus IV);\\
&for\ j = 2 \ldots n\\
&\quad C_j = CIPH_K\left(P_j \oplus C_{j-1}\right).
\end{aligned}
$$

Deciphering in the mode of cipher-blocks chaining:

$$
\begin{aligned}
&P_1 = CIPH_K^{-1}(C_1) \oplus IV;\\
&for\ j = 2 \ldots n\\
&P_j = CIPH_K^{-1}\left(C_j\right) \oplus C_{j-1}.
\end{aligned}
$$

The first input cipher-block is being formed as a result of the EXCLUSIVE OR (i.e. through the operation $\oplus$) of the first block of the open text with the initiation vector *IV*. The ciphering function is being applied to the first input block, its output is the first block of the ciphered text. This output block also is being added (through the operation $\oplus$) to the second data block of the open text, in order to form the second output block of ciphering. Further away the ciphering function is being applied again, which is forming the second output block, that is the second block of the ciphered text. The output block is being added through the EXCLUSIVE OR to the block of the usual text, in order to form a following input block, and the operations are being repeated. In such a manner, each consequent block of the open text is being added through the EXCLUSIVE OR to the anterior result of ciphering, in order to form a new input block. The ciphering function is being applied to each input block in order to form a block of the ciphered text. The ciphering and deciphering scheme in the mode of cipher-blocks chaining is presented in the Fig. 5.4.

While deciphering in the mode of cipher-blocks chaining, the deciphering function is being applied to the first block of the ciphered text, its result is an output block, which is being added through the operation of the EXCLUSIVE OR (the operation $\oplus$) with the initiation vector *IV*, in order to create the first block of the open text. The inverse function of the cipher is applied to the second block of the ciphered

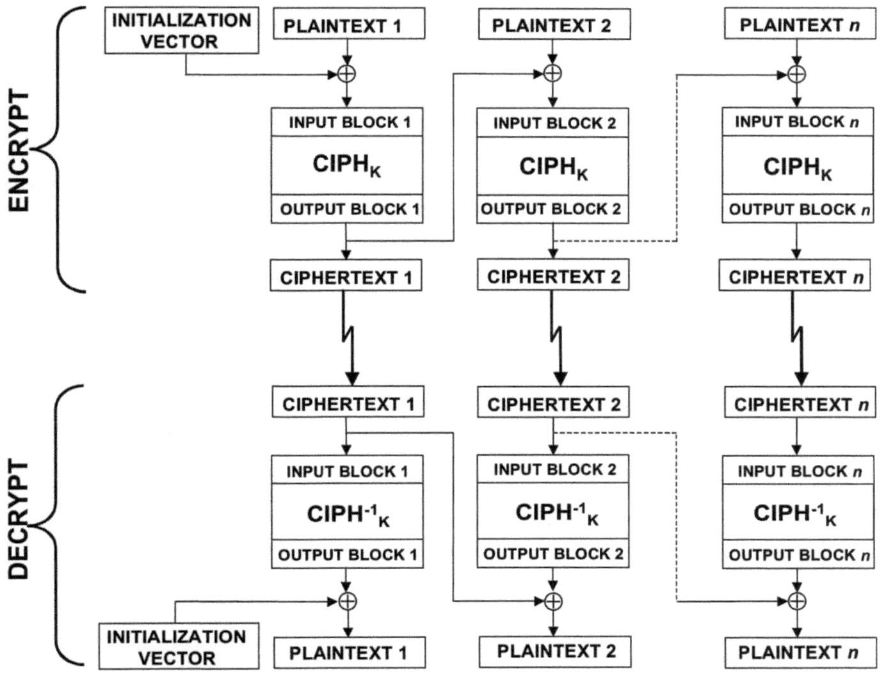

Fig. 5.4 Scheme of ciphering and deciphering in the Cipher block chaining mode

text, its output block through the operation of the EXCLUSIVE OR is being added to the first block of the ciphered text in order to form the second block of the open text. Generally, in order to form any block (besides the first) of the open text, the inverse cipher function is applied to the respective block of the ciphered text, and the resulting block is being added through the EXCLUSIVE OR with the anterior block of the ciphered text.

While ciphering in the mode of cipher-blocks chaining, each input ciphering block depends on the result of the anterior ciphering, for this reason the ciphering cannot be executed in parallel. However, while deciphering the input blocks (the blocks of the enciphered text) are accessible immediately, because of this the multiple deciphering can be executed in parallel.

Let us examine the specification of the cipher-blocks chaining, which is defined in the national standard DSTU 7624:2014 «Information technologies. Cryptographic information protection. Algorithm of the symmetric block conversion».

5.4.1 General Parameters

The mode ensures the confidentiality in the way of ciphering.

The ciphering executes a direct imaging of the message $M(|M| = l \cdot n$, where n—a positive integer) into a cipher-text C, $|C| = |M|$, and an inverse imaging of the cipher-text C into the message M.

The mode parameters are the ciphering key K, $|K| = k$, the synchro-reference S, $|S| = l$.

An additional requirement for the synchro-reference in this mode is the casualty (unpredictability of the value, which will be applied to any message).

The mode of the cipher-blocks chaining is designed as Kalyna-*l/k-CBC*.

5.4.2 Enciphering

The message M being supplied as a sequence of blocks $M = m_1\|m_2\| \ldots \|m_n, |m_i| = l$ for $i = 1, 2, \ldots, n$.

The value $c0 = S$ is established.

For $i = 1, 2, \ldots, n$ are being calculated $c_i = T_{l,k}^{(K)}(c_{i-1} \oplus m_i)$.

The result of enciphering is the cipher-text $C = c_1\|c_2\| \ldots \|c_n$.

5.4.3 Deciphering

The cipher-text $C(|C| = l \cdot n$, where n—a positive integer) is supplied as a sequence of blocks: $C = c_1\|c_2\| \ldots \|c_n, |c_i| = l$ for $i = 1, 2, \ldots, n$.

The value $c_0 = S$ is established.

For $i = 1, 2, \ldots, n$ are being calculated $m_i = c_{i-1} \oplus U_{l,k}^{(K)}(c_i)$.

The deciphering of the cipher-text result is the message $M = m_1\|m_2\| \ldots \|m_n$.

5.5 Output Feedback—OFB

The inhibition mode with feedback after the ciphering keystream, defined in the *NIST Special Publication 800-38* «*Block Cipher Modes*» and in the DSTU 7624 «Kalyna», corresponds to the fourth *OFB*-mode from *ISO/IEC* 10116 «*Information technology—Security techniques—Modes of operation for an n-bit block cipher*» and it has been destined for ensuring of confidentiality.

This mode is based on the initiation vector *IV* ciphering for generating of a sequence of output blocks (cipher-keystream), which are being added through the EXCLUSIVE OR operation (operation $\oplus$) to the usual text in order to form a ciphered text, and inversely, to the cipher-text for its deciphering.

The mode requires that the initiation vector *IV* should be unique for each appliance with the given (fixed) key.

The inhibition mode with feedback after the ciphering keystream is being defined in such a way.

The ciphering in the inhibition mode with feedback after the ciphering keystream:

$$
\begin{aligned}
& I_1 = IV; \\
& for\ j = 2 \ldots n \\
& \quad I_j = O_{j-1}; \\
& for\ j = 1, 2 \ldots n \\
& \quad O_j = CIPH_K(I_j); \\
& for\ j = 1, 2 \ldots n - 1 \\
& \quad C_j = P_j \oplus O_j; \\
& C_n^* = P_n^* \oplus MSB_u(O_n).
\end{aligned}
$$

Deciphering in the inhibition mode with feedback after the ciphering keystream:

$$
\begin{aligned}
& I_1 = IV; \\
& for\ j = 2 \ldots n \\
& \quad I_j = O_{j-1}; \\
& for\ j = 1, 2 \ldots n \\
& \quad O_j = CIPH_K(I_j); \\
& for\ j = 1, 2 \ldots n - 1 \\
& \quad P_j = C_j \oplus O_j; \\
& P_n^* = C_n^* \oplus MSB_u(O_n).
\end{aligned}
$$

While ciphering to the initiation vector *IV* is being applied the ciphering function for getting of the first output block (the first keystream block). The first keystream block through the operation of the EXCLUSIVE OR is being added to the first block of the open text, the result is the first block of the ciphered text. The ciphering function is being applied again at the first keystream block for getting the second keystream block. The obtained second keystream block through the operation of EXCLUSIVE OR is being added to the second block of the open text, as a result—the second block of the ciphered text. The ciphering function is being initiated again, and operations are being repeated. In such a way, the consequent output blocks (keystream blocks) are being generated through application of the ciphering function to the anterior keystream blocks, which through application of the EXCLUSIVE OR operation is being added to the respective blocks of the open text for the forming blocks of the ciphered text. To the last block of the open text, which is, probably, a partial filled block with u bits, are being added through the EXCLUSIVE OR operation the u most significant bits of the last keystream block; the b–u bits of the remaining last keystream block are being truncated.

While deciphering, to the initiation vector *IV* is applied the ciphering function for forming of the first keystream block (first output block). The first keystream block through the operation of EXCLUSIVE OR is being added to the first block of the ciphered text, the result is the first block of the open text. The first keystream block is being processed by the ciphering function for forming of the second keystream block. The second keystream block through the operation of EXCLUSIVE OR is being added to the second block of the ciphered text for getting of the second block of the open text. The second keystream block is being processed by the ciphering function again, and the operation is being repeated. In such a way, the consequent output blocks (keystream blocks) are being generated by applying of the ciphering function to the antarior keystream blocks, which through the operation of EXCLUSIVE OR are being added to the respective blocks of the ciphered text for forming of the open text blocks. To the last block of the ciphered text, which is, probably, a partially filled block from u bits, are being added the u most significant bits of the last keystream block; the remaining $b - u$ bits of the last keystream block are being truncated. The ciphering and deciphering scheme in the inhibition mode with feedback after the ciphering keystream is presented in the Fig. 5.5.

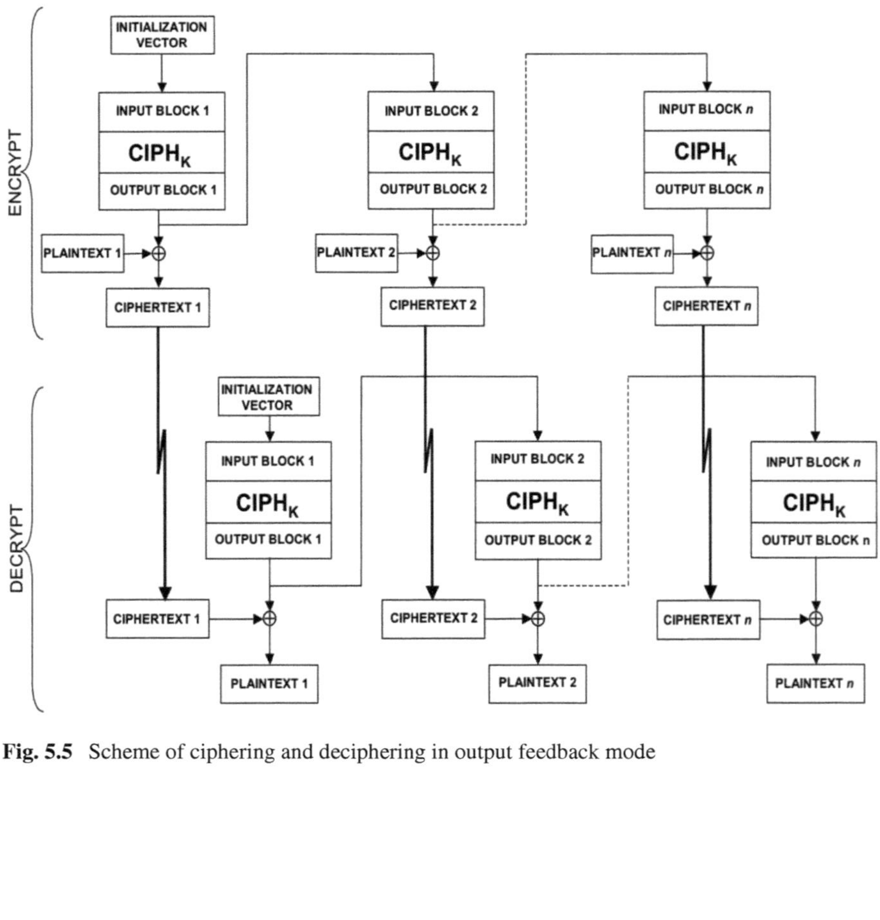

Fig. 5.5 Scheme of ciphering and deciphering in output feedback mode

Both while ciphering, and while deciphering in the inhibition mode after the ciphering keystream each result (besides the first) of the ciphering function depends on the results of the precedent ciphering. Because of this, the multiple initiations of the ciphering function cannot be executed in parallel. However, if the initiation vector *IV* is known, the keystream blocks can be generated by anticipation before obtaining of data of the open or of the ciphered text.

The inhibition mode with feedback after the ciphering keystream requires a unique value of the initiation vector *IV* for each message, which will be some enciphered by the given key. If this requirement will be violated, i.e. the initiation vector *IV* will be used for ciphering of more than one message, the confidentiality of such messages, possibly, will be endangered. In particular, if a block of the usual text of any from these messages is known, tell the *j*-th block of the usual text, then the *j*-th block of the keystream can be easily determined by the *j*-th block of the enciphered text of the message. This information enables to obtain the *j*-th block of the cipher-text for any other message, enciphered by using of the same initiation vector *IV*.

The confidentiality, possibly, can be endangered as well, if any of the keystream blocks is being used as an initiation vector while ciphering of another message with the given key.

Let us examine the specification of the inhibition mode with feedback after the ciphering keystream, which is defined in the national standard DSTU 7624:2014 «Information technologies. Cryptographic information protection. Algorithm of symmetric block conversion».

5.5.1 General Parameters

The mode ensures the confidentiality of the message be ciphering.

The ciphering executes a direct message imaging $M (M \geq 1)$ into a cipher-text C, $|C| = |M|$, and an inverse imaging of the cipher-text C into amessage M.

The requirements concerning multiplicity of the message length to the size of the basic conversion block are not imposed.

The mode parameters are the ciphering key K, $|K| = k$ and the synchro-reference S, $|S| = l$.

Additional requirements concerning the synchro-reference are not being imposed.

The inhibition mode is designed as Kalyna-*l*/*k*-*OFB*.

5.5.2 Enciphering

The message $M (|M| \geq 1)$ is being given as a sequence of blocks:

$$M = m_1 \| m_2 \| \ldots \| m_n, \; |m_i| = l \text{ for } i = 1, 2, \ldots, n-1, \; 1 \leq |m_n| \leq l.$$

The initial value of the keystream block $\gamma_0(|\gamma_0| = l)$ is calculated as $\gamma_0 = T_{l,k}^{(K)}(S)$.

Each of the blocks of the cipher-text is being calculated in accordance to the relationship: $c_i = m_i \oplus L_{l,|m_i|}(\gamma_{i-1})$ for $i = 1, 2, \ldots, n$ and $\gamma_i = T_{l,k}^{(K)}(\gamma_{i-1})$ for $i = 1, 2, \ldots, n-1$.

The result of the message enciphering is the cipher-text $C = c_1 \| c_2 \| \ldots \| c_n$.

5.5.3 *Deciphering*

The cipher-text $C(|C| \geq 1)$ is being supplied as a sequence of blocks:

$$C = c_1 \| c_2 \| \ldots \| c_n, \; |c_i| = l \text{ for } i = 1, 2, \ldots, n-1, \; 1 \leq |c_n| \leq l.$$

The initial value of the keystream block $\gamma_0(|\gamma_0| = l)$ is being calculated as $\gamma_0 = T_{l,k}^{(K)}(S)$.

Each of the message blocks is being calculated accordingly with the relationship $m_i = c_i \oplus L_{l,|m_i|}(\gamma_{i-1})$ for $i = 1, 2, \ldots, n$ and $\gamma_i = T_{l,k}^{(K)}(\gamma_{i-1})$ for $i = 1, 2, \ldots, n-1$.

The result of deciphering of the ciphered-text is the message $M = m_1 \| m_2 \| \ldots \| m_n$.

5.6 Conclusions and Recommendations

1. For the cryptographic protection of the information in contemporary ITS are being used the BSC, of which efficiency is determined not only by the properties of the basic conversion, but also by the properties of the selected mode of the BSC-using. Certain specificities of stream ciphering modes determine their utilizing for the cryptographic protection of important state information, privacies and other prescribed by law secrets. In truth, the level of credibility for stream modes is the highest, because in this case the data for being processed are supplied as a stream of interlinked symbols, and a cramming of false information or erroneous working modes of end user equipment is being complicated substantially.
2. The performed analysis of the valid regulations has shown that the most used and widespread have become the stream BSC-modes: Counter—CTR, Cipher Feedback—CFB, Cipher Block Chaining—CBC, Output Feedback—OFB. The specifications of shown modes are known, they are presented in numerous national and international standards, in special in the NIST Special Publication 800-38 «Block Cipher Modes», in the DSTU GOST 28147:2009 «Information processing systems. Cryptographic protection. Alhorithm of cryptographic conversion» and in the DSTU 7624 «Information technologies. Cryptographic information protection. Algorithm of symmetric block conversion».

3. The national standards of the USA and of Ukraine determine the largest list of BSC-modes using, i.e. by these normative documents is being provided the most complete branch for using of the cryptographic information protection by the aid of BSC:

    - ECB is the simplest one and consists in ensuring of confidentiality of separate blocks of the open text in the way of their ciphering after the introduced secret key. I.e. each block of the open text is being confronted a block of the cipher-text, what is some analogy with attribution of a codeword in the ciphering book;
    - CTR is destined for ensuring of confidentiality through a quick ciphering of the blocks of the open text with the possibility of parallel calculations applying. It consists in ciphering of the set of input blocks, which are the counter output, and in adding (with the aid of the EXCLUSIVE OR $\oplus$) of the resulting blocks (the keystream blocks) to the blocks of the open text. Each block on the counter output has to be distinct of another block. This property is not limited to one informational message, i.e, for all messages, which are liable to be ciphered with the same key, the blocks of the counter output must differ one from another;
    - CFB is destined for ensuring of confidentiality by adding (through the operation $\oplus$) to each block of the open text of the result of ciphering of precedent cipher-text. This mode repeats, in its essence, the anterior inhibition mode and differs from it by the keystream-applience rule—it is formed not as a result of ciphering of the output sequences of the counter, but as a result of ciphering of the precedent block of the cipher-text;
    - the cipher-blocks chaining (CBC) is destined for ensuring of confidentiality, in which the ciphering process is built by association («forming of a chain») of blocks of the open text with precedent blocks of the ciphered text;
    - OFB is based on the ciphering of the initiation vector *IV* for generation of the sequence of output blocks (cipher-keystream) which are being added through the operation of EXCLUSIVE OR (operation $\oplus$) to the usual text in order to form the enciphered text, and, vice-versa, to the cipher-text for its deciphering.

4. The performed analysis and the comparative researches of indicated stream BSC-modes have shown that their application enables to ensure a high level of cryptographic stability and velocity at ensuring of the information confidentiality in defined supplements. In particular, the national standards of the USA and of Ukraine have being determined as far as 10 BSC application modes, which enables to ensure a cryptographic conversion in various cryptographic appliances.

References

1. Dolmatov, V.: GOST 28147-89: Encryption, Decryption, and Message Authentication Code (MAC) Algorithms. RFC Editor (2010). https://doi.org/10.17487/rfc5830
2. ISO/IEC 10116:2004. Information technology—Security techniques—Modes of operation for an nbit block cipher [Electronic resource]. Access mode http://www.iso.org/iso/home/store/catalogue_tc/catalogue_detail.htm?csnumber=38761
3. DSTU 7624: 2014 Information technologies. Cryptographic information protection. Symmetric block transformation algorithm, 227 p. Ministry of Economic Development of Ukraine, Kyiv
4. A New Encryption Standard of Ukraine: The Kalina Block Cipher [Electronic resource]. Access mode https://eprint.iacr.org/2015/650.pdf
5. NIST Special Publication 80038. Recommendation for Block Cipher Modes of Operation [Electronic resource]. Access mode http://csrc.nist.gov
6. DSTU 7624: 2014. Information Technology. Cryptographic information protection. Symmetric block transformation algorithm. Enter. 01–07–2015. Ministry of Economic Development of Ukraine, Kyiv (2015)

Chapter 6
Comparison of Stream Modes in Block Symmetric Ciphers

For researching of statistical security of contemporary cryptographic algorithms the most frequently used are the testing methods *NIST STS*, *DIEHARD* and others [1–3]. Their essence lies in verification of the hypothesis about a incidental character of the output sequence of the researched cryptoalgorithm, i.e. in verification of the supposition that the formed data does not differ statistically from some hypothetical «incidental» sequence. Otherwise telling, a hypothesis is being made that the researched cryptoalgorithms is a generator, output sequences of which are a concrete realization of the random process. For instance, by the procedure of *NIST STS*, the given hypothesis is being verified after 15 independent statistical tests (taking into consideration various input parameters, 188 tests are being performed), after each of those the respective test passing probability P_j, $j = 1, \ldots 188$ is being calculated. This probability is being used by the rule of sentence acceptance (fitting criterion) about the hypothesis verity or falsity, i.e. if the probability value P_j of *j*-th test passing is not lower as some threshold value, for example, if $P_j \geq 0{,}99$, then the hypothesis H_0 is accepted (at the *j*-th test). Otherwise, is accepted an alternative hypothesis.

Consequently, in the result of the statistical testing of the output sequence of the cryptoalagorithm (generator) the vector $P = \{P_1, P_2 \ldots P_n\}$ is being formed of P_j, where n—the number of independent tests after the respective methodology (in *NIST STS* the value $n = 188$). The analysis of the elements P_j of the vector P enables to show the concrete defects of «randomness» of the tested sequence, i.e. the low P_j values show on the apparent difference of the researched sequence from the realization of the random process, discovered by the *j*-th statistical test. For ensuring a high authenticity of statistical testing results in this work has been estimated the mathematical expectancy of the number of the passed tests X_i by the researched generator (cryptoalgorithm). This methodology has been already applied in other searching works as well [4, 5].

The methods of statistical testing was being examined in detail while carrying out comparative researches in Chap. 10.

A. A. Kuznetsov et al., *Stream Ciphers in Modern Real-time IT Systems*,
Studies in Systems, Decision and Control 375,
https://doi.org/10.1007/978-3-030-79770-6_6

6.1 Statistical Researches of the BSC «Kalyna»

6.1.1 Statistical Researches of BSC «Kalyna» in the ECB-Mode

While executing of experimental researches let us concentrate attention on the properties of pseudorandom sequences, in particular on their statistical characteristics, and, in the first instance, on how these properties are being changed during increasing of ciphering cycles number.

Let us examine the pseudorandom sequences, which are being formed while using a certain number of ciphering modes with utilizing of the *ECB*-mode. The obtained results of experimental researches of the statistical security of the BSC «Kalyna» after the earlier developed methodology are summarized in Tables 6.1, 6.2, 6.3, 6.4 and 6.5, in particular:

- in Table 6.1 are presented the results of statistical researches of PRS, which are formed with using of the algorithm Kalyna-128/128;
- in Table 6.2 are presented the results of statistical researches of PRS, which are formed with using of the algorithm Kalyna-128/256;
- in Table 6.3 are presented the results of statistical researches of PRS, which are formed with using of the algorithm Kalyna-256/256;
- in Table 6.4 are presented the results of statistical researches of PRS, which are formed with using of the algorithm Kalyna-256/512;
- in Table 6.5 are presented the results of statistical researches of PRS, which are formed with using of the algorithm Kalyna-512/512.

The obtained results are generalizing the data for $N = 100$ realizations of statistical testing, i.e. the obtained after the results of processing of 100 pseudorandom sequences according to (6.1)–(6.5). In the first column of Tables 6.1, 6.2, 6.3, 6.4

Table 6.1 Results of researches of PRS with using of BSC Kalyna-128/128

	M099	D099	S099	P099	M096	D096	S096	P096	Min096
KALYNA_ECB_1	133.64	94.34	9.71	0.96	183.63	2.82	1.68	1.00	174
KALYNA_ECB_2	132.33	23.97	4.90	1.00	186.85	0.29	0.54	1.00	183
KALYNA_ECB_3	131.07	46.30	6.80	1.00	186.67	0.80	0.90	1.00	181
KALYNA_ECB_4	132.50	29.99	5.48	1.00	186.80	0.51	0.72	1.00	182
KALYNA_ECB_5	133.35	29.94	5.47	1.00	186.86	0.76	0.87	1.00	180
KALYNA_ECB_6	133.10	26.39	5.14	1.00	186.72	1.09	1.05	1.00	179
KALYNA_ECB_7	132.31	29.05	5.39	1.00	186.79	1.08	1.04	1.00	180
KALYNA_ECB_8	132.63	25.19	5.02	1.00	186.72	0.60	0.77	1.00	181
KALYNA_ECB_9	131.57	35.13	5.93	1.00	186.75	0.83	0.91	1.00	181
KALYNA_ECB_10	131.87	33.41	5.78	1.00	186.62	0.70	0.83	1.00	180

Table 6.2 Results of researches of PRS with using of BSC Kalyna-128/256

	M099	D099	S099	P099	M096	D096	S096	P096	Min096
KALYNA_ECB_1	132.83	94.57	9.72	0.96	183.74	1.26	1.12	1.00	176
KALYNA_ECB_2	132.04	25.14	5.01	1.00	186.81	0.28	0.53	1.00	183
KALYNA_ECB_3	132.29	27.86	5.28	1.00	186.70	0.49	0.70	1.00	182
KALYNA_ECB_4	131.81	41.57	6.45	1.00	186.77	0.88	0.94	1.00	179
KALYNA_ECB_5	133.52	35.85	5.99	1.00	186.84	0.52	0.72	1.00	182
KALYNA_ECB_6	131.25	21.54	4.64	1.00	186.64	0.12	0.35	1.00	184
KALYNA_ECB_7	134.41	23.37	4.83	1.00	186.99	0.31	0.56	1.00	182
KALYNA_ECB_8	133.32	19.13	4.37	1.00	186.86	0.53	0.73	1.00	182
KALYNA_ECB_9	133.27	25.76	5.08	1.00	186.87	0.38	0.62	1.00	182
KALYNA_ECB_10	131.86	26.70	5.17	1.00	186.65	0.26	0.51	1.00	183
KALYNA_ECB_11	131.99	32.25	5.68	1.00	186.83	0.52	0.72	1.00	182
KALYNA_ECB_12	133.54	35.26	5.94	1.00	186.81	0.62	0.79	1.00	181
KALYNA_ECB_13	132.55	30.39	5.51	1.00	186.53	1.20	1.10	1.00	179
KALYNA_ECB_14	132.47	35.73	5.98	1.00	186.73	0.13	0.36	1.00	184

Table 6.3 Results of researches of PRS with using of BSC Kalyna-256/256

	M099	D099	S099	P099	M096	D096	S096	P096	Min096
KALYNA_ECB_1	134.60	52.94	7.28	0.99	184.62	1.23	1.11	1.00	178
KALYNA_ECB_2	132.80	29.09	5.39	1.00	186.86	0.53	0.73	1.00	182
KALYNA_ECB_3	132.29	18.55	4.31	1.00	186.86	0.29	0.54	1.00	183
KALYNA_ECB_4	132.55	27.30	5.22	1.00	186.64	0.78	0.89	1.00	180
KALYNA_ECB_5	133.55	32.48	5.70	1.00	186.72	0.27	0.52	1.00	183
KALYNA_ECB_6	133.15	25.05	5.00	1.00	186.85	0.29	0.54	1.00	183
KALYNA_ECB_7	131.06	23.39	4.84	1.00	186.63	0.25	0.50	1.00	183
KALYNA_ECB_8	132.77	33.31	5.77	1.00	186.76	0.50	0.71	1.00	182
KALYNA_ECB_9	131.64	36.80	6.07	1.00	186.73	0.73	0.85	1.00	180
KALYNA_ECB_10	132.32	17.11	4.14	1.00	186.84	0.29	0.53	1.00	183
KALYNA_ECB_11	132.31	31.55	5.62	1.00	186.46	0.73	0.86	1.00	181
KALYNA_ECB_12	132.39	32.10	5.67	1.00	186.79	0.28	0.53	1.00	183
KALYNA_ECB_13	132.10	25.04	5.00	1.00	186.61	1.94	1.39	1.00	177
KALYNA_ECB_14	132.24	18.40	4.29	1.00	187.00	0.31	0.56	1.00	183

and 6.5 is demonstrated the number of ciphering cycles, which have being executed for forming of these pseudorandom sequences, for example: «*KALYNA_ECB_4*»—4 ciphering cycles in the *ECB*-mode have been executed.

Table 6.4 Results of researches of PRS with using of BSC Kalyna-256/512

	M099	D099	S099	P099	M096	D096	S096	P096	Min096
KALYNA_ECB_1	133.19	64.45	8.03	0.99	184.55	0.76	0.87	1.00	179
KALYNA_ECB_2	132.14	29.67	5.45	1.00	186.96	0.31	0.55	1.00	183
KALYNA_ECB_3	132.16	34.96	5.91	1.00	186.68	0.26	0.51	1.00	183
KALYNA_ECB_4	132.02	29.51	5.43	1.00	186.51	0.75	0.87	1.00	181
KALYNA_ECB_5	132.52	34.53	5.88	1.00	186.69	0.48	0.70	1.00	182
KALYNA_ECB_6	132.41	37.05	6.09	1.00	186.85	0.29	0.54	1.00	183
KALYNA_ECB_7	131.64	32.47	5.70	1.00	186.70	0.49	0.70	1.00	182
KALYNA_ECB_8	133.67	31.26	5.59	1.00	186.81	0.62	0.79	1.00	181
KALYNA_ECB_9	132.57	24.79	4.98	1.00	186.56	1.20	1.10	1.00	180
KALYNA_ECB_10	132.13	29.43	5.42	1.00	186.87	0.29	0.54	1.00	183
KALYNA_ECB_11	131.88	22.40	4.73	1.00	186.90	1.04	1.02	1.00	179
KALYNA_ECB_12	131.16	25.40	5.04	1.00	186.52	0.75	0.87	1.00	181
KALYNA_ECB_13	131.91	27.52	5.25	1.00	186.80	0.51	0.72	1.00	182
KALYNA_ECB_14	132.86	24.18	4.92	1.00	186.84	0.63	0.79	1.00	181
KALYNA_ECB_15	132.07	43.90	6.63	1.00	186.80	0.51	0.72	1.00	182
KALYNA_ECB_16	133.49	21.79	4.67	1.00	186.91	0.54	0.74	1.00	182
KALYNA_ECB_17	132.83	27.19	5.21	1.00	186.94	0.66	0.81	1.00	181
KALYNA_ECB_18	132.65	34.71	5.89	1.00	186.94	0.39	0.63	1.00	182

In Tables 6.1, 6.2, 6.3, 6.4 and 6.5 are used such significations:

- «KALYNA_ECB_n»—the obtained results of statistical researches while realization of BSC Kalyna in ECB-mode at executing of n ciphering rounds;
- «M096», «M099»—the estimation of mathematical expectation (selective averages) of the numbers of passed statistical test after the criterion $P_j \geq 0{,}96$ and after the criterion $P_j \geq 0{,}99$ respectively;
- «D096», «D099» («S096», «S099»)—estimations of the dispersion (root-mean-square deviation) of testing results of passed statistical tests after the criteria $P_j \geq 0{,}96$ and $P_j \geq 0{,}99$ respectively;
- «P099»—the calculated value of credible probability for the number of passed statistical tests after the criterion $P_j \geq 0{,}99$ and the faithfulness $\varepsilon = 2$;
- «P096»—the calculated value of credible probability for the number of passed statistical tests after the criterion $P_j \geq 0{,}96$ and the faithfulness $\varepsilon = 1$.

In the last column «Min096» of Tables 6.1, 6.2, 6.3, 6.4 and 6.5 are presented the minimal values of the number of passed statistical tests after the criterion $P_j \geq 0{,}96$.

The obtained results of performed researches demonstrate that the elaborated and accepted as the national standard DSTU 7624 the block cipher «Kalyna» ensures high indices of statistical security. In particular, in each of researched modes the output sequence complies with all statistical tests. Therefore, it can be affirmed that

Table 6.5 Results of researches of PRS with using of BSC Kalyna-512/512

	M099	D099	S099	P099	M096	D096	S096	P096	Min096
KALYNA_ECB_1	134.91	38.14	6.18	1.00	184.64	0.47	0.69	1.00	180
KALYNA_ECB_2	132.21	18.57	4.31	1.00	186.74	0.60	0.78	1.00	181
KALYNA_ECB_3	132.43	23.03	4.80	1.00	186.87	1.16	1.08	1.00	179
KALYNA_ECB_4	131.06	34.92	5.91	1.00	186.51	1.24	1.11	1.00	179
KALYNA_ECB_5	131.72	32.00	5.66	1.00	186.54	0.24	0.49	1.00	183
KALYNA_ECB_6	133.44	25.67	5.07	1.00	186.67	1.56	1.25	1.00	178
KALYNA_ECB_7	131.99	34.15	5.84	1.00	186.87	0.29	0.54	1.00	183
KALYNA_ECB_8	133.12	22.32	4.72	1.00	186.81	0.86	0.92	1.00	181
KALYNA_ECB_9	132.94	22.44	4.74	1.00	186.87	0.38	0.62	1.00	182
KALYNA_ECB_10	132.74	21.76	4.66	1.00	186.74	0.59	0.77	1.00	180
KALYNA_ECB_11	133.10	22.33	4.73	1.00	186.75	0.50	0.71	1.00	182
KALYNA_ECB_12	132.16	26.51	5.15	1.00	186.72	0.49	0.70	1.00	182
KALYNA_ECB_13	131.08	20.56	4.53	1.00	186.76	0.27	0.52	1.00	183
KALYNA_ECB_14	131.02	33.81	5.81	1.00	186.91	0.63	0.79	1.00	180
KALYNA_ECB_15	131.95	25.07	5.01	1.00	186.67	0.67	0.82	1.00	181
KALYNA_ECB_16	132.74	32.14	5.67	1.00	186.77	1.11	1.06	1.00	179
KALYNA_ECB_17	134.30	30.17	5.49	1.00	187.11	0.33	0.58	1.00	183
KALYNA_ECB_18	132.30	27.53	5.25	1.00	186.59	0.46	0.68	1.00	182

the BSC can be considered as a generator of sequences, which are scarcely different statistically from a random sequence. This conclusion can be interpreted even for the output sequences of the cipher just after the first ciphering round, what attests high characteristics of the cipher and a certain stability margin.

6.1.2 Statistical Researches of the BSC «Kalyna» in the CTR-Mode

The obtained results of experimental researches of the BSC «Kalyna» statistical security with using of the *CTR*-mode are summarized in Table 6.6. Herewith the full-round versions of the algorithm were researched.

All designations, used in Table 6.6 and further away, comply with earlier accepted designations, the researches results were being obtained at the same initial conditions.

The obtained results confirm the conclusion about high indices of the statistical security of the cipher «Kalyna». In the CTR application mode the output sequences do not differ statistically from random sequences.

Table 6.6 Results of PRS researches with using of the CTR-mode

	M099	D099	S099	P099	M096	D096	S096	P096	Min096
KALYNA_CTR-128-128	131.97	38.62	6.21	1.00	186.61	0.78	0.89	1.00	181
KALYNA_CTR-128-256	132.09	22.79	4.77	1.00	186.74	0.83	0.91	1.00	181
KALYNA_CTR-256-256	132.63	26.88	5.18	1.00	186.68	0.59	0.77	1.00	181
KALYNA_CTR-256-512	133.22	21.27	4.61	1.00	186.73	0.69	0.83	1.00	181
KALYNA_CTR-512-512	132.63	21.85	4.67	1.00	186.70	0.94	0.97	1.00	180

Table 6.7 The results of PRS researches with using of the CFB-mode

	M099	D099	S099	P099	M096	D096	S096	P096	Min096
KALYNA_CFB-128-128	132.70	33.74	5.81	1.00	187.07	0.33	0.57	1.00	183
KALYNA_CFB-128-256	132.50	25.79	5.08	1.00	186.50	0.44	0.66	1.00	182
KALYNA_CFB-256-256	133.67	29.83	5.46	1.00	186.90	0.30	0.54	1.00	183
KALYNA_CFB-256-512	132.04	20.19	4.49	1.00	186.82	0.71	0.84	1.00	181
KALYNA_CFB-512-512	132.18	26.46	5.14	1.00	186.83	0.86	0.93	1.00	181

6.1.3 Statistical Researches of the BSC «Kalyna» in the CFB-Mode

The obtained results of experimental researches of the statistical security of the BSC «Kalyna» at using of the *CFB*-mode are summarized in Table 6.7. Herewith the full-round versions of the algorithm were researched.

6.1.4 Statistical Researches of the BSC «Kalyna» in the CBC-Mode

The obtained results of experimental researches of the statistical security of the BSC «Kalyna» at using of the *CBC*-mode are summarized in Table 6.8. Herewith the full-round versions of the algorithm were researched.

Table 6.8 The results of PRS researches with using of the CBC-mode

	M099	D099	S099	P099	M096	D096	S096	P096	Min096
KALYNA_CBC-128-128	132.39	32.00	5.66	1.00	186.89	0.73	0.86	1.00	181
KALYNA_CBC-128-256	132.53	20.17	4.49	1.00	186.63	1.06	1.03	1.00	179
KALYNA_CBC-256-256	132.27	31.55	5.62	1.00	186.66	0.58	0.76	1.00	181
KALYNA_CBC-256-512	132.22	28.53	5.34	1.00	187.01	0.68	0.82	1.00	181
KALYNA_CBC-512-512	133.12	36.78	6.06	1.00	186.87	0.73	0.85	1.00	181

Table 6.9 The results of PRS researches with using of the OFB-mode

	M099	D099	S099	P099	M096	D096	S096	P096	Min096
KALYNA_OFB-128-128	132.64	28.88	5.37	1.00	186.71	0.27	0.52	1.00	183
KALYNA_OFB-128-256	132.35	20.53	4.53	1.00	186.63	0.57	0.76	1.00	181
KALYNA_OFB-256-256	132.08	34.86	5.90	1.00	186.74	0.27	0.52	1.00	183
KALYNA_OFB-256-512	132.36	33.58	5.79	1.00	186.77	1.11	1.06	1.00	179
KALYNA_OFB-512-512	131.12	37.12	6.09	1.00	186.50	0.44	0.66	1.00	182

6.1.5 Statistical Researches of the BSC «Kalyna» in the OFB-Mode

The obtained results of experimental researches of the statistical security of the BSC «Kalyna» at using of the *OFB*-mode are summarized in Table 6.9. Herewith the full-round versions of the algorithm were researched.

In the *OFB*-mode the cipher «Kalyna» demonstrates good statistical properties as well and can be considered a generator of random sequences.

Therefore, after the totality of obtained results of researches can be made the conclusion about a high statistical security of the new national block cipher of Ukraine, in all the tested modes it functions as a generator of random sequences.

6.2 Statistical Researches of BSC AES (FIPS-197) and GOST 28147-89 (DSTU GOST 28147:2009) in Different Using Modes

In Table 6.10 are summarized the results of the performed experimental researches of the statistical security of the BSC *AES (FIPS*-197*)* in different application modes. In Table 6.11 the respective data for GOST 28147-89 (DSTU GOST 28147:2009) are presented.

Table 6.10 Results of statistical researches of the BSC AES (FIPS-197) in various using modes

	M099	D099	S099	P099	M096	D096	S096	P096	Min096
AES_CBC_128_256	132.31	32.30	5.68	1.00	186.89	0.29	0.54	1.00	183
AES_CTR_128_256	132.73	30.31	5.51	1.00	186.83	1.70	1.31	1.00	176
AES_ECB_128_256	132.06	19.34	4.40	1.00	186.62	0.66	0.81	1.00	181
AES_OFB_128_256	131.98	29.01	5.39	1.00	186.89	0.29	0.54	1.00	183
AES_CFB_128_256	133.63	18.23	4.27	1.00	186.72	0.60	0.77	1.00	181

Table 6.11 Results of statistical researches of the BSC GOST 28147-89 (DSTU GOST 28147:2009) in various application modes

	M099	D099	S099	P099	M096	D096	S096	P096	Min096
GOST_CBC	131.76	21.74	4.66	1.00	186.67	0.34	0.59	1.00	182
GOST_CTR	130.62	23.71	4.87	1.00	186.88	0.88	0.94	1.00	181
GOST_ECB	131.72	34.17	5.85	1.00	186.77	1.28	1.13	1.00	178
GOST_OFB	131.71	31.68	5.63	1.00	186.64	2.38	1.54	1.00	176
GOST_CFB	132.07	28.30	5.32	1.00	187.02	0.15	0.39	1.00	184

All the significations, used in Tables 6.10 and 6.11, comply with the earlier accepted significations, the researches results have been obtained at the same initial conditions.

In the first column of Table 6.10 are indicated the ciphering mode and parameters, being applied while forming of the PRS. For example, «*AES_CFB*_128_256»—the BSC *AES* in inhibition mode with feedback after the cipher-text (*CFB*) was applied, the data blocks length is equal with 128 bits, the key length is equal with 256 bits. For the BSC the ciphering parameters are unchanged (after specification the data blocks length is equal with 64 bits, the key length constitutes 256 bits).

The obtained results, which are presented in Tables 6.10 and 6.11, testify high indices of the statistical security of the BSC *AES* (*FIPS*-197) and GOST 28147-89 (DSTU GOST 28147:2009). In each of the examined appliance modes the formed BSC have an aspect of pseudorandomness and does not differ statistically from a certain hypothetic «random sequence». The experimental data have been obtained with a high accuracy and veracity, so it can be affirmed that the researched algorithms of block symmetric cryptoconversion are statistically secure in any of the practical appliances.

6.3 Comparison of Statistical Security Indices of BSC

Let us make a comparison of statistical security indices for different appliance modes of the BSC «Kalyna», *AES* (*FIPS*-197) and GOST 28147-89 (DSTU GOST 28147:2009). For this aim let us make use of earlier presented results.

Let us examine the electronic codebook-mode (basic cryptoconversion). In Table 6.12 are presented the results of researches of the ciphers fullround versions.

The obtained results of the comparative analysis, presented in Table 6.12, testify high indices of statistical security of all examined ciphers. Practically each algorithm can be considered as a generator of pseudorandom sequences, the work of which is well approximated by the random process.

Concerning concrete values of statistical security, presented in Table 6.12, a small advantage of the BSC «Kalyna» would be mentioned. The most correct is the comparison at the same key length what equals 256 bits. The mathematical

Table 6.12 Results of researches of the ciphers fullround versions of the BSC «Kalyna», AES (FIPS-197) and GOST 28147-89 (DSTU GOST 28147:2009) in the mode of an electronic codebook

	M099	D099	S099	P099	M096	D096	S096	P096	Min096
KALYNA_ECB_128_128	131.87	33.41	5.78	1.00	186.62	0.70	0.83	1.00	180
KALYNA_ECB_128_256	132.47	35.73	5.98	1.00	186.73	0.13	0.36	1.00	184
KALYNA_ECB_256_256	132.24	18.40	4.29	1.00	187.00	0.31	0.56	1.00	183
KALYNA_ECB_256_512	132.65	34.71	5.89	1.00	186.94	0.39	0.63	1.00	182
KALYNA_ECB_512_512	132.30	27.53	5.25	1.00	186.59	0.46	0.68	1.00	182
AES_ECB_128_256	132.06	19.34	4.40	1.00	186.62	0.66	0.81	1.00	181
GOST_ECB	131.72	34.17	5.85	1.00	186.77	1.28	1.13	1.00	178

expectancy of passed statistical tests both after criterion $P_j \geq 0{,}99$, and after criterion $P_j \geq 0{,}96$ at the BSC «Kalyna», *AES* (*FIPS*-197) and GOST 28147-89 (DSTU GOST 28147:2009) is approximately the same, but the minimum value of the passed statistical tests after the criterion $P_j \geq 0{,}96$ at the BSC «Kalyna» is the highest (184 compared to 181 for *AES* and 178 for GOST 28147-89).

The analysis of presented in Tables 6.1, 6.2, 6.3, 6.4, 6.5, 6.6, 6.7, 6.8 and 6.9 data (concerning statistical security of the BSC «Kalyna» and their comparison with the respective testing results of the BSC *AES* (*FIPS*-197) and GOST 28147-89 (DSTU GOST 28147:2009), that are presented in Tables 6.10 and 6.11 after each application mode respectively, show that the examined BSCs, being compared, possess very like values both after mathematical expectancy, and after the minimum value of passed tests quantity. It testifies high values of BSC statistical security, i.e. they can be considered as efficient PRS generators in any application mode.

6.4 Substantiation of Recommendations Concerning BSC-Usage in Various Cryptographic Appliances

The obtained results of performed comparative analysis are evidence of high statistical indices of the formed PRS, and the BSC themselves can be considered as generators of pseudorandom sequences. Each of the modes forms some sequences, similar to the random ones, what is evidence of a high statistical security of the BSCs in various cryptographic appliances.

The standard DSTU «Information technologies. Cryptographic information protection. Algorithm of symmetric block conversion» provides for such recommendations and restrictions for using of the BSC.

1. Ciphering key and authentication key of the cryptographic algorithm, all the intermediate values. Formed while processing, the processing time (if it depends on the message content or key significance) have to be protected against divulging to an unauthorized part.

Table 6.13 Recommended restrictions for the summary messages length, defensible by using of a single key in stream application modes

Mode designation	Block size (l)		
	128	256	512
CTR, CFB	2^{60} (16 mln TB)	2^{124}	2^{251}

2. In the case, when the message length (in particular during anterior using of compression algorithms) or messages quantity per unit of time is connected with the information content, which is contained in the message itself, the developer of the cryptographic information protection mean must provide some supplementary measures.
3. The device, which executes cryptographic conversions, has to be protected from the external influence, what can cause the appearance of errors while processing.

The general summary messages length of defensible data by using of a single key is recommended to be restricted in dependence on the basic conversion block size. In compliance with the usage condition of the recommended parameters of working modes, these restrictions are summarized in Table 6.13.

In the case, when the summary messages length during the practical exploitation of the cryptographic information protection means for the block length $l = 128$ bits can potentially exceed the in Table 6.13 indicated restrictions, a larger block size is being applied (i.e. the lengths of a block are being enlarged till $l = 256$ or $l = 512$).

6.5 Conclusions and Recommendations

1. The obtained results of the experimental researches of the statistical security of the BSC «Kalyna» testify high characteristics of the researched cryptographic typical. The formed pseudorandom sequences do not differ statistically from a random sequence, i.e. the hypothesis that the researched cryptoalgorithm is a generator, the output sequences of which would be the realization of a random process, is not denied.
2. The statistical tests have been performed for each of the following modes: electronic codebook (*ECB*) or simple change (basic conversion) (*CTR*), inhibition with feedback after the cipher-text (*CFB*), cipher-blocks chaining (*CBC*), inhibition with feedback after the ciphering keystream (*OFB*). Herewith the modes *CTR, CFB, CBC, OFB* statistically were tested on the fullround algorithm version, the mode of electronic codebook or simple change (basic conversion) were tested both on the fullround version, and each round separately.
3. The results of researches of the mode of electronic codebook on each round separately have shown that the researched BSC (its basic conversion) possesses higher indices of statistical security. Already after the first ciphering round the output sequences look like as pseudorandom ones, it can be affirmed with a high probability that the generated sequence is statistically secure. This means that

the BSC «Kalyna» approximates well the generators of «random» sequences; the hypothesis that it is a generator the output sequences of which are realization of a random process, is not refutable.

4. The results of performed statistical researches of various stream ciphering modes (*CTR*, *CFB*, *CBC*, *OFB*) on the full algorithm version testify that the BSC «Kalyna» at any installed mode generates pseudorandom sequences, i.e. with high probability and exactitude can be affirmed that the researched algorithm of block symmetric cryptoconversion will be statistically secure in any practical application.
5. The obtained results of experimental researches of statistical properties of pseudorandom sequences, which are formed by the block symmetric ciphers *AES* (*FIPS*-197) and GOST 28147-89 (DSTU GOST 28147:2009) testify about high characteristics of examined cryptographic typicals. In each of the examined application modes (*ECB*, *CTR*, *CFB*, *CBC*, *OFB*) the formed PRSs look like as pseudorandom and statistically do not differ from the hypothetic «random sentence». The experimental data are obtained with high exactitude and credibility, i.e. it can be affirmed that the researched algorithms of block symmetric cryptoconversion are statistically secure in each of the practical applications.
6. The obtained estimations of the statistical security DSC *AES* (*FIPS*-197) and GOST 28147-89 (DSTU GOST 28147:2009) have been compared with the results of researches of the BSC «Kalyna». The results of comparative analysis testify about high indices of statistical security of all researched ciphers. Practically each cryptoalgorithm can be considered as a pseudorandom sequences generator, the function of which is well approximated by a random process. Concerning concrete values of the statistical security, it should be noted a small advantage of the BSC «Kalyna», because the minimum value of the number of the passed statistical tests at the BSC «Kalyna» is the highest among the researched cryptoalgorithms.
7. After the results of comparative researches, the recommendations concerning using modes of the BSC «Kalyna» in various cryptographic appliances have been substantiated. The general summary lengths of the messages, which are defensible by utilizing of a single key, is recommended to be restricted in dependence on the size of the basic conversion block. The obtained results of statistical researches demonstrate that each of the researched modes forms the sequences, similar to random ones, what speaks about a high statistical BSC security in various cryptographic appliances.

References

1. NIST Special Publication 80022: A Statistical Test Suite for Random and Pseudorandom Number Generators for Cryptographic Applications [Electronic resource]. Access mode http://csrc.nist.gov/publications/nistpubs/80022rev1a/SP80022rev1a.pdf

2. Dieharder: A Random Number Test Suite—Free Implementation of Statistical Tests (including diehard and NIST tests) in C. [Electronic resource]. Access mode http://www.phy.duke.edu/~rgb/General/dieharder.php
3. The Marsaglia Random Number CDROM Including the Diehard Battery of Tests of Randomness [Electronic resource]. Access mode http://stat.fsu.edu/pub/diehard
4. Research of modes of application of block symmetric ciphers: report on GDR (final)/KhNU. V.N. Karazin; ker. Kuznetsov OO; Ex .: Shlokin VM [etc., only 4 people]. H.: KhNU. V.N. Karazina, 89 p. (2014)
5. Kuznetsov, A.A., Mordvinov, R.I., Kolovanova, E.P., Samoilova, A.V.: Methods of statistical testing of cryptographic algorithms. Special Telecommunication Systems and Information Security, vol. 1(25), pp. 54–61. Kyiv (2014)

Chapter 7
Analysis of Standardized Algorithms for Streaming Cryptographic Convention, Defined in ISO/IEC 18033-4

7.1 General Models of Stream Codes, Defined in ISO/IEC 18033-4

7.1.1 Models of Key Stream Generators

7.1.1.1 Model of Synchronous Generator of Key Streams

The synchronous key sstream generator according to ISO/IEC 180334 is a finite state machine. It is defined as follows:

1. *An initialization function, Init*, which takes the K key and the IV initialization vector as input, and outputs the initial state S0 for the key stream generator. The initialization vector must be chosen such that no two messages can ever be encrypted using the same key and the same IV.
2. *Next state function*, ***Next***, which takes as input the current state of the S_i key stream generator and outputs the next state of the S_{i+1} key stream generator.
3. *The key stream function is* ***Strm***, which takes the status of the S_i key stream generator as input and outputs the key stream block Z_i.

When the synchronous key stream generator is first initiated, it will enter an initial state of S_0 defined as

$$S_0 = Init(IV, K).$$

Upon request, a synchronous key stream generator for i = 0, 1, … will execute:

1. Output of key stream block

$$Z_i = Strm(S_i, K);$$

A. A. Kuznetsov et al., *Stream Ciphers in Modern Real-time IT Systems*,
Studies in Systems, Decision and Control 375,
https://doi.org/10.1007/978-3-030-79770-6_7

2. Finite state machine status update

$$S_{i+1} = Next(S_i, K).$$

Therefore, to determine the synchronous key stream generator, all you need to do is describe the functions of *Init, Next, and Strm*, along with key lengths and alphabets, initialization vector, state, and source block.

7.1.1.2 Model of Asynchronous Key Stream Generator (Self-synchronizing Generator)

Generation of a key stream for an asynchronous stream cipher depends only on the previous encrypted texts, key and initialization vector.

ISO/IEC 18033-4 defines a common model for the key stream generator of an asynchronous stream cipher as follows.

1. *An initialization function,* ***Init***, which accepts as input *K* key and *IV* initialization vector and outputs internal input to the key stream generator *S* and *r* dummy $C_{-1}, C_{-2}, \ldots, C_{-r}$ ciphertext blocks.
2. *The key stream function,* ***Strm***, which accepts as input *S* and *r* ciphertext blocks $C_{-1}, C_{-2}, \ldots, C_{-r}$ and outputs the key stream block Z_i.

To determine the asynchronous key stream generator, it is only necessary to determine the number of *r* blocks and the functions *Init* and *Strm*.

7.1.2 Output Function Models

ISO/IEC 18033-4 describes two output functions of a stream cipher, that is, the methods used in a stream cipher to combine a key stream with plaintext to obtain ciphertext.

The output of the output function for synchronous or asynchronous stream cipher is an inverse function of *Out*, which combines the plaintext block P_i, the key stream block Z_i and, optionally, some other input *R*, to obtain the ciphertext block C_i ($i \geq 0$). We now define a general model for the output function of the streaming cipher.

The encryption of the P_i plaintext block using the Z_i key stream block is represented as follows:

$$C_i = Out(P_i, Z_i, R),$$

and decrypting the ciphertext block C_i using the key stream block Z_i is given as follows:

$$P_i = Out^{-1}(C_i, Z_i, R).$$

The output function must be such that for any block of the key stream Z_i, the plaintext block P_i and other input R, the relation is:

$$P_i = Out^{-1}(Out(P_i, Z_i, R), Z_i, R).$$

7.1.2.1 Model of Binary Additive Function

A binary additive stream cipher is called a stream cipher in which blocks of key stream, plaintext and ciphertext are submitted in binary digits and the operation of combining plaintext with key stream is submitted by bit *XOR*. The Out operation takes two input values and does not use any additional R information for its calculation.

Let n be a bit length P_i. This function is described as follows:

- set $R = 0^{(n)}$;
- $Out\ (P_i, Z_i, R) = P_i \oplus Z_i \oplus R$.

The notation $0^{(n)}$ is understood to mean an n-bit vector, where each bit is 0.

The Out–1 operation is described as follows:

- set $R = 0^{(n)}$;
- $Out^{-1}\ (C_i, Z_i, R) = C_i \oplus Z_i \oplus R$.

The binary additive ciphertext does not provide any integrity protection for the encrypted data. If data integrity is to be ensured, then in accordance with *ISO/IEC 180334*, it is recommended to use the *MULTIS01* output function or a separate integrity mechanism (for example, message authentication code according to *ISO/IEC 9797* international standard or DSTU 7564 cryptographic hashing standard: 2014 Information technology—Cryptographic information security—Hashing function [1].

7.1.2.2 Model of the Output Function MULTI-S01

According to *ISO/IEC 18033-4*, the output function *MULTI-S01* is defined for synchronous streaming cipher. The function supports both confidentiality and data integrity, it is intended for encryption in an operational environment. However, the *MULTI-S01* decryption operation can only be performed in an offline situation, since integrity checking is only performed after all blocks of ciphertext have been accepted.

MULTI-S01 has a security parameter n, only messages whose length is multiple of n are subject to processing. To encrypt messages whose length is not multiple of n, you need to fill in plain text. The supplement method is described below.

In addition to P and Z, the *MULTI-S01* function accepts some redundancy as other inputs. One possibility is to create an R fixed fixed $n$bit value. It must be accessible to both encryption and decryption processes.

The *MULTI-S01*-based encryption mechanism is considered secure as long as the primary key stream generator is protected. The data integrity protection level is usually as high as the key stream generator security level. However, the security level is always limited by the length of the fake message above. It is considered that a fake message with a length of $u \bullet n$ bits can be performed with a probability of no more than $(u - 2) \bullet 2 - n$, where n is a mode protection parameter. For practical purposes, the parameter n should be at least 64. For higher protection, the recommended value is $n = 128$.

The *Out* function operates in an iterative manner, processing n bits of plaintext for iteration.

Out algorithm

Input: $n \bullet bit$ plaintext P, key stream Z, $n$bit redundancy R.

Output: C. Ciphertext.

1. Let t be the leftmost value of $i (i \geq 0)$ such that $Zi \neq 0^{(n)}$.
2. Let $(P_0, P_1, \ldots, P_{u-1}) = P$, where P_i is a n-bit block.
3. Set $P_u = Z_{t+u+3}$.
4. Set $P_{u+1} = R$.
5. Set $W_{-1} = O^{(n)}$.
6. For all $i = 0, 1, \ldots, u + 1$ perform the following calculations:

    6.1 Let $W_i = P_i \oplus Z_{t+i+1}$;
    6.2 Let $X_i = Z_t \otimes W_i$ (in the finite field $GF(2_n)$);
    6.3 Let $C_i = X_i \oplus W_{i-1}$.

7. Set $C = C_0 \parallel C_1 \parallel \ldots \parallel C_{u+1}$.
8. Output C.

In Fig. 7.1 *Out* functions are shown schematically.

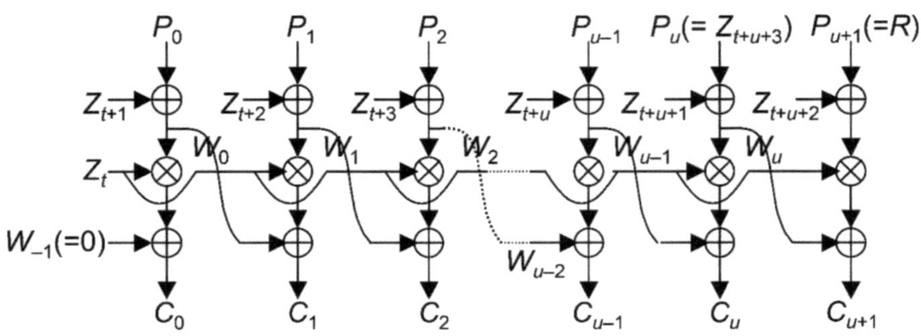

Fig. 7.1 *MULTI-S01* mode *Out* function calculation scheme

The irreducible polynomial used to determine the multiplication in the field $GF(2^n)$ (step 6.2) depends on n. For example, when $n = 64$ and 128, the polynomials $x^{64} + x^4 + x^3 + x + 1$ and $x^{128} + x^7 + x^2 + x + 1$ are used.

The Out^{-1} function does not output any text before checking integrity, the basic calculation of the Out^{-1} function is also incremental and has an internal variable W. To generate a pre-opened P_i value (*preplaintext*), the function requires a W_{i-1} value other than C_i, generating W_i for the next process.

Out^{-1} implementation algorithm.

Input: $n \bullet v$-bit ciphertext C, key stream Z, n-bit redundancy R.

Output: P plaintext or "reject".

1. Let t be the smallest value of $i (i \geq 0)$ such that $Z_i \neq 0^{(n)}$.
2. Let $(C_0, C_1, \ldots, C_{v-1}) = C$, where C_i is a n-bit block.
3. For each C_i, perform the following calculations ($i = 0, 1, \ldots, v - 1$):

   3.1 Let $X_i = C_i \oplus W_{i-1}$, where $W_{-1} = 0^{(n)}$;
   3.2 Let $W_i = Z_t^{-1} \otimes X_i$ (in the finite field $GF(2^n)$);
   3.3 Let $P_i = W_i \oplus Z_{t+i+1}$.

4. If $P_{v-2} = Z_{t+v+1}$ and $P_{v-1} = R$, print $P = P_0 || P_1 || \ldots || P_{v-3}$ as plaintext. Otherwise, display a special symbol for "rejection" without any text.

In Fig. 7.2 shows the operation of the Out^{-1} function.

The *ISO/IEC 18033-4* standard recommends a *Pad*(M) add-on mechanism designed to handle incoming messages of multiple length n.

Input: $(n \bullet v + c)$ bit string M, where v is an integer positive and $0 \leq c < n$.

Output: Supplemented plaintext P.

1. Add one bit "1" to the end of the message.
2. Add $(n - c - 1)$ bit line $0^{(n-c-1)}$ to the end of the line generated in step 1.
3. Output the entire data string $(n \bullet v + n)$ of bits.

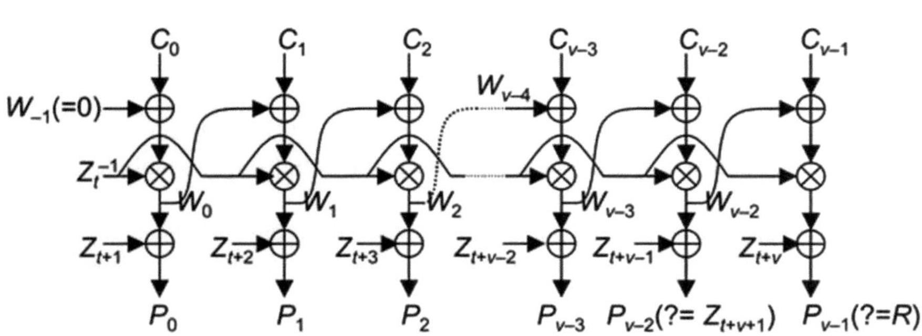

Fig. 7.2 Calculation scheme of the Out^{-1} function of *MULTI-S01* mode

If the length of the message is multiple of n in an environment where this situation is not guaranteed, it is recommended that this mechanism be used. To reproduce the message, delete consecutive 0 bits from the end of the data and delete the other bit "1".

7.2 Specialized Key Stream Generators

International standard *ISO/IEC 18033-4*, except for common models of streaming ciphers and modes of BSC application, also specifies five specialized key stream generators (*MUGI, SNOW 2.0, Rabbit, Decim*^{v2}, *KCipher-2*) for synchronous stream cipher, i.e. key stream generation methods with key and initialization vector.

Let's look at these generators, examine their structure and basic cryptoconversions.

7.2.1 *MUGI Key Stream Generator*

The *MUGI* algorithm is a key stream generator that uses a 128-bit K secret key, a 128-bit IV initialization vector, and a state variable $S_i (i \geq 0)$ consisting of 19 64-bit blocks (note that the term "block" is used throughout the *MUGI* specification for the 64-bit block) and outputs the key stream block Z_i at each iteration of the *Strm* function.

The S_i state variable is constructed from a combination of a 3-block variable:

$$a^{(i)} = \left(a_0^{(i)}, a_1^{(i)}, a_2^{(i)}\right),$$

where $a_j^{(i)}$ is a block (for $j = 0, 1, 2$), and 16 is a block variable:

$$b^{(i)} = \left(b_0^{(i)}, b_1^{(i)}, b_{15}^{(i)}\right),$$

where $b_j^{(i)}$ is a block (for $j = 0, 1, \ldots, 15$).

The *Init* function of the *MUGI* algorithm takes the 128 bit K key and the 128 bit IV initialization vector as input and produces the initial state variable $S_0 = (a^{(0)}, b^{(0)})$.

The *Next* function of the *MUGI* algorithm takes as the input 19-block state variable $S_i = (a^{(i)}, b^{(i)})$ and produces as the output the following state variable value: $S_{i+1} = (a^{(i+1)}, b^{(i+1)})$.

The *Strm* function of the *MUGI* algorithm takes the 19-block state variable $S_i = (a^{(i)}, b^{(i)})$ as input and produces the key stream block Z_i as output.

The *Next* function is defined in terms of the functions ρ_1 and λ_1, defined below.

There are three constants used in *MUGI*—D_0 in the *Init* initialization function and D_1, D_2 in the ρ_1 function. They are set as follows:

$$D_0 = 0x6A09E667F3BCC908,$$
$$D_1 = 0xBB67AE8584CAA73B,$$
$$D_2 = 0x3C6EF372FE94F82B.$$

7.2.1.1 *Init* Initialization Function

MUGI initialization takes eight steps. The blocks of the left and right halves of K are denoted by K_0 and K_1 respectively. IV_0 and IV_1 are defined in the same way. The *Init* initialization function is described as follows:

Input: 128-bit K key, 128-bit IV initialization vector.

Output: The initial vector of the state variable $S_0 = (a^{(0)}, b^{(0)})$.

1. Setting the K key to the state variable $a^{(-49)}$ as follows:

   1.1 Set $(K_0, K_1) = K$, where K_i – 64 bits for $i = 1, 2$;
   1.2 Set $a_0^{(-49)} = K_0$;
   1.3 Set $a_1^{(-49)} = K_1$;
   1.4 Set $a_2^{(-49)} = (K_0 <<<_{64} 7) \oplus D_0$.

2. For all $i = -49, -48, \ldots, -34$ set $a^{(i+1)} = \rho_1\left(a^{(i)}, 0^{(64)}, 0^{(64)}\right)$
3. For all $i = 0, 1, \ldots, 15$ set $b_{15-i}{}^{(-16)} = a_0{}^{(-48)}$.
4. Add IV initialization vector to state a as follows:

   4.1 Set $IV_0 \,||\, IV_1 = IV$, where IV_i – a block;
   4.2 Set $a_0{}^{(-32)} = a_0{}^{(-33)} \oplus IV_0$;
   4.3 Set $a_1{}^{(-32)} = a_1{}^{(-33)} \oplus IV_1$;
   4.4 Set $a_2{}^{(-32)} = a_2{}^{(-33)} \oplus (IV_0 <<<_{64} 7) \oplus (IV_1 <<<_{64} 7) \oplus D_0$.

5. For all $i = -32, -31, \ldots, -17$ set $a^{(i+1)} = \rho_1(a^{(i)}, 0^{(64)}, 0^{(64)})$.
6. Set $S_{-16} = (a^{(-16)}, b^{(-16)})$.
7. 16x iteration of the *Next* update feature:
   Set $S_0 = Next^{16}(S_{-16})$, where $Next^{16}$ is set for 16 iterations of the next *Next* function.
8. Output S_0.

The algorithm indicates:

$<<<_n t$ – t—bit cyclic left shift in n-bit.

$>>>_n t$ – t—bit cyclic shift to the right in n-bit.

7.2.1.2 *Next* State Function

Input: State variable $S_i = (a^{(i)}, b^{(i)})$.

Output: The next value of the state variable $S_{i+1} = (a^{(i+1)}, b^{(i+1)})$.

1. Set $a^{(i+1)} = \rho_1\left(a^{(i)}, b_4^{(i)}, b_{10}^{(i)}\right)$.
2. Set $b^{(i+1)} = \lambda_1\left(b^{(i)}, a_0^{(i)}\right)$.
3. Set $S_{i+1} = (a^{(i+1)}, b^{(i+1)})$.
4. Derive $S_{i+1} = (a^{(i+1)}, b^{(i+1)})$.

7.2.1.3 Function of Key Stream *Strm*

Input: S_i state variable.

Output: Z_i key stream unit.

1. Set $Z_i = a_2^{(i)}$.
2. Output Z_i.

7.2.1.4 The Function ρ_1

Input: State variable $a^{(i)}$, two 64-bit parameters W_1, W_2.

Output: The next value of the state variable $a^{(i+1)}$.

1. Set $a_0(i+1) = a_1(i)$.
2. Set ${a_1}^{(i+1)} = {a_2}^{(i)} \oplus F\left({a_1}^{(i)}, W_1\right) \oplus D_1$.
3. Set ${a_2}^{(i+1)} = {a_0}^{(i)} \oplus F\left({a_1}^{(i)}, (W_1 <<<\ _{64}17)\right) \oplus D_2$.
4. Derive $a^{(i+1)}$.

D_1, D_2 denote constants in the algorithm.
In Fig. 7.3 shows the operation ρ_1 of the function.
A detailed description of F is given below.

7.2.1.5 λ_1 Function

Input: Status variable $b^{(i)}$, 64-bit parameter a'.

Output: The next value of the state variable $b^{(i+1)}$.

1. Set ${b_j}^{(i+1)} = {b_{j-1}}^{(i)}$ for all $j \neq 0, 4, 10$.
2. Set ${b_0}^{(i+1)} = {b_{15}}^{(i)} + a'$.
3. Set ${b_4}^{(i+1)} = {b_3}^{(i)} \oplus b_7^{(i)}$.
4. Set $b_{10}(i+1) = {b_9}^{(i)} \oplus \left({b_{13}}^{(i)} <<<\ _{64}32\right)$.
5. Derive $b^{(i+1)}$.

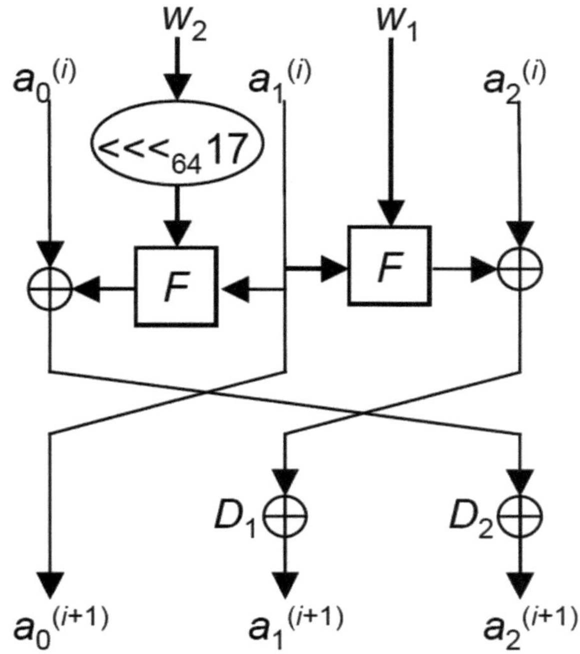

Fig. 7.3 Scheme of calculation of function ρ_1 of *MUGI* algorithm

7.2.1.6 *F* Function

The function F will use operations on a finite field $GF(2^8)$. In the polynomial representation, $GF(2^8)$ is implemented as $GF(2)[x]/m(x)$, where $m(x)$ is an irreducible polynomial of degree 8 defined over $GF(2)$. The *MUGI* key stream generator will use the following irreducible polynomial: $m(x) = x^8 + x^4 + x^3 + x + 1$.

Function F is a composition of key addition (compilation of part of state variable b), nonlinear transformation using S_R, function, linear transformation using matrix M and byte mixing (see Fig. 7.4).

We denote the input and output data for the function F as X and Y, respectively. Then function F is described as follows.

Input: Two 64-bit X and T strings.

Output: 64-bit string Y.

1. $X' = X \oplus T$.
2. Set $(X_0, X_1, X_2, X_3, X_4, X_5, X_6, X_7) = X'$, where X_i – 8bit string.
3. Set $P_i = S_R(X_i)$ for all $i = 0, 1, \ldots, 7$.
4. Set $P_L = P_0 \,||\, P_1 \,||\, P_2 \,||\, P_3$.
5. Set $P_R = P_4 \,||\, P_5 \,||\, P_6 \,||\, P_7$.
6. Set $Q_L = M(P_L)$.
7. Set $Q_R = M(P_R)$.

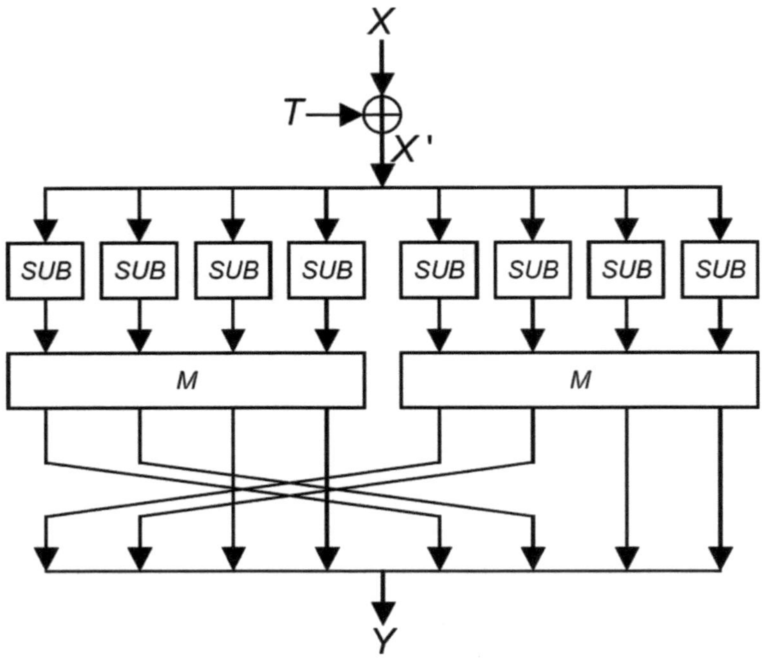

Fig. 7.4 Calculation scheme of function F of the *MUGI* algorithm

8. Set $(Q_0, Q_1, Q_2, Q_3) = Q_L$.
9. Set $(Q_4, Q_5, Q_6, Q_7) = Q_R$.
10. Set $Y = Q_0 \parallel Q_1 \parallel Q_2 \parallel Q_3 \parallel Q_4 \parallel Q_5 \parallel Q_6 \parallel Q_7$.
11. Derive Y.

7.2.1.7 S_R Function

The S_R function is an internal function F. The S_R function can be described using the substitution table. In this case, the S_R function is described as follows.

Input: 8-bit x string.

Output: 8-bit y string.

1. Set $y = SUB\,[x]$.
2. Derive y.

The *SUB* operation used to calculate the S_R function is the substitution table shown in Fig. 7.5.

The *SUB* substitution table is identical to the S-block defined in the *AES* [2].

```
SUB [256] = {
0x63, 0x7c, 0x77, 0x7b, 0xf2, 0x6b, 0x6f, 0xc5, 0x30, 0x01, 0x67, 0x2b, 0xfe, 0xd7, 0xab, 0x76,
0xca, 0x82, 0xc9, 0x7d, 0xfa, 0x59, 0x47, 0xf0, 0xad, 0xd4, 0xa2, 0xaf, 0x9c, 0xa4, 0x72, 0xc0,
0xb7, 0xfd, 0x93, 0x26, 0x36, 0x3f, 0xf7, 0xcc, 0x34, 0xa5, 0xe5, 0xf1, 0x71, 0xd8, 0x31, 0x15,
0x04, 0xc7, 0x23, 0xc3, 0x18, 0x96, 0x05, 0x9a, 0x07, 0x12, 0x80, 0xe2, 0xeb, 0x27, 0xb2, 0x75,
0x09, 0x83, 0x2c, 0x1a, 0x1b, 0x6e, 0x5a, 0xa0, 0x52, 0x3b, 0xd6, 0xb3, 0x29, 0xe3, 0x2f, 0x84,
0x53, 0xd1, 0x00, 0xed, 0x20, 0xfc, 0xb1, 0x5b, 0x6a, 0xcb, 0xbe, 0x39, 0x4a, 0x4c, 0x58, 0xcf,
0xd0, 0xef, 0xaa, 0xfb, 0x43, 0x4d, 0x33, 0x85, 0x45, 0xf9, 0x02, 0x7f, 0x50, 0x3c, 0x9f, 0xa8,
0x51, 0xa3, 0x40, 0x8f, 0x92, 0x9d, 0x38, 0xf5, 0xbc, 0xb6, 0xda, 0x21, 0x10, 0xff, 0xf3, 0xd2,
0xcd, 0x0c, 0x13, 0xec, 0x5f, 0x97, 0x44, 0x17, 0xc4, 0xa7, 0x7e, 0x3d, 0x64, 0x5d, 0x19, 0x73,
0x60, 0x81, 0x4f, 0xdc, 0x22, 0x2a, 0x90, 0x88, 0x46, 0xee, 0xb8, 0x14, 0xde, 0x5e, 0x0b, 0xdb,
0xe0, 0x32, 0x3a, 0x0a, 0x49, 0x06, 0x24, 0x5c, 0xc2, 0xd3, 0xac, 0x62, 0x91, 0x95, 0xe4, 0x79,
0xe7, 0xc8, 0x37, 0x6d, 0x8d, 0xd5, 0x4e, 0xa9, 0x6c, 0x56, 0xf4, 0xea, 0x65, 0x7a, 0xae, 0x08,
0xba, 0x78, 0x25, 0x2e, 0x1c, 0xa6, 0xb4, 0xc6, 0xe8, 0xdd, 0x74, 0x1f, 0x4b, 0xbd, 0x8b, 0x8a,
0x70, 0x3e, 0xb5, 0x66, 0x48, 0x03, 0xf6, 0x0e, 0x61, 0x35, 0x57, 0xb9, 0x86, 0xc1, 0x1d, 0x9e,
0xe1, 0xf8, 0x98, 0x11, 0x69, 0xd9, 0x8e, 0x94, 0x9b, 0x1e, 0x87, 0xe9, 0xce, 0x55, 0x28, 0xdf,
0x8c, 0xa1, 0x89, 0x0d, 0xbf, 0xe6, 0x42, 0x68, 0x41, 0x99, 0x2d, 0x0f, 0xb0, 0x54, 0xbb, 0x16}
```

Fig. 7.5 Tabular execution of *SUB* substitution of *MUGI* algorithm

7.2.1.8 *M* Function

Function M is an internal function of F function. The function M is described as follows.

Input: 32-bit X string.

Output: 32-bit Y string.

1. Set $(x_0, x_1, x_2, x_3) = X$, where x_i is an 8-bit string and GF element $GF(2^8)$.
2. Set

$$\begin{pmatrix} y_0 \\ y_1 \\ y_2 \\ y_3 \end{pmatrix} = \begin{pmatrix} 0x02 & 0x03 & 0x01 & 0x01 \\ 0x01 & 0x02 & 0x03 & 0x01 \\ 0x01 & 0x01 & 0x02 & 0x03 \\ 0x03 & 0x01 & 0x01 & 0x02 \end{pmatrix} \begin{pmatrix} x_0 \\ x_1 \\ x_2 \\ x_3 \end{pmatrix}.$$

   where 0x01, 0x02 and 0x03 are hexadecimal expressions of $GF(2^8)$ elements.
3. Set $Y = y_0 \,\|\, y_1 \,\|\, y_2 \,\|\, y_3$.
4. Derive Y.

7.2.2 SNOW *2.0 Key Stream Generator*

SNOW 2.0, later simply referred to as *SNOW*, is a key stream generator that uses 128 or 256-bit K secret key and 128-bit *IV* initialization vector as input. They are used to initialize the S_i ($i \geq 0$) state variable consisting of eighteen $n = 32$-bit blocks. The bit/byte order is *big-endian*, that is, if the initialization key and vector are provided as a sequence of bits/bytes, then the first/leftmost bit/byte is the most significant of the corresponding data.

For each iteration of the *Strm* 32-bit key stream Z is produced as output.

The S_i state variable of the *SNOW* generator consists of two components.

Firstly, sixteen 32-bit variables

$$a^{(i)} = \left(a_{15}{}^{(i)}, b_{14}{}^{(i)}, \ldots, b_0{}^{(i)}\right)$$

implement LFSR.

Secondly, two 32-bit variables

$$b^{(i)} = \left(b_2{}^{(i)}, b_1{}^{(i)}\right)$$

maintain the state of the finite state machine (FSM).

The *SNOW* algorithm is better understood with reference to Fig. 7.6, showing the operation of the generator at time i. The time dependency variable (i) is not specified.

The *SNOW* operation is defined as follows.

The *Init* function accepts 128 or 256-bit K and 128-bit IV as input, and produces the initial state variable $S_0 = (a^{(0)}, b^{(0)})$.

The *Next* function accepts the eighteen 32-bit state variables $S_i = (a^{(i)}, b^{(i)})$ as input and produces the next state variable value as the input $S_{i+1} = (a^{(i+1)}, b^{(i+1)})$. The *Next* function is executed in two modes, depending on the method of iteration execution—initialization mode or normal output generation mode. By default, they use normal mode.

The *Strm* function accepts eighteen 32-bit state variables $S_i = (a^{(i)}, b^{(i)})$ as input and produces a 32-bit key stream Z_i as input.

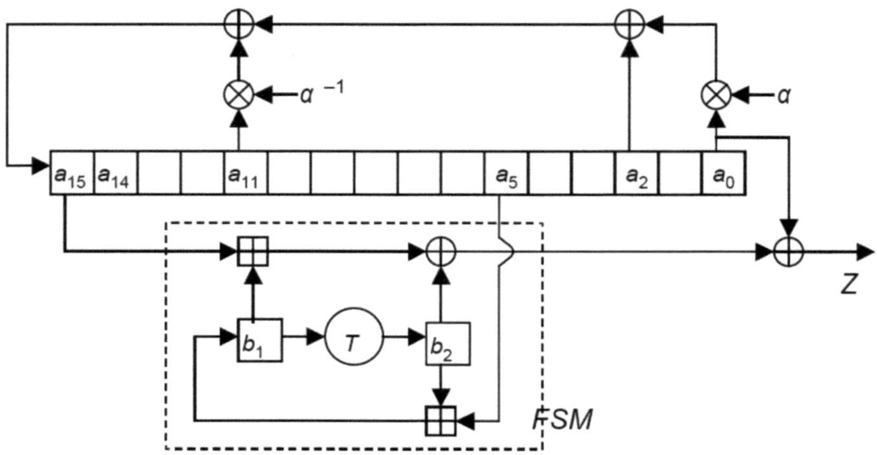

Fig. 7.6 Schematic representation of the *SNOW* generator

For *SNOW*, the maximum recommended number of key stream produced from a given *(K, IV)* is 2^{50} 32-bit blocks. This limit is chosen to provide a large margin of security against cryptanalysis and implies the absence of any practical limitations in the applicability of the algorithm.

7.2.2.1 *Init* Initialization Function

The *Init* implementation function is described as follows.

Input: 128 or 256-bit *K* key, 128-bit *IV* initialization vector.

Output: The initial value of the state variable $S_0 = (a^{(0)}, b^{(0)})$.

1. Initializing registers with key information.

   a. For 128-bit key
   $(K_3, K_2, K_1, K_0) = K$.

   For all $j = 0, 1, 2, 3$ to set

$$a_{15-j}{}^{(-34)} = a_{15-j-8}{}^{(-34)} = K_{3-j},$$
$$a_{15-j-4}{}^{(-34)} = a_{15-j-12}{}^{(-34)} = -K_{3-j}.$$

   b. For 256-bit key
   $(K_7, K_6, \ldots, K_0) = K$.

   For all $j = 0, 1, \ldots, 7$ to set

$$a_{15-j}^{(-34)} = K_{7-j},$$
$$a_{15-j-8}^{(-34)} = K_{7j}.$$

2. Set $S_{-33} = (a^{(-33)}, b^{(-33)})$:

   a. $(IV_3, IV_2, IV_1, IV_0) = IV$.
   b. Set $a_j^{(-33)} = a_j^{(-34)}$ for all $j = 0,1,2,3,4,5,6,7,8,11,13,14$.
   c. Set

$$a_{15}^{(-33)} = a_{15}^{(-34)} \oplus IV_0;$$
$$a_{12}^{(-33)} = a_{12}^{(-34)} \oplus IV_1;$$
$$a_{10}^{(-33)} = a_{10}^{(-34)} \oplus IV_2;$$
$$a_9^{(-33)} = a_9^{(-34)} \oplus IV_3.$$

   d. Set $b_1^{(-33)} = b_2^{(-33)} = 0^{(32)}$.

3. Set $S_{-1} = Next^{32}(S_{-33}, \text{INIT})$, where $Next^{32}$ denotes 32 iterations of the Next function.

4. $S_0 = Next\ (S_{-1})$.
5. Derive S_0.

7.2.2.2 *Next* State Function

SNOW has two modes for the *Next* function.

Input: State variable $S_i = (a^{(i)}, b^{(i)})$, mode = {*INIT*, *null*},

where *INIT* is the initialization mode;

null—default mode (normal mode).

Output: The next value of the state variable $S_{i+1} = (a^{(i+1)}, b^{(i+1)})$.

1. Set ${b_2}^{(i+1)} = T\left({b_1}^{(i)}\right)$.
2. Set ${b_1}^{(i+1)} = {b_2}^{(i+1)} +_{32} {a_5}^{(i)}$,

   where $+_m$ denotes adding integers module 2^m.
3. For all $j = 0, 1, \ldots, 14$ set ${a_j}^{(i+1)} = {a_{j+1}}^{(i)}$.
4. For *INIT* mode:
   Set ${a_{15}}^{(i+1)} = \left({a_0}^{(i)} \otimes \alpha\right) \oplus {a_2}^{(i)} \oplus \left({a_{11}}^{(i)} \otimes \alpha^{-1}\right) \oplus FSM\left({a_{15}}^{(i)}, {b_1}^{(i)}, {b_2}^{(i)}\right)$,
   Otherwise set ${a_{15}}^{(i+1)} = \left({a_0}^{(i)} \otimes \alpha\right) \oplus {a_2}^{(i)} \oplus \left({a_{11}}^{(i)} \otimes \alpha^{-1}\right)$.
5. $S_{i+1} = (a^{(i+1)}, b^{(i+1)})$.
6. Derive S_{i+1}.

Figure 7.7 shows a block diagram of the *Next* function in *INIT* mode.

7.2.2.3 Function of Key Stream Strm

The function of the key stream of *Strm* is described as follows.

Input: S_i state variable.

Output: 32-bit Z_i key stream.

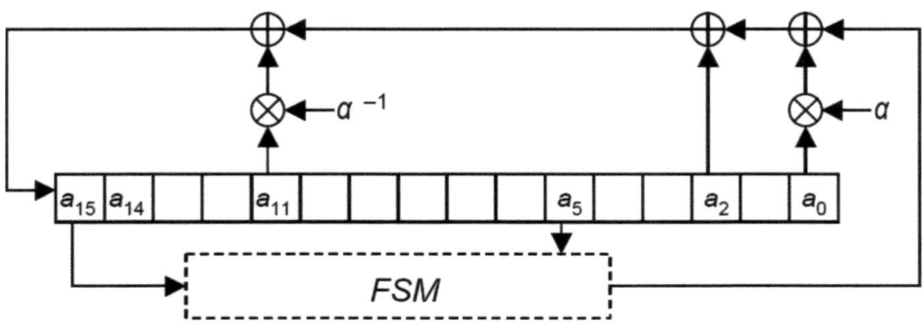

Fig. 7.7 Block diagram of *Next* function in *INIT* mode

1. Set $Z_i = FSM\left(a_{15}{}^{(i)}, b_1{}^{(i)}, b_2{}^{(i)}\right) \oplus a_0{}^{(i)}$.
2. Derive Z_i.

7.2.2.4 T Function

The function T is a substitution, in particular the permutation of $GF(2^{32})$ elements based on components from *AES* [2]. To this end, we used a finite field $GF(2^8)$, which is represented as $GF(2)[x]$ by the module of irreducible polynomial

$$f(x) = x^8 + x^4 + x^3 + x + 1$$

and the polynomial ring $GF(2^8)[y]$ module $(y^4 + 1)$.

Input: 32-bit string w.

Output: 32-bit string $q = T(w)$.

1. Set $(w_3, w_2, w_1, w_0) = w$, where each w_j is 8 bits.
2. Set $t_j = SUB[w_j]$ for all $j = 0, 1, 2, 3$.
3. Let $t(y)$ be a polynomial $t(y) = t_3y^3 + t_2y^2 + t_1y^1 + t_0$ in $GF(2^8)[y]$, where t_j is interpreted as an element of $GF(2^8)$ in the natural way: $t_j = t_{j,7}x^7 + \cdots + t_{j,2}x^2 + t_{j,1}x + t_{j,0}$ B $GF(2)$.
4. Set $q(y) = c(y) \cdot t(y)$ modulo $(y^4 + 1)$, where $c(y) = (x + 1)y^3 + y^2 + y + x$ B $GF(2^8)[y]$.
5. Associate the 32-bit string $q = (q_3, q_2, q_1, q_0)$ with the above result, $q(y) = q_3y^3 + q_2y^2 + q_1y^1 + q_0$.
6. Derive q.

Note that in step 3, the multiplication of two polynomials is performed, where the coefficient operations are performed in $GF(2^8)$ as defined by the function $f(x)$ above. Then the result is summed up by the module $y^4 + 1$.

7.2.2.5 Multiplication by A in Finite Field Arithmetic

Input: A 32-bit string representing the $GF(2^{32})$ element.

Output: A 32-bit string w', representing $\alpha \oplus w$ B $GF(2^{32})$.

1. Set $w' = (w \ll_{32} 8) \oplus \alpha_{MUL}(w \gg_{32} 24)$.
2. Derive w'.

In the algorithm, α_{MUL} denotes the table execution of the operation shown in Fig. 7.8.

α_{mul} [256] = {
```
0x00000000,0xE19FCF13,0x6B973726,0x8A08F835,0xD6876E4C,0x3718A15F,0xBD10596A,0x5C8F9679,
0x05A7DC98,0xE438138B,0x6E30EBBE,0x8FAF24AD,0xD320B2D4,0x32BF7DC7,0xB8B785F2,0x59284AE1,
0x0AE71199,0xEB78DE8A,0x617026BF,0x80EFE9AC,0xDC607FD5,0x3DFFB0C6,0xB7F748F3,0x566887E0,
0x0F40CD01,0xEEDF0212,0x64D7FA27,0x85483534,0xD9C7A34D,0x38586C5E,0xB250946B,0x53CF5B78,
0x1467229B,0xF5F8ED88,0x7FF015BD,0x9E6FDAAE,0xC2E04CD7,0x237F83C4,0xA9777BF1,0x48E8B4E2,
0x11C0FE03,0xF05F3110,0x7A57C925,0x9BC80636,0xC747904F,0x26D85F5C,0xACD0A769,0x4D4F687A,
0x1E803302,0xFF1FFC11,0x75170424,0x9488CB37,0xC8075D4E,0x2998925D,0xA3906A68,0x420FA57B,
0x1B27EF9A,0xFAB82089,0x70B0D8BC,0x912F17AF,0xCDA081D6,0x2C3F4EC5,0xA637B6F0,0x47A879E3,
0x28CE449F,0xC9518B8C,0x435973B9,0xA2C6BCAA,0xFE492AD3,0x1FD6E5C0,0x95DE1DF5,0x7441D2E6,
0x2D699807,0xCCF65714,0x46FEAF21,0xA7616032,0xFBEEF64B,0x1A713958,0x9079C16D,0x71E60E7E,
0x22295506,0xC3B69A15,0x49BE6220,0xA821AD33,0xF4AE3B4A,0x1531F459,0x9F390C6C,0x7EA6C37F,
0x278E899E,0xC611468D,0x4C19BEB8,0xAD8671AB,0xF109E7D2,0x109628C1,0x9A9ED0F4,0x7B011FE7,
0x3CA96604,0xDD36A917,0x573E5122,0xB6A19E31,0xEA2E0848,0x0BB1C75B,0x81B93F6E,0x6026F07D,
0x390EBA9C,0xD891758F,0x52998DBA,0xB30642A9,0xEF89D4D0,0x0E161BC3,0x841EE3F6,0x65812CE5,
0x364E779D,0xD7D1B88E,0x5DD940BB,0xBC468FA8,0xE0C919D1,0x0156D6C2,0x8B5E2EF7,0x6AC1E1E4,
0x33E9AB05,0xD2766416,0x587E9C23,0xB9E15330,0xE56EC549,0x04F10A5A,0x8EF9F26F,0x6F663D7C,
0x50358897,0xB1AA4784,0x3BA2BFB1,0xDA3D70A2,0x86B2E6DB,0x672D29C8,0xED25D1FD,0x0CBA1EEE,
0x5592540F,0xB40D9B1C,0x3E056329,0xDF9AAC3A,0x83153A43,0x628AF550,0xE8820D65,0x091DC276,
0x5AD2990E,0xBB4D561D,0x3145AE28,0xD0DA613B,0x8C55F742,0x6DCA3851,0xE7C2C064,0x065D0F77,
0x5F754596,0xBEEA8A85,0x34E272B0,0xD57DBDA3,0x89F22BDA,0x686DE4C9,0xE2651CFC,0x03FAD3EF,
0x4452AA0C,0xA5CD651F,0x2FC59D2A,0xCE5A5239,0x92D5C440,0x734A0B53,0xF942F366,0x18DD3C75,
0x41F57694,0xA06AB987,0x2A6241B2,0xCBFD8EA1,0x977218D8,0x76EDD7CB,0xFCE52FFE,0x1D7AE0ED,
0x4EB5BB95,0xAF2A7486,0x25228CB3,0xC4BD43A0,0x9832D5D9,0x79AD1ACA,0xF3A5E2FF,0x123A2DEC,
0x4B12670D,0xAA8DA81E,0x2085502B,0xC11A9F38,0x9D950941,0x7C0AC652,0xF6023E67,0x179DF174,
0x78FBCC08,0x9964031B,0x136CFB2E,0xF2F3343D,0xAE7CA244,0x4FE36D57,0xC5EB9562,0x24745A71,
0x7D5C1090,0x9CC3DF83,0x16CB27B6,0xF754E8A5,0xABDB7EDC,0x4A44B1CF,0xC04C49FA,0x21D386E9,
0x721CDD91,0x93831282,0x198BEAB7,0xF81425A4,0xA49BB3DD,0x45047CCE,0xCF0C84FB,0x2E934BE8,
0x77BB0109,0x9624CE1A,0x1C2C362F,0xFDB3F93C,0xA13C6F45,0x40A3A056,0xCAAB5863,0x2B349770,
0x6C9CEE93,0x8D032180,0x070BD9B5,0xE69416A6,0xBA1B80DF,0x5B844FCC,0xD18CB7F9,0x301378EA,
0x693B320B,0x88A4FD18,0x02AC052D,0xE333CA3E,0xBFBC5C47,0x5E239354,0xD42B6B61,0x35B4A472,
0x667BFF0A,0x87E43019,0x0DECC82C,0xEC73073F,0xB0FC9146,0x51635E55,0xDB6BA660,0x3AF46973,
0x63DC2392,0x8243EC81,0x084B14B4,0xE9D4DBA7,0xB55B4DDE,0x54C482CD,0xDECC7AF8,0x3F53B5EB}
```

Fig. 7.8 Tabular execution of α_{MUL} operation

7.2.2.6 Multiplication by α^{-1} in Finite-Field Arithmetic

Input: A 32-bit string representing the $GF(2^{32})$ element.

Output: A 32-bit string y', representing $\alpha \oplus w$ в $GF(2^{32})$.

1. Set $y' = (y \ll_{32} 8) \oplus \alpha_{INV\_MUL}(w \gg_{32} 24)$.
2. Derive y'.

The algorithm named α_{INV_MUL} denotes the table execution of the operation shown in Fig. 7.9.

```
αinv_mul [256]= {
0x00000000,0x180F40CD,0x301E8033,0x2811C0FE,0x603CA966,0x7833E9AB,0x50222955,0x482D6998,
0xC078FBCC,0xD877BB01,0xF0667BFF,0xE8693B32,0xA04452AA,0xB84B1267,0x905AD299,0x88559254,
0x29F05F31,0x31FF1FFC,0x19EEDF02,0x01E19FCF,0x49CCF657,0x51C3B69A,0x79D27664,0x61DD36A9,
0xE988A4FD,0xF187E430,0xD99624CE,0xC1996403,0x89B40D9B,0x91BB4D56,0xB9AA8DA8,0xA1A5CD65,
0x5249BE62,0x4A46FEAF,0x62573E51,0x7A587E9C,0x32751704,0x2A7A57C9,0x026B9737,0x1A64D7FA,
0x923145AE,0x8A3E0563,0xA22FC59D,0xBA208550,0xF20DECC8,0xEA02AC05,0xC2136CFB,0xDA1C2C36,
0x7BB9E153,0x63B6A19E,0x4BA76160,0x53A821AD,0x1B854835,0x038A08F8,0x2B9BC806,0x339488CB,
0xBBC11A9F,0xA3CE5A52,0x8BDF9AAC,0x93D0DA61,0xDBFDB3F9,0xC3F2F334,0xEBE333CA,0xF3EC7307,
0xA492D5C4,0xBC9D9509,0x948C55F7,0x8C83153A,0xC4AE7CA2,0xDCA13C6F,0xF4B0FC91,0xECBFBC5C,
0x64EA2E08,0x7CE56EC5,0x54F4AE3B,0x4CFBEEF6,0x04D6876E,0x1CD9C7A3,0x34C8075D,0x2CC74790,
0x8D628AF5,0x956DCA38,0xBD7C0AC6,0xA5734A0B,0xED5E2393,0xF551635E,0xDD40A3A0,0xC54FE36D,
0x4D1A7139,0x551531F4,0x7D04F10A,0x650BB1C7,0x2D26D85F,0x35299892,0x1D38586C,0x053718A1,
0xF6DB6BA6,0xEED42B6B,0xC6C5EB95,0xDECAAB58,0x96E7C2C0,0x8EE8820D,0xA6F942F3,0xBEF6023E,
0x36A3906A,0x2EACD0A7,0x06BD1059,0x1EB25094,0x569F390C,0x4E9079C1,0x6681B93F,0x7E8EF9F2,
0xDF2B3497,0xC724745A,0xEF35B4A4,0xF73AF469,0xBF179DF1,0xA718DD3C,0x8F091DC2,0x97065D0F,
0x1F53CF5B,0x075C8F96,0x2F4D4F68,0x37420FA5,0x7F6F663D,0x676026F0,0x4F71E60E,0x577EA6C3,
0xE18D0321,0xF98243EC,0xD1938312,0xC99CC3DF,0x81B1AA47,0x99BEEA8A,0xB1AF2A74,0xA9A06AB9,
0x21F5F8ED,0x39FAB820,0x11EB78DE,0x09E43813,0x41C9518B,0x59C61146,0x71D7D1B8,0x69D89175,
0xC87D5C10,0xD0721CDD,0xF863DC23,0xE06C9CEE,0xA841F576,0xB04EB5BB,0x985F7545,0x80503588,
0x0805A7DC,0x100AE711,0x381B27EF,0x20146722,0x68390EBA,0x70364E77,0x58278E89,0x4028CE44,
0xB3C4BD43,0xABCBFD8E,0x83DA3D70,0x9BD57DBD,0xD3F81425,0xCBF754E8,0xE3E69416,0xFBE9D4DB,
0x73BC468F,0x6BB30642,0x43A2C6BC,0x5BAD8671,0x1380EFE9,0x0B8FAF24,0x239E6FDA,0x3B912F17,
0x9A34E272,0x823BA2BF,0xAA2A6241,0xB225228C,0xFA084B14,0xE2070BD9,0xCA16CB27,0xD2198BEA,
0x5A4C19BE,0x42435973,0x6A52998D,0x725DD940,0x3A70B0D8,0x227FF015,0x0A6E30EB,0x12617026,
0x451FD6E5,0x5D109628,0x750156D6,0x6D0E161B,0x25237F83,0x3D2C3F4E,0x153DFFB0,0x0D32BF7D,
0x85672D29,0x9D686DE4,0xB579AD1A,0xAD76EDD7,0xE55B844F,0xFD54C482,0xD545047C,0xCD4A44B1,
0x6CEF89D4,0x74E0C919,0x5CF109E7,0x44FE492A,0x0CD320B2,0x14DC607F,0x3CCDA081,0x24C2E04C,
0xAC977218,0xB49832D5,0x9C89F22B,0x8486B2E6,0xCCABDB7E,0xD4A49BB3,0xFCB55B4D,0xE4BA1B80,
0x17566887,0x0F59284A,0x2748E8B4,0x3F47A879,0x776AC1E1,0x6F65812C,0x477441D2,0x5F7B011F,
0xD72E934B,0xCF21D386,0xE7301378,0xFF3F53B5,0xB7123A2D,0xAF1D7AE0,0x870CBA1E,0x9F03FAD3,
0x3EA637B6,0x26A9777B,0x0EB8B785,0x16B7F748,0x5E9A9ED0,0x4695DE1D,0x6E841EE3,0x768B5E2E,
0xFEDECC7A,0xE6D18CB7,0xCEC04C49,0xD6CF0C84,0x9EE2651C,0x86ED25D1,0xAEFCE52F,0xB6F3A5E2}
```

Fig. 7.9 Tabular operation of α_{INV_MUL} operation

7.2.2.7 FSM (x, y, z) Function

Input: Three 32-bit strings, x, y, z.

Output: 32-bit string q.

1. Set $q = (x +_{32} y) \oplus z$.
2. Derive q.

7.2.3 Rabbit Streaming Code

The *Rabbit* algorithm is a high-speed streaming cipher that was first introduced in February 2003 at the *10*th *FSE* Symposium [3]. In May 2005, this code was sent to the *eStream* contest. The developers of the *Rabbit* algorithm are Martin Boesgaard, Mette Vesterager, Thomas Pedersen, Jesper Christiansen та Ove Scavenius.

The algorithm uses a 128-bit key and a 64-bit initialization vector. The cipher was designed to be used in software with high speed encryption, that is, this cipher was nominated for competition on the first profile—software oriented algorithms. Encryption rates by project authors reach 3.7 cycles per byte for the *Pentium 3* processor and 10.5 cycles per byte for the *ARM7*. In addition, the *Rabbit* cipher also proved to be fast and compact to implement in hardware.

The main component of the cipher is the bit stream generator, which generates 128 bits of the iteration message. The advantage of the *Rabbit* cipher is the careful mixing of its internal states between two successive iterations. The shuffle function is entirely based on the arithmetic operations available on modern processors, that is, *Substitution Blocks* and Search Sheets are not required to implement the cipher.

The *Rabbit* algorithm can be briefly described as follows. As input, the algorithm takes the 128-bit secret key and the 64-bit *IV* initialization vector (optional) and generates an output block of 128 pseudorandom bits for each iteration from a combination of internal state bits. Encryption/decryption is performed by *XOR* pseudorandom data with open/encrypted text. The internal state size is 513 bits, which are split between eight 32-bit state variables, eight 32-bit counter variables, and one counter is used for the carry bit. Eight state variables are updated with eight related nonlinear functions. The counters provide the lower limit of the length of the period for the state variables.

The *Rabbit* algorithm was designed to provide high-speed performance, in particular to be faster than commonly used 128-bit key encryption for encrypting up to 2^{64} blocks of text. For an attacker who does not know the key, the ability to distinguish 2^{64} cipher blocks from the output of the generator of these random bits needs no less action than would be required to search 2^{128} keys.

7.2.3.1 Specification of *Rabbit* Stream Encryption

The *Rabbit* stream cipher specification uses the notation given in Table 7.1 [3].

During bit numbering, the lower bit is denoted by 0. For hexadecimal numbers, the notation with a prefixed «0x» is used.

The internal state of the streaming cipher consists of 513 bits. Of these, 512 bits are divided between the eight 32-bit state variables $x_{j,i}$ and the eight 32-bit variables of the counter $c_{j,i}$. In these designations $x_{j,i}$ is the variable of the state of its subsystem at

Table 7.1 The operations and designations used in the specification the *Rabbit* cipher

Symbol	Operation or notation
$\oplus$	The operation of a logical exclusive OR (XOR)
&	The operation of logical I (AND)
$\gg$	The shift operator to the right for the specified number of bits
$\ll$	The shift operator to the left for the specified number of bits
>>>	The cyclic shift operator to the right for the specified number of bits
<<<	The cyclic shift operator to the left for the specified number of bits
$\lozenge$	Concatenation of two bit sequences
$A^{[g \ldots h]}$	Designation of sequence of bits from *g* to *h* of the variable *A*

its i iteration, and $c_{j,i}$ is the variable of the counter of its j subsystem at its i iteration, respectively. One transfer bit counter, denoted as $\varphi_{7,i}$ should also be used, which should be stored between iterations. This transfer bit counter is set to zero. Eight state variables and eight variable counters are set from the key during initialization.

7.2.3.2 Scheme of Installation of the Key

The *Rabbit* encryption algorithm is initiated by the extension of a 128-bit key in both components: eight state variables and eight variable counters. There is a one-to-one correspondence between the key and the initial state of these variables.

Key installation can be divided into three steps: key extension, iteration system, counter modification.

The key extension stage guarantees a one-to-one correspondence between the status key and the counter key, which prevents excess keys. It also distributes key bits optimally for all iterations.

At this point, the *Rabbit* algorithm's current encryption key, denoted as $K^{[127\ldots0]}$, is divided into eight parts, the so-called subkeys:

$$k_0 = K^{[15\ldots0]}, k_1 = K^{[31\ldots16]}, \ldots, k_7 = K^{[127\ldots112]}.$$

The iteration system ensures that after one iteration of the next state function, each bit of the key will affect all eight state variables. This also ensures that after the second iteration of the next state function, all key bits will affect all state bits with a probability of 0.5.

The state variables and counters are initialized from these subkeys as follows:

$$x_{j,0} = \begin{cases} k_{(j+1 \bmod 8)} \overset{\wedge}{\vee} k_j, & for\ all\ even\ j; \\ k_{(j+5 \bmod 8)} \overset{\wedge}{\vee} k_{(j+4 \bmod 8)}, & for\ all\ odd\ j; \end{cases}$$

$$c_{j,0} = \begin{cases} k_{(j+4 \bmod 8)} \overset{\wedge}{\vee} k_{(j+5 \bmod 8)}, & for\ all\ even\ j; \\ k_j \overset{\wedge}{\vee} k_{(j+1 \bmod 8)}, & for\ all\ odd\ j; \end{cases}$$

The key installation system is repeated four times according to the next status function (defined below). This is to reduce the correlation between the key bits and the bits of the internal state variables.

Modification of the counter. Even if the counters are known to the attacker, modifying the counter will make it difficult to recover the key by inverting the system counter, since this will require information about the state variables. In addition, the modification violates the unambiguous relationship between the key and the counter.

The counter variables are modified according to the following rule:

$$c_{j,4} = c_{j,4} \oplus x_{(j+4 \bmod 8),4},$$

which is performed for all j.

7.2.3.3 Initialization Vector Installation Diagram

Let us denote the internal state after the key installation scheme as a master, and let a copy of this master be modified according to the scheme of the *IV* initialization vector. Installation Scheme IV functions by changing the meter status as a function of *IV*. This is done by performing an XOR operation on the 64-bit *IV* initialization vector on all 256 bits of the counter state. These 64 bits of the *IV* initialization vector are denoted as $IV^{[63...0]}$. The meters are modified as follows:

$$\begin{aligned}
c_{0,4} &= c_{0,4} \oplus IV^{[31...0]}, \\
c_{1,4} &= c_{l,4} \oplus \left(IV^{[63...48]} \diamond IV^{[31...16]}\right), \\
c_{2,4} &= c_{2,4} \oplus IV^{663...32]}, \\
c_{3,4} &= c_{3,4} \oplus \left(IV^{[47...32]} \diamond IV^{[15...0]}\right), \\
c_{4,4} &= c_{4,4} \oplus IV^{[31...0]}, \\
c_{5,4} &= c_{5,4} \oplus \left(IV^{[63...48]} \diamond IV^{[31...16]}\right), \\
c_{6,4} &= c_{6,4} \oplus IV^{[63...32]}, \\
c_{7,4} &= c_{7,4} \oplus \left(IV^{[47...32]} \diamond IV^{[15...0]}\right).
\end{aligned}$$

The above modification system is repeated four times to make all the status bits nonlinearly dependent on all bits of the *IV* initialization vector. Modifying the counter variables on the *IV* initialization vector ensures that the use of all 2^{64} different *IV* initialization vectors will lead to the formation of unique key streams.

7.2.3.4 The Next State Function

The core of the *Rabbit* algorithm is an iterative scheme, which is determined by the following equations:

$$
\begin{aligned}
x_{0,i+1} &= g_{0,i} + (g_{7,i} <<< 16) + (g_{6,i} <<< 16);\\
x_{1,i+1} &= g_{1,i} + (g_{0,i} <<< 8) + g_{7,i};\\
x_{2,i+1} &= g_{2,i} + (g_{l,i} <<< 16) + (g_{0,i} <<< 16);\\
x_{3,i+1} &= g_{3,i} + (g_{2,i} <<< 8) + g_{1,i};\\
x_{4,i+1} &= g_{4,i} + (g_{3,i} <<< 16) + (g_{2,i} <<< 16)(5);\\
x_{5,i+1} &= g_{5,i} + (g_{4,i} <<< 8) + g_{3,i};\\
x_{6,i+1} &= g_{6,i} + (g_{5,i} <<< 16) + (g_{4,i} <<< 16);\\
x_{7,i+1} &= g_{7,i} + (g_{6,i} <<< 8) + g_{5,i},
\end{aligned}
$$

where

$$
g_{j,i} = \left(\left(x_{j,i} + c_{j,i+1}\right)^2 \oplus \left(\left(x_{j,i} + c_{j,i+1}\right)^2 >> 32\right)\right) \bmod 2^{32}
$$

and all additions are made by Module 2^{32}.

These analytical expressions in conjunction with each other can be represented graphically, as shown in Fig. 7.10. Before iteration, the counters increase as described below.

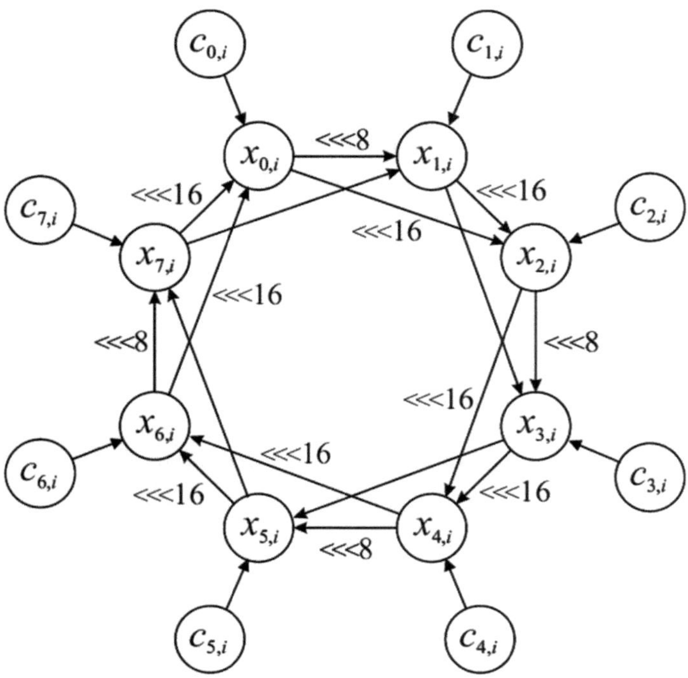

Fig. 7.10 Graphic illustration of the next state of the *Rabbit* algorithm

7.2.3.5 System of Meters

The dynamics of change of meters is defined as follows:

$$
\begin{aligned}
c_{0,i+1} &= c_{0,i} + a_0 + \phi_{7,i} \bmod 2^{32};\\
c_{l,i+1} &= c_{1,i} + a_I + \phi_{0,i+1} \bmod 2^{32};\\
c_{2,i+1} &= c_{2,i} + a_2 + \phi_{1,i+1} \bmod 2^{32}\\
c_{3,i+1} &= c_{3,i} + a_3 + \phi_{2,i+1} \bmod 2^{32};\\
c_{4,i+1} &= c_{4,i} + a_4 + \phi_{3,i+1} \bmod 2^{32};\\
c_{5,i+1} &= c_{5,i} + a_5 + \phi_{4,i+1} \bmod 2^{32};\\
c_{6,i+1} &= c_{6,i} + a_6 + \phi_{5,i+1} \bmod 2^{32};\\
c_{7,i+1} &= c_{7,i} + a_7 + \phi_{6,i+1} \bmod 2^{32},
\end{aligned}
$$

where the transfer bit counter $\varphi_{j,i+1}$ is given as

$$
\phi_{j,i+1} = \begin{cases} 1, & c_{0,i} + a_0 + \phi_{7,i} \geq 2^{32} \wedge j = 0 \\ 1, & c_{j,i} + a_j + \phi_{j-1,i+1} \geq 2^{32} \wedge j > 0; \\ 0 & otherwise. \end{cases}
$$

The constants a_j are defined as:

$$
\begin{aligned}
a_0 &= 0x4D34D34D;\\
a_1 &= 0xD34D34D3;\\
a_2 &= 0x34D34D34;\\
a_3 &= 0x4D34D34D;\\
a_4 &= 0xD34D34D3;\\
a_5 &= 0x34D34D34;\\
a_6 &= 0x4D34D34D;\\
a_7 &= 0xD34D34D3.
\end{aligned}
$$

7.2.3.6 Output Scheme

After each iteration, the output key stream is formed as follows:

$$
\begin{aligned}
s_i^{[15...0]} &= x_{0,i}^{[15...0]} \oplus x_{5,i}^{[31...16]};\\
s_i^{[31...16]} &= x_{0,i}^{[31...16]} \oplus x_{3,i}^{[15...0]};\\
s_i^{[47...32]} &= x_{2,i}^{[15...0]} \oplus x_{7,i}^{[31...16]};\\
s_i^{[63...48]} &= x_{2,i}^{[31...16]} \oplus x_{5,i}^{[15...0]};\\
s_i^{[78...64]} &= x_{4,i}^{[15...0]} \oplus x_{1,i}^{[31...16]};\\
s_i^{[95...80]} &= x_{4,i}^{[31...16]} \oplus x_{7,i}^{[15...0]};\\
s_i^{[111...96]} &= x_{6,i}^{[15...0]} \oplus x_{3,i}^{[31...16]};\\
s_i^{[127...112]} &= x_{6,i}^{[31...16]} \oplus x_{1,i}^{[15...0]},
\end{aligned}
$$

where s_i is the 128 bit key stream block at i iteration.

7.2.3.7 Encryption/Decryption Scheme

The generated key stream bits are added (by XOR operation) to the plain/encrypted text for encryption/decryption:

$$
\begin{aligned}
c_i &= p_i \oplus s_i,\\
p_i &= c_i \oplus s_i,
\end{aligned}
$$

where c_i and p_i denote the i 128-bit block of encrypted text and the block of plaintext, respectively.

7.2.3.8 Rabbit Stream Encryption Algorithm Security

Attacks on the key install function. Once the encryption key is specified, the bits of the counter and the status bits are strictly and very nonlinearly dependent on the bits of the key. This complicates the implementation of attacks based on guessing part of the key, even if the bits of the counter were known after modifying the counter. Of course, knowledge of meters facilitates other types of analysis.

Collision based attack. *Rabbit* cipher uses ambiguous mapping, different keys can potentially lead to the same gamut. This problem mainly boils down to the fact that different keys lead to the same counter values, since different counter values will almost certainly lead to different keystream generations. It should be noted that the key extension and iteration system have been designed in such a way that each key leads to unique counter values. However, modifying the counter can result in equal meter values for two different keys. Considering that after four initial iterations, the internal state is essentially random and does not correlate with the system of counters, the probability of collisions is given by the "paradox of birthdays", that is, for all 2^{128} keys, one collision is expected to be a 256-bit counter. Thus, the collision of the meter system should not lead to safety problems in practice.

Attack with a linked key. The attack is based on the use of symmetry properties in the next state functions and key installation. Consider, for example, two K and $K*$ keys related by the relation $K^{[i]} = K*^{[i+2]}$ for all.

This leads to the ratios:

$$x_{j,0} = x^*{}_{j+2,0}$$

and

$$c_{j,0} = c^*{}_{j+2,0}$$

Now consider if K and $K*$ are the same key. If the condition $a_{\mathrm{j},0} \neq a^*_{j+2,0}$. is satisfied, then the function of the following state would preserve the symmetry property. But it is easy to check that the constants a_j, are chosen such that $a_{j,0} \neq a^*_{j+2,0}$. Thus, the next state function is not prone to attacks based on related keys.

Attack based on partial guess key. *Guess and Verify Attack* attack will be possible if the output bits can be predicted by a small set of internal state bits. The attacker should guess the appropriate part of the state variables, predict the output bits and compare them directly with the output bits to make sure in the correctness of his guess. The attacker should guess at least 2×12 input bytes for different *g*-functions to check against one byte. This is identical to guessing 192 bits, which is more difficult than a complete search of all keys.

The essence of *the Guess and Determine Attack* method is that it is necessary to guess several unknown variables of a cipher and, using them, to derive the unknown remaining ones. After that, the system is banished several times, and what is output is compared to the real output of the cipher to check the assumption. The attacker tries to reproduce 512 bits of the internal state, that is, he watches 4 consecutive 128-bit data at the output of the cipher and does the following:

- divides 32-bit counter and 32-bit state variable into 8-bit variables;
- draws up a system of equations that simulates changes in counters and state variables, as well as outputs. As a result, 160 equations and 160 unknowns are obtained;
- solves this system, guessing as many unknowns as possible.

The effectiveness of this approach depends on the number of variables mentioned. This number is limited to the bottom by the 8-bit subsystem with the least number of input variables. If counters are neglected, each byte of the next state function depends on 12 input bytes. Taking into account the counters, each byte at the output of the subsystem already depends on 24 input bytes. Therefore, the attacker must guess more than 128 bits, which makes the attack impossible.

Algebraic attacks. For the *Rabbit* cipher function, the algebraic normal forms for 32 Boolean subfunctions have at least an algebraic degree of 30. The number of monomials varies from $2^{24.5}$ до $2^{30.9}$, and for a random function it must be 2^{31}. The distribution of the number of monomials as a function of degree is given in Fig. 7.11.

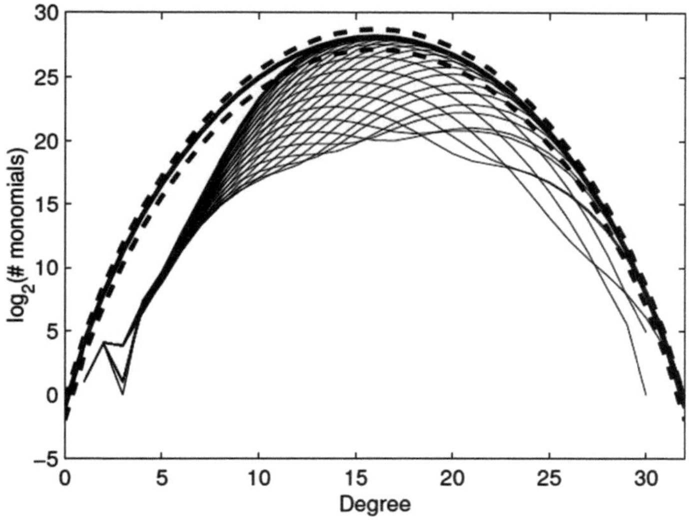

Fig. 7.11 The distribution of the number of monomials of each degree in each of the 32 Boolean subfunctions of g-function. The solid solid line and two dotted lines indicate the mean and the variance for an ideal random function [3]

Ideally, the major part of the number of monomials in each of the 32 Boolean subfunctions of g-function should be in the space between the dashed lines, which show deviations from the mean for an ideal random function. Some Boolean functions differ significantly from the results for a random function, but they all have a large number of high-degree monomials.

In addition, the developers of the algorithm investigated the partial coincidence of 32 functions. The total number of monomials that occur once is $2^{26{,}03}$, while the number of monomials that does not occur at all is $2^{26{,}2}$. Compared to a random function, these are quite large deviations. This information may be useful for analyzing the cipher in the future.

Correlation Attacks. The linear approximation attack implies a better linear approximation between the bits at the input of the next state function and the bits at the output of the output circuit. To do this, use the Walsh–Adamar transform, assuming that all input data is linearly independent. The cipher developers have found that the best linear correlation has a correlation coefficient of about $2^{-57{,}8}$, which involves generating output from 2^{114} iterations for comparison with a random function.

The second-order approximation can be much better. This can be achieved by eliminating from the algebraic normal form of g-function that is higher than the second degree. Denote by $f^{[j]}$ the second-order approximation of the algebraic normal form of the function $g^{[j]}$. According to the developers' experiments, the correlation coefficient between f and g is less than $2^{-9{,}5}$, and the correlation coefficient between $f^{[j]} \oplus f^{[j+1]}$ and $g^{[j]} \oplus g^{[j+1]}$ is approximately equal to $2^{-2{,}72}$. This means that some higher-order members are cut off when two adjacent bits are added by module 2. Building on this data a second order approximation cipher gives, at best, a correlation coefficient of about $2^{-26.4}$. This correlation coefficient value is insufficient to attack. If you consider the meters, the analysis is much more complicated.

7.2.4 Decim *Key Stream Generator*

The hardware-specific *Decim* SSC algorithm was developed by Burbaine et al. [4]. This is a hardware-oriented 80-bit key and 64-bit initialization vector *(IV)* stream cipher that was submitted to the *eSTREAM* streaming encryption project (no further than the third stage of the competition).

The *Decim* design is based on a nonlinear linear feedback filter (*LFSR*) and an irregular pseudorandom sequence dilution mechanism called *ABSG*. As a consequence, *Decim* has a low hardware complexity. Following the identification of certain shortcomings in [5], this algorithm was refined. The new version of *Decim*, called Decim^{v2}, looks more secure and also has less hardware complexity than the previous *Decim* version.

7.2.4.1 General Design of SSC $Decim^{v1}$

Decim [4] is a hardware-oriented current cipher that was submitted to the *ECRYPT* stream encryption project [6]; it is now called $Decim^{v1}$. It was developed within the framework of the *ABSG* mechanism, which provides a method for the uneven rarefaction of pseudorandom sequences. The overall operation of $Decim^{v1}$ (as well as the new $Decim^{v2}$ cipher) consists in generating a binary sequence in the usual way with the help of LFSR, which is filtered by a Boolean function. The resulting sequence is filtered by the *ABSG* mechanism.

In their work [5], Hongjun Wu and Bart Preneel describe two identified deficiencies of the $Decim^{v1}$ stream cipher.

The first drawback concerns the initialization phase, i.e. the calculation of the initial internal state to start generating the key stream.

In a nutshell, the $Decim^{v1}$ initialization mechanism works like this.

1. Filling the LFSR with an 80-bit secret key and 64-bit public initialization vector *(IV)*.
2. 192 *LFSR* updates. One update consists of the following three steps:
   - calculating the feedback value (non-linearly);
   - applying one of two permutations over 7 elements of the current state of the *LFSR*; the choice of permutation is controlled by the output of the *ABSG*;
   - move to one *LFSR* position.

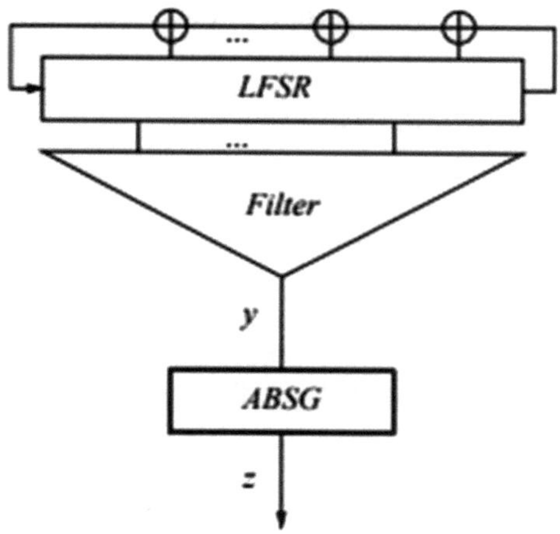

Fig. 7.12 Generating the *Decim*^{v1} key stream

The purpose of permutations is to ensure high nonlinearity during the initialization phase. However, a side effect of permutations is that a large number of *LFSR* elements (after initial fill) with high probability can never be updated during the initialization process. This shortcoming allowed Hongjun Wu and Bart Preneel to simulate an effective key recovery attack on *Decim*^{v1}. For *Decim*^{v2} it is offered a simpler and more secure initialization procedure than for *Decim*^{v1} (including the permissions associated with the *Decim*^{v1} initialization procedure, which implies a significant increase in hardware cost, were removed in *Decim*^{v2}).

The main drawback pointed out by Wu and Preneel [5] is the key flow generation algorithm shown in Fig. 7.12.

More precisely, the drawback is the generation of the sequence *y*, which is the filter output (the sequence processed by the *ABSG* mechanism). In short, this drawback is due to the fact that the sequence y is the direct input of a symmetric Boolean function that is not first-order correlative. There is a correlation between the outputs of a function associated with two inputs of vectors that have one common element. Using this weakness, Hongjun Wu and Bart Preneel demonstrate correlation between some bits of the key stream sequence and then show that the *Decim*^{v1} key stream is too predictable.

Decim^{v2} offers a simpler and more secure filter than *Decim*^{v1} by selecting a filter that is first order correlative. Let us briefly review the proposed improvements, following the appropriate terminology and designations [5].

According to the specification provided by the Ecrypt Stream Encryption Project, *Decim*^{v2} accepts an 80-bit secret key and a 64-bit public initialization vector as input.

7.2.4.2 Specification of the *Decim*^{v2} Algorithm

Generation of key stream

The size of the internal state of *Decim*^{v2} does not change, i.e. 192 bits. The key stream generation mechanism is shown in Fig. 7.13. The *LFSR* internal state bits are numbered from 0 to 191 and are denoted x_0, ..., x_{191}. The sequence of linear feedback values *LFSR* is denoted by $s = (s_t)_{t \geq 0}$.

The Boolean function is a 13-digit balanced quadratic symmetric function.

Let x_{i1}, ..., x_{i14} denote the 14 initial *LFSR* internal states, which are the input parameters of the filter. The sequence *y*, which is the filter output, is defined as:

$$y_t = f(s_{i1+t}, \ldots, s_{i13+t}) \oplus s_{i14+t}$$

ABSG accepts as the input sequence $y = (y_t)_{t \geq 0}$. The buffer mechanism guarantees the constant throughput of the key stream; 32-bit buffer length and 1-bit buffer output are selected for every 4 offsets by one *LFSR* position (see [4] for more details).

Remark 1 To generate the key stream, the difference between *Decim*^{v1} and *Decim*^{v2} is the choice of filter. The *Decim*^{v1} filter has a vector function defined as

$$F : F_2^{14} \rightarrow F_2^2; x_{i1}, \ldots, x_{i14} \mapsto (f(x_{i1}, \ldots, x_{i7}), f(x_{i8}, \ldots, x_{i14}))$$

where *f* is a 7-digit balanced symmetric Boolean function, but it is not first-order correlation stable.

Setting Key/Vector Initialization (IV)

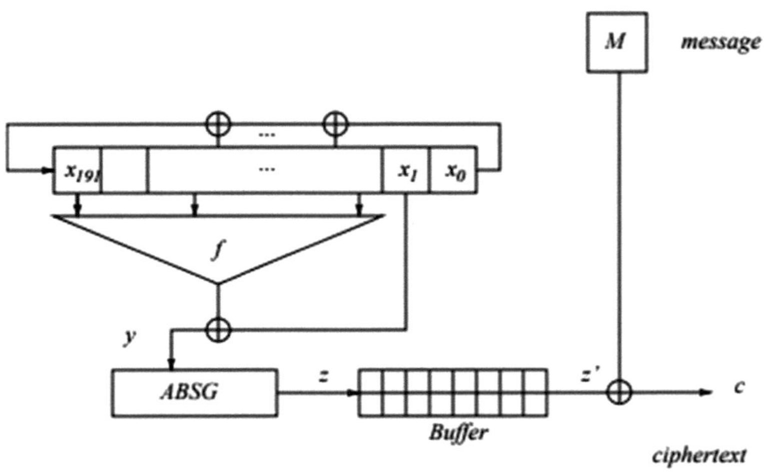

Fig. 7.13 Generating the *Decim*^{v2} key stream

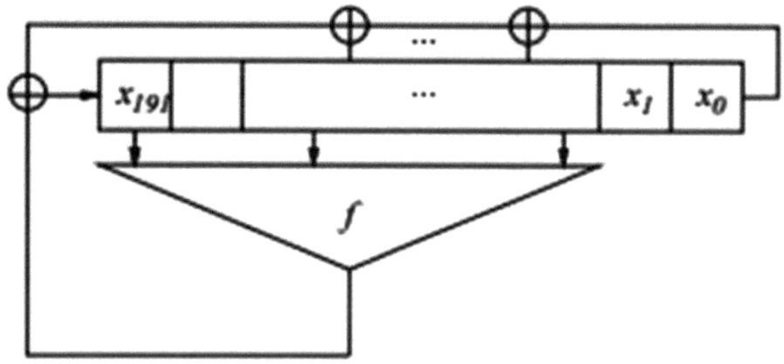

Fig. 7.14 Initialization key/vector setting mechanism

The initial *LFSR* fill from the key and the initialization vector was modified in $Decim^{v2}$ compared to $Decim^{v1}$. The mechanism of setting the key/vector of initialization (*IV*) is to clock $4 \times 192 = 768$ times the *LFSR* using the nonlinear feedback described in Fig. 7.14.

Remark 2 For the initialization stage, the main differences between $Decim^{v1}$ and $Decim^{v2}$ are filling in the *LFSR* that has been modified, deleting permutations, and selecting a filter. As a result, the number of cycles at the initialization stage increases from 192 to 768.

Filtered LFSR

Consider a filtered *LFSR* that generates sequences of y (sequences of y are input parameters of the *ABSG* mechanism).

LFSR (unchanged). The main *LFSR* is the maximum length *LFSR*, with a length of 192 over F_2. It is defined by the following primitive feedback polynomial:

$$P(X) = X^{192} + X^{189} + X^{188} + X^{169} + X^{156} + X^{155} + X^{132} + X^{131} + X^{94} + X^{77} + X^{46} + X^{17} + X^{16} + X^5 + 1$$

Filter (modified). The filter function is a 14-digit Boolean function defined as:

$$F : F_2^{14} \to F_2; a_1, \ldots, a_{14} \mapsto F(a_1, \ldots, a_{13}) \oplus a_{14}$$

where f is a symmetric quadratic Boolean function, defined as follows:

$$f(a_1, \ldots, a_{13}) = \oplus_{1 \le i \le j \le 13} a_i a_j \oplus_{1 \le i \le 13} a_i$$

Filter position labels:

$$191 - 186 - 178 - 172 - 162 - 144 - 111 - 104 - 65 - 54 - 45 - 28 - 13 - 1$$

and input ABSG state t is

$$y_t = f(s_{t+191}, s_{t+186}, s_{t+178}, s_{t+172}, s_{t+62}, s_{t+144}, s_{t+111}, s_{t+104}, s_{t+65}, s_{t+54}, s_{t+45}, s_{t+28}, s_{t+13}) \oplus s_{t+1}$$

Fig. 7.15 ABSG algorithm

Input: $(y_0, y_1, \ldots)$
Set: $i \leftarrow 0$; $j \leftarrow 0$;
Repeat the following steps:
1. $e \leftarrow y_i$, $z_j \leftarrow y_{i+1}$;
2. $i \leftarrow i+1$;
3. **while** $(y_i = \overline{e})$ $i \leftarrow i+1$;
4. $i \leftarrow i+1$;
5. **output** z_j
6. $j \leftarrow j+1$

Decimation (unchanged)

This section describes how the key stream sequence z is obtained from the sequence y. The ABSG algorithm is shown in Fig. 7.15.

Buffer mechanism (unchanged)

The speed of the ABSG mechanism is not uniform, so they use a buffer to guarantee constant bandwidth. A buffer of length 32 is selected, and for each of the 4 bits present on the ABSG input, the buffer must output exactly 1 bit. With these parameters, the probability that the buffer is empty when it should output 1 bit is less than 2^{-89}. If ABSG outputs one bit when the buffer is full, then the new calculated bit is not queued, that is, it drops. Assuming that the initial internal state is calculated (it is denoted $z_0, \ldots, z_{191}$), the ABSG mechanism starts from the beginning of the cycle and the buffer is empty. Generation of the key stream begins when the buffer is full.

7.2.4.3 Key/Initialization Vector Setting (IV)

This section describes how to calculate the initial internal state to generate a key stream. It should be noted that the *ABSG* mechanism is no longer used during the initialization phase.

LFSR Initial Fill (modified)

The *K* secret key is an 80-bit key labeled $K = K_0, \ldots, K_{79}$, and the *IV* initialization vector is 64-bit, *IV* is denoted $IV_0, \ldots, IV_{63}$.

The initial filling of the *LFSR* is as follows:

$$x_i = \begin{cases} K_i & 0 \le i \le 79 \\ K_{i-80} \oplus IV_{i-80} & 80 \le i \le 143 \\ K_{i-80} \oplus IV_{i-144} \oplus IV_{i-128} \oplus IV_{i-112} \oplus IV_{i-96} & 144 \le i \le 159 \\ IV_{i-160} \oplus IV_{i-128} \oplus 1 & 160 \le i \le 191 \end{cases}$$

The number of possible initial states of the *LFSR* is $2^{89+64} = 2^{144}$.

LFSR status update

The *LFSR* is clocked $4 \times 192 = 768$ times using nonlinear feedback. Let y_t, denoted as output f at time t, and l_{vt} denote the linear feedback at $t > 0$. Then x_{191} at time t is calculated using the equation:

$$x_{191} = l_{v_t} \oplus y_t$$

Note that this step does not have an *LFSR* status output bit.

7.2.4.4 Justification of the Design

The rationale behind the *Decim*^{v2} design is that the basic ideas of *Decim*^{v1}, namely filtering and then destroying the *LFSR* output using *ABSG*, are undeniable. Thus, the *Decim*^{v2} kernel is the only filter based on the Boolean function and then the decimation, which in turn is based on *ABSG*.

Filter

In *Decim*^{v2} (as well as *Decim*^{v1}), the Boolean function is used to filter the *LFSR*, whereas the linear complexity of the *LFSR* output sequences with a primitive feedback polynomial is the length of the *LFSR*. The interest of the filter is to greatly increase the linear complexity of the output sequences that is input to the *ABSG* mechanism. This occurs to increase the minimum length equivalent to the *LFSR*, which produces the same sequence as the outputs of the filtered *LFSR*.

Choosing a filter is very important as the filter does not need to detect any weaknesses of the streaming cipher (as is the case with *Decim*^{v1}). An important feature of the filter is that its output must be evenly distributed. In *Decim*^{v1} 7 digit Boolean function f used in the filter is balanced, that is, the value of f is evenly distributed over $\{0, 1\}$, whereas the estimate of f is uniformly over $\{0, 1\}^7$. *Decim*^{v1} is a hardware-oriented streaming cipher, and the filter should have an inexpensive hardware implementation. In *Decim*^{v1}, the filter is a symmetric Boolean function f (i.e. the value of f depends only on the Hamming weight of the input signal) in order to reduce the cost of the equipment, and the function f is balanced.

The attack made by Wu and Preneel [5] demonstrated that it is important to choose a Boolean function f that is first order correlation stable, that is, a function that has no relationship to the outputs of functions associated with two input vectors, which have 1 common element. Since the function f must also be balanced, this means that f must be 1-elastic. In *Decim*^{v1}, the Boolean function is balanced but not 1-elastic. The *Decim*^{v2} filter is constructed with a balanced 13-digit symmetric function (which is not first-order correlation stable) and the entire filter F is a 1-elastic Boolean function.

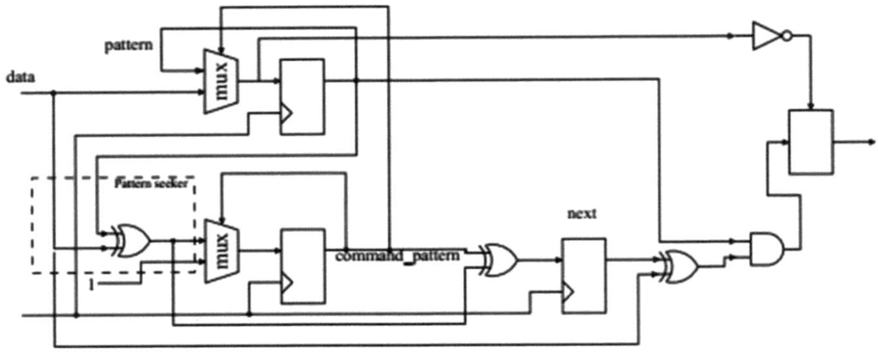

Fig. 7.16 Hardware implementation of *ABSG*

Position switch: filter and feedback log

Assuming the key stream z is known, the attacker must guess some bits of the sequence y to attack f. Knowing the bits of y gives the equation in bits of the initial state of the *LFSR*. Thus, the number of monomial bits of the initial state of the *LFSR* involved in these equations should be maximized. In addition, this number should increase rapidly during the first *LFSR* cycles. There are two conditions in mind:

1. Each difference between the two positions of the bits entered in f must appear only once.
2. Some inputs f must be received at bit positions near one of the feedback bits (which means that some inputs should be leftmost in Fig. 7.16). Finally, the switches of the Boolean function f inputs and the feedback inputs must be independent.

Key/Initialization Vector Setting (*IV*)

The key stream generation components are reused to configure the Initialization Key (*IV*); no new components introduced. Using the 80-bit key and 64-bit initialization vector, the number of possible initial states may exceed 2^{144}, which occurs in the case of $Decim^{v2}$, whereas the number of possible initial states of $Decim^{v1}$ is 2^{136}. The first attack given in [5] is based on a permutation effect π_1 and π_2, used in the initialization process. Indeed, some *LFSR* bits are not updated properly. Then the attack is to track some bits during the initialization process. In $Decim^{v2}$, the permutations are removed and the number of clock cycles is increased in order to provide sufficient nonlinearity of initialization steps.

7.2.4.5 Hardware Implementation

The number of gateways involved in hardware implementation can be estimated as follows: based on the evaluation of the elementary components given in [7], i.e. 12

gateways for triggers, 2.5 for *XOR*, 1.5 gateways for *AND*, and 5 gateways for *MUX*. Here we have the following values for each component of the chain:

- *LFSR*: 2339 gateways corresponding to 192 triggers and 14 *XOR*s (instead of 3334 gateways for $Decim^{v1}$).
- Filtering function: 86.5 gateways corresponding to 6 complete adders and 7 *XOR* (instead of 74 gateways for $Decim^{v1}$; details of the hardware implementation of the quadratic symmetric function are given in [4]).
- *1 ABSG* input as shown in Fig. 7.16*:* 67 gateways corresponding to 2 *MUX*, 3 *XOR*, 1 *AND* and 4 triggers.

Remark 3 For the proposed hardware implementation, the major differences between $Decim^{v1}$ and $Decim^{v2}$ are that the *LFSR* will now be clocked 4 times instead of 2 before outputting bits, i.e. $Decim^{v2}$ is half that of $Decim^{v1}$. In addition, the throughput of the generator can be doubled in the case of low cost of implementation using a simple acceleration mechanism. This can be done using a circuit that calculates 2 bits of feedback for the *LFSR* at a time, as shown in ([4], Sect. 7.1). This dual-clocked *LFSR* can be implemented with 192 triggers and 28 *XOR*s. One additional copy of the filtering function is also required, and a 2-input *ABSG* mechanism must be used (see [4] for details).

7.2.4.6 Security Features

The discussion given in [4] about random and deterministic attacks, excellent attacks, and side-channel attacks also holds true for $Decim^{v2}$. Tactically controlled linear feedback registers, that is, *LFSR*s that are irregularly synchronized according to the destructive sequence that determines the number of characters to be deleted before the next input character, are resistant to rapid correlation attacks. In [8] Golik developed a theory of rapid correlation attacks on irregularly synchronized *LFSR*s based on linear statistical weakness. This attack may be realistic in special cases, but $Decim^{v2}$ maintains its security in such attacks. Indeed, in order to increase the linear complexity of the sequence (that is, the minimum equivalent length of the *LFSR* that produces the same sequence), which is reduced by the *ABSG* mechanism, we use the *LFSR* filtered by the Boolean function. Similarly, the expected linear complexity of the original sequence using the Boolean function is 18528, i.e. the expected minimum length of the *LFSR* that generates the same sequence as that generated by the filtered *LFSR Decim*-18528.

Thus, the $Decim^{v2}$ streaming encryption scheme is based on $Decim^{v1}$ stream encryption and eliminates any deficiencies found in the original scheme. The $Decim^{v2}$ stream encryption is particularly suitable for hardware applications with limited resources, such as limited memory or gateways. For programs requiring higher bandwidth, special acceleration mechanisms can be used to accelerate $Decim^{v2}$ due to the higher complexity of the equipment.

7.2.5 *Key Stream Generator* KCipher-2

The *KCipher-2* stream encryption is based on several independent linear shift registers (LFSR), which are used in conjunction with nonlinear functions to produce the key stream. Some stream ciphers use a generic nonlinear function that allows one or more LFSRs to be uneven. Currently, different types of generators based on clock control are offered, and there are various cryptographic attacks against them. An example of such algorithms is A5, a well-known streaming algorithm designed to ensure privacy in mobile communications.

The stream encryption control mechanism typically controls the timing of LFSR or shortens the output sequence. In the latter case, the performance of the stream cipher is reduced because some elements of the output sequence are discarded. If you use the reduction of the output sequence obtained from the word-oriented algorithm when controlling the clock signal, the performance in this case is markedly reduced. A clock control mechanism for updating the state of a bit-oriented generator is also ineffective when the mechanism is used to control each register. On the other hand, dynamic feedback control for LFSR is an effective way to improve the security of streaming encryption.

KCipher2 is a streaming word-based cipher that has an effective dynamic feedback management mechanism used as irregular synchronization. The basic idea of the algorithm is to change the mixing operation when updating the internal state of the generator.

The inverse polynomials for the word-oriented LFSR are described by certain coefficients; multiplying the input word by a factor means mixing words. A typical example of such a generator is the linear shift registers used in *SNOW 2.0*. As a rule, the coefficients are chosen so that the feedback polynomial is primitive over the field. *KCipher-2* uses irregular synchronization for this mixing operation, and the modification causes only a minimal decrease in the encryption/decryption rate. In other words, at least one feedback shift register (FSR), which has an irregular internal refresh rate, will dynamically modify the feedback function for dynamic feedback control of the output sequences of other FSS. For example, the feedback function is defined as

$$s_{t+a} = \alpha_0^{\{0,1\}} s_{t+b} \oplus \alpha_1^{\{0,1\}} s_{t+c} \oplus \alpha_2^{\{0,1\}} s_{t+d}$$

where $\{0,1\}$ is selected by the dynamic feedback control mechanism. This is called a dynamic shift register (DFSR). The dynamic feedback control mechanism improves the security of the stream cipher because it changes the deterministic linear output sequence of some registers to a probable output sequence. This property allows the algorithm to effectively counteract some attacks. For a successful cryptographic attack, an attacker must obtain a linear output sequence derived from the linear sequence of some registers. In the case of irregular synchronization, the possibility of obtaining such a linear sequence is unlikely. In order to carry out the attack, the attacker must guess some inputs to the nonlinear function; however, irregular controls make this impossible: the attacker must first guess the inputs of the dynamic feedback controller. Thus, irregular modification of the feedback function improves the security of the streaming cipher. Dynamic Feedback Management potentially enhances effectiveness against multiple attacks, both existing and new. First of all, *KCipher-2* provides not only the high performance inherent in most stream ciphers based on LFSR, but also high cryptoconversion.

The *KCipher-2* key stream generation rate is 4.77 cycles/byte on a *Pentium 4* computing platform. Thus, the cipher competes with existing *CRYPTREC* stream ciphers. In addition, *KCipher-2* was designed to take into account the implementation of two attacks on SNOW 2.0 SSG: algebraic attack and differential attack. *SNOW 2.0* is vulnerable to them; however, *KCipher-2* has a larger amount of cryptoconversions against these two attacks. In fact, *KCipher-2* has not yet detected any attacks that would require less than 2^{256} operations. Details on crypto-stability assessments are given below.

7.2.5.1 General Description of the Structure

The following specifications are used for the stated specification of the cipher [9].

Marking

$\oplus$—bit exclusive *XOR*
$\boxplus$—adding an integer to the module $2^{32} - 1$
0x—prefix for hexadecimal values
$GF(2^n)$ is a finite field to which 2^n values belong
$\gg n$—n-bit bit shift to the right of the 32-bit register
$\ll n$—n-bit bit shift left 32-bit register.

Data structure

Data word size in *KCipher-2* is 32 bits. In the algorithm, the order of bits/bytes in the word is "*big-endian*", that is, if the key and the initialization vector are given as a sequence of bits/bytes, then the first/leftmost bit/byte is the most significant.

7.2.5.2 *KCipher-2* Components and Basic Features

The *KCipher-2* consists of two feedback shift registers (*FSR*), *FSR-A* and *FSR-B*, a nonlinear function with four internal registers R_1, R_2, L_1 and L_2 and dynamic feedback control, as shown in Fig. 7.17.

The *FSR-B* is a dynamic feedback shift register. The length of each register is 32 bits.

Feedback shift register

The *FSR-A* consists of five cells, and the *FSR-B* includes eleven cells. Let β be the root of the generating polynomial $x^8 + x^7 + x^6 + x + 1 \in GF(2)[x]$. The byte string

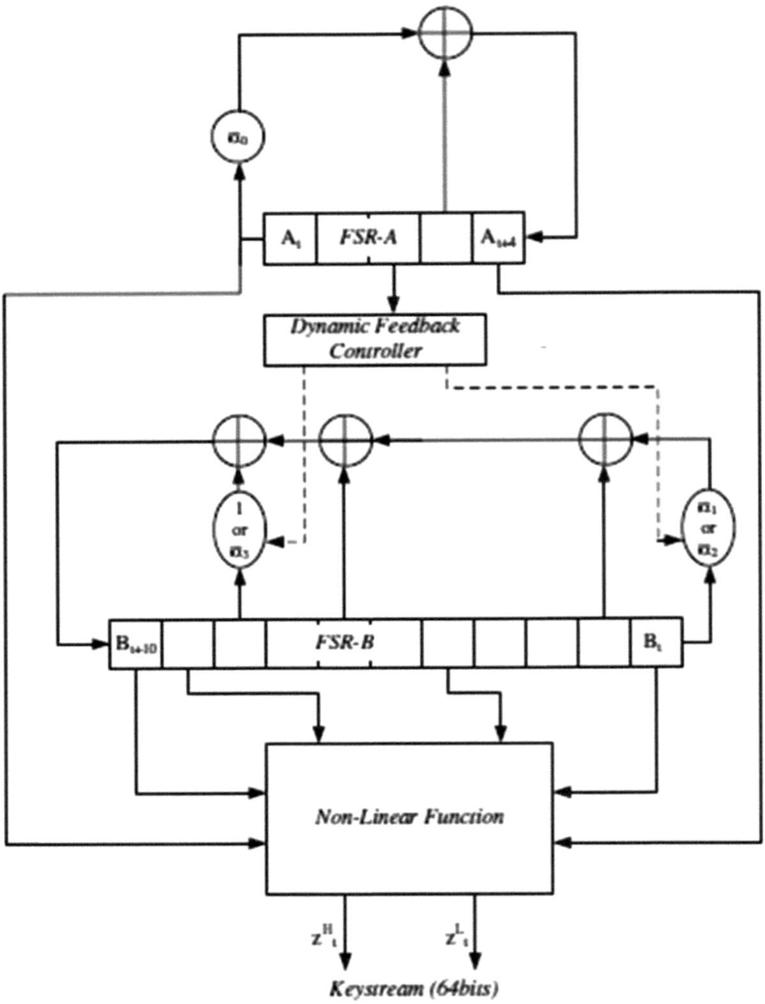

Fig. 7.17 Schematic representation of KCipher-2

is denoted as $(y_7, y_6, \ldots, y_1, y_0)$, where y_7 is the most significant bit, y_0 is the least significant. y is given as $y = y_7\beta^7 + y_6\beta^6 + \cdots + y_1\beta + y_0$.

Likewise, let λ, δ, ζ be the roots of primitive polynomials

$$x^8 + x^5 + x^3 + x^2 + 1 \in GF(2)[x],$$
$$x^8 + x^6 + x^3 + x^2 + 1 \in GF(2)[x],$$
$$x^8 + x^6 + x^5 + x^2 + 1 \in GF(2)[x]$$

in accordance. Let α_0 be the root of an irreducible fourth degree polynomial

$$x^4 + \beta^{24}x^3 + \beta^3x^2 + \beta^{12}x + \beta^{71} \in GF(2^8)[x]$$

We denote a row of 32-bit Y-long lines as (Y_3, Y_2, Y_1, Y_0), where Y_1 is the byte of the string and Y_3 is the most significant byte. In this case Y is given as… Let $\alpha_1, \alpha_2, \alpha_3$ correspond to the roots of irreducible fourth-degree polynomials

$$\begin{gathered}
x^4 + \gamma^{230}x^3 + \gamma^{156}x^2 + \gamma^{93}x + \gamma^{29} \in GF(2^8)[x], \\
x^4 + \delta^{34}x^3 + \delta^{16}x^2 + \delta^{199}x + \delta^{248} \in GF(2^8)[x], \\
x^4 + \zeta^{157}x^3 + \zeta^{253}x^2 + \zeta^{56}x + \zeta^{16} \in GF(2^8)[x],
\end{gathered}$$

respectively.

Polynomial feedback $f_A(x)$ and $f_B(x)$ for the *FSR-A* and *FSR-B* registers remains as follows:

$$\begin{aligned}
f_A(x) &= \alpha_0 x^5 + x^2 + 1 \\
f_B(x) &= \left(\alpha_1^{cl1_t} + \alpha_2^{cl1_t} - 1\right)x^{11} + x^{10} + x^5 + \alpha_3^{cl2_t}x^3 + 1
\end{aligned}$$

We denote by $cl1$ and $cl2$ the sequences that describe the output of the dynamic feedback controller. The output sequences at a certain time t are defined in terms of some bits of *FSR-A*. Let A_x be the output of *FSR-A* at x time.

$A_x[y] = \{0, 1\}$ determines the yth bit of the variable A_x, where A_x is the most significant bit of A_x. Then $cl1$ and $cl2$ (the so-called clock control bits) are described as $cl1_t = A_{t+2}$, $cl2_t = A_{t+2}$. Both $cl1$ and $cl2$ are binary variables; more precisely $cl1_t = \{0, 1\}$ and $cl2_t = \{0, 1\}$. The stopwatch generator algorithm is efficient in terms of computational capabilities since no computation is required when variables are 0. However, the feedback function does not have transformations for feedback registers with a probability of 1/4, when all registers are clocked. So we use two types of clock control for the feedback function. *FSR-B* is defined by a primitive polynomial in which $cl2 = 0$.

Nonlinear Conversion Function

The nonlinear *KCipher-2* function receives the input of two *FSR-A* cells and four *FSR-B* cells and the internal registers $R1$, $R2$, $L1$, $L2$, and 64 bits of key stream are generated at each output. In Fig. 7.18 shows the nonlinear function of *KCipher-2*. The nonlinear function involves four replacement operations marked with the *Sub* symbol.

$$Sub\colon A \in GF(2^{32}) \mapsto B = Sub(A) \in GF(2^{32})$$

In the first step of the replacement operation, the input 32-bit string is divided into four byte substrings, each of which uses a non-linear $8 \rightarrow 8$ substitution, and then a 32-bit linear permutation is applied. Replacing $8 \rightarrow 8$ bits corresponds to

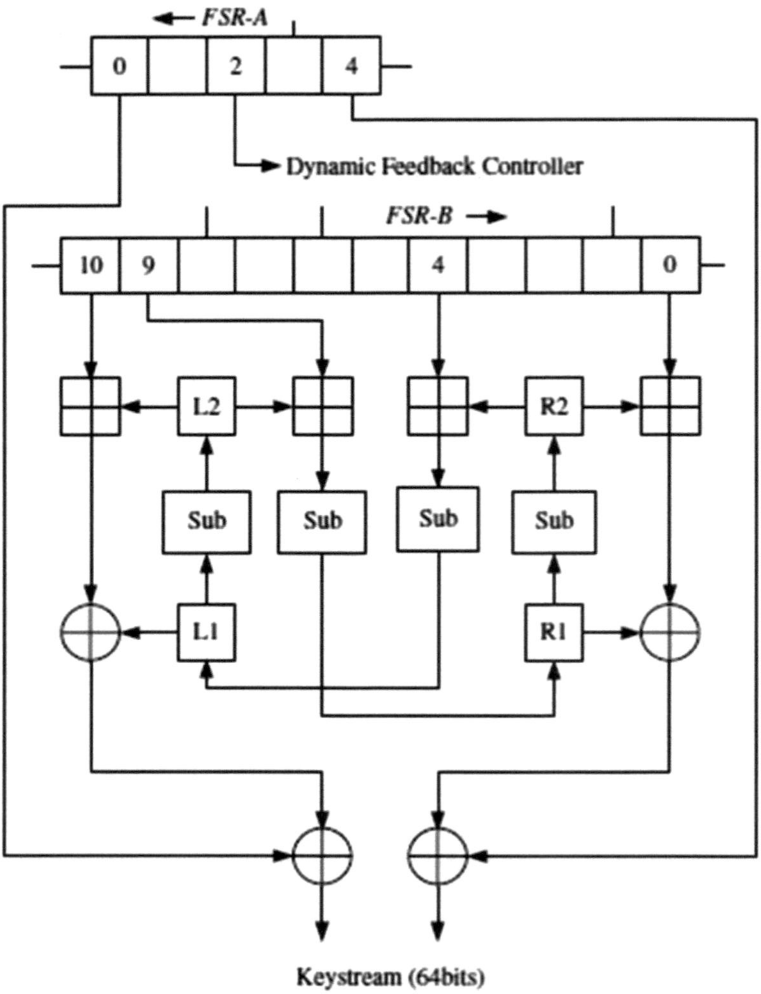

Fig. 7.18 *KCipher-2* nonlinear function

the substitution in *AES* [10], and the linear permutation is the same as the *AES Mix Column* operation.

Substitution $8 \to 8$ in *Sub* function. The *8*-bit substitution consists of two functions *g* and *f* and is determined by the composition function $f \circ g$. The function *g* calculates the inverse multiplicative element modulo irreducible polynomial $m(x) = x^8 + x^4 + x^3 + x + 1$ without 0x00, since 0x00 transforms itself:

$$g : a \in GF(2^8) \mapsto b = a^{-1} \in GF(2^8)$$

On the other hand, we have an affine transformation $f : a \in GF(2^8) \mapsto b \in GF(2^8)$ defined as:

$$\begin{vmatrix} b_7 \\ b_6 \\ b_5 \\ b_4 \\ b_3 \\ b_2 \\ b_1 \\ b_0 \end{vmatrix} = \begin{vmatrix} 1\,1\,1\,1\,1\,0\,0\,0 \\ 0\,1\,1\,1\,1\,1\,0\,0 \\ 0\,0\,1\,1\,1\,1\,1\,0 \\ 0\,0\,0\,1\,1\,1\,1\,1 \\ 1\,0\,0\,0\,1\,1\,1\,1 \\ 1\,1\,0\,0\,0\,1\,1\,1 \\ 1\,1\,1\,0\,0\,0\,1\,1 \\ 1\,1\,1\,1\,0\,0\,0\,1 \end{vmatrix} \times \begin{vmatrix} a_7 \\ a_6 \\ a_5 \\ a_4 \\ a_3 \\ a_2 \\ a_1 \\ a_0 \end{vmatrix} \oplus \begin{vmatrix} 0 \\ 1 \\ 1 \\ 0 \\ 0 \\ 0 \\ 1 \\ 1 \end{vmatrix}$$

where $a = (a_7, \ldots, a_0)$ is the input and $b = (b_7, \ldots, b_0)$ is the output a_0 and b_0 are the least significant bits (*LSB*).

Linear permutation $32 \rightarrow 32$ in *Sub*. Let C be an input 32-bit string consisting of four 8-bit values (c_3, c_2, c_1, c_0). And denote D as an output sequence of 32 bits given by four 8-bit variables (d_3, d_2, d_1, d_0). The linear permutation $p : C \in GF(2^{32}) \mapsto D = p(C) \in GF(2^{32})$ is described as follows:

$$\begin{pmatrix} d_0 \\ d_1 \\ d_2 \\ d_3 \end{pmatrix} = \begin{pmatrix} 02 & 03 & 01 & 01 \\ 01 & 02 & 03 & 01 \\ 01 & 01 & 02 & 03 \\ 03 & 01 & 01 & 02 \end{pmatrix} \begin{pmatrix} c_0 \\ c_1 \\ c_2 \\ c_3 \end{pmatrix}$$

$GF(2^8)$ with irreducible polynomial $m(x) = x^8 + x^4 + x^3 + x + 1$.

7.2.5.3 *KCipher-2* Operations

This section provides a detailed description of the basic operations in *KCipher-2*. *KCipher-2* has two operations, the keystream encryption operation and the generator internal state initialization operation. Initialization is performed first and then a generation operation.

Operation of generation of a key stream

We define the key stream at time t as $Z_t = (z_t^H, z_t^L)$ (each variable z_t^x is a 32-bit value and z_t^H is the highest row). The key sequence z_t^H, z_t^L is calculated as follows:

$$z_t^L = B_t + R2_t \oplus R1_t \oplus A_{t+4}$$

where A_x та B_x denote the outputs of *FSR-A* and *FSR-B* at time x, $R1_x$, $R2_x$, $L1_x$ and $L1_x$ denote the inner cells at moment x. Finally, the inner cells are updated as follows:

$$R1_{t+1} = Sub(L2_t + B_{t+9}),\quad R2_{t+1} = Sub(R1_t)$$
$$L1_{t+1} = Sub(R2_t + B_{t+4}),\quad L2_{t+1} = Sub(L1_t)$$

$$A_{t+i+1} = \begin{cases} A_{t+i+1} & (i = 0, 1, 2, 3) \\ A_{t+3} \oplus \alpha_0 A_t & (i = 4) \end{cases}$$

$$B_{t+i+1} = \begin{cases} B_{t+i+1} & (i = 0, 1, \ldots, 9) \\ \left(\alpha_1^{cl1_t} + \alpha_2^{cl1} - 1\right) B_t \oplus B_{t+1} \oplus B_{t+6} \oplus \alpha_3^{cl2} B_{t+8} & (i = 10) \end{cases}$$

where *Sub*(X) is the output of *Sub* for step X. The set $\{B_t, B_{t+3}, B_{t+8}, B_{t+10}\}$ is the set of complete positive differences (*FPDS*) [11]. Section 7.2.5.4 describes in detail the *Sub* function and finite-field arithmetic for fixed multiplicative elements α_0, α_1, α_2 and α_3 to implement the *KCipher-2* algorithm.

Initialization process

The initialization process for the *KCipher-2* SSC consists of two steps: the key download step and the internal state initialization step of the generator.

The key download step. The initial internal state is formed from the 128-bit initial key and the 128-bit initialization vector using the key allocation algorithm. The key allocation algorithm is similar to the generation of round keys in *AES BSC*; in *KCipher-2*, the algorithm extends the 128-bit input key to 384 bits. The key distribution algorithm for a 128-bit long key is defined as:

$$K_i = \begin{cases} IK_i & (0 \le i \le 3) \\ K_{i-4} \oplus Sub\big((K_{i-1} << 8) \oplus (K_{i-1} >> 24)\big) \oplus Rcon[i/4 - 1] & (i = 4n) \\ K_{i-4} \oplus K_{i-1} & (i \ne 4n) \end{cases}$$

where $IK = (IK_0, IK_1, IK_2, IK_3)$ is the initial key, i is a positive integer $0 \le i \le 11$ and n is a positive integer. The function of *Sub (X)* key distribution algorithm is the same as in the nonlinear function. This function is different from the round key generation function in *AES*, but the other part of the key distribution algorithm is similar to the round key generation function in *AES*. $Rcon[i]$ is defined as $\left(x^i \bmod x^8 + x^4 + x^3 + x + 1, 0x00, 0x00, 0x00\right)$ and x is $0x02$. Therefore, $Rcon[0] = (0x01, 0x00, 0x00, 0x00)$ for $i = 4$ and $Rcon[1] = (0x02, 0x00, 0x00, 0x00)$ for $i = 8$. The internal state is initialized by the key K_i and the initialization vector $IV = (IV_0, IV_1, IV_2, IV_3)$ as follows:

$$A_m = K_{4-m} (m = 0, \ldots, 4),$$
$$B_0 = K_{10}, B_1 = K_{11}, B_2 = IV_0, B_3 = IV_1,$$
$$B_4 = K_8, B_5 = K_9, B_6 = IV_2, B_7 = IV_3,$$
$$B_8 = K_7, B_9 = K_5, B_{10} = K_6.$$

The internal registers $R1$, $R2$, $L1$ and $L2$ are set as 0x00.

Stage of initialization of the internal state. When the initial data is set, the internal state of the generator is formed by performing the clocking operation of the generator 24 times ($j = 1, \ldots, 24$). The stage of initialization of the internal state is shown in Fig. 7.19.

The internal state of y is updated as follows:

$$R1_j = Sub\big(L2_{j-1} + B_{j+8}\big), \quad R2_j = Sub\big(R1_{j-1}\big)$$
$$L1_j = Sub\big(R2_{j-1} + B_{j+3}\big), \quad L2_j = Sub\big(L1_{j-1}\big)$$

$$A_{j+i} = \begin{cases} A_{j+i} & (i = 0, 1, 2, 3) \\ \alpha_0 A_{j-1} \oplus A_{j+2} \oplus z^L_{j-1} & (i = 4) \end{cases}$$

$$B_{j+i} = \begin{cases} B_{j+i} & (i = 0, 1, \ldots, 9) \\ \left(\alpha_1^{cl1_{j-1}} + \alpha_2^{cl1_{j-1}} - 1\right) B_{j-1} \oplus B_j \oplus B_{j+5} \oplus \alpha_3^{cl2_{j-1}} B_{j+7} \oplus z^L_{j-1} & (i = 10) \end{cases}$$

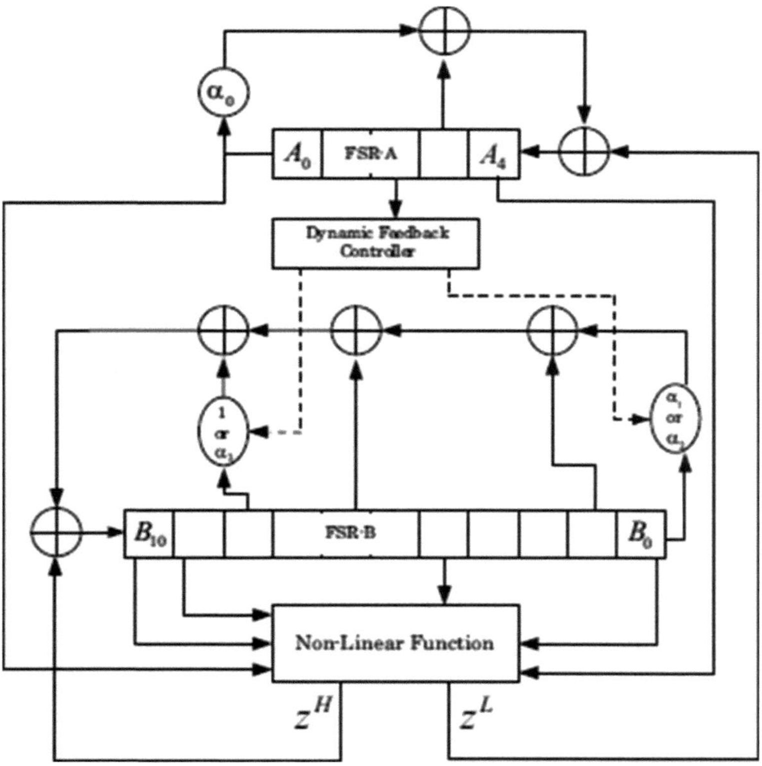

Fig. 7.19 Initializing *KCipher-2*

7.2.5.4 Implementation of *KCipher-2*

Quick implementation using conversion tables

Multiplicative operations with α_0, α_1, α_2 та α_3. Multiplicative operations with 32-bit variables and fixed elements α_0, α_1, α_2 та α_3 are performed in *FSR-A* and *FSR-B*. Operations can be calculated quickly when recalculation tables are used. Let *w* be a 32-bit value and *alpha_i*[*j*] corresponds to the *j*th element ($0 \leq j \leq 255$) *alpha_i* of the recalculation table ($i = 0, 1, 2, 3$).

The multiplication $\alpha_i w$ is defined as

$$a_i w = (w << 8) \oplus alpha\_i[(w >> 24)].$$

For example, the result of multiplication α_0 та $w =$ 0x0123467 will be

$$\begin{aligned}\alpha_0 \cdot (0x01234567) &= ((0x01234567) \ll 8) \oplus alpha\_0[0x01] \\ &= 0x23456700 \oplus 0xB6086D1A \\ &= 0x954D0A1A\end{aligned}$$

Conversion tables for Sub. The *Sub* operation $32 \to 32$ of a nonlinear function can be performed fairly quickly if the conversion tables are used. Let 32-bit *A* be fed to *Sub* input. But let *T_i*[*j*] be the *jth* element ($0 \leq j \leq 255$) in the table *T_i* ($i = 0, 1, 2, 3$). In this case, the result of the *Sub (A)* function would be:

$$\begin{aligned}Sub(A) = T\_0[A\,\&\,0xFF] \oplus T\_1[(A \gg 8)\&0xFF] \oplus \\ \oplus T\_2[(A \gg 16)\&0xFF] \oplus T\_3[A \gg 24]\end{aligned}$$

where & denotes the bitwise *AND*. We have the following result of function execution for $A = 0x01234567$:

$$\begin{aligned}Sub(0x01234567) &= T\_0[0x67] \oplus T\_1[0x45] \oplus T\_2[0x23] \oplus T\_3[0x01] \\ &= 0x94858511 \oplus 0x6E6EDCB2 \oplus 0x264C6A26 \oplus 0xF8847C7C \\ &= 0x24234FF9\end{aligned}$$

Compact implementation without conversion tables

Next, let's consider a compact implementation with no conversion tables for α_i and *Sub*.

Multiplications by multiplicative elements α_0, α_1, α_2 *ma* α_3. Let $x = (x_{31}, \ldots, x_0)$—is a variable of 32 bits long. And $y(i) = (y(i)_{31}, \ldots, y(i)_0)$ is a variable that is the result of multiplying $y(i) = \alpha_i x$ for $i = 0,1,2,3$. In this case, each bit for $y(i)$ can be calculated as follows (see pages 132–133 below):

– for $y(0)$.	– for $y(1)$
$y(0)_0 = x_{28} \oplus x_{30}$,	$y(1)_0 = x_{27} \oplus x_{29} \oplus x_{31}$,
$y(0)_1 = x_{24} \oplus x_{28} \oplus x_{29} \oplus x_{30} \oplus x_{31}$,	$y(1)_1 = x_{24} \oplus x_{28} \oplus x_{30}$,
$y(0)_2 = x_{25} \oplus x_{29} \oplus x_{30} \oplus x_{31}$	$y(1)_2 = x_{24} \oplus x_{25} \oplus x_{27}$,
$y(0)_3 = x_{24} \oplus x_6 \oplus x_{30} \oplus x_{31}$,	$y(1)_3 = x_{24} \oplus x_{25} \oplus x_{26} \oplus x_{27} \oplus x_{28} \oplus x_2 \oplus x_{31}$,
$y(0)_4 = x_{24} \oplus x_{25} \oplus x_{27} \oplus x_{31}$,	$y(1)_4 = x_{25} \oplus x_{26} \oplus x_{27} \oplus x_{28} \oplus x_{20} \oplus x_{30}$,
$y(0)_5 = x_{25} \oplus x_{26} \oplus x_{28}$,	$y(1)_5 = x_{24} \oplus x_{26} \oplus x_{28} \oplus x_{30}$,
$y(0)_6 = x_{26} \oplus x_{27} \oplus x_{28} \oplus x_{29} \oplus x_{30}$,	$y(1)_6 = x_{25} \oplus x_{27} \oplus x_{20} \oplus x_{31}$,
$y(0)_7 = x_{27} \oplus x_{29} \oplus x_{31}$,	$y(1)_7 = x_{26} \oplus x_{28} \oplus x_{30}$,
$y(0)_8 = x_0 \oplus x_{24} \oplus x_{26} \oplus x_{28}$,	$y(1)_8 = x_0 \oplus x_{25} \oplus x_{26} \oplus x_{27} \oplus x_{30}$,
$y(0)_9 = x_1 \oplus x_{25} \oplus x_{26} \oplus x_{27} \oplus x_{28} \oplus x_{29}$,	$y(1)_9 = x_1 \oplus x_{26} \oplus x_{27} \oplus x_{28} \oplus x_{31}$,
$y(0)_{10} = x_2 \oplus x_{24} \oplus x_{26} \oplus x_{27} \oplus x_{28} \oplus x_{29} \oplus x_{30}$,	$y(1)_{10} = x_2 \oplus x_{24} \oplus x_{25} \oplus x_{26} \oplus x_{28} \oplus x_{29} \oplus x_{30}$,
$y(0)_{11} =$ $x_3 \oplus x_{24} \oplus x_{25} \oplus x_{27} \oplus x_{28} \oplus x_{20} \oplus x_{30} \oplus x_{31}$,	$y(1)_{11} = x_3 \oplus x_{24} \oplus x_{20} \oplus x_{3,}$,
$y(0)_{12} = x_4 \oplus x_{25} \oplus x_{26} \oplus x_{28} \oplus x_{20} \oplus x_{30} \oplus x_{31}$,	$y(1)_{12} = x_4 \oplus x_{24} \oplus x_{25} \oplus x_{30}$,
$y(0)_{13} = x_5 \oplus x_{24} \oplus x_{26} \oplus x_{27} \oplus x_{29} \oplus x_{30} \oplus x_{31}$,	$y(1)_{13} = x_5 \oplus x_{24} \oplus x_{27} \oplus x_{30} \oplus x_{31}$,
$y(0)_{14} = x_6 \oplus x_{24} \oplus x_{25} \oplus x_{26} \oplus x_{27} \oplus x_{30} \oplus x_{31}$,	$y(1)_{14} = x_6 \oplus x_{24} \oplus x_{25} \oplus x_{28} \oplus x_{31}$,
$y(0)_{15} = x_7 \oplus x_{25} \oplus x_{27} \oplus x_{31}$,	$y(1)_{15} = x_7 \oplus x_{24} \oplus x_{25} \oplus x_{26} \oplus x_{20}$,
$y(0)_{16} = x_8 \oplus x_2 \oplus x_{30}$,	$y(1)_{16} = x_8 \oplus x_{24} \oplus x_{25} \oplus x_{26} \oplus x_{27} \oplus x_{20} \oplus x_{30}$,
$y(0)_{17} = x_9 \oplus x_{29} \oplus x_{31}$,	$y(1)_{17} = x_9 \oplus x_{25} \oplus x_{26} \oplus x_{27} \oplus x_{28} \oplus x_{30} \oplus x_{31}$,
$y(0)_{18} = x_{10} \oplus x_{30}$,	$y(1)_{18} = x_{10} \oplus x_{24} \oplus x_{25} \oplus x_{28} \oplus x_{30} \oplus x_{31}$,
$y(0)_{19} = x_{11} \oplus x_{24} \oplus x_{31}$,	$y(1)_{19} = x_{11} \oplus x_{27} \oplus x_{30} \oplus x_{31}$,
$y(0)_{20} = x_{12} \oplus x_{25}$,	$y(1)_{20} = x_{12} \oplus x_{24} \oplus x_{28} \oplus x_{31}$,
$y(0)_{21} = x_{13} \oplus x_{26}$,	$y(1)_{21} = x_{13} \oplus x_{24} \oplus x_{26} \oplus x_{27} \oplus x_{30}$,
$y(0)_{22} = x_{14} \oplus x_{27} \oplus x_{20} \oplus x_{30}$,	$y(1)_{22} = x_{14} \oplus x_{24} \oplus x_{25} \oplus x_{27} \oplus x_{28} \oplus x_{31}$,
$y(0)_{23} = x_{15} \oplus x_{28} \oplus x_{29} \oplus x_{31}$,	$y(1)_{23} = x_{15} \oplus x_{24} \oplus x_{25} \oplus x_{26} \oplus x_{28} \oplus x_{29}$,
$y(0)_{24} = x_{16} \oplus x_{25} \oplus x_{26} \oplus x_{27} \oplus x_{28} \oplus x_{31}$,	$y(1)_{24} = x_{16} \oplus x_{25} \oplus x_{27} \oplus x_{28}$,
$y(0)_{25} = x_{17} \oplus x_{24} \oplus x_{25} \oplus x_{20} \oplus x_{31}$,	$y(1)_{25} = x_{17} \oplus x_{26} \oplus x_{28} \oplus x_{22}$,
$y(0)_{26} = x_{18} \oplus x_{25} \oplus x_{26} \oplus x_{30}$,	$y(1)_{26} = x_{18} \oplus x_{25} \oplus x_{28} \oplus x_{29} \oplus x_{30}$,
$y(0)_{27} = x_{19} \oplus x_{25} \oplus x_{26} \oplus x_{27} \oplus x_{31}$,	$y(1)_{27} =$ $x_{19} \oplus x_{25} \oplus x_{26} \oplus x_{27} \oplus x_{28} \oplus x_{29} \oplus x_{30} \oplus x_{31}$,
$y(0)_{28} = x_{20} \oplus x_{24} \oplus x_{26} \oplus x_{27} \oplus x_{28}$,	$y(1)_{28} = x_{20} \oplus x_{26} \oplus x_{27} \oplus x_{28} \oplus x_{29} \oplus x_{30} \oplus x_{31}$,
$y(0)_{29} = x_{21} \oplus x_{24} \oplus x_{25} \oplus x_{27} \oplus x_{28} \oplus x_{28}$,	$y(1)_{29} = x_{21} \oplus x_{24} \oplus x_{25} \oplus x_{29} \oplus x_{30} \oplus x_{31}$,
$y(0)_{30} = x_{22} \oplus x_{27} \oplus x_{20} \oplus x_{30} \oplus x_{31}$,	$y(1)_{30} = x_2 \oplus x_{25} \oplus x_{26} \oplus x_{30} \oplus x_{31}$,
$y(0)_{31} = x_{23} \oplus x_{24} \oplus x_{25} \oplus x_{26} \oplus x_{27} \oplus x_{30}$.	$y(1)_{31} = x_{23} \oplus x_{24} \oplus x_{26} \oplus x_{27} \oplus x_{31}$,
– for $y(2)$	– for cases $y(3)$

(continued)

(continued)

– for $y(0)$.	– for $y(1)$
$y(2)_0 = x_{24} \oplus x_{25} \oplus x_{27} \oplus x_{28} \oplus x_{29} \oplus x_{31}$,	$y(3)_0 = x_{24} \oplus x_2 \oplus x_{27} \oplus x_{28} \oplus x_{31}$,
$y(2)_1 = x_{24} \oplus x_{25} \oplus x_{26} \oplus x_{28} \oplus x_2 \oplus x_{30}$,	$y(3)_1 = x_{24} \oplus x_{25} \oplus x_{26} \oplus x_{28} \oplus x_{29}$,
$y(2)_2 = x_{26} \oplus x_{28} \oplus x_{30}$,	$y(3)_2 = x_{26} \oplus x_{28} \oplus x_{29} \oplus x_{30} \oplus x_{31}$,
$y(2)_3 = x_{25} \oplus x_{28}$,	$y(3)_3 = x_{24} \oplus x_{27} \oplus x_{29} \oplus x_{30} \oplus x_{31}$,
$y(2)_4 = x_{24} \oplus x_{26} \oplus x_{29}$,	$y(3)_4 = x_{25} \oplus x_{28} \oplus x_{30} \oplus x_{31}$,
$y(2)_5 = x_{25} \oplus x_{27} \oplus x_{30}$,	$y(3)_5 = x_{25} \oplus x_{26} \oplus x_{27} \oplus x_{28} \oplus x_{20}$,
$y(2)_6 = x_{25} \oplus x_{26} \oplus x_{27} \oplus x_{20}$,	$y(3)_6 = x_{25} \oplus x_{26} \oplus x_{20} \oplus x_{30} \oplus x_{31}$,
$y(2)_7 = x_{24} \oplus x_{26} \oplus x_{27} \oplus x_{28} \oplus x_{30}$,	$y(3)_7 = x_{24} \oplus x_{26} \oplus x_{27} \oplus x_{30} \oplus x_{31}$,
$y(2)_8 = x_0 \oplus x_{24} \oplus x_{26} \oplus x_{27} \oplus x_{30}$,	$y(3)_8 = x_0 \oplus x_{26} \oplus x_{29} \oplus x_{30}$,
$y(2)_9 = x_1 \oplus x_{24} \oplus x_{25} \oplus x_{27} \oplus x_{28} \oplus x_{31}$,	$y(3)_9 = x_1 \oplus x_{24} \oplus x_{27} \oplus x_{30} \oplus x_{31}$,
$y(2)_{10} = x_2 \oplus x_{24} \oplus x_{25} \oplus x_{27} \oplus x_{28} \oplus x_{20} \oplus x_{30}$,	$y(3)_{10} =$ $x_2 \oplus x_{24} \oplus x_{25} \oplus x_{26} \oplus x_{28} \oplus x_2 \oplus x_{30} \oplus x_{31}$,
$y(2)_{11} = x_3 \oplus x_{24} \oplus x_{25} \oplus x_{27} \oplus x_{28} \oplus x_{20} \oplus x_{31}$,	$y(3)_{11} = x_3 \oplus x_{25} \oplus x_{26} \oplus x_{27} \oplus x_{29} \oplus x_{30} \oplus x_{31}$,
$y(2)_{12} = x_4 \oplus x_{24} \oplus x_{25} \oplus x_{26} \oplus x_{28} \oplus x_2 \oplus x_{30}$,	$y(3)_{12} = x_4 \oplus x_{24} \oplus x_{26} \oplus x_{27} \oplus x_{28} \oplus x_{30} \oplus x_{31}$,
$y(2)_{13} =$ $x_5 \oplus x_{24} \oplus x_{25} \oplus x_{26} \oplus x_{27} \oplus x_{29} \oplus x_{30} \oplus x_{31}$,	$y(3)_{13} = x_5 \oplus x_{25} \oplus x_{26} \oplus x_{27} \oplus x_{28} \oplus x_{30} \oplus x_{31}$,
$y(2)_{14} = x_6 \oplus x_{24} \oplus x_{25} \oplus x_{28} \oplus x_{31}$,	$y(3)_{14} = x_6 \oplus x_{24} \oplus x_{27} \oplus x_{28} \oplus x_{30} \oplus x_{31}$,
$y(2)_{15} = x_7 \oplus x_{25} \oplus x_{26} \oplus x_{20}$,	$y(3)_{15} = x_7 \oplus x_{25} \oplus x_{28} \oplus x_2 \oplus x_{31}$,
$y(2)_{16} = x_8 \oplus x_{25} \oplus x_{26} \oplus x_{29} \oplus x_{30} \oplus x_{31}$,	$y(3)_{16} = x_8 \oplus x_{24} \oplus x_{26}$,
$y(2)_{17} = x_9 \oplus x_{26} \oplus x_{27} \oplus x_{30} \oplus x_{31}$,	$y(3)_{17} = x_9 \oplus x_{25} \oplus x_{27}$,
$y(2)_{18} = x_{10} \oplus x_{25} \oplus x_{25} \oplus x_{27} \oplus x_{28} \oplus x_{29} \oplus x_{30}$,	$y(3)_{18} = x_{10} \oplus x_{28}$,
$y(2)_{19} = x_{11} \oplus x_{24} \oplus x_{25} \oplus x_{27} \oplus x_{28}$,	$y(3)_{19} = x_{11} \oplus x_{24} \oplus x_{29}$,
$y(2)_{20} = x_{12} \oplus x_{24} \oplus x_{25} \oplus x_{26} \oplus x_{28} \oplus x_{29}$,	$y(3)_{20} = x_{12} \oplus x_{24} \oplus x_{25} \oplus x_{30}$,
$y(2)_{21} = x_{13} \oplus x_{24} \oplus x_{25} \oplus x_{26} \oplus x_{27} \oplus x_{20} \oplus x_{30}$,	$y(3)_{21} = x_{13} \oplus x_{25} \oplus x_{31}$,
$y(2)_{22} = x_{14} \oplus x_{24} \oplus x_{27} \oplus x_{28} \oplus x_{22}$,	$y(3)_{22} = x_{14} \oplus x_{24}$,
$y(2)_{23} = x_{15} \oplus x_{24} \oplus x_{25} \oplus x_{28} \oplus x_{29} \oplus x_{30}$,	$y(3)_{23} = x_{15} \oplus x_{25}$,
$y(2)_{24} = x_{16} \oplus x_{24} \oplus x_{26} \oplus x_{29} \oplus x_{31}$,	$y(3)_{24} = x_{16} \oplus x_{24} \oplus x_{26} \oplus x_{28} \oplus x_2$,
$y(2)_{25} = x_{17} \oplus x_{24} \oplus x_{25} \oplus x_{27} \oplus x_{30}$,	$y(3)_{25} = x_{17} \oplus x_{25} \oplus x_{27} \oplus x_{29} \oplus x_{30}$,
$y(2)_{26} = x_{18} \oplus x_{25} \oplus x_{28} \oplus x_{20}$,	$y(3)_{26} = x_{18} \oplus x_{24} \oplus x_{20} \oplus x_{30} \oplus x_{31}$,
$y(2)_{27} = x_{19} \oplus x_{24} \oplus x_{30} \oplus x_{31}$,	$y(3)_{27} = x_{19} \oplus x_{25} \oplus x_{30} \oplus x_{31}$,
$y(2)_{28} = x_{20} \oplus x_{24} \oplus x_{25} \oplus x_{31}$,	$y(3)_{28} = x_{20} \oplus x_{26} \oplus x_{31}$,
$y(2)_{29} = x_{21} \oplus x_{25} \oplus x_{26}$,	$y(3)_{29} = x_{21} \oplus x_{26} \oplus x_{27} \oplus x_{28} \oplus x_{29}$,
$y(2)_{30} = x_{22} \oplus x_{24} \oplus x_{27} \oplus x_{29} \oplus x_{31}$,	$y(3)_{30} = x_{22} \oplus x_{24} \oplus x_{26} \oplus x_{27} \oplus x_{30}$,
$y(2)_{31} = x_{23} \oplus x_{25} \oplus x_{28} \oplus x_{30}$,	$y(3)_{31} = x_{23} \oplus x_{25} \oplus x_{27} \oplus x_{28} \oplus x_{31}$,

Sub function. The $32 \to 32$ *Sub* substitution function in a nonlinear function can also be implemented without conversion tables. The *Sub* function consists of two stages: the substitution phase and the permutation phase. In the substitution phase, operations in the finite field must be used.

First, let's describe the operations in the subfield.

- Multiplication in the $GF(2^4)$ field. Let $x = (x_3, x_2, x_1, x_0)$ and $y = (y_3, y_2, y_1, y_0)$ be 4-bit variables. We also define a variable $z = (z_3, z_2, z_1, z_0)$ with a length of 4 bits as a result of multiplication $z = x \times_4$, where $\times_4$ is a multiplication operator in the field $GF(2^4)$. Then

  $z = x \times_4$ is obtained as follows:

1. Let t_6, ... t_0 be local 1-bit variables, which are calculated as follows:

$$
\begin{aligned}
t_6 &= (x_3 \& y_3), \\
t_5 &= (x_3 \& y_2) \oplus (x_2 \& y_3), \\
t_4 &= (x_3 \& y_1) \oplus (x_2 \& y_2) \oplus (x_1 \& y_3), \\
t_3 &= (x_3 \& y_0) \oplus (x_2 \& y_1) \oplus (x_1 \& y_2) \oplus (x_0 \& y_3), \\
t_2 &= (x_2 \& y_0) \oplus (x_1 \& y_1) \oplus (x_0 \& y_2), \\
t_1 &= (x_1 \& y_0) \oplus (x_0 \& y_1), \\
t_0 &= (x_0 \& y_0).
\end{aligned}
$$

where & is denoted by the *AND* bitwise operation.

2. From t_6, ... t_0 we get z:

$$
\begin{aligned}
z_3 &= t_3 \oplus t_6, \\
z_2 &= t_2 \oplus t_5 \oplus t_6, \\
z_1 &= t_1 \oplus t_4 \oplus t_5, \\
z_0 &= t_0 \oplus t_4.
\end{aligned}
$$

– *Multiplicative feedback element in the $GF(2^4)$ field.* Let's define $x = (x_3, x_2, x_1, x_0)$ as the 4-bit variable, and $y = (y_3, y_2, y_1, y_0)$ as the 4-bit output variable, which is $y = x^{-1} \in GF(2^4)$. Then y is obtained from the following mappings:

$x \mapsto y$	$x \mapsto y$
$(0, 0, 0, 0) \mapsto (0, 0, 0, 0)$	$(1, 0, 0, 0) \mapsto (1, 1, 1, 1)$
$(0, 0, 0, 1) \mapsto (0, 0, 0, 1)$	$(1, 0, 0, 1) \mapsto (0, 0, 1, 0)$
$(0, 0, 1, 0) \mapsto (1, 0, 0, 1)$	$(1, 0, 1, 0) \mapsto (1, 1, 0, 0)$
$(0, 0, 1, 1) \mapsto (1, 1, 1, 0)$	$(1, 0, 1, 1) \mapsto (0, 1, 0, 1)$
$(0, 1, 0, 0) \mapsto (1, 1, 0, 1)$	$(1, 1, 0, 0) \mapsto (1, 0, 1, 0)$
$(0, 1, 0, 1) \mapsto (1, 0, 1, 1)$	$(1, 1, 0, 1) \mapsto (0, 1, 0, 0)$
$(0, 1, 1, 0) \mapsto (0, 1, 1, 1)$	$(1, 1, 1, 0) \mapsto (0, 0, 1, 1)$
$(0, 1, 1, 1) \mapsto (0, 1, 1, 0)$	$(1, 1, 1, 1) \mapsto (1, 0, 0, 0)$

With the above operations, the following operations are performed in the $GF(2^4)$ field during the substitution and permutation steps.

– *Substitution stage.* The substitution step is a function of 8 → 8. We denote the input sequence as $x = (x_7, \ldots, x_0)$ and $y = (y_7, \ldots, y_0)$ as the output sequence. Further, this step is implemented as follows.
– Define $a = (a_7, \ldots, a_0)$ as an 8-bit local variable. Calculate:

$$a_7 = x_7 \oplus x_5,$$
$$a_6 = x_7 \oplus x_5 \oplus x_3 \oplus x_2,$$
$$a_5 = x_7 \oplus x_6 \oplus x_4 \oplus x_1,$$
$$a_4 = x_6 \oplus x_5 \oplus x_4,$$
$$a_3 = x_7 \oplus x_6 \oplus x_2 \oplus x_1,$$
$$a_2 = x_6 \oplus x_4 \oplus x_1,$$
$$a_1 = x_3 \oplus x_1,$$
$$a_0 = x_7 \oplus x_6 \oplus x_4 \oplus x_3 \oplus x_2 \oplus x_0.$$

2. Let (b_3, b_2, b_1, b_0) be a 4-bit local variable defined as:

$$b_3 = a_7 \oplus a_3,$$
$$b_2 = a_6 \oplus a_2,$$
$$b_1 = a_5 \oplus a_1,$$
$$b_0 = a_4 \oplus a_0.$$

3. Set the 4-bit local variable as (c_3, c_2, c_1, c_0). We multiply $(c_3, c_2, c_1, c_0) = (a_3, a_2, a_1, a_0) \times_4 (b_3, b_2, b_1, b_0)$, where $\times_4$ is a multiplicative operation on the field $GF(2^4)$.
4. Determine a local variable (d_3, d_2, d_1, d_0) with a length of 4 bits. We perform the multiplication $(d_3, d_2, d_1, d_0) = (a_7, a_6, a_5, a_4) \times_4 (a_7, a_6, a_5, a_4)$, where $\times_4$ is a multiplicative operation on the field $GF(2^4)$.
5. Let (e_3, e_2, e_1, e_0) be a 4-bit local variable. Let's multiply $(e_3, e_2, e_1, e_0) = (1, 0, 0, 1) \times_4 (d_3, d_2, d_1, d_0)$, where $\times_4$ is a multiplicative operation over the field $GF(2^4)$.
6. A local variable (f_3, f_2, f_1, f_0) of 4 bits long is calculated as follows:

$$f_3 = c_3 \oplus e_3,$$
$$f_2 = c_2 \oplus e_2,$$
$$f_1 = c_1 \oplus e_1,$$
$$f_0 = c_0 \oplus e_0.$$

7. Define (g_3, g_2, g_1, g_0) as a local 4-bit variable. We calculate the multiplicative inverse element $(g_3, g_2, g_1, g_0) = (f_3, f_2, f_1, f_0)^{-1} \in GF(2^4)$.
8. Let (h_3, h_2, h_1, h_0) be a 4-bit local variable. We multiply $(h_3, h_2, h_1, h_0) = (b_3, b_2, b_1, b_0) \times_4 (g_3, g_2, g_1, g_0)$, where $\times_4$ is a multiplicative operation on the field $GF(2^4)$.
9. Specify a local variable (h_7, h_6, h_5, h_4) with a length of 4 bits. Calculate $(h_7, h_6, h_5, h_4) = (a_7, a_6, a_5, a_4) \times_4 (g_3, g_2, g_1, g_0)$, where $\times_4$ is a multiplicative operation over the field $GF(2^4)$.
10. Determine the 8-bit local variable $(i_7, ..., i_0)$. We get:

$$
\begin{aligned}
i_7 &= h_3 \oplus 1,\\
i_6 &= h_6 \oplus h_4 \oplus h_2 \oplus 1,\\
i_5 &= h_7 \oplus h_5 \oplus h_1,\\
i_4 &= h_5 \oplus h_2 \oplus h_0,\\
i_3 &= h_7,\\
i_2 &= h_5 \oplus h_2,\\
i_1 &= h_7 \oplus h_3 \oplus h_1 \oplus 1,\\
i_0 &= h_4 \oplus h_2.
\end{aligned}
$$

11. Obtain the following initial sequence for the substitution step:

$$
\begin{aligned}
y_7 &= i_5 \oplus i_2,\\
y_6 &= i_7 \oplus i_6 \oplus i_5 \oplus i_3 \oplus i_2 \oplus i_1,\\
y_5 &= i_7 \oplus i_5 \oplus i_2,\\
y_4 &= i_6 \oplus i_4 \oplus i_3 \oplus i_1,\\
y_3 &= i_7 \oplus i_5 \oplus i_4 \oplus i_1,\\
y_2 &= i_6 \oplus i_5 \oplus i_4 \oplus i_1,\\
y_1 &= i_7 \oplus i_5 \oplus i_4,\\
y_0 &= i_6 \oplus i_4 \oplus i_0.
\end{aligned}
$$

- *Permutation stage.* The permutation step is called the function $32 \to 32$. We define $x = (x_{31}, \ldots, x_0)$ as the input sequence and $y = (y_{31}, \ldots, y_0)$ as the output sequence. This stage is implemented as follows:

$$
\begin{aligned}
y_{31} &= x_{15} \oplus x_{23} \oplus x_7 \oplus x_6 \oplus x_{30}, & y_{15} &= x_{31} \oplus x_7 \oplus x_{23} \oplus x_{22} \oplus x_{14},\\
y_{30} &= x_{14} \oplus x_{22} \oplus x_6 \oplus x_5 \oplus x_{29}, & y_{14} &= x_{30} \oplus x_6 \oplus x_{22} \oplus x_{21} \oplus x_{13},\\
y_{29} &= x_{13} \oplus x_{21} \oplus x_5 \oplus x_4 \oplus x_{28}, & y_{13} &= x_{29} \oplus x_5 \oplus x_{21} \oplus x_{20} \oplus x_{12},\\
y_{28} &= x_{12} \oplus x_{20} \oplus x_4 \oplus x_3 \oplus x_7 \oplus x_{27} \oplus x_{31}, & y_{12} &= x_{28} \oplus x_4 \oplus x_{20} \oplus x_{19} \oplus x_{23} \oplus x_{11} \oplus x_{15},\\
y_{27} &= x_{11} \oplus x_{19} \oplus x_3 \oplus x_2 \oplus x_7 \oplus x_{26} \oplus x_{31}, & y_{11} &= x_{27} \oplus x_3 \oplus x_{19} \oplus x_{18} \oplus x_{23} \oplus x_{10} \oplus x_{15},\\
y_{26} &= x_{10} \oplus x_{18} \oplus x_2 \oplus x_1 \oplus x_{25}, & y_{10} &= x_{26} \oplus x_2 \oplus x_{18} \oplus x_{17} \oplus x_9,\\
y_{25} &= x_9 \oplus x_{17} \oplus x_1 \oplus x_0 \oplus x_7 \oplus x_{24} \oplus x_{31}, & y_9 &= x_{25} \oplus x_1 \oplus x_{17} \oplus x_{16} \oplus x_{23} \oplus x_8 \oplus x_{15},\\
y_{24} &= x_8 \oplus x_{16} \oplus x_0 \oplus x_7 \oplus x_{31}, & y_8 &= x_{24} \oplus x_0 \oplus x_{16} \oplus x_{23} \oplus x_{15},\\
y_{23} &= x_7 \oplus x_{15} \oplus x_{31} \oplus x_{30} \oplus x_{22}, & y_7 &= x_{23} \oplus x_{31} \oplus x_{15} \oplus x_{14} \oplus x_6,\\
y_{22} &= x_6 \oplus x_{14} \oplus x_{30} \oplus x_{29} \oplus x_{21}, & y_6 &= x_{22} \oplus x_{30} \oplus x_{14} \oplus x_{13} \oplus x_5,\\
y_{21} &= x_5 \oplus x_{13} \oplus x_{29} \oplus x_{28} \oplus x_{20}, & y_5 &= x_{21} \oplus x_{29} \oplus x_{13} \oplus x_{12} \oplus x_4,\\
y_{20} &= x_4 \oplus x_{12} \oplus x_{28} \oplus x_{27} \oplus x_{31} \oplus x_{19} \oplus x_{23}, & y_4 &= x_{20} \oplus x_{28} \oplus x_{12} \oplus x_{11} \oplus x_{15} \oplus x_3 \oplus x_7,\\
y_{19} &= x_3 \oplus x_{11} \oplus x_{27} \oplus x_{26} \oplus x_{31} \oplus x_{18} \oplus x_{23}, & y_3 &= x_{19} \oplus x_{27} \oplus x_{11} \oplus x_{10} \oplus x_{15} \oplus x_2 \oplus x_7,\\
y_{18} &= x_2 \oplus x_{10} \oplus x_{26} \oplus x_{25} \oplus x_{17}, & y_2 &= x_{18} \oplus x_{26} \oplus x_{10} \oplus x_9 \oplus x_1,\\
y_{17} &= x_1 \oplus x_9 \oplus x_{25} \oplus x_{24} \oplus x_{31} \oplus x_{16} \oplus x_{23}, & y_1 &= x_{17} \oplus x_{25} \oplus x_9 \oplus x_8 \oplus x_{15} \oplus x_0 \oplus x_7,\\
y_{16} &= x_0 \oplus x_8 \oplus x_{24} \oplus x_{31} \oplus x_{23}, & y_0 &= x_{16} \oplus x_{24} \oplus x_8 \oplus x_{15} \oplus x_7.
\end{aligned}
$$

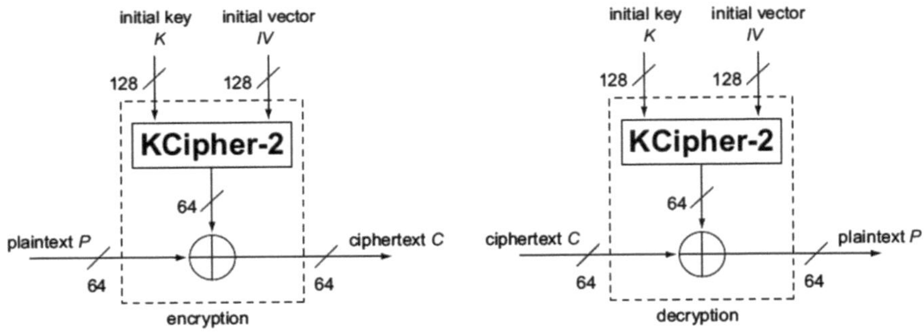

Fig. 7.20 Encryption and decryption using KCipher-2 SSC

7.2.5.5 Usage Notes

How to use initialization keys and vectors

The maximum number of cycles that can be calculated for the *KCipher-2* SSC without re-initializing and changing the key is 2^{58} cycles (or 2^{64} bits of the key stream). After completing 2^{58} key sequence production cycles, you must change the key and the initialization vector.

Encryption and decryption

To perform encryption and decryption with *KCipher-2*, we first split the plaintext data into 64-bit blocks. Then we perform an exclusive OR operation on the plaintext block and the 64 bit key stream block generated by the key and the initialization vector, as shown in Fig. 7.20. The decryption can be done in the same way.

Thus, five specialized key strem generators are specified by *ISO/IEC 18033-4*: *MUGI*, *SNOW 2.0*, *Rabbit*, *Decimv2*, *KCipher-2*.

The following is an analysis of the installed levels of protection of stream encryption with *ISO/IEC 18033-4*, in particular the binary additive mode, MULTIS01 mode discussed above, BSC streaming modes and specialized *MUGI* and *SNOW 2.0* key stream generators.

7.3 Fluid Code Security Levels with ISO/IEC 18033-4

The levels of protection of streaming ciphers in binary additive mode and MULTIS01 mode are given in Table. 7.2.

The installed protection levels of the key flow generators are given in Table. 7.3.

A trade-off between security and performance in MULTI-S01

Let n be the block size of the output function *MULTI-S01*. For the n–$u$bit message, the *MULTI-S01* output function performs a $(u + 2)$-multiple iteration of the block

Table 7.2 The security levels of the streaming cipher modes are set

Function	Data confidentiality	Data integrity
Binary additive mode	An encryption mechanism based on these source functions is considered safe as long as it is a secure key stream generator This is mathematically proven	This mode does not provide data integrity protection
MULTIS01 mode		Data integrity protection is generally as secure as a key stream generator However, the security level is always limited by the length of the fake message above. Fake messages whose length is *un* bits can be received with probability of success $(u-2)2^{-n}$, where *n* is the mode protection parameter

process. In both software and hardware implementation, the multiplication in the finite field $GF(2^n)$ is the dominant computation.

When comparing two implementations with different sizes of block *n*, implementation with smaller *n* is usually faster than the other, although implementation with smaller value *n* is iterated more than once, each multiplication can be calculated in less time and using less space. The factor of calculation acceleration depends on the algorithm or platform used to implement it. Table 7.4 shows some experimental results of *MULTI-S01* mode implementations with $n = 64$. Table 7.5 shows examples of implementation of specialized generators.

Thus, not listed in Tables 7.2 and 7.3 The data indicate that streaming cryptoconversions, as standardized in the international normative document *ISO/IEC 18033-4*, provide the necessary level of security. In addition, as the results from Tables 7.4 and 7.5, appropriate transformations can be efficiently implemented both in hardware using (*Application Specific Integrated Circuit ASIC*) and software through the development of appropriate software.

7.4 Conclusions and Recommendations

1. Cryptographic protection of information, including models and methods of symmetric transformation, is an important component in ensuring the security of modern information communication systems and technologies. Stream cryptoconversions are most trusted in users of critical and secure systems, in particular, they are used in systems with high and rigid security and performance requirements, as the streaming method is more reliable and the corresponding cryptographic information security tools are faster and less demanding for computing resources.
2. World-famous crypto-algorithms, internationally or nationally standardized, have gained the most trust and distribution at this time: ISO/IEC 180334

Table 7.3 Levels of protection for BSC streaming and specialized modes key stream generators

Key flow generators	Claims of protection	Key length	Computational complexity for the best known attack
Streaming *CFB*, *OFB*, *CTR* modes with *n*-bit *BSC* with *k*-bit key length	Assume that the basic block cipher is secure, then the *PRS* generated by the *BCS* in *CFB*, *OFB*, and *CTR* modes is usually indistinguishable from a random sequence. However, if the number of blocks processed exceeds approximately 2*n*/2, then the likelihood of a difference between such a key stream and a random sequence becomes significant. The protection of such a key stream is restricted from above by the protection of the underlying *BSC* depending on the length of the key and the degree of cryptanalysis of the cipher	*k* bits	$2^{n/2}$
MUGI	There is no weakness against *MUGI*, which is faster than exhaustive key search. *MUGI* has a 128-bit key length	128 bits	2^{128}
SNOW 2.0	For 256-bit secret key mode, differentiating algorithms were developed to distinguish *SNOW 2.0* output from a random sequence. A better rated attack requires approximately 2^{231} bits of known key stream. It hardly breaks the theoretical level of protection expected from a 256-bit key length. However, any attack associated with known differentiating algorithms (distinguishers) will not be a threat to practical applications	128 or 256 bits	2^{231}

Table 7.4 Experimental results of *MULTI-S01* mode implementation without key stream generation ($n = 64$)

	Speed/size	Platform specification
High-speed ASIC	5.1 Gb/s (80 MHz)/25.7 thousand valves	ASIC using the Hitachi HG73C Element Library (0.35 μm)
Small-scale ASIC	100 Mbps (100 MHz)/2.0 thousand valves	
Software	10 cycles/byte or, which is equivalent, 520 Mbps (650 MHz)	Intel Pentium III 650 МГц (Coppermine), Windows98 SE, RAM 64 MB, Visual C++ Ver.6.0. Service Pack 3

Table 7.5 Examples of implementation of specialized key streams generators

	Software	ASIC
MUGI	6.5 cycles/byte on Pentium III or equivalent, 800 Mbps (650 MHz)	– 14.4 Gbps (225.2 MHz)/52.4 Kilo Valves – 3.61 Gbps (56.4 MHz)/3.61 Kilo Valves (using 0.13 μm standard element library)
SNOW 2.0	3.13 cycles/byte on Pentium III or equivalent, 1661 Mbps (650 MHz)	No implementation results are known

"Information technology—Security techniques—Encryption algorithms—Part 4: Stream ciphers"; ISO/IEC 10116 «Information technology—Security techniques—Modes of operation for an nbit block cipher»; NIST Special Publication 80038 "Block Cipher Modes"; DSTU 7624: 2014 «Information technologies. Cryptographic protection of information. Symmetric Block Conversion Algorithm"; NIST Special Publication 80090A «Recommendation for Random Number Generation Using Deterministic Random Bit Generators»; DSTU GOST 28147: 2009 «Information processing systems. Cryptographic protection. Cryptographic conversion algorithm"; GOST R ISO/IEC 1011693 «Information technology. Modes of operation for n-bit block encryption algorithm"; eSTREAM—an international project to identify new current ciphers; ISO/IEC 291923 «Information technology—Security techniques—Lightweight cryptography».

3. The analysis of standardized algorithms of stream cryptoconversions, specification of which is given in ISO/IEC 180334, showed that the considered stream ciphers (algorithms MUGI, SNOW 2.0, Rabbit, DEcim, KCipher-2) are constructed according to the scheme of key stream generators, which are built using linear techniques. Shift registers and cryptographic elements of block ciphers. In particular, these generators use nonlinear complication functions that are constructed using nonlinear AES replacement units (FIPS197). This, on the one hand, can greatly simplify the analysis of the stability of the cipher,

using a proven mathematical apparatus and known statements about certain cryptographic properties; on the other hand, it allows the use of pre-existing crypto-primitives and possibly their software implementations, which greatly accelerates the implementation of algorithms on different computing systems.

4. In addition to the specialized MUGI and SNOW 2.0 key stream generators, the international ISO/IEC 180334 standard also contains common PSS models and BSS streaming modes for the construction of synchronous (OFB and CTR) and asynchronous (CFB) cryptoconversions. The standard also specifies the levels of protection of stream cryptoconversions and indicates a possible compromise between the level of protection and the effectiveness of software or hardware implementation. This chapter is the most interesting from the point of view of possible implementation of streaming cryptoconversions in different computer systems, because it allows to adapt certain protection mechanisms taking into account the peculiarities of environment construction for the application and functional purpose of information and telecommunication systems.
5. The above results indicate that streaming cryptoconversions, as standardized in ISO/IEC 180334, provide the required level of security. In addition, the appropriate transformations can be efficiently implemented both hardware—using Application-Specific Integrated Circuit (ASIC) and software—by developing appropriate software (software).

References

1. DSTU 7564: 2014: Information Technology. Cryptographic information protection. Hashing function [Text]. Enter. 01–07–2015. Ministry of Economic Development of Ukraine, Kyiv (2015)
2. FIPS197: Advanced Encryption Standard (AES). National Institute of Standards and Technology [Electronic resource] (2001). Access mode http://csrc.nist.gov/publications/fips/fips197/fips197.pdf
3. The eSTREAM Project—eSTREAM Phase 3. Rabbit (Portfolio Profile 1) [Electronic resource]. Access mode http://www.ecrypt.eu.org/stream/rabbitpf.html
4. Berbain, C., Billet, O., Canteaut, A., Courtois, N., Debraize, B., Gilbert, H., Goubin, L., Gouget, A., Granboulan, L., Lauradoux, C., Minier, M., Pornin, T., Sibert, H.: Decim—A New Stream Cipher for Hardware Applications. In: ECRYPT Stream Cipher Project Report 2005/004. Available at http://www.ecrypt.eu.org/stream/
5. Wu, H., Preneel, B.: Cryptanalysis of Stream Cipher Decim. Available at http://www.ecrypt.eu.org/stream/
6. eStream, Stream cipher project of the European Network of Excellence in Cryptology ECRYPT. http://www.ecrypt.eu.org/stream/
7. Batina, L., Lano, J., Ors, S.B., Preneel, B., Verbauwhede, I.: Energy, perfomance, area versus security tradeoffs for stream ciphers. In: The State of the Art of Stream Ciphers: Workshop Record, pp. 302–310. Brugge, Belgium, October (2004)
8. Golic, J.: Towards fast correlation attacks on irregularly clocked shift registers. In: Proceedings of Eurocrypt'95, Lecture Notes in Computer Science (1995)
9. ISO/IEC 180334:2011: Information technology—Security techniques—Encryption algorithms—Part 4: Stream ciphers [Electronic resource]. Access mode http://www.iso.org/iso/home/store/catalogue_ics/catalogue_detail_ics.htm?csnumber=54532

10. Daemen, J., Rijmen, V.: The Design of Rijndael: AES—The Advanced Encryption Standard, Information Security and Cryptography. SpringerVerlag (2002)
11. Golic, J.D.: On security of nonlinear filter generators. In: Proceeding of FSE 1996, Lecture Notes in Computer Science, vol. 1039, pp. 173–188 (1996)

Chapter 8
Comparative Analysis of Determined Generators for Random Bits, Defined in the Nist Special Publication 800-90A

In the standard NIST Special Publication 800-90A are defined:

- functional models of DGRB;
- conceptual mechanisms and base requirements for DGRB;
- mechanisms of functions DGRB;
- specifications of algorithms, which realize certain DGRB, and namely:
  - based on the using of the hashing function (Hash_DRBG and HMAC_DRBG);
  - based on the using of BSC (CTR_DRBG);
  - based on the using of the number-theoretical problems (Dual Elliptic Curve Deterministic RBG—Dual_EC_DRBG).

It should be noted that the generator Dual Elliptic Curve Deterministic RBG was contained only in the precedent version of the USA national standard NIST Special Publication 800-90A [1]. In the renewed version of the standard [2] this generator has been excluded due to certain deficiencies, but in this chapter is researched nevertheless its structure and its main properties. This direction is noteworthy and worth of detailed study, because the utilizing of numeric-theoretical enables to build generators with high indices of statistical security [3, 4].

Let us examine more detailed these algorithms, let us research the particularities of their building up and the possibilities concerning information protection.

In Fig. 8.1 is presented the functional model of the generator of pseudorandom sequences according to the NIST Special Publication 800-90A [1].

The entropy introducing is executed for a seed, i. e. for generator initiation. The entropy introducing and the so-called seed (Engl. Seed) must be held in security. The security classification of this information ensures a base for the generator.

Another (complement) information can be introduced into the generator, its security is not correlated with the security classification of the complement information. The complement information is recommended to be checked for authenticity, when

A. A. Kuznetsov et al., *Stream Ciphers in Modern Real-time IT Systems*,
Studies in Systems, Decision and Control 375,
https://doi.org/10.1007/978-3-030-79770-6_8

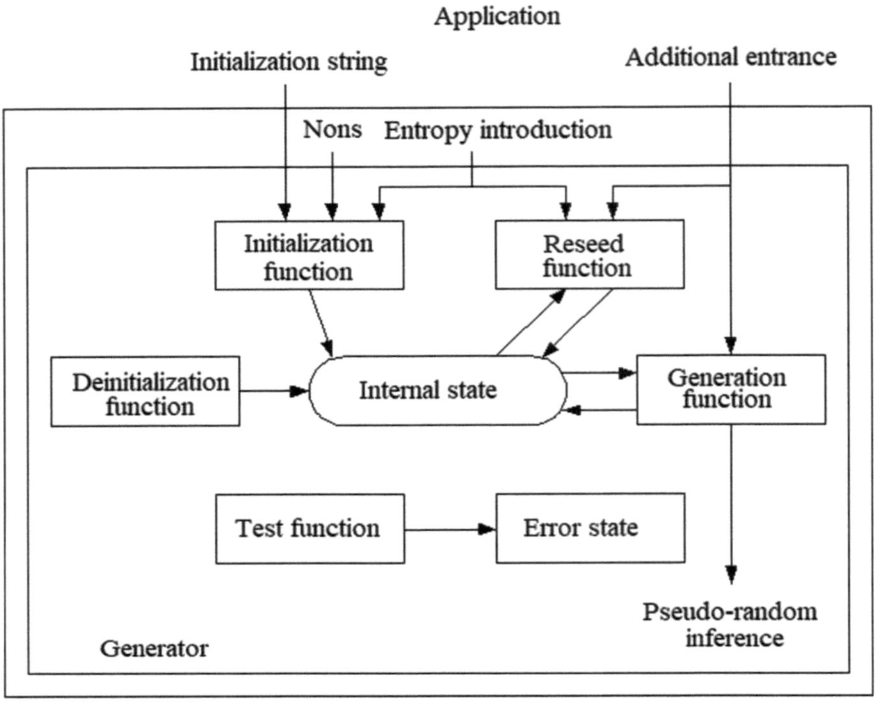

Fig. 8.1 Functional model of PRS-generator

possible, for example, if for introducing is utilized the time, the time format and reasoning can be verified.

During the generator initiation the so-called nonce (Engl. Nonce) can be needed, and if it is used, so it is combined with the entropy introducing for creating of the generator initial seed. The nonce has to be:

- an unpredictable value with at least (½ key length) bits of entropy;
- an inimitable value, with expected repetitions not more frequent as (½ key length)-bit of a random line.

For the first case the nonce can be obtained from the same source and in the same time as the entropy introducing.

The nonce grants a bigger certitude that the generator assures the target security level for the user. When the generator is initiated many-fold without nonce, the compromise becomes more verisimilar. In some applications the single compromise of generator can discover the long-time secrecy (for example, the secret sense compromise for a message in the standard of digital signature discovers the signature key).

The internal state is the generator memory, which includes all parameters variables and other stored values, which the generator is utilizing or performs operations with them. The internal state includes both the administrative data (for instance, security level), and the data it operates with or customizes them while generating pseudorandom bits (i. e. working state).

The generator functions perform the processing of the internal state. Individual mechanisms have five separate functions:

1. The initiation function accepts the entropy introducing and can combine it with the nonce and with the personalization line for the seed (seed) creation, from which is created the primary internal state.
2. The generation function generates pseudorandom bits after request by using of the current internal state, and generates a new internal state for the next request.
3. The reseed function accepts the new entropy introducing and combines it with its current internal state and with any complement introducing, which is assured for creating a new seed and a new internal state.
4. The deinitiation function zeroise (i. e. erases) the internal state.
5. The functioning test function determines that the generator continue working correctly.

The generators, which are defined in the standard NIST Special Publication 800-90A, support four security levels: 112, 128, 192 or 256 bits. The security level is commanded by the application supplement during initiation, and the initiation function obtains the respective entropy quantity for the demanded security level. Any security level can be commanded (up to 256 bits), but the generator will be initiated to one of the indicated security levels. The necessary security level, if it is lower than 112 bits or lies between the enumerated 4 levels, will be initiated up to next higher level (i. e. for the demanded level of 80 bits will be initiated the 112-bit security level). The real security level, supported by the given initiation, depends on the generator realization and on the entropy quantity, transmitted into the initiation function.

For convenience, there is utilized in the standard the entity of «generator margin» for explication of operations and their interaction. The generator margin includes all functions and internal states, needed for the generator working. The data are introduced into the generator margin through the public interfaces.

In the generator margin:

- the internal state and the operations functions must be determined only according to the specification;
- the internal state must exist exclusively in the generator margin and may not be accessible for the non-generator functions;
- the information about secret components of the generator internal state and of the calculations intermediate values, which involve these secret components, may not influence any information that escape from the generator margins, except for determined in output pseudorandom bits.

Every generator mechanism includes one or more cryptographic typicals (for instance, hashing function). Other applications can utilize the same cryptographic typical, till it does not influence the internal state and the functions of the generator.

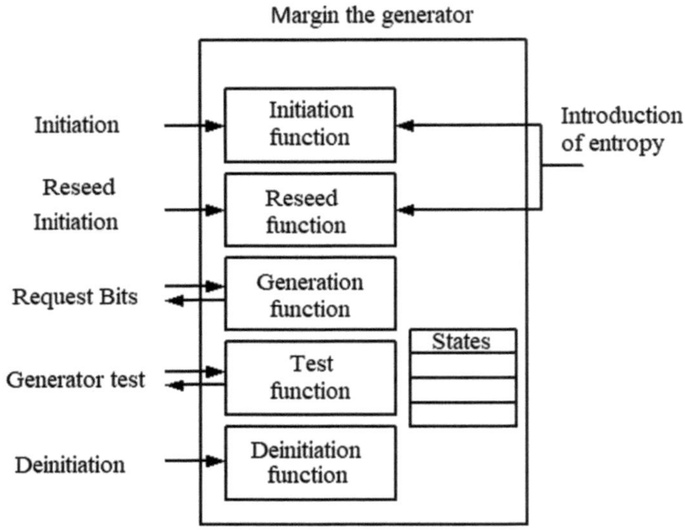

Fig. 8.2 Functions of the generator in one single device

The generator functions can be realized in one device, or can be distributed in several devices. In Fig. 8.2 is drawn the generator, all functions of which are placed in one single device. The margin of the whole generator must include the totality of submargins that ensure its functionality. The using of distributed functions of the generator can be efficient for restricted environments (circumstances) (for instance, annexes for smartcards), in which the first-priority usage of the generator do not require an iterative using of initiation functions or of reseed.

According to the main statements of the standard NIST Special Publication 800-90A, there are determined several approaches to the building up of the generation function and, respectively, to the generators of pseudorandom sequences. The main generators, which are presented in the specification, are the following:

- mechanisms of determined forming of pseudorandom sequences, which use the mathematical apparatus of the hashing functions;
- mechanisms of determined forming of pseudorandom sequences, which use the block symmetric ciphering;
- mechanisms of determined forming of pseudorandom sequences, which are based on the numeric-theoretical problems.

The general classification of methods of forming of pseudorandom sequences and of the respective generators is presented in Fig. 8.3.

Let us examine the main conversions, which are used by these three types of generators, let us research their particularities from the point of view of ensuring of the certain properties of the formed PRSs.

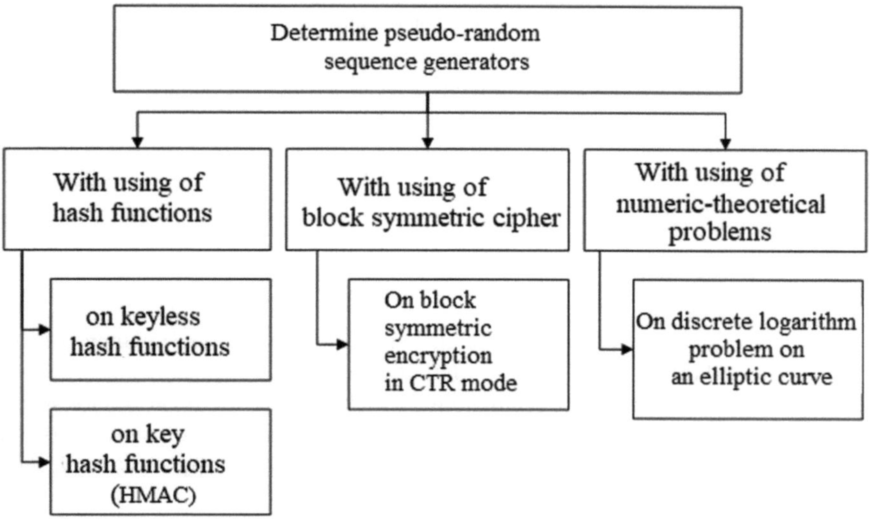

Fig. 8.3 General classification of the PRS-generators, which are specified in the NIST Special Publication 800-90A

8.1 Analysis and Research of the Main Conversions in DGRB with Using of the Hashing Funcions

The PRS-generators, according to the standard NIST Special Publication 800-90A, can be based on the using of the hashing function, which are irreversible or unilateral functions of conversions [5].

Based on hashing, the PRS-generators must use any approved, i. e. validated by the respective standard, the hashing function:

- on keyless functions of hashing;
- on key functions of hashing, built up by the scheme HMAC.

The maximum security, which can be supported by such a generator, corresponds to security level of the utilized hashing function. As the main hashing algorithm the standard NIST Special Publication 800-90A proposes the using of the hash-function SHA, which is specified in the respective USA-standard. However, for the PRS-generators only four security levels are supported: 112, 128, 192 and 256 bits. Table 8.1 concretizes the values, which must be used for building up of the generators for each approved hashing function.

In Fig. 8.4 is drawn the scheme of PRS-forming with the using of the keyless hashing function.

The internal structure of the generator consists of following elements:

Table 8.1 Main parameters of generators, based on hashing

	SHA-1	SHA-224 та SHA-512/224	SHA-256 та SHA-512/256	SHA-384	SHA-512
Length of the output block (outlen)	160	224	256	384	512
Length of the seed (seedlen)	440	440	440	880	880

1. Working states:
   - The value (V) *seedlen* bits (the bit-length of the seed, *seed*), which is being renewed during each generator initiation.
   - The constant value (C) *seedlen* bits, which depends on the generator seed.
   - Counter (*reseed_counter*), which indicates the number of requests of the pseudorandom bits since that time, as a new input entropy was introduced.
2. Administrative information:
   - *Security_strength*—the initiated level of the generator security.
   - *Prediction_resistance_flag*—the flag of the stability prognostication, which indicates the generator ability to support the forecasted security level.

The values V and C—are the critical values concerning the internal state, on which is based the security of this mechanism (i. e. V and C—are the «secret values» of the internal state).

The generator scheme on the key hashing functions, which are built up after the scheme *HMAC* on the approved (validated by the respective standards) keyless hashing functions, are drawn in Fig. 8.5.

Figure 8.5 demonstrates the generator working at the three stages with the using of the internal function (*HMAC_DRBG_Update*). This function is destined for regulation of the internal state. The actions in the upper part of the figure are executed, only if the additional input is not valid.

8.2 Analysis and Research of the Main Conversions in DGRB with Using of Symmetric Ciphering

The PRS-generators can be based on the using of the BSC-algorithms, for instance, *DES* or *AES* [6, 7]. They must utilize every approved, i. e. the validated by the respective standard, BSC-algorithm. The maximum security, which can be supported by such a generator, corresponds to the security level of the utilized cipher. As the main BSC-algorithms the standard *NIST Special Publication 800-90A* proposes to use the 3-key *TDEA*, or various variants of the algorithm *AES*: *AES*-128, *AES*-196, *AES*-256 [6, 7].

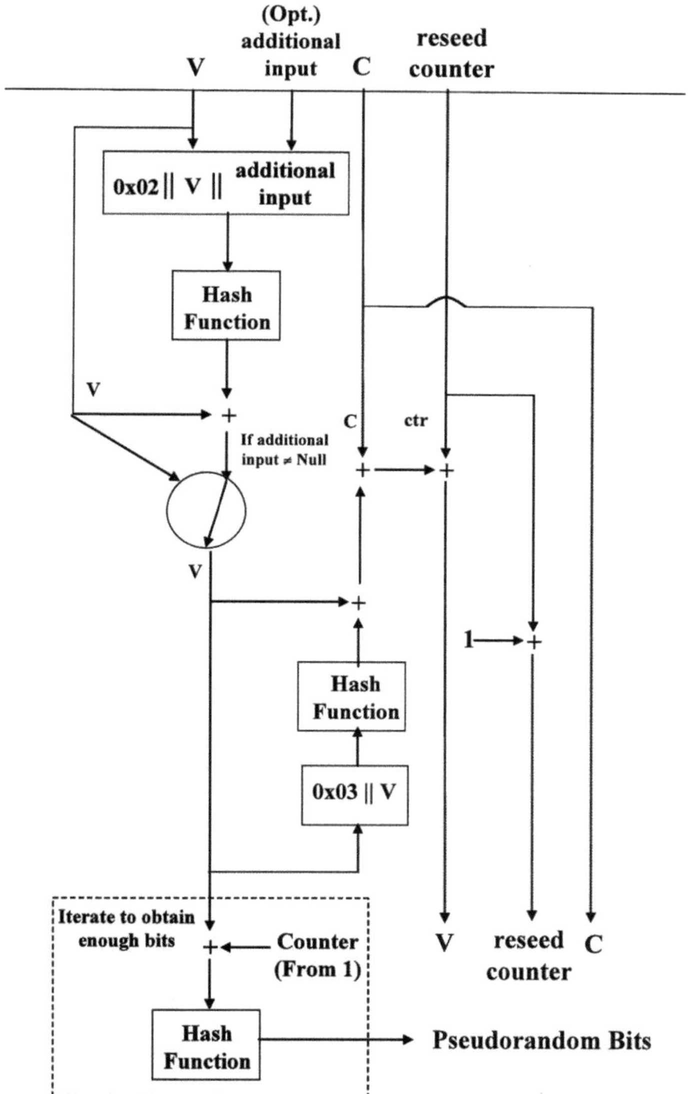

Fig. 8.4 Scheme of PRS forming with the using of the keyless hashing function

The PRS-generators, which are based on utilizing of the symmetric ciphers, apply the counter-mode (*CTR*), what is specified in the standard *NIST Special Publication 800-90A* [8]. This and others bitstreamed BSC-modes have been discussed above. The PRS-generators utilize also the internal function (*CTR_DRBG_Update*). This function is destined for regulating of the internal state.

Figure 8.6 demonstrates the DRGB-work with using of symmetric ciphering at three stages. The acts in the upper part of the figure are executed only if the additional input is not valid.

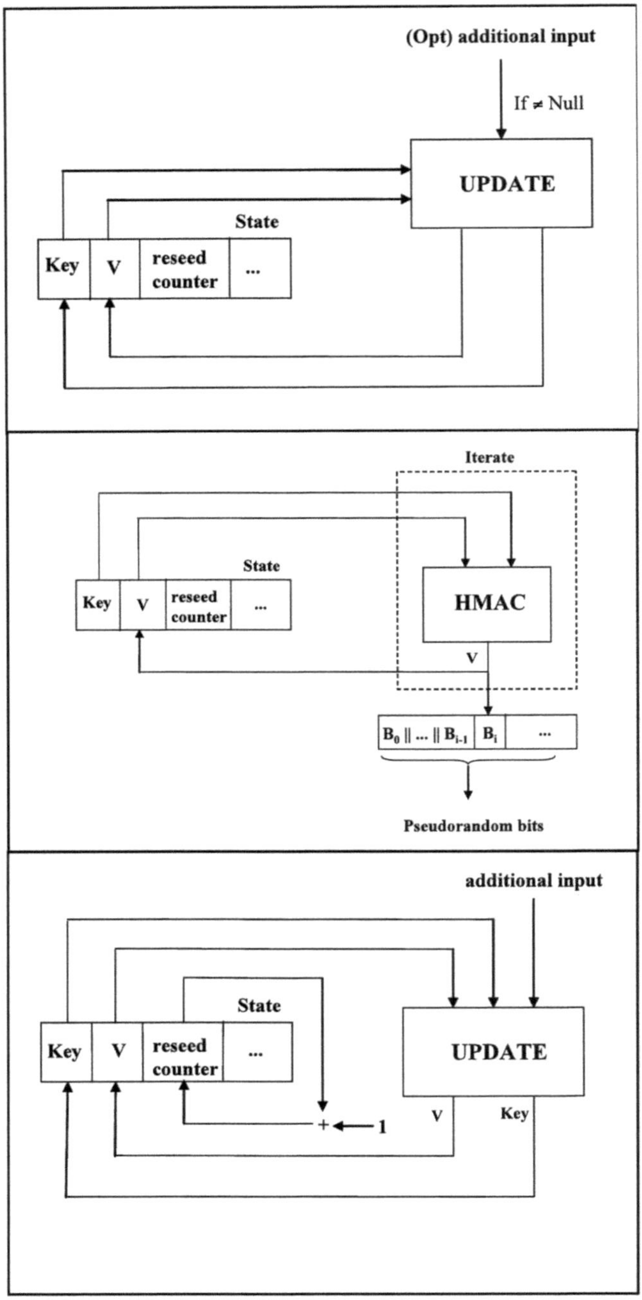

Fig. 8.5 Scheme of the generator on the key hashing functions built up after the scheme HMAC

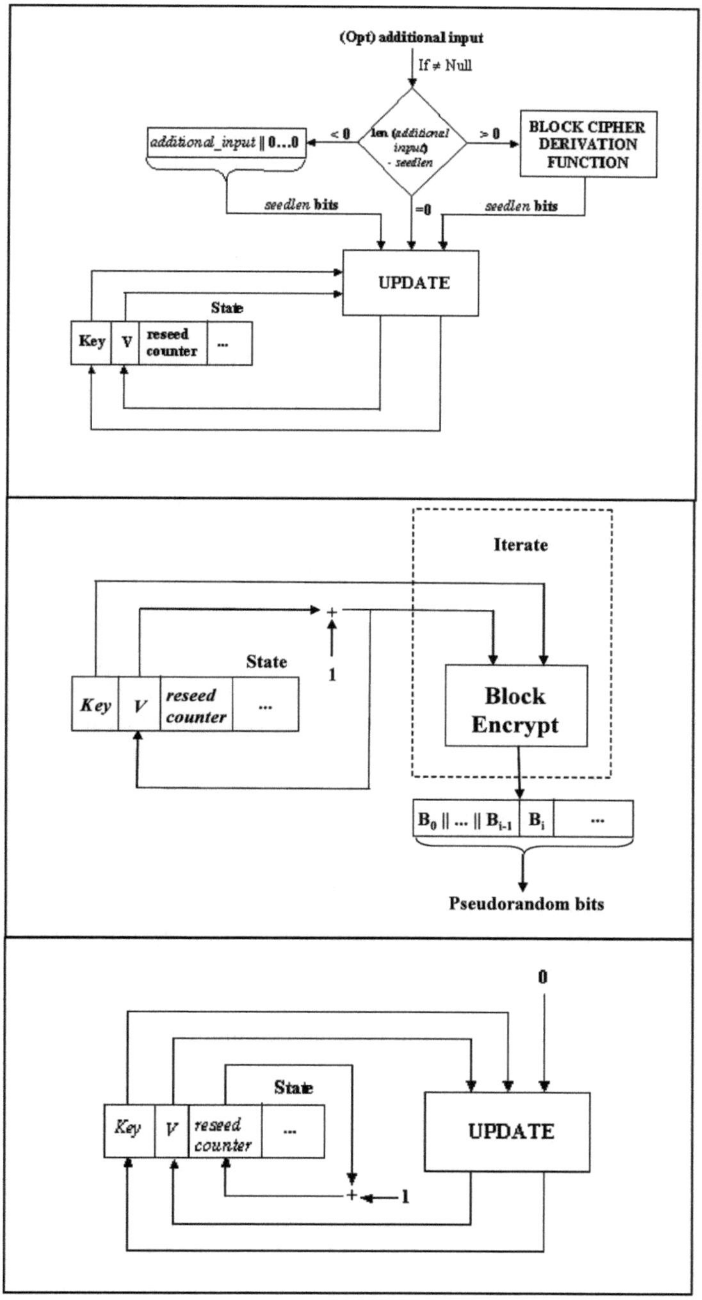

Fig. 8.6 Scheme of DGRB working with the BSC-using

Table 8.2 Main parameters of the generator on block ciphers

	TDEA	AES-128	AES-196	AES-256
Length of the output block (outlen)	64	128	128	128
Key length (keylen)	168	128	196	256
Length of seed (seedlen = outlen + keylen)	232	256	320	384

Table 8.2 concretizes the values which must be used for the considered generator.

8.3 Analysis and Researches of Main Conversions in DRGB, Based on Numeric-Theoretical Problems

The PRS-generators can be based on numeric-theoretical problems, for example, on the problem of discrete logarithmation [1]. If the generator is built up correctly, then its properties concerning randomness and/or unpredictability of the formed PRS will be reduced to the finding of the solution of this problem.

In the standard NIST Special Publication 800-90A the generator is concretized, which is based on the problem of discrete logarithmation on elliptic curve. This problem is formulated in such a way [9]: the points P and Q are given on the elliptic curve of the order n, such an a is needed to be found that $Q = aP$.

The generator uses the initial seed (seed) of the length 2*security_strength bits (it means of the double bit-stability) for initiation of generation of the outlen-bits of pseudorandom lines, executing scalar multiplications of two points of the point group of the elliptic curve, where the curve is defined over the field approximatively by 2^m as size.

For all curves the value m corresponds to the at least double bit-stability (2*security_strength bits) and never less than 256. The internal structure of the generator uses for initiation the seed (seed) of the length seedlen. In Fig. 8.7 is drawn the structural scheme of such a generator.

The generator initiation requires a corresponding elliptic curve for the wanted security level. The seed is utilized in order to determine the initial state (s) of the

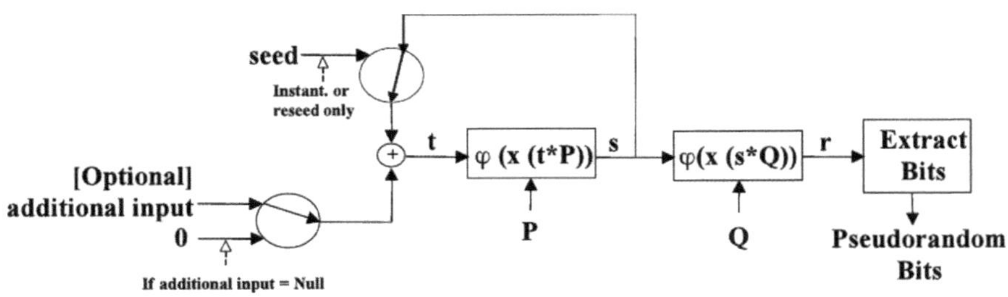

Fig. 8.7 Scheme of DGRB working on elliptic curves

Table 8.3 Main DGRB-parameters on the elliptic curves

	P-256	P-384	P-521
Size of the base field (in bits), analogue *seedlen*	256	384	521
Length of the output block (*outlen*)	240	368	504

generator, it must at least the *security_strength* bits of entropy (i. e. the entropy must not be less as the bit-level of security).

The maximum security level, which can be supported by the analyzed generator, is the security level of the utilized curve. Table 8.3 concretizes the values, which have to be used for each curve.

Let us perform some comparative researches of the discussed methods of PRS-forming, let us analyze their possibilities from the point of view of eventual realization in the schemes of bitstreamed cryptoconversion.

8.4 Comparative DGRB-Researches, Defined in the Standard NIST Special Publication 800-90A

Because of the main statements of the theory of information protection while estimated the security level of the protection mechanisms, the so-called undoubted and conventional efficiency indices, which were discussed in [10, 11]. To undoubted are attributed the indices, the executing of requirements after which is unconditioned, i.e. a necessary criterion of generator using, because namely they determine the achievement of an objective—operation of system. According to this, to unconditioned are attributed the partial indices of stability(security) of generator. The indices of the generator velocity are attributed to conventional ones, because the raise of this index is wanted, but not a necessary criterion of the generator utilizing.

To informal indices of the efficiency of generators must be attributed the appartenence of their security to one of the stability models, which were defined in [10, 11], and namely:

- model of unconditional security (unconditional secrecy), which was introduced by C. Shannon for designing of system stability, which does not reduce its properties even by admission of unrestricted computing resources at violator [12]. This theoretical presumption requires superlarge volumes of key informations, which are formed incidentally, equiprobably and independently from each other, because of this the model of unconditional security is considered, most commonly, exclusively within theoretical plane [12, 13];
- model of argumentative security, which is introduced, when the task of security violation (connected, for instance, with probability increasing for the rule discovering of the forming of a pseudorandom sequence) equivalently to the solution of the known numeric-theoretical task, for example of factorization, discrete logarithmation and so on. If during extended time (several centuries) this task was not

resolved, it is scarcely probable that a solution will be found and, respectively, the algorithm security will be reduced;

- model of practical security, when the algorithm stability is being considered according to the needed computing resources, that are demanded by the known analysis methods in order to reduce the security. It is a very practical model, because the verification is performed against the known attacks, but, according to new approaches, an existence of more efficient methods of algorithm analysis is possible.

The results of the performed analysis of the discussed above PRS-generators are presented in Table 8.4.

The performed analysis demonstrates that different generators, of which specification is presented in the standard NIST Special Publication 800-90A, have nearly identical security indices. It concerns, firstly, the generators on the hashing functions and on the BSC. From the point of view of the assured security, for they has not been found any effective analysis method, because of this it is considered that the probability of discovering of the rule of forming of pseudorandom sequences does not exceed the lower limit, i.e. $P \leq 2^{-k}$. The same concerns the estimation of the period of the formed pseudorandom sequences L, but from the point of view that there are no confirmed statements regarding periodical properties of the formed sequences, in the table is presented the value $L \approx 2^2$. However, as after indices of statistical security, the generators built up with utilizing of numeric-theoretical problems have a certain advantage, because for them the part of statistical tests after which the formed PRS is undistinguishable from the realization of an incidental process, is the highest. For generators on the hashing funcions and on the BSC there are statistical tests, after which the formed PRS (with a probability >1–0.96) can be distinguished from the realization of a random process. Obviously, it is demonstrated in the Figs. 8.8, 8.9 and 8.10, in which are presented the statistical portraits of respective DGRB.

From the point of view of ensured security, the generators, based on reducing of the task of stability violation to the resolving of an numeric-theoretical problem,

Table 8.4 Results of comparative researches of PRS-generators properties

	DGRB on hash functions	DGRB on BSC	DGRB on numerical theoretical problems
Discovering probability of *PB*	$P \leq 2^{-k}$, satisfies the requirements	$P \leq 2^{-k}$, satisfies the requirements	$P \leq 2^{-k}$, satisfies the requirements
Period of sequences *L*	$L \approx 2^2$	$L \approx 2^2$	$L \approx 2^2$
Statistical security	Satisfies the requirements	Satisfies the requirements	Satisfies the requirements
Forming velocity of V	10^7–10^9 bit/s	10^6–10^8 bit/s	10^3–10^4 bit/s
Model of generator security	Practical security	Practical security	Argumentative security

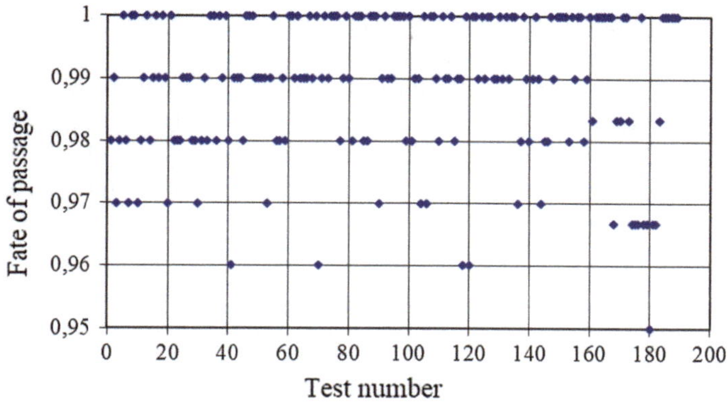

Fig. 8.8 Statistical portrait of the generator on Hash-function (SHA-1)

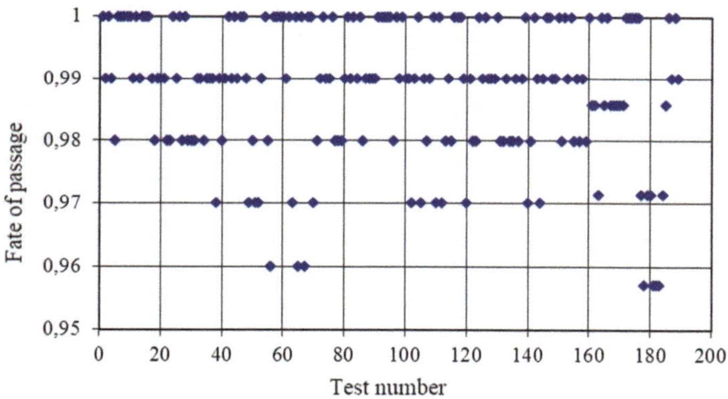

Fig. 8.9 Statistical portrait of the generator on BSC (3-*DES*)

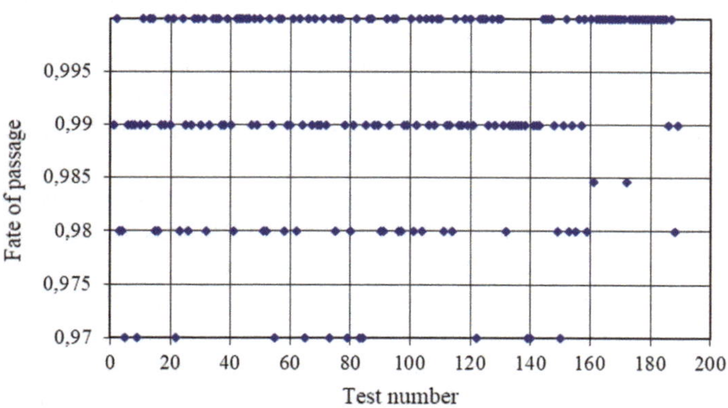

Fig. 8.10 Statistical portrait of the generator, which is based on a numeric-theoretical problem

correspond to the model of argumentative security, and the confidence to their security indices is higher. By the way, it should be noted that the increasing of security indices of such generators is conjugated with a drastic velocity fall. In reality, after the velocity of the PRS-forming the generators on numeric-theoretical problems dice away substantially from other constructions, for example, on hashing functions and on BSC, moreover this underrun is very large—three and more orders of magnitude.

8.5 Conclusions and Recommendations

1. The recommendations of the national institute of standards and technologies of the USA concerning the building up of generators of pseudorandom sequences with using of various methods and of respective mathematical apparatus are reflected in the national standard NIST Special Publication 800-90A «Recommendation for Random Number Generation Using Deterministic Random Bit Generators». In it are defined: the functional DGRB models; the conceptual mechanisms and basic requirements to DGRB; the mechanisms DGRB-functions; the specifications of algorithms realizing certain DGRB.
2. The performed analysis has demonstrated that all algorithms of PRS-forming, defined in the NIST Special Publication 800-90A, are detached generally in several main groups, and namely: based on using of the hashing functions (Hash_DRBG and HMAC_DRBG).
3. The performed researches demonstrated that the greatest velocity while PRS-forming ensures the generators, built up with using of the hashing functions and of the BSC. It is explained by the simplicity of mathematical operations and by the possibility of their optimization for realizing on contemporary calculating systems. However, from the point of view of statistical security the better indices are demonstrated by the PRS-generators, built up on utilizing of numeric-theoretical problems.
4. The researches of the PRS-generators, built up on utilizing of numeric-theoretical problems, demonstrated that after indices of statistical security they get the highest values. In particular, the statistical portrait of the generator Dual Elliptic Curve Deterministic RBG – Dual_EC_DRBG, built up with utilizing of mathematic apparatus of elliptic curves, is the best, all statistical testings demonstrates with a high probability that the formed PRS does not differ statistically from the random process realization. In addition, such generators correspond to the model of argumentative security, and the confidence to their indices is higher.
5. The performed researches of the mathematic conversions totality, on which are based the respective PRS-generators, demonstrated that the highest operating velocity is ensured by the algorithms, based on symmetric cryptoconversions (with using of hashing functions or of BSC). The most expensive are the generators, based on utilizing of numeric-theoretical problems, for instance, on elliptic curves. Thus, the election of a concrete conversion is determined by

way of compromise between the expected level of statistical security and the admissible expenditure for realization of the PRS-forming algorithms.

References

1. NIST Special Publication 80090A: Recommendation for Random Number Generation Using Deterministic Random Bit Generators. http://csrc.nist.gov/publications/nistpubs/80090A/SP80090A.pdf
2. Barker, E., Kelsey, J.: Recommendation for Random Number Generation Using Deterministic Random Bit Generators. National Institute of Standards and Technology, 101 p (2015). https://doi.org/10.6028/NIST.SP.80090Ar1
3. Luby, Rackoff, C.: How to construct pseudorandom permutations from pseudorandom functions. SIAM J. Comput. **17**(2), 373–386 (1988)
4. NIST Special Publication 80022: A Statistical Test Suite for Random and Pseudorandom Number Generators for Cryptographic Applications. http://csrc.nist.gov/publications/nistpubs/80022rev1a/SP80022rev1a.pdf
5. Seberry J.: Relationships among nonlinearity criteria. Seberry, J., Zhang, X.M., Zheng, Y.: Advances in Cryptology, EUROCRYPT'94, vol. 950, pp. 376–388.
6. FIPS197: Advanced Encryption Standard (AES): National Institute of Standards and Technology. http://csrc.nist.gov/publications/fips/fips197/fips197.pdf (2001)
7. FIPS46: Data Encryption Standard. National Institute of Standards and Technology (NIST), National Security Agency (NSA). http://www.itl.nist.gov/fipspubs/fip462.htm (1977)
8. NIST Special Publication 80038A: Recommendation for Block Cipher Modes of Operation. http://csrc.nist.gov/publications/nistpubs/80038a/sp80038a.pdf
9. Menezes, A.J., van Oorschot, P.C., Vanstone, S.A.: Handbook of Applied Cryptography, 794 p. CRC Press (1997)
10. NESSIE Call for Cryptographic Primitives, Version 2.2. http://cryptonessie.org (2000)
11. NESSIE Security Report, Deliverable D21, Version 1. http://www.cryptonessie.org (2002)
12. Shannon, K.: Works on Information Theory and Cybernetics, 830 c. M.: ИЛ (Section—Theory of Communication in Secret Systems) (1963)
13. Feistel, H.: Cryptography and Computer Privacy, vol. 228, No. 5, pp. 15–23. Scientific American (1973)

Chapter 9
Analysis of Bitstreamed Cryptographic Conversion Algorithms from the International Project eSTREAM

9.1 Analysis of Software-Oriented Stream Ciphers, Recognized as Winners of the International Project eSTREAM

The international project eSTREAM [1] was performed in the years 2004–2008 by the European Union and was destined for discovering of a new stream cipher, able for wide using. It was initiated because of failure of all six stream ciphers, presented at one time to the project NESSIE. The project eSTREAM consisted of several phases and was aimed on discovering of algorithms, able for various using variants.

The presentations for the eSTREAM found themselves into one or both profiles:

- Profile 1: Stream cipher for software with high requirements to throughput capacity;
- Profile 2: Stream cipher for hardware with restricted resources, for instance, restriction of memory, of transistors number or other devices, or energy consumption.

Both the profiles contain subcategory (1A and 2A) with the ciphers that ensure authentication with enciphering. In the third phase no cipher, which would ensure the authentication, was not examined.

As for today, such ciphers were selected, which correspond to the established requirements of the project eSTREAM (see Table 9.1) [1–7].

The presented in the Table 9.1 stream cipher have not as for today restrictions concerning their utilizing, inclusive patent ones or any others [2–8].

Let us examine more in detail each of the finalists of the international contest of the stream ciphers *eSTREAM*, let us analyze the particularities of their building up and of their appliance efficiency.

A. A. Kuznetsov et al., *Stream Ciphers in Modern Real-time IT Systems*,
Studies in Systems, Decision and Control 375,
https://doi.org/10.1007/978-3-030-79770-6_9

Table 9.1 Stream ciphers, which were selected during the international project *eSTREAM*

Profile 1 (software-oriented ciphers)	Profile a (hardware-oriented ciphers)
HC-128	Grain
Rabbit	MICKEY
Salsa20/12	Trivium
SOSEMANUK	

9.1.1 Stream Cipher HC-128

The stream cipher HC-128 is a simplified version of the cipher HC-256 [8] for 128-bit level of security. The SSC HC-128 is simple, secure, software-oriented cipher with an efficient realization and can be free used.

The stream cipher HC-128 consists of two secret tables, each of those contains 512 of 32-bit elements. On each iteration is executed the renewing of one table element with non-linear feedback function. All elements of the two tables are renewed each 1024-iterations. On each iteration one 32-bit output is generated from the non-linear function of input filtration.

Consequently, the SSC HC-128 suits for contemporary (and future) superscalar microprocessors. Interdependence between the the operations in the HC-128 is very small: three consequent iterations can be calculated in parallel; on each iteration the feedback functions and the instructions output can be calculated in parallel. The high degree of parallelness allows the cipher HC-128 to be extremely effective while working on a contemporary processor. In the author realization of the SSC HC-128 in the language C the ciphering velocity achieved 3,05 cycles/byte on the processor Intel Pentium M. Herewith the cipher HC-128 is secure. The performed while eSTREAM contest analysis showed that the renewal of the HC-128-key is such sophisticated as an exhaustive key retrieval. In order to distinguish the keystream from the incidental bits, are necessary, expectedly, over 2^{64} bits of the ciphering keystream.

9.1.1.1 Specification of the HC-128-Cipher

The stream cipher HC-128 is destined for the keystream forming with the length till 2^{64} bits, which is elaborated from the 128-bit key and the 128-bit initiation vector.

For realization of SSC HC-128 are used the following operations (Table 9.2).

The two tables, used in the SSC HC-128, are denoted as P and Q. The key and the initiation vector of the cipher HC-128 are denoted as K and IV. The keystream, generated by the cipher, is denoted as s. These notations are reduced to the Table 9.3.

The stream cipher *HC*-128 uses the six functions in the same way, as the cipher *HC*-256. The functions $f_1(x)$ and $f_2(x)$ coincide with the functions $\sigma_0^{\{256\}}(x)$ and $\sigma_1^{\{256\}}(x)$, which are used in the hashing algorithm *SHA*-256. For the function $h_1(x)$

Table 9.2 Operations used for the realization of the stream cipher HC-128

Designation	Operation
$+$	Operation of addition after 2^{32}
$-$	Operation of subtraction after the modulus 512
$\oplus$	Operation of bitwise exclusive OR (XOR)
$\|\|$	Concatenation
>>	Operator of right shift by the set amount of bits
<<	Operator of left shift by the set amount of bits
>>>	Operator of cyclic right shift by the set amount of bits
<<<	Operator of cyclic left shift by the set amount of bits

Table 9.3 Notations used for realization of the stream cipher HC-128

Designation	Contents of the designation
P	Table with 512 32-bit elements. Each table element is denoted as *P[i]*, $0 \leq i \leq 511$
Q	Table with 512 32-bit elements. Each table element is denoted as *Q[i]*, $0 \leq i \leq 511$
K	128-bit key of the stream cipher *HC*-128
IV	128-bit vector of initiation of the stream cipher *HC*-128
S	Keystream, generated by the stream cipher *HC*-128 On each iteration a 128-bit output is generated, which is denoted as s_i, $s = s_0 \|\| s_1 \|\| s_2 \|\| \ldots$

as *S*-block is utilized the table *Q*. For the function $h_2(x)$ as *S*-block is utilized the table *P*.

The functions are written as:

$$
\begin{aligned}
f_1(x) &= (x{>}{>}{>}7) \oplus (x{>}{>}{>}18) \oplus (x{>}{>}3);\\
f_2(x) &= (x{>}{>}{>}17) \oplus (x{>}{>}{>}19) \oplus (x{>}{>}10);\\
g_1(x,\ y,\ z) &= ((x{>}{>}{>}10) \oplus (z{>}{>}{>}23)) + (y{>}{>}{>}8);\\
g_2(x,\ y,\ z) &= ((x <<< 10) \oplus (z <<< 23)) + (y <<< 8);\\
h_1(x) &= Q[x_0] + Q[256 + x_2];\\
h_2(x) &= P[x_0] + P[256 + x_2],
\end{aligned}
$$

where $x = x_3||x_2||x_1||x_0$, x—is a 32-bit word; x_0, x_1, x_2, x_3—the four bytes. The bytes x_1 and x_2 are denoted as the most significant and the junior byte x respectively.

9.1.1.2 Initiation Algorithm

The initiation process of the stream cipher *HC*-128 consists of the extension of key and of initiation vector in *P* and *Q*, (analogically with the messages establishing in *SHA*-256) and of ciphering start by 1024 times without generation of output result (the outputs are used for renewing of *P* and *Q*).

1. Let be $K = K_0||K_1||K_2||K_3$ and $IV = IV_0||IV_1||IV_2||IV_3$ where each *Ki* and *IVi* denote a 32-digit number. Let be $Ki + 4 = Ki$ and $IVi + 4 = IVi$ for $0 \le i < 4$. The key and the initiation vector are extended into the batch $W_i (0 \le i \le 1279)$ in such a way:

$$W_i = \begin{cases} K_i, & 0 \le i \le 7; \\ IV_{i-8}, & 8 \le i \le 15; \\ f_2(W_{i-2}) + W_{i-7} + f_1(W_{i-15}) + W_{i-16} + i, & 16 \le i \le 1279. \end{cases}$$

2. Renewing of tables *P* and *Q* from the batch *W*.

$$P[i] = W_{i+256} \text{ для } 0 \le i \le 511;$$
$$Q[i] = W_{i+768} \text{ для } 0 \le i \le 511.$$

3. Executing of 1024 iterations of ciphering and using of outputs for changing of the table elements as follows:

For all from $i = 0$ till 511:

$$P[i] = (P[i] + g_1(P[i-3],\ P[i-10],\ P[i-511])) \oplus h_1(P[i-12]);$$

For all from $i = 0$ till 511:

$$Q[i] = (Q[i] + g_2(Q[i-3], Q[i-10], Q[i-511])) \oplus h_2(Q[i-12]).$$

After executing of written iterations the initiation process ends, and the cipher is ready to generate the keystream.

9.1.1.3 Algorithm of Keystream Generating

In each iteration one element of the table is renewed and one 32-bit output is generated. Each *S*-block is used for generation of only 512 outputs, it is being renewed during the next 512 iterations.

The algorithm of keystream of the cipher *HC*-128 is presented on the Fig. 9.1, with the symbol «⊟» is denoted «–» by the modulus 512.

```
i = 0;
repeat until enough keystream bits are generated.
{
    j = i mod 512;
    if (i mod 1024) < 512
    {
        P[j] = P[j] + g1(P[j ⊟ 3], P[j ⊟ 10], P[j ⊟ 511]);
        s_i = h1( P[j ⊟ 12] ) ⊕ P[j];
    }
    else
    {
        Q[j] = Q[j] + g2(Q[j ⊟ 3], Q[j ⊟ 10], Q[j ⊟ 511]);
        s_i = h2( Q[j ⊟ 12] ) ⊕ Q[j];
    }
    end-if
    i = i + 1;
}
end-repeat
```

Fig. 9.1 Algorithm of keystream generation of the cipher *HC*-128

9.1.1.4 Security Analysis of the Stream Cipher HC-128

After the results of the security analysis the stream cipher *HC*-128 is similar with the cipher *HC*-256. The input and the inverse functions of the *HC*-128 are not non-linear, because of this the fast correlation *S*-boxes, which are renewed while generating of the keystream, and this complicates substantially the clarification of relationships, which connect the linear input and output bits of the *S*-block.

Period. The 32778-bit state of the cipher *HC*-128 guarantees that the keystream period is extremely large. But the exact period of the keystream of *HC*-128 is hardly to determine. The average period of the keystream, by estimates, is much more than 2256. A large quantity of states eliminates the time-memory-data objection on stream ciphers [8].

Security of the secret key. Let us design that the output function and the feedback function of the cipher *HC*-128 are non-linear. The non-linear output function displays a small quantity of the partial data concerning state on each iteration. The non-linear feedback function guarantees that the secret key can not be taken away from any partial information. Consequently, as is reported by the algorithm developers, the secret key of the stream cipher *HC*-128 can not determined faster as an exhaustive search of the key.

Security of initialization process. The initialization process of the *HC*-128 is composed of the two above discussed stages. The tables *P* and *Q* are converted with the aid of the key *K* and of the initialization vector *IV*. Each bit of *K* and of *IV* influences all bits of the two tables, and each change in them produces uncontrolled changes in the tables. The ciphering is initiated by 1024 times without generating of the output result, thus the tables *P* and *Q* become more random. After the initialization process the changes in *K* and *IV* do not produce changes in the key sequence.

The stream cipher *HC*-128 is convenient for the contemporary microprocessors, moreover the dependence between the operations can be reduced to minimum: the three consequent iterations of algorithm can be calculated in parallel. The paralleling ability enables this cipher to be very efficient on contemporary processors.

9.1.2 Stream Cipher Salsa20

The algorithm *Salsa20* is a system of stream ciphering, which was developed by *Daniel J. Bernstein* (Engl.). The algorithm became a winner of the contest *eSTREAM* in the first profile (the stream ciphers for software using with large throughput capacity) [3].

The algorithms *Salsa20/8* and *Salsa20/12* that are some ciphersystems, which utilize the same mechanism as also *Salsa20*, but with 8 and 12 rounds of ciphering respectively instead of the 20 original at *Salsa20*. The output algorithm *Salsa20* was made with a large stability margin, while the *Salsa20/8* demonstrates good results concerning operating velocity, advancing in majority of cases many other ciphersystems, inclusive *AES* and *RC*4 [3]. The algorithm utilizes a hash-function with the 20 cycles. Its main cryptoconversions resemble the algorithm *AES*.

For exposing the specification of the stream cipher Salsa20 are used the notations, presented in the Table 9.4 [3].

A word for describing of the algorithm *Salsa20* is named the element of the multitude $\{0, 1, \ldots, 2^{32}-1\}$ and is written in hexadecimal notation with the prefix 0x.

Table 9.4 Operations and notations, which are used in the specification of the cipher Salsa 20

Designation	Operation or designation
$\oplus$	Operation of bitwise addition by the modulus 2 (*XOR*)
+	Operation of addition of 32 bit-numbers by the modulus 2^{32}
<<<	Operator of cyclic left shift by the indicated number of bits

9.1.2.1 Function *Quarterround*

The main functional block of the ciphering system quarterround(y) above the 4 words. From this conversion are built up and described more general and more complicated conversions.

The essence of the conversion *quarterround(y)* lies in the fact that for each word are added the precedent two, the obtained sum is cyclically shifted by certain quantity of bits, and the result is bitwise summarised with the selected word. The next operations are executed with new meanings of words.

The scheme of computing of the function $quarterround(y) = (z_0, z_1, z_2, z_3)$ is presented on the Fig. 9.2, with the symbol «⊞» is denoted the operation «+» addition (from Table 9.4).

If y is the sequence of $_4$ words, i. e. $y = (y_0, y_1, y_2, y_3)$, then the computing of the function $quarterround(y) = (z_0, z_1, z_2, z_3)$ lies in the following:

$$z_1 = y_1 \oplus ((y_0 + y_3) <<< 7),$$
$$z_2 = y_2 \oplus ((z_1 + y_0) <<< 9),$$

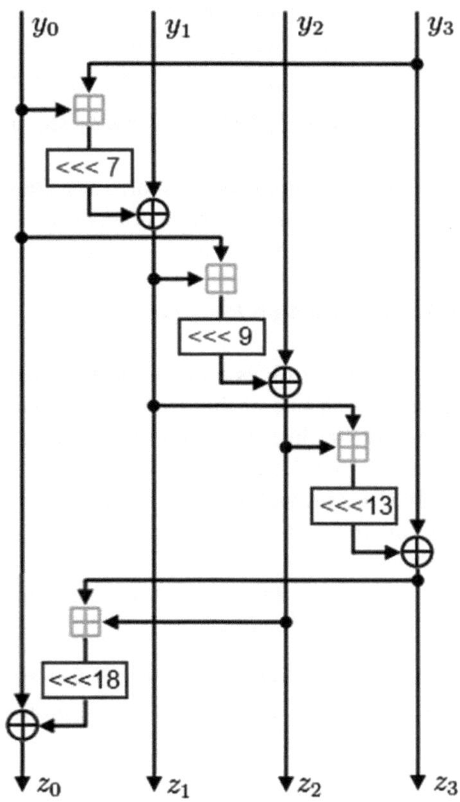

Fig. 9.2 Scheme of computing of the function $quarterround(y) = (z_0, z_1, z_2, z_3)$

$$z_3 = y_3 \oplus ((z_2 + z_1) <<< 13),$$
$$z_0 = y_0 \oplus ((z_3 + z_2) <<< 18).$$

For instance, the executing of the function $quarterround(y) = (z_0, z_1, z_2, z_3)$ for some values will be written as follows:

$$quarterround(0x00000000, 0x00000000, 0x00000000, 0x00000000)$$
$$= (0x00000000, 0x00000000, 0x00000000, 0x00000000),$$
$$quarterround(0x00000001, 0x00000000, 0x00000000, 0x00000000)$$
$$= (0x08008145, 0x00000080, 0x00010200, 0x20500000),$$
$$quarterround(0x00000000, 0x00000001, 0x00000000, 0x00000000)$$
$$= (0x88000100, 0x00000001, 0x00000200, 0x00402000),$$
$$quarterround(0x00000000, 0x00000000, 0x00000001, 0x00000000)$$
$$= (0x80040000, 0x00000000, 0x00000001, 0x00002000),$$
$$quarterround(0x00000000, 0x00000000, 0x00000000, 0x00000001)$$
$$= (0x00048044, 0x00000080, 0x00010000, 0x20100001),$$
$$quarterround(0xe7e8c006, 0xc4f9417d, 0x6479b4b2, 0x68c67137)$$
$$= (0xe876d72b, 0x9361dfd5, 0xf1460244, 0x948541a3),$$
$$quarterround(0xd3917c5b, 0x55f1c407, 0x52a58a7a, 0x8f887a3b)$$
$$= (0x3e2f308c, 0xd90a8f36, 0x6ab2a923, 0x2883524c).$$

This function can be estimated as a conversion of the words y_1, y_2, y_3 and y_4. Each of these conversions is reversible, as the function in whole as well.

9.1.2.2 Function *Rowround*

In this conversion as input are utilized the 16 words, which are supplied as a 4×4 matrix. Each line of this matrix is processed by the function *quarterround*(y). The words of the line are taken by the order, beginning with the i-th for the i-th line, where $i = \{0, 1, 2, 3\}$.

If $y = (y_0, y_1, y_2, ..., y_{15})$ is a sequence of the 16 words, then the calculating of the function $rowround(y) = (z_0, z_1, z_2, \ldots, z_{15})$ lies in the following:

$$(z_0, z_1, z_2, z_3) = quarterround(y_0, y_1, y_2, y_3),$$
$$(z_5, z_6, z_7, z_4) = quarterround(y_5, y_6, y_7, y_4),$$
$$(z_{10}, z_{11}, z_8, z_9) = quarterround(y_{10}, y_{11}, y_8, y_9),$$
$$(z_{15}, z_{12}, z_{13}, z_{14}) = quarterround(y_{15}, y_{12}, y_{13}, y_{14}).$$

For example, the execution of the function $rowround(y) = (z_0, z_1, z_2, \ldots, z_{15})$ for some values will be as follows:

$$\begin{aligned}
&rowround(0x00000001, 0x00000000, 0x00000000, 0x00000000,\\
&\quad 0x00000001, 0x00000000, 0x00000000, 0x00000000,\\
&\quad 0x00000001, 0x00000000, 0x00000000, 0x00000000,\\
&\quad 0x00000001, 0x00000000, 0x00000000, 0x00000000)\\
&\quad = (0x08008145, 0x00000080, 0x00010200, 0x20500000,\\
&\quad 0x20100001, 0x00048044, 0x00000080, 0x00010000,\\
&\quad 0x00000001, 0x00002000, 0x80040000, 0x00000000,\\
&\quad 0x00000001, 0x00000200, 0x00402000, 0x88000100),\\
&rowround(0x08521bd6, 0x1fe88837, 0xbb2aa576, 0x3aa26365,\\
&\quad 0xc54c6a5b, 0x2fc74c2f, 0x6dd39cc3, 0xda0a64f6,\\
&\quad 0x90a2f23d, 0x067f95a6, 0x06b35f61, 0x41e4732e,\\
&\quad 0xe859c100, 0xea4d84b7, 0x0f619bff, 0xbc6e965a)\\
&\quad = (0xa890d39d, 0x65d71596, 0xe9487daa, 0xc8ca6a86,\\
&\quad 0x949d2192, 0x764b7754, 0xe408d9b9, 0x7a41b4d1,\\
&\quad 0x3402e183, 0x3c3af432, 0x50669f96, 0xd89ef0a8,\\
&\quad 0x0040ede5, 0xb545fbce, 0xd257ed4f, 0x1818882d).
\end{aligned}$$

9.1.2.3 Function *Columnround*

For realization of this function as input are used the columns of the same matrix as at the function *rowround*. The conversion is also analogical with using of the the function *quarterround*(*y*), the argument of which are the values of columns of the matrix, beginning with the *j*-th for the *j*-th column, where $j = \{0, 1, 2, 3\}$.

The function $columnround(y) = (z)$ operates with the sequence of the 16 words as well, so that:

$$\begin{aligned}
(z_0, z_4, z_8, z_{12}) &= quarterround(y_0, y_4, y_8, y_{12}),\\
(z_5, z_9, z_{13}, z_1) &= quarterround(y_5, y_9, y_{13}, y_1),\\
(z_{10}, z_{14}, z_2, z_6) &= quarterround(y_{10}, y_{14}, y_2, y_6),\\
(z_{15}, z_3, z_7, z_{11}) &= quarterround(y_{15}, y_3, y_7, y_{11}).
\end{aligned}$$

The equivalent formula:

$$\begin{aligned}
&(z_0, z_4, z_8, z_{12}, z_1, z_5, z_9,\\
&z_{13}, z_2, z_6, z_{10}, z_{14}, z_3, z_7, z_{11}, z_{15})
\end{aligned}$$

$$= rowround(y_0, y_4, y_8, y_{12}, y_1, y_5, y_9, y_{13}, y_2, y_6, y_{10}, y_{14}, y_3, y_7, y_{11}, y_{15}).$$

For instance, for executing of the function $columround(y) = (z_0, z_1, z_2, \ldots, z_{15})$ for some values will be written as follows:

$$\begin{aligned}
&columnround(0x00000001, 0x00000000, 0x00000000, 0x00000000,\\
&\quad 0x00000001, 0x00000000, 0x00000000, 0x00000000,\\
&\quad 0x00000001, 0x00000000, 0x00000000, 0x00000000,\\
&\quad 0x00000001, 0x00000000, 0x00000000, 0x00000000)\\
&\quad\quad = (0x10090288, 0x00000000, 0x00000000, 0x00000000,\\
&\quad 0x00000101, 0x00000000, 0x00000000, 0x00000000,\\
&\quad 0x00020401, 0x00000000, 0x00000000, 0x00000000,\\
&\quad 0x40a04001, 0x00000000, 0x00000000, 0x00000000),\\
&columnround(0x08521bd6, 0x1fe88837, 0xbb2aa576, 0x3aa26365,\\
&\quad 0xc54c6a5b, 0x2fc74c2f, 0x6dd39cc3, 0xda0a64f6,\\
&\quad 0x90a2f23d, 0x067f95a6, 0x06b35f61, 0x41e4732e,\\
&\quad 0xe859c100, 0xea4d84b7, 0x0f619bff, 0xbc6e965a)\\
&\quad\quad = (0x8c9d190a, 0xce8e4c90, 0x1ef8e9d3, 0x1326a71a,\\
&\quad 0x90a20123, 0xead3c4f3, 0x63a091a0, 0xf0708d69,\\
&\quad 0x789b010c, 0xd195a681, 0xeb7d5504, 0xa774135c,\\
&\quad 0x481c2027, 0x53a8e4b5, 0x4c1f89c5, 0x3f78c9c8).
\end{aligned}$$

9.1.2.4 Function *Doubleround*

The function *doubleround*(y) is a consequent application of the functions *columnround*and then *rowround*, i. e.:

$$doubleround(y) = rowround(columnround(y)).$$

For example, the executing of the function $doubleround(y) = (z_0, z_1, z_2, \ldots, z_{15})$ for some values will be written as follows:

$$\begin{aligned}
&doubleround(0x00000001, 0x00000000, 0x00000000, 0x00000000,\\
&\quad 0x00000000, 0x00000000, 0x00000000, 0x00000000,\\
&\quad 0x00000000, 0x00000000, 0x00000000, 0x00000000,\\
&\quad 0x00000000, 0x00000000, 0x00000000, 0x00000000)\\
&\quad\quad = (0x8186a22d, 0x0040a284, 0x82479210, 0x06929051,
\end{aligned}$$

$0x08000090,\ 0x02402200,\ 0x00004000,\ 0x00800000,$
$0x00010200,\ 0x20400000,\ 0x08008104,\ 0x00000000,$
$0x20500000,\ 0xa0000040,\ 0x0008180a,\ 0x612a8020),$

$$doubleround(0xde501066,\ 0x6f9eb8f7,\ 0xe4fbbd9b,\ 0x454e3f57,$$
$$0xb75540d3,\ 0x43e93a4c,\ 0x3a6f2aa0,\ 0x726d6b36,$$
$$0x9243f484,\ 0x9145d1e8,\ 0x4fa9d247,\ 0xdc8dee11,$$
$$0x054bf545,\ 0x254dd653,\ 0xd9421b6d,\ 0x67b276c1)$$
$$= (0xccaaf672,\ 0x23d960f7,\ 0x9153e63a,\ 0xcd9a60d0,$$
$$0x50440492,\ 0xf07cad19,\ 0xae344aa0,\ 0xdf4cfdfc,$$
$$0xca531c29,\ 0x8e7943db,\ 0xac1680cd,\ 0xd503ca00,$$
$$0xa74b2ad6,\ 0xbc331c5c,\ 0x1dda24c7,\ 0xee928277).$$

9.1.2.5 Function *Littleendian*

The algorithm *Salsa20* utilizes the word writing which begins with the junior byte. For that is introduced a separate conversion *littleendian*(b).

Let be $b = (b_0,\ b_1,\ b_2,\ b_3)$ a 4-byte sequence, then the function *littleendian*(b) restores the word, such that

$$littleendian(b) = b_0 + 2^8 b_1 + 2^{16} b_2 + 2^{24} b_3.$$

For instance, the executing of function $littleendian(b_0,\ b_1,\ b_2,\ b_3)$ for some values will be written as follows:

$$littleendian(0,\ 0,\ 0,\ 0) = 0x00000000,$$
$$littleendian(86, 75, 30, 9) = 0x091e4b56,$$
$$littleendian(255, 255, 255, 250) = 0xfaffffff.$$

It should be noted that this function is revertible.

9.1.2.6 Hash-Function *Salsa20(X)*

This function uses as an input the 64-byte sequence and converts the 64-byte sequence. The calculating of the function *Salsa*20(x) begins with the vector

$$x = (x[0], x[1], \dots, x[63]),$$

where

$$x_0 = littleendian(x[0],\ x[1],\ x[2],\ x[3]),$$
$$x_1 = littleendian(x[4],\ x[5],\ x[6],\ x[7]),$$
$$x_2 = littleendian(x[8],\ x[9],\ x[10],\ x[11]),$$
$$\dots$$
$$x_{15} = littleendian(x[60],\ x[61],\ x[62],\ x[63]).$$

Here are utilized the following notations: $x[i]$ bytes of the input sequence x; x_j the words which are used in the above-described functions.

Let us determine $(z_0, z_1, \dots, z_{15}) = doubleround^{10}(x_0, x_1, \dots, x_{15})$, then the value of the function *Salsa20(x)* is calculated as follows:

$$littleendian^{-1}(z_0 + x_0);$$
$$littleendian^{-1}(z_1 + x_1);$$
$$littleendian^{-1}(z_2 + x_2);$$
$$\dots$$
$$littleendian^{-1}(z_{15} + x_{15}).$$

For example, the executing of the function *Salsa20(x)* for some values will be written as follows:

$$\begin{aligned} &Salsa20(0,\ 0,\ 0,\ 0,\ 0,\ 0,\ 0,\ 0,\ 0,\ 0,\ 0,\ 0,\ 0,\ 0,\ 0,\ 0,\\ &0,\ 0,\ 0,\ 0,\ 0,\ 0,\ 0,\ 0,\ 0,\ 0,\ 0,\ 0,\ 0,\ 0,\ 0,\ 0,\\ &0,\ 0,\ 0,\ 0,\ 0,\ 0,\ 0,\ 0,\ 0,\ 0,\ 0,\ 0,\ 0,\ 0,\ 0,\ 0,\\ &0,\ 0,\ 0,\ 0,\ 0,\ 0,\ 0,\ 0,\ 0,\ 0,\ 0,\ 0,\ 0,\ 0,\ 0,\ 0)\\ &= (0,\ 0,\ 0,\ 0,\ 0,\ 0,\ 0,\ 0,\ 0,\ 0,\ 0,\ 0,\ 0,\ 0,\ 0,\ 0,\\ &0,\ 0,\ 0,\ 0,\ 0,\ 0,\ 0,\ 0,\ 0,\ 0,\ 0,\ 0,\ 0,\ 0,\ 0,\ 0,\\ &0,\ 0,\ 0,\ 0,\ 0,\ 0,\ 0,\ 0,\ 0,\ 0,\ 0,\ 0,\ 0,\ 0,\ 0,\ 0,\\ &0,\ 0,\ 0,\ 0,\ 0,\ 0,\ 0,\ 0,\ 0,\ 0,\ 0,\ 0,\ 0,\ 0,\ 0,\ 0). \end{aligned}$$

$$\begin{aligned} &Salsa20(211, 159,\ 13, 115,\ 76,\ 55,\ 82, 183,\ 3, 117, 222,\ 37, 191, 187, 234, 136,\\ &49, 237, 179,\ 48,\ 1, 106, 178, 219, 175, 199, 166,\ 48,\ 86,\ 16, 179, 207,\\ &31, 240,\ 32,\ 63,\ 15,\ 83,\ 93, 161, 116, 147,\ 48, 113, 238,\ 55, 204,\ 36,\\ &79, 201, 235,\ 79,\ 3,\ 81, 156,\ 47, 203,\ 26, 244, 243,\ 88, 118, 104,\ 54)\\ &= (109,\ 42, 178, 168, 156, 240, 248, 238, 168, 196, 190, 203,\ 26, 110, 170, 154,\\ &29,\ 29, 150,\ 26, 150,\ 30, 235, 249, 190, 163, 251,\ 48,\ 69, 144,\ 51,\ 57,\\ &118,\ 40, 152, 157, 180,\ 57,\ 27,\ 94, 107,\ 42, 236,\ 35,\ 27, 111, 114, 114,\\ &219, 236, 232, 135, 111, 155, 110,\ 18,\ 24, 232,\ 95, 158, 179,\ 19,\ 48, 202),\\ &Salsa20(88, 118, 104,\ 54,\ 79, 201, 235,\ 79,\ 3,\ 81, 156,\ 47, 203,\ 26, 244, 243,\\ &191, 187, 234, 136, 211, 159,\ 13, 115,\ 76,\ 55,\ 82, 183,\ 3, 117, 222,\ 37, \end{aligned}$$

86, 16, 179, 207, 49, 237, 179, 48, 1, 106, 178, 219, 175, 199, 166, 48,
238, 55, 204, 36, 31, 240, 32, 63, 15, 83, 93, 161, 116, 147, 48, 113)
= (179, 19, 48, 202, 219, 236, 232, 135, 111, 155, 110, 18, 24, 232, 95, 158,
26, 110, 170, 154, 109, 42, 178, 168, 156, 240, 248, 238, 168, 196, 190, 203,
69, 144, 51, 57, 29, 29, 150, 26, 150, 30, 235, 249, 190, 163, 251, 48,
27, 111, 114, 114, 118, 40, 152, 157, 180, 57, 27, 94, 107, 42, 236, 35),

$Salsa20^{1000000}$(6, 124, 83, 146, 38, 191, 9, 50, 4, 161, 47, 222, 122, 182, 223, 185,
75, 27, 0, 216, 16, 122, 7, 89, 162, 104, 101, 147, 213, 21, 54, 95,
225, 253, 139, 176, 105, 132, 23, 116, 76, 41, 176, 207, 221, 34, 157, 108,
94, 94, 99, 52, 90, 117, 91, 220, 146, 190, 239, 143, 196, 176, 130, 186)
= (8, 18, 38, 199, 119, 76, 215, 67, 173, 127, 144, 162, 103, 212, 176, 217,
192, 19, 233, 33, 159, 197, 154, 160, 128, 243, 219, 65, 171, 136, 135, 225,
123, 11, 68, 86, 237, 82, 20, 155, 133, 189, 9, 83, 167, 116, 194, 78,
122, 127, 195, 185, 185, 204, 188, 90, 245, 9, 183, 248, 226, 85, 245, 104).

9.1.2.7 Extention Function *Salsa20k(n)*

The function *Salsa*20*k*(n) utilizes as input a 32-byte or a 16-byte sequence k and a 16-byte sequence n and inverts the 64-byte sequence. For this reason are used the constants:

$$\begin{aligned}
\sigma_0 &= (101, 120, 112, 97),\\
\sigma_1 &= (110, 100, 32, 51),\\
\sigma_2 &= (50, 45, 98, 121),\\
\sigma_3 &= (116, 101, 32, 107).
\end{aligned}$$

If k_0, k_1, n are 16-byte sequences, then

$$Salsa20_{k0,k1}(n) = Salsa20(\sigma_0,\ k_0, \sigma_1,\ n, \sigma_2,\ k_1, \sigma_3).$$

Let us denote

$$\begin{aligned}
\tau_0 &= (101, 120, 112, 97),\\
\tau_1 &= (110, 100, 32, 49),\\
\tau_2 &= (54, 45, 98, 121),\\
\tau_3 &= (116, 101, 32, 107).
\end{aligned}$$

If k, n are 16-byte sequences, then

$$Salsa20_k(n) = Salsa20(\tau_0,\ k,\ \tau_1,\ n,\ \tau_2,\ k).$$

For example, for the input values

$$\begin{aligned} k_0 &= (1,\ 2,\ 3,\ 4,\ 5,\ \ldots,\ 16), \\ k_1 &= (201,\ 202,\ 203,\ 204,\ 205,\ \ldots,\ 216), \\ n &= (101,\ 102,\ 103,\ 104,\ 105,\ \ldots,\ 116) \end{aligned}$$

we have:

$$\begin{aligned} Salsa20_{k0,k1}(n) = Salsa20(&101, 120, 112, 97, 1, 2, 3, 4, 5, 6, 7, 8, 9, 10, 11, 12, \\ &13, 14, 15, 16, 110, 100, 32, 51, 101, 102, 103, 104, 105, 106, 107, 108, \\ &109, 110, 111, 112, 113, 114, 115, 116, 50, 45, 98, 121, 201, 202, 203, 204, \\ &205, 206, 207, 208, 209, 210, 211, 212, 213, 214, 215, 216, 116, 101, 32, 107) \\ = (&69, 37, 68, 39, 41, 15, 107, 193, 255, 139, 122, 6, 170, 233, 217, 98, \\ &89, 144, 182, 106, 21, 51, 200, 65, 239, 49, 222, 34, 215, 114, 40, 126, \\ &104, 197, 7, 225, 197, 153, 31, 2, 102, 78, 76, 176, 84, 245, 246, 184, \\ &177, 160, 133, 130, 6, 72, 149, 119, 192, 195, 132, 236, 234, 103, 246, 74), \end{aligned}$$

and

$$\begin{aligned} Salsa20_{k0}(n) = Salsa20(&101, 120, 112, 97, 1, 2, 3, 4, 5, 6, 7, 8, 9, 10, 11, 12, \\ &13, 14, 15, 16, 110, 100, 32, 49, 101, 102, 103, 104, 105, 106, 107, 108, \\ &109, 110, 111, 112, 113, 114, 115, 116, 54, 45, 98, 121, 1, 2, 3, 4, \\ &5, 6, 7, 8, 9, 10, 11, 12, 13, 14, 15, 16, 116, 101, 32, 107) \\ = (&39, 173, 46, 248, 30, 200, 82, 17, 48, 67, 254, 239, 37, 18, 13, 247, \\ &241, 200, 61, 144, 10, 55, 50, 185, 6, 47, 246, 253, 143, 86, 187, 225, \\ &134, 85, 110, 246, 161, 163, 43, 235, 231, 94, 171, 51, 145, 214, 112, 29, \\ &14, 232, 5, 16, 151, 140, 183, 141, 171, 9, 122, 181, 104, 182, 177, 193). \end{aligned}$$

The widening belongs also to the extention (k, n) in *Salsa*20$k(n)$. It belongs also to the extention k in the long stream of outputs довгому потоку виходів *Salsa*20k for different variables.

9.1.2.8 Ciphering Function *Salsa20*

The cipher-text of the l-byte sequence m for $l \in \{0, 1, \ldots, 2^{70}\}$ will be

$$Salsa20_k(v) \oplus m$$

where ν —the unique 8-byte number (*nonce*).

The cipher-text has the size of $l$bytes, the same as the initial text.

$Salsa20k(v)$ is the key stream of $_2{}^{70}$ bytes:

$$Salsa20_k(v, \underline{0}), Salsa20_k(v, \underline{1}), Salsa20_k(v, \underline{2}), \ldots, Salsa20_k(v, \underline{2^{64} - 1}),$$

where $\underline{i}$ is the unique sequence of 8 bytes (i_0, i_1, …, i_7), such that

$$i = i_0 + 2^8 i_1 + \cdots + 2^{56} i_7.$$

Respectively

$$Salsa20_k(v) \oplus (m[0],\ m[1], \ldots,\ m[l-1]) = (c[0],\ c[1], \ldots,\ c[l-1]),$$

where

$$c\,[i] = m\,[i] \oplus Salsa20_k(v,\ \lfloor i/64 \rfloor)\,[imod64].$$

The definitions from the algorithm *Salsa20* can be easily transferred from the sequences of bytes into bit sequences, taking in account the bytes-coding as a sequence of bits.

9.1.2.9 Realization Aspects of *Salsa20*

Thanks to the fact that the conversions of each column and of each line do not depend one on another, the calculations, necessary for ciphering, are easily parallelizable. It gives a substantial advantage in operating velocity for majority of contemporary platforms.

The algorithm Salsa20 practically has not any unprofitable computing, necessary for start of ciphering cycle. In the same way this is related to the key change. Many ciphering systems are counting on antecedent calculations, the results of which must be stored in the first level cash (L1) of the processor. Inasmuch as they depend on the key, the calculations have to be executed again. But at Salsa20 it is sufficient to download the key into memory.

9.1.3 Stream Cipher SOSEMANUK

SOSEMANUK is a new synchronous software-oriented stream cipher, which corresponds to the profile 1 of the international contest eECRYPT [4]. Its key length can be selected between 128 and 256 bits. The cipher works with the 128-bit initial value, herewith, as algorithm developers affirm [4], any length of the key achieve a 128-bit protection.

The algorithm SOSEMANUK utilizes some main principles of the bitstreamed cipher Snow 2.0 [9] and some conversions, obtained from the bitstreamed cipher SERPENT [10]. The SOSEMANUK is directed onto enhancing of the Snow 2.0 both in relation to security and in relation to the realization efficiency. In particular, the SOSEMANUK utilizes a quick procedure of installing of the initialization vector IV. It requires a reduced quantity of statistical data, what gives a higher productivity on several architectures.

9.1.3.1 Specification

SERPENT and its derivatives. The SERPENT [10] is a BSC which corresponded to the conditions of the contest AES for defining of the new BSC standard of the USA, adopted in the year 2001. The algorithm SERPENT works above the blocks of 128 bits which are divided into four 32-bit words and which are jointed into so-called «bitslice» («slice» Engl. «part/particle») bit-sections. Consequently, the SERPENT can be defined as a cipher which works with the four-groups of 32-bit words. Let us number the input and the output the four-groups of the algorithm SERPENT from 0 till 3 and write them in the order: (Y_3, Y_2, Y_1, Y_0). Y_0 is the minimum significant word and contains the minimum significant bytes of 324-bit words, which log in into the S-block of the cipher *SERPENT*. While the output of the algorithm *SERPENT* is written in 16 bytes, the values Y_i are written from the junior to the senior (the first the least significant byte), Y_0 is derived as the first, further Y_1 and so on.

From the algorithm SERPENT are defined the two typicals Serpent1 and Serpent24 [10].

Serpent1 The rounds BSC *SERPENT* are compiled in such an order:

- bitwise operation *XOR*;
- utilizing of S-block (what is expressed as a set of bit-permutation between the four utilized 32-bit words in «*bitslice*» mode);
- linear bijective transformation (which is composed of several operations *XOR*, shifts and returns into the «*bitslice*» mode).

Thus, the typical *Serpent*1 is on round of the algorithm BSC *SERPENT* without adding of the key and of the linear transformation. The *SERPENT* utilizes the 8 unique S-blocks (presented below), enumerated, respectively, from *S*0 till *Sl* on 4-bit words. The *Serpent*1 is defined as utilizing of *S*2 in «*bitslice*» mode. It is the third S-block of the algorithm *SERPENT*. The *Serpent*1 receives the four 32-bit words to the input and supplies the four 32-bit words to the output.

Serpent24 The *Serpent*24 is the algorithm *SERPENT*, reduced to 24 rounds instead of 32 rounds of the full version *SERPENT*. The *Serpent*24 is equivalent to the first 24 rounds of *SERPENT*, where the last, the 24-th round, is full one and includes the full round with linear transformation and with the operation *XOR* with 25 subkeys. Put in other words, the 24-th round of *Serpent*24 is equivalent to the thirty second round of the algorithm *SERPENT*, except that it contains the linear transformation and that here are utilized the 24-th and 25-th subkeys (the 32-th and 33-th subkeys in the algorithm *SERPENT*). Consequently, the last round equation is:

$$R_{23}(X) = L\left(S_{23}(X \oplus K_{23})\right) \oplus K_{24}.$$

The typical *Serpent*24 uses only 25 128-bit subkeys, which are the first 25 subkeys of the algorithm *SERPENT*. The *Serpent*24 in *SOSEMANUK* is utilized on the initialization stage, only in ciphering mode. Deciphering is not utilized.

LFSR For the great part of the bitstreamed ciphers the internal state is stored and renewed in SRLFB. In the cipher *SOSEMANUK* is used the SRLFB which contains 10 elements of the field $GF(2^{32})$, i.e. of the finite field with 2^{32} elements.

Basic Finite Field The elements of the field $GF(232)$ are presented in the same way, as in the specification of the cipher *SNOW* 2.0. Let us present here briefly these notations.

Let be $GF(2)$ the sign of the finite field with two elements. Let be β is the root of the primitive polynomial:

$$Q(X) = X^8 + X^7 + X^5 + X^3 + 1$$

in the ring of polynomials *GF(2)[X]*. Let us determine the field $GF(2^8)$ as a ring *GF(2)[X]/Q(X)*. Each element in the field $GF(2^8)$ we can determine by utilizing the basis $\beta^7, \beta^6, \ldots, \beta,\ 1)$. In as much as the selected polynomial is primitive one, then β is a multiplicative generating element of the field $GF(2^8)$: each non-zero element from $GF(2^8)$ is equal to β^k for certain non-negative integer $(0 \leq k \leq 254)$. Every element from $GF(2^8)$ is identified with a 8-bit number by the following bijective function:

$$\phi : GF\left(2^8\right) \rightarrow \{0, 1, \ldots, 255\}, x = \sum_{i=0}^{7} x_i \beta^i \mapsto \sum_{i=0}^{7} x_i 2^i,$$

where each x_i is either 0 or 1. For instance, β^{23} is presented as an integer number $\varphi(\beta^{23}) = 0xE1$ (in hexadecimal system). Thus, the addition of two elements in $GF(2^8)$ corresponds to the bit-operation *XOR* between respective integer-valued representations. The multiplication by β is equivalent to left shift by one bit of an integer-valued representation. It is executed by the operation *XOR* with a fixed mask, if the highest order bit, which was eliminated by the shift, is equal to 1.

Let α be the root of the primitive polynomial:

$$P(X) = X^4 + \beta^{23}X^3 + \beta^{245}X^2 + \beta^{48}X + \beta^{239}$$

from $GF(2^8)[X]$. The field $GF(2^{32})$ further is defined as the ring $GF(2^8[X]/P(X)$, i. e. its elements are represented in the basis $(\alpha^3, \alpha^2, \alpha, 1)$. Every element from the $GF(2^{32})$ is identified with the $_{32}$-bit integer number by the following bijective function:

$$\psi : GF(2^{32}) \to \{0, 1, \ldots, 2^{32} - 1\}, x = \sum_{i=0}^{3} y_i \alpha^i \mapsto \sum_{i=0}^{3} \phi(y_i) 2^{8i}.$$

Consequently, the addition of two elements in $GF(2^{32})$ corresponds to the bitwise operation *XOR* between its integer-valued representations. Thereafter this operation will be denoted as $\oplus$. *SOSEMANUK* utilizes the operations of multiplication and division of elements in the $GF(2^{32})$ by α. The multiplication $z \in GF(2^{32})$ by α corresponds to the left shift by 8 bits of $\psi(z)$, which follows after the operation *XOR* with a 32-bit mask, what depends only on the most significant bytes of $\psi(z)$.

Definition of of LFSR The *LFSR* works with the elements from $GF(2^{32})$. The initial state in the time moment $t = 0$ corresponds to the ten 32-bit values of s_1–s_{10}. On each stage, the new value is calculated by the following recursion:

$$s_{t+10} = s_{t+9} \oplus \alpha^{-1} s_{t+3} \oplus \alpha s_t, \forall t \geq 1$$

and the register is shifted.

For *LFSR*, the scheme of which is presented in the Fig. 9.3, is utilized such a feedback polynomial:

$$\pi(X) = \alpha X^{10} + \alpha^{-1} X^7 + X + 1 \in GF(2^{32})[X].$$

Inasmuch as *LFSR* is nonsingular and π is a simple polynomial, so the sequence of the 32-bit words $(s_t)t \geq 1$ is periodical one and has a maximum period, which is equal to 2^{320}–1.

Finite State Machine (***finite state automaton***). The finite state machine (*FSM*) is a component with 64 bits of memory, which corresponds to the two 32-bit registers

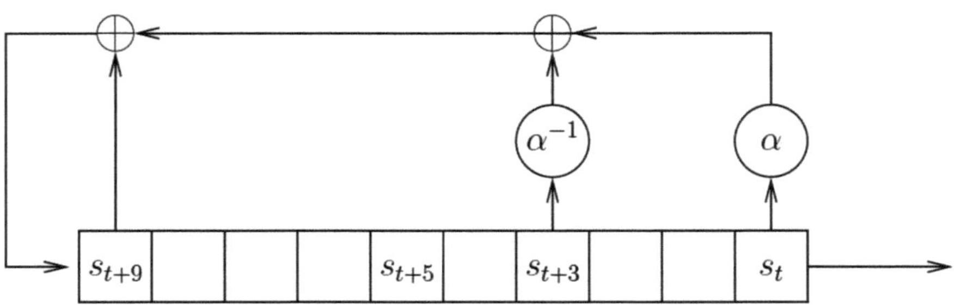

Fig. 9.3 Structural scheme of *LFSR* what is utilized in the algorithm *SOSEMANUK*

*R*1 and *R*2. In each stage the *FSM* accepts to the input several words from the state *LFSR;* it renews the memory bits and elaborates 32 bits to output. The *FSM* operates over the states of the *LFSR* at the time moment $t \geq 1$, as shown below:

$$FSM_t : (R1_{t-1},\ R2_{t-1},\ s_{t+1},\ s_{t+8},\ s_{t+9})1 \to (R1_t,\ R2_t,\ f_t),$$

where

$$\begin{aligned} R1_t &= (R2_{t-1} + mux\,(lsb\,(R1_{t-1}),\ s_{t+1},\ s_{t+1} \oplus s_{t+8}))\ mod\ 2^{32}, \\ R2_t &= Trans\ (R1_{t-1}), \\ f_t &= \left(s_{t+9} + R1_t\ mod\ 2^{32}\right) \oplus R2_t, \end{aligned}$$

where *lsb*(*x*) is the least significant bit *x*, *mux*(*c*, *x*, *y*) is equal to *x*, if $c = 0$, or *y*, if $c = 1$.

The internal transitive function Trans on GF(232) is defined as follows:

$$Trans(z) = \left(M \times z\,mod 2^{32}\right) <<< 7,$$

where *M* is the constant value 0x54655307 and <<< means a bit shift of the 32-bit value (in this case by 7 bits).

Output Conversion The outputs of *FSM* are grouped in four, and the *Serpent*1 is applied to each group. Further the result is combined by the operation *XOR* with the respective values of *LFSR*-state for obtaining at output of values *zt*:

$$(z_{t+3},\ z_{t+2},\ z_{t+1},\ z_t) = Serpent1(f_{t+3},\ f_{t+2},\ f_{t+1},\ f_t) \oplus (s_{t+3},\ s_{t+2},\ s_{t+1},\ s_t).$$

The four consequent rounds of *SOSEMANUK* are schematically drawn in the Fig. 9.4.

Working Process of SOSEMANUK The cipher *SOSEMANUK* combines in itself the *FSM* and the *LFSR* for obtaining of output values *zt*. The time moment $t = 0$ denotes the internal state after initialization; the first output value $z1$. In the Fig. 9.5 is given the graphical notation of the cipher *SOSEMANUK*.

At the time moment $t \geq 1$ the following operations are executed:

- the *FSM* is renewed: *R*1*t*, *R*2*t*, the intermediary value *ft* is calculated from $R1_{t-1},\ R2t - 1,\ s_{t+1}, s_{t+8}$ and s_{t+9};
- the *LFSR* is renewed: s_{t+10} is calculated on account of $s_t,\ s_{t+3}$ and s_{t+9}. The value *st* is sent into the internal buffer, and the *LFSR* is shifted.

Once for four iterations the output values $z_t,\ z_{t+1},\ z_{t+2},\ z_{t+3}$ arise from the accumulated values $f_t,\ f_{t+1},\ f_{t+2},\ f_{t+3}$ and $s_t,\ s_{t+1},\ s_{t+2},\ s_{t+3}$. Consequently, the algorithm *SOSEMANUK* elaborates $_{32}$-bit values. It is recommended to encipher them in groups by four bytes, using the order of bytes from the junior to the senior, because

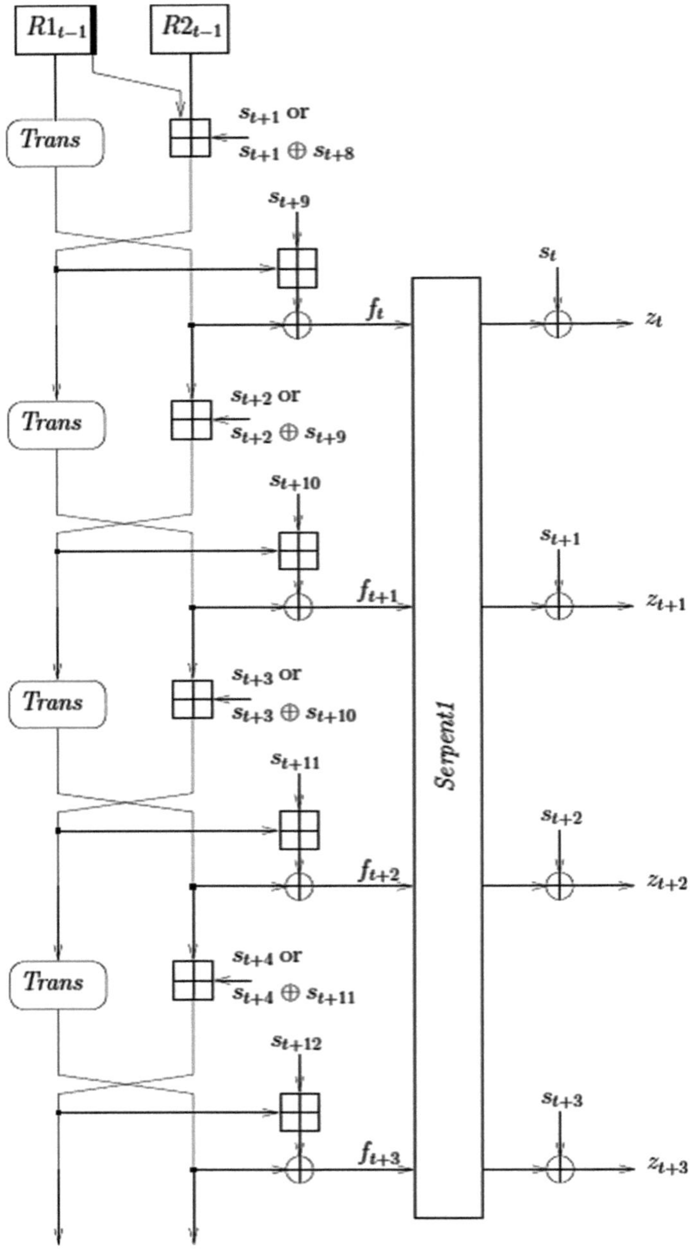

Fig. 9.4 Scheme of the output conversion with four consequent rounds of *SOSEMANUK*

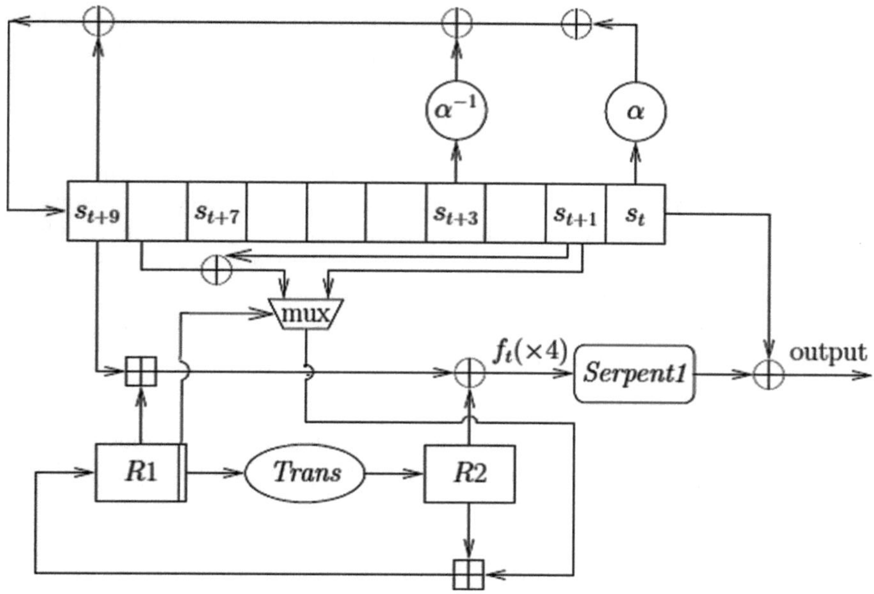

Fig. 9.5 Scheme of the working process of the cipher *SOSEMANUK*

the last method has a greater velocity on widely used software platforms of high class (x86-compatible personal computers). In addition, this method utilizes the algorithm *SERPENT*.

Consequently, the first four iterations of *SOSEMANUK* are the following:

- the initial state of *LFSR* contains the values s_1–s_{10}; s_0 contains no value. The initial state of *FSM* contains $R1_0$ and $R2_0$;
- during the first stage the $R1_1$, $R2_1$ and $f1$ are calculated arising from $R1_0$, $R2_0$, s_2, s_9 and s_{10};
- from the first stage the buffered intermediary values of $s1$ and f_1 are obtained;
- during the first stage the inverse word s_{11} is calculated from s_{10}, s_4 and s_1, then the internal state of *LFSR* is renewed, what leads to a new state, which consists of s_2–s_{11};
- the first four output values are z_1, z_2, z_3 and z_4, and are calculated by using the *Serpent*1 above (f_4, f_3, f_2, f_1), where the output is combined by the operations *XOR* with (s_4, s_3, s_2, s_1).

Initialization Key of the Algorithm and Injection IV The initialization process of the algorithm *SOS-EMANUK* is divided in two stages:

- generation of keys, which processes the secret key, but is independent on *IV*;
- injection of *IV*, which utilizes the result of the precedent operation and *IV*. It initiates the internal state of the stream cipher.

Generation of Keys The generation of keys corresponds to the generation of keys *Serpent*24, which elaborates the 25 128-bit subkeys, as well the 100 32-bit words. These 25 128-bit subkeys are identical to the first 25 128-bit subkeys, which elaborates the generation of keys of the algorithm *SERPENT*.

The algorithm *SERPENT* accepts every length of the key from 1 till 256 bits, consequently, the *SOSEMANUK* can work with exactly the same keys. However, inasmuch as the *SOSEMANUK* is directed to the 128-bit ciphering, its lengths of keys can be at least of 128 bits. Thus, 128 bits are standard length of a key. Supported is every length of the key from 128 to 256 bits. But the security level, as before, is the 128-bit one. Alternatively saying, the using of a longer secret key does not guarantee the same protection level as the key length.

Injection of IV The initialization vector *IV* is a 128-bit value. It is utilized as input data of the block cipher *Serpent*24, which is initiated by the previous operation (generation of keys). The algorithm *Serpent*24 consists of 24 rounds. Let us denote these outputs in such a way:

- $\left(Y_3^{12}, Y_2^{12}, Y_1^{12}, Y_0^{12}\right)$: output of the 12-th round;
- $\left(Y_3^{18}, Y_2^{18}, Y_1^{18}, Y_0^{18}\right)$: output of the 18-th round;
- $\left(Y_3^{24}, Y_2^{24}, Y_1^{24}, Y_0^{24}\right)$: output of the 24-th round.

The output of each round consists of four 32-bit words immediately after the linear conversion, except the 24-th round, for which the output is valid only after addition with the twenty fifth subkey.

These values are utilized for initialization of the internal state *SOSEMANUK* with following values:

$$(s_7, s_8, s_9, s_{10}) = \left(Y_3^{12}, Y_2^{12}, Y_1^{12}, Y_0^{12}\right);$$

$$(s_5, s_6) = \left(Y_1^{18}, Y_3^{18}\right);$$

$$(s_1, s_2, s_3, s_4) = \left(Y_3^{24}, Y_2^{24}, Y_1^{24}, Y_0^{24}\right);$$

$$R1_0 = Y_0^{18};$$

$$R2_0 = Y_2^{18}.$$

9.1.3.2 Substantiation of Construction

Initialization Key and Injection of IV The first improved property of the initialization process lies in the fact that it is divided into two separate stages: generation of keys, which does not depend on *IV*, and injection of *IV*, which generates the initial state of generator from the initialization vector *IV* and from the output of the scheme of generating of keys. Then, the adjusting of *IV* for a fixed key is less expensive than a full key installation. It improves the construction, because the change of the initialization vector *IV* occurs oftener than the change of the secret key.

Another improved characteristic is the fact that in the algorithm *SOSEMANUK* the process of installation of the initialization vector *IV* is obtained by using of an block cipher above *IV*. If we shall examine the function, which reflects F_K n-bit initialization vector *IV* in the first n bits of the output stream that is generated from the key K and *IV*, then F_K has to be calculatingly indistinguishable from a random function. Consequently, the calculation of F_K can not be «morally» faster than the most known PRS-generator above n-bit blocks. It happens that the fastest PRS-generator utilizes the same realization methods as the known fast pseudorandom permutations, which are the block ciphers, and with an equivalent productivity.

9.2 Analysis of Hardware-Oriented Stream Ciphers, Recognised as Winners of the International Project eSTREAM

9.2.1 Stream Cipher Grain

The new stream cipher *Grain* was proposed in [5]. The design of cipher is aimed into hardware media with a very limited number of gates (*gate*), restricted power consumption and memory, i.e. this cipher is hardware-oriented and was nominated to the international contest *eSTREAM* after another profile.

The ciphering algorithm *Grain* is based on the two shift registers and a non-linear output function. The cipher has an additional function, whereupon the operating velocity can be increased at the cost of additional equipment. The key size makes up 80 bits and, as was defined, for today is known no faster attack than a full key exhaustion. The complexity of hardware realization and operating velocity compares favourable this algorithm from other hardware-oriented stream ciphers, for instance such as *E*o and *A*5/1.

9.2.1.1 Substantiation of Construction

While designing of the cryptographic typical there are a lot of different properties, which must be resolved. They incorporate, for example, velocity, security and realization simplicity. Comparing several ciphers, it is quite probable that the result can be obtained, at which one algorithm is faster on 32-bit processors, another is faster on 8-bit processors, and one more is faster on hardware facilities. The simplicity of construction is one more factor, which is necessarily to be taken in consideration, but moreover that the software realization can be very simple, the hardware realization can be quite sophisticated. The necessity of cryptographic typicals exists, which have a very low complicity of hardware features. In particular, the *RFID*-mark is a typical example of product, where the memory capacity and the facilities are very limited. These are the microchips, capable to transmit the identifying sequence on the request from the reader unit. The renewal of the *RFID*-tag can have devastating consequences, if the tag is used, for example, in electronic payments and, consequently, there is a necessity in cryptographic typicals realized in these marks. Today, for example, the hardware realization *AES* on the radio-frequency identification tag appears not possible due to the great number of necessary gates (*gates*). The algorithm *Grain* is a primitive stream cipher, destined for realization on very simple and limited hardware facilities.

Many stream ciphers based on *SRLFB* ensure not only good statistical properties of PRS, but also the simplicity and rapidity of their hardware realization. There are several proposals of stream ciphers based on *SRLFB*, for instance, in [11, 12]. The utilizing of the word-oriented *SRLFB* enables to obtain a more efficient realization in software, but it enables also to enlarge the through-pass capacity, because during one time the blocks of bits are put out, but not separate bits. In the hardware realization the word-oriented cipher will be, apparently, more sophisticated than the bit-oriented one. The developers of the cipher *Grain* researched this problem. The user can decide how will be the cipher speed depending on the quantity of hardware.

The algorithm *Grain* is a synchronous stream cipher, in which the key stream is generated independently on the text. The construction is based on the two shift-registers: one with a linear feedback (*FLSR*) and one shift-register with a non-linear feedback (*NFLSR*). The usage of the *NFLSR* guarantees a minimum period for the keystream, and also ensures the balanced state on the output. The usage of *NFLSR* together with the non-linear output function introduces the non-linearity to the cipher. The *NFLSR*-input is masked from the *FLSR* S-output in such a way, that the *NFLSR*-state becomes balanced. Consequently, in the algorithm *Grain* is used the notation *NFLSR*, although in fact this is a filter. That, what is known about cyclic structures of the shift registers with non-linear feedback, can not be applied immediately for the cipher analysis. Both shift registers have a size of 80 bits. The key size makes up 80 bits, and the size of the initialization vector *IV* makes up 64 bits. The cipher is developed in such a way that no attack should not be faster than an exhaustive key search; consequently, it is better that the attack should require a calculating complexity substantially lower than 2^{80}.

9.2.1.2 Specification of Construction

Let us examine the specification of the cipher *Grain*. The structural scheme of the cryptoconversion is schematically drawn in the Fig. 9.6.

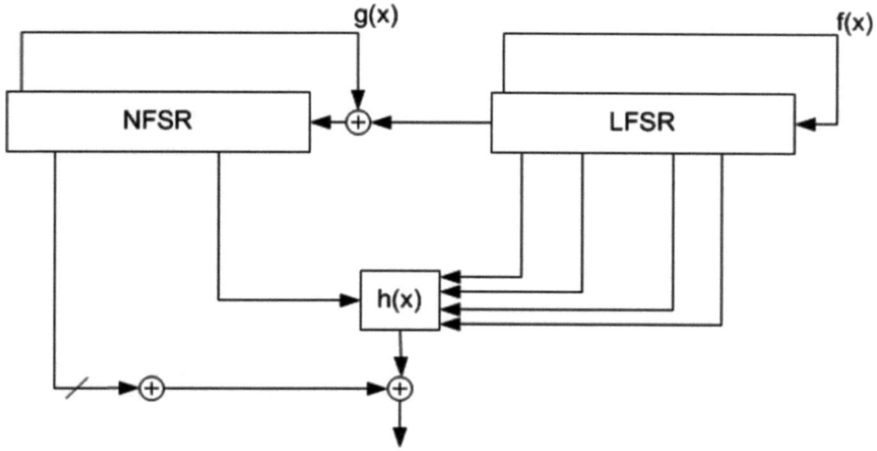

Fig. 9.6 Structural scheme of the stream cipher *Grain*

The cipher consists from three main blocks, and namely: *FLSR*, *NFLSR* and an output function. The content of the *FLSR* is denoted as

$$s_i, s_{i+1}, \ldots, s_{i+79},$$

the content of *NFLSR* is denoted as

$$b_i, b_{i+1}, \ldots, b_{i+79}.$$

The feedback polynomial of *FLSR* $f(x)$ is a primitive polynomial of the degree 80. It is written in the following way:

$$f(x) = 1 + x^{18} + x^{29} + x^{42} + x^{57} + x^{67} + x^{80}.$$

In order to exclude any possible ambiguity in the specification of algorithm the ciphering function of renewal of *FLSR* is defined as follows:

$$s_{i+80} = s_{i+62} + s_{i+51} + s_{i+38} + s_{i+23} + s_{i+13} + s_i.$$

The polynomial of feedback *NFLSR* $g(x)$ is defined as

$$\begin{aligned} g(x) = 1 &+ x^{18} + x^{20} + x^{28} + x^{35} + x^{43} + x^{47} + x^{52} + x^{59} + x^{66} + x^{71} + x^{80} \\ &+ x^{17}x^{20} + x^{43}x^{47} + x^{65}x^{71} + x^{20}x^{28}x^{35} + x^{47}x^{52}x^{59} + x^{17}x^{35}x^{52}x^{71} \\ &+ x^{20}x^{28}x^{43}x^{47} + x^{17}x^{20}x^{59}x^{65} + x^{17}x^{20}x^{28}x^{35}x^{43} + x^{47}x^{52}x^{59}x^{65}x^{71} \\ &+ x^{28}x^{35}x^{43}x^{47}x^{52}x^{59}. \end{aligned}$$

In order to exclude any ambiguity, the renewal function of *NFLSR* we shall write as follows:

$$\begin{aligned} b_{i+80} = {} & s_i + b_{i+62} + b_{i+60} + b_{i+52} + b_{i+45} + b_{i+37} + b_{i+33} + b_{i+28} + b_{i+21} \\ & + b_{i+14} + b_{i+9} + b_i + b_{i+63}b_{i+60} + b_{i+37}b_{i+33} + b_{i+15}b_{i+9} \\ & + b_{i+60}b_{i+52}b_{i+45} + b_{i+33}b_{i+28}b_{i+21} + b_{i+63}b_{i+45}b_{i+28}b_{i+9} \\ & + b_{i+60}b_{i+52}b_{i+37}b_{i+33} + b_{i+63}b_{i+60}b_{i+21}b_{i+15} \\ & + b_{i+63}b_{i+60}b_{i+52}b_{i+45}b_{i+37} + b_{i+33}b_{i+28}b_{i+21}b_{i+15}b_{i+9} \\ & + b_{i+52}b_{i+45}b_{i+37}b_{i+33}b_{i+28}b_{i+21}. \end{aligned}$$

It is noteworthy that the bit s_i is masked with the input of the renewal function.

The content of the two shift registers *FLSR* and *NFLSR* are the cipher state. The five variables from this state are taken as the input of the Boolean function, $h(x)$. This function is a non-linear filter, it is selected in such a way in order to be balanced one, to ensure a correlative immunity of the first order and to have the algebraic degree 3. The maximum possible non-linearity of such a function can be 12. The input is received both from *FLSR* and *NFLSR*. The Boolean function is defines as

$$\begin{aligned} h(x) == {} & x_1 + x_4 + x_0x_3 + x_2x_3 + x_3x_4 + x_0x_1x_2 \\ & + x_0x_2x_3 + x_0x_2x_4 + x_1x_2x_4 + x_2x_3x_4, \end{aligned}$$

where the variables x_0, x_1, x_2, x_3 and x_4 correspond to the positions $s_{i+3}, s_{i+25}, s_{i+46}, s_{i+64}$ and b_{i+63} respectively.

The output function is written as follows:

$$z_i = \sum_{k \in A} b_i + k + h(s_{i+3}, s_{i+25}, s_{i+46}, s_{i+64}, b_{i+63}),$$

where $A = \{1, 2, 4, 10, 31, 43, 56\}$.

Initialization of Key Before the generation of key stream the cipher has to be initiated by the key and by the initialization vector. Let be the key bits k denoted as *ki*, herewith $0 \leq i \leq 79$, and the bits of initialization vector will be denoted as *IVi*, $0 \leq i \leq 63$. The initialization of key happens in following way.

First, the *NFLSR* is downloaded by the key bits:

$$b_i = k_i, \ 0 \leq i \leq 79,$$

further the first 64 *FLSR*-bits are downloaded by the key bits:

$$s_i = IV_i, \ 0 \leq i \leq 63.$$

The remaining *FLSR*-bits are filled in by figures one:

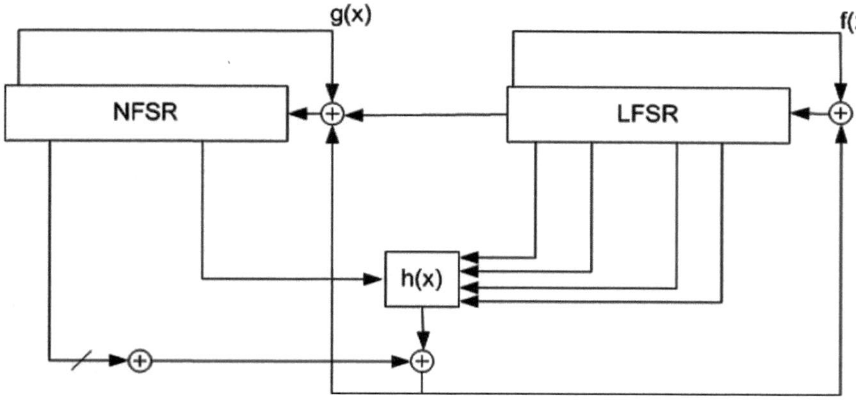

Fig. 9.7 Scheme of key initialization of the cipher *Grain*

$$s_i = 1,\ 64 \le i \le 79.$$

Due to this, the *FLSR* can not be initiated by the zero-state (zero-values only). Further the cipher operates 160 times without outputting of the key stream. Instead of this the output function is supplied backwards, and the operation *XOR* is executed with the output both for the *FLSR*, and for the *NFLSR*. Schematically it is given on the Fig. 9.7.

9.2.1.3 Criteria of Cipher Developing

The cipher construction is directed to the more simple as possible hardware realization. The security requirements correspond to the calculating complexity 2^{80}, what is equivalently to the exhaustive key search. For satisfying of this requirement it is necessary to build a cipher with a memory of 160 bits. The realization of the 160-bit-memory on the hardware medium is considered below, additionally an estimation of such a realization complexity is given. For utilizing of little hardware resources is necessary a concentrating on minimizing of functions, which are utilized in common with this memory. The functions utilized while developing of the cipher must be small in order to fulfil the restrictions concerning gates number (*gate*), but still large enough to ensure a high security level.

It is known that the LFSR with a primitive polynomial of the feedback of the degree d gives out an output signal with the period $2^d - 1$. The LFSR in the cipher *Grain* has a size of 80 bits, and the feedback is preset by a primitive polynomial, i. e. there is guaranteed that the PRS-period with the probability of з імовірністю $1–2^{-80}$ will be equal at least to $2^{80} - 1$. Because of utilizing of the NLFSR and of that fact that its input is masked with the output of the LFSR, the exact period will depend on the key and on the initialization vector. The NLFSR-input is masked with the LFSR-output in order to be convinced that the state of LFSR is balanced one.

The non-linear feedback has a doubled stability, because the terms $x80$, $x18$ and $x66$ are set only by a linear law.

The filter function is quiet little, it possesses only 5 variables and the non-linearity 12. However it is partially compensated by the fact that one of the inputs of this function is taken from the NLFSR. The input bit from the NLFSR will linearly depend on other bits in the state both on LFSR and on NLFSR. The little filter function is compensated as well by the addition of 7 bits after the linear law from NLFSR in respective positions with creating of the output function.

In the phase of the key initialization, the aim lies in separation of the content of the shift registers before beginning the generation of the key stream. The quantity of shifts is substantiated by the compromise between security and operating velocity. If the cipher must be initiated repeatedly, and this event happens very frequently with the renewal of the initialization vector, then the cipher efficiency will be reduced, because in this case the initialization will be a «bottle-neck». After initialization the LFSR contains an initialization vector and 16 units. For initialization with two different initialization vectors, which distinguish only by one bit, the probability of that fact that the register motion gives an equal bit for both initializations has to be near to 0.5. The modelling shows that it is achieved after 160 shifts. Further this fact will be disputed in more detail.

Outputting Velocity The both shift registers are shifted regularly, i.e. one shifting will output to forming with the cipher of one bit. The cipher velocity can be increased on the cost of more powerful hardware. It can be made through the increasing of the quantity of the realizations of feedback functions $f(x)$ and $g(x)$. Because of this, the outputting velocity will be increased by several times. In order to simplify this realization, the last 15 bits of shift registers s_i, $65 \leq i \leq 79$ and b_i, $65 \leq i \leq 79$, are not utilized in the feedback functions or on the input into the filter function. It enables to multiply easily the velocity up to 16 times, sure, if there is accessible a sufficient amount of hardware resources. An example of cipher realization when the velocity is doubled can be seen in the Fig. 9.8.

Naturally, the shift registers also must be realized in such a way, that each bit should be shifted by t steps instead of one, when the velocity is increased by a coefficient t. In the case of increasing of velocity by the coefficient 16, the cipher puts out 16 bits during one shifting. Inasmuch in the key initialization of the cipher 160 shiftings are realized, the possibilities for increasing of velocity are limited by the index $t \leq 16$, herewith t must be a divisor of 160. The number of shiftings, which are utilized for the key initialization, will be equal to $160/t$. Inasmuch the filter and feedback functions are quite small, it may well be to increase the velocity in such a way.

9.2.1.4 Strengths and Weaknesses of the Cipher *Grain*

The design of the cipher is concentrated on some specific properties. It is impossible to have a design which should ideally fit to all purposes, e. g., for processors of all word

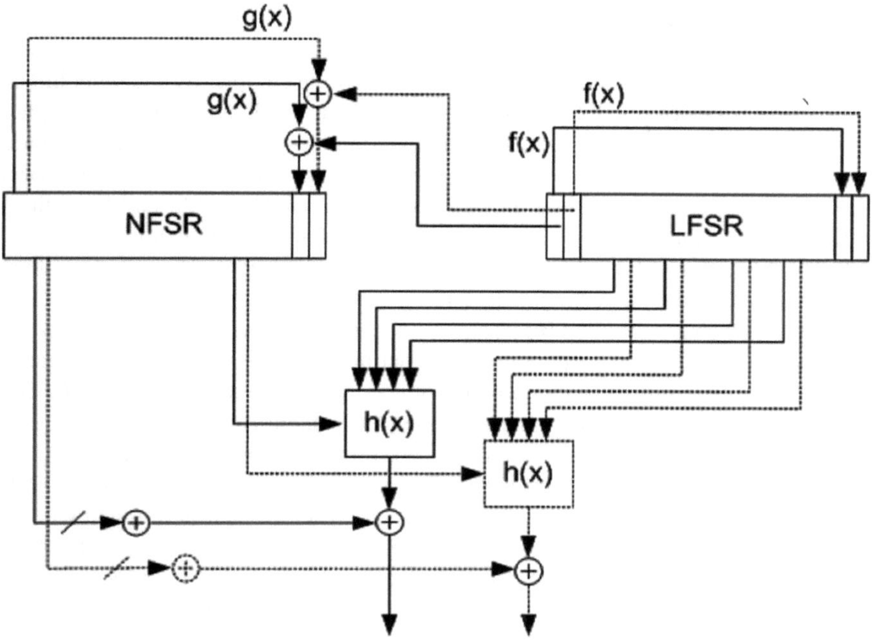

Fig. 9.8 Structural scheme of the stream cipher *Grain* with the doubled velocity

lengths, for all hardware appliances, for memory restrictions etc. The cipher *Grain* is developed for hardware-oriented realization with utilizing of minimum possible amount of gates (*gate*), as far as possible, by saving of high security requirements.

The cipher is destined for using in the media, where the amounts of gates, of energy consumption and of memory are very small, i. e. the algorithm is oriented on the hardware realization with limited resources. The cipher *Grain* can be used in a common applied software, however there are several ciphers developed namely software-oriented realization. For this reason there is no sense to compare the software productivity for the cipher *Grain* and that of the software-oriented ciphers.

The base realization has the velocity of 1 bit/time (shifting). The velocity of the word-oriented ciphers is usually higher, in particular it is equal to 1 word/time (shifting). The cipher *Grain* is oriented to the realization on little hardware means after complexity, thus its disadvantages were recovered by the possibility to increase the velocity at the cost of more powerful hardware. It allows to supplier to select, how fast the cipher has to form a PRS and what amount of technical means will be necessary for it to him.

9.2.1.5 Equipments Complexity

In order to obtain some practical instructions concerning complexity and other important particularities of possible equipment for hardware realization of stream cipher, the algorithm developers performed the realization based on the standard architecture of the programmed by the user gate matrix (Engl. *Field-Programmable Gate Array FPGA*). In addition to the normal working mode (Fig. 9.6) and key initialization (Fig. 9.7), the developers introduced the third mode downloading of key bits in the NLFSR and of bits of initialization vector in the LFSR. This whole scheme was described by the language for describing of the integrated circuit devices (*Very high speed integrated circuits*) *Hardware Description Language VHDL*) (nearby 300 lines code), depending on the parameter t, as defined above. For algorithm realization was selected the family of programmable logical integrated circuits (PLIC) *ALTERA MAX 3000A*. With the using of the designing instrument *Altera Quartus* a logical synthesis was performed, after which was determined, that at $t \leq 4$ the realization is possible on the microcircuit *EPM*3256, what produces an utilization of nearby 90% from existing 256 microelements. The maximum synchronization frequency lies in the band 35–50 MHz, dependently on the working mode and on the output interface. The number of output bits per second is t after clock frequency. In order to make a correct comparison between different ciphers, the realizations bust be tested on the same PLIC. For highlighting the productivity difference between different PLICs, the realization on two additional PLIC families was simulated, and namely *ALTERA MAX II* and *ALTERA Cyclone*. These two cipher realizations are directed to the working with an enhanced clock frequency and a higher velocity of PRS-forming: the velocity was increased by 16 times, i. e. $t = 16$. It is noteworthy that there are also other PLIC-producers (e. g. *Actel, Xilinx*), which can propose the devices that will satisfy all requirements and at lower prices. Some products, including security mechanisms, deny the inverse designing of a programmed microcircuit. The gates number for functions changes depending on complexity and functionality. In various realizations the gates number was calculated for $t = 1, 2, 4, 8$ and 16. The Table 9.5 shows the gates number and the respective velocity of the PRS-forming for three different PLIC.

Table 9.5 Gates number and velocity of PRS-forming by the cipher Grain

T	Number of gates	Velocity of PRS-forming		
		MAX 3000A (Mbit/s)	MAX II (Mbit/s)	Cyclone (Mbit/s)
1	1450	49	200	282
2	1637	98.4	422	576
4	2010	196	632	872
8	2756	–	1184	1736
16	4248	–	2128	3136

9.2.1.6 Cryptoanalysis

Let us analyse some common attacks against stream ciphers and let us research, how far they can be applied to the cipher *Grain*. The stability against all known cryptoanalytical attacks is the most important property of a cipher. The complexity of no of the attacks may be less than an exhaustive search of keys. The initial attempts of cryptoanalytical attacks against the cipher *Grain* demonstrate the following.

Correlational Analysis Because of statistical properties of the LFSR of maximum length, the bits in the registers will be (almost) exactly balanced. The non-balanced state can be observed in case of NLFSR, when it is started autonomously. However, in the feedback for the function $g(x)$ is applied the operation *XOR* with the LFSR-state, for this reason the bits in the NLFSR will be balanced. In addition to that, $g(x)$ is a balanced function. Consequently, it is safe to assume that the bits in NLFSR will be non-correlated with the LFSR-bits.

The function $h(x)$ is selected under condition of the correlative immunity of the first order. This does not exclude the fact that there exists a correlation of the output $h(x)$ with the sum of inputs. But the correlations of the generator output with the sums of SRLFB-bits will be quite little, to prevent any attack against the cipher *Grain* [5].

Algebraic Attacks The filter-generator with the output function $h(x)$ of the degree three could be sensitive to algebraic attacks. The algebraic attacks will not work for determination of the initial 160-bit state of the full generator. Inasmuch as the NLFSR-renewal is non-linear, and the next bits of the NLFSR-state depend as a function from the initial bits of state. This function will have a various, but very large algebraic degree. By utilizing the initialization key, the generator output can be expressed depending on the state of bits only in LFSR. The filter function $h(x)$ has one input, which comes in from the NLFSR, and $h(x)$ *XOR*-ing with the linear combination of bits of the NLFSR-state. Respectively, the algebraic degree of the input bits expressed as a function from the bits of the LFSR-state will be large as a whole and various timely. It is a defeat of algebraic attacks.

Attack on the Base of a Compromise Time-Memory-Data The complexity of the attacks based on the compromise time-memory-data against the stream ciphers sets up $O\ (2^{n/2})$, where n is the number of internal states of a stream cipher. Applying this estimation, $n = 160$ was selected. Thus, the stability is finally quite large, and the attacks based on the compromise time-memory-data must have a complexity not lower than $O\ (2^{80})$.

Attack on the Initialization Vector A necessary condition for stability against differential or statistical attacks on the initialization vector is the fact that the initial states for the every two selected *IV* should be algebraically and statistically unrelated. The quantity of the key initializations cycles was selected so that the Heming weight for the two differences of full initial $_{160}$-bit states after initialization should be near to the random one. It may prevent attacks against the initialization vector.

9.2.2 *Stream Cipher* MICKEY

In the work [6] there is presented an upgraded version 2.0 of the stream cipher *MICKEY* (deciphered as *Mutual Irregular Clocking KEYstream generator* key stream generator with the mutual nonuniform motion), which is destined for hardware platforms with limited resources. Consequently, the stream cipher *MICKEY* was developed with the profile 2, i. e. as a hardware-oriented cipher.

The ciphering algorithm *MICKEY* has a simple hardware realization and herewith ensures a high security level. In it is used an irregular motion of the shift registers, and also new methods, which ensure a quite large PRS-period and a stability against some cryptoanalitical attacks.

9.2.2.1 Input and Output Parameters

The stream cipher *MICKEY* 2.0 accepts the two input parameters:

- the key K of the length 80 bits, its bits are denoted k_0, k_1 ..., k_{79};
- the initialization vector IV of the length from 0 till 80 bits, its bits are denoted $iv_0, \ldots, iv_{LENGTH\ IV-1}$.

The consequence of the key stream bits, that form the cipher *MICKEY* 2.0, are denoted z_0, z_1. The cipher text is elaborated from the uncodified (open) text by the aid of a bitwise operation *XOR* with the key bitstream, as in the majority of stream ciphers.

9.2.2.2 Limitations for Applying

The maximum length of the key stream, obtained by the aid of one pair (K, IV) sets up 2^{40} bits. It is acceptable for generating of similar sequences with one key K, but with various values IV. It is inadmissible to use two initialization vectors of different length with one and the same key K. Certainly, it is not acceptable to use repeatedly one and the same value IV with a fixed K.

9.2.2.3 Component Parts of a Generator of Sequences of Keys

Registers The generator consists of two registers R and S. The length of each register is equal to 100 digits, each digit contains 1 bit.

The bits of these registers are denoted r_0, r_1, ..., r_{99} and s_0, s_1, ..., s_{99} respectively. Generally, in the algorithm specification is indicated that R is a «linear register», but S «non-linear register».

Register Shift R The inputs set of feedback of the register R.

$$\begin{aligned} RTAPS = \{&0, 1, 3, 4, 5, 6, 9, 12, 13, 16, 19, 20, \\ &21, 22, 25, 28, 37, 38, 41, 42, \\ &45, 46, 50, 52, 54, 56, 58, 60, 61, 63, \\ &64, 65, 66, 67, 71, 72, 79, 80, 81, 82, 87, 88, \\ &89, 90, 91, 92, 94, 95, 96, 97\} \end{aligned}$$

The register shift operation

$$CLOCK\_R(R, INPUT\_BIT\_R, CONTROL\_BIT\_R)$$

is determined by such a sequence of actions:

- let be r_0, r_1, ..., r_{99} state of register R before shifting, and $r'_0, r'_1, \ldots, r'_{99}$ after shifting;
- *FEEDBACK_BIT* $= r_{99} \oplus$ *INPUT_BIT_R*;
- for all $1 \leq i \leq 99$, $r\prime i = r\prime i - 1$, $r'_0 = 0$;
- for all $0 \leq i \leq 99$, if $i \in$ *RTAPS*, $r\prime i = r\prime i$ *FEEDBACK_BIT*;
- if *CONTROL_BIT_R* $= 1$, then for all $0 \leq i \leq 99$, $r\prime i = r\prime i \ \oplus ri$.

Shifting of register S. Let us define the four sequences

$$\begin{aligned} &COMP0_1 \ldots COMP0_{98}, COMP1_1 \ldots COMP1_{98}, \\ &FB0_0 \ldots FB0_{99}, FB1_0 \ldots FB1_{99} \end{aligned}$$

as it is presented in the Table 9.6.
Function of register shifting

$$CLOCK\_S(S, INPUT\_BIT\_S, CONTROL\_BIT\_S)$$

is defined by such a sequence of actions:

- let be $s0$, $s1$, ..., $s99$ state of the register S before shifting, $s\prime 0, s\prime 1, \ldots, s\prime 99$ after shifting. For simplicity we shall denote the intermediary register state as $s_0^\wedge, s_1^\wedge, \ldots, s_{99}^\wedge$;
- *FEEDBACK_BIT* $= s99 \oplus$*INPUT_BIT_S;*
- for all $1 \leq i \leq 98$, $\hat{s}_i = s_{i-1} \oplus ((s_i \oplus COMP0_i) \bullet (s_{i+1} \oplus COMP1_i))$; $\hat{s}_0 = 0$; $\hat{s}_{99} = s_{98}$;
- if *CONTROL_BIT_S* $= 0$, then for all $0 \leq i \leq 99$, $s'_i = \hat{s}_i \oplus (FB0_i \bullet FEEDBACK\_BIT)$;
- if *CONTROL_BIT_S* $= 1$, then for $0 \leq i \leq 99$, $s'_i = \hat{s}_i \oplus (FB1_i \bullet FEEDBACK\_BIT)$.

The motion control of the whole generator. Let us define the function
CLOCK_KG (R, S, MIXING, INPUT_BIT)
consequently:

Table 9.6 Defining of register sequences

I	0	1	2	3	4	5	6	7	8	9	10	11	12	13	14	15	16	17	18	19	20	21	22	23	24
$COMP0_i$		0	0	0	1	1	0	0	0	1	0	1	1	1	1	0	1	0	0	1	0	1	0	1	0
$COMP1_i$		1	0	1	1	0	0	1	0	1	1	1	1	0	0	1	0	1	0	0	0	1	1	0	1
$FB0_i$	1	1	1	1	0	1	0	1	1	1	1	1	1	1	1	0	0	1	0	1	1	1	1	1	1
$FB1_i$	1	1	1	0	1	1	1	0	0	0	0	1	1	1	0	1	0	0	1	1	0	0	0	1	0
i	25	26	27	28	29	30	31	32	33	34	35	36	37	38	39	40	41	42	43	44	45	46	47	48	49
$COMP0_i$	1	0	1	0	1	1	0	1	0	0	1	0	0	0	0	0	0	0	1	0	1	0	1	0	1
$COMP1_i$	0	1	1	1	0	1	1	1	1	0	0	0	1	1	0	1	0	1	1	1	0	0	0	0	1
$FB0_i$	1	1	1	1	0	0	1	1	0	0	0	0	0	0	1	1	1	0	0	1	0	0	1	0	1
$FB1_i$	0	1	1	0	0	1	0	1	1	0	0	0	1	1	0	0	0	0	0	1	1	0	1	1	0
i	50	51	52	53	54	55	56	57	58	59	60	61	62	63	64	65	66	67	68	69	70	71	72	73	74
$COMP0_i$	0	0	0	0	1	0	1	0	0	1	1	1	1	0	0	1	0	1	0	1	1	1	1	1	1
$COMP1_i$	0	0	0	1	0	1	1	1	0	0	0	1	1	1	1	1	1	0	1	0	1	1	1	0	1
$FB0_i$	0	1	0	0	1	0	1	1	1	1	0	1	0	1	0	1	0	0	0	0	0	0	0	0	0
$FB1_i$	0	0	1	0	0	0	1	0	0	1	0	0	1	0	1	1	0	1	0	1	0	0	1	0	1
i	75	76	77	78	79	80	81	82	83	84	85	86	87	88	89	90	91	92	93	94	95	96	97	98	99
$COMP0_i$	1	1	1	0	1	0	1	1	1	1	1	1	0	1	0	1	0	0	0	0	0	0	1	1	
$COMP1_i$	1	1	1	0	0	0	1	0	0	0	0	1	1	1	0	0	0	1	0	0	1	1	0	0	
$FB0_i$	1	1	0	1	0	0	0	1	1	0	1	1	1	0	0	1	1	1	0	0	1	1	0	0	0
$FB1_i$	0	0	0	1	1	1	1	0	1	1	1	1	1	0	0	0	0	0	0	1	0	0	0	0	1

- $CONTROL\_BIT\_R = s_{34} \oplus r_{67}$;
- $CONTROL\_BIT\_S = s_{67} \oplus r_{33}$;
- if *MIXING* = *TRUE*, then $INPUT\_BIT\_R = INPUT\_BIT \oplus s_{50}$, but if *MIXING* = *FALSE*, then *INPUT_BIT_R* = *INPUT_BIT*;
- *INPUT_BIT_S* = *INPUT_BIT;*
- *CLOCK_R (R, INPUT_BIT_R, CONTROL_BIT_R);*
- *CLOCK_S (S, INPUT_BIT_S, CONTROL_BIT_S).*

9.2.2.4 Downloading of Key and Initialization

The initialization of the registers after input parameters is performed in such a way:

- filling out of registers *R* and *S* with zeros;
- (downloading into *IV*). For all $0 \leq i \leq$ *IV*LENGTH – 1, *CLOCK_KG* (*R, S, MIXING* = *TRUE, INPUT_BIT* = *ivi*);
- (downloading into *K*). For all $0 \leq i \leq 79$, *CLOCK_KG* (*R, S, MIXING* = *TRUE, INPUT_BIT* = *ki*);
- (before shifting). For all $0 \leq i \leq 99$, *CLOCK_KG* (*R, S, MIXING* = *TRUE, INPUT_BIT* = 0).

9.2.2.5 Generating of the Key Stream

After downloading and initialization the key stream *z*0, …, *zL* − 1 can be generated in following way:

- for all $0 \leq i \leq L-1$,
- $zi = r0 \oplus s0$;
- *CLOCK_KG (R, S, MIXING* = *FALSE, INPUT_BIT* = *0).*

9.2.2.6 Principles of Designing

The variable of Motion for the Register R (Destination) When *CONTROL_BIT_R* = 0, *R* presents an usual shift register with linear feedback (with a feedback of the type Galois with a primitive characteristic polynomial $\boldsymbol{C_R(x)} = \boldsymbol{x}^{100} + \sum_{\boldsymbol{i \in RTAPS}} \boldsymbol{x}^{\boldsymbol{i}}$, hereat *INPUT_BIT_R* are added to the feedback by the operation *XOR*). It is presented illustratively in the Fig. 9.9.

If the elements of the field $GF(2^{100})$ should be presented as polynomials after modulus $C_R(x)$, then the shifting of the register corresponds to the multiplication by *x* in this field (Fig. 9.9).

When *CONTROL_BIT_R* = 1, then, besides of right shifting of each bit, the operation *XOR* is applied with the preliminary bit value, as shown in the Fig. 9.10. It corresponds to the multiplication by $x + 1$ in the same field.

The characteristic polynomial $C_R(x)$ is selected in such a way that it is a divisor of the polynomial $x^J + x + 1$ (that is $C\,(x\,)|x^J + x + 1$), where $J = 2^{50} - 157$.

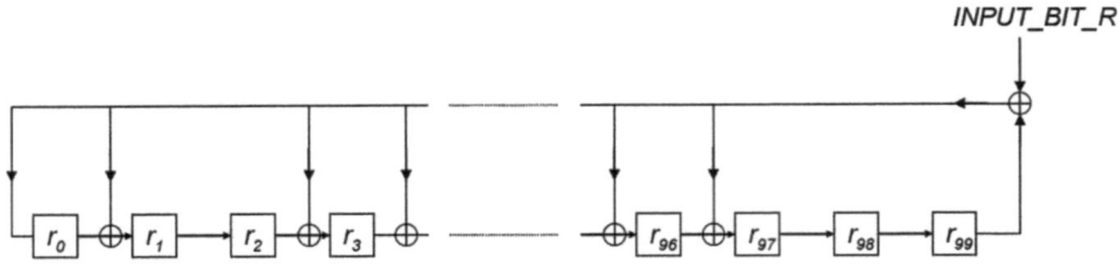

Fig. 9.9 Motion of the register R з $CONTROL\_BIT\_R = 0$

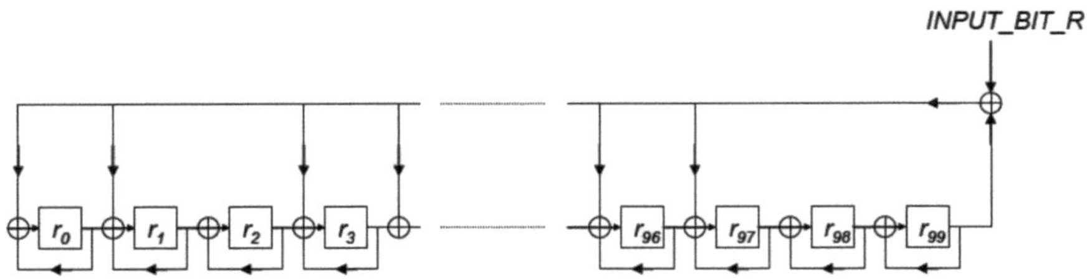

Fig. 9.10 Motion of the register $R3$ $CONTROL\_BIT\_R = 1$

Consequently, the clocktime of the register R in the case $CONTROL\_BIT\_R = 1$ corresponds to its shifting by J times.

This method is a simple operation, connected with the standard operation of linear register shifting, but it is equivalent insofar as the register makes a «jump» in shifting by J times. The method is applicable to the shift register with linear feedback of the type Fibonacci, but it is clear that such an approach influences the registers of the type Galois as well.

The Causes of Using of Irregular Motion The stream ciphers, which use the irregular motion, are frequently prone to statistical attacks. In such ciphers the premise is used how many times the register was shifted for a certain period of time. There is a number of characteristics of the cipher construction, which are responsible for the possible realization of such an attack.

For illustration of these possible characteristics let us examine the stream of the cipher *LILI*-128 [6]. *LILI*-128 utilizes two shift registers with linear feedback, with a length 39 and 89; the 89-digit register is shifted by 1, 2, 3 or 4 times during each clocktime of the whole generator, based on two control bits of the 39-digit register. The attacks based on guessing of probable clocktimes of the 89-digit register is possible, because:

1. The register motion for initially m times, and then n times gives the same result as the initially triple shifting, and then double shifting. Different variants of shiftings are commutated. So, for example, the attacker can get wind that after

ten clocktimes of the whole generator the 89-digit register had two simple clocktimes, three doubled clocktimes, three tripled clocktimes and 2 fourfold clocktimes. He does not need to guess in which order various shiftings have been produced;

2. In addition, a one-time-movement and then a fourfold one give the same result as a motion initially doubled and then tripled. There are many combinations which give, for instance, the register clocktime after 10 clocktimes of the whole generator; the attacker can determine one general probability of this event from the lot of various shifting combinations, which should lead to it. It facilitates additionally the efficiency of a statistical attack;
3. And, finally, the 25 clocktimes of a 89-digit register can be obtained after the 10 clocktimes of the generator or after the 9 clocktimes of the generator, or after the 11 clocktimes of the generator, ... Once more, it can be used in order, for instance, to make the attacks more efficient [6].

Principles, which lie in the base of creating the cipher MICKEY 2.0:

- to utilize all advantages of an irregular shifting for protection against many types of attacks;
- to ensure a great period and a local randomness;
- to reduce as much as possible the probability of statistical attacks, to which are inclined the ciphers of this type.

What concerns the given properties 1–3, which make worse the cryptostability, the cipher *MICKEY* behaves as follows:

1. Correct for the register *R*, because $clock^J clock^1 = clock^1 \bullet clock^J$, but incorrect for the register *S*, because its shift operation do not commutate.
2. Incorrect for both registers. For *R* for every values $t \leq 2^{40}$ and u exist not more than one possible pair of values $n1$ and nJ such as $0 \leq n_1, n_j \leq t, n_1 + n_j = t$ and $n_1 + n_j J = u$, where n_1 and $n_j J$ correspond to a singlefold and a *J*-fold shifting.
3. Incorrect for both registers. For *R* for every preset u exists not more than one triad $t, n1$ and nJ such as $t \leq 2^{40}, 0 \leq n_1, n_j \leq t, n_1 + n_j = t$ and $n_1 + n_j J = u$.

In the cipher *MICKEY* 2.0 the register *R* presents itself as an «engine», guaranteeing that the generator state does not repeat during generation of one key sequence, and good local statistical properties are ensured. The influence of the *R* on the motion of the *S* prevents the *S* for the cycling in the short cycle. If the «jump index» $J \leq 2^{60}$, than the register state *R* will not repeat during the generation of a key sequence with a length through to maximum length of 2^{40} bits, and if $J \geq 2^{40}$, than the 3-d property is fulfilled. The «jump index» *J* is selected by the cipher developers so as to have the greatest possible value, which is less than 250 ($J < 250$).

The architecture of the irregular motion, applied in the cipher *MICKEY* 2.0, is schematically demonstrated in the Fig. 9.11.

Selection of the Bits of the Motion Control The developers of the cipher *MICKEY* selected consciously the bits of the motion control of the each register, so that they

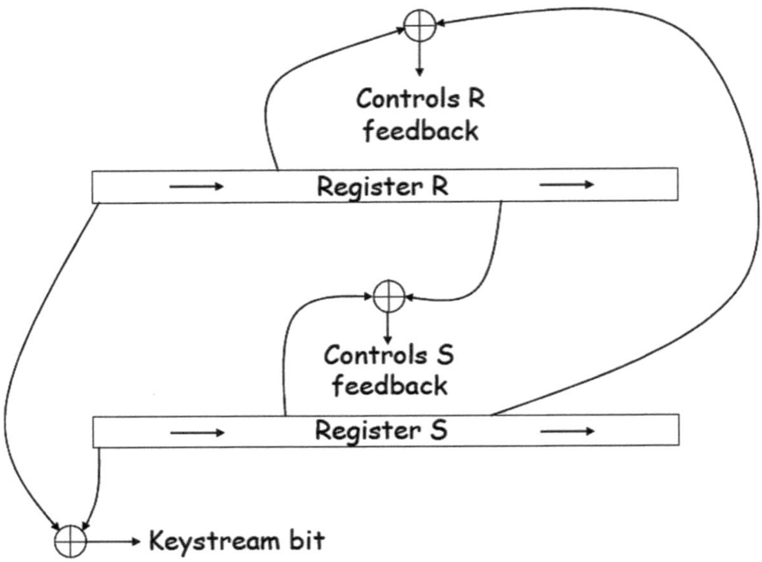

Fig. 9.11 Architecture of the irregular motion in the cipher *MICKEY*

should not depend on the values of both registers. Consequently, if the attacker will know the state of one from the registers, it will be not sufficient for determining of the next motion register. It helps at protecting against the attack based on guess and next determining («*guess and determine attack*»), or attack based on the principle «divide and conquer» («*divide and conquer attack*»).

Feedback of the Register S For every fixed value of the variable *CONTROL_BIT_S*, the function of motion of the register *S* is revertible (so as the area of possible register values does not descend as result of the motion of the register *S*). The scheme of control of motion of the register *S* is shown in the Fig. 9.12.

The aim of creating of the motion function of the register *S* can be formulated as follows. Let us suppose that the initial state of the register *S* is selected in a random manner and that the sequence of values of *CONTROL_BIT_S*, applied in the motion control of the register *S*, selected in a random manner as well. Let us examine the sequence ($s0(i)$: $i = 0,1,2$ …). (Under $s0(i)$ is meant the value $s0$ after the generator has been eliminated by i times). The aim consists in avoidance of every strong affine relations in this sequence, i. e. it is necessary to eliminate the existence of the countless number of I's which with a great probability will be equal to 0 (or 1), and for avoiding that *CONTROL_BIT_S* should take on all possible values.

The construction utilized in the cipher *MICKEY* keeps its local randomness in those sense that the initial state is distributed uniformly in a random manner, thus the sequence of $_{100}$ consequent bits $s0(i)$ also will be distributed uniformly in a random manner. So, for instance, there is no sum, less as $_{101}$ consequent bit $s0(i)$, which is equal to $_{0}$ with a probability not identical to ½. On the base of an empiric analysis

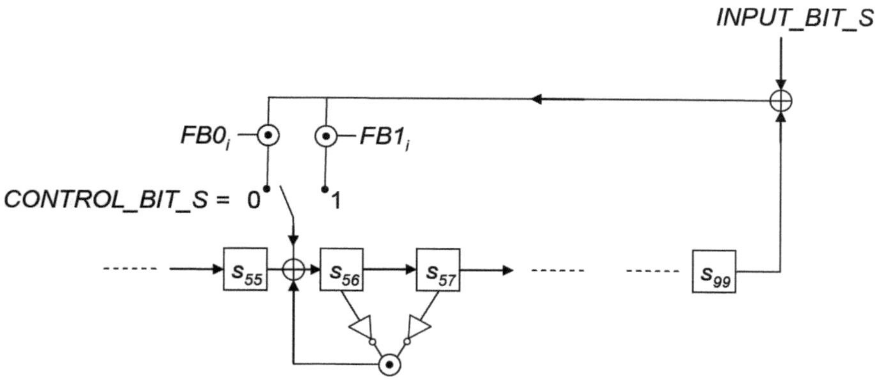

Fig. 9.12 Scheme of the motion control of the register *S*

by the algorithm authors is stated that a substantial shifting from ½ will take place only if the combination will be selected after 100 consequent bits $s0(i)$.

Key Downloading In the algorithm *MICKEY* is utilized a non-linear mechanism for downloading for protection against desynchronising attacks.

Algebraic Attacks The algebraic attacks become obviously possible, when the key sequence correlates with one or more linearly shifted registers, of which motions was either fully predictable, or they could be guessed. At the developing of the cipher *MICKEY*, the attention was concentrated on the fact, that the attacker should not be able to eliminate the indeterminacy concerning motion of one or other register by guessing of a small number of values. Among other things, each key sequence of bits, obtained thanks to *MICKEY* $_{2.0}$, does not correlate with the content of anyone register.

Entropy Quantity The irregular motion is controlled by the bits of the generator itself. It leads to the entropy reduction in the generator: some generator states after motion have two or more possible pre-images For the *MICKEY* version $_1$ is indicated [6] that the lost of entropy can lead to convergency of different key streams in the limits of a normal cipher utilizing. For example, if the key stream *V* with the length 2^{40} is generated by different (*K*, *IV*) pairs, then for a quite great *V* will be possible state collisions and, naturally, the identical states will be achieved once, respectively, all next key streams will be identical. It seems difficult to perform an exact analysis, but the key stream *V* cannot be much greater than 2^{22}, otherwise collisions will arise. This unfavourable property is based on the fact that after the generator actuation for the elaborating of a 2^{40}-bit sequence the quantity of entropy can descend almost by 40 bits: since the initial value 2^{60} to 2^{120}. Inasmuch as 120 is less than the double key length, the collisions start to appear at a less data quantity than the key length. At the *MICKEY* 2.0 the state size is equal to 200 bits, but the maximum admissible length of one key sequence equals to 2^{40} bits. Even though the generator has worked quite a long time, elaborating a 2^{40}-bit sequence, the entropy nevertheless will be slightly greater than 160 bits. Inasmuch as it double exceed the key size, so the problem is resolved.

The utilizing of the operation *XOR* and of the bits from the two registers for obtaining of the control bit ensures the absence of correlation between this control bit and the controlled register, and this was a main particularity for designing of the cipher *MICKEY*. Thanks to this, the entropy does not descend depending on the state of any one register of the generator.

Output Function The cipher *MICKEY* 2.0 utilizes a very simple output function ($r0 \oplus s0$) for calculating of the key stream bits from the registers' state.

9.2.2.7 Estimated Stability of Algorithm

The cipher *MICKEY* 2.0, being utilized according to the exposed above rules, can resist every attack. There is no attack more efficient than an exhaustive search of keys. The algorithm developers did not add purposely any hidden vulnerabilities into the algorithm *MICKEY*.

9.2.2.8 Productivity of the Algorithm

The stream cipher *MICKEY* 2.0 is not destined for using in particular rapid software, although it is very simply realizable and quite effective calculatively. The own realization by the algorithm developers is quite efficient (but without optimization), it generates 10^8 bits of key stream for 3,81 s, by using the processor Pentium 4 with the frequency 3.4 GHz [6]. Naturally, there is a possibility for more efficient software realizations, which elaborate several bits of the key sequence for one time, by utilizing the table realization of the control register and of creating of the output key stream.

9.2.3 Stream Cipher Trivium

The algorithm *Trivium* is a hardware-oriented parallel bitstreamed cipher [7]. It was designed as an example for research of dependence of the stream cipher simplifying without damage for security, velocity and flexibility. Taking in account that simple constructions are more vulnerable to simple and, probably, destroying attacks (and namely because of this reason the developers do not recommend the *Trivium* for these purposes), they, although, inspire confidence than sophisticated schemes, taking in account that they are being studied for long time, in spite of its simplified nature.

In [7] is mentioned a full description of the bitstreamed cipher *Trivium*, researched the particularities of the cipher and the aspects of the algorithm security.

Table 9.7 Parameters of the cipher Trivium

Parameters of the cipher Trivium	
Key size	80 bits
Initialization vector size	80 bits
Internal state	288 bits

After the specification [7] the algorithm *Trivium* is a parallel bitstreamed cipher, destined for the generation of 2^{64} bits of the key stream from the 80 bits of the initialization vector (Table 9.7).

As in majority of the other bitstreamed ciphers, the ciphering process consists of two phases: initially the internal state of the cipher is initiated by the aid of the key and of the initialization vector, further the state is renewed by several times and is used for the generation of the key stream bits.

Let us analyse initially the second phase.

9.2.3.1 Generating of the Key Stream

The ciphering scheme contains 288 bits of the internal state and is denoted (s_1, ..., s_{28}). The generation process of the key stream is iterative one and derives the values of $_{15}$ special states bits and uses all them for the renewal of the $_3$ state bits and for the calculation of the $_1$ bit of the key stream z_i.

Further the bits of state are converted and the process is repeated again, till will be generated $N \leq 2^{64}$ bits of the key stream. The full description is given as a simple pseudocode (Fig. 9.13).

Let us remark that the operations «+» and «−» denote addition and multiplication in the field $GF(2)$, i.e. these are respectively the operations *XOR* and *AND*. A graphical representation of the process of a key stream generating is presented in the Fig. 9.14.

for $i = 1$ to N **do**

$t_1 \leftarrow s_{66} + s_{93}$

$t_2 \leftarrow s_{162} + s_{177}$

$t_3 \leftarrow s_{243} + s_{288}$

$z_i \leftarrow t_1 + t_2 + t_3$

$t_1 \leftarrow t_1 + s_{91} \cdot s_{92} + s_{171}$

$t_2 \leftarrow t_2 + s_{175} \cdot s_{176} + s_{264}$

$t_3 \leftarrow t_3 + s_{286} \cdot s_{287} + s_{69}$

$(s_1, s_2, \ldots, s_{93}) \leftarrow (t_3, s_1, \ldots, s_{92})$

$(s_{94}, s_{95}, \ldots, s_{177}) \leftarrow (t_1, s_{94}, \ldots, s_{176})$

$(s_{178}, s_{279}, \ldots, s_{288}) \leftarrow (t_2, s_{178}, \ldots, s_{287})$

end for

Fig. 9.13 Pseudocode of the function of the key stream generating

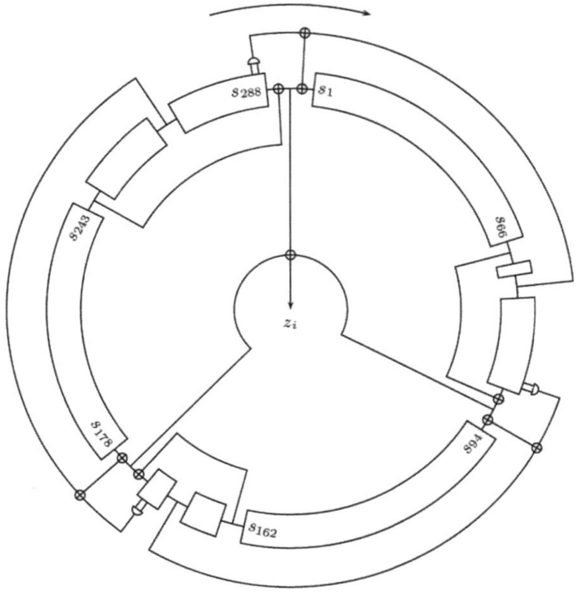

Fig. 9.14 Scheme of the calculation of the function of the key stream generation

9.2.3.2 Setting of the Key and of the Initialization Vector

The algorithm begins with the setting of a 80-bit key and of a 80-bit initialization vector into the 288-bit internal state, all other bits are being set into 0, excluding the s_{286}, s_{287}, s_{288}. Later on, the state value is being mixed while the full 4 cycles in the same way, as were described above, however without generation of the key bits stream. All the above told is described by the pseudocode, presented in the Fig. 9.15.

9.2.3.3 Realization Aspects

Trivium is a hardware-oriented cipher, the construction of which is focused on its flexibility. It is aimed on the compactness in an environment with limited input parameters, on the power efficiency on the platforms with small power sources, and on the operations velocity in the appliances which require a high-speed ciphering as well.

The necessity of a compact realization foresees a bit-oriented approach. This encourages the usage of the non-linear internal state for not spending all the built non-linearity to the output of the key stream generator. In order to ensure a powerful and fast realization, the construction must also ensure an operations paralleling. In the case of *Trivium* it will be made on the cost of ensuring that every bit value is not used at least in 64 iterations after it was changed. Consequently, before these 64 iterations the value can be calculated one time, by the condition that the 3 inputs of the adding element (*AND*, logical conjunction) and the 11 inputs of the element

$$(s_1, s_2, \ldots, s_{93}) \leftarrow (K_1, \ldots, K_{80}, 0, \ldots, 0)$$
$$(s_{94}, s_{95}, \ldots, s_{177}) \leftarrow (IV_1, \ldots, IV_{80}, 0, \ldots, 0)$$
$$(s_{178}, s_{279}, \ldots, s_{288}) \leftarrow (0, \ldots, 0, 1, 1, 1)$$
for $i = 1$ **to** $4 \cdot 288$ **do**
$$t_1 \leftarrow s_{66} + s_{91} \cdot s_{92} + s_{93} + s_{171}$$
$$t_2 \leftarrow s_{162} + s_{175} \cdot s_{176} + s_{177} + s_{264}$$
$$t_3 \leftarrow s_{243} + s_{286} \cdot s_{287} + s_{288} + s_{69}$$
$$(s_1, s_2, \ldots, s_{93}) \leftarrow (t_3, s_1, \ldots, s_{92})$$
$$(s_{94}, s_{95}, \ldots, s_{177}) \leftarrow (t_1, s_{94}, \ldots, s_{176})$$
$$(s_{178}, s_{279}, \ldots, s_{288}) \leftarrow (t_2, s_{178}, \ldots, s_{287})$$
end for

Fig. 9.15 Pseudocode of the function of setting of the key and of the initialization vector

Table 9.8 Estimation of the input values for 1-bit ... 64-bit hardware realization

Value	1-bit	8-bit	16-bit	32-bit	64-bit
Flip-flops	288	288	288	288	288
AND-inputs	3	24	48	96	192
XOR-inputs	11	88	176	352	704
NAND-inputs	3488	3712	3968	4480	5504

XOR (disjunction) are doubled in the original scheme by the respective number of times. It allows to divide the clock-time frequency by 64 without disadvantage for the throughput capability.

The results of the calculations of the number of inputs, which are needed for different degree of paralleling, are presented in the Table 9.8 [7].

9.2.3.4 Hardware Realization

In spite of the fact that the *Trivium* is developed not for the software realization, the cipher is efficient also on the standard personal computers. The yields of the cipher executing with the usage of the language C on the processor *Xeon* with the frequency of 1,5 GHz are presented in the Table 9.9.

Table 9.9 Cipher characteristics for the software realization with the using of the processor *Intel Xeon CPU* 1,5 GHz

Operations	
Stream generation	12 cycles/bit
Key setting	55 cycles
Initialization vector setting	2050 cycles

9.2.3.5 Security

The security requirement, that is imposed to the cipher, lies in the fact that every cryptographical attack must not be substantially easier realized on the cipher *Trivium*, as well as on every other possible bitstreamed cipher with the same output parameters (2^{64} bits of the key stream from a $_{80}$-bit secret key and from a 80-bit initialization vector).

Correlation Analysing the protection of the synchronous bitstreamed cipher, the cryptoanalyst examines obviously the two different correlation (interconnection) types. The first type that is an interconnection between a linear bit combination of a key sequence and the bits of an input sequence, which can potentially cause a full renewal of the state value. Another type, used by some attacks, that is an interconnection between the bits of the key sequence itself.

Obviously, the linear correlations between the bits of the key sequence and the bits of the input sequence is easier to find: as soon as z_i will be equal to $s_{66} + s_{93} + s_{162} + +s_{177} + s_{243} + s_{288}$. However, for the cipher *Trivium*, which is based on the shift register with the non-linear feedback, the value of the cipher state is developed by the non-linear way, and it is unclear that how the attacker has to comprise this equation for an effective discovering of state values.

An easy method to find the correlation of another type is to follow the linear way of the cipher and to approach to 0 the output values, which come to the input of all *AND*s. However the outputs disposition in the *Trivium* is selected in such a way that to open every of outputs is necessary to reduce at least the values of 72 *AND*-outputs. For example, the correlative linear combination of the key sequence bits will be written as follows:

$$z_1 + z_{16} + z_{28} + z_{43} + z_{46} + z_{55} + z_{61} + z_{73} + z_{88}$$
$$+ z_{124} + z_{133} + z_{202} + z_{211} + z_{220} + z_{289}.$$

By supposing that the correlation of this linear combination is fully explained by a specific method, then the correlation coefficient will be 2^{-72}. In order to discover a similar correlation, are needed at least 2^{144} bits of the key sequence, what is substantially less than the requirements of the cipher protection.

More sophisticated species of the linear connections with a greater correlation can exist as well, however in this case it will be unacceptable, because such a correlation will exceed 2^{-40}.

Period Inasmuch as the internal state of the *Trivium* is developed non-linearly, its period is hardly to determine. And although several observations can be made.

First, by omitting the *AND*-inputs (as a result of the fully linear scheme), can be discovered that every pair Key/Initialization vector, which will be generated, will have the period $2^{96-3} - 1$. It has no immediate consequences for the cipher, but it reflects the sign that the utilized ramifications are selected correctly.

Second, the state of the cipher *Trivium* is refreshed reversibly, but the initialization of the values $(s_{178}, \ldots, s_{288})$ prevents the state from repetitions at least while 111 iterations. If assuming that the *Trivium* transforms into a random transposition after a certain number of iterations, so all the cycles with a length till 2^{288} will be equiprobable, and consequently the possibility of obtaining the pair Key/Initialization vector of cycles less as 2^{80} will be equal to 2^{-208}.

Attacks of Anticipation and Determining In each iteration of the cipher *Trivium* are used only several state bits, in spite of the main rule that the rare renewal functions must be avoided in the scheme. As a result, it offers the possibility of the existence such attacks against the cipher, as guessing and determining attacks. The first attack will calculate the bits $(s_{25}, \ldots, s_{93})$, $(s_{97}, \ldots, s_{177})$ and $(s_{244}, \ldots, s_{288})$, all together $_{195}$ bits, after that the other bits can be found immediately from the key sequence. The further researches must be performed for detecting of a more updated attack, which should reduce this number.

Algebraic Attacks The cipher *Trivium* presents an exceptional interest for algebraic attacks. The full scheme can be easily described by an interminable number of separate equations of a lower degree. However, its states are being developed by a non-linear way, and, consequently, the effectiveness of the linearization techniques, which is utilized for resolving of equations systems, generated by the schemes, based on the shift register with linear feedback, are quite labour-consuming in realization. And in the sane time, other techniques can be utilized, and their effectiveness for resolving of such equations systems must be researched necessarily.

Attacks of Repeated Synchronization Another kind of attacks are the attacks of repeated synchronization, where the enemy can change the value of the initialization vector and tries to obtain the information about keys by studying the respective key sequence. The cipher *Trivium* prevents the realization of such sort of attack, the internal state is initiated several times within the scheme before supplying it to the output. It demonstrates that each state bit appears in each key and initialization vector after two full cycles (i.e. 2 • 288 iterations). It is expected that these two cycles will be sufficient for protecting the cipher against an attack of repeated synchronization.

Consequently, the cipher *Trivium* is the simple parallel bitstreamed cipher, which matches to using in appliances with a compact hardware realization. It is evidently in practice, although the further researches will demonstrate whether it will correspond to the security requirements, or not.

9.3 Stream Cipher *CryptMT* Version 3

The SSC *CryptMT* was presented to the contest *eSTREAM* after the profile of software-oriented ciphers. But, after the researches results, it did not pass the third stage of tests and is taken off from the contest. In this chapter the structure and the main conversions of this SSC are analysed [13, 14].

The third version of the algorithm *CryptMT* was proposed for the accelerating of generation and initialization, by utilizing the parallelism, for example, the containerization operation and the operations with several control streams and several data streams (*SIMD*), which are included into contemporary processors, in the filter, on the way of increasing of the relation between the quantity of input bits and the output bits from ¼ till ½.

As result the *CryptMT* of the version 3 has the following indices: the PRS-generation by 1.8 times and the initialization by 48.7 times faster than at the version 1, concurrently the security level remains comparable with the first version. The generation/initialization velocity is comparable with such fast bitstreamed ciphers, as *SNOW* 2.0. And what is more, it is proved that the *CryptMT* of the version 3 has a period factor of $2^{19937} - 1$ and resists reliably against the most frequent attacks.

9.3.1 Specification of the Cipher

The idea of the bitstreamed cipher *CryptMT* consists in uniting of a linear generator of states (so-called mother generator) and of a memory filter, as shown in the Fig. 9.16.

9.3.1.1 Notations

Definition 1 (*Generator with a filter with memory*) Let be *X* a finite multitude (obviously it is a multitude of integer numbers). The mother generator *G* generates the

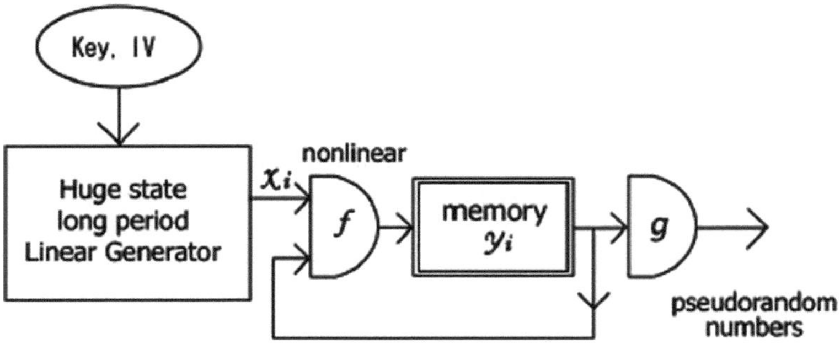

Fig. 9.16 Combined generator = linear generator + memory filter

sequence $x_0, x_1, \ldots \in X$ Let be Y a finite multitude which represents the multitude of all possible states of the filter memory. Let us take $y_0 \in Y$. Let be $f : Y \times X \to Y$ the function of state transition of the filter memory, i. e. filling out of the variable y_1 of the memory is executed by the recursion

$$y_{i+1} := f(y_i, x_i).$$

The output in the i-th iteration is given by $g(y_i)$, where $g : Y \to O$ this is an initial function, that transforms the memory content into the output symbol, which belongs to the multitude O.

In the precedent works as the mother generator was selected the Mersenne twister (BM) [15], which generates a sequence of 32-bit integer numbers by the aid of the F_2-linear recursion. The filter is defined as

$$f(y, x) := y \times (x|1) \bmod 2^{32},\ g(y) := 8MSBs \tag{9.1}$$

where $(x|1)$ means x with the least significant bit, set in $_1$, but 8 *MSBs* mean the 8 most significant bits of the integer y. The initialization of the internal memory is performed by the variable y_0, what represents an odd integer number.

Let us define a conventional notation for 128-bit integer numbers. By the bold italic $\boldsymbol{x}$ is denoted the 128-bit integer number. This is a concatenation of 32-bit registers, of which each is denoted as $x[3], x[2], x[1], x[0]$ respectively.

The notation $x[3][2]$ denotes the 64-bit integer number, obtained on the way of unification of two 32-bit integers $x[3]$ and $x[2]$ in this order. Analogically $x[0][3][2][1]$ denotes a 128-bit integer number, obtained on the way of permutation (cyclic shifting) of the four 32-bit integer numbers into x.

The operation on the 128-bit registers, which is executed for each 32-digit integer, is denoted by the symbol 32. For example

$$x +_{32} y := [(x[3] + y[3]), (x[2]) + y[2], (x[1] + y[1]), (x[0], y[0])]$$

that is the first 32-bit part represents the addition of $x[3]$ and $y[3]$ after the modulus 2^{32}, other 32-bits there are the bits $x[2]$ and $y[2]$ (without transferring from the second 32-bit part into the first 32-bit part, different from the adding of the 128-bit integer numbers). External majority $[\cdot]$ in the right part highlights that they are unified for obtaining of a 128-bit integer number.

Analogically, for the integer number S

$$x{>>}_{32}S := [(x[3]{>>}S), (x[2]{>>}S), (x[1]{>>}S), (x[0]{>>}S)]$$

means the shifting, which is applied for each of the two $_{64}$-bit integer numbers. Further we frequently use such functions as

$$x{>>}_{64}S := [(x[3][2]{>>}S), (x[1][0]{>>}S)]$$

which we call *perm-shift*. $\oplus$ here means a bitwise exclusive OR. The transmutation [2] [1] [0] [3] can be an arbitrary transmutation with a left shifting.

The function of the type

$$x \mapsto x[i_3][i_2][i_1][i_0] \oplus (x >>_{32} S)$$

also is called *perm-shift*, where i_3, i_2, i_1, i_0 is the transmutation 3, 2, 1, 0. The *Perm-shift* is an $F2$-linear conversion, but if $S \geq 1$, so this is a bijection.

Let be n a natural number, and x a 32-bit integer number. And n most significant bits of x are denoted $MSB^n(x)$. An analogical notation $LSB^n(x)$ is used as well. For the 128-bit integer x we define

$$MSB_{32}^n(x) := \left[MSB^n(x[3]), MSB^n(x[2]), MSB^n(x[1]), MSB^n(x[0])\right]$$

what is an $(n \times 4)$-bit integer number.

The function $f : Y \times X \rightarrow Z$ is bi-bijective, if for every fixed $x \in X$ the reflection $Y \rightarrow Z, y \mapsto f(y, x)$ is bijective, and for every fixed $y \in Y$ the reflection $X \rightarrow Z, x \mapsto f(y, x)$ is bijective. It is necessary that the powers of multitudes should converge: $\#(X) = \#(Y) = \#(Z)$.

9.3.1.2 Simple Fast BM

The version 3 accepts the mother generator, a so-called Mersenne Twister (*Simple Fast Mersenne Twister*, SFMT) [16] as a basic generator.

Let be N an integer number, but $x_0, x_1, \ldots, x_{N-1}$ – N 128-bit integer numbers, assumed as an initial state. The SFMT are used for the generation of a sequence from 128-bit integer numbers with a following наступною $F2$-linear recursion:

$$\begin{aligned} x_{N+j} &:= \left(x_{N+j-1} \& 128 - bitMASK\right) \oplus \\ &(x_{M+j} \gg_{64} S) \oplus (x_{M+j}[2][0][3][1]) \oplus (x_j[0][3][2][1]). \end{aligned} \tag{9.2}$$

Here and denotes the bitwise logical «*AND*», as a consequence the first term is the result of the bit-mask x_{N+j-1} of an immutable 128-bit *MASK*. The second term represents the concatenation of two $_{64}$-bit integer numbers $(x_{M+j}[3][2] \gg S)$ and $(x_{M+j}[1][0] \gg S)[0] \gg S)$, as was explained above. The third term that is the shifting of four 32-bit integer numbers x_{M+j}, but the last term is a shifting of those that are in the x_j. Consequently, the SFMT based on the linear recursion of the N-th order above 128-dimensional vectors $\mathbb{F}_2^{128}$. The Fig. 9.17 describes the structure of the SFMT.

permute:$y \mapsto y[0][3][2][1]$.

perm-shift: $y \mapsto y[2][0][3][1] \oplus (y \gg 64^3)$.

бітова маска: ffdfafdf f5dabfff ffdbffff ef7bffff

During a computer search we found the parameters

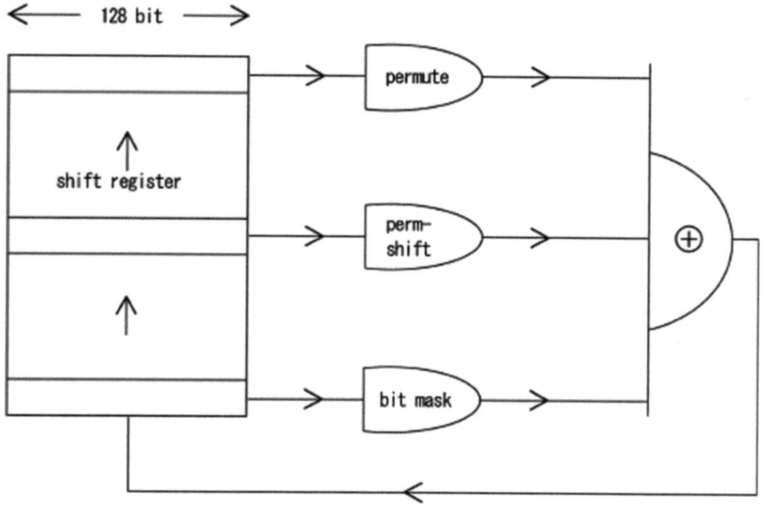

Fig. 9.17 Mother generator: simple fast Mersenne Twister

$$N = 156, \quad M = 108, \quad S = 3$$

$$\text{MASK} = \text{ffdfafdf f5dabfff ffdbffff ef7bffff}$$

in hexadecimal. Such a mask is necessary for an abruption of symmetry (i. e. without introduction of asymmetry, in the case if each 128-bit integer x_i in the array of the initial state complies with the expression $x_i[3] = x_i[2] = x_i[1] = x_i[0]$, so this equality will be fulfilled steadily while further forming of numbers by the generator). The mask is selected with a more quantity of ones than of zeros so that much information will not be lost.

It is proved that if $x_0[3] = 0x4d734e48$, so the period of the generated sequence SFMT is multiple to the simple Mersenne number $2^{19937} - 1$, and the output is 155-dimensionally evenly distributed.

These operations are selected so that they should correspond to the instructions *SIMD* in contemporary processors, such as the *Pentiun IV*. Let us mention that even for the processors without *SIMD* the calculation of such recursive formulas occurs rapidly, because this fits in the conveyor working mode.

9.3.1.3 New Filter

The earlier proposed filter (9.1) utilizes the integer-valued multiplication in the ring $\mathbb{Z}/2^{32}$. In order to avoid degeneration we limit the multiplication by the multitude of the odd integer numbers $\mathbb{Z}/2^{32}$, setting the *LSB* equal to 1 in (9.1)

In the version 3 is used such a binary operation $\tilde{\times}$ by $\mathbb{Z}/2^{32}$ instead of $\times$: for $x, y \in \mathbb{Z}/2^{32}$, let us define

$$x\tilde{\times}y := 2xy + x + y \bmod 2^{32}$$

which is, in its essence, the multiplication of 33-bit odd integer numbers. Let be S a countless amount of odd integer numbers $\mathbb{Z}/2^{32}$. We have a bijection

$$\varphi : \mathbb{Z}/2^{32} \to S, x \mapsto 2x + 1.$$

Then $\tilde{\times}$ is defined by the formula:

$$x \tilde{\times} y := \varphi^{-1}(\varphi(x) \times \varphi(y)),$$

where x denotes the multiplication in S. Consequently, $\tilde{\times}$ there is set taken in account only the upper 32 bits of the multiplication in S. So, $\tilde{\times}$ is bi-bijective.

The majority of contemporary processors possess a 32-bit integer-number multiplication, but not a 64-bit or 128-bit multiplication. Hence, the most simple solution of the multiplication problem of numbers of greater digit capacity than 32 should be the paralleling (8.1) of the operation performing in such a way: $X = Y = \left(\mathbb{Z}/2^{32}\right)^4$, and

$$f(x, y) := y\tilde{\times}32x,$$

(it means $f(y, x)[i] := y[i] \tilde{\times} x[i]$ for $i = 3, 2, 1, 0$), and

$$g(y) := MSB_{32}^{8}(x)$$

represents by itself an output of 8×4-bit integer numbers.

In the version 3 a modified filter is utilized (see the Fig. 9.18) in such a way. For

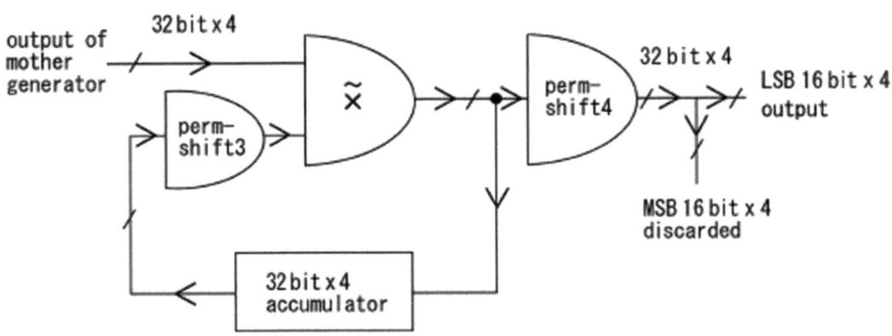

Fig. 9.18 Filter *CryptMT Ver.*3

a given pair of the 128-bit integer numbers x, y let us define

$$f(y, x) := (y \oplus (y[0][3][2][1]>>_{32}1)) \underset{32}{\tilde{\times}} x. \tag{9.3}$$

The operation, applied to y in the right part, is an operation *perm-shift*, hence, is a bijective one. Insomuch it is a bi-bijective, then f is bi-bijective as well. The aim of the *perm-shift* introduction is unification of information between the four 32-bit memory centres in the filter and sending of information about upper bits into the junior bits. It completes the multiplication.

perm-shift3: $y \mapsto y \oplus (y[0][3][2][1] \gg_{32} 1)$.

perm-shift4: $y \mapsto y \oplus (y \gg_{32} 16)$.

$\tilde{\times}$: multiplication of 33-bit odd integer numbers.

Output function

$$g(y) := LSB_{32}^{16}(y \oplus (y>>_{32}16)).$$

Thus, the new filter has a 128-bit memory, accepts a 128-bit integer number and leads out the integer number (16×4). The compression coefficient of this filter is equal to (128:64), what is less than (32:8), as it was in the earlier proposed tfilter. This correlation change is necessary for increasing of the calculating velocity, but such an approach can weaken the security. In order to equilibrate this, the leading-out function assumes the result of executing of the exclusive OR operation above the 16 most significant bits and the 16 least significant bits $y[i]$, $i = 3, 2, 1, 0$.

9.3.1.4 Conversion into 8-bit Integer Numbers

Since the outputs of the filter are (16×4) bit integers, and the required specification is 8-bit integer outputs, we need to convert them to 8-bit integers. Due to the nature of 128-bit SIMD instructions, the following strategy is used to increase computing speed. Let

UPPER16:=(0xffff0000, 0xffff0000, 0xffff0000, 0xffff0000)
LOWER16:=(0x0000ffff, 0x0000ffff, 0x0000ffff, 0x0000ffff)

will be 128-bit masks.

Let $y_0, y_1, \ldots, y_{2i}, y_{2i+1}, \ldots$ be the contents of the memory cells in the filter at each step generated by the operation $y_{i+1} := f(y_i, x_i)$ in (9.3). Then y_{2i} and y_{2i+1} is used to generate igo 128 integer output z_i by the formula:

$$z_i := [(y_{2i} \oplus (y_{2i}>>_{32}16))\&LOWER16] | [(y_{2i+1} \oplus (y_{2i+1} <<_{32} 16))\&UPPER16]$$

where |—denotes bitwise OR. Z_i is then divided into 16 8-bit integers from lower to higher bits and used as 8-bit integers.

9.3.1.5 New Initialization Operation

SFMT requires N = 156 of 128-bit integers as the initial state. We need to extend the key and IV to the initial state during initialization, but this is costly if the message length is much less than N = 128 bits. The strategy presented in [17] is to use a smaller PRS generator called a loader. Its role is to extend the key and IV to a sequence of 128-bit integers. The loader output signal is transmitted to the above-described filter to generate a pseudo-random integer stream and at the same time is used to fill the SFMT state. Once the SFMT state is filled, the generation switches from bootloader to SFMT. Such a bootloader is described in Fig. 9.19.

perm-shift1:$x \mapsto (x[2][1][0][3]) \oplus (x>>_{32}13)$.

perm-shift2:$x \mapsto (x[1][0][2][3]) \oplus (x>>_{32}11)$.

$\tilde{\times}$: multiplication of 33-bit odd integers

Choose an integer h according to the size of the key and IV. The boot state space is a shift register consisting of H 128-bit integers. Select the initial state $x_0, x_1, \ldots, x_{H-1}$ and the initial value a0 of the drive (128-bit memory), as described below. Then the transition from state to state is set by recursion

$$a_j := \left(a_{j-1}\tilde{x}_{32} perm - shift2\left(x_{H+j-1}\right)\right)$$
$$x_{H+j} := perm - shift1\left(x_j +_{32} x_{H+j-2}\right) -_{32} a_j,$$

where

$$perm - shift1(x) := (x[2][1][0][3] \oplus (x>>_{32}13))$$
$$perm - shift2(x) := (x[1][0][2][3] \oplus (x>>_{32}11))$$

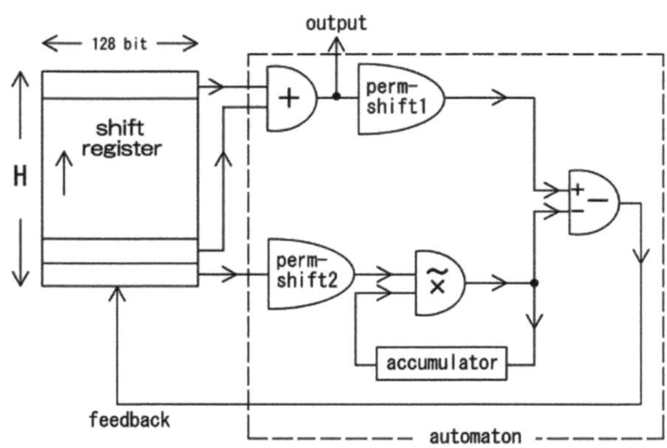

Fig. 9.19 *CryptMT Ver.3* loader

Similar to the notation +32, –32 denotes the subtraction modulo 2^{32} for each of the four 32-bit integers. The output of j step is equal to $x_j + x_{32}x_{H+j-2}$.

As shown in Fig. 9.19, the loader consists of a machine with three inputs and two outputs of 128-bit integers with a shift register. In the implementation of the developers, the shift register is an array of 128-bit integers with a length of $2H+2+N$. The reasons for choosing such a loss of length for idling are explained below.

9.3.1.6 Installing the Key and IV

It is assumed that both IV and K are given as arrays of 128-bit integers with a length of 1 to 16 for each. Thus, the key size can be flexibly selected from 128 to 2048 bits, as well as size IV. It is claimed that the security level is similar to the size of the key.

During IV and key installation, these arrays are combined and copied twice to the array, as described in Fig. 9.20. To eliminate symmetry, a constant 128-bit integer (846264, 979323, 265328, 314159) (four 32-bit integers in decimal notation) is added to the bottom line of the second copy of the key. Now the size H of the shift register in the loading means is equal to 2 × (IV size + key size (in bits))/128; namely, doubling the number of 128-bit integers contained in IV and K. For example, if the size IV and the size of the key are 128 bits, then H = 2 × (1 + 1) = 4. The machine in the bootloader, described in Fig. 9.19, uses this array, as shown in Fig. 9.20. The battery of the bootloader is set to (top row of the key array) | (1,1,1,1), ie the top line is copied to the drive, and then the LSB of each of the 32-bit integers in the battery is set to 1.

Key and vector arrays are concatenated and copied to the array twice. Then a constant is added to the bottom of the second copy of the key to eliminate possible symmetry.

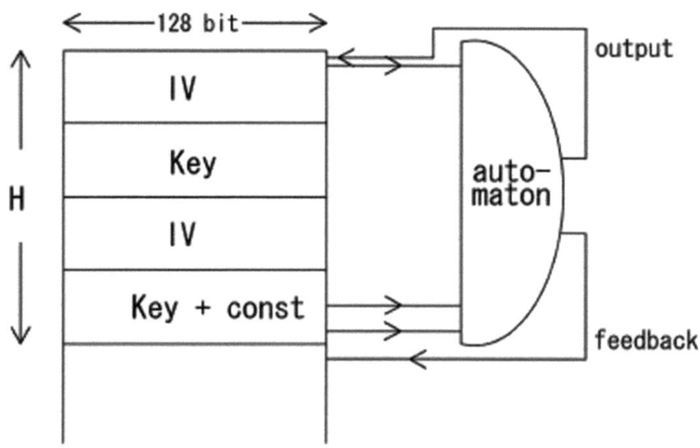

Fig. 9.20 Setting the initialization key and vector

During the first generation, the machine reads three 128-bit integers from the array and writes the original 128-bit integer at the top of the array. Feedback to the shift register is written to the record (H + 1) of the array. For the next generation, the machine is clocked once and then do the same.

For idling, H + is repeated 2 times. Then the last modified string in the array is (2H + 2) and the string is copied to 128-bit memory in the filter. The upper records of the H + 2 array are discarded. This completes the installation of the initialization key and vector. In Fig. 9.21 shows the state after installation.

After installation, the bootloader produces 128-bit integer output numbers no more than N times. Let L be the number of 8-bit integers in the message. If $L \times 8 \leq N \times 64$, then no mother generator is required. The required number of 128-bit integers is generated by the loader and transmitted to the filter to obtain the required outputs. If $L \times 8 \geq N \times 64$, then N 128-bit integers are generated by the loader and transmitted to the filter to obtain N 64-bit integers that are used as the first output numbers. At the same time, these N 128-bit integers are written to the array, and they are passed to SFMT as the initial state.

To exclude the possibility of a shorter period than 2^{19937} - 1, 32 MSB of the first line of the SFMT state array is set as the magic number 0 × 4d734e48 in hexadecimal. This is illustrated in Fig. 9.22. That is, we begin the recursion (8.2) of SFMT with $x_0, x_1, \ldots, x_{N-1}$ - an array of length N, shown in Fig. 9.22, and we get $x_N, x_{N+1}, \ldots$. Since x_N can be easier to guess through the constant part in the initial state, skip it and pass 128-bit integers $x_{N+1}, x_{N+2}, \ldots$, to the filter.

The cryptanalysis developed in [14] for CryptMT can also be applied to Version 3. We discuss some of the properties of SFMT required in the following cryptanalysis.

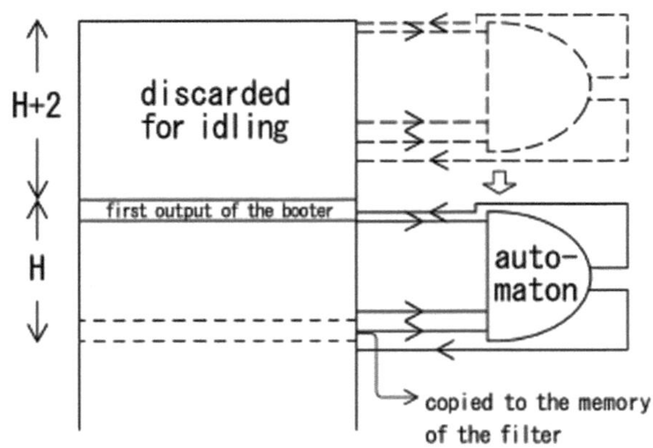

Fig. 9.21 After setting the key and initialization vector

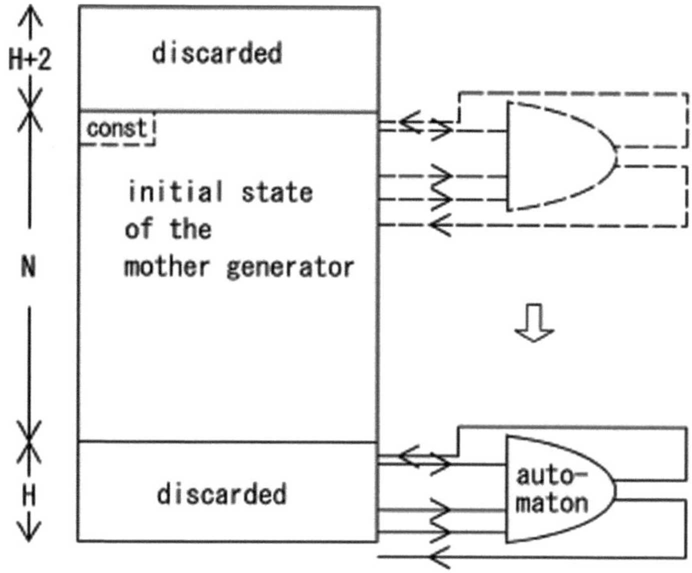

Fig. 9.22 Initialization of the parent SFMT generator

9.3.2 CryptMT *Cipher Security*

Proposition 1. SFMT is an automaton with state space S, which is an array of 128-bit integers with a length of 156 (therefore, has 19968 = 128 × 156 bits).

1. The transition function h SFMT is an F_2-linear bijection, the characteristic polynomial of which is factorized as

$$\chi_h(t) = \chi_{19937}(t) \times \chi_{31}(t),$$

   where $\chi_{19937}(t)$ is a primitive polynomial of degree 19,937, and $\chi_{31}(t)$ is a polynomial of degree 31.
2. The state S is uniquely divided into the direct sum of h-invariant subspaces of degrees 19,937 and 31

$$S = V_{19937} + V_{31},$$

   where the characteristic polynomial h bounded by V_{19937} is equal to $\chi_{19937}(t)$.
3. From any initial state $\mathbf{S}_0$, which is not contained in V_{31}, the period P of the state transition is a multiple of the 24th Mersenne prime number $2^{19937} - 1$, namely $P = (2^{19937} - 1)q$ it holds for some $1 \le q \le 2^{31} - 1$ (q may depend on $\mathbf{S}_0$). The period of the original sequence is also equal to P.

In this case, in addition, the original sequence of 128-bit SFMT 155 integers is evenly distributed with the defect q [14].
4. Let $\mathbf{S}_0$ be the initial state of SFMT, i.e. an array of 128-bit integers of length 156. If 32 MSB of the first 128-bit integer in $\mathbf{S}_0$ is equal to $0 \times$ 4d734e48, then $s_0 \notin V_{31}$. During SFMT initialization, the corresponding 32 bits are set to $\mathbf{S}_0$.
5. $\mathrm{n}\chi_h(t)$ has 8928 non-zero members (which is much more than 135 in the case of MT19937), and $\chi_{19937}(t)$ has 9991 non-zero members.

9.3.2.1 Period

Proposition 2 Any bit of the 8-bit integer stream generated by CryptMT Ver.3 has a period multiple of $2^{19937} - 1$.

Proof Put $Q := 2^{19937} - 1$. Suppose that there is an inverse element, then there is one bit (out of 8 possible bits) of an 8-bit integer stream, the period of which is not a multiple of Q. Denote by $h_0, h_1, h_2, \ldots$ the original 8-bit the whole sequence of CryptMT Ver.3.

If we consider CryptMT Ver.3 as a 64-bit integer generator, then its $z_0, z_1, z_2, \ldots$, are determined by $h_0, h_1, h_2, \ldots$ from

$$
\begin{aligned}
z0 &= (h_{13}, h_{12}, h_9, h_8, h_5, h_4, h_1, h_0) \\
z1 &= (h_{15}, h_{14}, h_{11}, h_{10}, h_7, h_6, h_3, h_2) \\
z2 &= (h_{29}, h_{28}, h_{25}, h_{24}, h_{21}, h_{20}, h_{17}, h_{16}) \\
z3 &= (h_{31}, h_{30}, h_{27}, h_{26}, h_{23}, h_{22}, h_{19}, h_{18})
\end{aligned}
$$

It can be seen that the corresponding bits in $z_0, z_2, z_4, \ldots$ (there are 8 bits for each) have a period that is not a multiple of Q (because they are obtained with each 16th h).

This means that each of the corresponding 8 bits in $z_0, z_2, z_4, \ldots$ has a period not a multiple of Q.

We use Theorem A.1 from [14] to show that any two bits of 64 bits in z_i have a period multiple of Q (as a 2-bit integer sequence), which proves the assumption. *CryptMT Ver.3* is considered as a 64-bit integer stream generator. Then he satisfies the conditions of the theorem, where $n = 155$, $Q = 2^{19937} - 1$, $q < 2^{31}$ and $Y = \mathbb{F}_2^{128}$. If we define the mapping in $g : Y \to B$ in Theorem A.1, assuming $B := \mathbb{F}_2^2$ and

$$
g : y \mapsto \text{ any fixed two bits in } LSB_{32}^{16}(y \oplus (y \gg_{32} 16)),
$$

then $r = 1/4$ and inequality

$$
r^{-156} = 2^{312} > q \times \#(Y)^2 \left(< 2^{31} < 2^{256}\right)
$$

implies that any pair of bits in 64 bits has a period of at least Q according to Theorem A.1.

9.3.2.2 An Attack on the Compromise of Time and Memory

This attack takes a time approximately equal to the square root of the size of the state space, which is equal to $O\left(\sqrt{2^{19968+128}}\right) = O\left(2^{10048}\right)$ for Version 3.

9.3.2.3 Estimation of Uniform Distribution

Proposition 1 shows that SFMT satisfies all conditions, has the period $P = \left(2^{19937} - 1\right)q\left(1 \leq q \leq 2^{31}\right)$ and $n = 155$-dimensional equi-distribution with defect $d = q$. It follows from the Proposition that the original 64-bit integer sequence of *CryptMT Version 3* is 156-dimensional evenly distributed with a defect $q \cdot 2^{128} < 2^{159}$, and therefore, 1248-dimensional are evenly distributed as 8-bit integers.

9.3.2.4 Correlation Attacks

Considering a simple attack on *CryptMT Ver.3* of order ≤ 155, we determine that its security level will be equal to $2^{19937 \times 2}$, since $P/d = 2^{19937} - 1$.

Due to the 156-dimensional property of uniform distribution, correlation attacks should not be used.

9.3.2.5 Algebraic Degree of the Filter

The statement belongs to the multiplicative filter, so it does not apply to *CryptMT Ver.3* in its pure form. However, because the *Version 3* filter introduces more bit mixing than the original multiplier filter, it is assumed that each *CryptMT Ver.3* output bit will have a high algebraic degree close to the upper limit based on the number of variables.

9.3.3 Comparison of Productivity

A performance testing tool with *eSTREAM* was used [18] to compare the speed of Phase 2 candidate algorithms with *CryptMT Ver.3* on two different platforms, ie a 1.4 GHz *Pentium-M* processor and an *AMD* Athlon 64 processor with a 3.4.4 gcc compiler.

The proposed SSC *CryptMT Ver.3* is 1.8 times faster than the first version (faster than SNOW 2.0 on AMD Athlon on platforms with quick SIMD instructions), while the period $\geq 2^{19937} - 1$ and 1248-dimensional property of uniform distribution (in the form of an 8-bit integer generator). The key size and size IV can be flexibly selected from 128 to 2048 bits. The size of the state and the length of the period make it impossible to attack the compromise of time and memory, and the high nonlinearity introduced by integer multiplication will make it impossible to use algebraic attacks and the Berlecamp Messi attack. *CryptMT* has no tables and is therefore resistant to cache attacks. *CryptMT Ver.3* uses up to 2.6 KB of memory.

9.4 Conclusions and Recommendations

1. The eSTREAM project, organized by the European Union, was a multi-year competition to identify new streaming ciphers suitable for widespread use. The main reason for initiating such studies was the failure of all six stream ciphers submitted for the NESSIE project. The eSTREAM project ran from 2004 to 2008 to help develop efficient and compact streaming ciphers suitable for widespread use. As a result of the project, a portfolio of new streaming ciphers was announced in April 2008. The eSTREAM portfolio was revised in September 2008 and currently contains seven streaming ciphers (HC128; Rabbit; Salsa20/12; SOSEMANUK; Grain; MICKEY and Trivium). All ciphers have undergone extensive exploration and careful analysis of both cryptographic properties and performance (PRS rate) on various computing platforms. Therefore, the winners of this project are de facto (individual algorithms and deure) world standards for streaming symmetric cryptographic transformation.
2. The analysis of streaming ciphers, selected as the winners of the international project eSTREAM, showed that according to their specification, the corresponding algorithms are divided into two functional groups. The first group of algorithms are software-oriented stream ciphers, which include: HC128; Rabbit; Salsa20/12; SOSEMANUK. The second group of hardware-oriented ciphers includes Grain algorithms; MICKEY and Trivium.
3. Analysis of the structure and main applied functions of conversion of software-oriented stream ciphers (HC128, Rabbit, Salsa20/12, SOSEMANUK) showed that the corresponding cryptocurrency algorithms use mainly simple and computationally efficient bitwise addition operations modulo 2, AND, cyclic shifts functions with the ability to apply all the benefits of parallelization of the processes of formation of the key flow. The state of the generator is usually presented either in words or in tabular form, which greatly simplifies the implementation of calculations during software implementation.
   An exception is the SOSEMANUK algorithm, which uses a shift register of length 10 to represent the state of the generator, but the small length of this register does not significantly reduce the performance even during software implementation. It should also be noted that almost all algorithms use elements of BSS constructions, in particular, the SOSEMANUK algorithm uses nonlinear replacement nodes constructed using Sblock.
   BSC AES (FIPS197).
4. Analysis of hardware-oriented ciphers, which include algorithms

Grain, MICKEY and Trivium, showed that long shift registers are usually used in their structure. This allows during the hardware implementation for one time cycle of the device to change the state of the key stream generator, which allows to achieve extremely high rates of PRS formation rate. To ensure cryptographic stability, a nonlinear function is also added to the generator structure, which is implemented either by nonlinear feedback in the shift registers or, for example, through the use of a cryptographic Boolean function with high nonlinearities. The complexity of cryptanalysis also increases due to the use of uneven movement of registers, which complicates the original PRS.

5. All ciphers researched in the eSTREAM project are the result of careful research of the author's teams of developers and certify the level of the relevant scientific cryptographic school. Unfortunately, Ukrainian researchers did not take part in the competition, but the experience of organizing and conducting the eSTREAM project and the results obtained, including the structure of selected cryptocurrencies, basic primitives used, requirements and restrictions for the key SSC system, etc., must be taken into account. time of development of new stream ciphers in Ukraine.

References

1. eStream, Stream cipher project of the European Network of Excellence in Cryptology ECRYPT. http://www.ecrypt.eu.org/stream/
2. The eSTREAM Project—eSTREAM Phase 3. Rabbit (Portfolio Profile 1). Access mode http://www.ecrypt.eu.org/stream/rabbitpf.html.
3. The eSTREAM Project—eSTREAM Phase 3. Salsa20 (Portfolio Profile 1). Access mode http://www.ecrypt.eu.org/stream/salsa20pf.html
4. The eSTREAM Project—eSTREAM Phase 3. SOSEMANUK (Portfolio Profile 1). Access mode http://www.ecrypt.eu.org/stream/sosemanukpf.html
5. The eSTREAM Project—eSTREAM Phase 3. Grain (Portfolio Profile 2). Access mode http://www.ecrypt.eu.org/stream/grainpf.html
6. The eSTREAM Project—eSTREAM Phase 3. MICKEY (Portfolio Profile 2). Access mode http://www.ecrypt.eu.org/stream/mickeypf.html
7. The eSTREAM Project—eSTREAM Phase 3. Trivium (Portfolio Profile 2). Access mode http://www.ecrypt.eu.org/stream/triviumpf.html
8. The eSTREAM Project—eSTREAM Phase 3. HC (Portfolio Profile 1). Access mode http://www.ecrypt.eu.org/stream/hcpf.html
9. Schafheutle, M.: A First Report on the Stream Cipher SNOW. http://www.cryptonessie.org
10. Anderson, R.J.: Serpent: A Candidate Block Cipher for the Advanced Encryption Standard. University of Cambridge Computer Laboratory. Retrieved 20130114. Access mode http://www.cl.cam.ac.uk/~rja14/serpent.html
11. NESSIE Call for Cryptographic Primitives, Version 2.2. March, 2000. Access mode http://cryptonessie.org
12. NESSIE Security Report, Deliverable D21, Version 1.0. 30 Oct 2002. Access mode http://www.cryptonessie.org
13. Matsumoto, M., Saito, M., Nishimura, T., Hagita, M.: CryptMT Version 2.0: a large state generator with faster initialization. In: SASC2006 Conference Volume http://www.ecrypt.eu.org/stream/

14. Matsumoto, M., Saito, M., Nishimura, T., Hagita, M.: Cryptanalysis of CryptMT: effect of huge prime period and multiplicative filter. In: SASC2006 Conference Volume http://www.ecrypt.eu.org/stream/
15. Matsumoto, M., Nishimura, T.: Mersenne twister: a 623dimensionally equidistributed uniform pseudorandom number generator. ACM Trans. Model. Comput. Simul. **8**, 3–30 (1998)
16. Saito, M., Matsumoto, M.: Simple and fast mersenne twister. In: To be presented at Monte Carlo and Quasi Monte Carlo Method (2006)
17. Gorbenko, I.D., Dolgov, V.I., Oleynikov, R.V., Ruzhentsev, V.I., Mikhailenko, M.S., Gorbenko Y.I., Neyvanov A.V.: Principles of construction and properties of block symmetric cipher "Kalina". In: Applied Radio Electronics, issue 2 (2007)
18. Batina, L., Lano, J., Ors, S.B., Preneel, B., Verbauwhede, I.: Energy, perfomance, area versus security tradeoffs for stream ciphers. In: The State of the Art of Stream Ciphers: Workshop Record, pp. 302–310, Brugge, Belgium, Oct 2004

Chapter 10
Analysis of Stream Cryptographic Transfer Algorithms for Light (Less-Resource) Cryptographies Defined in ISO/IEC 29192

10.1 Stream Cipher Enocoro

The *Enocoro* cipher belongs to a family of PRS generators with 11 parameters. Its general specification was published in [1, 2]. The previous version of the algorithm recommended certain parameters for 80-bit and 128-bit versions, which were defined as Enocoro80v1 та Enocoro128v1. Later, the recommended settings for the 128-bit version have changed (i.e. *Enocoro-128v1* is outdated), and the new cipher is called *Enocoro-128v1.1.*

The overall part of the PRS generation algorithm, which will be discussed later, is slightly different from the first version of *Enocoro*, that is, we focus solely on the updated general version of the *Enocoro v2* algorithm, which was described in detail in [1]. The difference between the *Enocoro-128v2* and *Enocoro-128v1.1* algorithms lies only in the characteristic of the polynomial φ_8 over the finite field $GF(2^8)$ and the initialization process.

To describe the specifications of the stream cipher Enocoro use the designations given in Table 10.1 [1, 2].

10.1.1 Data Structure

The unit of data size of the *Enocoro* algorithm is 8 bits, or one byte.

Submission of data. The *Enocoro* cipher uses operations on the finite field $GF(2^8)$ and $GF(2^4)$. Elements of the binary expanded field are represented in polynomial form, with coefficients 0 and 1. The polynomial is represented as a bit string. For example, bit string 0 x2 is represented as a polynomial x. For a simple explanation, consider the notation for the field $GF(2^8)$. The field element $GF(2^8)$ is supplied by a polynomial whose degree is less than 8. Such a polynomial as

A. A. Kuznetsov et al., *Stream Ciphers in Modern Real-time IT Systems*,
Studies in Systems, Decision and Control 375,
https://doi.org/10.1007/978-3-030-79770-6_10

Table 10.1 Operations and designations used in the Enocoro cipher specification

Designation	Operation or designation
$\oplus$	Operation bitwise addition modulo 2 (XOR)
$\wedge$	Bitwise operation AND
||	Concatenation operation
$+$	Operation of adding 32-bit numbers modulo 2^{32}
$<<<_m n$	Left shift operator by n bits (above m-bit register)
$>>>_m n$	Right shift bit by n bits (above m-bit register)
0x	Hexadecimal representation

$$b_7x^7 + b_6x^6 + b_5x^5 + b_4x^4 + b_3x^3 + b_2x^2 + b_1x + b_0$$

given as

$$b_7||b_6||b_5||b_4||b_3||b_2||b_1||b_0$$

where b_j is 0 or 1.

For example, the polynomial $x^6 + x^4 + x^2 + x + 1$ is given as 0 x7 = 01010111.

Addition The sum of two polynomials over a finite field is a polynomial whose coefficients are the sum of the corresponding coefficients modulo 2. On the other hand, the addition is given by a bit addition *(XOR)* of two rows. For example, the sum of 0x7 and 0xa3 is calculated as follows:

$$\begin{aligned} 0x57 + 0x87 &= (x^6 + x^4 + x^2 + x + 1) + (x^7 + x^5 + x + 1) \\ &= x^7 + x^6 + x^5 + x^4 + x^2 \leftrightarrow 0xf4 \end{aligned}$$

Multiplication For the multiplication operation, it is first necessary to determine the polynomial φ_8 of the eighth degree. In the *Enocoro v2* specification, the following polynomial is used:

$$\varphi_8(x) = x^8 + x^4 + x^3 + x^2 + 1.$$

In bit format, $\varphi_8(x)$ is given as 0 x11d[3].

Multiply polynomial $f(x) = \sum a_i x^i$ by *x*:

$$x \cdot f(x) = \sum a_i x^{i=1} mod \varphi_8(x).$$

For example

$$\begin{aligned} 0x02 \cdot 0x87 &= x \cdot \left(x^7 + x^2 + x + 1\right) = x^8 + x^3 + x^2 + x \\ &= (x^4 + x^3 + x^2 + 1) + (x^3 + x^2 + x) \\ &= x^4 + x + 1 = 0x13. \end{aligned}$$

The multiplication of $f(x)$ by x^i for any positive integer i occurs by induction. Multiplication of any two elements is as follows:

$$f(x) = \sum a_i x^i, g(x) = \sum b_i x^i$$

Definition of the GF (2^4). The *Enocoro* cipher uses multiplication in the $GF(2^4)$ field as it does in the $GF(2^8)$. The representation of elements and operations is performed in the same way as in $GF(2^8)$. The $GF(2^4)$ field element is fed by a 4-bit string $b_3||b_2||b_1||b_0$, which are given in polynomial form as $b_3x^3 + b_2x^2 + b_1x + b_0$.. The following are the characteristics of polynomial φ_4 over the finite field $GF(2^4)$:

$$\varphi_4(x) = x^4 + x + 1$$

Generator of Pseudorandom Numbers The PRS generator contains the state of the finite state machine (*FSM*), the *Init* initialization function, and the *Out* function. *FSM*, in turn, consists of an internal state (or register) of $S^{(t)}$, which depends on the meter, and the function of its *Next* update. The initialization function sets the initial internal state $S^{(0)}$ from the initial input, that is, from the secret key K and the initialization vector *IV*. The output function generates output bits $Z^{(t)}$ from the internal state $S^{(t)}$ at each instant of time t. So we have:

$$\begin{aligned} S^{(0)} &= Init(K, I), \\ Z^{(t)} &= Out\left(S^{(t)}\right), \\ S^{(t+1)} &= Next\left(S^{(t)}\right). \end{aligned}$$

The block diagram of the *Enocoro* key flow generator is shown in Fig. 10.1.

PANAMA-Like Key Stream Generator *PANAMA-like* Key Stream Generator is a class of *PRNG* generators that defines *PANAMA*-oriented *PRNG* software [1, 2]. Internally, the state of such a generator is divided into two: state $a^{(t)}$ and buffer $b^{(t)}$. The upgrade functions are denoted as ρ and λ respectively, and both functions

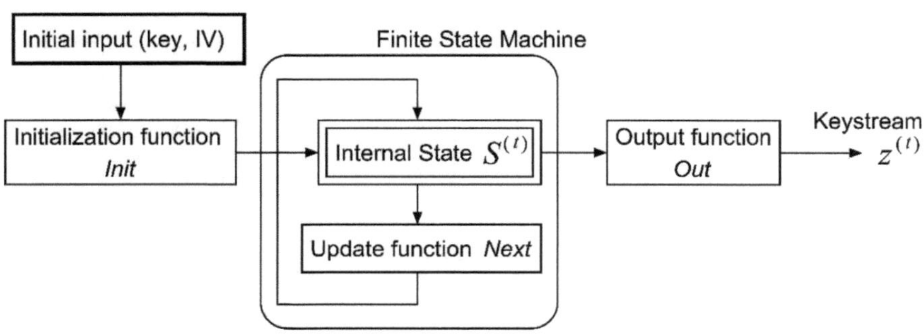

Fig. 10.1 Block diagram of the *Enocoro* key stream generator

assume a different intermediate internal state as a parameter. Completely the entire *Next* function is a composition of ρ and λ:

$$\left(a^{(t+1)}, b^{(t+1)}\right) = Next\left(S^{(t)}\right) = \left(\rho\left(a^{(t)}, b^{(t)}\right), \lambda\left(a^{(t)}, b^{(t)}\right)\right).$$

10.1.2 The General Specification of Enocoro V2

Consider the *Enocoro v2* Key Stream Generator family specification.

This cipher has 11 parameters. Let the size of the cipher buffer be n_b bytes and the input from the buffer in ρ function to be

$$b_{k_1}, b_{k_2}, b_{k_3}, b_{k_4}.$$

The parameters that define its λ function will be

$$q_1, p_1, q_2, p_2, q_3, p_3.$$

If you detail the parameters, then *Enocoro* is denoted as

$$(n_b; k_1, \ldots, k_4, q_1, p_1 \ldots, q_3, p_3).$$

A schematic representation of the *Enocoro* cipher is shown in Fig. 10.2.

Internal State Condition a consists of two bytes. Higher bytes are denoted as a_0,, younger bytes are designated a_1. Buffer b consists of n_b bytes $b_0, b_1, \ldots, b_{nb-1}$.

Function ρ The status update function ρ in *Enocoro v2* cipher uses as input parameters. Function ρ consists of corresponding S-boxs, linear transformation L over

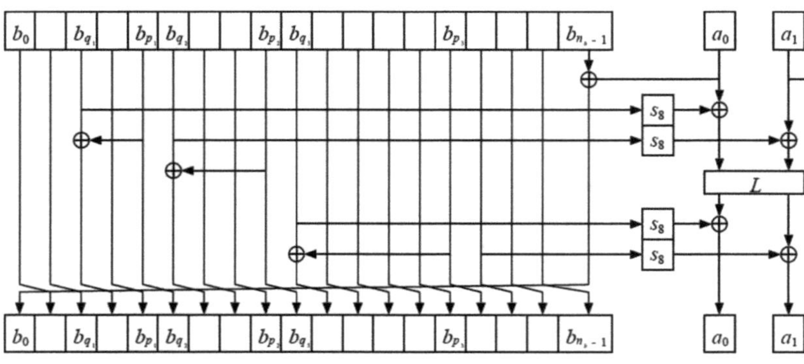

Fig. 10.2 Schematic representation of the *Enocoro* cipher

the $GF(2^8)$ field, and bit addition operations (*XOR*). In particular, the conversion is defined as follows:

$$\begin{aligned} u_0 &= a_0^{(t)} \oplus s_8\left[b_{k_1}^{(t)}\right], \\ u_1 &= a_1^{(t)} \oplus s_8\left[b_{k_2}^{(t)}\right], \\ (v_0, v_1) &= L(u_0, y_1), \\ a_0^{(t+1)} &= v_0 \oplus s_8\left[b_{k_3}^{(t)}\right], \\ a_1^{(t+1)} &= v_1 \oplus s_8\left[b_{k_4}^{(t)}\right]. \end{aligned}$$

Linear Conversion The transformation L of the cipher is a linear transformation of the matrix 2×2 over the field $GF(2^8)$ and is defined as follows:

$$\begin{pmatrix} v_0 \\ v_1 \end{pmatrix} = L(u_0, u_1) = \begin{pmatrix} 1 & 1 \\ 1 & d \end{pmatrix}\begin{pmatrix} u_0 \\ u_1 \end{pmatrix}, \ d \in GF(2^8),$$

where $d = 0\text{x}02$ is a constant used in the *Enocoro v2* cipher.

S-box. S-box (wildcard) s_8 is an 8-bit wildcard input into 8-bit output. It also contains an *SPS*-structure and includes 4 small S-boxes s_4 (4-bit input to 4-bit output) and a linear conversion l (matrix 2×2 over the $GF(2^4)$ field). The s_4 S-box looks like this:

$$s_4[16] = \{1, 3, 9, 10, 5, 14, 7, 2, 13, 0, 12, 15, 4, 8, 6, 11\}.$$

The linear transformation of l is as follows:

$$l(x, y) = \begin{pmatrix} 1 & e \\ e & 1 \end{pmatrix}\begin{pmatrix} x \\ y \end{pmatrix}, \ x, y, e \ \in GF(2^8).$$

In Figs. 10.3 and 10.4 show the construction of 8x8 Sboks s_8, which contains the following:

$$\begin{aligned} y_0 &= s_4[s_4[x_0] \oplus e \cdot s_4[x_1] \oplus 0x\text{a}], \\ y_1 &= s_4[e \cdot s_4[x_0] \oplus s_4[x_1] \oplus 0x5]. \end{aligned}$$

where $e = 0\text{x}04$ is the constant used in *Enocoro v2*. The output value is shifted 1 bit to the left:

$$s_8[x] = (y_0 \| y_1) <<<_8 1.$$

A table view of *S-box* s_8 is shown in Fig. 10.3. A schematic representation of the structure of *S-box* s_8 is shown in Fig. 10.4.

$s_8[256] = \{$
```
99, 82, 26, 223, 138, 246, 174, 85, 137, 231, 208, 45, 189, 1, 36, 120,
27, 217, 227, 84, 200, 164, 236, 126, 171, 0, 156, 46, 145, 103, 55, 83,
78, 107, 108, 17, 178, 192, 130, 253, 57, 69, 254, 155, 52, 215, 167, 8,
184, 154, 51, 198, 76, 29, 105, 161, 110, 62, 197, 10, 87, 244, 241, 131,
245, 71, 31, 122, 165, 41, 60, 66, 214, 115, 141, 240, 142, 24, 170, 193,
32, 191, 230, 147, 81, 14, 247, 152, 221, 186, 106, 5, 72, 35, 109, 212,
30, 96, 117, 67, 151, 42, 49, 219, 132, 25, 175, 188, 204, 243, 232, 70,
136, 172, 139, 228, 123, 213, 88, 54, 2, 177, 7, 114, 225, 220, 95, 47,
93, 229, 209, 12, 38, 153, 181, 111, 224, 74, 59, 222, 162, 104, 146, 23,
202, 238, 169, 182, 3, 94, 211, 37, 251, 157, 97, 89, 6, 144, 116, 44,
39, 149, 160, 185, 124, 237, 4, 210, 80, 226, 73, 119, 203, 58, 15, 158,
112, 22, 92, 239, 33, 179, 159, 13, 166, 201, 34, 148, 250, 75, 216, 101,
133, 61, 150, 40, 20, 91, 102, 234, 127, 206, 249, 64, 19, 173, 195, 176,
242, 194, 56, 128, 207, 113, 11, 135, 77, 53, 86, 233, 100, 190, 28, 187,
183, 48, 196, 43, 255, 98, 65, 168, 21, 140, 18, 199, 121, 143, 90, 252,
205, 9, 79, 125, 248, 134, 218, 16, 50, 118, 180, 163, 63, 68, 129, 235
```
$\};$

Fig. 10.3 Tabular presentation of *S-box* s_8

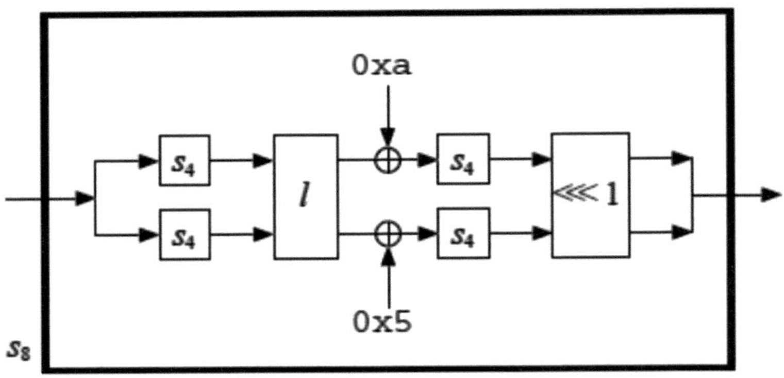

Fig. 10.4 Schematic representation of *S-box* s_8 structure

Function λ The *Enocoro v2* encryption function consists of three feedbacks (*XOR*), bitwise additions (*XOR*) a_0 with the rightmost byte of the b_{nb-1} buffer of vector b and a buffer byte offset. This process is described in detail below:

$$
\begin{aligned}
b_i^{(t+1)} &= b_{i-1}^{(t)}, \; i \neq 0, q_1+1, q_2+1, q_3+1, \\
b_0^{(t+1)} &= b_{n_b-1}^{(t)} \oplus a_0^{(t)}, \\
b_{q_j+1}^{(t+1)} &= b_{q_j}^{(t)} \oplus b_{p_j}^{(t)}, i = 1, 2, 3,
\end{aligned}
$$

where $p_i - q_i \neq p_j - q_j$ if i $\neq$ j.

Out Function The *Enocoro v2* encryption function outputs lower status bytes:

$$Out\left(S^{(t)}\right) = a_1^{(t)}.$$

Introduction and Initialization Function Setting the input (key and vector of initialization) and defining the initialization function for each algorithm are described in more detail below.

10.1.3 Enocoro-128v2 *Stream Encryption Specification*

Parameters *Enocoro-128v2* is a PRS generator with 128-bit input key, 64-bit initialization vector, and with the following specific parameters:

$$\begin{aligned} &n_b = 32, \\ &k_1 = 2, \quad k_2 = 7, \quad k_3 = 16, \quad k_4 = 29, \\ &p_1 = 6, \quad p_2 = 15, \quad p_3 = 28, \\ &q_1 = 2, \quad q_2 = 7, \quad q_3 = 16. \end{aligned}$$

Initialization Function At the beginning of the work, the initialization function sets in the registers the values of the key K, the vector of initialization IV and the internal state C as follows:

$$\begin{aligned} b_i^{(-96)} &= K_i, \quad 0 \le i < 16, \\ b_{i+16}^{(-96)} &= I_i, \quad 0 \le i < 8, \\ b_{24}^{(-96)} &= C_0 = 0x66, \\ b_{25}^{(-96)} &= C_1 = 0xe9, \\ b_{26}^{(-96)} &= C_2 = 0x4b, \\ b_{27}^{(-96)} &= C_3 = 0xd4, \\ b_{28}^{(-96)} &= C_4 = 0xef, \\ b_{29}^{(-96)} &= C_5 = 0x8a, \\ b_{30}^{(-96)} &= C_6 = 0x2c, \\ b_{31}^{(-96)} &= C_7 = 0x3b, \\ a_0^{(-96)} &= C_8 = 0x88, \\ a_{1a}^{(-96)} &= C_9 = 0x4c. \end{aligned}$$

Then the state is updated with 96 iterations of two functions: one of them is a bitwise addition of the counter value from b_{31}, the other is an update of the *Next* function. The size of the counter takes bytes. Initially, it is initialized as 0 x01 and then incremented by multiplying by 0 x02 in the finite field $GF(2^8)$.

```
Init (a[2], b[32], K[16], I[8]){
  // set initial values
  for (i = 0; i < 16; i++){ b[i] = K[i]; }
  for (i = 0; i < 8; i++){ b[i+16] = I[i]; }
  for (i = 0; i < 8; i++){ b[i+24] = C[i]; }
  a[0] = C[8]; a[1] = C[9];
  ctr = 1;

  // update the state 96 times
  for (r = 0; r < 96; r++){
      b[31] ^= ctr;
      ctr = gf256multiplication(ctr, 2);
      Next(a, b);
  }
}
```

Fig. 10.5 The pseudocode initialization function

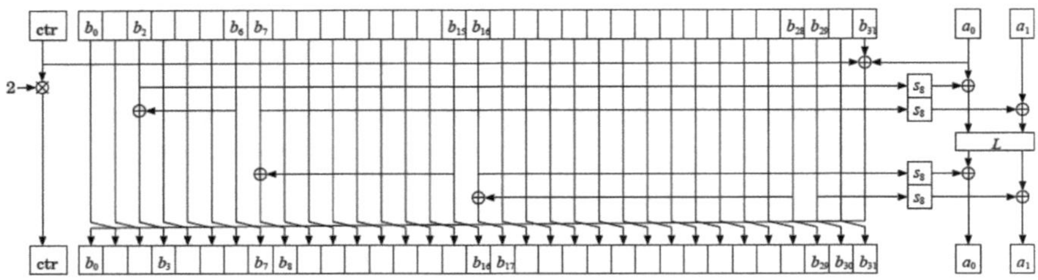

Fig. 10.6 Status update process during *Enocoro-128v2* initialization

To avoid confusion, the initialization function is shown in the pseudocode below (Fig. 10.5). The process of updating the state during initialization of Enocoro-128v2 is schematically shown in Fig. 10.6.

10.1.4 Data Encryption Using Enocoro-128v2

Choice of Key and Vector of Initialization Preferably, the output sequence generated by the PRS generator is uniquely determined by the combination of the secret key K and the initialization vector IV. For this reason, it is forbidden to use the identification combination twice. Especially if a key stream is generated on the same key, it is imperative to use a different initialization vector.

Encryption and Decryption The binary assembly module provides a data encryption mechanism using a PRS generator, and it simply combines the bitstream (XOR) of the key stream and the input. The decryption happens the same way. Let $p^{(t)}$, $c^{(t)}$, $z^{(t)}$ be

Fig. 10.7 Test vectors for the *Enocoro-128v2* encryption algorithm

key[16] = {0}
iv[8] = {0}
output =
`0x63 0xd7 0xda 0x6b 0x55 0x73 0x7f 0xcf`
`0x57 0x34 0xb6 0x77 0x3a 0xe7 0x72 0xe8`
...

key[10] =
{`0x00, 0x01, 0x02, 0x03, 0x04, 0x05, 0x06, 0x07`
`0x08, 0x09, 0x0a, 0x0b, 0x0c, 0x0d, 0x0e, 0x0f`}
iv[8] =
{`0x00, 0x10, 0x20, 0x30, 0x40, 0x50, 0x60, 0x70`}
output =
`0xc8 0xc8 0xee 0x43 0x3b 0x0d 0xc0 0x40`
`0xe5 0x3b 0xc5 0x06 0xea 0x21 0xad 0x82`
...

input, ciphertext, and output byte at time t, respectively. Then bit-add in encryption and decryption operations will look like this:

$$c^{(t)} = p^{(t)} \oplus z^{(t)},$$
$$p^{(t)} = c^{(t)} \oplus z^{(t)}.$$

Finally, here are the test vectors (see Fig. 10.7), which make sure that the *Enocoro-128v2* encryption algorithm is implemented correctly.

10.1.5 Key Stream Statistics Tests

In a study [3], statistical tests were performed to check the randomness of a key stream. A set of statistical *NIST* was selected for the studies [4]. This is a set of 15 statistical tests that can be applied to bit PRS. In the context of pseudorandom generators, it is very important for cryptographic purposes to understand the applicability of these types of tests. If the generator fails one or more tests, it gives a clear idea that the key stream is not accidental and susceptible to appropriate attacks (and possibly a key recovery attack). If, on the other hand, the generator passes all the tests, it does not indicate the safety of the generator. The reason for this is that statistical tests investigate the properties of PRS and do not take into account the features of the actual design of the generator.

In [3], 200 key streams generated by a pair of random key *K*/initialization vector *IV* were tested. Each sequence was 10^6 bits long. The test results are shown in Table 10.2 [3].

Table 10.2 Results of statistical testing of PRS, which are generated by the *Enocoro* stream cipher

Statistical test	Proportion	*P*-values	C1	C2	C3	C4	C5	C6	C7	C8	C9	C10
Frequency	198/200	0.196920	23	12	18	24	17	19	17	17	22	31
BlockFrequency	198/200	0.678686	20	18	23	26	20	19	25	13	18	18
CumulativeSums	198/200	0.428095	23	18	17	23	14	13	20	21	24	27
Runs	199/200	0.064822	24	18	31	18	14	15	23	27	17	13
LongestRun	197/200	0.689019	19	25	19	20	21	24	11	20	22	19
Rank	195/200	0.096578	32	17	18	28	17	14	17	22	16	19
FFT	197/200	0.080519	17	31	18	18	16	24	19	26	10	21
NonOverlappingTemplate	197/200	0.419021	20	18	29	15	24	21	18	20	13	22
OverlappingTemplate	199/200	0.202268	31	23	21	14	15	21	21	23	16	15
Universal	196/200	0.334538	18	20	18	24	24	17	21	29	12	17
ApproximateEntropy	199/200	0.534146	22	29	24	20	17	16	19	16	21	16
RandomExcursions	128/129	0.973388	12	13	14	16	11	13	11	15	12	12
RandomExcursionsVariant	128/129	0.182384	11	8	12	12	10	21	12	12	12	19
Serial	194/200	0.162606	22	28	16	25	22	20	23	19	16	9
LinearComplexity	196/200	0.437274	18	19	30	20	22	19	14	23	20	15

According to the results obtained, the *P*-meaning can be interpreted as the probability that an ideal random number generator would generate a sequence less random than the one tested. Each row of Table 10.2 states: the name of the test, the proportion of sequences that passed the appropriate test, the P-Values and the distribution of *200 P*-values for the individual tests. As can be seen, the *Enocoro-128v2* cipher has not been tested for all 200 tests (PRS molded). However, obtained $\alpha = 0,01$, as expected. A detailed description of the tests can be found in [3].

Summarizing the results of the statistical testing, we can conclude that the key stream *Enocoro-128v2* has passed all the statistical tests of the *NIST* test kit by the appropriate method. But it should be noted that this does not say anything about the security of the streaming cipher.

10.1.6 Attacks Based on a Compromise of Time-Memory-Data

Consider common attacks based on time-memory-data compromise on *Enocoro-128v2*, in particular focusing on two main types of compromises for stream ciphers, namely the Babbage-Golich compromise and the Biriukov-Shamir compromise. Denote by: N is the search space, t is the time complexity of the online attack phase, D is the amount of data required to attack, P is the time complexity offline (or during pre-calculation) and M is the amount of memory (storage) required for attacks. Consider both attacks on the internal state and the secret key.

In the overall discussion, we denote the key sizes *K* and the initialization vector *IV* as *k* and *iv*, respectively.

Attacks on the Internal State

The *Enocoro-128v2* algorithm has a 272-bit internal state and a 128-bit key. That is, we can assume that *Enocoro-128v2* can resist attacks based on a compromise of time-data-memory-internal state. The evaluation of the complexity of different options for such attacks is summarized in Table 10.3.

Table 10.3 shows the complexity of an attack on *Babbage-Golic (BG)* and two variants with different parameters for *Biryukov-Shamir* (BS1 and BS2) attack on the internal state. Other compromises are also possible to implement this attack.

Attacks on Initialization Functions

The attacks discussed above can also be applied to initialization functions [3]. The one-way function f is considered in these attacks to be a reflection of the $(k + iv)$ bits of the key and the initialization vector into the first $(k + iv)$ bits of the key stream, i.e.:

$$f : \{0, 1\}^{k+iv} \rightarrow \{0, 1\}^{k+iv}.$$

In this case, $N = 2^{k+iv}$. When using *Enocoro-128v2*, the key consists of 128 bits and the initialization vector is 64 bits.

Table 10.4 shows the complexity of the *Babbage-Golic (BG)* attack and the two options with different parameters for the *Biryukov-Shamir* attack (BS_1 and BS_2) on the initialization function. There are also other options for compromising this attack.

Other Notes about Attacks Based on Compromise of Time—Memory—Data

In the selected scenario, the initialization vector can be fixed to build tables for the attacks under consideration. In this case, $N = 2^k$. Even if the *IV* initialization vector cannot be fully determined, it may be that some of the *IV* will be detected and/or known to the attacker in advance. This will greatly improve the effectiveness of the attacks. For comparison, Table 10.4 shows the corresponding figures for *P, M, T* and *D*, when both halves *IV* and whole *IV* can be selected or predicted by the attacker

Table 10.3 Assessing the complexity of a compromise attack time—memory—data to the internal state

Time-memory-data compromise attacks on the internal state						
	General			Enocoro-128v2 (N = 272)		
	BG	BS1	BS2	BG	BS1	BS2
P	$N^{1/2}$	$N^{3/4}$	$N^{2/3}$	2^{136}	2^{204}	2^{181}
T	$N^{1/2}$	$N^{1/2}$	$N^{2/3}$	2^{136}	2^{136}	2^{181}
M	$N^{1/2}$	$N^{1/2}$	$N^{1/3}$	2^{136}	2^{136}	2^{91}
D	$N^{1/2}$	$N^{1/4}$	$N^{1/3}$	2^{136}	2^{68}	2^{91}

Table 10.4 Difficulty ratings for attack based on time-memory-data compromise on the initialization function, when the attacker can control 0, 50 and 100% of the initialization vector

Time-memory-data compromise attacks on the initialization function												
	General			Enocoro-128v2 (N = 192)								
				No IV Control			50% IV Control			Full IV Control		
	BG	BS1	BS2	BG	BS1	BS2	BG	BS1	BS2	BG	BS1	BS2
P	$N^{1/2}$	$N^{3/4}$	$N^{2/3}$	2^{96}	2^{144}	2^{128}	2^{80}	2^{120}	2^{107}	2^{64}	2^{96}	2^{85}
T	$N^{1/2}$	$N^{1/2}$	$N^{2/3}$	2^{96}	2^{96}	2^{128}	2^{80}	2^{80}	2^{107}	2^{64}	2^{64}	2^{85}
M	$N^{1/2}$	$N^{1/2}$	$N^{1/3}$	2^{96}	2^{96}	2^{64}	2^{80}	2^{80}	2^{53}	2^{64}	2^{64}	2^{43}
D	$N^{1/2}$	$N^{1/4}$	$N^{1/3}$	2^{96}	2^{48}	2^{64}	2^{80}	2^{40}	2^{53}	2^{64}	2^{32}	2^{43}

[3]. The applicability of these numbers, of course, depends on the use case and how the *IV* initialization vectors are selected.

It is important to note that there are important fundamental differences between attacks on the internal state and attacks on initialization functions. Internal state attacks can be used to break a stream cipher that uses one specific key and one *IV* vector of initialization. A key stream from this key and *IV* is used in this type of attack. If the status update function and the initialization function are reversible, it will also be possible to recover the *IV* key and vector. Attacks on initialization functions recover the key and *IV* for the specific key stream used. Only one key stream is used for each *IV* initialization key/vector pair. Thus, this attack requires a key stream from different key/*IV* pairs and recovers only one such pair.

Conclusions on the Possibility of Using a Compromise Attack Time—Memory—Data

The internal state of *Enocoro-128v2* is 272 bits. Thus, there is no attack on the basis of time-memory-data compromise on the internal state, which takes time, memory and data less than 2^{128}. The key search will always be more effective than the online phase of such an attack.

The relatively short value of the *IV* initialization vector (64 bits) makes the *Enocoro-128v2* cipher more vulnerable to attack based on a time-to-memory compromise on the initialization function. There are several options for performing these attacks with time, memory and data costs of less than 2^{128}. However, the preliminary computation time is equal to or more than the time spent on exhaustive key search, even if data and/or memory costs are reduced. In addition, only one of the several *IV* key/vector pairs can be recovered. If an attacker can choose an *IV*, then the difficulty to recover one of the 2^{64} keys will be quite moderate.

It has nothing to do with the length of the *IV* initialization vector, but depends on the 128-bit key.

10.1.7 Initialization Vector Recovery Attacks

Under the *IV* Initialization vector recovery attacks, we mean cryptographic attacks that aim to initialize the cipher. These attacks typically use some of the key stream sequences obtained from the same key, but when using different *IV* values [3]. We may also look at several key stream sequences generated by different but related keys and possibly by different *IV* values. This gives relevant key attacks.

Denote by $i_0, i_1, \ldots, i_{m-1}$ bits *IV*, and $k_0, k_1, \ldots, k_{n-1}$—key bits, and let z_i be the key bit of the key stream. Note that bits are indicated here, not bytes, as in the cipher specification. Then you can write z_i as a function of the key bits and the initialization vector:

$$z_i = f_i(i_0, i_1, \ldots, i_{m-1}, k_0, k_1, \ldots, k_{n-1}), \; i_j \in I, \; k_j \in \mathrm{K}.$$

The basic assumption is that we know $Z = z_0, z_1, \ldots$ for a number of different values of $IV(i_0, i_1, \ldots, i_{m-1}) \in I$ for some set I.

Consider a few possible attacks for this assumption. One is differential attack.

Selective Differential Attack on Initialization Vector The Differential Initialization Attack treats the *Enocoro-128v2* as 96 round functions and examines how input differences, in the case of *IV* bits and possibly key bits, extend through 96 rounds and eventually appear in the key bit stream.

Referring to the *Enocoro-128v2* initialization, it can be seen that the key bits set the state of bytes b_0–b_{15}, and the bits of the initialization vector are *IV* – b_{16}–b_{23}.. Introduce the notation

$$Y(t) = (b_0(t), b_1(t), \ldots, b_{31}(t)|a_0(t), a_1(t))$$

for full condition after t rounds (offsets). In addition, consider the differences between the generation of any two sequences. We have a sequence of values of the state $Y(t)$ for the generation of the first key stream and another sequence of values of the state $Y'(t)$ for the generation of the second key stream. In addition, denote by *XOR* the difference between them $\delta Y(t)$, that is

$$\delta Y(t) = Y(t) + Y'(t).$$

A promising differential attack may be if we start from one differential in the second byte *IV*, that is, in the state variable b_{17}. Then we will have a differential spread like

$$\delta Y(0) = (0000000000000000a000000000000000 \mid 00),$$

then

$$\delta Y(12) = (00000000000000000000000000000a000 \mid 00),$$
$$\delta Y(13) = (0000000000000000a00000000000a00100).$$

Further, the difference in b_{29} goes through S-block to a_1 also as a one-step transition. The difference after S-block is now the new zero value of b. That is, we reach

$$\delta Y(14) = (00000000000000000a000000000000a0 \mid 0b).$$

Further, the difference $a_0, a_1, \delta(a_0, a_1) = (0, b)$ turns to $\delta(a_0, a_1) = (b, c = d * b)$,

$$\delta Y(15) = (000000000000000000a000000000000a \mid bc).$$

This approach would lead to a successful attack if we could eliminate the differences b_{31}, a_0, a_1 in the next step. However, since the updates a_0, a_1 are linear, it is impossible to reach $\delta(a_0, a_1) = (0, 0)$ in the next step, and the probability of reaching

$$\delta Y(16) = (000000000000000000a000000000000 \mid 00)$$

is zero. This shows that the linearity in the update (a_0, a_1) is critical. If the update is non-linear, there will probably be some positive probability of reaching $Y(16)$, as described above. The procedure can be repeated to reach a large number of rounds, possibly all 96.

Please note that this type of analysis is closely related to the search for collisions for hash functions.

Statistical Testing of Attacks on Initialization Vector The key statistical tests of the key stream sequence are very simple and, as a rule, not very successful if the initialization is carefully designed and investigated.

Again, consider a few bits of the key stream $Z = z_0, z_1, \ldots$ for a number of different values of initialization vectors $(i_0, i_1, \ldots, i_{m-1}) \in I$ for some set I. Consider the n bit lines of the key stream generated for different IV like

$$Z_0, \ Z_1, \ldots, \ Z_{m-1}.$$

This can be considered as m implementations taken from the distribution D as bit strings of length n. If we can find a statistical test that is very likely to show that D is not an even distribution, then we have a delimiter.

We perform several simulations of generating m of different output rows of length n for a fixed key, and then use a set of statistical tests provided by *NIST* at the output [4]. Table 10.5 summarizes the results of one such test [3]. This test received 1 byte $(n = 8)$ of 1,000,000 initialization using a fixed key and a random *IV* initialization vector. This was done for 200 random keys [3]. The table is prepared by analogy to Table 10.2.

Note that some tests run with different parameters. Only one result for each test was included in the table. The *RandomExcursions* and *RandomExcursionsVariant* tests were interrupted in some cases (the number of zero crossings is too small to give a good result) [3].

Conclusions on the Possibility of Using an Initialization Vector Recovery Attack Neither selective differential attacks nor statistical analysis provide attacks that could be potentially more effective than a complete key search. A differential attack will be powerful if the differential (a_0, a_1) of the update can be canceled, but this is not possible due to the linearity (a_0, a_1) of the update. General statistical tests using the *NIST* test suite [4] do not indicate any deviation for chance when considering multiple key streams from initialization with a random *IV* and a fixed key.

Table 10.5 The results of the NIST statistical test run for the Enocoro-128v2 key stream

Statistical test	Proportion	*P*-value	C1	C2	C3	C4	C5	C6	C7	C8	C9	C10
Frequency	197/200	0.118812	21	29	26	12	12	19	16	23	23	19
BlockFrequency	196/200	0.296834	19	15	13	27	23	23	14	18	24	24
CumulativeSums	195/200	0.036352	24	26	24	16	11	16	18	15	18	32
Runs	198/200	0.118812	15	22	15	19	25	30	18	21	24	11
LongestRun	199/200	0.242986	16	22	26	26	14	26	20	21	16	13
Rank	199/200	0.112047	13	21	33	18	16	21	19	16	18	25
FFT	197/200	0.249284	21	28	19	20	13	28	16	18	22	15
NonOverlappingTemplate	198/200	0.960198	21	22	19	21	17	21	25	16	20	18
OverlappingTemplate	195/200	0.0 57146	24	30	20	28	17	12	19	22	14	14
Universal	200/200	0.055361	20	14	24	20	15	9	29	26	24	19
ApproximateEntropy	197/200	0.202268	18	24	28	19	25	18	16	15	12	25
RandomExcursions	173/174	0.679903	19	20	21	19	17	14	18	21	10	15
RandomExcursionsVariant	174/174	0.716158	14	13	15	14	19	22	16	19	20	22
Serial	197/200	0.788728	20	18	22	13	21	26	22	18	22	18
LinearComplexity	199/200	0.875539	22	22	18	21	13	21	24	17	21	21

10.1.8 The Maximum Degree of Monomials Test

The maximum degree of monomials test can be used to test if all bits of *IV* initialization keys/vectors are correctly added to the state at the initialization stage. The initialization procedure is considered as a Boolean function. The input for the Boolean function is a key set/*IV*. Exit is the first bit of the key stream. Other bits of the key stream can be considered, but the more key streams are generated, the more bits of the *IV* initialization key/vector are added to the state. Let I be the set of all bits of the *IV* initialization vector, and K is the set of all key bits, and let z_0 be the first bit of the key stream.

Consider the function

$$z_i = f_i(i_0, i_1, \ldots, i_{m-1}, k_0, k_1, \ldots, k_{n-1}), \; i_j \in I, \; k_j \in K.$$

The idea behind the test is to look at the coefficient of the higher-degree monomial in the Boolean function, that is, the monomial $i_0, i_1, \ldots, i_{m-1}, k_0, k_1, \ldots, k_{n-1}$.

In a random Boolean function, this factor must be 1 with probability 0.5. The idea is to find a subset of *IV* initialization key/vector bits for which this coefficient is 0 with a very high probability. If only *IV* bits are used, this will make it possible to selectively attack the *IV* resolver. If key bits are used then this cannot be associated with any attack as the attacker will generally not be able to select the value of specific key bits. However, this will still indicate the non-randomness of the streaming encryption, which could potentially be used in other attacks.

Finding a coefficient for the maximum degree of a monomial Boolean function can be very easily accomplished by a simple *XOR* operation for all records of the truth table, that is, the coefficient is set

$$\sum_{i_0,i_1,\ldots,i_{m-1},k_0,k_1,\ldots,k_{n-1}\in\{0,1\}^{n+m}} f(i_0, i_1, \ldots, i_{m-1}, k_0, k_1, \ldots, k_{n-1}),$$

over F_2.

The complexity of determining the coefficient of a monotone of the maximum degree depends exponentially on the number of bits selected for the key/*IV*. Finding the best subset of the $n + m$ bits of key/vector initialization is an open problem. One method is to use the greedy algorithm. The best subset of a bits is used as a starting point to determine the subset of $a + b$ bits. The complexity of this algorithm for selectively distinguishing *IV* is specified as

$$\binom{|I|}{i} + \sum_{j=0}^{n+m} \binom{|I| - i - j*b}{b}$$

and for a non-random detector the complexity is defined as

$$\binom{|I| + |K|}{i} + \sum_{j=0}^{n+m} \binom{|I| + |K| - i - j*b}{b}.$$

The greedy algorithm gives very good results, for example, for the analysis of the *Grain-128* and *Trivium* algorithms. This definition algorithm was also applied to *Enocoro-128v2*. In [3], $i = 1$ was selected and the maximum degree of the monomial was tested using the greedy algorithm for $b = 1$, $b = 2$, and $b = 3$. Testing for $n = 0$ was also carried out, which corresponds to a selective attack of resolution *IV* and $n = 128$, which corresponds to the randomness test.

The tests were conducted as follows. The *Enocoro-128v2* initialization procedure was modified to form 8 bits at each initialization step. These bits are stored in the vector x. Thus, the first 8 bits in the vector corresponded to the first key byte stream if 0 was used to initialize the digits, the next 8 bits of the vector x corresponded to the first output bytes if 1 was used for initialization, and so on. All vectors x, each corresponding to one bit of a given sequence of $a + b$ bits key/vector initialization, are subjected to the *XOR* operation and the number of initial zeros in the result vector is calculated. Full initialization of *Enocoro-128v2* is 96 steps, i.e. $96 \bullet 8 = 768$ bits are subject to initialization. The initialization key/vector bits that are not used in the test are fixed at 0. The results are shown in Table 10.6. In Figs. 10.8 and 10.9 show graphs of test performance.

Table 10.6 shows the maximum number of prevailing zeros in the *XOR* of all vectors x, the size of the set of bits that gave the maximum number of prevailing zeros, the bits in the bit set, and the maximum size of the tested bit sets.

Table 10.6 The results of the test of the maximum degree of monomial

		Leading zeroes (%)	Bit-set size	Bit-set	Max size
Dist	b = 1	178/768 (23.2%)	20	12, 7, 23, 18, 16, 30, 19, 38, 31, 39, 20, 32, 37, 17, 29, 22, 21, 57, 41, 14	30
	b = 2	180/768 (23.4%)	19	12, 7, 23, 28, 46, 17, 22, 36, 38, 16, 18, 19, 43, 20, 21, 5, 15, 29, 55	23
	b = 3	174/768 (22.7%)	19	12, 18, 34, 46, 9, 17, 30, 20, 23, 31, 11, 14, 25, 21, 26, 27, 16, 19, 22	19
	b = 1	161/768 (21.0%),	26	Key bits: 2, 69, 77, 44, 31, 126, 62, 61, 5, 122 IV bits: 12, 32, 28, 25, 16, 21, 10, 8, 26, 9, 23, 30, 42, 38, 39, 11	26
NR	b = 2	162/768 (21.09%)	3	Key bits: 61, 115 IV bits: 12	23
	b = 3	166/178 (21.61%)	4	Key bits: 35, 122 IV bits: 12, 1	16

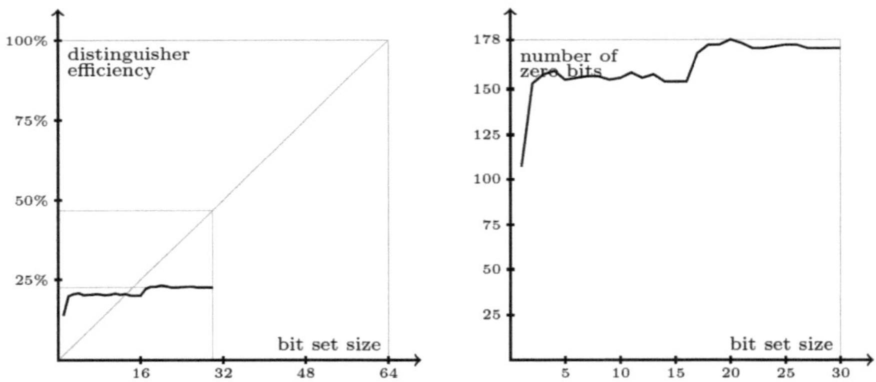

Fig. 10.8 Number of initialization steps that failed the maximum monomial test for a given bit size

Figure 10.8 corresponds to the case where only bits *IV* are allowed in the greedy algorithm (corresponds to a selective distinguish *IV* attack). The 178 prevailing zero bits correspond to a possible selective distinction attack on the 21 rounds of *Enocoro-128v2* initialization, as after round 21 bits 169 - 176 are output.

Figure 10.9 corresponds to the case when both key bits and bits *IV* are allowed. It should be noted that this estimate is worse than when only bits *IV* are used. This is a property of the greedy algorithm. If all variants of bit selection were tested, then the result would be at least as good as if only *IV* bits were selected.

Conclusions on the Possibility of Using a Selective Discrimination Attack The *Enocoro-128v2* initialization function has high resistance to the maximum monomial

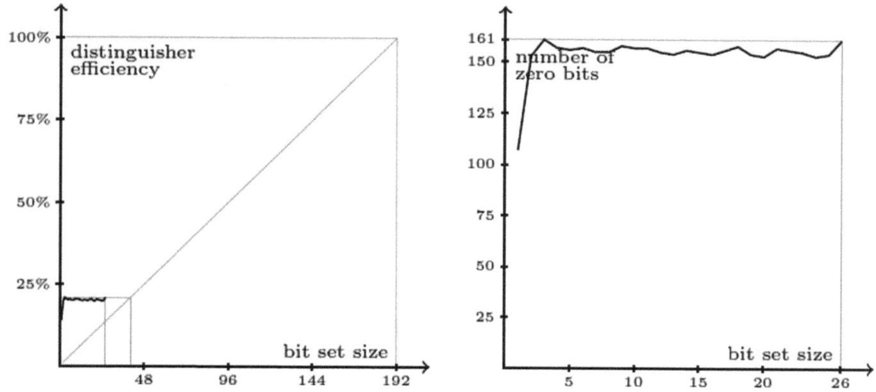

Fig. 10.9 Number of initialization steps that failed the maximum monomial test for a given bit size

test. A selective attack of distinction *IV* was detected against 21 initiating rounds [3]. It should be noted that the key bits and the *IV* bits were set to 0 in this test. Preliminary results on *Grain-128* and *Trivium* indicate that for an accidental key the attack should be slightly worse.

There are several other options for this attack. Instead of looking at the full 8-bit word, you can only consider one bit in the word. In addition, any linear combination of bits in a byte can be the target. Depending on the strategy, it is likely that the attack can be improved by several rounds. However, the full initialization of 96 rounds made the algorithm resistant to this attack.

10.2 *RC4* Stream Cipher

The *RC4* stream cipher was created in 1987 by Ronald Rivest, an employee of *RSA Security*. The abbreviation "*RC4*" officially stands for "*Rivest cipher 4*" or "*Rivest cipher*" ("4"—version number) [5].

For seven years, the code was kept secret and an accurate description of the algorithm was provided only after the non-disclosure agreement was signed, but in September 1994 its description was anonymously sent to the *Cypherpunks* mailing list [6]. Soon a description of *RC4* was published in the usenet newsgroup "*sci.crypt*". From there the source code got to many sites on the Internet. The published algorithm on the output issued ciphertexts that match the ciphertexts that appear to be true *RC4*. Holders of legal copies of *RC4* source code have confirmed the identity of the algorithms for differences in designation and structure of the program.

As this algorithm is known, it is no longer a trade secret. However, the name "*RC4*" is a trademark of "*RSA Security*". To avoid possible claims by the trademark owner, the code is sometimes referred to as "*ARCFOUR*" or "*ARC4*", meaning in English *"alleged RC4"—"predicted"RC4* (since *RSA Security* did not officially publish the algorithm).

The *RC4* encryption algorithm is used in some common standards and encryption protocols (such as *WEP, WPA, SSL*, and *TLS*).

RC4 became popular thanks to:

- simplicity of its hardware and software implementation;
- high speed of algorithm operation in both cases.

10.2.1 Specification of the Algorithm

The kernel of the current cipher algorithm consists of a function - a pseudorandom bit generator (keystream), which outputs a key bit stream (key stream, keystream, pseudorandom bit sequence).

RC4 is a class of algorithms defined by block size (hereinafter S-block). The parameter n specifies the word size for the algorithm and determines the length of the S-block. Usually $n = 8$, but for analysis purposes it can be reduced. To increase security, you need to increase this value. There is no contradiction in the algorithm for increasing the size of the S-block. In case of increase, for example, up to 16 bits in S-block will be 65,536 elements and accordingly the time of initial iteration will be increased. However, the encryption rate will increase [7].

The internal state of the *RC4* is supplied as a 2^n array and two counters. The array is known as S-block, and will be referred to as *S*. It always contains a permutation of 2^n possible word meanings. Two counters are designated as *i* and *j*.

RC4 initialization consists of two parts:

- initialization of the S-block;
- the generation of the pseudorandom word *K*.

10.2.1.1 Initializing the *S*-Block

The algorithm is also known as "key planning algorithm" or "*KSA*". This algorithm uses a user-supplied key stored in the key and a length of *L* bytes. Initialization starts with filling in the array *S*, then this array is mixed by permutations defined by the key.

S always contains one set of values that was given for the initial initialization ($S[i]: = i$).

10.2.1.2 Generation of the Pseudorandom Word *K*

This part of the algorithm is called the PRS generator. Key Stream Generator (see Fig. 10.10) *RC4* rearranges the value stored in S. In one *RC4* loop, one *n*-bit word *K* from the key stream is determined. In the future, the key word will be composed

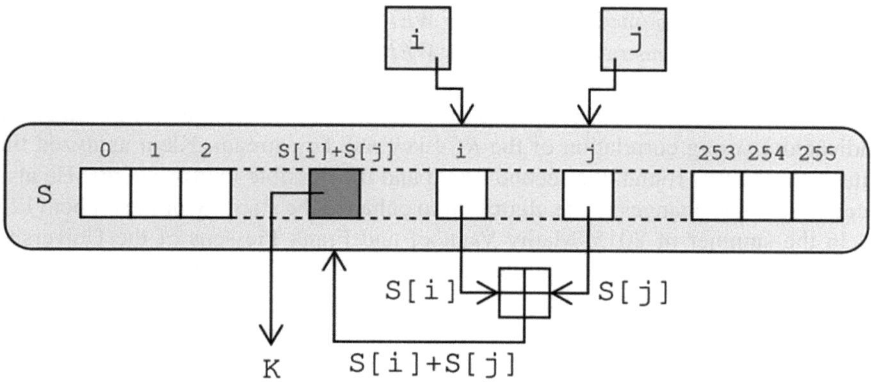

Fig. 10.10 RC4 key stream generator

of module two with source text that the user wants to encrypt and receive encrypted text.

10.2.2 RS4 *SSC Security*

Unlike modern ciphers (such as those offered in *eSTREAM* and discussed in the previous section), the *RC4* SSC does not use the *nonce* parameter with the key. This means that if a single key is to be used for a long time to encrypt multiple threads, the crypto system itself using RC4 must combine a nonce and a long-term key to obtain a streaming key for *RC4*. One way out is to generate a new *RC4* key using long-range keyhole and nonce. However, many applications that use *RC4* simply concatenate (unite) the key and nonce. Because of this, and also because of the poor key layout used in *RC4*, a system using SSC can be vulnerable [8, 9]. That is why many of the companies are obsolete with *RC4* SSC today. For example, *Microsoft's.NET Framework* already lacks *RC4* implementation.

In 1995, Andrew Roos experimentally observed that the first byte of a key stream is correlated with the first three bytes of a key, and the first few bytes of permutations after a keyhole algorithm with some linear combination of key bytes [10]. Further studies proved the correlation of key and key stream as well as the correlation of permutation and key. Later, a lot of work was published about restoring the key from the internal state of *RC4*.

In 2001, Florer, Mantinee, and Shamir published work on the vulnerability of a key *RC4* schedule. They showed that among all the possible keys, the first few bytes of the key stream are quite random. From these bytes, it is highly likely to obtain information about the key used. And if the long key and nonce are simply concatenated to generate the *RC4* key, then this long key can be obtained by processing a sufficient number of messages encrypted using this key [11]. This vulnerability and some of its related effects have been exploited when hacking *WEP* encryption on *IEEE 802.11* wireless networks. This demonstrated the need for *WEP* replacement as soon as possible and a new *WPA* wireless security standard was developed.

In 2005, Andreas Klein presented an analysis of the *RC4* cipher, in which he indicated a strong correlation of the *RC4* key and key stream. Klein analyzed the attacks in the first round, the second round and the possible improvements. He also suggested some changes to the algorithm to enhance the stability of the cipher [12].

In the summer of 2015, Mathy Vanhoef and Frank Piessens of the University of Leuven in Belgium demonstrated a real attack on *TLS* using *RC4* to encrypt transmitted data [13].

The list of publications on *RC4* SSC vulnerabilities can be continued, so far this crypto algorithm is considered to be unstable.

10.2.3 Realization

The work of many current ciphers is based on LFSR. This allows to achieve high efficiency of implementations of the cipher in the form of an integrated circuit (hardware implementation), but complicates the software implementation of such ciphers. Since the *RC4* cipher does not use LFSR and is based on byte operations, it is convenient to implement it programmatically. A typical implementation executes 8–16 machine commands per byte of text, so software implementation of the cipher must work very quickly [7–13].

10.3 Conclusions and Recommendations

1. Modern research in the field of high-performance cryptographic information security has shown that the implementation of reliable cryptographic methods and algorithms is extremely difficult in devices with low computing power, low volume and low power consumption. This is especially due to the development of the so-called "Internet of Things", objects of communication which may include various home appliances, vehicles, intelligent sensors, radio frequency identification tags and more. Limited physical parameters, low power consumption, low computing power and other characteristic attributes of green IT engineering are forcing new approaches to the design of cryptographic security tools focused on low-resource systems and technologies.
2. The International *ISO/IEC 291923 Project Information Technology - Security Techniques - Lightweight cryptography* is dedicated to the presentation of streaming algorithms for lightweight cryptography, the implementation of which is focused on low-cost and low-performance devices and memory. Among the specific algorithms, the *Trivium* streaming cipher is also considered as the winner of the international *eSTREAM* project, in which its properties are studied

in detail and all the advantages and disadvantages of the algorithm are noted. The second *Enocoro* streaming cipher is very similar to the *MUGI* cipher, which is specified in the international standard *ISO/IEC 180334*, and is, in fact, its simplified version for simple systems with small computing resources. The application of the non-linear complication function in *Enocoro*, built with reference to the AES S-block Board BSC (FIPS197), also aims at providing the required stability indicators with high requirements for the speed of formation of the original PRS.

3. Studies have shown that lightweight (lightweight) streaming crypto-algorithms "*Enocoro*" and "*Trivium*" are hardware-oriented streaming ciphers, the design of which is focused on flexibility. Their main goal is compactness in an environment with limited inputs, energy efficiency on low-energy platforms, and speed in applications requiring high-speed encryption.
4. Additionally, the *RC4* SSC algorithm, developed by *RSA Security* employee Ronald Rivest in *1987*, was investigated. Unlike many other SSC, the *RC4* algorithm does not use LFSR and is based on byte operations. This feature of its construction allows you to conveniently implement SSC programmatically, i.e. according to the classification of the *eSTREAM* project *RC4* algorithm should be attributed to program-oriented ciphers. According to the algorithm specification, a typical implementation executes from *8* to *16* machine commands for each byte of text, so software implementation of the cipher will be computationally efficient.
5. There are some drawbacks to the security of the *RC4* algorithm. In particular, *RC4* SSC does not use the nonce parameter, which is not safe from the perspective of the current SSC paradigm: if a single key is to be used for a long time to encrypt multiple threads, the system using *RC4* will be compromised. A way out of this inconvenience is the combination of nonce and the long-term key that must be performed from the outside of the SSC. But today there are many publications that point out the shortcomings of the key cipher list, the correlation of the key and the key stream *RC4*, and more. So today, the *RC4* algorithm is considered dangerous and, as a consequence, is gradually being phased out.

References

1. ISO/IEC 291923: Information technology—Security techniques—Lightweight cryptography—Part 3: Stream ciphers. Access mode http://www.iso.org/iso/home/store/catalogue_tc/catalogue_detail.htm?csnumber=56426 (2012).
2. Pseudorandom Number Generator Enocoro. Access mode http://www.cryptrec.go.jp/english/cryptrec_13_spec_cypherlist_files/PDF/23_00espec.pdf.
3. Hell M., Johansson T.: Security Evaluation of Stream Cipher Enocoro128v2. Access mode https://lup.lub.lu.se/luur/download?func=downloadFile&recordOId=2433384&fileOId=2433492
4. NIST Special Publication 80022. A Statistical Test Suite for Random and Pseudorandom Number Generators for Cryptographic Applications. Access mode http://csrc.nist.gov/publications/nistpubs/80022rev1a/SP80022rev1a.pdf.

5. Frequently Asked Questions. Access mode http://people.csail.mit.edu/rivest/faq.html#Ron
6. Thank you Bob Anderson. Cypherpunks. Access mode http://cypherpunks.venona.com/date/1994/09/msg00304.html (1994)
7. Schneier, B.: Applied Cryptography, 2nd edn. Wiley (1996)
8. Klein, A.: Attacks on the RC4 stream cipher. Access mode http://www.networklife.net/images/weprc4/RC4.pdf
9. McKague, M.E.: Design and analysis of RC4like stream ciphers. A thesis presented to the University of Waterloo in fulfillment of the thesis requirement for the degree of Master of Mathematics in Combinatorics and Optimization Waterloo, Ontario, Canada, Access mode https://uwspace.uwaterloo.ca/bitstream/handle/10012/1141/memckagu2005.pdf;jsessionid=E279D5A0119BA568F834582DC8D95BE0?sequence=1 (2005)
10. Weak Keys in RC4. VERÖFFENTLICHT AM 19950922 VON MARCEL WALDVOGEL. Access mode https://netfuture.ch/1995/09/weakkeysinrc4/
11. Fluhrer, S.R., Mantin, I., Shamir, A.: Weaknesses in the key scheduling algorithm of RC4. In: Lecture Notes in Computer Science, vol. 2259, p. 124. Access mode http://www.math.psu.edu/mathnet/mdoc/md15/rc4_ksaproc1.pdf (2001)
12. Klein A.: Attacks on the RC4 stream cipher. Designs Codes Cryptogr **48**(3), 269286. Access mode http://www.networklife.net/images/weprc4/RC4.pdf (2008)
13. RC4 NOMORE. Numerous Occurrence Monitoring & Recovery Exploit. By Mathy Vanhoef and Frank Piessens, iMindsDistriNet, KU Leuven. Access mode http://www.rc4nomore.com/

Chapter 11
Comparative Research of Fluid Cryptographic Transformation Algorithms

11.1 Comparative Studies of Statistical Security of Stream Cryptographic Transformation Algorithms

Since the original sequences are generated using deterministic algorithms, sometimes there is a high probability of distinguishing such sequences from truly random sequences. In this case, the encryption algorithm is vulnerable to many cryptographic attacks. The simplest way to implement such attacks is to perform a statistical analysis of the source sequences produced by the encryption algorithm. As a rule, the internal structure of the cipher is not taken into account in statistical testing. Therefore, this type of testing can be considered as black box testing because the internal structure, the initial data used to initialize the internal state, or the steps of initializing the internal state of the cipher are not taken into account [1–4].

When testing the statistical properties of stream ciphers, the general approach is to create a large number of key streams and apply some statistical tests to them. If the test results do not meet the requirements of the test, it is not considered that such a cipher is not resistant to attacks full recovery of key data or plaintext, but for such an algorithm encryption can be accurately asserted that the key sequence differs in its properties from the truly random.

When conducting statistical analysis, it is important to consider the relationship between the key, the initialization vector, the internal state, and the key stream, since if the cryptanalyst performs a key sequence analysis, he should not obtain any information about the generator's internal state or the secret key. Therefore, the encryption algorithm must form a key sequence of sufficiently long length (greater than the probable length of plaintext) and having pseudo-random sequence properties [1, 2, 4].

Usually the statistical results of sequence testing are compared with the theoretical distributions of truly random sequences. If there is a systematic deviation from the distribution of a truly random sequence, then the decision is made about the danger of the algorithm and its vulnerability to cryptographic attacks. It is

A. A. Kuznetsov et al., *Stream Ciphers in Modern Real-time IT Systems*,
Studies in Systems, Decision and Control 375,
https://doi.org/10.1007/978-3-030-79770-6_11

the generator-generated key sequence that determines the cryptoconversion of the encrypted sequence. Here is evidence [4–6].

Let the input text $M = (m_1, m_2, \ldots, m_N)$ look like a binary sequence of length N. The probability distribution $P_m(m_i)$ of occurrence of symbols of the input sequence can be any. The key sequence (gamut of encryption) $KS = (ks_1, ks_2, \ldots, ks_N)$ is represented by a binary sequence of length N. The key sequence must necessarily have a uniform distribution, valid for a pseudorandom sequence.

$$P_{ks}(ks_i) = \frac{1}{2^N}.$$

For stream symmetric encryption algorithms, the plaintext encryption operation is performed by summing the input sequence and key sequence modulo 2. The result is a cipher C:

$$C = M \oplus KS = (m_1 \oplus ks_1, m_2 \oplus ks_2, \ldots, m_N \oplus ks_N).$$

Prove that the probabilistic distribution of ciphertext elements is uniform, since the sum of two random variables is a random variable with a uniform distribution if one of the additives has a uniform distribution. We get:

$$\begin{aligned} P(c = a) &= P_m(m \oplus ks = a) \\ &= \sum_m P(m) \cdot P(m \oplus ks = a|m) \\ &= \sum_m P_m(m) \cdot P(ks \oplus m) = \sum_m P_m(m) \cdot \frac{1}{2^N} = \frac{1}{2^N}. \end{aligned}$$

Let us prove that for a crypto-encryption algorithm, the amount of information that a cryptanalyst can obtain from ciphertext relative to an input sequence must be *0*. For this we write an expression for the joint distribution of two sequences of input text and the key:

$$P(m = a, c = b) = P(m = a) \cdot P(c = b|m = a).$$

Therefore, for interdependent random variables, the distribution condition is:

$$\begin{aligned} P(c = b|m = a) &= P(m \oplus ks = b|m = a) \\ = P(ks = b \oplus a|m = a) &= P(ks = b \oplus a) = \frac{1}{2^N}. \end{aligned}$$

Finally we get:

$$P(m = a, c = b) = P(m = a) \cdot \frac{1}{2^N}.$$

Based on these principles of statistical analysis, we will conduct experimental studies of the source sequences produced by the encryption algorithms.

In this work, statistical testing of the source sequences (key stream or encryption keystream) was used to conduct experimental studies of the cryptographic properties of PSS. The most famous sets of statistical tests include [1–4, 7]:

- DIEHARD. The earliest and best known set of tests. It contains 12 statistical tests;
- NIST Statistical Test Suite (NIST STS), developed by the National Institute of Standards and Technology of the United States. It consists of 15 statistical tests.

Below there is a generalized characteristic of statistical testing. Statistical tests are used to compare the generator-generated pseudorandom sequence with a truly random sequence. The probable result of the statistical tests applied to a truly random sequence is known a priori and can be used to compare similar properties of a pseudorandom sequence. Since there are so many tests that evaluate whether a sequence is truly random, and because of a number of factors that are related to the probabilistic nature of the result, it is appropriate to indicate that a separate final set of tests does not provide a comprehensive assessment of the cryptographic properties of the sequence generator. In addition, the results of the statistical test should be interpreted with some caution and carefulness to avoid making erroneous conclusions about a particular generator.

A statistical test is formed to test a certain null hypothesis H_0, which assumes that the sequence under study is random. As a result, a null hypothesis is accepted or rejected for each test performed, depending on the sequence properties generated by the sequence generator.

Based on the assumption that the sequence is random, the statistical results of its testing have a certain probability distribution. The theoretical reference distribution of these statistics for the null hypothesis is determined by mathematical methods. A critical value is determined from this reference distribution. The test statistic is calculated during the test. It compares with the critical value. If the test statistic value exceeds the critical value, the null hypothesis for chance is rejected.

Otherwise, the null hypothesis is accepted. Testing of statistical hypotheses is effective because the reference distribution and the critical value are dependent and generated according to the previous assumption of chance. If the sequence randomness assumption is true, then the resulting test statistic will have a very low probability of exceeding the critical value.

On the other hand, if the calculated value of the test statistic exceeds the critical value (i.e., an event with a low probability), then, from the point of view of testing the statistical hypothesis, the event with a low probability cannot occur naturally. Therefore, when the estimated value of the test statistic exceeds the critical value, it is concluded that the initial contingency assumption is suspicious or erroneous. In this case, the conclusion is made about the rejection of H_0 (chance) and the hypothesis that the sequence is not coincidental.

Testing a statistical hypothesis is a procedure for deciding on a random sequence that has two possible results: either adopting a hypothesis of randomness H_0 or not adopting H_0, in which case it is decided that the sequence is not accidental. Table 11.1

Table 11.1 Possible options for hypothesizing about chance and the randomness of the sequence

Sequence	Acceptance of the hypothesis	
	About randomness	About non-randomness
Really random sequence (the H_0 contingency hypothesis is correct)	No mistake	Mistake of the 1st kind
The sequence is not random (the H_0 contingency hypothesis is incorrect)	Mistake of the 2nd kind	No mistake

lists the possible choices. If the sequence is in fact random, then the conclusion that the null hypothesis is rejected (that is, the non-coincidence hypothesis) can be rarely accepted. This conclusion is called first-order error. If the data is not random, then the conclusion that the null hypothesis (i.e., the data is random) is called a second kind of error. The conclusions about the acceptance of H_0 (when the data is truly random) and the rejection of H_0 (when the data is non-random) are correct [4].

The test statistic uses a calculation of a certain probability value P_j (where j is the test number) because a random number generator would generate a sequence less random than the one investigated. If this probability for the test is *1*, then the sequence is completely random. A value of *0* indicates that the sequence is absolutely not random. For each test, a certain level of significance should be chosen. If the value of the probability obtained is greater than or equal to the level of significance, then the null hypothesis is accepted, i.e. the sequence is considered random. If the probability value is less than α, then the null hypothesis is rejected, i.e. the sequence is considered non-random.

For example, according to the *NIST STS* method, this hypothesis about the random nature of the original sequence of the crypto algorithm under test is tested by *15* independent statistical tests (taking into account different input parameters, *188* tests are performed), each of which calculates the corresponding probability P_j, $j = 1, \ldots, 188$ passing the test. This probability is used by the rule of acceptance (criterion of consent) of the truth or falsehood of the hypothesis, that is, if the value of the probability of P_j passing the test does not fall below a certain threshold, for example, if, $p_j \geq 0, 99$ then the hypothesis H_0 is accepted (on his test). Otherwise, an alternative hypothesis is accepted.

11.1.1 Methods of Analysis of Passing Statistical Tests

Analysis of passing statistical tests begins with the analysis of test statistics. There are three options for evaluating test statistics.

1. *Limit values.* If the test statistic is more (less) than the threshold, the sequence is considered non-random.
2. *Fixed intervals.* If the test statistic exceeds the specified interval, the sequence is considered non-random.

3. *Probabilistic values.* For the test statistic, P_j. is calculated. The significance level α is selected for the test. If P_j is greater than or equal to α, then the sequence is considered random.

Because the first two methods require thresholds and fixed intervals to be calculated in advance, calculating P_j can be an effective way to evaluate test statistics.

In the papers [8–12], a technique for statistical testing of cryptographic algorithms was proposed, which consists in estimating the mathematical expectation of the number of tests passed by the investigated generator (crypto algorithm). It is with this approach of probabilistic value estimation that experimental studies of SSC cryptographic properties were carried out in this work.

Thus, as a result of statistical testing of the original sequence of the crypto-algorithm (generator), the vector $P = \{P_1, P_2, \ldots, P_n\}$ of P_j, values is formed, where n is the number of independent tests according to the appropriate procedure (in *NIST STS*, the value is $n = 188$). The analysis of the elements P_j of the vector P allows us to point to specific defects in the "randomness" of the tested sequence, that is, the low values of P_j, indicate a clear difference between the studied sequence and the implementation of the random process revealed by the statistical test.

Most modern crypto algorithms have P_j, values that exceed the threshold and use only the number of tests passed as a test result, i.e. the number of probabilities $P_j \geq 0,99$ from the set $P = \{P_1, P_2, \ldots P_n\}$. Let us denote the number of tests passed for a specific i sample by the symbol X_i, $0 \leq X_i \leq n$, $i = 1, \ldots, N$, where N is the number of cryptographic algorithm output sequences tested. It should be noted that the values of X_i, as well as the values of the set $P = \{P_1, P_2 \ldots P_n\}$, characterizing the statistical security of the crypto algorithm under study, are stochastic in nature. These values depend directly on the properties of the generator being tested and on the initial data during the experimental studies. In other words, the values of the elements P_j are determined for a specific i sample $i = 1, \ldots, N$, that is, for a specific output sequence of a given length crypto algorithm. Different initial data (different initial sequences of a given length in the experiment) can give different values of the elements P_j, and differences in eigenvalues can be significant. In practice, this means that the results of statistical testing of different sequences of the same crypto algorithm with different likelihood will indicate the detected defects of "randomness" of the tested sequences.

11.1.2 NIST STS Statistical Test Package

The *NIST STS* Statistical Testing Package was developed during the AES competition for the study of random or pseudorandom number generators and is the most common tool for assessing the statistical security of cryptographic primitives. The use of this package allows us to estimate how closely the crypto algorithms under study approximate the generators of "random" sequences, that is, with a high probability to state whether the generated sequence is statistically safe.

The order of testing of a single binary sequence S is as follows [13]:

- the null hypothesis H_0 is put forward—the assumption that this binary sequence S is random;
- for the sequence S the statistics of the test $c(S)$ are calculated;
- using the special function and test statistics, the probability value $P = f(c(S))$ is calculated;
- the probability value P is compared with the threshold value $\alpha \in [0,96; 0,99]$. If $P \geq \alpha$, then hypothesis H_0 is accepted. Otherwise, an alternative hypothesis is accepted.

The package contains *15* statistical tests, but, in fact, depending on the input parameters, *188* values of probability P are calculated, which can be considered as the result of individual tests.

1. *Frequency bit test*. Aims to determine the relation between zeros and ones in a binary sequence of a certain length. For a truly random binary sequence, the number of zeros and ones is almost the same. The test estimates how close the unit is to *0.5*.
2. *Frequency block test*. The essence of the test is to determine the fraction of units within the block with a length of m bits, i.e. it is necessary to find out whether the frequency of units in the block with a length of m bits is approximately equal to *m/2* , as might be assumed in the case of random sequence.
3. *Test for the sequence of identical bits*. This test searches for rows, that is, continuous sequences of identical bits. A series (series) of length k bits consists of k absolutely identical bits, beginning and ending with a bit containing the opposite value. In this test, you need to find out if the number of such rows really matches the number in a random order. Specifically, it determines whether units and zeros alternate quickly or slowly in the initial sequence.
4. *Test for the longest sequence of units in the block*. This test determines the longest row of units inside the block with a length of m bits. It is necessary to find out whether the length of such a line actually meets the expectation of the length of the longest row of units in the case of an absolutely random sequence.
5. *Rank test of binary matrices*. This is where rankings are calculated continuous sub arrays constructed from the initial binary sequence. The purpose of this test is to test the linear dependence of the fixed length substrings that make up the initial sequence. This test is also available in the DIEHARD package [2, 3].
6. *Spectral test*. The essence of the test is to estimate the peak height of the discrete Fourier transform of the initial sequence. The purpose is to identify periodic properties of the input sequence, such as closely spaced repeating plots. The idea is that the number of peaks exceeding the *95%* amplitude threshold is much greater than *5%*.
7. *Match test for non-overlapping templates*. This test calculates the number of predefined templates found in the original sequence. It is necessary to identify

random or pseudorandom number generators that form too often non-periodic patterns. As in Test # 8 on matching overlapping patterns, a window with a length of m bits is also used to search for specific patterns with a length of m bits. If no pattern is found, the window shifts one bit. If a pattern is found, then the window moves to the bit that follows the pattern found, and the search continues.

8. *Test overlapping overlapping patterns.* The essence of this test is to calculate the number of predefined templates found in the original sequence. As in Test # 7 on matching non-overlapping patterns, a window with a length of m bits is used to search for specific patterns with a length of m bits. The search itself is conducted in a similar way. If no pattern is found, the window shifts one bit. The difference between this test and test # 7 is that when the pattern is found, the window moves only one bit forward, and then the search continues.
9. *Maurer Universal Statistical Test.* This determines the number of bits between the same patterns in the original sequence (a measure that is directly related to the length of the compressed sequence). It is necessary to find out whether this sequence can be significantly compressed without loss of information. If possible, it is not truly random.
10. *Linear Complexity Test.* The test is based on the principle of the linear shift register feedback. You need to find out if the input sequence is complex enough to be considered completely random. Absolutely random sequences are characterized by long linear shift registers. If such a register is too short, then it is assumed that the sequence is not completely random.
11. *Frequency test.* This test is to calculate the frequency of all possible overlaps of patterns of length m bits during the initial sequence of bits. The purpose is to determine whether the number of occurrences of $2m$ overlapping patterns with a length of m bits is really the same as in the case of a completely random input bit sequence. The latter is known to be unique, that is, each template with a length of m bits appears in sequence with equal probability. It should be noted that in the case of $m = 1$, the periodicity test passes into the frequency bit test *(# 1)*.
12. *Test of approximate entropy.* As with the periodicity test, this test focuses on calculating the frequency of all possible overlaps of patterns of length m bits during the initial sequence of bits. It is necessary to compare the overlap frequencies of two consecutive blocks of the initial sequence with the lengths m and $m + 1$ with the overlap frequencies of similar blocks in a completely random sequence.
13. *The cumulative sum test.* The test is the maximum deviation (from zero) at an arbitrary bypass determined by the cumulative sum of the given $(-1, +1)$ digits in the sequence. It is necessary to determine whether the cumulative sum of the partial sequences occurring in the input sequence is too large or too small compared to the expected behavior of such a sum for a completely random input sequence. Thus, the cumulative amount can be considered as a bypass. For a random sequence, the deviations from the bypass should be about zero.

14. *Random deviation test*. The essence of this test is to calculate the number of cycles that have strictly k visits with an arbitrary bypass of the cumulative amount. The arbitrary crawl of a cumulative sum begins with partial sums after the sequence (*0, 1*), translated into the corresponding sequence (*–1, +1*).An arbitrary bypass cycle consists of a series of single-length steps performed in random order. In addition, such a crawl begins and ends on the same element. The purpose of this test is to determine whether the number of visits to a particular state within a cycle differs from a similar number in the case of a completely random input sequence. In fact, this test is a set consisting of eight tests that are performed for each of the eight cycle states: *–4, –3, –2, –1*, and *+1, +2, +3, +4*.
15. *Another test for arbitrary deviations*. This test calculates the total number of visits to a given condition when you randomly bypass the cumulative amount. The purpose is to determine deviations from the expected number of visits to different states at random bypass. In fact, this test consists of 18 tests performed for each state: *–9, –8,…,* –1 and *+1, +2,…, +9*.

Thus, as a result of testing the binary sequence, the vector $P = \{P_1, P_2, \ldots, P_{188}\}$ of probability values is formed. The analysis of the constituents P_j of this vector allows us to point to specific defects in the randomness of the tested sequence.

Passing each of the *15* statistical tests is an important criterion for evaluating a pseudorandom generator. Therefore, even a mismatch by one or more criteria means that the key stream cannot withstand cryptanalysis at a high level. If, on the other hand, the generator passes all the tests, it does not indicate the safety of the generator, since such tests do not take into account the features of the actual design of the generator.

The accumulated experience of conducting statistical testing shows that the number of tests passed by the generator under study depends directly on the selected cryptographic algorithm output sequence. To ensure the reliability of the results of statistical testing in [8–12], it is proposed to estimate the mathematical expectation of the number of tests passed X_i by the investigated generator (crypto algorithm), considering each test as a single observation (experience), i.e. as a specific implementation X.

11.1.3 Results of Experimental Studies of SSC Statistical Properties Using the NIST STS Statistical Testing Package

In accordance with the methodology of statistical testing [8–12], experimental studies of cryptographic properties of different SSC were carried out. The work has tested the key streams of modern streaming ciphers: AES, CryptMT, DECIM, Enocoro, Grain, HC, KCipher, Mickey2, MUGI, Rabbit, RC4, Salsa20, Snow 2, Sosemanuk, Trivium, and proposed in [8–12] cipher "STRUMOK". The structure and basic transformations of the SSC "STRUMOK" are described in more detail in Chap. 15, which also

Table 11.2 Results of statistical testing of PRS of modern AES, CryptMT, DECIM, Enocoro, Grain, HC, KCipher, Mickey 2, MUGI, Rabbit, RC4, Salsa20, SNOW 2, SOSEMANUK, Strumok, Trivium stream ciphers

Algorithm	M099	D099	S099	P099	M096	D096	S096	P096	MIN
AES-128_K_IV	127.07	20.456	4.438	1.00	186.63	0.3191	0.554	1.00	185
CryptMT_K_IV	130.89	52.988	7.279	1.00	158.56	5.284	2.299	1.00	181
DECIM_K_IV	132.44	19.358	4.399	1.00	186.44	0.9136	0.956	1.00	185
Enocoro_K_IV	132.92	51.22	7.157	1.00	187.17	0.79	0.89	1.00	185
Grain_K_IV	132.36	57.32	7.571	1.00	186.92	1.414	1.185	1.00	182
HC_256_K_IV	133.75	36.44	6.04	1.00	186.66	1.93	1.381	1.00	182
KCipher_K_IV	131.29	11.061	3.3258	1.00	186.71	0.489	0.699	1.00	186
Mickey_2_K_IV	133.53	61.65	7.85	1.00	186.6	2.302	1.51	1.00	179
MUGI_K_IV	132.23	53.721	7.329	1.00	186.5	0.978	0.989	1.00	185
Rabbit_K_IV	132.65	16.87	4.017	1.00	187.22	0.451	0.657	1.00	185
RC4_K_IV	133.7	67.01	8.186	1.00	186.3	1.61	1.269	1.00	184
Salsa20_K_IV	134.16	28.055	5.27	1.00	187.001	1.01	0.99	1.00	183
Snow2.0_K_IV	132.78	23.93	4.89	1.00	186.79	0.43	0.656	1.00	183
Sosemanuk_K_IV	131.73	49.36	6.991	1.00	186.8	2.240	1.49	1.00	184
Strumok_256	130.01	23.6	4.86	1.00	186.45	1.4555	1.206	1.00	184
Strumok_512	132.83	56.516	7.518	1.00	186.90	0.802	0.896	1.00	185
Trivium_K_IV	130.24	99.683	9.935	1.00	187.15	1.49	1.214	1.00	182

presents the results of studies of the rate of transformations in comparison with other crypto algorithms. The sequences for statistical testing were generated by the "random key K/random initialization vector IV" pair. Each sequence is 106 bits long. The test results are summarized in Table 11.2.

Table 11.2 provides the following data:

- "*M096*" and "*M099*" are the estimates of the mathematical expectation (sample averages) of the number of statistical tests passed by the criteria $P_j \geq 0, 96$ and $P_j \geq 0, 99$ respectively;
- "*D096*" and "*D099*" ("*S096*" and "*S099*")—estimates of variances (root mean square deviations) of the results of testing the number of passed statistical tests by the criteria $P_j \geq 0, 96$ and $P_j \geq 0, 99$ respectively;
- "*P099*"—the value of the confidence probability for the number of statistical tests passed by the criterion $P_j \geq 0, 99$ and to the accuracy of $\varepsilon = 2$;
- "*P096*"—the value of the confidence probability for the number of passed statistical tests according to the criterion $P_j \geq 0, 96$ and to the accuracy $\varepsilon = 1$;
- "*Min096*"—the minimum values of the number of passed statistical tests by the criterion $P_j \geq 0, 96$.

This is where we investigate a previous version of the "STRUMOK" algorithm with *512* and *1024* bit keys, which was later placed and refined.

The test results of various stream cryptoconversions confirm the high cryptographic performance. All the ciphers studied showed a high number of successfully passed tests: *130–134* on the criterion $P_j \geq 0, 99$ and 186–187 on the criterion $P_j \geq 0, 96$ (except *CryptMT*, which received a score of *158*). These estimates were obtained with very high confidence ($P_\partial = 0, 99$ for criterion $P_j \geq 0, 99$ and $P_\partial \approx 1$ for criterion $P_j \geq 0, 96$). The minimum values of the number of passed statistical tests (Min096) range from 179 to 186 tests. The highest number of tests passed showed KCipher.

It should be noted the high performance of the statistical safety of the algorithm of SSC "Strumok", which revealed certain properties of the random bit generator. In particular, the results of the table show that the formed PRS in its properties are not inferior to the PRS, which is formed by the world-known streaming cryptographic algorithms, in particular *HC-256* and *SNOW 2.0* ciphers. In addition, for SSC "STRUMOK", the minimum values of the number of passed statistical tests by the criterion $P_j \geq 0, 96$ are higher than in these algorithms, which indicates a slight superiority of the statistical safety indicators of the developed algorithm.

11.1.4 Characteristics of DIEHARD Statistical Tests

George Marsalgia introduced the *DIEHARD* statistical test suite in 1995. *DIEHARD* tests are considered as a set of tests with the most stringent criteria for sequence properties. Despite being introduced more than *20* years ago, the system has not lost its relevance, and even today it can be used as one of the main tools for evaluating the statistical properties of a sequence. The *DIEHARD* tests are intended to characterize the randomness (or lack thereof) in a sequence of integers formed by a particular pseudorandom sequence generator.

A specific feature of the *DIEHARD* system is the practical orientation of the tests, that is, some tests are not based on theoretical calculations of the statistical safety assessment, but rather on the evaluation of the results based on practical tests carried out by the author. Due to the specific construction of test algorithms, this package is characterized by the fact that it is difficult for it to form a bad sequence that does not meet the requirements of chance, but would successfully pass all the tests.

In order to conduct a complete evaluation of cryptographic properties, it is necessary to form a sufficiently large number of test sequences of sufficient length. Because it is wrong to make a conclusion about the quality of the generator based on the results of the sequence analysis for any one initial fill. Similarly, it is incorrect to conclude that a random number generator is based on the results of the analysis of one sequence fragment produced by it. A *108*-bit sequence is desirable for testing.

DIEHARD includes 12 testing algorithms. The DIEHARD software implementation includes the following tests, which total 215 tests:

1. Checking birthdays.
2. Check for permutations of intersections.

3. Checking the ranks of matrices.
4. Bitstream tests.
5. Counting units.
6. Parking test.
7. The minimum distance test.
8. Random Sphere Test.
9. Compression test.
10. Test the intersecting sums.
11. Test sequences.
12. The dice game test.

Some tests use an aggregate probability estimate to evaluate the results of some tests using the Kolmogorov—Smirnov criterion, that is, the values of several Pj are reduced to a final estimate using KStests. The Kolmogorov—Smirnov criterion can be used when the test results are infinite. Its essence is as follows.

Let n be the result of the values obtained $X_i, i = \overline{1, n}$.

We construct an empirical distribution function

$$F_n(x) = \frac{1}{n}\sum_{i-1}^{n} X_i \leq x$$

The Kolmogorov—Smirnov criterion determines how different the empirical distribution function $F_n(x)$ is from the distribution function $F(x)$, which determines the probability that a random variable X will have a value less than or equal to x for a given distribution. However, calculations of this criterion are quite time-consuming, so replace the Kolmogorov—Smirnov test in mathematical statistics using the Anderson—Darling test.

According to the Anderson—Darling criterion, it is understood that n tests were performed and statistics obtained $X_i, i = \overline{1, n}$. In the *DIEHARD* test, the test result is the probability P_j.. Placing these values in ascending order, we get:

$$P_1 \leq P_2 \leq P \leq \ldots \leq P_n.$$

In this case, the test statistic becomes meaningful:

$$A^2 = -n - \sum_{j-1}^{n} \frac{2j-1}{n}\left[\ln P_j + \ln P_{n+1-j}\right].$$

The final probability estimate is defined as:

$$P_{jkS} = \begin{cases} 0, & A^2 < \alpha \\ 2e^{-\frac{12337}{A^2}}\left(1 + \frac{A^2}{2^3} - \frac{0.04958\left(A^2\right)^2}{1.325+A^2}\right), & \alpha \le A^2 < 2 \\ 1 - 0.6621361 \cdot e^{-1.091638A^2} - 0.95095 \cdot e^{-2005138A^2}, & 2 \le A^2 < 4 \\ 1 - 0.4938691 \cdot e^{-1.000321A^2} - 0.5946336 \cdot e^{-1.527198A^2} & A^2 \ge 4 \end{cases}$$

where α is the level of significance.

Thus, the probability of P_j or the probability obtained by the Anderson-Darling P_{jKS}. criterion are used to evaluate the passage of a particular *j*-test.

A general description of the tests is given below.

1. ***Birthday Spacings***. Random points are selected at a large interval. The distances between the points must be asymptotically distributed by Poisson. Named this test based on the paradox of birthdays.
2. ***Overlapping Permutations***. Sequences of five consecutive random numbers are analyzed. The 120 possible permutations should meet with a statistically equivalent probability.
3. *Bitstream tests* ***(Monkey Tests, Bitstream test)***. Sequences with a number of bits are interpreted as words. Crossing words in the stream are counted. The number of "words" that do not appear should satisfy the known distribution. The test was called *Monkey Test* because it is based on the infinite number of monkeys theorem.
4. ***Ranks of matrices***. A few bits of a number of random numbers are selected to form a matrix above *{0,1}*, then the rank of the matrix is determined.
5. ***Count the 1's***. Single bits are counted in each of the following or selected bytes.
6. ***Parking Lot Test***. The individual circles are randomly placed in a square 100 × 100. If the circle intersects the existing one, an attempt is made to re-position the circle. After *12,000* attempts, the number of successfully "parked" circles must be normally distributed.
7. ***Minimum Distance Test***. *8000* points are randomly placed in a square of 10 000 × 10 000, then determine the minimum distance between any pairs. The square of this distance should be exponentially distributed with some median.
8. ***Random Spheres Test***. Randomly selected *4000* points in a cube with an edge of *1000*. At each point is placed a sphere, the radius of which is a minimum distance to another point. The minimum volume of a sphere must be exponentially distributed with some median.
9. **The Squeeze Test**. 2^{31} is multiplied by random real numbers in the range [0,1] until it yields 1. Repeats 100,000 times. The number of real numbers required to reach 1 must be distributed in a certain way.
10. **Overlapping Sums Test**. A long sequence of real numbers from the interval [0,1] is generated. It sums up every 100 consecutive numbers. The amounts should be normally distributed with characteristic mean and variance.
11. ***Runs Test***. A long sequence at the interval [0,1] is generated. Ascending and descending sequences are calculated. The numbers must satisfy some distribution.

12. The Craps Test. 200,000 dice games are played, victories and the number of shots in each game are counted. Each number received must satisfy some distribution.

11.1.5 Results of Experimental Statistical Studies SSC Properties Using the DIEHARD Package

It is known that in the case of testing the statistical properties of a truly random sequence, the characteristic property is a uniform distribution of the probabilities of passing the tests. That is, among all obtained values of probabilities P_j,, their distribution must be uniform at a single interval. Deviations from the even distribution indicate that some of the *DIEHARD* tests found ordered patterns in the sequence being tested.

If the obtained *215* probability values of P_j, in ascending order, are presented as a graph, then any, even minor, deviations from a uniform distribution can be clearly defined.

An example of the results of bad testing, in a statistical sense, is shown in Fig. 11.1. This sequence was formed by the standard random number generator for the *C++* language. It is clear from the distribution that the sequence generated by such a generator cannot be used as part of the implementation of the cryptographic encryption algorithm.

Probability distribution at a single interval allows you to roughly estimate the randomness of the sequence that was tested, to match the distribution of a truly random sequence. DIEHARD tests were carried out for SSC "Strumok", the results of which are shown in Figs. 11.2 and 11.3. It can be seen that the probability distribution, for both the 256-bit version of the "STRUMOK" key and the 512-bit version, corresponds to a uniform distribution with small fluctuations.

Also, a slightly different approach is used to evaluate the statistical properties of the tests. In addition to the fact that the probability distribution should be uniform in

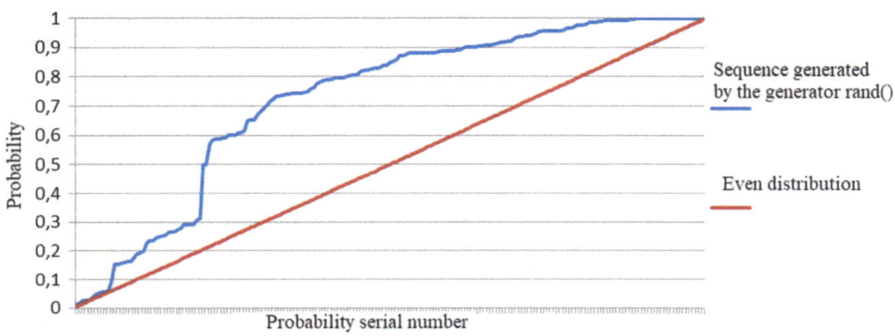

Fig. 11.1 Comparison of the probability distribution of sequence testing with even distribution

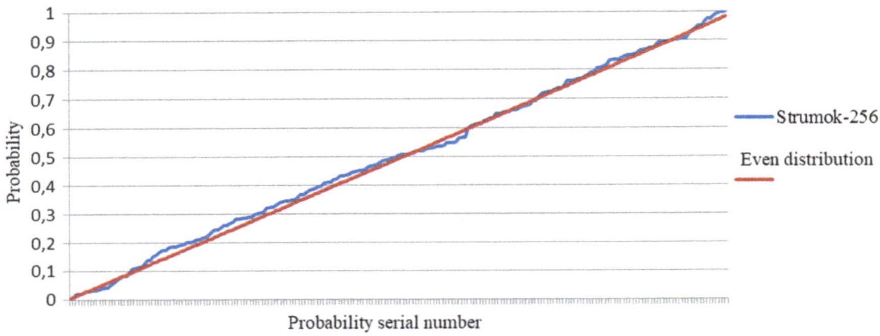

Fig. 11.2 *DIEHARD* test probability distribution for "STRUMOK-256"

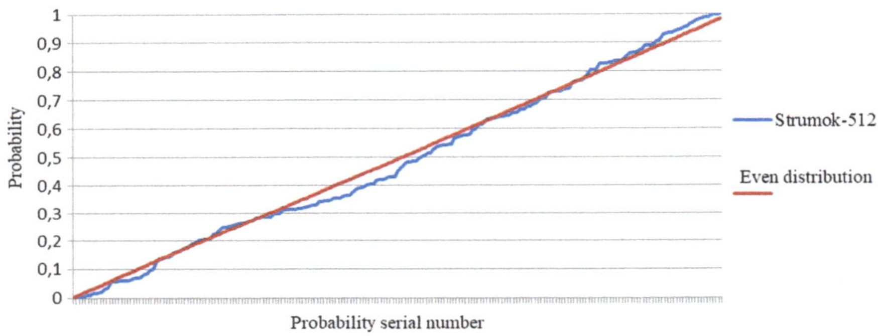

Fig. 11.3 *DIEHARD* test probability distribution for "STRUMOK-512"

a single interval, the probabilities of the test results should be normally distributed for a truly random sequence. Since, under normal distribution, a large number of results having an asymptotic approximation to *0* or *1* means that the sequence has poor statistical properties.

Proceeding from this, we conditionally divide a single segment by several intervals, namely *3*. If the probability belongs to a certain interval, it characterizes the passing of its *j*-test as:

1. test failure, with probability $P_j \leq 0.1$ or $P_j > 0.9$;
2. the test results are doubtful if the probability belongs to one of the intervals $0.1 < P_j \leq 0.25$ or $0.75 < P_j \leq 0.9$;
3. the test is passed if the probability is $0.25 < P_j \leq 0.75$

Graphically, such a characteristic is shown in Fig. 11.4.

Obviously, when the majority of results belong to the interval (*0.25; 0.75*], the sequence can be considered to have sufficient statistical properties compared to the truly random one. And if the vast majority of results belong to the intervals (0; 0.25)

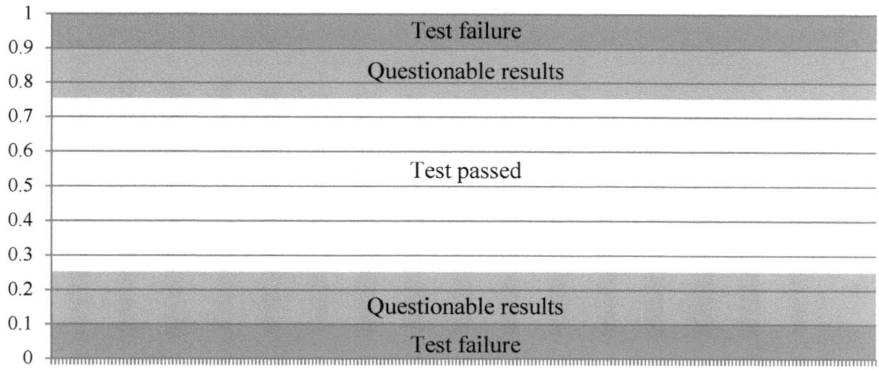

Fig. 11.4 Characterization of single interval intervals

and [0.9; 1), the sequence under test can be considered to be bad in the statistical sense.

The calculated number of test probabilities for each of the above codes is summarized in Table 11.3.

Table 11.3 Number of test scores distributed at five intervals

Cipher	Probability value				
	$P_j \le 0.1$	$0.1 < P_j \le 0.25$	$0.25 < P_j \le 0.75$	$0.75 < P_j \le 0.9$	$P_j > 0.9$
AES-128	16	40	106	29	28
AES-256	15	33	119	31	21
CryptMT	21	33	99	36	25
DECIM	25	29	92	42	26
Enocoro	18	36	102	44	16
Grain	17	27	124	32	19
HC-128	17	39	103	36	24
HC-256	25	36	110	30	18
KCipher-2	25	43	90	43	25
Mickey2	23	36	104	33	23
MUGI	25	35	107	27	25
RC4	23	21	103	35	32
Sasla20	17	34	107	27	34
Snow 2	21	31	112	33	18
Sosemanuk	13	25	125	34	17
Струмок-256	20	29	113	35	18
Струмок-512	26	26	113	28	22
Trivium	24	27	107	35	22

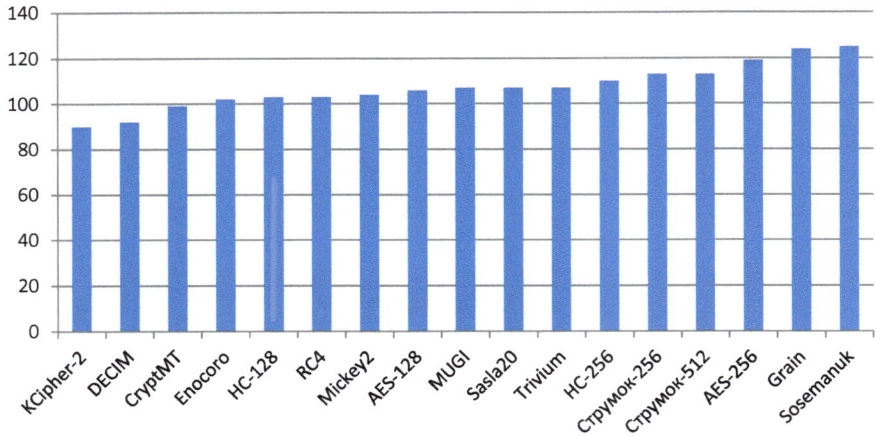

Fig. 11.5 Number of *DIEHARD* statistical tests successfully completed

All ciphers showed a sufficient level of statistical certainty. For the sake of clarity, the number of tests passed, that is, in which the result belongs to the interval $0.25 < P_j \leq 0.75$ is reduced to the histogram in Fig. 11.5.

Despite the fact that the *Sosemanuk* cipher according to the results of matching the probability distribution to the uniform showed a slight, but slightly larger, unlike other ciphers, deviations from the uniform distribution, in the results of this test it occupies a leading position. It should also be noted that in the number of statistical tests of the *NIST STS* package, the *KCipher* cipher took the first position, but in the *DIEHARD* testing it was in last place.

11.2 Comparative Researches of Computing Complexity of the FLOW-Cryptographic Transformation Algorithms

11.2.1 The Essence of the Method of Testing Speed Characteristics Stream Ciphers

An important characteristic of modern cryptographic information security is performance indicators that characterize the ability of the cryptosystem to process large amounts of data in a fixed time. In this sense, it is important to study the velocity characteristics of streaming ciphers, to compare them against different metrics that reflect the ability of crypto-primitives to quickly process different volumes of data.

In order to achieve this goal, it was decided to analyze the stream ciphers according to the method proposed at the world famous eSTREAM competition. This technique is that when testing streaming encryption, you need not dwell solely on investigating the speed of encryption of large volumes of data and the time of initialization/generation of key parameters; That is why in this work in accordance with the specified methodology simulate different conditions of use of stream ciphers. It is this approach that demonstrates the weaknesses and strengths of the elements of the cipher structure.

The eSTREAM methodology provides the following criteria:

1. The criterion for encryption of long streams. Stream ciphers have the most potential advantage over block ciphers when encrypting long streams;
2. The criterion for encryption of short streams. This metric reflects the speed of encryption of packets of different lengths;
3. Criterion of initialization/generation of key parameters. This feature individually reflects the efficiency of key installation and initialization vector.

The following ciphers were involved in the experiment—the winners of the eSTREAM competition: HC, MICKEY, RABBIT, Salsa, SOSEMANUK, TRIVIUM; SNOW 2.0, AES and the promising stream "Stream" cipher, which was later adopted as the national standard of Ukraine by DSTU 8845 [8–12]. For the sake of clarity of the results of the study, it was decided to add to the general list of selected ciphers other versions of the ciphers: AES-128, AES-256, SNOW 2.0–128, SNOW 2.0–256, "Stream" 512, in order to demonstrate the relationship between speed and block size.

It should be noted that this chapter explores a previous version of the "STRUMOK" algorithm, which was subsequently refined and improved. As a result, a new version of the "STRUMOK" algorithm was obtained, the two modes of which (C-256 and C-512) are explored in Chap. 15.

All ciphers were implemented using a computer with an Intel (R) Pentium (R) CPU P6200 @ 2.13 GHz, 2128, RAM: 2 to 2 GB (1333 MHz), Cache: 1st level (128 Kb); 2nd level (512 Kb); Level 3 (3072 Kb) in Windows 8.1 Professional 64bit, on the Microsoft Visual Studio 2012 32bit compiler version 11.00.50727.1.

After the implementation of the selected ciphers, all ciphers were run on different processors in power/performance, on different platforms, implemented in different programming languages and compilers. Test results are given below.

11.2.2 Investigation of the Characteristics of Ciphers on Different Processors

11.2.2.1 Testing Symmetric Stream Encrypts by the Long Stream Encryption Criterion

One area where stream ciphers can be used is long stream encryption. For example, this problem occurs when encrypting hard disk data. There are many reasons for encrypting data on your hard drive, first and foremost keeping your personal information private, but paying for keeping your personal data is a slower operating system.

To determine the speed of modern ciphers, a preliminary experiment was conducted, using VeraCrypt and TrueCrypt disk encryption software, and running standard encryption speed tests for all possible algorithms. The main reason for choosing these programs is that when they are used, the data must be encrypted on the fly, so special attention is given to the speed of algorithms in such programs.

The experiment was conducted under the following conditions [8–10]: 1 GB of data was generated and ciphers: AES, Twofish, AES-Twofish, SERPENT, SERPENTAES, AES-Twofish-SERPENT, SERPENT-Twofish-AES, Twofish-SERPENT, encrypted and decrypted. The experimental results are shown in Figs. 11.6 and 11.7.

The criterion of encryption of long streams is a characteristic feature of streaming ciphers, so attention is paid to these indicators in the first place. The research method randomly generates 1 GB of data, then randomly generates one key and sets the random state of the registers using the initialization vector. The next step is directly encrypting 1 GB of data, measuring the time of the encryption process and the algorithm's throughput (the number of bytes encrypted in one microsecond). The results are summarized in Table 11.4. Based on the obtained results, we conclude that

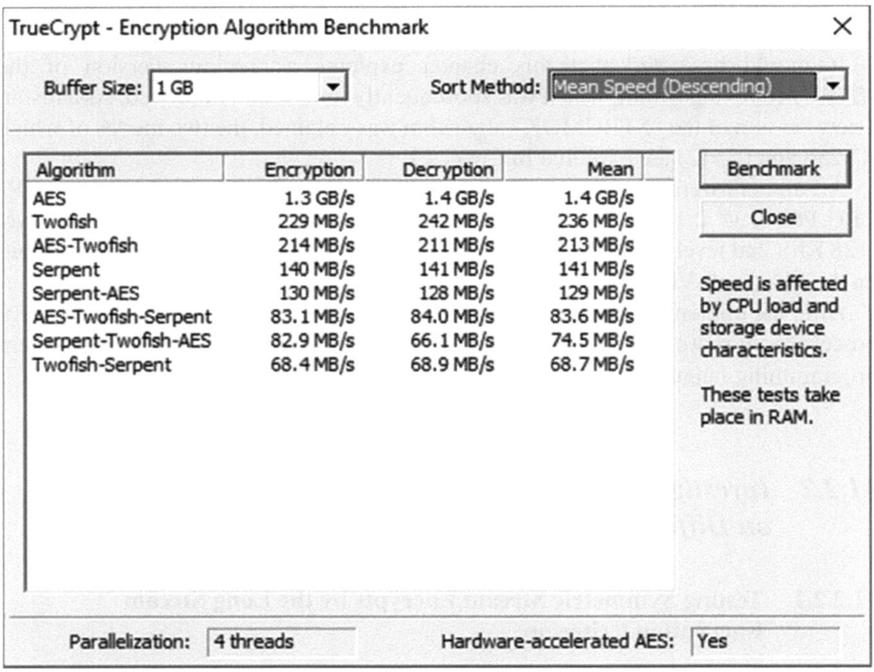

Fig. 11.6 Results of experimental studies of the speed characteristics of *TrueCrypt* ciphers

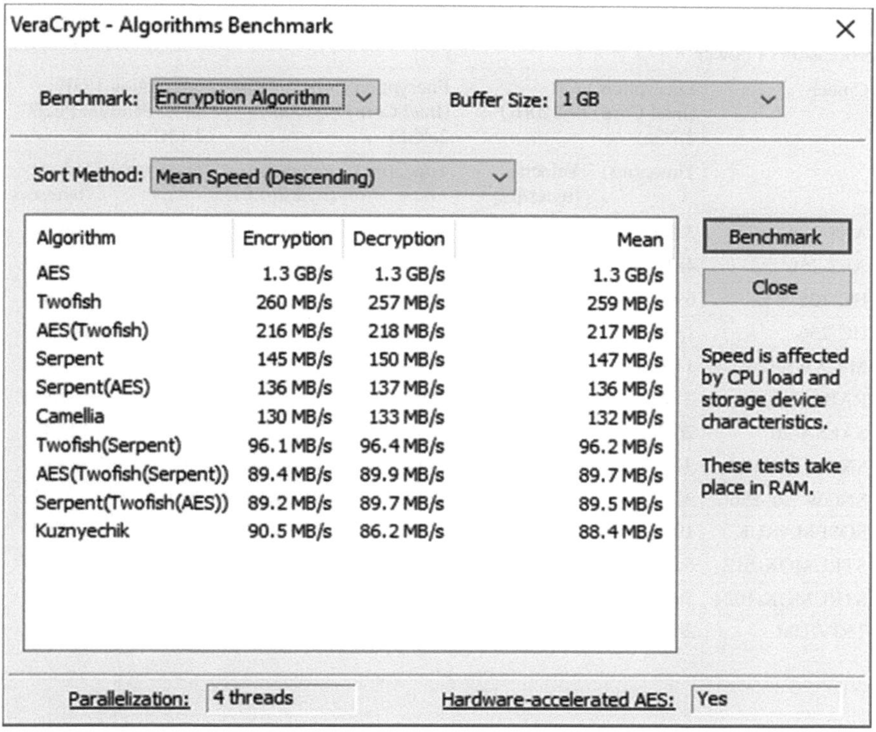

Fig. 11.7 The results of experimental studies of the velocity characteristics of *VeraCrypt* ciphers

the best indicators for the criterion of long message encryption were “STRUMOK”, HC-128 and “SNOW 2.0” (see Figs. 11.8 and 11.9).

Taking into account the results of experimental studies of the speed characteristics of ciphers on different processors, it can be seen that on low-frequency processors cipher “STRUMOK” loses in speed, but on more powerful processors speed increases. This loss is, first of all, due to the larger block size than other algorithms, which takes more time to calculate by the mathematical operations processor. An improved version of the “STRUMOK” algorithm is more productive. This is evidenced by the results in Chap. 15.

11.2.2.2 Testing Symmetric Streaming Ciphers by the Criterion of Encryption of Short Streams

The speed of encrypting long streams is a basic feature of block ciphers, which reflects the ability of crypto primitives to function for their intended purpose. Therefore, when

Table 11.4 Results of experimental velocity studies characteristics of ciphers on different processors of power

Cipher	Encryption 1 GB (Intel Core i7-6820HQ 2.7Gh)		Encryption 1 GB (Intel Core i7-5500u 2.4Gh)		Encryption 1 GB (Intel Pentium P6200 2.13Gh)	
	Time, (ms)	Velocity, (bytes/μs)	Time, (ms)	Velocity, (bytes/μs)	Time, (ms)	Velocity, (bytes/μs)
AES-128	3229	332.510165	4570	234.969872	9787	109.717
AES-256	4819	236.830435	6766	158.692002	14,133	75.976
HC-128	698	1537.651187	1040	1032.245553	2073	518.090
HC-256	1559	688.825907	2061	521.082124	4465	240.474
MICKEY-128	116,786	9.194097	164,304	6.535092	265,471	4.045
RABBIT	2190	490.270684	2893	371.215842	5656	189.848
SALSA-20	2650	405.262059	3885	276.402766	8428	127.402
SNOW 2.0–128	913	1176.187780	1474	728.503850	2924	367.179
SNOW 2.0–256	917	1170.801247	1445	742.868288	2989	359.219
SOSEMANUK	1967	545.822399	3124	343.685367	4973	215.906
STRUMOK-512	601	1788.079640	797	1347.060374	3648	294.329
STRUMOK-1024	584	1839.543985	821	1308.005633	3677	292.032
TRIVIUM	2058	521.715089	2879	372.930614	5016	214.051

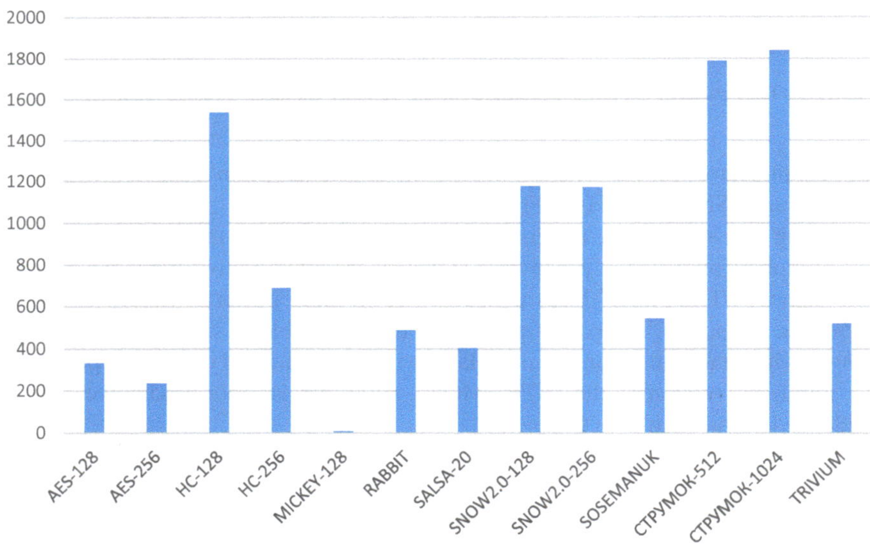

Fig. 11.8 Test results of long-range encryption streaming encryption, *Intel Core i76820HQ 2.7Gh*

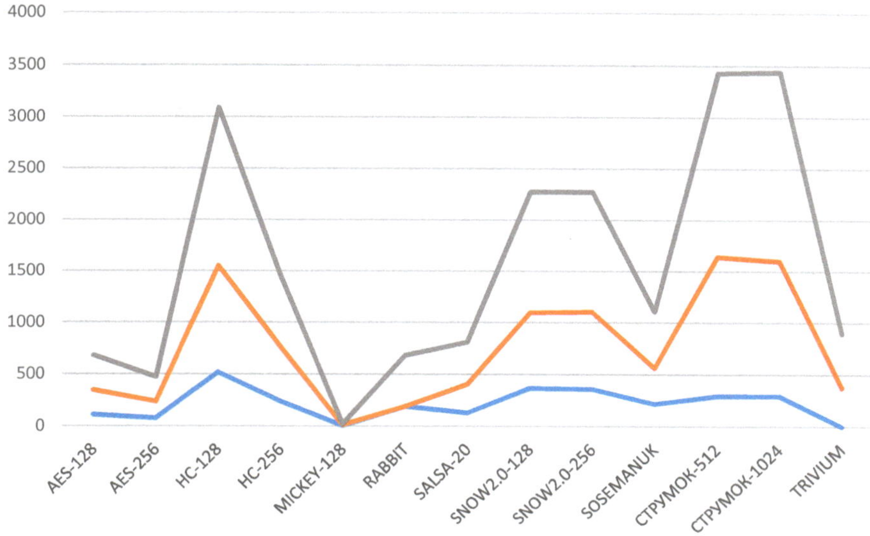

Fig. 11.9 Test results for long-range encryption streaming encryption, *Intel Core i75500u 2.4Gh*

testing streaming ciphers, this criterion is explored to determine the behavior of algorithm structures, identify weaknesses, and identify the advantages and disadvantages of certain ciphers.

The criterion of short-stream encryption according to the research methodology reflects the encryption speed of packets of various lengths (40, 576 and 1500 bytes), which represent the traffic of a telecommunication transmission channel. That is, these studies are a certain imitation of the functioning of streaming cipher when used to encrypt the traffic of modern packet-switched telecommunication systems.

According to the methodology, packet data is generated randomly. The test measures packet encryption time, the rate of encrypted bytes per microsecond, and the rate of encrypted packets per microsecond.

The test results for the short message encryption test are summarized in Tables 11.5, 11.6 and 11.7.

Summarizing these data, we can draw the following conclusions:

- the best performance by the criterion of short-stream encryption was obtained for the ciphers "STRUMOK", "SNOW 2.0" and "SOSEMANUK" respectively;
- the obtained figures show that the SNOW 2.0 and SOSEMANUK stream ciphers show better performance in the case of packets that are large but small in size;
- We have the opposite results with regard to the "STRUMOK" stream code: the code shows better performance when there are few packages, but they are large in size.

Table 11.5 Experimental results of the study on the criterion of short message encryption (*Intel Core i76820HQ 2.7Gh*)

Cipher	Packet encryption								
	50 packets by 1500 bytes			120 packets by 576 bytes			350 packets by 40 bytes		
	Time, (ms)	Velocity, (bytes/µs)	Velocity, (packets/µs)	Time (ms)	Velocity, (bytes/µs)	Velocity, (packets/µs)	Time, (ms)	Velocity, (bytes/µs)	Velocity, (packets/µs)
AES-128	230	6.524576	0.217486	211	2.736342	0.570071	60	0.667780	5.843072
AES-256	323	4.639654	0.154655	300	1.917443	0.399467	81	0.491400	4.299754
HC-128	257	5.845674	0.194856	546	1.054559	0.219700	1492	0.026804	0.234537
HC-256	1420	1.056115	0.035204	3413	0.168766	0.035160	9605	0.004165	0.036440
MICKEY-128	7819	0.191840	0.006395	7542	0.076372	0.015911	3532	0.011325	0.099094
RABBIT	154	9.771987	0.325733	158	3.654822	0.761421	88	0.453001	3.963760
SALSA-20	190	7.886435	0.262881	175	3.298969	0.687285	58	0.687285	6.013746
SNOW 2.0–128	70	21.337127	0.711238	71	8.089888	1.685393	64	0.624025	5.460218
SNOW 2.0–256	71	21.246459	0.708215	71	8.089888	1.685393	64	0.627943	5.494505
SOSEMANUK	91	16.574586	0.552486	102	5.658153	1.178782	79	0.505689	4.424779
STRUMOK-512	49	30.425963	1.014199	60	9.616027	2.003339	76	0.528402	4.623514
STRUMOK-1024	50	30.241935	1.008065	61	9.473684	1.973684	75	0.530504	4.641910
TRIVIUM	153	9.791123	0.326371	168	3.428571	0.714286	128	0.311526	2.725857

Table 11.6 Experimental results of the short message encryption criteria study (*Intel Core i75500u 2.4Gh*)

Cipher	Packet encryption								
	50 packets by 1500 bytes			120 packets by 576 bytes			350 packets by 40 bytes		
	Time (ms)	Velocity (bytes/μs)	Velocity, (packets/μs)	Time (ms)	Velocity, (bytes/μs)	Velocity, (packets/μs)	Time (ms)	Velocity, (bytes/μs)	Velocity, (packets/μs)
AES-128	313	4.796930	0.159898	291	1.982788	0.413081	91	0.438596	3.837719
AES-256	432	3.473830	0.115794	406	1.417323	0.295276	113	0.353669	3.094607
HC-128	412	3.637245	0.121242	844	0.682303	0.142146	2241	0.017849	0.156180
HC-256	2148	0.698259	0.023275	4916	0.117178	0.024412	14,225	0.002812	0.024605
MICKEY-128	10,651	0.140832	0.004694	10,462	0.055056	0.011470	4963	0.008060	0.070522
RABBIT	210	7.156489	0.238550	209	2.762590	0.575540	105	0.380952	3.333333
SALSA-20	262	5.723006	0.190767	233	2.471042	0.514801	77	0.517464	4.527814
SNOW 2.0–128	118	12.722646	0.424088	117	4.906303	1.022147	101	0.397219	3.475670
SNOW 2.0–256	110	13.599275	0.453309	120	4.792013	0.998336	103	0.387973	3.394762
SOSEMANUK	150	10	0.333333	171	3.376319	0.703400	122	0.328677	2.875924
STRUMOK-512	66	22.692890	0.756430	82	7.041565	1.466993	108	0.371058	3.246753
STRUMOK-1024	69	21.802326	0.726744	84	6.840855	1.425178	126	0.318471	2.786624
TRIVIUM	211	7.105637	0.236855	238	2.419152	0.503990	175	0.228311	1.997717

Table 11.7 Experimental results of the short message encryption criteria study (*Intel Pentium P6200 2.13Gh*)

Cipher	Packet encryption								
	50 packets by 1500 bytes			120 packets by 576 bytes			350 packets by 40 bytes		
	Time (ms)	Velocity (bytes/µs)	Velocity (packets/µs)	Time (ms)	Velocity, (bytes/µs)	Velocity, (packets/µs)	Time (ms)	Velocity, (bytes/µs)	Velocity, (packets/µs)s
AES-128	691	2.170	0.072	621	0.927	0.193	179	0.224	1.957
AES-256	983	1.526	0.051	911	0.632	0.132	245	0.163	1.429
HC-128	657	2.282	0.076	1363	0.423	0.088	3673	0.011	0.095
HC-256	4210	0.356	0.012	9658	0.060	0.012	27,350	0.001	0.013
MICKEY-128	19,010	0.079	0.003	18,628	0.031	0.006	8423	0.005	0.042
RABBIT	409	3.670	0.122	409	1.408	0.293	211	0.190	1.662
SALSA-20	601	2.496	0.083	541	1.065	0.222	170	0.236	2.062
SNOW 2.0–128	220	6.812	0.227	220	2.618	0.545	177	0.226	1.974
SNOW 2.0–256	221	6.790	0.226	223	2.582	0.538	177	0.226	1.976
SOSEMANUK	285	5.258	0.175	297	1.938	0.404	186	0.215	1.883
STRUMOK-512	293	5.125	0.171	340	1.693	0.353	394	0.102	0.889
STRUMOK-1024	293	5.125	0.171	343	1.682	0.350	396	0.101	0.884
TRIVIUM	376	3.986	0.133	404	1.427	0.297	313	0.128	1.119

11.2.2.3 Testing Streaming Ciphers by Criterion Initialization/Generation of Key Parameters

This criterion separately characterizes such an element of the cipher structure as the key set and the initialization vector. These two components of the encryption structure are the least critical for displaying the speed of the algorithm, as it takes little time to install the key and the initialization vector when compared to the encryption process.

According to the method, the following was determined: *7000* key installations are required to evaluate the key deployment scheme—*700* installations per key (*10* keys per *700* installations). To evaluate the initial vector initialization process, the following has been determined: *500* key installations are *50* installations per vector (*10* vectors of *50* installations).

To estimate the parameters described above, the total operation time, the number of installation cycles spent, and the number of settings that can be done per second will be recorded.

The results are summarized in Tables 11.8, 11.9 and 11.10.

Based on the obtained criteria of the initialization/generation of key parameters, we can conclude that the worst values are in the codes *HC-128, HC-256, MICKEY-128*. This explains the poor performance in *HC-128* and *HC-256* by the criterion of short message encryption—it takes a lot of time in the algorithm to update new key parameters. *AES-128, AES-256* and *Salsa20* have the best performance.

Table 11.8 Experimental results of cipher study by the criterion of initialization/generation of key parameters (*Intel Core i76820HQ 2.7Gh*)

Cipher	Setting key parameters			
	key		Initialization vector	
	Time (ms)	Installation speed (ms)	time (ms)	Installation speed (ms)
AES-128	483	14.489754	0.09	5494.505495
AES-256	1075	6.510417	0.06	8474.576271
HC-128	22	325.581395	2101.60	0.237914
HC-256	339	20.679468	13,347.60	0.037460
MICKEY-128	7	1044.776119	2987.80	0.167347
RABBIT	886	7.899786	58.30	8.576329
SALSA-20	25	277.777778	0.09	5787.037037
SNOW 2.0–128	33	210.210210	54.90	9.107468
SNOW 2.0–256	64	108.695652	55.10	9.074410
SOSEMANUK	1084	6.456373	70.20	7.122507
-512	33	209.580838	62.40	8.012821
-1024	34	207.100592	61.30	8.156607
TRIVIUM	46	152.173913	138.00	3.623188

Table 11.9 Experimental results of the study of ciphers by the criterion of initialization/generation of key parameters (*Intel Core i75500u 2.4Gh*)

Cipher	Setting key parameters			
	key		Initialization vector	
	Time (ms)	Installation speed (ms)	time (ms)	Installation speed (ms)
AES-128	657	10.652869	0.08	6112.469438
AES-256	1436	4.874312	0.09	5617.977528
HC-128	29	241.379310	2982.80	0.167628
HC-256	441	15.873016	22,889.10	0.021844
MICKEY-128	9	744.680851	4257.90	0.117429
RABBIT	1215	5.761791	80.80	6.188119
SALSA-20	36	196.078431	0.12	4101.722724
SNOW 2.0–128	50	140.845070	91.70	5.452563
SNOW 2.0–256	100	70.281124	90.30	5.537099
SOSEMANUK	1550	4.516712	101.60	4.921260
-512	44	159.453303	84.30	5.931198
-1024	44	159.090909	105	4.761905
TRIVIUM	66	105.900151	219.90	2.273761

Table 11.10 Experimental results of the study of ciphers by the criterion of initialization/generation of key parameters (*Intel Pentium P6200 2.13Gh*)

Cipher	Setting key parameters			
	key		Initialization vector	
	Time (ms)	Installation speed (ms)	Time (ms)	Installation speed (ms)
AES-128	1482	4.724	0.16	3060
AES-256	2109	3.319	0.23	2159
HC-128	38	183.727	5150	0.097
HC-256	787	8.897	38,976	0.013
MICKEY-128	10	686.275	7166	0.070
RABBIT	2387	2.933	175	2.864
SALSA-20	40	175	0.20	2495
SNOW 2.0–128	105	66.414	151	3.318
SNOW 2.0–256	174	40.299	152	3.283
SOSEMANUK	2824	2.479	163	3.054
-512	191	36.688	345	1.449
-1024	111	63.234	345	1.449
TRIVIUM	115	61.082	336	1.488

11.3 Conclusions and Recommendations

1. The studied symmetric encryption algorithms provide high cryptographic performance. Statistical security requirements are fulfilled, that is, the sequence generated by the generators does not differ in its statistical properties from the truly random sequence.
2. The statistical research methodology discussed in this monograph and its results can be considered as a primary analysis of the cryptographic properties of generators, since such statistical tests do not take into account the generator's own structure. The test results of various stream cryptoconversions confirm their high cryptographic performance. All the ciphers studied showed a large number of successfully passed tests: 130–134 on the criterion $P_j \geq 0,99$ and 186–187 on the criterion $P_j \geq 0,96$. It is worth noting the high performance of the statistical safety of the algorithm of the "Strumok" SSC (this chapter explored the previous version of the "Strumok" algorithm with 512 and 1024 bit keys), which revealed certain properties of the random bit generator. In particular, the properties of the generated sequences are not inferior to the PRS, formed by world-famous streaming cryptographic algorithms, in particular HC256 and SNOW 2.0 ciphers. In addition, for "Strumok" SSC the minimum values of the number of passed statistical tests by the criterion $P_j \geq 0,96$ are higher than in these algorithms, which indicates a slight preference for the statistical safety indicators of the "Strumok" algorithm.
3. The calculated number of probabilities of passing the tests for each of the above ciphers is summarized according to the methodology considered. All ciphers showed a sufficient level of statistical certainty. Despite the fact that the SOSEMANUK cipher in the results of matching the probability distribution to the uniform showed a slight, but slightly larger, unlike other ciphers, deviations from the uniform distribution, in the results of this test, it occupies a leading position.
4. The results of DIEHARD statistical testing for symmetric AES, Enocoro, Grain, HC, MICKEY, MUGI, Salsa20, SOSEMANUK, Trivium, and "Strumok" SSC show that by most indicators all streaming crypto algorithms have comparable statistical security. For "Strumok" SSC, additional DIEHARD tests were performed for probability distribution, which show that the probability distribution corresponds to a uniform law with small fluctuations. This additionally demonstrates the benefit of the conclusion regarding the high statistical safety performance of the "Strumok" algorithm. A promising area of further research is to evaluate the robustness of a generator against various cryptographic attacks.
5. The Stream algorithm is capable of delivering high performance and operating efficiently across different computing platforms. In particular, the results of the research found that compared to the best of the world analogs (*Grain, HC-128, HC-256, MICKEY, Rabbit, Salsa20, SNOW 2.0, SOSEMANUK, Trivium* streaming crypto algorithms) "Strumok" crypto algorithm provides the highest speeds (after *HC-128* and *HC-256*). It should be noted that these estimates

are not marginal, that is, a promising direction for further development is the optimization of software implementation to increase throughput, for example, due to parallelization or acceleration due to the attraction of additional hardware.

6. The test results showed that by the criterion of long message encryption, the best indicators were the codes "Strumok", *HC-128* and *SNOW 2.0*. And the algorithm "Strumok" gives better results than *HC-128* by 30%, and better than *SNOW 2.0* almost twice.
7. Experimental results on the criterion of short message encryption have shown that the best indicators are the codes "Strumok", *SNOW 2.0* and *SOSEMANUK* respectively. *SNOW 2.0* and *SOSEMANUK* ciphers perform better when packets are large but small in size. In the stream "Stream" cipher, the results are reversed: the code shows better performance when packets are large in size.
8. According to the indicators of the criterion of initialization/generation of key parameters, the worst performance in the codes *HC-128, HC-256, MICKEY-128*. This explains the poor performance in *HC-128* and *HC-256* by the criterion of short message encryption—it takes a lot of time in algorithms to update new key parameters. The *AES-128*, *AES-256* and *Salsa20* ciphers have the best performance.
9. The results of experimental studies of the speed characteristics of stream ciphers indicate that the algorithm "Strumok" is the most prudent solution, it is able to produce huge rates of encryption speed. It is practically proved that the current rate of encryption algorithm "Strumok" on modern computing systems can reach 10–15 Gbit/s.

References

1. NIST Special Publication 80090A. Recommendation for Random Number Generation Using Deterministic Random Bit Generators. (Electronic resource). Access mode: http://csrc.nist.gov/publications/nistpubs/80090A/SP80090A.pdf
2. Dieharder: a random number test suite–free implementation of statistical tests (including diehard and NIST tests) in C. (Electronic resource). Access mode: http://www.phy.duke.edu/~rgb/General/dieharder.php
3. The Marsaglia Random Number CDROM including the Diehard Battery of Tests of Randomness. (Electronic resource). Access mode: http://stat.fsu.edu/pub/diehard
4. HMRAE 1040403. The Evaluation of Randowness of RPG100 by Using NIST and DIEHARD Tests, 6 p. FDK Corp (2003)
5. Gorbenko, I.D., Gorbenko, Y.I.: Applied Cryptology: Monograph, 1st and 2nd edition, 868 p. Kharkiv Fort, KNURE (2012)
6. Gorbenko, I.D., Gorbenko, Y.I.: Applied Cryptology: Textbook, 1st and 2nd editions. 878 p. Kharkiv, KNURE, Fort (2012)
7. Nyffeler, P.: Binare Automation und ihre linearen Rekursionen. Ph.D thesis. University of Berne (1975)
8. Kuznetsov, A., Gorbenko, Y., Andrushkevych, A., Belozersev, I.: Analysis of block symmetric algorithms from international standard of lightweight cryptography ISO/IEC 29192-2. In: 2017 4th International Scientific-Practical Conference Problems of Infocommunications. Science

and Technology (PIC S T), pp. 203–206 (2017). https://doi.org/10.1109/INFOCOMMST.2017.8246380
9. Gorbenko, I., Kuznetsov, A., Gorbenko, Y., Vdovenko, S., Tymchenko, V., Lutsenko, M.: Studies on statistical analysis and performance evaluation for some stream ciphers. Int. J. Comput. **18**, 82–88 (2019)
10. Gorbenko, I., Kuznetsov, A., Tymchenko, V., Gorbenko, Y., Kachko, O.: Experimental studies of the modern symmetric stream ciphers. In: 2018 International Scientific-Practical Conference Problems of Infocommunications. Science and Technology (PIC S T), pp. 125–128 (2018). https://doi.org/10.1109/INFOCOMMST.2018.8632058
11. Kuznetsov, A., Frolenko, V., Eremin, E., Zavgorodnia, O.: Research of cross-platform stream symmetric ciphers implementation. In: 2018 IEEE 9th International Conference on Dependable Systems, Services and Technologies (DESSERT), pp. 300–305 (2018). https://doi.org/10.1109/DESSERT.2018.8409148
12. Gorbenko, I., Kuznetsov, A., Lutsenko, M., Ivanenko, D.: The research of modern stream ciphers. In: 2017 4th International Scientific-Practical Conference Problems of Infocommunications. Science and Technology (PIC S T), pp. 207–210 (2017). https://doi.org/10.1109/INFOCOMMST.2017.8246381
13. Babbage, S.: A Space/Time TradeOff in Exhaustive Search Attacks on Stream Ciphers. Vodafone Ltd, Newbury, UK. 09.04.1996

Chapter 12
Areas of Application for Nonlinear Shift Registers in PRS Generators

12.1 Key Issues During the Design of Streaming Encryption Algorithms

Common cryptographic algorithms that are built using the LFSR are: *A5/1* streaming encryption, which is used to ensure confidentiality in *GSM* cellular telephone communications [1]; E0 stream cipher used in *Bluetooth* protocol [2] and others. LFSR is one of the most popular devices for generating pseudorandom binary sequences. They have many applications, including for the detection and correction of errors [3], data compression [4], testing and simulation [5], cryptography [6].

One of the first to discuss the generation of PRS using shift registers was *Golomb* [7]. Contemporary interpretation of the issue is contained in the work of *Golomb* and *Gong* [8].

However, in spite of the numerous advantages of using LFSR in streaming encryption, they have a number of significant disadvantages due to their linearity [9]. In addition, the security of some streaming algorithms based on LFSR is at risk of algebraic attacks [10]. Recent studies have shown that PRS generators based on LFSR are prone to various threats, such as interception of transmitted information, surveillance, and information exchange. This problem is particularly relevant in wireless data technologies where the cryptographic security mechanisms used are not sufficiently reliable [11]. It has been found that streaming ciphers can be prone to various network attacks [12], serious vulnerabilities have been found in A5X [13] and E0 [14] algorithms.

The Berlekamp-Massey algorithm can be used to construct the smallest LFSR that generates a given binary sequence. It was originally invented by Berlekamp to decode Bose (Bose-Chaudhuri-Hocquenghem) codes [15]. Massey linked Berlekamp's algorithm to the synthesis of RZLZZ and simplified it [16]. In the future, there were many modifications and improvements to the algorithm, such as Mandelbaum developed its arithmetic analogue [17], Imamura and Yoshida showed an alternative and simpler

A. A. Kuznetsov et al., *Stream Ciphers in Modern Real-time IT Systems*,
Studies in Systems, Decision and Control 375,
https://doi.org/10.1007/978-3-030-79770-6_12

derivation [18], Fitzpatrick found polynomials for symmetry [19] and Fleischmann modified the algorithm to extend the sequence in both directions around any given data bit [20]. In addition, it has been shown that results similar to those of the BerlekampMassey algorithm can be obtained using the Euclid algorithm [21] and continuous fractions [22].

The BerlekampMassey algorithm builds LFSR that generate one output bit per beat. A number of methods have been developed for the construction of a radar with a degree of parallelization *p*, among which there are two main approaches:

1. Synthesis of subsequences representing p thinning of some phase shift of the original LFSR sequence [23];
2. Calculate the set of states achievable from any state in *p* steps.

The latter is usually done by calculating the degree of coherence of the LFSR matrix [6]. High parallelization LFSR are used in applications where high data rates are required and are important, for example, for cyclic redundancy code (CRC), which is widely used for data transmission and storage to detect packet errors [3].

As noted in the first sections, there are several methods of designing PRS generators that destroy the linear properties and thus make the following systems cryptographically more stable [24]:

- the use of a nonlinear function that combines the outputs of several LFSR (such as the Geffen generator);
- the use of a nonlinear filtering function for the contents of each cell of a single LFSR;
- using the output of the LFSR to control the clock signal of one or more LFSR (algorithm A5);
- dynamic change of recurrent parameters (register length and feedback coefficients), etc.

One promising approach is to use a design based on shift registers with nonlinear feedback (SRNLF or NLFSR) [25–29]. LFSR represent the generalized case of LFSR, but, unlike the last, in NLFSR the current condition is a nonlinear function of the previous states [7]. As shown in [27], at this time NLFSR are quite important in random number generation systems.

The advantages of systems in which cryptographic primitives based on NLFSR are introduced, in comparison with LFSR, are:

- higher stability (in particular linear complexity);
- the original sequence has statistical characteristics no worse than the well-studied LFSR (passing tests for randomness of the generated sequence);
- more complex, in comparison with LFSR, cryptanalysis, the lack of a well-developed mathematical apparatus;
- structure (amount of memory, number of operations per output bit, production architecture), almost identical to LFSR;

- nonlinearity is already entered in the register, which does not require additional complication (and as a consequence, additional RAM and additional computational operations), ie the time/memory ratio is similar to LFSR;
- Significantly more feedback combinations (including those that generate a sequence with the maximum possible period). This allows you to implement schemes based on one basic model (fixed number of collar shift registers) and different long-term keys, or when you need to perform a large number of encoders (for example, for serial production), but it is necessary to ensure incompatibility of different batches;
- the lack of simple algorithms for restoring the structure of the NLFSR using the sequence generated by them, such as the Burlackamp-Messi algorithm for LFSR;
- simplicity of software and hardware implementation.

There are many theoretical results that demonstrate the benefits of using nonlinear sequences in the field of wireless communication. For example, complementary sequences can solve the known problem of power control in systems with orthogonal frequency division multiplexing (OFDM), maintaining the ratio of peak power to average [30]. In [31] it was shown that the use of systems using several carriers in the spectrum complement each other and extend the Legendre sequences, ahead of the systems with code division multiplexing (CDMA). However, due to the lack of effective hardware methods for generating nonlinear sequences, their theoretical advantages cannot be used at this time.

Similar to the case of LFSR, NLFSR can be reused to generate p bits of the sequence in one clock cycle. This is usually done by duplicating the feedback functions from the NLFSR p times, as in [32–34]. This method requires that the leftmost cells of the NLFSR not be used as input for feedback functions or output functions. In this case, the size of the circuits that implement the first degree of parallelization can grow much more than p times [35].

Studies of standardized streaming cryptographic conversion algorithms show that there is a steady trend in the construction of PRS generator circuits constructed using shift registers [25], including nonlinear feedback. Stream ciphers based on NLFSR are included in Achterbahn [36], Dragon [37], Grain [38], Trivium [39], VEST [34]. In [40–43], it was shown that NLFSR are more resistant to cryptanalytic attacks than LFSR.

12.2 General Second-Order NLFSR Model

The general design of the NLFSR for the register, consisting of $L = 4$ cells, is shown in Fig. 12.1. The registers use the product of only two cells, such NLFSR will be called NLFSR of the second order. In what follows, unless otherwise specified, we will understand by NLFSR such constructions NLFSR of the second order in $GF(2)$.

The following notation is introduced in Fig. 12.1: $a_{ij} \in \{0, 1\}$ is the feedback coefficient, which corresponds to the presence ($a_{ij} = 1$) or absence ($a_{ij} = 0$) of

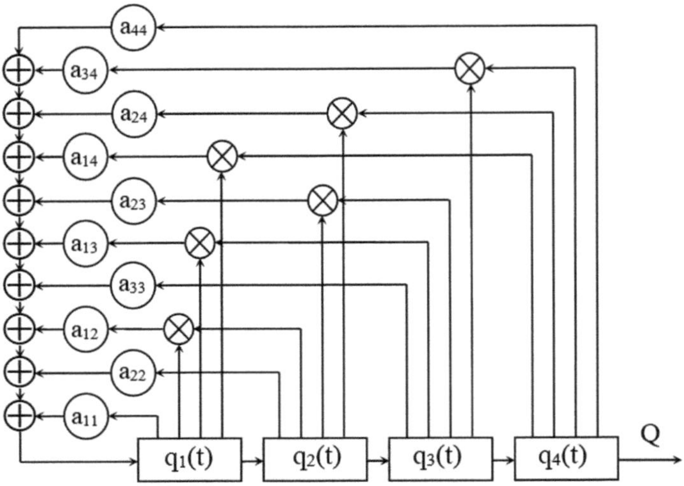

Fig. 12.1 General design of the NLFSR

feedback from the product of the i and j cells of the register ($i, j = 1, \ldots, L$) $q_i(t) \in \{0, 1\}$ is the value of cell i at time *t; Q* is the sequence of bits generated. The sign $\otimes$ indicates a nonlinear multiplication function (corresponding to the logical operation "and"—*AND*) and the sign $\oplus$ - a linear addition function (corresponding to the logical operation of exclusion "OR"—*XOR*).

In the general case, with the introduced notation system (see Fig. 12.1), the feedback for the NLFSR can be set as follows:

$$q_1(t+1) = \sum_{i=1}^{L} a_{ii} q_i(t) + \sum_{i=1}^{L-1} \sum_{j=i+1}^{L} a_{ij} q_i(t) q_j(t) \tag{12.1}$$

which will correspond to the feedback function *f (x)*, written in algebraically normal form:

$$\begin{aligned} f(x) = a_0 + a_{11} \cdot x_1 + a_{22} \cdot x_2 + \cdots + a_{LL} \cdot x_L + a_{12} \cdot x_1 \cdot x_2 \\ + a_{13} \cdot x_1 \cdot x_3 + \cdots + a_{(L-1)L} \cdot x_{L-1} \cdot x_L \end{aligned}$$

We call the coefficients a_{ij} *linear coefficients* if $i = j$ and *nonlinear coefficients*—if $i \neq j$. In this case, the state of LFSR is a special case of NLFSR at all nonlinear coefficients equal to zero, and is calculated by the ratio:

$$q_1(t+1) = \sum_{i=1}^{L} a_{ii} q_i(t) \tag{12.2}$$

For convenience of consideration and perception of designations of feedback coefficients a_{ij} in NLFSR we arrange them in the form of a matrix:

$$\begin{matrix} a_{11} & a_{12} & a_{13} & \dots & a_{1L} \\ & a_{22} & a_{23} & \dots & a_{2L} \\ & & a_{33} & \dots & a_{3L} \\ & & & \dots & \dots \\ & & & & a_{LL} \end{matrix}$$

In this case, the diagonal elements, i.e. a_{ii},, correspond to a special case—LFSR.

Let us denote the number of feedback coefficients a_{ij}, used in the construction by the symbol n_L. For NLFSR of the second order n_L is defined unambiguously for the set L by the formula:

$$n_L = \frac{L \cdot (L + 1)}{2}.$$

Since any NLFSR is an automaton of finite states, then, as noted in [44], the PRS generated by it will inevitably be periodic and will have a period T. A characteristic indicator T is the maximum possible period ($T_{\max}$) for a given value L. Since the zero state of all registers is forbidden, the maximum possible value of the device states, and accordingly ($T_{\max}$) is defined as:

$$T_{\max} = 2^L - 1. \tag{12.3}$$

A sequence that has a maximum period for a given value of the registers L will be called M-sequence. NLFSR, which generate M-sequence, call M- NLFSR.

12.3 Research of the Basic Methods of Construction of NLFSR

The theory of LFSR is studied well enough [45], NLFSR are studied much less [46]. The first algorithm for constructing the smallest NLFSR generating a given binary sequence was introduced by Jansen in 1991 [47, 48]. Alternative algorithms were presented in [49], as well as in [50] and [51].

It is known how to construct LFSR with the maximum period, they correspond to primitive polynomials over $GF(2)$. In the general case, it is unknown how to build all NLFSR with the maximum period. The main method is to find such registers with the appropriate properties. Typically, all nonlinear feedback registers in the literature have a rather complex structure and do not consist of a single structural element. If the NLFSR has a simple structure, then it is a register with a small size.

There are no general methods for designing NLFSR maximum period [45, 52]. The construction of a special class of NLFSR with the maximum period was submitted by Mykkeltveit and others. Mykkeltveit et al. [53]. The maximum period of NLFSR to the order of $L = 64$ was presented in [54], however, the given NLFSR have a very complex algebraic normal form. Dubrova [52] gives an example of a shift of the Galois register with size $L = 100$, which generates a sequence with a maximum period, but this sequence has no power de Bruijn, i.e. some L-bit models appear in the sequence more than once.

In [45], the results for second-order NLFSR generating a sequence of the maximum period with a nonlinear feedback function for the following register sizes are presented:

$$L = 22, x_{22} + x_{19} + x_{15} + x_{13} + x_9 + x_8 + x_4 + x_3 + x_2 + x_{11} \cdot x_{17}$$

$$\begin{aligned} L = 23, x_{23} &+ x_{17} + x_{15} + x_{14} + x_{12} + x_{10} \\ &+ x_9 + x_8 + x_6 + x_4 + x_3 + x_2 + x_1 + x_{13} \cdot x_{21} \end{aligned}$$

$$L = 24, x_{24} + x_{15} + x_{14} + x_{10} + x_8 + x_7 + x_5 + x_9 \cdot x_{14}$$

$$\begin{aligned} L = 25, x_{25} &+ x_{24} + x_{22} + x_{20} + x_{17} + x_{15} + x_{13} \\ &+ x_{10} + x_9 + x_7 + x_6 + x_3 + x_2 + x_{13} \cdot x_{20} \end{aligned}$$

$$L = 26, x_{26} + x_{24} + x_{22} + x_{21} + x_{18} + x_{11} + x_7 + x_5 + x_4 + x_{10} \cdot x_{20}$$

$$\begin{aligned} L = 27, x_{27} &+ x_{26} + x_{25} + x_{23} + x_{19} \\ &+ x_{17} + x_{16} + x_{13} + x_{10} + x_8 + x_6 + x_{17} \cdot x_{21} \end{aligned}$$

$$\begin{aligned} L = 28, x_{28} &+ x_{24} + x_{23} + x_{22} + x_{20} + x_{17} \\ &+ x_{14} + x_{10} + x_9 + x_7 + x_6 + x_2 + x_1 + x_1 \cdot x_{20} \end{aligned}$$

$$L = 29, x_{29} + x_{26} + x_{24} + x_{23} + x_{18} + x_{17} + x_{13} + x_{10} + x_7 + x_6 + x_2 + x_1 \cdot x_9$$

In addition, [52] presents the results of the found nonlinear feedback functions for higher-order nonlinearity:

$$L = 26, x_{26} + x_{13} + x_{12} + x_7 + x_3 + x_2 \cdot x_3 \cdot x_4 \cdot x_8$$

$$L = 27, x_{27} + x_{24} + x_{22} + x_{18} + x_{17} + x_{16} + x_9 + x_7 + x_3 + x_1 \cdot x_4 \cdot x_{16} \cdot x_{18}$$

$$L = 28, x_{28} + x_{27} + x_{20} + x_{11} + x_{10} + x_9 + x_7 + x_6 + x_5 + x_{10} \cdot x_{14} \cdot x_{17} \cdot x_{24}$$

$$L = 29, x_{29} + x_{28} + x_{26} + x_{25} + x_{16} + x_{13} + x_{11} + x_{10} + x_9 + x_7 + x_6 + x_2 + x_2 \cdot \overline{x}_3 \cdot x_5 \cdot x_8 \cdot \overline{x}_9 \cdot \overline{x}_{10} \cdot \overline{x}_{19} \cdot x_{23} + x_2 \cdot x_3 \cdot \overline{x}_5 \cdot x_8 \cdot x_9 \cdot \overline{x}_{10} \cdot \overline{x}_{19} \cdot \overline{x}_{23}$$

$$L = 30, x_{30} + x_{26} + x_{23} + x_{22} + x_{21} + x_{20} + x_{19} + x_{18} + x_{17} + x_{16} + x_{15} + x_8 + x_3 + x_2 + x_3 \cdot x_5 \cdot \bar{x}_6 \cdot x_{10} \cdot \bar{x}_{11} \cdot \bar{x}_{15} \cdot x_{17} \cdot \bar{x}_{19} \cdot \bar{x}_{21} \cdot x_{23} \cdot \bar{x}_{27} \cdot x_{29} + \bar{x}_3 \cdot \bar{x}_5 \cdot x_6 \cdot \bar{x}_{10} \cdot x_{11} \cdot x_{15} \cdot \bar{x}_{17} \cdot \bar{x}_{19} \cdot x_{21} \cdot x_{23} \cdot \bar{x}_{27} \cdot \bar{x}_{29}$$

$$L = 31, x_{31} + x_{29} + x_{25} + x_{24} + x_{23} + x_{22} + x_{21} + x_{17} + x_{16} + x_{15} + x_{11} + x_{10} + x_5 + x_2 + \overline{x}_2 \cdot \overline{x}_5 \cdot x_6 \cdot x_7 \cdot x_9 \cdot \overline{x}_{10} \cdot \overline{x}_{12} \cdot x_{19} \cdot x_{22} \cdot x_{24} \cdot x_{26} \cdot x_{27} + x_2 \cdot x_5 \cdot \overline{x}_6 \cdot \overline{x}_7 \cdot \overline{x}_9 \cdot x_{10} \cdot x_{12} \cdot \overline{x}_{19} \cdot \overline{x}_{22} \cdot \overline{x}_{24} \cdot x_{26} \cdot \overline{x}_{27}$$

where $\overline{x}_i = x_i + 1$.

One of the leading researchers in this field, Dubrova, argues [55] that modern computing power allows to model NLFSR only up to $L < 35$, in the work of the same author [56] NLFSR with a dimension up to $L < 25$.

The maximum M- NLFSR at this time is the register represented in the *Achterbahn-128* streaming encryption algorithm [36], where the maximum register length is $L < 33$ and fourth-order nonlinearity is used, and the algorithm contains primitive NLFSR with nonlinearity up to the sixth order.

12.3.1 De Bruijn's Sequence

The de Bruijn sequence of order $n(a_0, \ldots, a_2{}^L - {}_2)$ of elements from the binary field $GF(2)$ is a sequence of period 2^L, in which all different L-courts appear only once.

It was proved [57] in 1894, and independently by *de Bruijn* [58] in 1946, that the number of cyclically equivalent sequences satisfying the de Bruijn sequences is equal to

$$B_L = 2^{2^{L-1}-L}$$

The modified de Bruijn sequence of order $L(a_0, \ldots, a_{2^L-2})$ is a sequence of period $2^L - 1$, obtained from the sequence de Bruijn of order L by removing the tuple L of consecutive zeros.

In 1990, Mayhew and Golomb [59] investigated sequences that satisfied de Bruijn's modified sequences and their linear complexity. These sequences were called by Hammel and others in [36] primitive sequences. In the case of LFSR, these sequences are generated by primitive polynomials, and their theory has been studied quite well [60]. Primitive sequences are very important in cryptographic applications because [61]:

1. They exist. There are B_L primitive sequences in general (linear and nonlinear). The number of primitive LFSR is equal to

$$\frac{\varphi(2^L - 1)}{L}, \frac{\varphi(2^L - 1)}{L}, \frac{\varphi(2^L - 1)}{L},$$

where φ denotes the Euler function, therefore, the NLFSR is much larger than the LFSR.

2. Primitive sequences have good statistical properties. They satisfy the basic postulates of Golomb. The linear complexity of the NLFSR is much larger than L, and most of the generated sequences have the maximum possible linear complexity of $2^L - 2$. Recall that the linear complexity of the primitive LFSR of the order of L is simply L.
3. There are primitive NLFSR, for which the algebraic normal form of Boolean feedback functions is quite simple; it has a low algebraic degree and possibly a small number of terms. Since there are different Boolean functions from $L - 1$ variables, therefore, the probability that the randomly selected feedback function is a primitive NLFSR is equal to

$$\frac{2^{2^{L-1}-L}}{2^{2^{L-1}}} = \frac{1}{2^L}$$

and decreases with increasing L.

The task is to find primitive NLFSR with the simplest algebraic form. The method of constructing such primitive NLFSR is not known. Hammel and others [36] found simple primitive NLFSR up to 33 orders, which were used in the construction of the Achterbahn stream cipher. But no method gives a good average search time needed to find the maximum period.

It is also an open problem to prove the lower estimate of the linear complexity of the NLFSR. Mayhew and Golomb [59] investigated all modified de Bruijn sequences of size *5* and *6*, which exist $2^{11} = 2048$ and $2^{26} = 67108864$, respectively. It turns out that there are very few such sequences with low linear complexity. In the case when $L = 5$ there is no NLFSR with a linear complexity of *10*, and there are only *10* sequences with a linear complexity of *15*. It is possible to hypothesize that NLFSR of dimension L, being a prime number, has a lower limit of the corresponding linear complexity, which is equal to 3 • L.

This follows from the more general hypothesis formed in the article [62] and the results [59]. The upper limit of the linear complexity of NLFSR $2^L - 2$, this estimate

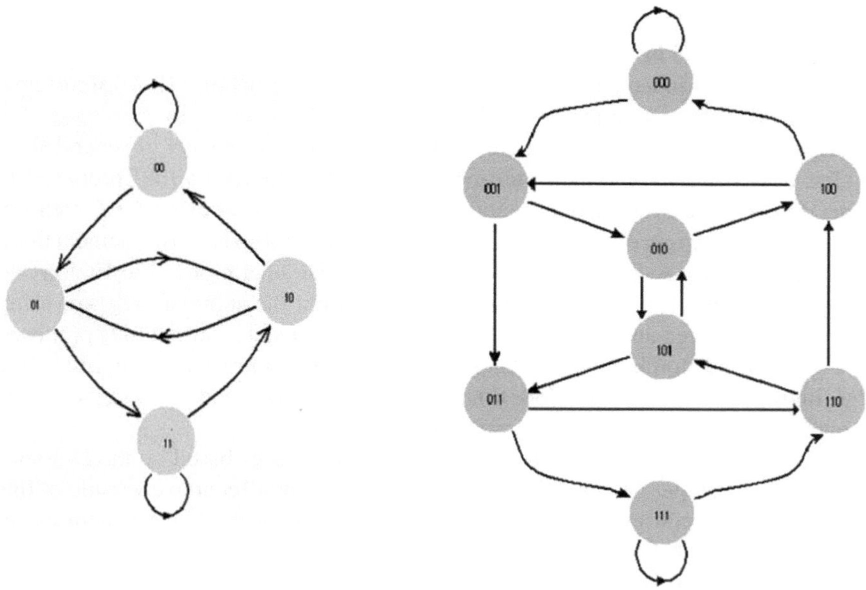

Fig. 12.2 Counts de Bruijn with dimensions 2 and 3, respectively

is rigid. To date, there is great interest in finding and constructing primitive NLFSR suitable for cryptographic applications, see [56, 63, 64] as an example.

Visually, the de Bruijn sequence can be represented by so-called graphs [65]. De Bruijn graphs of order L have the same set of vertices as a hypercube, i.e. 2^L binary values, but the edges are different. Graphs de Bruijn with dimensions *2* and *3* are shown in Fig. 12.2.

There are various methods for generating de Bruijn sequences, but the most popular is to find linear recurrent relations, which can be almost a complete period of NLFSR, based on primitive polynomials in *GF (2)*.

12.3.2 Algorithms for Generating De Bruijn Sequences

The number of publications discussing methods for generating de Bruijn sequences is quite large, for example [46, 53, 66–70].

In most methods, the de Bruijn sequence is obtained by combining many small loops, but such a procedure requires additional memory to store information about the states for further connection of loops, or the feedback function must contain many coefficients in order to get one sequence. Most existing methods are not effective for obtaining de Bruijn sequences of period 2^L for $L \geq 30$.

Finding the de Bruijn sequence is equivalent to solving the problem of finding the Hamilton cycle in the shift register graph [71]. An overview of algorithms for generating de Bruijn sequences using shift registers is given in [46, 72]. All algorithms can be divided into two main groups [55].

Algorithms in the first group are obtained using the shift register of several short cycles, and then combine them into one large cycle. Fredriksen [68] proposed to generate complete cycles from several cycles of the L register using $6 \bullet L$ memory bits and L time steps to obtain the next state from the current state. To implement these calculations, you need to perform ~ $O(L \cdot 2^L)$ bitwise operations. The Etzion and Lempel algorithm [73] generates complete loops from the summary register using $L^2/4$ bits for storage and L time steps to obtain the next state. Jansen [47] proposed an algorithm for combining the states of cycles of an arbitrary shift register. This algorithm generates complete loops using $3 \bullet L$ storage bits and $4 \bullet L$ steps in time to obtain the next state from the current one.

A recursive algorithm for generating de Brain sequences based on the D homomorphism of Lempel's works was presented in [74]. A non-recursive version of this algorithm was proposed by Chang etc. [75]. In both algorithms, the generation of L variable Boolean functions of sequences where the de Bruijn order of order L is constructed from Boolean functions with fewer variables. In [71] a method and algorithm for finding nonlinear Boolean feedback functions that generate de Bruijn sequences in the shift register based on the properties of homomorphism of graphs of shift register states, construction of automatic models of main trees of the shift register graph, and formation of rules of transitions of the states of the automaton forming a Hamilton cycle in the shift register.

In the second group, the method of algorithms is to use cross-connections of pairs between shift registers with a known period to obtain another shift register with the same period. There are different approaches to choosing a crossover pair. Fredriksen [76] proposed the "best unit" ("prefer one") algorithm for generating NLFSR from de Bruijn sequences. In the algorithm of the "best unit" of the de Bruijn sequence, the L-tuple $(1, a_1, a_2, \ldots, a_{L-1})$ precedes the tuple L $(0, a_1, a_2, \ldots, a_{L-1})$ for all $(L - 1)$ tuples $(a_1, a_2, \ldots, a_{L-1})$, except for zero. Dubrova [52] presents a method for constructing NLFSR with a period of $2^L - 1$ from the maximum LFSR period, in which the cross-coupling of pairs is determined by an additional nonlinear function to NLFSR. The number of possible pairs of cross-compounds in the LFSR of the maximum period was obtained by Helleseth and Kløve [77].

Dubrova [55] presents a method for generating complete cycles for NLFSR, but their original sequences are not de Bruijn sequences. According to [78], the feedback shift register may need up to L steps to generate a binary sequence of length L. For example, the smallest LFSR and NLFSR generating binary sequence

$$\underbrace{0, 0, \ldots, 0, 1}_{L-1}$$

have L and $L - 1$ degrees, respectively [47].

On average, in this technique, LFSR require *L/2* stages to generate a binary sequence of length L [79], and NLFSR requires $2 \bullet \log_2 L$ stages to create such a sequence [47]. These estimates reflect the size of the stages only, they do not take into account the size of the circuits of the computational feedback functions. Because the nonlinear feedback function in NLFSR is generally larger than the linear feedback function in LFSR, the NLFSR $k$stage can be much larger than the $k$stage for LFSR.

But, despite all the previously mentioned works, the synthesis of NLFSR with a guaranteed duration of the period remains an open problem [80]. A systematic algorithm for the synthesis of NLFSR has not been discovered yet. In addition to the approaches described above, we can note the solutions in some individual cases listed in [81–84].

The disadvantage of the described methods for the synthesis of NLFSR by searching for de Bruijn sequences with the period $2^L - 1$ and the subsequent restoration of the feedback structure is that it is impossible to determine the order of nonlinearity of the synthesized NLFSR. As noted in [45, 61], in the polynomials found for $L = 30.31$ there are several terms of multiplication of *12* register cells, including inverted ones. In practice, such a complex structure leads to a significant reduction in the speed of generators containing such large structures, and increase the resources spent to ensure their efficiency.

12.4 Problems of SPS Synthesis and Analysis on the Basis of NLFSR

In order to build streaming ciphers, which are based on schemes with NLFSR, which can be applied in practice, it is necessary to take into account a number of factors:

1. The generated sequence must be long enough to replace the key [85].

Thus, with modern technologies, it is necessary to generate streams of encryption to close the Ethernet channel in real time. That is, to generate an encryption range at a speed of 1 GB/s, while reloading key information should not be at least a day, and it is desirable to ensure continuous operation during the year. To meet the requirement of no scale overlap, the scale period must be at least $T > 10^9 \cdot 60 \cdot 60 \cdot 24 \cdot 365 \approx 32 \times 10^{15}$ (operation at 1 GB/s during the year). For these reasons, we have the minimum length of the register, which can generate a sequence with a period $L = 55$, while $T = 2^{55} - 1 = 36 \times 10^{15}$. After that, in order to avoid repetition of the encryption sequence, it is necessary to change the feedback coefficients. Or take a system with T that is much higher than the minimum specified, and change the initial state of the registers.

To achieve the required length of the period when solving such a problem, it is possible to use two approaches:

- construction of one M- NLFSR the size, for example 2^{128}, or another rather big M- NLFSR;
- construction of several M- NLFSR (with possible use of some M- NLFSR) of smaller length, and to receive the initial sequence of all generator as the sum (or more difficult function) from outputs of all M- NLFSR.

At the same time, it is undesirable to use a significant number of short NLFSR, as their dimensions must meet certain requirements to obtain a given period. Computational complexity (for software implementation) will increase using each additional register. Thus, it is desirable to reduce the number of NLFSR used in the system to the minimum necessary.

For example, to generate a sequence equivalent in length to the sequence generated by the NLFSR with $L = 120$, it is possible to use *4* NLFSR size *23, 29, 31, 37* and the function of simple summation of their sequences.

The first problem directly follows another, which must be solved when designing streaming encryption systems.

2. The forming polynomial for the sequences must form M- NLFSR, otherwise we can get $T = 1$ or a period less than the maximum possible.

There is a certain difficulty in solving this problem, because the algorithms for finding large M- NLFSR, as mentioned earlier, are not known. And solving this problem in terms of complexity is equivalent to solving the problem of finding prime numbers. But the larger the size of the NLFSR, the more difficult this task becomes.

An unambiguous answer to the question: "Does this NLFSR give the maximum period or not?"—can give only the analysis of the generated scale. The complexity of such an analysis is proportional to the length of the registers and is defined as 2^{L+1} (for the analysis you need to get a scale of length $2 \bullet T$).

In this case, the proportion of polynomials that generates the M sequence from the total number of possible polynomials decreases with increasing L as will be shown below. An approximate formula for estimating the upper limit of the number of M-NLFSR:

$$M_0 \leq 2^{\frac{L(L-1)}{2} - L + 2}$$

The total number of different polynomials for NLFSR is defined as

$$\Lambda_{\max} = 2^{\frac{L(L+!)}{2}}$$

Thus, the share of M-NLFSR can be determined by the ratio:

$$\frac{M_0}{\Lambda_{\max}} \leq \frac{4}{2^{2-L}}$$

For $L = 55$ the part of polynomials that generate the M-sequence will be less than 3×10^{-33}, and for $L = 128$—less than 3×10^{-77} of the total number possible. For L

= *128*, the total number of possible NLFSR will be 2^{8256}. At the same time, even at $L = 16$, the analysis of the period of only one polynomial takes about a minute (on a middle-class personal computer).

Thus, for large values of *L*, the time spent on solving the problem of finding all possible M- NLFSR or at least sufficient to solve the problem of synthesis, becomes unacceptably important if you use only known and published at this time methods and techniques. Moreover, with increasing L, the probability that an arbitrarily taken NLFSR will form a sequence will decrease with degree dependence.

3. The third aspect to pay attention to during design streaming encryption systems is an analysis of possible attacks and the complexity of solving hacking problems by force.

As will be shown below, most polynomials can be discarded by analyzing only their general form. The complexity of analyzing the form of a polynomial is incomparably less than the complexity of analyzing the sequence generated by this polynomial. As a result, the complexity of the power attack tasks is significantly reduced as the total number of polynomials that can be used in the encryption generator is reduced.

By approximating the obtained result to LFSR, which is a special case of NLFSR, it is possible to exclude in advance a significant number of polynomials that deliberately cannot generate a sequence with a maximum period. Thus, it is possible to reduce the complexity of solving computational problems for the search for generators LFSR for encoders by several orders of magnitude of the stated complexity. Of course, this is the case when the structure of the encryption system and the polynomial forming itself are not known to us at first.

12.5 Cryptographic Stability of NLFSR

Accurate assessment of the reliability of streaming encryption systems remains an open question. It is known that to generate a sequence of a given length, you can use the sum of sequences from shorter NLFSR second order. However, which sources can be used in this second-order NLFSR and how long the total sequence will be, are not given in the available sources.

The literature indicates that systems that include NLFSR have good statistical properties, but there are no quantitative estimates of the tests performed. In the studied sources there is no analysis of the sequence obtained by summing the initial values from several NLFSR, and it is on this principle that many modern generators using NLFSR (e.g. Achterbahn128/80) are built.

In the works where generators using summation of sequences from several NLFSR are described, the description of one of important parameters—the period of the resulting sequence is not given. As will be shown in the following sections, the equivalent length sequence from the shift register with $L = 120$ can be used *4* NLFSR length *23, 29, 31, 37* and the function of simple summation of their sequences. This approach significantly expands the possibilities of application of already found NLFSR and synthesis by means of these registers of generators with the set characteristics.

The question of the linear and quadratic equivalent complexity of the sum of the sequence generated by the NLFSR of sufficiently large size, which can significantly simplify the cryptanalysis of systems that use NLFSR, has not been studied.

When designing streaming encryption systems, an important characteristic is the speed of the system, which is limited by the generation speed in the NLFSR. The following sections provide estimates of the speed characteristics of the NLFSR in software implementation and analyze the possibilities for the generation rate for hardware implementation.

An open question is to compare the capabilities of cryptographic systems based on LFSR and NLFSR. In the available literature, there is almost no comparison of performance, i.e. the amount of data generated per unit time using constructions based on LFSR and NLFSR.

These indicators can be considered as components of a comprehensive assessment of the cryptographic stability of NLFSR in the case of NLFSR as a basic element of the cryptographic primitive.

12.5.1 System of Criteria and Indicators of Cryptographic Evaluation Stability of NLFSR

The problem of evaluating the effectiveness of streaming encryption schemes can be solved by taking into account various factors of different nature. The methodological basis for the preparation and justification of decisions on the possibility of using the studied cryptosystems is a system analysis [86]. The involvement of methodological tools of systems analysis is primarily due to the fact that the decision has to be made in conditions of uncertainty caused by the presence of factors that can not be strictly quantified. Techniques and methods of system analysis are aimed at proposing alternative solutions to this problem, identifying the scale of uncertainty for each option and comparing options for their effectiveness. Obviously, this approach will be effective for the development of streaming encryption systems.

Under *cryptographic stability* we understand the ability of a cryptographic primitive to resist cryptanalysis. Algorithm is considered stable, which for a successful attack requires from the enemy unattainable computing resources, unattainable amount of intercepted open and encrypted messages, or such a time of disclosure that at the end of the protected information will no longer be relevant. In most cases, cryptographic stability cannot be proved mathematically, it is only possible to prove the vulnerabilities of the cryptographic algorithm.

Criteria for cryptographic stability—rules that allow you to evaluate and select indicators that characterize the cryptographic stability of streaming encryption schemes in accordance with the required level of reliability. The assessment includes a quantitative assessment obtained analytically or empirically and eligibility criteria ("yes" or "no").

Evaluation of efficiency is carried out using the criterion W, which reflects the degree of performance of the scheme of the functional task. Since the provision of cryptographic stability is achieved only by fulfilling many different conditions, the result of the system is described by many conditions, and the efficiency index is vector, i.e. contains many values of W_i individual performance indicators, namely

$$W = |W_1, W_2, \ldots W_n|,$$

where n is the number of individual criteria by which the study is conducted.

For cryptography of practical interest are M- NLFSR, i.e. NLFSR, generating *M-sequence*. By M sequence or sequence of maximum length we mean a pseudo-random binary sequence, which is generated by the shift register and has a maximum period.

NLFSR are investigated only as a basic element of the streaming encryption scheme, as an alternative to LFSR. Cryptographic security assessment is performed only by the NLFSR as part of the streaming encryption scheme, and in no case as a sufficient element of PRS generation.

In accordance with the main methods of cryptanalysis of algorithms based on NLFSR, the study of stability will be conducted on the basis of a system of unconditional and conditional criteria and indicators on which the selected criteria are based.

12.5.2 Unconditional Criteria of Cryptographic Stability

Unconditional criteria will be those criteria that are mandatory, i.e. unconditional. The indicators that belong to the unconditional criteria include the following:

- Linear complexity of PRS, which generates a single NLFSR.

The linear complexity (*Li*) of PRS is the shortest shift register that generates a given periodic sequence, provided that the first L values of the sequence are the initial filling of the register. Estimation of linear complexity is one of the main parameters of the system. Any sequence that can be generated automatically (linear or nonlinear) over a finite field has finite linear complexity. Thus, it is possible to build an algorithm that will determine the linear complexity of any sequence, regardless of the method of its generation.

- The quadratic complexity of PRS, which generates a single NLFSR.

The quadratic complexity (Li^2) is the smallest length of the second-order NLFSR, with which it is possible to reproduce the original sequence.

- Linear and quadratic complexity of the total sequence.

Under *the total sequence* we mean the sequence obtained as a result of the operation of bitwise addition of two or more sequences from M-NLFSR. The generation of sequences in each of the M- NLFSR involved in the addition is carried out independently of each other. The total linear (Li_{sum}) and total quadratic () complexity are determined from the total sequence.

- The period of the total sequence formed as a result of summation of sequences from two or more M- NLFSR.

The sequence period (T) is the number of bits generated by the register to the value when the generated sequence begins to repeat. *The total period* (T^{sum}) is the smallest period obtained, which is determined from the total sequence. To generate a sequence of a given length quite often use the sum of sequences from shorter M-NLFSR second order.

- Statistical properties of sequences generated from M- NLFSR.

The statistical properties of PRS are one of the components that determine the stability of conversion schemes. The stability of the scheme depends on how closely it approximates the random number generator, i.e. how PRS will be computationally unpredictable and indistinct in comparison with a truly random sequence.

12.5.3 Conditional Criteria of Cryptographic Stability

Conditional criteria include criteria that are met only under certain conditions and are desirable but not mandatory. The indicators related to the conditional criteria include:

- Profile of the linear complexity of PRS, which is formed by a separate M- NLFSR.

The criterion for checking the quality of PRS, based on calculating the linear complexity of its segments and comparing the sample distribution of the obtained values with their distribution for a random sequence, which is ideal.

- Profile of the linear complexity of the total sequence from different M- NLFSR.

The same as the profile of the linear complexity of PRS, but calculated from the total sequence.

- The possibility of applying the decimation of the sequence formed by M- NLFSR.

Decimation of the M-sequence by the index $k(k = 1, 2, 3, \ldots)$ is a sample of k elements of this M-sequence. If the period M-sequence and the decimation coefficients k are mutually prime values, then the decimation is considered proper or normal.

For LFSR, the result of any normal decimation is also M-sequence. If the decimation coefficients belong to one cyclotomic class, they will correspond to the same sequence with an accuracy of a certain shift. If the decimation coefficients belong to different cyclotomic classes, then the decimation results in a sequence other than the derivative. Moreover, if you go through all the decimation coefficients from *1* to $2^L - 1$ (then it makes no sense, because due to its cyclicity the sequences will be repeated), it is possible to obtain all possible M-sequences for LFSR for a given *L*, and using the Burleckamp-Messi algorithm, it is possible to restore the polynomial LFSR.

– The ability to apply the property of group addition M-sequence, formed by M-NLFSR.

Property of group addition if you add (modulo *2*) some bits of LFSR, then the result of such an operation will be M-sequence, the same as at the output of NLFSR, only with some offset. Thus, adding different bits of LFSR, it is possible to receive all $2^L - 1$ shift. Using the properties of group addition, knowing the forming polynomial and performing preliminary calculations, it is possible to obtain a sequence that will be generated by LFSR through a certain number of iterations, without running all register states to the desired iteration, which is impossible with large *L* values.

– The volume of possible different structures of the second-order RZNZZ for a fixed value of L (M).

Under the specified volume we understand the full set of different NLFSR of the second order at the set value of the size of the register, i.e. quantitative value of separate structures which can be created, using the register from L cells.

– The volume of the ensemble M- NLFSR (M_0) in *GF(2)*.

The term ensemble volume will mean the number of different M-sequences that can generate M- NLFSR second order of fixed size *L*.

– Carrying out a comparative assessment of the performance of M- NLFSR and M-NLFSR on the software implementation of the algorithm.

Under the productivity of software implementation we understand the amount of data generated per unit time by constructions based on M- NLFSR second order (M-LFSR). When designing streaming encryption systems, an important characteristic is the speed of the system, which is limited by the generation rate in the NLFSR. Testing did not use acceleration methods in the form of generating several bits per clock cycle or optimizing calculations for a specific feedback structure.

Criteria and indicators for assessing the resilience of the NLFSR, in addition to those listed, may be: balance; dissemination criterion; avalanche criterion; correlation immunity; function stability; algebraic immunity; level of affinity; differential uniformity; properties of completeness; criterion of computational complexity and others. However, most of them make sense if you use NLFSR as a filtering or combining function.

12.5.4 Estimation of Cryptographic Stability of NLFSR as Filtering or Combining Function

Consider the general block diagram of the combining generator (Fig. 12.3a) and the filter generator (Fig. 12.3b) PRS using several shift registers with linear or nonlinear feedback—SR_i ($i = 1, \ldots, L$). In this case, f is considered as a combining or filtering function of L variables.

Consider some of the main indicators of sustainability assessment in this case:

– *Balance.*

A Boolean function f of L variables is said to be balanced if the function takes values 0 and 1 equally often. This is one of the most natural essential properties imposed on Boolean functions used in streaming ciphers [87].

If the Boolean function is balanced, then the probability that it will take the value *0* or *1* is the same and equal to *1/2*. This allows you to reduce the statistical relationships between the input of the function and its output. Otherwise, the cryptanalyst has the opportunity, using the distribution of all relations, to conduct cryptanalysis of the cipher.

– *The presence of prohibitions.*

In the case of PRS analysis generated by a filter generator, there is another concept—the prohibition of the Boolean function, i.e. the presence of combinations of the output sequence, which can not occur with any combination of the input sequence.

It is intuitive that the presence of prohibitions in the filter function of the generator makes it "weaker", this prohibition will never appear in the original sequence of the generator, which degrades its statistical properties.

– *Correlation immunity.*

The requirement of correlated immune function is associated with resisting a correlation attack, the idea of which is as follows [88]. Consider a combining PRS

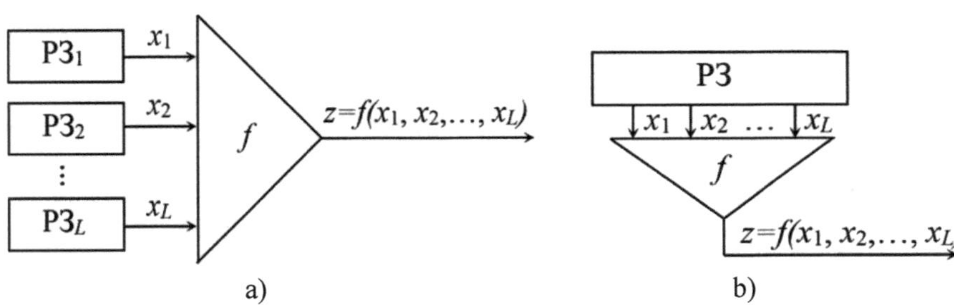

Fig. 12.3 Block diagram of combining (**a**) and filtering (**b**) PRS generator

generator (Fig. 12.3 a). The key of the generator is the initial state of all registers. The volume of the key is equal to $2^{l_1+\ldots+lL}$, where l_i is the length of SR_i for $i = 1, \ldots, L$.

Each of the SR generates a sequence $x_i = x_i^1 x_i^2 \ldots$ that is usually close in its properties to the random one. In particular, for a sufficiently long sequence length for a randomly selected bit x_i^j, there is a probability of a random event $x_i^j = 0$: $\mathrm{P}\left[x_i^j = 0\right] \approx 1/2$. Therefore, if $y = y^1 y^2 \ldots$ is an arbitrary sequence that does not depend on x_i, then.

$$\mathrm{P}\left[x_i^j = y^j\right] = \mathrm{P}[x_i^j = 0] \cdot \mathrm{P}\left[y^j = 0\right] + \mathrm{P}[x_i^j = 1] \cdot \mathrm{P}\left[y^j = 1\right]$$
$$\approx \frac{1}{2}(\mathrm{P}[y^j = 0] + \mathrm{P}[y^j = 1]) = \frac{1}{2}.$$

Assume that $\mathrm{P}[f = x_1] \neq 1/2$ (in which case we say that *the function f correlates with the variable x_I*). Using the correlation attack, we find the initial state s_I SR_1. To do this, we will go through all possible 2^{l1} states SR_1, for each of them we construct the sequence $z' = z_1' z_2' \ldots$ and count the number of coincidences with PRS $z_i' = z_i$. For all sequences except one (generated by s_I), the proportion of matches will be $\approx 1/2$. Thus we determine that part of the key is the state s_I. If the function f has a correlation with all its variables (or with all but one—then the state of the register corresponding to this variable, knowing the state of all other registers we find the last), then we find the generator key for $2^{l1} + \cdots + 2^{lL}$ tests that much less the complexity of a brute force attack.

- *Nonlinearity.*

Practice shows [89] that cryptographic transformations, which have properties close to the properties of linear functions, in many cases lead to a significant reduction in the stability of ciphers. For this reason, functions in cryptography are important, the properties of which exclude weaknesses inherent in functions close to linear. Thus, the desired quality of a function is its nonlinearity, which is understood, in a broad sense, as a negation of linearity. In block and stream ciphers, the use of a function with high nonlinearity helps to increase the stability of ciphers to linear and differential methods of cryptanalysis.

The relationship between the various cryptographic properties has been little described in the literature. Practice shows [87] that as a component of the cipher it is necessary to choose "good on all sides" functions, which is actually a very difficult task, because many properties contradict each other. Although theoretical results show that in a random function most cryptographic parameters are close to optimal. The question of how to choose it, random?

Let's limit ourselves to the consideration of only the above indicators, which, in our opinion, are the most appropriate in the study of RHP. Due to the fact that these indicators are especially relevant in the study of filtering or combining functions, we will consider them separately from others.

The issue of cryptographic stability and the study of the above indicators are discussed in more detail in the fourteenth section.

12.6 Conclusions

1. The lack of a well-developed mathematical apparatus for the analysis and synthesis of streaming encryption algorithms based on NLFSR, which significantly inhibits their potential development.
2. It is determined that the available methods have a high computational complexity, which does not allow to search for M-NLFSR with a size of more than 33 cells. In addition, the available methods do not allow you to search for M-NLFSR with the specified design characteristics (maximum number and location of feedback coefficients), and thus perform the search for M-NLFSR under the specified technical conditions.
3. A very small amount of research on the quality of PRS formed NLFSR and its cryptographic stability, in connection with which the unconditional and conditional indicators and criteria of stability NLFSR, according to which it is desirable to conduct research.

References

1. Biham, E., Dunkelman, O.: Cryptanalysis of the A5/1 GSM stream cipher. In: INDOCRYPT '00: Proceedings of the First International Conference on Progress in Cryptology, (London, UK), Springer, pp. 43–51 (2001)
2. Shaked, O., Wool, A.: Cryptanalysis of the Bluetooth E0 cipher using OBDD's. (Електронний ресурс) (2006). URL: http://citeseer.ist.psu.edu/viewdoc/download?doi=10.1.1.60.6279&rep=rep1&type=pdf (дата звернення: 01.08.2017)
3. McCluskey, J.: High speed calculation of cyclic redundancy codes. In: Proceedings of the 1999 ACM/SIGDA Seventh International Symposium on Field Programmable Gate Arrays, FPGA '99, (New York). ACM, 250–256 (1999)
4. Mrugalski, G., Rajski, J., Tyszer, J.: Ring generators—new devices for embedded test applications. Trans. Comput. Aided Des. Integr. Circ. Syst. **23**(9), 1306–1320 (2004)
5. David, R.: Random Testing of Digital Circuits. Marcel Dekker, New York (1998)
6. Mukhopadhyay, S., Sarkar, P.: Application of LFSRs for parallel sequence generation in cryptologic algorithms. In: Computational Science and Its Applications—ICCSA 2006, vol. 3982 of Lecture Notes in Computer Science, pp. 436–445. Springer Berlin / Heidelberg (2006)
7. Golomb, S.W.: Shift Register Sequences. San Francisco, HoldenDay, 1967, revised edition. Aegean Park Press, Laguna Hills (1982)
8. Golomb, S.W., Gong, G.: Signal Design for Good Correlation. For Wireless Communication, Cryptography, and Radar. Cambridge University Press (2005)
9. Schneier, B.: A selfstudy course in blockcipher cryptanalysis. Cryptologia **XXIV**(1), 18–33 (2000)
10. Courtois, N.: Higher order correlation attacks, XL algorithm and cryptanalysis of toyocrypt. In: Lee, P., Lim, C. (eds) Information Security and Cryptology—ICISC 2002, 5th International Conference Seoul, Korea, November 28–29, 2002, Revised Papers, volume 2587 of Lecture Notes in Computer Science, 182–199. Berlin Heidelberg, SpringerVerlag (2003)
11. Garland, M., Le Grand, S., Nickolls, J., Anderson, J., Hardwick, J., Morton, S., Phillips, E., Zhang, Y., Volkov, V. (eds) Parallel Computing Experiences with CUDA. IEEE Micro, vol. 28, 13–27 (2008)

12. Preneel, B., Paar, C., Pelzl, J.: Understanding Cryptography: A Textbook for Students and Practitioners. Springer (2009)
13. Barkan, E., Biham, E., Keller, N.: Instant Ciphertextonly Cryptanalysis of GSM Encrypted Communication. Crypto (2003)
14. Armknecht, F.: A Linearization Attack on the Bluetooth Key Stream Generator. Eprint (2002)
15. Berlekamp, E.R.: Nonbinary BCH decoding. In: International Symposium on Information Theory, (San Remo, Italy) (1967).
16. Massey, J.: Shiftregister synthesis and BCH decoding. IEEE Trans. Inf. Theory **15**, 122–127 (1969)
17. Mandelbaum, D.: An approach to an arithmetic analog of Berlekamp's algorithm. IEEE Trans. Inf. Theory **30**(5), 758–762 (1984)
18. Imamura, K., Yoshida, W.: A simple derivation of the Berlekamp Massey algorithm and some applications. IEEE Trans. Inf. Theory **33**(1), 146–150 (1987)
19. Fitzpatrick, P.: New time domain errors and erasures decoding algorithm for bch codes. Electron. Lett. **32**(2), 110–111 (1994)
20. Fleischmann, M.: Modified Berlekamp Massey algorithm for two sided shift register synthesis. Electron. Lett. **31**(8), 605–606 (1995)
21. Dornstetter, J.: On the equivalence between Berlekamp's and Euclid's algorithms. IEEE Trans. Inf. Theory **33**(3), 428–431 (1987)
22. Welch, L., Sholtz, R.: Continued fractions and Berlekamp's algorithm. IEEE Trans. Inf. Theory **25**(1), 19–27 (1979)
23. Lempel, A., Eastman, W.L.: High speed generation of maximal length sequences. IEEE Trans. Comput. **20**, 227–229 (1971)
24. Torba, A.A., Bobukh, V.A., Torba, M.O., Tobra, A.O.: Deterministic pseudo-random sequence generators for DLRR-based streaming encryption. Appl. Radio Electron. **15**(3), 191–194 (2016)
25. Karazin, V.N., Kuznetsov, O.O., Malakhov, S.V.: Research of flow symmetric ciphers and flow modes of block symmetric ciphers: report on GDR (intermediate), code "Stream". Analysis and comparative studies of modern algorithms for streaming cryptocurrency/KhNU. (etc., a total of 11 people). H.: KhNU. V.N. Karazina, 233 p (2015)
26. The main trends in the development of open cryptography (review commissioned by cryptography.ru). Published: geo.com.ru. (Electronic resource) (2001). URL: http://images.geo.web.ru/pubd/2001/10/10/ 0001161293/tend.pdf. Access date: 01.08.2017
27. Asoskov, A.V., Ivanov, M.A., Mirsky, A.A., Ruzin, A.V., Slanin, A.V., Tyutvin, A.N.: Current ciphers. M .: KUDITSOBRAZ 336 p (2003).
28. Gammel, B.M., Gottfert, R., Kniffler, O.: An NLFSRBased Stream Cipher. Infineon Technologies AG, Munich, Germany. (Electronic resource). URL: https://www.researchgate.net/publication/224647778 (date: 01.08.2017).
29. Hell, M., Johansson, T., Meier, W.: Grain—A Stream Cipher for Constrained Environments, eSTREAM submission, (Electronic resourse). URL: http://www.ecrypt.eu.org/stream/p3ciphers/grain/Grain_p3.pdf (date: 01.08.2017)
30. Davis, J., Jedwab, J.: Peak to mean power control in OFDM, Golay complementary sequences, and ReedMuller codes. IEEE Trans. Inf. Theory. **45**(7), 2397–2417 (1999)
31. Popovic, B.: Spreading sequences for multicarrier CDMA systems. IEEE Trans. Commun. **47**, 918–926 (1999)
32. Canniere, C., Preneel, B.: Trivium. New Stream Cipher Designs: The eSTREAM Finalists, LNCS, vol. 4986, 244–266 (2008)
33. Hell, M., Johansson, T., Maximov, A., Meier, W.: The Grain family of stream ciphers. New Stream Cipher Designs: The eSTREAM Finalists, LNCS 4986, 179–190 (2008)
34. Gittins, B., Landman, H.A., O'Neil, S., Kelson, R.: A presentation on VEST hardware performance, chip area measurements, power consumption estimates and benchmarking in relation to the AES, SHA256 and SHA512. Cryptology ePrint Archive, Report 415. (Electronic resource) (2005). URL: http://eprint.iacr.org/2005/415 (date: 01.08.2017)

35. Dubrova, E., Mansouri, S.: A BDDbased approach to constructing LFSRs for parallel CRC encoding. In: Proceedings of International Symposium on MultipleValued Logic, 128–133 (2012)
36. Gammel, B., Gottfert, R., Kniffler, O.: Achterbahn128/80: Design and analysis. In: SASC'2007: Workshop Record of The State of the Art of Stream Ciphers, 152–165 (2007)
37. Chen, K., Henricken, M., Millan, W., Fuller, J., Simpson, L., Dawson, E., Lee, H., Moon, S.: Dragon: a fast word based stream cipher. In: eSTREM, ECRYPT Stream Cipher Project. Report 2005/006 (Electronic resources) (2005). URL https://cr.yp.to/streamciphers/dragon128/desc.pdf. Access date: 01.08.2017
38. Hell, M., Johansson, T., Meier, W.: Grain a stream cipher for constrained environments (Electronic resource) (2005). URL: http://citeseer.ist.psu.edu/viedoc/summary?doi=10.1.1.107.9707 (date: 01.08.2017)
39. Canniere, C., Preneel, B.: TRIVIUM specifications. (Electronic resources) (2006). URL: http://citseer.ist.psu.edu/viewdoc/summary? doi=10.1.1.59.9030 (date: 07.10.2016)
40. Canteaut, A.: Open problems related to algebraic attacks on stream ciphers. In: WCC, 120–134 (2005)
41. Preneel, B.: A survey of recent developments in cryptographic algorithms for smart cards. Comput. Netw. **51**(9), 2223–2233 (2007)
42. Derbunovich, L.V., Derbunovich, L.V., Karaman, D.G., Osipenko, A.N.: Key sequence generators in streaming cryptographic ciphers. Abstracts of reports at the 3rd International Scientific-Practical Conference "Information Technologies and Computer Engineering". VNTU, Vinnytsia, 138–139 (2012).
43. Zeng, K., Yang, C., Wei, D., Rao, T.R.N.: Pseudorandom bit generators in streamcipher cryptography. Computer (1991)
44. Games, R.A., Cham, A.H.: A fast algorithm for determining the complexity of a binary sequence with period 2n. IEEE Trans. Inf. Theory, IT29–1983, pp. 144–146
45. Dabrowski, P., Łabuzek, G., Rachwalik, T., Szmidt, J.: Searching for Nonlinear Feedback Shift Registers with Parallel Computing. (Electronic resource) (2013). URL: https://eprint.iacr.org/2013/542.pdf. Access date: 07.10.2016
46. Fredricksen, H.: A survey of full length nonlinear shift register cycle algorithms. SIAM Rev. **24**(2), 195–221 (1982)
47. Jansen, C.J.: Investigations on Nonlinear Streamcipher Systems: Construction and Evaluation Methods. Ph.D. Thesis, Technical University of Delft (1989)
48. Jansen, C.J.: The maximum order complexity of sequence ensembles. Lecture Notes in Computer Science, Adv. Cryptology Eupocrypt'1991, Berlin, Germany, vol. 547, 153–159 (1991)
49. Linardatos, D., Kalouptsidis, N.: Synthesis of minimal cost nonlinear feedback shift registers. Signal Process. **82**(2), 157–176 (2002)
50. Rizomiliotis, P., Kalouptsidis, N.: Results on the nonlinear span of binary sequences. IEEE Trans. Inf. Theory **51**(4), 1555–5634 (2005)
51. Limniotis, K., Kolokotronis, N., Kalouptsidis, N.: On the nonlinear complexity and Lempel Ziv complexity of finite length sequences. IEEE Trans. Inf. Theory **53**(11), 4293–4302 (2007)
52. Dubrova, E.: A scalable method for constructing Galois NLFSRs with period 2n1 using crossjoin pairs. IEEE Trans. Inf. Theory **59**(1), 703709 (2013)
53. Mykkeltveit, J., Siu, M.K., Tong, P.: On the cyclic structure of some nonlinear shift register sequences. Inform. Control. **43**, 202215 (1979)
54. Mandal, K., Gong, G.: Cryptographically strong de Bruijn sequences with large periods. In: Knudsen, L.R., Wu, K. (eds) Selected Areas in Cryptography. LNCS, Springer, vol. 7707, 104118 (2012)
55. Dubrova, E.: A Method for Generating Full Cycles by a Composition of NLFSRs. Designs, Codes and Cryptography. ISSN 09251022, EISSN 1573–7586, November 2014, 73, 2, 469–486 (2014)
56. Dubrova, E.: A list of maximumperiod NLFSRs. Cryptology ePrint Archive, Report 2012/166. (Electronic resource) (2012). URL: http://eprint.iacr.org/2012/166. Access date: 07.10.2016

57. Flye, S.M.C.: Solution to question nr. 48. L'Intermédiaire des Mathématiciens 1, 107–110 (1894)
58. De Bruijn, N.G.: A combinatorial problem. Indag. Math. **8**, 461–467 (1946)
59. Mayhew, G.L., Golomb, S.W.: Linear spans of modified de Bruijn sequences. IEEE Trans. Inform. Theory, **36**(5), 1166–1167 (1990)
60. Lidl, R., Niederreiter, H.: Introduction to Finite Fields and their Applications (Revisited Edition). Cambridge University Press, Cambridge, (Electronic resource) (1994). URL: http://math.boisestate.edu/~liljanab/MATH508/FiniteFields_and_Applications.pdf. Access date: 07.10.2016
61. Rachwalik, T., Szmidt, J., Wicik, R., Zabłocki, J.: Generation of Nonlinear Feedback Shift Registers with Special Purpose Hardware. (Electronic resource) (2012). URL: https://eprint.iacr.org/2012/314.pdf. Access date: 07.10.2016
62. Kyureghyan, G.M.: Minimal polynomials of the modified de Bruijn sequences. Discr. Appl. Math. **156**, 1549–1553 (2008)
63. Mandal, K., Gong, G.: Probabilistic Generation of Good Span n Sequences from Nonlinear Feedback Shift Registers. University of Waterloo, Preprint (2012)
64. Turan, M.S.: On the nonlinearity properties of maximumlength NFSR feedbacks. Cryptology ePrint Archive, 2012/112. (Electronic resource) (2012). URL: www.iacr.org. Access date: 07.10.2016
65. Alhakim, A.: A simple combinatorial algorithm for de Bruijn sequences. Am. Math. Monthly. **117**(8), 728–732 (2010). (Electronic resource). URL: https://www.mimuw.edu.pl/~rytter/TEACHING/TEKSTY/Prefer Opposite.pdf. Access date: 07.10.2016
66. Chan, A.H., Games, R.A.: On the quadratic spans of de Bruijn sequences. IEEE Trans. Inf. Theory **36**(4), 822–829 (1990)
67. Etzion, T., Lempel, A.: Construction of de Bruijn sequences of minimal complexity. IEEE Trans. Inf. Theory **30**(5), 705–709 (1984)
68. Fredricksen, H.: A class of nonlinear de Bruijn cycles. J. Comb. Theory Ser. A **19**(2), 192–199 (1975)
69. Jansen, C.J.A., Franx, W.G., Boekee, D.E.: An efficient algorithm for the generation of de Bruijn Cycles. IEEE Trans. Inf. Theory **37**(5), 1475–1478 (1991)
70. Lempel, A.: On a homomorphism of the de Bruijn graph and its applications to the design of feedback shift registers. IEEE Trans. Comput. **C19**(12), 1204–1209 (1970)
71. Derbunovich, L.V., Karaman, D.G., Osipenko, A.N., De Brain sequence generators on shift registers with nonlinear feedback. (Electronic resource) (2013). URL: http://irbisnbuv.gov.ua/cgibin/irbis_nbuv/cgiirbis_64.exe?C21COM=2&I21DBN=UJRN&P21DBN=UJRN&IMAGE_FILE_DOWNLOAD=1&Image_file_name=PDF/201kh3tus
72. De, R.A.: Bruijn sequences—a model example of interaction of discrete mathematics and computer science. Am. Math. Monthly **55**(3), 131–143 (1982)
73. Etzion, T., Lempel, A.: Algorithms for the generation of fulllength shift register sequences. IEEE Trans. Inf. Theory **3,** 480–484 (1984)
74. Annexstein, F.S.: Generating de Bruijn sequences: an efficient implementation. IEEE Trans. Comput. **46**, 198–200 (1997)
75. Chang, T., Park, B., Kim, Y.H., Song, I.: An efficient implementation of the Dhomomorphism for generation of de Bruijn sequences. IEEE Trans. Inf. Theory., **45**, 1280–1283 (1999)
76. Fredricksen, H.M.: Disjoint cycles from de Bruijn graph. Tech. Rep. 225, USCEE (1968)
77. Helleseth, T., Klove, T.: The number of crossjoin pairs in maximum length linear sequences. IEEE Trans. Inf. Theory, **31**, 1731–1733 (1991)
78. Li, N., Dubrova, E.: An Algorithm for Constructing a Smallest Register with NonLinear Update Generating a Given Binary Sequence. (Electronic resources) (2013). URL: https://arxiv.org/pdf/1306.5596v1.pdf. Access date: 07.10.2016
79. Rueppel, R.: Linear complexity and random sequences. In: Pichler, F. (ed) Advances in Cryptology EUROCRYPT'85. Springer Berlin Heidelberg, vol. 219 of Lecture Notes in Computer Science, 167–188 (1986)

80. Dubrova, E., Teslenko, M., Tenhunen, H.: On Analysis and Synthesis of (n,k)NonLinear Feedback Shift Registers. (Electronic resource) (2008). URL: https://www.dateconference.com/proceedingsarchive/PAPERS/2008/DATE08/PDFFI LES/10.4_3.PDF. Access date: 07.10.2016
81. Ronce, C.A.: Feedback Shift Registers, vol. 169 (1984)
82. Robshaw, M.J.B.: On Binary Sequences with Certain Properties. Ph.D. Thesis, University of London (1992)
83. Ahmad, A., AlMushrafi, M.J., AlBusaidi, S.: Design and study of a strong cryptosystem model for ecommerce. In: ICCC '02: Proceedings of the 15th International Conference on Computer Communication, (Washington, DC, USA), International Council for Computer Communication, 619–630 (2002)
84. Janicka Lipska, J.S.I.: Boolean feedback functions for fulllength nonlinear shift registers. Telecommun. Inf. Technol. **5**, 28–29 (2004)
85. Karazin, V.N., Kuznetsov, O.O.: Research of flow symmetric ciphers and flow modes of block symmetric ciphers: report on GDR (final), code "Stream", Volume 1. Research of opportunities to increase the efficiency of streaming encryption procedures, KhNU. Malakhov SV (etc., only 10 people). H.: KhNU. V.N. Karazina 104 p (2015)
86. Reliability and Efficiency in Technology: Handbook: In: 10 vols. Ed. advice: V.S. Avduevsky (pred.) and others M.: Mechanical Engineering, Vol. 3 (1986)
87. Gorodilova, A.A.: From crypto analysis of a cipher to a cryptographic property of a Boolean function. Appl. Discr. Math. **3**(33), 16–44 (2016)
88. Pankratova, A.: Boolean Functions in Cryptography: Textbook. Allowance. Publishing House of Tomsk State University, Tomsk, 88 p (2014)
89. Mukhachev, V.A., Khoroshko, V.A.: Methods of Practical Cryptography. K.: Polygraph Consulting LLC. 215 p (2005)

Chapter 13
Requirements for Feedback Coefficients, What Should the M-Nlfsr Respond to?

13.1 Requirements 1–4

13.1.1 Formulation and Substantiation of Requirements 1–4

We formulate the requirements that must be met by a combination of coefficients a_{ij} (i.e. the type of feedback), during which a binary sequence with a maximum period is generated.

Requirement 1. To ensure the maximum period, the linear feedback factor from the last cell must always be present:

$$a_{LL} = 1.$$

Indeed, if the period of the sequence is equal to T_{max}, then there must be a register state in which $q_1(t) = 0, q_2(t) = 0, \ldots, q_{L-1}(t) = 0, q_L(t) = 1$, in this case $q_1(t+1) = 1$ is possible only if $a_{LL} = 1$.

Consider the case when all cells of the register correspond to a single state. In this case, any product of the cells will give one, and the sum of any odd number of units (modulo 2) will also be equal to one. Given that any initial fill (for M-NLFSR) must pass a state when all the cells in the register correspond to one, we obtain a loop with $T = 1$ with an odd number of a_{ij}.. Taking into account the above, we formulate the following requirement, which must meet M-NLFSR.

Requirement 2. The sum of all non-zero feedback coefficients a_{ij} must be an even number, i.e.:

$$\sum_{ij} a_{ij} - \text{even number.}$$

A. A. Kuznetsov et al., *Stream Ciphers in Modern Real-time IT Systems*, Studies in Systems, Decision and Control 375, https://doi.org/10.1007/978-3-030-79770-6_13

Half of the allowed (subject to Requirement 1) combinations will give a period $T = 1$, because the sum of every second combination will be an odd value.

Consider in more detail the sum of the coefficients a_{ij}.. We perform the calculation of all NLFSR for $L = 6$, taking into account Requirements 1 and 2 for $L = 7$. We sum all the coefficients $a_{ij} \neq 0$ only for the set of combinations a_{ij}. that correspond to M-NLFSR. The results of the calculations are summarized in Table 13.1

From the Table 13.1 we see that there are coefficients that do not participate in the generation of the maximum period. All other coefficients (except $a_{LL} = 1$ of Requirement 1) occur with approximately the same number. A similar pattern of distribution will be for $L = 4.5$. We formulate the following requirement, which must be met by the feedback coefficients, for M-NLFSR.

Requirement 3. There must be no nonlinear feedback received from the source cell and any other, i.e.:

$$\forall a_{iL} = 0, i = \overline{\{1, L-1\}}.$$

Consider how the state of NLFSR, which does not satisfy Requirement *3*, changes, and show that Requirement 3 is valid for any value of *L*.

For certainty, we will consider only those NLFSR, in which the forming polynomial is always present $a_{LL} = 1$. From Requirement 1, we may not consider other NLFSR, because they clearly will not form M-sequence.

Consider the NLFSR length $L = 3$, which is given by different polynomials.

(see Fig. 13.1). Option (a)—the polynomial is primitive and generates a sequence; (b)—the polynomial is not M-NLFSR, but meets the Requirements 3 (does not meet the Requirements 2); (*c*) and (*d*)—polynomials do not meet Requirement 3.

The presence of the coefficient $a_{iL} = 1(i \in \{1, L-1\})$ leads to the appearance not of circles, i.e. a series of state tuples that pass through the NLFSR, passing from one state to another, (the number of such states is limited and equal to *T*), and loops, i.e. different initial states of the register lead to one final state.

Consider the sequences of the whole set of polynomials of length $L = 3$, in which at least one coefficient $a_{iL} = 1$. Install the following:

1. If $a_{13} = 1$, then the pair of register states (1,1,1) and (1,1,0), as well as (1,0,1) and (1,0,0) always gives the same the state of the register, and the pair (0,1,1) and (0,1,0) are different, independently of the others $a_{ij} = 1(j \neq L)$;
2. If $a_{23} = 1$, then the pair of register states (1,1,1) and (1,1,0), as well as (0,1,1) and (0,1,0) always gives the same the state of the register, and the pair (1,0,1) and (1,0,0) are different, independently of the others $a_{ij} = 1(j \neq L)$;
3. If $a_{13} = a_{23} = 1$, then the pair of states of the register (1,0,1) and (1,0,0), as well as (0,1,1) and (0,1,0) always gives one and the same state of the register, and the pair (1,1,1) and (1,1,0) are different, independently of the others $a_{ij} = 1(j \neq L)$.

For clarity, we summarize the data obtained in Table 13.2. A similar summary table (taking into account Requirement 1) can be compiled for $L = 4$ (Table 13.3).

Thus, we see that when $a_{iL} = 1$, then the corresponding coefficients of states (i.e. $x_1, x_2, \ldots, x_{i-1}, 1, x_{i+1}, \ldots, x_{L-1}, 1$) will always give the value that is zero,

Table 13.1 Distribution of the sum of the coefficients for M-NLFSR for $L = 6$ and $L = 7$

i / j	1	2	3	4	5	6	i / j	1	2	3	4	5	6	7
1	984	1000	1236	1000	984	1952	1	31,996	32,116	32,266	32,266	32,116	31,996	64,056
2	968	968	968	968	0		2	31,768	32,122	31,924	32,122	31,768	0	
3	996	1016	996	0			3	32,068	31,890	31,890	32,068	0		
4	972	972	0				4	31,942	32,048	31,942	0			
5	908	0					5	32,070	32,070	0				
6	0						6	32,160	0					
							7	0						

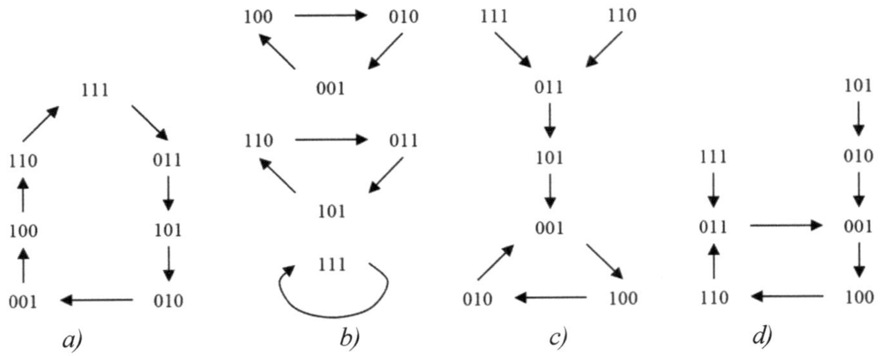

Fig. 13.1 Diagram of states of NLFSR with forming polynomials **a** $\left(x^3+x^1+1\right)$, б) $\left(x^3+1\right)$, **b** $\left(x^3+x^3x^2+1\right)$ та г) – $\left(x^3+x^3x^2+x^2x^1+x^1+1\right)$

Table 13.2 Combinations of second-order NLFSR states for $L=3$, which will give the same resultant states depending on the presence coefficients $a_{iL}=1$ and for any other $a_{ij}(j \neq L)$

States of registers / a_{ij}	(1,0,0)	(0,1,0)	(1,1,0)
$a_{13}=1$	(1,0,1)		(1,1,1)
$a_{23}=1$		(0,1,1)	(1,1,1)
$a_{13}=a_{23}=1$	(1,0,1)	(0,1,1)	

Table 13.3 Combinations of NLFSR states for L = 3, which will give the same resulting states

States of registers / a_{ij}	(1,0,0,0)	(0,1,0,0)	(0,0,1,0)	(1,1,0,0)	(0,1,1,0)	(1,0,1,0)	(1,1,1,0)
$a_{14}=1$	(1,0,0,1)			(1,1,0,1)		(1,0,1,1)	(1,1,1,1)
$a_{24}=1$		(0,1,0,1)		(1,1,0,1)	(0,1,1,1)		(1,1,1,1)
$a_{34}=1$			(0,0,1,1)		(0,1,1,1)	(1,0,1,1)	(1,1,1,1)
$a_{14}=a_{24}=1$	(1,0,0,1)	(0,1,0,1)			(0,1,1,1)	(1,0,1,1)	
$a_{24}=a_{34}=1$		(0,1,0,1)	(0,0,1,1)	(1,1,0,1)		(1,0,1,1)	
$a_{14}=a_{34}=1$	(1,0,0,1)		(0,0,1,1)	(1,1,0,1)	(0,1,1,1)		
$a_{14}=a_{24}=a_{34}=1$	(1,0,0,1)	(0,1,0,1)	(0,0,1,1)				(1,1,1,1)

independently of the others $x_j(j \neq i, L)$. The result will also not depend on other coefficients a_{ij}. Let's show it.

Consider what values are generated by a fixed combination of nonlinear relations of the form $a_{iL}=1$, $a_{LL}=1$, which corresponds to the polynomial representation:

x_i	x_L	$x^L + x^L x^i$
1	0	$0 + 0{\cdot}1 = 0$
1	1	$1 + 1{\cdot}1 = 0$

That is, if $x_i = 1$, then regardless of x_L the result will be zero, because the nonlinear combination will compensate for the linear component.

Consider the case when two L coefficients are equal to one, i.e. $a_{iL} = 1 (i \in \{1, L-1\})$, and let it be combinations $a_{kL} = a_{tL} = 1$, all others $a_{iL} = 0$. Then, if $x_k = 1$, and $x_t = 0$ (or $x_k = 0$, $x_t = 1$), which will correspond to the case described above. When $x_k = 1$ and $x_t = 1$, then at different values of x_L (0 or 1) different output values will be generated. Indeed, if $x_L = 0$, then the combination of nonlinear terms $x^L + x^L x^k + x^L x^t$ always gives zero, and for $x_L = 1$ the same combination will always give one. Note that all of the above does not depend on the values of the other $x_L = 1$ and x_i.

If we consider the case when the three coefficients $a_{iL} = 1 (i \in \{1, L-1\})$, i.e. $x^L + x^L x^k + x^L x^t + x^L x^p$, then $(x_k = x_t = 0,\ x_p = 1)$, or $\left(x_k = x_p = 0, x_t = 1\right)$, or $\left(x_p = x_t = 0, x_k = 1\right)$, or $\left(x_k = x_t = x_p = 1\right)$ will always give a resultant sum equal to zero regardless of the value of x_L (0 or 1).

Summarizing the obtained results, it is possible to formulate the following rule for the formation of the resulting value in the sum of nonlinear combinations of individual registers.

If $a_{LL} = a_{iL} = a_{jL} = a_{kL} = \ldots = 1$ (*i, j, k* are not equal to each other and less than *L*), which corresponds to the state of the registers x_L, x_i, x_j, x_k, then:

- if $\sum_{i,j,k,\ldots} x$—an odd number, the result of the sum $x^L + x^L x^i + x^L x^j + x^L x^k + \cdots$ will be equal to zero regardless of the value of x_L;
- if $\sum_{i,j,k,\ldots} x$—an even number, then at $x_L = 0$ the result is also equal to zero, and at $x_L = 1$ the result is equal to one.

All of the above is true for any other values of x_i and a_{ij}.

In Table 13.4 there is a certain pattern, which is subject to the coefficients $a_{ij} = 1$ of the whole set of M-NLFSR, and there is a symmetry for *i* relative to the average values. We introduce the following definition of a symmetric polynomial.

Definition: Two polynomials with maximum degree *L* will be called *symmetric polynomials* if they generate a symmetric sequence. That is, if two polynomials produce two different sequences Q_1 and Q_2 with the same period *T*, then for any reference point in the sequence Q_1 there is a reference point in the sequence Q_2 such that in the case of right-to-left (left to right) sequence Q_1, it will be completely identical to the sequence Q_2 in the case of passing from left to right (right to left) along the entire period *T*.

Table 13.4 shows the whole set a_{ij} for $L = 4$, for which $T = T_{max}$, and the sequence of bits generated by the above combinations. It should be noted that the sequences have an identical appearance when they are passed in reverse order (mirror image). It is easy to see that each combination a_{ij} corresponds to a symmetric one (for example: $x^1 + x^1 x^2 + x^2 + x^4 \leftrightarrow x^2 + x^2 x^3 + x^3 + x^4$).

Table 13.4 The complete set of feedback coefficients a_{ij} and their sequences (Q) with T_{max} for $L = 4$

$a_{11}a_{12}a_{13}a_{14}a_{22}a_{23}a_{24}a_{33}a_{34}a_{44}$	$Q(t) \rightarrow$	$a_{11}a_{12}a_{13}a_{14}a_{22}a_{23}a_{24}a_{33}a_{34}a_{44}$	$\leftarrow Q(t)$
1000000001	111101011001000	0000000101	000100110101111
1100100001	111100101101000	0000110101	000101101001111
1010100001	111100101000110	0010100101	011000101001111
0110100001	111100110100010	0010110001	010001011001111
1100010001	111100100011010	0100010101	010110001001111
1000110001	111101000110010	0100100101	010011000101111
1110110001	111101100101000	0110110101	000101001101111
1110100101	111101001011000	1010110101	000110100101111

Table 13.5 Combinations a_{ij} for $L = 4$, meeting the requirements of 1–3, and their respective periods

a_{ij}	$a_{(L-j)(L-i)}$	T	a_{ij}	$a_{(L-j)(L-i)}$	T
1110000001	0010010101	5	1000000001	0000000101	15
0100110001		5	1100100001	0000110101	15
1000100101		5	1010100001	0010100101	15
0000100001		6	0110100001	0010110001	15
0110010001		6	1100010001	0100010101	15
0010000001		7	1000110001	0100100101	15
1010000101		7	1110110001	0110110101	15
1110010101		7	1110100101	1010110101	15
1100110101		7			
0100000001	0000010001	9			
1010010001	0110000101	10			
1100000101	1000010101	10			

It is possible to obtain coefficients that are symmetric with respect to the coefficients a_{ij}, using the following relation:

$$a_{ij}^{1} = a_{(L-j)(L-i)}.$$

Let's go back to the feedback factors and consider them in terms of symmetry. Table 13.5 shows the coefficients a_{ij}, corresponding to Requirements 1–3 for $L = 4$, and their symmetrical coefficients. The sequences that generate the maximum period are located on the right in Table 13.5, and the other sequences are on the left.

From Table 13.5 we see that some combinations are symmetric to themselves and have a period less than the maximum. This statement is also true for $L = 5$, 6 and 7, calculated without taking into account Requirements 1–3. Thus, we can formulate the following requirement.

Requirement 4. A polynomial based on the second-order NLFSR, which generates a sequence of the maximum period, should not be symmetric to itself, i.e.

$$a_{ij} \neq a_{(L-j)(L-i)}.$$

13.1.2 Calculation of the Number of NLFSR that Meet the Requirements 1–4. Empirical Expression for the Calculation of M-NLFSR

Denote by Λ_0 the complete set of all different combinations that can be taken by the feedback coefficients a_{ij} in NLFSR:

$$\Lambda_0 = 2^{n_L}. \tag{13.1}$$

Λ_0 is a fairly large number even for small values of L (see Table 13.6), so the process of finding all M-NLFSR (or estimating their number) is a very time-consuming process. Given the above requirements, it is possible to significantly reduce the set of combinations a_{ij}, which can potentially generate M-sequence. Let Λ denote the set of admissible values a_{ij}, i.e. those that meet the requirements and

Table 13.6 Combined results of calculations for NLFSR

	Length of NLFSR (L)				
	4	5	6	7	8
Number of feedback coefficients (n_L)	10	15	21	28	36
The maximum possible PRS period (T_{max})	15	31	63	127	255
Number possible feedback combinations for NLFSR (Λ_0)	1 024	32,768	2,097,152	268,435,456	68,719,476,736
The number of admissible (in the presence of T = T_{max}) feedback combinations taking into account Requirements 1–4 $\Lambda_0 - \Lambda^{1-4}$)	24	480	16,128	1,046,528	134,184,960
$\frac{\Lambda_0 - \Lambda^{1-4}}{\Lambda_0} \cdot 100\%$	2.34%	1.46%	0.77%	0.39%	0.20%
The amount of M-NLFSR obtained empirically $\left(\Lambda^1_{max}\right)$	16	128	2048	65,536	4,194,304
Experimental value of M-NLFSR (Λ_{max})	16	128	1952	64,056	4,017,998

can generate PRS with M-period. Fulfillment of Requirement 1 reduces the number of required combinations by 2^1 times, because the coefficient a_{LL} can take only one state. If we denote by Λ^1 the set of combinations that do not meet Requirement 1, then it can be determined by the following formula:

$$\Lambda^1 = 2^{n_L - 1}. \tag{13.2}$$

Requirement 2 reduces the remaining (and/or initial) set by 2^1 times, because the sum of only every second combination will be even. Denoting by Λ^2 the set of combinations that do not meet only Requirement 2, we obtain:

$$\Lambda^2 = 2^{n_L - 1}. \tag{13.3}$$

To determine the number of polynomials that satisfies Requirement 3 (denote it by Λ^3), consider the feedback coefficients in matrix form:

$$\begin{matrix} a_{11} & a_{12} & a_{13} & \ldots & a_{1L} \\ & a_{22} & a_{23} & \ldots & a_{2L} \\ & & a_{33} & \ldots & a_{3L} \\ & & & \ldots & \ldots \\ & & & & a_{LL} \end{matrix}$$

Then the complete set of coefficients a_{ij}, their total n_L, can be divided into two parts: the number of coefficients $a_{iL} (i \in \{1, L-1\})$ (the number of such coefficients is denoted by $n^{3.1}$) and the whole set of other coefficients. In this case, the number of possible combinations can be represented as the product of the number of combinations from the first set to the number of combinations from the second set:

$$\Lambda = \Lambda^{3.1} \cdot \Lambda^{3.2},$$

where $\Lambda^{3.1} = 2^{n^{3.1}}$ is the set of combinations composed of the coefficients of the first set;

$\Lambda^{3.2} = 1$ is the set of combinations composed of the coefficients of the second set (since, according to Requirement 3, all $a_{1L} = a_{2L} = \ldots = a_{(L-1)L} = 0$).

The number of coefficients in the first set can be determined by considering the coefficients a_{ij} in matrix form without the last column, but adding the coefficient a_{LL}:

$$n^{3.1} = \frac{(L-1) \cdot (L-1+1)}{2} + 1 = \frac{L \cdot (L-1)}{2} + 1.$$

Thus, the total number of polynomials that meet only Requirement 3 will be defined as:

$$\Lambda^3 = 2^{[n_L]} - 2^{\left[\frac{L\cdot(L-1)}{2}+1\right]} \tag{13.4}$$

To determine the set of polynomials that meet Requirements 1–3, it is necessary to subtract from the whole set of possible combinations $\square_0$ those combinations that do not meet the requirements. To do this, it is necessary to exclude sets that intersect and do not meet Requirements 1–3. Denote by ${}^{-1}\square^2$ the set of combinations that meet only Requirement 2 and do not intersect with the set of combinations that do not meet Requirement 1. From Formulas (13.1) *and* (13.2) such a set is determined by the equation:

$${}^{-1}\Lambda^2 = \frac{1}{2}\cdot\Lambda^2 = \frac{1}{2}\cdot 2^{n_L-1} = 2^{n_L-2}. \tag{13.5}$$

To determine the set of polynomials that meet Requirement 3 and do not meet Requirements 1 and/or 2, it is also necessary to exclude intersecting parts. Denote by ${}^{-1,2}\Lambda^3$ the set of polynomials that meet Requirement 3 and meet Requirements 1 and 2. From (13.3) it is not difficult to see that in the set (13.4) the coefficient a_{LL} is included (according to Requirement 1, the coefficient a_{LL} takes only one value), and the sum of the remaining half of the combinations of coefficients will be even. Thus, there is equality:

$${}^{-1,2}\Lambda^3 = \Lambda^3/4 = 2^{[n_L-2]} - 2^{\left[\frac{L\cdot(L-1)}{2}-1\right]} \tag{13.6}$$

From Formulas (13.1), (13.5) and (13.6) it is possible to obtain the number of combinations of coefficients that meet Requirements 1–3 (denote this set as Λ^{1-3}):

$$\begin{aligned}\Lambda^{1-3} &= \Lambda^1 +^{-1}\Lambda^2 +^{-1,2}\Lambda^3\\ &= 2^{[n_L]} - 2^{\left[\frac{L\cdot(L-1)}{2}-1\right]}\\ &= 2^{[n_L]}\cdot\begin{pmatrix} & -[L+1] \\ 1-2 & \end{pmatrix}\end{aligned}$$

The number of symmetric combinations a_{ij},, denoted by the symbol Λ^4, is uniquely determined by L. The dependence of the values of Λ^4 on L can be determined by considering the coefficients a_{ij} in matrix form. Symmetrical are those combinations a_{ij} that have axial symmetry along the diagonal that goes from the upper right corner (excluding the column with coefficients a_{iL}) to the lower left corner. For example, for $L = 4$, all possible combinations of coefficients a_{11}, a_{12}, a_{13}, a_{22} have symmetric coefficients a_{33}, a_{23}, a_{13}, a_{22} (visually, these are all coefficients on one side of the diagonal, including the diagonal itself, except for the column with coefficients a_{iL}) and are symmetrical pairs. It is possible to determine all the coefficients, the variations of which will provide a set of symmetric coefficients, for any L:

$$\Lambda^4 = 2^{n^4},$$

where $n^4 = \frac{L^2}{4} + L$ is the number of coefficients that meet Requirement 4 for even L;

$n^4 = \frac{L^2-1}{4} + L$—the number of coefficients that meet the Requirement 4 for odd L.

Let${}^{-1,2,3}\Lambda^4$ be a set of combinations of coefficients a_{ij}, that does not meet Requirement 4 and does not intersect with the set of combinations that do not meet Requirements 1–3. Requirement *1* will reduce Λ^4 2^1 times, Requirement 2 also 2^1 times, and Requirement 3—in 2^{L-1} time. Thus, ${}^{-1,2,3}\Lambda^4$ can be determined by the following equation:

$${}^{-1,2,3}\Lambda^4 = {}^{-1,2}\Lambda^3 + \frac{\Lambda^4}{2^{L+1}}$$

After the calculations, it becomes possible to write a set of logs which satisfies Requirements 1–4 $\left(\Lambda^{1-4}\right)$:

$$\Lambda^{1-4} = \Lambda^{1-3} + {}^{-1,2,3}\Lambda^4 = 2^{[n_L]} - 2^{\left[\frac{L\cdot(L-1)}{2}-1\right]} + 2^{[n^4]},$$

where $n^4 = \frac{L^2}{4} - 1$—for even L;

$$n^4 = \frac{L^2-1}{4} - 1\text{-for odd } L.$$

Or, if you make a factor of 2^{nL}, the ratio can be rewritten as follows:

$$\Lambda^{1-4} = 2^{[n_L]}\cdot\left(1 - 2^{-(L+1)} + 2^{-n^4}\right), \tag{13.7}$$

where $n^4 = \left(\frac{L}{2}\right)^2 + \frac{L}{2} + 1$—for even L;

$$n^4 = \left(\frac{L+1}{2}\right)^2 + 1 - \text{for odd } L.$$

Of some interest is the quantitative assessment of the result. From the expression for Λ^{1-4}, determine as a percentage the number of polynomials that meet Requirements 1–4 for the whole set of possible feedback combinations depending on L:

$$\frac{\Lambda_0 - \Lambda^{1-4}}{\Lambda_0}\cdot 100\% = \left(\frac{1}{2^{[L+1]}} - \frac{1}{2^{[n^4]}}\right)\cdot 100\%, \tag{13.8}$$

where n^4 take the value of Formula (13.7).

Based on the above relations, the calculations for $L = 1 - 8$, the results of the calculations are shown in Table 13.6.

We see that the percentage of polynomials that can theoretically generate M-sequence in relation to the whole set of polynomials decreases in degree as the degree of NLFSR of the second order increases. In the Table 13.6 shows that already at $L = 6$ the number of such polynomials is less than 1%. Also from (13.8) it can be stated that the contribution of Requirement 4 decreases in quadratic dependence with increasing L.

As can be seen from Table 13.6, the number of combinations a_{ij}, for which $T = T_{max}$ (denote them by $\Lambda_{\max}$) is less than the value of Λ^{1-4}, obtained by Formula (13.7), by several orders of magnitude.

Analysis of the results allows us to propose the following empirical expression for $\Lambda_{\max}$:

$$\Lambda^1_{\max} \leq 2^{\frac{L(L-1)}{2} - L + 2}. \tag{13.9}$$

The use of expression (13.9) gives an error for $L = 8$ not higher than 4.3%, and for $L = 9$ the error is 3.4%.

It should be noted that when searching for all second-order NLFSR with a maximum period, you can halve the number of combinations studied by implementing Requirement 4. Thus, if you find a combination a_{ij}, that gives a sequence with a maximum period, then from Requirement 4 maximum period, without spending additional resources to study its period.

13.1.3 Conclusions on the Application of Requirements 1–4

1. This section formulates four requirements that must be met in the M-NLFSR. Based on the formulated requirements, a strict estimate of the upper limit of the number of possible polynomials (13.7) for a given length M-NLFSR.
2. This result becomes especially relevant for the search for polynomials with large values of L that generate M-sequence, as well as for estimating the number of possible such solutions when designing PRS generators.
3. It was possible to significantly reduce the set of combinations a_{ij}, for which the system will generate M-sequence (see Table 13.6). As an example of the effectiveness of the formulated requirements, we can cite the fact that during the finding of all M-NLFSR for $L = 6$ and with a complete search of all possible values was spent 5 h 12 min of machine time (calculations were performed on a *Pentium* 4 1.6 GHz). Approximately 240 times less machine time (only 1 min 18 s) was spent on the same search to enter the constraints defined only by Requirements 1–3 into the search algorithm.

13.2 Requirements 5–6

13.2.1 Formulation and Substantiation of Requirements 5–6

Consider a number of polynomials in which of all terms there is only one linear term (from Requirement 1 it must be a term of the highest order, i.e. $a_{LL} = 1$) and any number of nonlinear terms. For example, take $L = 4$ and polynomials of the form $x^4+x^3x^1+x^2x^1+1; x^4+x^4x^1+x^3x^1+x^2x^1+1; x^4+x^2x^1+1; x^4+x^3x^2+1$ etc. If we study the sequences generated by such polynomials at different initial states, then from the whole set of circles of states (and from all possible periods), in all cases the same ring will take place: (1000 → 0100 → 0010 → 0001 → 1000) with a period equal to the length of the NLFSR.

If the product of any two different cells of the considered registers at initial filling of cells of the register with one unit and all zeros, the result will always be zero. Therefore, any considered NLFSR at the specified initial filling will be equivalent (to generate identical sequence) to the polynomial $x_L + 1$. On the basis of the stated above we will formulate the next requirement.

Requirement 5. A polynomial based on a second-order NLFSR that can generate an M-sequence must have more than one ($a_{LL} = 1$ of Requirement 1) linear feedback coefficient, i.e.:

$$\sum_{i=1}^{L-1} a_{ii} > 1.$$

We obtain an estimate of the number of NLFSR that do not meet Requirement 5 and as a result can be excluded from consideration during the search of M-NLFSR only from the analysis of the coefficients a_{ij}.

The remaining NLFSR with $L = 4$, which meet the Requirements 1–5, but do not generate the M-sequence, will generate rings of the form (1010) → (0101) → (1010). The creation of circles of this kind will be provided that the present linear and nonlinear coefficients a_{ij} (in these states) will compensate each other. For example, coefficients of the form $\underline{\underline{1}}0\underline{\underline{1}}0$ 010 00 1; $\underline{\underline{1}}1\underline{\underline{1}}0$ 000 00 1 and symmetric coefficients $01\underline{\underline{1}}0$ 000 $\underline{\underline{1}}0$ 1; $00\underline{\underline{1}}0$ $\overline{010}$ $\underline{\underline{1}}0$ 1 (hereafter in this record, the coefficients a_{ij} from expression (13.3) will be given in linear form, i.e. $(a_{11}a_{12}a_{13}a_{14}a_{22}a_{23}a_{24}a_{33}a_{34}a_{44})$. The underlined combinations (in these states) of the register compensate for each other. And it does not matter what other nonlinear combinations, because their sum will always (in these states) be equal to zero.

The same will be in the case of mutual absence of underlined terms.

Linear feedback can also be compensated by linear combinations; as an example: $\underline{\underline{1}}100$ 000 $\underline{\underline{1}}0$ 1; $\underline{\underline{1}}000$ 010 $\underline{\underline{1}}0$ 1.

In order for the registers to accept the initial state, it is necessary that a unit be fed to the input as a result of the second cycle. This can be obtained only if $a_{22} \neq 1$ and

$$\begin{array}{cccccc}
(a_{11}) & a_{12} & (a_{13}) & a_{14} & (a_{15}) & a_{16} \\
 & [a_{22}] & a_{23} & [a_{24}] & a_{25} & [a_{26}] \\
 & & (a_{33}) & a_{34} & (a_{35}) & a_{36} \\
 & & & [a_{44}] & a_{45} & [a_{46}] \\
 & & & & (a_{55}) & a_{56} \\
 & & & & & [a_{66}]
\end{array}$$

a)

$$\begin{array}{ccccccc}
(a_{11}) & a_{12} & (a_{13}) & a_{14} & (a_{15}) & a_{16} & (a_{17}) \\
 & [a_{22}] & a_{23} & [a_{24}] & a_{25} & [a_{26}] & a_{27} \\
 & & (a_{33}) & a_{34} & (a_{35}) & a_{36} & (a_{37}) \\
 & & & [a_{44}] & a_{45} & [a_{46}] & a_{47} \\
 & & & & (a_{55}) & a_{56} & (a_{57}) \\
 & & & & & [a_{66}] & a_{67} \\
 & & & & & & (a_{77})
\end{array}$$

b)

$$(101010) \rightarrow (010101) \rightarrow (101010)$$

c)

$$(1010101) \rightarrow (0101010) \rightarrow (1010101)$$

d)

Fig. 13.2 Feedback coefficients: **a** for $L = 6$, **b** for $L = 7$, and sequences capable of generating the following systems: **c** for $L = 6$, **d** for $L = 7$, respectively

$a_{24} \neq 1$ or $a_{22} = a_{24} = 1$ provided that $a_{44} = 1$ or $a_{24} = 1$, $a_{22} = 1$ provided that $a_{44} \neq 1$.

Consider now $L = 5$. In this case, there will also be a ring of the form: (10101) $\rightarrow$ (01010) $\rightarrow$ (10101). But unlike the previous case, to obtain such a ring, it is necessary that the feedback in the first bar does not compensate for each other. In the second cycle, a necessary condition for the repetition of the state of the NLFSR is the presence of $a_{22} = 1$ or $a_{44} = 1$, or a nonlinear coefficient $a_{24} = 1$, as well as the last of the possible combinations, when all $a_{22} = a_{44} = a_{24} = 1$.

It is possible to extend the obtained result to higher values of L. Figure 13.2 shows the coefficients for an even number of registers $L = 6$, and for an odd number of $L = 7$, as well as similar rings with period $T = 2$. In parentheses are coefficients that can change the value, generated on the first cycle, in square—coefficients that affect the original value only on the second cycle. The rest of the coefficients a_{ij} can be ignored, because at any rate they will give zero.

It should be noted that the coefficients a_{ij}, in which the indices i and j are odd numbers, are taken in parentheses, and in square brackets, only those coefficients a_{ij}, are taken, in which the indices i and j are even numbers. Thus, based on the above, we can formulate the following requirement.

Requirement 6. A necessary condition for the formation of M-NLFSR second order is the simultaneous non-fulfillment of the following pair of conditions:

- parity of the number of coefficients $a_{ij} = 1$, the indices i and j of which are odd numbers (in Fig. 13.2 the coefficients in parentheses);
- odd number of coefficients $a_{ij} = 1$, the indices i and j of which are even numbers (in Fig. 13.2 coefficients in square brackets), or the same in the formulaic expression:

$$\sum_{i,j\in A} a_{ij} \begin{cases} \text{an even number, where } A \text{ is the whole set of odd numbers from 1 to } L; \\ \text{an odd number, where } A \text{ is the whole set of even numbers from 1 to } L. \end{cases}$$

13.2.2 *Calculation of the Set of NLFSR that Do not Meet the Requirements 1–6*

Denote by Λ^5 the total number of NLFSR of length L, which do not meet Requirement 5. Arrange the feedback coefficients in the form of a matrix:

$$\begin{matrix} a_{11} & a_{12} & a_{13} & \dots & a_{1L} \\ & a_{22} & a_{23} & \dots & a_{2L} \\ & & a_{33} & \dots & a_{3L} \\ & & & \dots & \dots \\ & & & & a_{LL} \end{matrix}.$$

Then Λ^5 will correspond to the set of possible combinations of values of the coefficients a_{ij}, except for the main diagonal (because it contains only the coefficients of linear feedback) and the coefficient a_{LL}.. The number of such combinations will be 2^{t+1}, where t is the number of nonlinear feedback coefficients, 2^1 is the number of variations with the coefficient a_{LL}. The number t can be determined from the considered triangle of coefficients aij by the expression $\frac{L\cdot(L-1)}{2}$. Thus, the total number of NLFSR length L that do not meet Requirement 5 is equal to:

$$\Lambda^5 = 2^{\frac{L\cdot(L-1)}{2}+1}.$$

Or, given that the above equation $n_L - L = \frac{L\cdot(L+1)}{2} - L = \frac{L\cdot(L-1)}{2}$ can be written as:

$$\Lambda^5 = 2^{n_L-(L-1)}.$$

It should be noted that the set of polynomials corresponding to Requirement 5 intersects with the set of polynomials truncated by Requirements 1, 2, 3 and 4.

Let Λ be the total number of admissible polynomials formed by NLFSR of the second order, ie those which from the full set of all possible polynomials correspond to the requirements put forward to them, therefore, can generate M-sequences.

Taking into account the requirements 1–5, Λ will be determined by the formula:

$$\Lambda = 2^{\left[\frac{L\cdot(L-1)}{2}-1\right]}\cdot\left(1-2^{[1-L]}\right) - 2^{\left[\frac{L^2-k}{4}-1\right]}\cdot\left(1-2^{\left[\frac{k-1}{2}\right]}\right)$$

where k = 0—for even L;

k = 1—for odd L.

The number of NLFSR that meet the Requirement 6 can be obtained by considering the record of the coefficients aij in matrix form (Fig. 13.2a, b). The coefficients, taken in parentheses and square brackets, if taken separately, form a triangle equivalent to the derived triangle, but with twice smaller sides. This gives a method for calculating the number of possible combinations of these coefficients.

Denote by Λ^6 the number of combinations aij that do not meet Requirement 6. Λ^6 will be defined as

$$\Lambda^6 = 2^{\left[\frac{L\cdot(L+1)}{2}-2\right]}$$

Excluding the set of intersecting polynomials, while not satisfying several of the above Requirements, we obtain an expression for the exact calculation of the number of polynomials that do not meet Requirements 1–6, for $L \geq 4$:

$$\Lambda^{1,2,3,4,5,6} = 2^{n_L-1} + \frac{1}{2}\cdot 2^{n_L-1}$$
$$+\frac{1}{4}\cdot\Big\{2^{n_L-(L-1)}\cdot\left(2^{L-1}-1\right)+\frac{1}{2^{L-1}}\cdot\Big(2^{\frac{L^2-k}{4}+L}$$
$$+2^{n_L-(L-1)}+2^{n_L-2}-)\}$$
$$\left[2^{\frac{L^2-k}{4}+L-\frac{L-k}{2}}+k\cdot 2\cdot 2^{\frac{L^2-k}{4}+L-2}+2^{n_L-L-1}-k\cdot 2\cdot 2^{\frac{L^2-k}{4}-\frac{L-k}{2}+L-2}\right]\Big)\Big\} \quad (13.10)$$

where $k = 0$—for even L;

$k = 1$—for odd L.

Expression (13.10) gives the exact quantitative value of the number of polynomials that do not meet the requirements of 1–6 and, as a consequence, are not able to generate a sequence with a maximum period.

13.2.3 *Quantitative Evaluation of the Obtained Results*

Thus, two more Requirements for the form of polynomials have been formulated, which additionally excluded polynomials that are not M-polinomials. Consider the influence of each of the Requirements on the search for M-polynomials.

To speed up the calculations, including computer, it makes sense to assess which part of the whole set can be excluded by analyzing the form of the polynomial-forming NLFSR in terms of requirements, as well as—in what sequence it is desirable to test to more likely cut off not M-polinomials at the initial stage of verification.

Figure 13.3 shows the calculated number of polynomials: the complete set of possible polynomials (Λ_0) for a given L; the number of polynomials that meet Requirements 1–6 $\left(\Lambda_0 - \Lambda^{1,2,3,4,5,6}\right)$ and can potentially generate M-Sequence; the

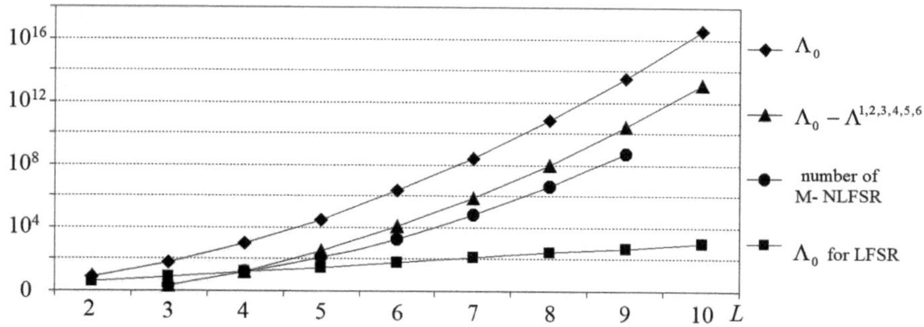

Fig. 13.3 Number of polynomials depending on the length of the NLFSR

number of M-NLFSR (established experimentally); for comparison, we give the complete set of possible polynomials for LFSR. Recall that LFSR is a special case of NLFSR, in which all nonlinear feedback coefficients are equal to zero.

The data shown in Fig. 13.3 make it possible to qualitatively estimate the set of polynomials that can be used in streaming cipher generators, as well as the tendency to increase with increasing L. Note that when designing streaming ciphers based on NLFSR for LFSR (including not even M-LFSR) for the same L.

We give the number of generating polynomials of NLFSR, which meet the requirements of 1–6. For quantitative estimation we will use standardized sizes. The obtained data are summarized in Table 13.7.

Table 13.7 Relative number of polynomials that do not meet Requirements 1–6

L	Λ_0	$\frac{\Lambda^1}{\Lambda_0}$	$\frac{\Lambda^2}{\Lambda_0}$	$\frac{\Lambda^3}{\Lambda_0}$	$\frac{\Lambda^4}{\Lambda_0}$	$\frac{\Lambda^5}{\Lambda_0}$	$\frac{\Lambda^6}{\Lambda_0}$	$\frac{\Lambda^{1,2,3,4,5,6}}{\Lambda_0}$
2	8	0.5	0.5	0.5	1	0.5	0.25	1.000000
3	64	0.5	0.5	0.75	0.5	0.25	0.25	0.968750
4	1024	0.5	0.5	0.875	0.25	0.125	0.25	0.984375
5	32,768	0.5	0.5	0.9375	0.0625	0.0625	0.25	0.989380
6	2,097,152	0.5	0.5	0.96875	0.015625	0.03125	0.25	0.994431
7	2.7×10^8	0.5	0.5	0.984375	0.001953	0.015625	0.25	0.997119
8	6.9×10^{10}	0.5	0.5	0.992188	0.000244	0.007813	0.25	0.998547
9	3.5×10^{13}	0.5	0.5	0.996094	1.5×10^{-5}	0.003906	0.25	0.999270
10	3.6×10^{16}	0.5	0.5	0.998047	9.5×10^{-7}	0.001953	0.25	0.999635
11	7.4×10^{19}	0.5	0.5	0.999023	3.0×10^{-8}	0.000977	0.25	0.999817
12	3.0×10^{23}	0.5	0.5	0.999512	9.3×10^{-10}	0.000488	0.25	0.999908
13	2.5×10^{27}	0.5	0.5	0.999756	1.5×10^{-11}	0.000244	0.25	0.999954
14	4.1×10^{31}	0.5	0.5	0.999878	2.3×10^{-13}	0.000122	0.25	0.999977

Table 13.7 denotes by Λ^m the set of polynomials that do not correspond to one of the six given Requirements, where m is the number of the extended Requirement for the form of the polynomial.

As can be seen from Table 13.7, when presenting the polynomial Requirements 1–6, already at $L = 9$ more than 99.9% of the whole set of polynomials is cut off and with increasing value of L, as can be seen from (13.10), this percentage will also be increase in degree dependence.

Also from Table 13.7 it is seen that the maximum contribution to the rejected polynomials that do not generate M-sequence, gives Requirement 3. Thus, when searching for polynomials from the whole set of possible for a given L must first check it for compliance with Requirement 3, then Requirements 1, 2 and 6, followed by Requirements 5, and only lastly Requirements 4. In addition, the fact that the verification of compliance with Requirements 3 and 1 is easy to implement in a software way and the computation time is less than when checking for compliance with Requirements 2, 4 and 6.

However, despite the small number of polynomials truncated by Requirement 4 compared to the number truncated by Requirement 3, Requirement 4 should not be neglected. Because at large values of L Requirement 4 still cuts off a considerable set not M-NLFSR, and the complexity of verification by analyzing each of the generated sequence is much higher than the analysis of the form of the polynomial.

As an example, the following quantitative calculations can be given: for $L = 9$ the number of possible polynomials $\Lambda_0 = 35, 184, 372{,}088{,}832$ (assuming 100%), Requirement 3 cuts off 35,046,933,135,360 polynomials (which will be 99.6%), and when this will leave another 137,438,953,472 polynomials (0.39%). When Polynomials of Requirements 1–6 are presented, 25,668,894,720 polynomials (0.073%) remain, which is 5.4 times less than the previous number. The number of M-NLFSR obtained experimentally will be only 519,239,794, which is 0.0015% of the total set or 2% of the set of polynomials that meet Requirements 1–6.

13.3 Requirements 7–8

13.3.1 Formulation and Substantiation of Requirements 7–8

Thus, Requirements 1–6 exclude all polynomials that are not M-NLFSR for $L = 4$. Consider polynomials with $L = 5$. Of all possible combinations, consider only those combinations that meet Requirements 1–6 and do not generate M-Sequence. There will be 220 such combinations. Consider these combinations in more detail.

Let the initial filling of all cells of the register be equal to 1, then we will see 16 combinations a_{ij},, in which, among other periods, will be $T = L + 1 = 6$ (see Table 13.8).

Table 13.8 Coefficients of NLFSR and symmetrical to them $a_{(L-j)(L-i)}$, that meet Requirements 1–6, but do not generate M-Sequence

No.	a_{ij}	$a_{(L-j)(L-i)}$
1.	111001010000001	001000010110101
2.	101100100100001	000101110000101
3.	100000010100001	001001000000101
4.	101101110010001	011100110100101
5.	111000000110001	010001010010101
6.	110100110110001	011101100010101
7.	111001100100101	100001110110101
8.	110101010100101	101101000110101

On the example of the first combination, consider how the state of the register changes when initially filled with all *1* in these combinations a_{ij}. All the above calculations are valid for other polynomials listed in Table 13.8. For ease of perception, we place the coefficients in matrix form:

$$111001010000001 \Rightarrow \begin{matrix} 1&1&1&0&0 \\ &1&0&1&0 \\ &&0&0&0 \\ &&&0&0 \\ &&&&1 \end{matrix}$$

The first number generated will be zero because $\sum a_{ij}$ it is an even number.

The state of the first register is zero, equivalent to deleting the first row (counting from top to bottom and from left to right), because there is a combination a_{1j}, which for any other values of $q_j (j \neq 1)$ will always give zero. Given that all other $q_j (j \in \{2, L\})$ at this stage is equal to *1*, the initial value will be determined by the parity of the sums remaining $a_{ij} (j \neq 1)$. We see that this sum will be an odd number, therefore, the initial value will be *1*.

At the beginning of the second cycle of the considered NLFSR there will be the following state of registers—(10,111). $q_2 = 0$ is equivalent to excluding the second row and the second column. Since all other cells of the register are equal to one, the initial value is also determined by the parity of the sum of non-zero coefficients a_{ij}, except $i = 2$ and $j = 2$. Calculate this number—it will be odd for all cases, i.e. at the output of the second cycle we always get unit.

A similar situation develops for the third, fourth and fifth cycles of the work, excluding respectively for the third cycle the 3rd line and the 3rd column, for the fourth—the 4th line and the 4th column, for the fifth—the 5th line and 5 and column, and in all these cases the residual number a_{ij} will be an odd number.

At the beginning of the sixth cycle we get the state of all cells of the register, which will be equal to one, i.e. the initial state. Thus, for all these polynomials, one of the possible circles will be a ring with period $T = L + 1$, when all or $(L - 1)$ cells of the register are initially filled with states equal to one. Obviously, this statement is true for any other value of L. Based on the results obtained, we can formulate such a statement.

Assertion. If the total number of elements $a_{ij} = 1$ is an even number, and the number of elements $a_{ij} = 1$ on all minors of the main diagonal is an odd number, then the whole set of generated sequences must have one ring with period $T = L+1$.

Let's take advantage of the fact that when you subtract an odd number from any even number, the result will definitely be an odd number. From a practical point of view, this would be an equivalent approach, as it is easier to count the number of elements that are excluded during the formation of a minor. Thus, we formulate the following requirement.

Requirement 7. A necessary condition for the existence of M-NLFSR second order, in which the number of all $a_{ij} = 1$ is an even number, is the parity of at least one sum for the elements $a_{ij} = 1$ in their rows and columns of any i from 1 to L.

Otherwise.

If the number $\sum_{i,j=1}^{L} a_{ij}$ is even and if $\sum_{i=1}^{k-1} a_{ik} + \sum_{j=k+1}^{L} a_{ik} + a_{kk}$ it is odd for all $k \in \{1, L\}$, then the NLFSR will not generate the M-sequence, and one of the possible periods will necessarily be $T = L + 1$.

Requirement 8. Let T be any number in the range from 1 to L inclusive. The condition for creating a NLFSR ring with period T will be the fulfillment of the condition:

$$\sum_{i=0}^{k+i\cdot T\le L} \sum_{j=a}^{k+j\cdot T\le L} a_{(k+i\cdot T)(k+j\cdot T)} \begin{cases} \text{even number, for all } k \in [1, T-1] \\ \text{odd number, for } k = T. \end{cases}$$

Requirement 8 includes:

- Requirement 2, which is a special case of Requirement 8 at $T = 1$;
- Requirement 6, which is also a special case of Requirement 8 at $T = 2$.

For clarity, consider an example of constructing many combinations that meet Requirement 8 in the example $L = 6$:

$$L = 6 \; T = 1$$

$$\begin{array}{cccccc}
\underline{a_{11}} & \underline{a_{12}} & \underline{a_{13}} & \underline{a_{14}} & \underline{a_{15}} & \underline{a_{16}} \\
 & \underline{a_{22}} & \underline{a_{23}} & \underline{a_{24}} & \underline{a_{25}} & \underline{a_{26}} \\
 & & \underline{a_{33}} & \underline{a_{34}} & \underline{a_{35}} & \underline{a_{36}} \\
 & & & \underline{a_{44}} & \underline{a_{45}} & \underline{a_{46}} \\
 & & & & \underline{a_{55}} & \underline{a_{56}} \\
 & & & & & \underline{a_{66}}
\end{array}$$

	k = 1		
$k+i\cdot T$	$=1$	1 1 1 1 1 2 2	2 2 2 3 3 3 3 4 4 4 5 5 6
i	$=0$	0 0 0 0 0 1 1	1 1 1 2 2 2 2 2 2 2 3 3 4
j	$=0$	1 2 3 4 5 1 2	3 4 5 2 3 4 5 3 4 5 4 5 5
$k+j\cdot T$	$=1$	2 3 4 5 6 2 3	4 5 6 3 4 5 6 4 5 6 5 6 6

$L = 6\ T = 2$

$$\begin{array}{cccccc}
\underline{a_{11}} & a_{12} & \underline{a_{13}} & a_{14} & \underline{a_{15}} & a_{16} \\
 & \underline{\underline{a_{22}}} & a_{23} & \underline{\underline{a_{24}}} & a_{25} & \underline{\underline{a_{26}}} \\
 & & \underline{a_{33}} & a_{34} & \underline{a_{35}} & a_{36} \\
 & & & \underline{\underline{a_{44}}} & a_{45} & \underline{\underline{a_{46}}} \\
 & & & & \underline{a_{55}} & a_{56} \\
 & & & & & \underline{\underline{a_{66}}}
\end{array}$$

		k = 1	k = 2
$k+i\cdot T$	$=1$	1 1 3 3 5 2	2 2 4 4 6
i	$=0$	0 0 1 1 2 0	0 0 1 1 2
j	$=0$	1 2 1 2 2 0	1 2 1 2 2
$k+j\cdot T$	$=1$	3 5 3 5 5 2	4 6 4 6 6

$L = 6T = 3$

$$\begin{array}{cccccc}
\underline{a_{11}} & a_{12} & a_{13} & \underline{a_{14}} & a_{15} & a_{16} \\
 & \underline{\underline{a_{22}}} & a_{23} & a_{24} & \underline{\underline{a_{25}}} & a_{26} \\
 & & \underline{\underline{\underline{a_{33}}}} & a_{34} & a_{35} & \underline{\underline{\underline{a_{36}}}} \\
 & & & \underline{a_{44}} & a_{45} & a_{46} \\
 & & & & \underline{\underline{a_{55}}} & a_{56} \\
 & & & & & \underline{\underline{\underline{a_{66}}}}
\end{array}$$

		k = 1	k = 2	k = 3
$k+i\cdot T$	$=1$	1 4 2	2 5 3	3 6
i	$=0$	0 1 0	0 1 0	0 1
j	$=0$	1 1 0	1 1 0	1 1
$k+j\cdot T$	$=1$	4 4 2	5 5 3	6 6

$L = 6\ T = 4$

$$\begin{array}{cccccc}
\underline{a_{11}} & a_{12} & a_{13} & a_{14} & \underline{a_{15}} & a_{16} \\
 & \underline{\underline{a_{22}}} & a_{23} & a_{24} & a_{25} & \underline{\underline{a_{26}}} \\
 & & \underline{\underline{\underline{a_{33}}}} & a_{34} & a_{35} & a_{36} \\
 & & & \utilde{a_{44}} & a_{45} & a_{46} \\
 & & & & \underline{a_{55}} & a_{56} \\
 & & & & & \underline{\underline{a_{66}}}
\end{array}$$

		k = 1	k = 2	k = 3	k = 4
$k+i\cdot T$	$=1$	1 5 2	2 6 3		4
i	$=0$	0 1 0	0 1 0		0
j	$=0$	1 1 0	1 1 0		0
$k+j\cdot T$	$=1$	5 5 2	6 6 3		4

$L = 6\ T = 5$

$$\begin{array}{llllll} \underline{a_{11}} & a_{12} & a_{13} & a_{14} & a_{15} & \underline{a_{16}} \\ & \underline{\underline{a_{22}}} & a_{23} & a_{24} & a_{25} & a_{26} \\ & & \underline{\underline{\underline{a_{33}}}} & a_{34} & a_{35} & a_{36} \\ & & & \underset{\sim\sim\sim}{a_{44}} & a_{45} & a_{46} \\ & & & & \underline{\underline{\underline{\underline{a_{55}}}}} & a_{56} \\ & & & & & \underline{a_{66}} \end{array} \qquad \begin{array}{rlllll} & k=1 & k=2 & k=3 & k=4 & k=5 \\ k+i\cdot T & =1 & 1\,6\,2 & 3 & 4 & 5 \\ i & =0 & 0\,1\,0 & 0 & 0 & 0 \\ j & =0 & 1\,1\,0 & 0 & 0 & 0 \\ k+j\cdot T & =1 & 6\,6\,2 & 3 & 4 & 5 \end{array}$$

$$L = 6\; T = 6$$

$$\begin{array}{llllll} \underline{a_{11}} & a_{12} & a_{13} & a_{14} & a_{15} & a_{16} \\ & \underline{\underline{a_{22}}} & a_{23} & a_{24} & a_{25} & a_{26} \\ & & \underline{\underline{\underline{a_{33}}}} & a_{34} & a_{35} & a_{36} \\ & & & \underset{\sim\sim\sim}{a_{44}} & a_{45} & a_{46} \\ & & & & \underline{\underline{\underline{\underline{a_{55}}}}} & a_{56} \\ & & & & & \underset{\cdots\cdots}{a_{66}} \end{array} \qquad \begin{array}{rllllll} & k=1 & k=2 & k=3 & k=4 & k=5 & k=6 \\ k+i\cdot T & =1 & 2 & 3 & 4 & 5 & 6 \\ i & =0 & 0 & 0 & 0 & 0 & 0 \\ j & =0 & 0 & 0 & 0 & 0 & 0 \\ k+j\cdot T & =1 & 2 & 3 & 4 & 5 & 6 \end{array}$$

The above feedback combinations will always generate rings with periods $T \leq L$ when all cells of the register are initially filled with zeros, except for cells multiples of T. For example, for $L = 6$ and $T = 2$ it will correspond to the initial filling with a combination of 0, 1, 0, 1, 0, 1, and for for $L = 6$ and $T = 5$ initial combination 0, 0, 0, 0, 1, 0.

The complete set of combinations a_{ij}, which does not satisfy Requirement *8*, (Λ^8) is determined by the relation:

$$\Lambda^8 = 2^n L \cdot \left(\sum_{i=1}^{L} 2^{-i} \sum_{i=2}^{L} \sum_{j=1}^{(i-1)\leq \frac{L}{2}} 2^{-i-j} + \sum_{i=3}^{L} \sum_{j=2}^{(i-1)\leq \frac{L}{2}} \sum_{k=1}^{(j-1)} 2^{-i-j-k} - \sum_{i=4}^{L} \sum_{j=3}^{(i-1)\leq \frac{L}{2}} \sum_{k=2}^{(j-1)} \sum_{t=1}^{(k-1)} 2^{-i-j-k-t} + \cdots \right)$$

Obtaining this formula is given in the next section.

13.3.2 The Method of Calculating the Set of Combinations, that Do not Meet the Requirement 8

Let's name combinations which at a certain filling will influence formation of an initial bit by significant combinations. Significant combinations consist of *significant coefficients* a_{ij}.

Denote by Θ the set of possible combinations (for a given L), which at a certain initial state will form a sequence ring with period T, i.e. will not meet Requirement 8.

Note that the symbol Θ will be used only in Sect. 13.3. In other sections, we will use the symbol Λ^8 to denote the complete set of combinations a_{ij} that does not satisfy Requirement 8. This double notation was introduced to avoid confusion with the symbol Λ, which will also be used in other sections with the corresponding index notation.

The set Θ will consist of subsets Θ_T, intersecting for all possible considered periods $T \in [1, L]$.

Denote by Θ_k the subset that forms significant coefficients for each k, where k is a certain set of significant coefficients for a given L and T. any restrictions Requirement 8.

For different k, T and L, the number of significant combinations will be determined by the following relationship:

$$n_k = \frac{\alpha \cdot (\alpha + 1)}{2}, \tag{13.11}$$

where n_k is the number of significant coefficients for a given k;

$\alpha$is the value of the rounding function to the lesser side (*floor*), calculated by the following ratio: $\alpha = floor\left(\frac{L+(T-k)}{T}\right) = floor\left(\frac{L-k}{T} + 1\right)$.

Given that for predetermined L and T the total number of coefficients will be determined by the ratio $n_L = \sum_{k=1}^{T} n_k + n_{\mathcal{H}3}$ determine the number of insignificant coefficients as

$$n_{\mathcal{H}3} = n_L - \sum_{k=1}^{T} n_k, \tag{13.12}$$

With the introduced notation system, we obtain that the subset Θ_k will be defined as half (even for $k \in [1, T-1]$ or odd for $k = T$ from Requirement 8) of all possible combinations of coefficients n_k. A subset of insignificant coefficients is taken in full. Or in the formulaic expression:

$$\Theta_k = \frac{1}{2} \cdot 2^{n_k} = 2^{n_k - 1}, \Theta_{\mathcal{H}3} = 2^{n_{\mathcal{H}3}}, \tag{13.13}$$

Given that all the numbers Θ_k and Θ_{H3} (for given L and T) according to the principles of construction of these sets, have no intersections, then for each Θ_T it will be true

$$\Theta_T = \prod_{k=1}^{T} \Theta_k \cdot \Theta_{\mathcal{H}3}, \tag{13.14}$$

Alternatively, given (13.12) and (13.13), we obtain:

$$\Theta_T = \prod_{k=1}^{T} 2^{n_k - 1} \cdot 2^{n_{\mathcal{H}3}} = \prod_{k=}^{T} 2^{n_k} \cdot \prod_{k=1}^{T} 2^{-1} \cdot 2^{n_{\mathcal{H}3}} = \prod_{k=1}^{T} 2^{n_k} \cdot 2^{-T} \cdot 2^{n_{\mathcal{H}3}}$$
$$= 2^{\sum_{k=1}^{T} n_k - T + n_{\mathcal{H}3}} = 2^{\sum_{k=1}^{T} n_k - T + \left(n_L - \sum_{k=1}^{T} n_k\right)} = 2^{n_L - T} \tag{13.15}$$

In order to find the complete set Θ, it is necessary to determine the sets that intersect for Θ_T at different values of T.

13.3.3 Inclusion and Exclusion Formula

To solve the problem of finding the complete set, we use the principles of combinatorics, namely, the formula of inclusions and exclusions. Suppose there is a finite set A, the number of elements of this set is called (in terms of combinatorics) *power* and is denoted as $| A |$. The inclusion and exclusion formula makes it possible to find the union power of any finite set of sets.

Thus, the formula of inclusions and exclusions for two finite sets $|A|$ and $|B|$ has the form: $|A \cup B| = |A| + |B| - |A \cap B|$.

Inclusion and exclusion formula for three finite sets $|A|$, $|B|$ and $|C|$: $|A \cup B \cup C| = |A| + |B| + |C| - |A \cap B| - |A \cap B| - |B \cap C| + |A \cap B \cap C|$.

In General, for finite sets A_1,A_2,…,A_n the formula of inclusions and exclusions has the form:

$$\left|\bigcup_{i=1}^{n} A_i\right|$$
$$= \sum_i |A_i| - \sum_{i<j} \left|A_i \cap A_j\right| + \sum_{i<j<k} \left|A_i \cap A_j \cap A_k\right| - \cdots$$
$$+ (-1)^{n-1} |A_1 \cap A_2 \cap \ldots \cap A_n| \tag{13.16}$$

The formula looks cumbersome, but its essence is simple: first we sum the powers of all sets, then we subtract the powers of all pairwise sections, then we add the powers of all triple sections, and so on, alternating signs, to the last term—the power of all sets.

13.3.4 Solve the Problem of Computing a Non-intersecting Set for Small Values of L

Solution for L = 2. In this case, the significant coefficients a_{ij} will be:

$$\begin{array}{cc} T = 1 & T = 2 \\ \underline{a_{11}a_{12}} & \underline{a_{11}}a_{12} \\ \underline{a_{22}} & \underline{\underline{a_{22}}} \end{array}$$

One line underlines the significant coefficients for $k = 1$ in (13.11), two dashes underline the significant coefficients for $k = 2$ for the corresponding T. The formula of inclusions and exclusions (13.16) in this case will look like:

$$\Theta^{L=2} = \Theta_1 + \Theta_2 - \Theta_{2//1}, \tag{13.17}$$

where Θ_1, Θ_2 are subsets calculated from (13.15) for $T = 1$ and $T = 2$, respectively, $\Theta_{2//1}$ is a set formed from the intersection of sets of combinations for subsets Θ_1 and Θ_2.

Using (13.15) and the definition for n_L, we obtain:

$$n_L = \frac{2 \cdot (2+1)}{2} = 3, \Theta_1 = 2^{3-1} = 2^2, \Theta_2 = 2^{3-2} = 2^1.$$

To determine $\Theta_{2//1}$ consider all combinations a_{ij} that are formed for $L = 2$.

$$\begin{array}{cccccccc} 0 & 0 & 1 & 0 & 0 & 1 & 1 & 1 \\ 0 & & \underline{0} & & \underline{0} & & 0 & \\ 0 & 0 & 1 & 0 & 0 & 1 & 1 & 1 \\ \underline{\underline{1}} & & 1 & & \underline{1} & & \underline{1} & \end{array}$$

One line underlines significant combinations a_{ij}, which meet the Requirements and belongs to the subset Θ_1, double line—to the subset Θ_2.

As we see, $\Theta_{2//1} = 1$ and corresponds to the combination 0, 0, 1.

Substituting the results obtained in (13.17). we have a complete set of combinations that do not meet Requirement 8 for $L = 2$:

$$\Theta^{L=2} = 2^2 + 2^1 - 2^0 = 5$$

Solution for L = 3. In this case, the significant coefficients a_{ij} will be:

$$\begin{array}{ccccccccc} & T=1 & & & T=2 & & & T=3 & \\ \underline{a_{11}} & \underline{a_{12}} & \underline{a_{13}} & \underline{a_{11}} & a_{12} & \underline{a_{13}} & \underline{a_{11}} & a_{12} & a_{13} \\ & \underline{a_{22}} & \underline{a_{23}} & & \underline{\underline{a_{22}}} & a_{23} & & \underline{\underline{a_{22}}} & a_{23} \\ & & \underline{a_{33}} & & & \underline{a_{33}} & & & \underline{\underline{\underline{a_{33}}}} \end{array}$$

One line emphasizes significant coefficients for $k = 1$ in (13.11), two dashes—for $k = 2$, three—for $k = 3$, with the corresponding T.

The formula of inclusions and exclusions (13.16) in this case takes the form:

$$\Theta^{L=3} = \Theta_1 + \Theta_2 + \Theta_3 - \left(\Theta_{2//1} + \Theta_{3//1} + \Theta_{3//2}\right) + \Theta_{3//2//1},$$

where Θ_1, Θ_2, Θ_3—subsets calculated from (13.15) for $T = 1$, $T = 2$ and $T = 3$, respectively;

$\Theta_{2//1}$, $\Theta_{3//1}$, $\Theta_{3//2}$—an intersecting set for combinations of subsets Θ_2 and Θ_1, Θ_3 and Θ_1, Θ_3 and Θ_2, respectively;

$\Theta_{3//2//1}$—a set that intersects for all three subsets Θ_1, Θ_2 and Θ_3 simultaneously.

From the Formula (13.15), given that $n_L = \frac{3\cdot(3+1)}{2} = 6$, we obtain the number of coefficients included in the subsets significant for a given value of T:

$$\begin{aligned} T &= 1, \Theta_1 = 2^{6-1} = 2^5, \\ T &= 2, \Theta_2 = 2^{6-2} = 2^4, \\ T &= 3, \Theta_3 = 2^{6-3} = 2^3. \end{aligned}$$

From the Formula (13.14) and significant combinations for the given L, T and k we calculate the sets of the obtained periton subsets. According to the definition of Requirement 8, the sums of all significant coefficients for $k < L$ must be even numbers, and for $k = L$—odd. Given that the number of significant coefficients for each k at $T = 3$ is equal to one, we obtain that the prerequisite: $a_{11} = 0; a_{22} = 0; a_{33} = 1$ (for $T = 3$). The number of significant coefficients for $k = 2$ at $T = 2$ is equal to one, i.e. the prerequisite for $T = 2$ will be $a_{22} = 1$. Therefore, for the intersections of sets of combinations for Θ_2 and Θ_3 we obtain (given the impossibility of simultaneous execution conditions of equality $a_{22} = 1$ and $a_{22} = 0$) an empty set. That is $\Theta_{3//2} = 0$ and $\Theta_{3//2//1} = 0$.

Considering the whole set of significant coefficients for $T = 2$, from (13.14) we see that it consists of three subsets $\Theta_2 = 2^{n1} \cdot 2^{n2} \cdot 2^{n_{\text{H3}}}$. In this case, $n_2 = 1$ and corresponds to the significant coefficient a_{22},$n_1 = 3$ and corresponds to the significant coefficients a_{11},a_{13},a_{33};$n_{nz} = 2$ corresponds to insignificant coefficients a_{12},a_{23}. The set $\Theta_1 = 2^{n1}$, where $n_1 = 6$ corresponds to all coefficients a_{ij}.

Since the sum of even and odd numbers will always be an even number, and from $T = 2$ it follows that the combination with a_{22} must be an odd number, and the combination with a_{11}, a_{13}, a_{33}—even, the sum of combinations of significant coefficients a_{11}, a_{13}, a_{22}, a_{33} will always be an odd number. The sum of two more insignificant coefficients must be an even number, then the sum of all coefficients will be an odd number, and will meet Requirement 8 for the set Θ_1. This condition is satisfied for every second combination of insignificant numbers, i.e. all such combinations will be $2^{n_{nz}-1}$. Hence and from (13.13), (13.14) we obtain:

$$\Theta_{2//1} = 2^{n_1} \cdot 2^{n_2} \cdot 2^{n_{\varkappa 3}-1} = 2^{3-1} \cdot 2^{1-1} \cdot 2^{2-1} = 2^2 \cdot 2^0 \cdot 2^1 = 2^3.$$

Similarly, we obtain an intersecting set for $\Theta_{3//1}$. Combinations of significant coefficients remain unchanged, as for many Θ_3, but the set of combinations that are formed from insignificant coefficients is halved:

$$\Theta_{3//1} = 2^{1-1} \cdot 2^{1-1} \cdot 2^{1-1} \cdot 2^{3-1} = 2^2.$$

Now we have everything we need to calculate the number of combinations that do not meet Requirement *8* for $L = 3$.

$$\begin{aligned} \Theta^{L=3} &= \Theta_1 + \Theta_2 + \Theta_3 - \left(\Theta_{2//1} + \Theta_{3//1} + \Theta_{3//2}\right) + \Theta_{3//2//1} \\ &= 2^5 + 2^4 + 2^3 - \left(2^3 + 2^2 + 0\right) + 0 = 32 + 16 + 8 - 8 - 4 = 44 \end{aligned}$$

Solution for L = 4. Significant coefficients a_{ij} will be:

$$\begin{array}{cccc} \multicolumn{4}{c}{T=1} \\ \underline{a_{11}} & \underline{a_{12}} & \underline{a_{13}} & \underline{a_{14}} \\ & \underline{a_{22}} & \underline{a_{23}} & \underline{a_{24}} \\ & & \underline{a_{33}} & \underline{a_{34}} \\ & & & \underline{a_{44}} \end{array} \qquad \begin{array}{cccc} \multicolumn{4}{c}{T=2} \\ \underline{a_{11}} & a_{12} & \underline{a_{13}} & a_{14} \\ & \underline{\underline{a_{22}}} & a_{23} & \underline{\underline{a_{24}}} \\ & & \underline{a_{33}} & a_{34} \\ & & & \underline{\underline{a_{44}}} \end{array} \qquad \begin{array}{cccc} \multicolumn{4}{c}{T=3} \\ \underline{a_{11}} & a_{12} & a_{13} & \underline{a_{14}} \\ & \underline{\underline{a_{22}}} & a_{23} & a_{24} \\ & & \underline{\underline{\underline{a_{33}}}} & a_{34} \\ & & & \underline{a_{44}} \end{array}$$

$$\begin{array}{cccc} \multicolumn{4}{c}{T=4} \\ \underline{a_{11}} & a_{12} & a_{13} & a_{14} \\ & \underline{\underline{a_{22}}} & a_{23} & a_{24} \\ & & \underline{\underline{\underline{a_{33}}}} & a_{34} \\ & & & \underset{\bullet\bullet\bullet\bullet}{a_{44}} \end{array}$$

One line underlines the significant coefficients for $k = 1$ in (13.11), two dashes for $k = 2$, three—for $k = 3$, points—for $k = 4$, with the corresponding T. The formula of inclusions and exclusions (13.16) in this case will type:

$$\begin{aligned} \Theta^{L=4} &= \Theta_1 + \Theta_2 + \Theta_3 + \Theta_4 - \left(\Theta_{2//1} + \Theta_{3//2} + \Theta_{3//1} + \Theta_{4//3} + \Theta_{4//2} + \Theta_{4//1}\right) \\ &\quad + \Theta_{3//2//1} + \Theta_{4//3//2} + \Theta_{4//3//1} + \Theta_{4//2//1} - \Theta_{4//3//2//1}, \end{aligned}$$

where Θ_1, Θ_2, Θ_3, Θ_4—subsets calculated from (13.15) for $T = 1$, $T = 2$, $T = 3$ and $T = 4$, respectively;

$\Theta_{m//n}$—the intersecting set for the subset Θ_m and Θ_n, , for $m, n \in [1, T]$ under the condition $m \neq n$;

$\Theta_{m//n//v}$—intersecting set for three subsets Θ_m, Θ_n and Θ_ν simultaneously for $m, n, v \in [1, T]$ under the condition $m \neq n \neq v$;

$\Theta_{4//3//2//1}$—intersecting set for four subsets Θ_1, Θ_2, Θ_3 and Θ_4 simultaneously.

Write down the values of all sets and their intersections:

$$\Theta_1 = 2^{10-1} = 2^9,$$
$$\Theta_2 = 2^{10-2} = 2^8, \quad \Theta_{2//1} = \frac{1}{2} \cdot \Theta_2 = 2^{8-1} = 2^7,$$
$$\Theta_3 = 2^{10-3} = 2^7, \quad \Theta_{3//1} = \frac{1}{2} \cdot \Theta_3 = 2^{7-1} = 2^6, \quad \Theta_{3//2} = 2^5,$$
$$\Theta_4 = 2^{10-4} = 2^6, \quad \Theta_{4//1} = \frac{1}{2} \cdot \Theta_4 = 2^{6-1} = 2^5,$$
$$\Theta_{4//2} = 2^4, \quad \Theta_{4//3} = 0,$$
$$\Theta_{4//2//1} = \frac{1}{2} \cdot \Theta_{4//2} = 2^{4-1} = 2^3,$$
$$\Theta_{4//3//1} = \frac{1}{2} \cdot \Theta_{4//3} = 0, \quad \Theta_{3//2//1} = \frac{1}{2} \cdot \Theta_{3//2} = 2^{5-1} = 2^4,$$
$$\Theta_{4//3//2} = 0, \quad \Theta_{4//3//2//1} = \frac{1}{2} \cdot \Theta_{4//3//2} = 0.$$

Θ_1, Θ_2, Θ_3, Θ_4 are calculated from (13.15), given that $n_L = \frac{4 \cdot (4+1)}{2} = 10$.

Any sets intersecting with Θ_1, are calculated as $\Theta_{...//1} = \frac{1}{2}\Theta_{...}$.

This follows from the fact that at $T = 1$ all coefficients are significant. And for $T > 1$ all values of n_1,n_2,n_3, … in (13.11), as well as their intersections, are a fixed value, and only n_{nz} can change so that every second combination with n_{nz} will satisfy the condition of oddness of all combinations for Θ_1.

The set that intersects between the sets Θ_3 and Θ_4,, i.e. $\Theta_{4//3}$, as well as their intersections with the other sets Θ_1 and Θ_2, will be an empty set. This follows from the fact that the prerequisite for $T = 4$ is $a_{33} = 0$, because the sum of all significant coefficients at $k = 3$ must be an even number, and for $T = 3$ the prerequisite is $a_{33} = 1$. Therefore, the coefficient a_{33} for these two sets will always be different.

Remains $\Theta_{4//2}$, $\Theta_{3//2}$.

Consider what are the requirements for the coefficients at $T = 4$ and $T = 2$:

$a_{44} = 1$—from the oddness of the sum of the coefficients at $T = 4$ for $k = 4$;

$a_{33} = 0$—from the parity of the sum of the coefficients at $T = 4$ for $k = 3$;

$a_{22} = 0$—from the parity of the sum of the coefficients at $T = 4$ for $k = 2$;

$a_{11} = 0$—from the parity of the sum of the coefficients at $T = 4$ for $k = 1$;

$a_{13} = 0$—from the parity of the sum of the coefficients $a_{11} + a_{13} + a_{13}$ at $T = 2$ for $k = 1$, taking into account the above requirements;

$a_{42} = 0$—from the oddness of the sum of the coefficients $a_{22} + a_{24} + a_{44}$ at $T = 2$ for $k = 2$, taking into account the above requirements;

no requirements are imposed on the coefficients a_{12}, a_{14}, a_{23}, a_{24}.

Therefore, $\Theta_{4//2}$ is formed as follows:

$$\begin{array}{llllllll} \Theta_{4//2} = 2^{1-1} \cdot & 2^{1-1} \cdot & 2^{1-1} \cdot & 2^{1-1} \cdot & 2^{1-1} \cdot & 2^{1-1} \cdot & 2^4 & = 2^4 \\ (a_{11}) & (a_{22}) & (a_{33}) & (a_{44}) & (a_{13}) & (a_{24}) & (a_{12}, a_{14}, a_{23}, a_{34}) & \end{array}$$

In 2^{m-1}—degree "-1" indicates that we take only half of the combinations (even or odd) of those combinations of coefficients that are indicated in parentheses.

Requirements for coefficients at $T = 3$ and $T = 2$:

$a_{33} = 1$—from the oddness of the sum of the coefficients at $T = 3$ for $k = 3$;

$a_{22} = 0$—from the parity of the sum of the coefficients at $T = 3$ for $k = 2$;

$a_{11} + a_{13} =$ an odd number, because with $T = 3$ for $k = 1$ the sum of the coefficients.

$a_{11} + a_{13} + a_{33}$—even, taking into account $a_{33} = 1$;

$a_{24} + a_{44} =$ an odd number, because with $T = 2$ for $k = 2$ the sum of the coefficients.

$a_{22} + a_{24} + a_{44}$—odd, taking into account $a_{22} = 0$;

to a_{14} also apply restrictions of parity or oddness depending on the selected a_{11}, a_{14} at $T = 3$ for $k = 1$, with the condition of oddness of the sums $a_{11} + a_{13}$ and $a_{24} + a_{44}$;

the coefficients a_{12},a_{23},a_{24} are not subject to any requirements.

Therefore, $\Theta_{3//2}$ is formed as follows:

$$\begin{array}{llllllll} \Theta_{3//2} = 2^{1-1} + & 2^{1-1} + & 2^{2-1} + & 2^{2-1} + & 2^{1-1} + & 2^3 & = 2^5 \\ (a_{22}) & (a_{33}) & (a_{11}, a_{13}) & (a_{24}, a_{44}) & (a_{14}) & (a_{12}, a_{23}, a_{34}). & \end{array}$$

Using the obtained values, we write Θ for $L = 4$ in numerical form:

$$\Theta^{L=4} = 2^9 + 2^8 + 2^7 + 2^6 - \left(2^7 + 2^5 + 2^6 + 0 + 2^4 + 2^5\right) + 2^4 + 0 + 0 + 2^3 - 0 = 712.$$

Solution for L = 5. In this case, the significant coefficients a_{ij} will be:

$$\begin{array}{ccccc} \multicolumn{5}{c}{T = 1} \\ \underline{a_{11}} & \underline{a_{12}} & \underline{a_{13}} & \underline{a_{14}} & \underline{a_{15}} \\ & \underline{a_{22}} & \underline{a_{23}} & \underline{a_{24}} & \underline{a_{25}} \\ & & \underline{a_{33}} & \underline{a_{34}} & \underline{a_{35}} \\ & & & \underline{a_{44}} & \underline{a_{45}} \\ & & & & \underline{a_{55}} \end{array} \quad \begin{array}{ccccc} \multicolumn{5}{c}{T = 2} \\ \underline{a_{11}} & a_{12} & \underline{a_{13}} & a_{14} & \underline{a_{15}} \\ & \underline{\underline{a_{22}}} & a_{23} & \underline{\underline{a_{24}}} & a_{25} \\ & & \underline{a_{33}} & a_{34} & \underline{a_{35}} \\ & & & \underline{\underline{a_{44}}} & a_{45} \\ & & & & \underline{a_{55}} \end{array} \quad \begin{array}{ccccc} \multicolumn{5}{c}{T = 3} \\ \underline{a_{11}} & a_{12} & a_{13} & \underline{a_{14}} & a_{15} \\ & \underline{\underline{a_{22}}} & a_{23} & a_{24} & \underline{\underline{a_{25}}} \\ & & \underline{\underline{\underline{a_{33}}}} & a_{34} & a_{35} \\ & & & \underline{a_{44}} & a_{45} \\ & & & & \underline{\underline{a_{55}}} \end{array}$$

$$
\begin{array}{ccccc}
\multicolumn{5}{c}{T=4}\\
\underline{a_{11}} & a_{12} & a_{13} & a_{14} & \underline{a_{15}}\\
 & \underline{\underline{a_{22}}} & a_{23} & a_{24} & a_{25}\\
 & & \underline{\underline{\underline{a_{33}}}} & a_{34} & a_{35}\\
 & & & \underset{\bullet\bullet\bullet\bullet}{a_{44}} & a_{45}\\
 & & & & \underline{a_{55}}
\end{array}
\qquad
\begin{array}{ccccc}
\multicolumn{5}{c}{T=15}\\
\underline{a_{11}} & a_{12} & a_{13} & a_{14} & a_{15}\\
 & \underline{\underline{a_{22}}} & a_{23} & a_{24} & a_{25}\\
 & & \underline{\underline{\underline{a_{33}}}} & a_{34} & a_{35}\\
 & & & \underset{\bullet\bullet\bullet\bullet}{a_{44}} & a_{45}\\
 & & & & \underset{\bullet-\bullet-}{a_{55}}
\end{array}
$$

One line $\left(\underline{a_{ij}}\right)$ emphasizes significant coefficients for $k=1$ in (13.11), two dashes $\left(\underline{\underline{a_{ij}}}\right)$—for $k=2$, three $\left(\underline{\underline{\underline{a_{ij}}}}\right)$—for $k=3$, dots $\left(\underset{\bullet\bullet\bullet\bullet}{a_{ij}}\right)$—for $k=4$, dotted line $\left(\underset{\bullet-\bullet-}{a_{ij}}\right)$—for $k=5$, with the corresponding T. inclusions and exclusions (13.16) in this case will take the form:

$$
\begin{aligned}
\Theta^{L=5} &= \Theta_1+\Theta_2+\Theta_3+\Theta_4+\Theta_5\\
&-\left(\Theta_{2//1}+\Theta_{3//2}+\Theta_{3//1}+\Theta_{4//3}+\Theta_{4//2}\right.\\
&\left.+\Theta_{4//1}+\Theta_{5//4}+\Theta_{5//3}+\Theta_{5//2}+\Theta_{5//1}\right)\\
&+\Theta_{3//2//1}+\Theta_{4//3//2}+\Theta_{4//3//1}+\Theta_{4//2//1}+\Theta_{5//4//3}\\
&+\Theta_{5//4//2}+\Theta_{5//4//1}+\Theta_{5//3//2}+\Theta_{5//3//1}+\Theta_{5//2//1}\\
&-\left(\Theta_{4//3//2//1}+\Theta_{5//4//3//2}+\Theta_{5//4//3//1}+\Theta_{5//4//2//1}+\Theta_{5//3//2//1}\right)\\
&+\Theta_{5//4//3//2//1},
\end{aligned}
$$

The given designations are similar to the designations considered in the previous cases. The sets Θ_m and $\Theta_{...//1}$ are calculated in the same way as in the previous cases.

As in the previous calculations, there are empty intersections of subsets: $\Theta_{4//3}$, $\Theta_{5//3}$, $\Theta_{5//4}$ and their intersections with other sets. Indeed, for $T=5$ for $k=3,4$ the necessary condition will be $a_{33}=0$ and $a_{44}=0$. For $T=4$ for $k=3$ the necessary condition will be $a_{33}=0$, and for $k=4a_{44}=1$, which is incompatible with $T=5$ for $k=4$. For $T=3$ for $k=3$, the necessary condition will be $a_{33}=1$, which is incompatible with $T=5$ and $T=4$ for $k=3$.

Let us dwell in more detail on the other intersecting sets. To reduce the record, we will indicate in parentheses those coefficients a_{ij}, which are combinations and conditions that affect the required sets.

$$
\begin{aligned}
&\Theta_{3//2}\\
&=\begin{pmatrix} 2^{1-1} \\ a_{33} \\ T=3\,k=3\end{pmatrix}\cdot\begin{pmatrix} 2^{3-1} \\ a_{22}a_{24}a_{44} \\ T=2\,k=2\end{pmatrix}\cdot\begin{pmatrix} 2^{2-1} \\ a_{35}a_{55} \\ T=3\,k=2\end{pmatrix}\cdot\begin{pmatrix} 2^{2-1} \\ a_{11}a_{14} \\ T=3\,k=1\end{pmatrix}\cdot\begin{pmatrix} 2^{3-1} \\ a_{13}a_{15}a_{35} \\ T=2\,k=1\end{pmatrix}\cdot\begin{pmatrix} 2^{4} \\ a_{12}a_{23}a_{34}a_{34} \end{pmatrix}\\
&=2^{10}
\end{aligned}
$$

$\Theta_{4//2}$

$$= \underset{\begin{pmatrix} a_{22} \\ T=4k=2 \end{pmatrix}}{2^{1-1}} \cdot \underset{\begin{pmatrix} a_{33} \\ T=4k=3 \end{pmatrix}}{2^{1-1}} \cdot \underset{\begin{pmatrix} a_{44} \\ T=4k=4 \end{pmatrix}}{2^{1-1}} \cdot \underset{\begin{pmatrix} a_{24} \\ T=2k=2 \end{pmatrix}}{2^{1-1}} \cdot \underset{\begin{pmatrix} a_{11}a_{15}a_{35} \\ T=4k=1 \end{pmatrix}}{2^{3-1}}$$

$$\cdot \underset{\begin{pmatrix} a_{13}a_{35} \\ T=4k=2 \end{pmatrix}}{2^{2-1}} \cdot \underset{\begin{pmatrix} a_{12}a_{14}a_{23}a_{25}a_{34}a_{45} \end{pmatrix}}{2^{6}} = 2^{9}$$

$\Theta_{5//2}$

$$= \underset{\begin{pmatrix} a_{11} \\ T=5k=1 \end{pmatrix}}{2^{1-1}} \cdot \underset{\begin{pmatrix} a_{22} \\ T=5k=2 \end{pmatrix}}{2^{1-1}} \cdot \underset{\begin{pmatrix} a_{33} \\ T=5k=3 \end{pmatrix}}{2^{1-1}}$$

$$\cdot \underset{\begin{pmatrix} a_{44} \\ T=5k=4 \end{pmatrix}}{2^{1-1}} \cdot \underset{\begin{pmatrix} a_{55} \\ T=5k=4 \end{pmatrix}}{2^{1-1}} \cdot \underset{\begin{pmatrix} a_{24} \\ T=2k=2 \end{pmatrix}}{2^{1-1}}$$

$$\cdot \underset{\begin{pmatrix} a_{13}, a_{15}, a_{34} \\ T=2k=1 \end{pmatrix}}{2^{3-1}} \cdot \underset{\begin{pmatrix} a_{12}, a_{14}, a_{23}, a_{25}, a_{34}, a_{45} \end{pmatrix}}{2^{6}} = 2^{8}$$

We group all the obtained values of the sets.

$$\Theta_1 = 2^{14},$$
$$\Theta_2 = 2^{13}, \quad \Theta_{2//1} = 2^{12},$$
$$\Theta_3 = 2^{12}, \quad \Theta_{3//1} = 2^{11},$$
$$\Theta_{3//2} = 2^{10}, \quad \Theta_{3//2//1} = 2^{9},$$
$$\Theta_4 = 2^{11}, \quad \Theta_{4//1} = 2^{10},$$
$$\Theta_{4//2} = 2^{9}, \quad \Theta_{4//2//1} = 2^{8},$$
$$\Theta_5 = 2^{10}, \quad \Theta_{5//1} = 2^{9},$$
$$\Theta_{5//2} = 2^{8}, \quad \Theta_{5//2//1} = 2^{7},$$
$$\Theta_{4//3} = 0, \quad \Theta_{4//3//2} = 0,$$
$$\Theta_{4//3//1} = 0, \quad \Theta_{4//3//2//1} = 0,$$
$$\Theta_{5//3} = 0, \quad \Theta_{5//3//2} = 0,$$
$$\Theta_{5//3//1} = 0, \quad \Theta_{5//3//2//1} = 0,$$
$$\Theta_{5//4} = 0, \quad \Theta_{5//4//3} = 0, \quad \Theta_{5//4//2} = 0,$$
$$\Theta_{5//4//1} = 0, \quad \Theta_{5//4//3//2} = 0, \quad \Theta_{5//4//3//1} = 0,$$
$$\Theta_{5//4//3//2//1} = 0.$$

Thus, the total number of combinations of coefficients a_{ij}, which meet the Requirement *8*, for $L = 5$ will be:

$$\begin{aligned}\Theta^{L=5} &= 2^{14} + 2^{13} + 2^{12} + 2^{11} + 2^{10} \\ &\quad - \left(2^{12} + 2^{10} + 2^{11} + 0 + 2^{9} + 2^{10} + 0 + 0 + 2^{8} + 2^{10}\right) \\ &\quad + 2^{9} + 0 + 0 + 2^{8} + 0 + 0 + 0 + 0 + 0 + 2^{7} \\ &\quad - (0 + 0 + 0 + 0 + 0) + 0 = 23168\end{aligned}$$

Solution for $L = 6$. The formula of inclusions and exclusions (13.16) in this case takes the form:

$$\begin{aligned}\Theta^{L=6} &= \Theta_1 + \Theta_2 + \Theta_3 + \Theta_4 + \Theta_5 + \Theta_6 \\ &- (\Theta_{2//1} + \Theta_{3//2} + \Theta_{3//1} + \Theta_{4//3} + \Theta_{4//2} + \Theta_{4//1} \\ &+ \Theta_{5//4} + \Theta_{5//3} + \Theta_{5//2} + \Theta_{5//1} \\ &+ \Theta_{6//5} + \Theta_{6//4} + \Theta_{6//3} + \Theta_{6//2} + \Theta_{6//1}) \\ &+ \Theta_{3//2//1} + \Theta_{4//3//2} + \Theta_{4//3//1} + \Theta_{4//2//1} \\ &+ \Theta_{5//4//3} + \Theta_{5//4//2} + \Theta_{5//4//1} + \Theta_{5//3//2} + \Theta_{5//3//1} + \Theta_{5//2//1} \\ &+ \Theta_{6//5//4} + \Theta_{6//5//3} + \Theta_{6//5//2} + \Theta_{6//5//1} + \Theta_{6//4//3} \\ &+ \Theta_{6//4//2} + \Theta_{6//4//1} + \Theta_{6//3//2} + \Theta_{6//3//1} + \Theta_{6//2//1} \\ &- (\Theta_{4//3//2//1} + \Theta_{5//4//3//2} + \Theta_{5//4//3//1} + \Theta_{5//4//2//1} + \Theta_{5//3//2//1} \\ &+ \Theta_{6//5//4//3} + \Theta_{6//5//4//2} + \Theta_{6//5//4//1} + \Theta_{6//5//3//2} + \Theta_{6//5//3//1} + \Theta_{6//5//2//1} \\ &+ \Theta_{6//4//3//2} + \Theta_{6//4//3//1} + \Theta_{6//4//2//1} + \Theta_{6//3//2//1}) + \Theta_{5//4//3//2//1} \\ &+ \Theta_{6//5//4//3//2} + \Theta_{6//5//4//3//1} + \Theta_{6//5//4//2//1} + \Theta_{6//5//3//2//1} + \Theta_{6//4//3//2//1} \\ &- \Theta_{6//5//4//3//2//1}.\end{aligned}$$

The empty intersection of subsets will be:
$\Theta_{5//4} = 0$, since $a_{44} = 1$ for $T = 4$ for $k = 4$ and $a_{44} = 0$ for $T = 5$ for $k = 4$, and the corresponding sets intersecting with other subsets:

$$\begin{aligned}\Theta_{5//4//3} &= \Theta_{5//4//2} \\ &= \Theta_{5//4//1} = \Theta_{5//4//3//2} = \Theta_{5//4//3//1} \\ &= \Theta_{5//4//2//1} = \Theta_{5//4//3//2//1} = 0;\end{aligned}$$

$\Theta_{6//4} = 0$ since $a_{44} = 1$ for $T = 4$ for $k = 4$ and $a_{44} = 0$ for $T = 6$ for $k = 4$, and the corresponding sets intersecting with other subsets:

$$\begin{aligned}\Theta_{6//4//3} &= \Theta_{6//4//2} = \Theta_{6//4//1} \\ &= \Theta_{6//4//3//2} = \Theta_{6//4//3//1} = \Theta_{6//4//2//1} \\ &= \Theta_{6//4//3//2//1} = 0; \Theta_{6//5} = 0,\end{aligned}$$

since $a_{55} = 1$ for $T = 5$ for $k = 5$ and $a_{55} = 0$ for $T = 6$ for $k = 5$, and the corresponding sets intersecting with other subsets:

$$\begin{aligned}\Theta_{6//5//4} &= \Theta_{6//5//3} = \Theta_{6//5//2} = \Theta_{6//5//1}\\ &= \Theta_{6//5//4//3} = \Theta_{6//5//4//2} = \Theta_{6//5//4//1}\\ &= \Theta_{6//5//3//2} = \Theta_{6//5//3//1} = \Theta_{6//5//2//1}\\ &= \Theta_{6//5//4//3//2} = \Theta_{6//5//4//3//1} = \Theta_{6//5//4//2//1}\\ &= \Theta_{6//5//3//2//1} = \Theta_{6//5//4//3//2//1} = 0.\end{aligned}$$

The sets Θ_m and $\Theta_{...//1}$ are calculated in the same way as in the previous cases. Here is the calculation of other intersecting sets. For a more compact record, in parentheses, we will also indicate the coefficients a_{ij} that make up the combinations and the conditions that affect these combinations.

$$\Theta_{3//2} = \underset{\begin{pmatrix} a_{33}, a_{36}, a_{66} \\ T=3\,k=3 \end{pmatrix}}{2^{3-1}} \cdot \underset{\begin{pmatrix} a_{22}, a_{25}, a_{55} \\ T=3\,k=2 \end{pmatrix}}{2^{3-1}} \cdot \underset{\begin{pmatrix} a_{11}, a_{14}, a_{44} \\ T=3\,k=1 \end{pmatrix}}{2^{3-1}} \cdot \underset{\begin{pmatrix} a_{24}, a_{26}, a_{46} \\ T=2\,k=2 \end{pmatrix}}{2^{3-1}} \cdot \underset{\begin{pmatrix} a_{23}, a_{25}, a_{35} \\ T=3\,k=1 \end{pmatrix}}{2^{3-1}} \cdot \underset{\begin{pmatrix} a_{22}, a_{23}, a_{34} \\ a_{45}, a_{56}, a_{16} \end{pmatrix}}{2^{6}} = 2^{16}$$

$$\Theta_{4//2} = \underset{\begin{pmatrix} a_{44} \\ T=4\,k=4 \end{pmatrix}}{2^{1-1}} \cdot \underset{\begin{pmatrix} a_{33} \\ T=4\,k=3 \end{pmatrix}}{2^{1-1}} \cdot \underset{\begin{pmatrix} a_{22}a_{26}a_{66} \\ T=4\,k=2 \end{pmatrix}}{2^{3-1}} \cdot \underset{\begin{pmatrix} a_{11}a_{15}a_{55} \\ T=4\,k=1 \end{pmatrix}}{2^{3-1}} \cdot \underset{\begin{pmatrix} a_{13}a_{35} \\ T=2\,k=1 \end{pmatrix}}{2^{2-1}} \cdot \underset{\begin{pmatrix} a_{24}a46 \\ T=2\,k=2 \end{pmatrix}}{2^{2-1}} \cdot \underset{\begin{pmatrix} a_{12}a_{14}a_{16}a_{23}a_{25} \\ a_{34}a_{36}a_{45}a_{56} \end{pmatrix}}{2^{9}} = 2^{15}$$

$$\Theta_{5//2} = \underset{\begin{pmatrix} a_{55} \\ T=5\,k=5 \end{pmatrix}}{2^{1-1}} \cdot \underset{\begin{pmatrix} a_{44} \\ T=5\,k=4 \end{pmatrix}}{2^{1-1}} \cdot \underset{\begin{pmatrix} a_{33} \\ T=5\,k=3 \end{pmatrix}}{2^{1-1}} \cdot \underset{\begin{pmatrix} a_{22} \\ T=5\,k=2 \end{pmatrix}}{2^{1-1}} \cdot \underset{\begin{pmatrix} a_{11}a_{16}a_{66} \\ T=5\,k=1 \end{pmatrix}}{2^{3-1}} \cdot \underset{\begin{pmatrix} a_{13}a_{15}a_{35} \\ T=2\,k=1 \end{pmatrix}}{2^{3-1}} \cdot \underset{\begin{pmatrix} a_{24}a_{26}a_{46} \\ T=2\,k=2 \end{pmatrix}}{2^{3-1}} \cdot \underset{\begin{pmatrix} a_{12}a_{14}a_{23}a_{25} \\ a_{34}a_{36}a_{45}a_{56} \end{pmatrix}}{2^{8}} = 2^{14}$$

$$\Theta_{6//2} = \underset{\begin{pmatrix} a_{66} \\ T=6k=6 \end{pmatrix}}{2^{1-1}} \cdot \underset{\begin{pmatrix} a_{55} \\ T=6k=5 \end{pmatrix}}{2^{1-1}} \cdot \underset{\begin{pmatrix} a_{44} \\ T=6k=4 \end{pmatrix}}{2^{1-1}} \cdot \underset{\begin{pmatrix} a_{33} \\ T=6k=3 \end{pmatrix}}{2^{1-1}} \cdot \underset{\begin{pmatrix} a_{22} \\ T=6k=2 \end{pmatrix}}{2^{1-1}} \cdot \underset{\begin{pmatrix} a_{11} \\ T=6k=1 \end{pmatrix}}{2^{1-1}} \cdot \underset{\begin{pmatrix} a_{13}a_{15}a_{35} \\ T=2\,k=1 \end{pmatrix}}{2^{3-1}}$$

$$\cdot \overset{2^{3-1}}{\begin{pmatrix} a_{24}2_{26}a_{46} \\ T=2k=2 \end{pmatrix}} \cdot \overset{2^{9}}{\begin{pmatrix} a_{12}a_{14}a_{26}a_{23}a_{25} \\ a_{34}a_{36}a_{45}a_{56} \end{pmatrix}} = 2^{13}$$

$$\begin{aligned}\Theta_{4//3} &= \overset{2^{1-1}}{\begin{pmatrix} a_{44} \\ T=4k=4 \end{pmatrix}} \cdot \overset{2^{1-1}}{\begin{pmatrix} a_{33} \\ T=4k=3 \end{pmatrix}} \cdot \overset{2^{3-1}}{\begin{pmatrix} a_{22}a_{26}a_{66} \\ T=4k=2 \end{pmatrix}} \cdot \overset{2^{3-1}}{\begin{pmatrix} a_{11}a_{15}a_{35} \\ T=4k=1 \end{pmatrix}} \\ &\cdot \overset{2^{1-1}}{\begin{pmatrix} a_{25} \\ T=3k=2 \end{pmatrix}} \cdot \overset{2^{1-1}}{\begin{pmatrix} a_{36} \\ T=3k=3 \end{pmatrix}} \cdot \overset{2^{10}}{\begin{pmatrix} a_{12}a_{13}a_{23}a_{24}a_{34} \\ a_{35}a_{45}a_{46}a_{56}a_{16} \end{pmatrix}} = 2^{14}\end{aligned}$$

$$\begin{aligned}\Theta_{5//3} &= \overset{2^{1-1}}{\begin{pmatrix} a_{55} \\ T=5k=5 \end{pmatrix}} \cdot \overset{2^{1-1}}{\begin{pmatrix} a_{44} \\ T=5k=4 \end{pmatrix}} \cdot \overset{2^{1-1}}{\begin{pmatrix} a_{33} \\ T=5k=3 \end{pmatrix}} \cdot \overset{2^{1-1}}{\begin{pmatrix} a_{22} \\ T=5k=2 \end{pmatrix}} \\ &\cdot \overset{2^{3-1}}{\begin{pmatrix} a_{11}a_{16}a_{66} \\ T=5k=1 \end{pmatrix}} \cdot \overset{2^{1-1}}{\begin{pmatrix} a_{14} \\ T=3k=1 \end{pmatrix}} \cdot \overset{2^{1-1}}{\begin{pmatrix} a_{25} \\ T=3k=2 \end{pmatrix}} \cdot \overset{2^{1-1}}{\begin{pmatrix} a_{36} \\ T=3k=3 \end{pmatrix}} \\ &\cdot \overset{2^{11}}{\begin{pmatrix} a_{12}a_{13}a_{23}a_{24}a_{34} \\ a_{35}a_{45}a_{46}a_{56}a_{15}a_{16} \end{pmatrix}} = 2^{13}\end{aligned}$$

$$\begin{aligned}\Theta_{6//3} &= \overset{2^{1-1}}{\begin{pmatrix} a_{66} \\ T=6k=6 \end{pmatrix}} \cdot \overset{2^{1-1}}{\begin{pmatrix} a_{55} \\ T=6k=5 \end{pmatrix}} \\ &\cdot \overset{2^{1-1}}{\begin{pmatrix} a_{44} \\ T=6k=4 \end{pmatrix}} \cdot \overset{2^{1-1}}{\begin{pmatrix} a_{33} \\ T=6k=3 \end{pmatrix}} \cdot \overset{2^{1-1}}{\begin{pmatrix} a_{22} \\ T=6k=2 \end{pmatrix}} \cdot \overset{2^{1-1}}{\begin{pmatrix} a_{11}a_{16}a_{66} \\ T=6k=1 \end{pmatrix}} \\ &\cdot \overset{2^{1-1}}{\begin{pmatrix} a_{14} \\ T=3k=1 \end{pmatrix}} \cdot \overset{2^{1-1}}{\begin{pmatrix} a_{25} \\ T=3k=2 \end{pmatrix}} \cdot \overset{2^{1-1}}{\begin{pmatrix} a_{36} \\ T=3k=3 \end{pmatrix}} \cdot \overset{2^{12}}{\begin{pmatrix} a_{12}a_{13}a_{23}a_{24}a_{34}a_{35} \\ a_{45}a_{46}a_{56}a_{15}a_{16}a_{26} \end{pmatrix}} = 2^{12}\end{aligned}$$

$$\begin{aligned}\Theta_{4//3//2} &= \overset{2^{1-1}}{\begin{pmatrix} a_{44} \\ T=4k=4 \end{pmatrix}} \cdot \overset{2^{1-1}}{\begin{pmatrix} a_{33} \\ T=4k=3 \end{pmatrix}} \\ &\cdot \overset{2^{3-1}}{\begin{pmatrix} a_{22}a_{26}a_{66} \\ T=4k=2 \end{pmatrix}} \cdot \overset{2^{3-1}}{\begin{pmatrix} a_{11}a_{15}a_{55} \\ T=4k=1 \end{pmatrix}} \cdot \overset{2^{1-1}}{\begin{pmatrix} a_{14} \\ T=3k=1 \end{pmatrix}} \cdot \overset{2^{1-1}}{\begin{pmatrix} a_{25} \\ T=3k=2 \end{pmatrix}} \\ &\cdot \overset{2^{1-1}}{\begin{pmatrix} a_{36} \\ T=3k=3 \end{pmatrix}} \cdot \overset{2^{2-1}}{\begin{pmatrix} a_{13}a_{35} \\ T=3k=1 \end{pmatrix}} \cdot \overset{2^{2-1}}{\begin{pmatrix} a_{24}a_{46} \\ T=2k=2 \end{pmatrix}} \cdot \overset{2^{6}}{\begin{pmatrix} a_{12}a_{23}a_{34} \\ a_{45}a_{56}a_{16} \end{pmatrix}} = 2^{12}\end{aligned}$$

$$\Theta_{5//3//2} = \overset{2^{1-1}}{\begin{pmatrix} a_{55} \\ T=5k=5 \end{pmatrix}} \cdot \overset{2^{1-1}}{\begin{pmatrix} a_{44} \\ T=5k=4 \end{pmatrix}} \cdot \overset{2^{1-1}}{\begin{pmatrix} a_{33} \\ T=5k=3 \end{pmatrix}}$$
$$\cdot \overset{2^{1-1}}{\begin{pmatrix} a_{22} \\ T=5k=2 \end{pmatrix}} \cdot \overset{2^{3-1}}{\begin{pmatrix} a_{11}a_{16}a_{16} \\ T=5k=1 \end{pmatrix}} \cdot \overset{2^{1-1}}{\begin{pmatrix} a_{14} \\ T=3k=1 \end{pmatrix}} \cdot \overset{2^{1-1}}{\begin{pmatrix} a_{25} \\ T=3k=2 \end{pmatrix}}$$
$$\cdot \overset{2^{1-1}}{\begin{pmatrix} a_{36} \\ T=3k=3 \end{pmatrix}} \cdot \overset{2^{3-1}}{\begin{pmatrix} a_{13}a_{15}a_{35} \\ T=2k=1 \end{pmatrix}} \overset{2^{3-1}}{\begin{pmatrix} a_{24}a_{26}a_{46} \\ T=2k=2 \end{pmatrix}} \cdot \overset{2^{5}}{\begin{pmatrix} a_{12}a_{23}a_{34} \\ a_{45}a_{46} \end{pmatrix}} = 2^{11}$$

$$\Theta_{6//3//2}$$
$$= \overset{2^{1-1}}{\begin{pmatrix} a_{66} \\ T=6k=6 \end{pmatrix}} \cdot \overset{2^{1-1}}{\begin{pmatrix} a_{55} \\ T=6k=5 \end{pmatrix}} \cdot \overset{2^{1-1}}{\begin{pmatrix} a_{44} \\ T=6k=4 \end{pmatrix}} \cdot \overset{2^{1-1}}{\begin{pmatrix} a_{33} \\ T=6k=3 \end{pmatrix}} \cdot \overset{2^{1-1}}{\begin{pmatrix} a_{22} \\ T=6k=2 \end{pmatrix}}$$
$$\cdot \overset{2^{1-1}}{\begin{pmatrix} a_{11} \\ T=6k=1 \end{pmatrix}} \cdot \overset{2^{1-1}}{\begin{pmatrix} a_{14} \\ T=3k=3 \end{pmatrix}} \cdot \overset{2^{1-1}}{\begin{pmatrix} a_{25} \\ T=3k=2 \end{pmatrix}} \overset{2^{1-1}}{\begin{pmatrix} a_{36} \\ T=3k=1 \end{pmatrix}}$$
$$\cdot \overset{2^{3-1}}{\begin{pmatrix} a_{13}a_{15}a_{35} \\ T=2k=1 \end{pmatrix}} \cdot \overset{2^{3-1}}{\begin{pmatrix} a_{24}a_{26}a_{46} \\ T=2k=2 \end{pmatrix}} \cdot \overset{2^{6}}{\begin{pmatrix} a_{12}a_{23}a_{34} \\ a_{45}a_{56}a_{16} \end{pmatrix}} = 2^{10}$$

We group all the obtained nonzero values of sets.

$$\Theta_1 = 2^{14},$$
$$\Theta_2 = 2^{13}, \quad \Theta_{2//1} = 2^{12},$$
$$\Theta_3 = 2^{12}, \quad \Theta_{3//1} = 2^{11}, \quad \Theta_{3//2} = 2^{10}, \quad \Theta_{3//2//1} = 2^{9},$$
$$\Theta_4 = 2^{11}, \quad \Theta_{4//1} = 2^{10}, \quad \Theta_{4//2} = 2^{9}, \quad \Theta_{4//2//1} = 2^{8},$$
$$\Theta_5 = 2^{10}, \quad \Theta_{5//1} = 2^{9}, \quad \Theta_{5//2} = 2^{8}, \quad \Theta_{5//2//1} = 2^{7},$$
$$\Theta_{4//3} = 0, \quad \Theta_{4//3//2} = 0, \quad \Theta_{4//3//1} = 0, \quad \Theta_{4//3//2//1} = 0,$$
$$\Theta_{5//3} = 0, \quad \Theta_{5//3//2} = 0, \quad \Theta_{5//3//1} = 0, \quad \Theta_{5//3//2//1} = 0,$$
$$\Theta_{5//4} = 0, \quad \Theta_{5//4//3} = 0, \quad \Theta_{5//4//2} = 0, \quad \Theta_{5//4//1} = 0,$$
$$\Theta_{5//4//3//2} = 0, \quad \Theta_{5//4//3//1} = 0,$$
$$\Theta_{5//4//3//2//1} = 0.$$

Using the obtained results, we calculate the total number of combinations of coefficients a_{ij} that do not meet Requirement *8*, for $L = 6$.

$$\Theta^{L=6} = 2^{20} +^{19} + 2^{18} + 2^{17} + 2^{16} + 2^{15}$$
$$- \left(2^{18} + 2^{16} + 2^{17} + 2^{14} + 2^{15} + 2^{16} + 0 + 2^{13} + 2^{14} + 2^{15} + 0 + 0 + 2^{12} + 2^{13} + 2^{14}\right)$$
$$+ 2^{15} + 2^{12} + 2^{13} + 2^{14} + 0 + 0 + 0 + 2^{11} + 2^{12} + 2^{13}$$
$$+ 0 + 0 + 0 + 0 + 0 + 0 + 0 + 2^{10} + 2^{11} + 2^{12}$$

$$-\left(2^{11}+0+0+0+2^{10}+0+0+0+0+0+0+0+0+0+2^{9}\right)$$
$$+0+0+0+0+0+0-0=1,484,288$$

13.3.5 Calculating the Complete Set of Feedback Combinations that Do not Meet Requirement 8

Let us pay attention to some regularities when constructing the complete set Θ.

Each set Θ^{L} contains all the coefficients of the formula of inclusion and exclusion, which make up the set Θ^{L-1}. And the value of the coefficients for L will correspond to the values of the coefficients for $L-1$, multiplied by 2^{L}. That is $\Theta_1^{L}=\Theta_1^{L-1}\cdot 2^{L}$, $\Theta_2^{L}=\Theta_2^{L-1}\cdot 2^{L}$, $\Theta_{2//1}^{L}=\Theta_{2//1}^{L-1}\cdot 2^{L}$ and beyond. Indeed, increasing the dimension of the coefficient matrix by one is equivalent to adding the elements $a_{1L}, a_{2L},\ldots,a_{LL}$, which is equivalent to adding L elements to all possible combinations of significant and non-significant sets. This property makes it possible to reduce the record, using for calculations the already calculated set of combinations for the previous value of L. We present this property in formulaic form, using (13.16).

For $L=n$:

$$\left|\bigcup_{i=1}^{n}A_i^n\right|=\sum_i\left|A_i^n\right|$$
$$-\sum_{i<j}\left|A_i^n\cap A_j^n\right|+\sum_{i<j<k}\left|A_i^n\cap A_j^n\cap A_k^n\right|-\cdots+(-1)^{n-1}\left|A_1^n\cap A_2^n\cap\ldots\cap A_n^n\right|$$

For $L=n+1$:

$$\left|\bigcup_{i=1}^{n+1}A_i^{n+1}\right|$$
$$=\sum_i\left|A_i^{n+1}\right|-\sum_{i<j}\left|A_i^{n+1}\cap A_j^{n+1}\right|$$
$$+\sum_{i<j<k}\left|A_i^{n+1}\cap A_j^{n+1}\cap A_k^{n+1}\right|-\cdots+(-1)^{(n+1)-1}\left|A_1^{n+1}\cap A_2^{n+1}\cap\ldots\cap A_n^{n+1}\right|$$
$$=\left|\bigcup_{i=1}^{n}A_i^{n+1}\right|+\left|A_{n+1}^{n+1}\right|-\sum_{i<n+1}\left|A_i^{n+1}\cap A_{n+1}^{n+1}\right|$$
$$+\sum_{i<j<n+1}\left|A_i^{n+1}\cap A_j^{n+1}\cap A_{n+1}^{n+1}\right|-\cdots+(-1)^{(n+1)-1}\left|A_1^{n+1}\cap A_2^{n+1}\cap\ldots\cap A_{n+1}^{n+1}\right|$$
$$=\left|\bigcup_{i=1}^{n}A_i^{n}\right|\cdot 2^{n+1}+\left|A_{n+1}^{n+1}\right|-\sum_{i<n+1}\left|A_i^{n+1}\cap A_{n+1}^{n+1}\right|$$
$$+\sum_{i<j<n+1}\left|A_i^{n+1}\cap A_j^{n+1}\cap A_{n+1}^{n+1}\right|-\cdots+(-1)^{(n+1)-1}\left|A_1^{n+1}\cap A_2^{n+1}\cap\ldots\cap A_{n+1}^{n+1}\right|$$

The following pattern.

There will be empty sets that intersect between the subsets Θ_i and Θ_k (i.e., the sets $\Theta_{i//k}$), provided that $i > k > L/2$. Indeed, for $k > L/2$ there will be only one significant coefficient for $T = k$, which according to the oddness of Requirement 8 must be equal to one, i.e. $a_{kk} = 1$, but the same significant coefficient (because it is unique for $k > L/2$) has be even (i.e. $a_{kk} = 0$) for $T = i$. Therefore, these two conditions are simultaneously unfeasible. Note that there will also be empty sets that intersect this set with all the others, i.e. $\Theta_{i//k//\ldots} = 0$.

This case was discussed in detail above. The above feature allows you to reduce the formula of inclusions and exclusions, removing from it the empty intersecting subsets:

$$\left|\bigcup_{i=1}^{L} A_i^L\right| = \left|\bigcup_{i=1}^{L-1} A_i^{L-1}\right| \cdot 2^L + \left|A_L^L\right| - \sum_{i\le\frac{L}{2}} \left|A_i^L \bigcap A_L^L\right| + \sum_{i<j\le\frac{L}{2}} \left|A_i^L \bigcap A_j^L \bigcap A_L^L\right| - \cdots + (-1)^{L-1} \left|A_1^L \bigcap A_2^L \bigcap \ldots \bigcap A^L_{floor\left(\frac{L}{2}\right)} \bigcap A_L^L\right|.$$

Or in the entered notation:

$$\Theta^L = \Theta^{L-1} \cdot 2^L + \Theta_L - \sum_{i\le\frac{L}{2}} \Theta_{L//i} + \sum_{i<j\le\frac{L}{2}} \Theta_{L//i//j} - \ldots + (-1)^{L-1}\Theta_{L//floor\left(\frac{L}{2}\right)//\ldots//1}$$

In the given expression Θ_L will be calculated by the relation (13.15).

After analyzing the numerical values of intersecting sets, it is possible to establish a pattern in the ratios of their quantities. Here is it:

$$\begin{aligned}
\Theta_{L//1} &= \Theta_L \cdot 2^{-1} = 2^{nL-T-1}, \\
\Theta_{L//2} &= \Theta_{L//1} \cdot 2^{-1} = \Theta_L \cdot 2^{-2} = 2^{nL-T-2}, \\
\Theta_{L//3} &= \Theta_{L//2} \cdot 2^{-1} = 2^{nL-T-3}, \\
\Theta_{L//i} &= 2^{nL-T-i}, \\
\Theta_{L//2//1} &= \Theta_{L//2} \cdot 2^{-1} = 2^{nL-T-2-1}, \\
\Theta_{L//3//1} &= \Theta_{L//3} \cdot 2^{-1} = 2^{nL-T-3-1}, \\
\Theta_{L//3//2} &== \Theta_{L//3//1} \cdot 2^{-1} = 2^{nL-T-3-2}, \\
\Theta_{L//i//j} &= 2^{nL-T-i-j}.
\end{aligned}$$

Or summarizing, we obtain a formula for quantifying the sets of intersecting *i, j, …,k* subsets:

$$\Theta_{i//j//...//k} = 2^{nL-i-j-...-k}.$$

The above formula will represent a complete set using the coefficients of all possible combinations for a given L:

$$\Theta^L = 2^{n_L} \cdot \left(\sum_{i=1}^{L} 2^{-i} - \sum_{i=2}^{L} \sum_{j=1}^{(i-1)\leq\frac{L}{2}} 2^{-i-j} + \sum_{i=3}^{L} \sum_{j=2}^{(i-1)\leq\frac{L}{2}} \sum_{k=1}^{(j-1)} 2^{-i-j-k} - \sum_{i=4}^{L} \sum_{j=3}^{(i-1)\geq\frac{L}{2}} \sum_{k=2}^{(j-1)} \sum_{t=1}^{(k-1)} 2^{-i-j-k-t} + \cdots \right).$$

Thus, we obtained a formula for calculating the complete set of combinations a_{ij} that do not satisfy Requirement *8* (Λ^8).

$$\begin{aligned}
\Theta^{L=7} = 2^{28} \cdot [&2^{-1} + 2^{-2} + 2^{-3} + 2^{-4} + 2^{-5} + 2^{-6} + 2^{-7} \\
&- (2^{-2-1} + 2^{-3-1} + 2^{-3-2} + 2^{-4-1} \\
&+ 2^{-4-2} + 2^{-4-3} + 2^{-5-1} + 2^{-5-2} + 2^{-5-3} + 0^{-5-4} + 2^{-6-1} \\
&+ 2^{-6-2} + 2^{-6-3} + 0^{-6-4} + 0^{-6-5} + 2^{-7-1} \\
&+ 2^{-7-2} + 2^{-7-3} + 0^{-7-4} + 0^{-7-5} + 0^{-7-6}) \\
&+ (2^{-3-2-1} + 2^{-4-2-1} + 2^{-4-3-1} + 2^{-4-3-2} + 2^{-5-2-1} \\
&+ 2^{-5-3-1} + 2^{-5-3-2} + 0^{-5-4-1} + 0^{-5-4-2} + 0^{-5-4-3} \\
&+ 2^{-6-2-1} + 2^{-6-3-1} + 2^{-6-3-2} + 0^{-6-4-1} + 0^{-6-4-2} \\
&+ 0^{-6-4-3} + 0^{-6-5-1} + 0^{-6-5-2} + 0^{-6-5-3} + 0^{-6-5-4} \\
&+ 2^{-7-2-1} + 2^{-7-3-1} + 2^{-7-3-2} + 0^{-7-4-1} + 0^{-7-4-2} \\
&+ 0^{-7-4-3} + 0^{-7-5-1} + 0^{-7-5-2} + 0^{-7-5-3} + 0^{-7-5-4} \\
&+ 0^{-7-6-1} + 0^{-7-6-2} + 0^{-7-6-3} + 0^{-7-6-4} + 0^{-7-6-5}) \\
&-(2^{-4-3-2-1} + 2^{-5-3-2-1} + 0^{-5-4-2-1} + 0^{-5-4-3-1} \\
&+ 0^{-5-4-2-2} + 2^{-6-3-2-1} + 0^{-6-4-2-1} + 0^{-6-4-3-1} \\
&+ 0^{-6-4-3-2} + 0^{-6-5-2-1} + 0^{-6-5-3-1} + 0^{-6-5-3-2} \\
&+ 0^{-6-5-4-1} + 0^{-6-5-4-2} + 0^{-6-5-4-3} + 0^{-7-3-2-1} \\
&+ 0^{-7-4-2-1} + 0^{-7-4-3-1} + 0^{-7-4-3-2} + 0^{-7-5-2-1} \\
&+ 0^{-7-5-3-1} + 0^{-7-5-3-2} + 0^{-7-5-4-1} + 0^{-7-5-4-2} \\
&+ 0^{-7-5-4-3} + 0^{-7-6-2-1} + 0^{-7-6-3-1} + 0^{-7-6-3-2} \\
&+ 0^{-7-6-4-1} + 0^{-7-6-4-2} + 0^{-7-6-4-3} + 0^{-7-6-5-1}
\end{aligned}$$

$$+0^{-7-6-5-2} +^{-7-6-5-3} + 0^{-7-6-5-4}) + (0^{-5-4-3-2-1}$$
$$+0^{-6-4-3-2-1} + 0^{-6-5-3-2-1} + 0^{-6-5-4-2-1}$$
$$+0^{-6-5-4-3-1} + 0^{-6-5-4-3-2} + 0^{-7-4-3-3-1}$$
$$+0^{-7-5-3-2-1} + 0^{-7-5-4-2-1} + 0^{-7-5-4-3-1} + 0^{-7-5-4-3-2}$$
$$+0^{-7-6-3-2-1} + 0^{-7-6-4-2-1} + 0^{-7-6-4-3-1}$$
$$+0^{-7-6-4-3-2} + 0^{-7-6-5-2-1} + 0^{-7-6-5-3-1} + 0^{-7-6-5-3-2}$$
$$+0^{-7-6-5-4-1} + 0^{-7-6-5-4-2} + 0^{-7-6-5-4-3})$$
$$-(0^{-6-5-4-3-2-1} + 0^{-7-5-4-3-2-1} + 0^{-7-6-4-3-2-1} + 0^{-7-6-5-3-2-1}$$
$$+0^{-7-6-5-4-2-1} + 0^{-7-6-5-4-3-1} + 0^{-7-6-5-4-3-2}) + 0^{-7-6-5-4-3-2-1}$$

The number of combinations for each set consisting of intersecting subsets is determined by the binomial coefficient formula.

Recall that in combinatorics, a *combination* of n through k is a set of k elements selected from a set containing n different elements. Sets that differ only in the order of passage of the elements (but not the composition) are considered the same.

The number of combinations from n to k is equal to the binomial coefficient:

$$C_n^k = \frac{n!}{k!(n-k)!}.$$

Thus, in our case for $L = 7$ the number of sets that intersect and participate in the formula of inclusion and exclusion can be calculated by the formula of the binomial coefficient:

$C_7^1 = \frac{7!}{1!(7-1)!} = \frac{5040}{1\cdot 720} = 7$ the number of subsets that do not intersect;

$C_7^2 = \frac{7!}{2!(7-2)!} = \frac{5040}{2\cdot 120} = 21$ the number of combinations for two intersecting subsets;

$C_7^3 = \frac{7!}{3!(7-3)!} = \frac{5040}{6\cdot 24} = 35$ the number of combinations for three intersecting subsets;

$C_7^4 = \frac{7!}{4!(7-4)!} = \frac{5040}{24\cdot 6} = 35$ the number of combinations for four intersecting subsets;

$C_7^5 = \frac{7!}{5!(7-5)!} = \frac{5040}{120\cdot 2} = 21$ the number of combinations for five intersecting subsets;

$C_7^6 = \frac{7!}{6!(7-6)!} = \frac{5040}{720\cdot 1} = 7$ number of combinations for six intersecting subsets;

$C_7^7 = \frac{7!}{7!(7-7)!} = \frac{5040}{5040\cdot 1} = 1$ the number of combinations for seven intersecting subsets.

Exception of empty sets of intersecting subsets. Empty (see remarks above) will be sets whose indices correspond to the condition $i > k > L/2$, i.e. for $L = 7$ it will be 7, 6, 5, 4 (only 4 sets).

Binomial coefficient $C_4^2 = \frac{4!}{2!(4-2)!} = \frac{24}{2\cdot 2} = 6$ calculation is the number of combinations that have an empty intersecting set for two subsets. In this case, the non-empty subsets that intersect will remain $21 - 6 = 15$.

The case of three intersecting subsets. Empty intersecting sets (from sets 4, 5, 6 and 7) will total $C_4^2 = 6$ and they can intersect with sets 1, 2 and 3, as well as give an empty intersecting set, i.e. all of them will be $6 \bullet 3 = 18$ To this number must be added $C_4^3 = \frac{4!}{3!(4-3)!} = \frac{24}{6\cdot 1} = 4$ the number of combinations that have empty sets that intersect for three subsets (4, 5, 6, and 7). Thus, only $18 + 4 = 22$ intersecting empty sets for three subsets, or $35 - 22 = 13$ non-empty ones.

The case of four intersecting subsets. Empty two sets (of sets 4, 5, 6 and 7) that intersect will have a total of $C_4^2 = 6$, and they can intersect with $C_3^2 = \frac{3!}{2!(3-2)!} = \frac{6}{2\cdot 1} = 3$ different combinations of sets 1, 2 and 3 and also give an empty intersecting set, i.e. all of them will be $6 \bullet 3 = 18$. To this number must be added $C_4^3 = \frac{4!}{3!(4-3)!} = \frac{24}{6\cdot 1} = 4$ the number of combinations that have an empty intersecting set for three subsets (4, 5, 6 and 7), which can also intersect with $C_3^1 = \frac{3!}{1!(3-1)!} = \frac{6}{1\cdot 2} = 3$ different combinations of 1, 2 and 3 sets and also give an empty intersecting set, i.e. there will be $4 \bullet 3 = 12$ in total. And it is necessary to add $C_4^4 = \frac{4!}{4!(4-4)!} = \frac{24}{24\cdot 1} = 1$ the number of combinations that have an empty intersecting set for four subsets (4, 5, 6 and 7). Thus, only $18 + 12 + 1 = 31$ intersecting empty sets for four subsets or $35 - 31 = 4$ not empty.

The case of five intersecting subsets. Calculate similarly.

$C_4^2 = 6$ (from sets *4, 5, 6* and *7*), $C_3^3 = \frac{3!}{3!(3-3)!} = \frac{6}{6\cdot 1} = 1$ (from sets 1, 2 and 3), total $6 \bullet 1 = 6$. $C_4^3 = 4$ (from sets 4, 5, 6 and 7), $C_3^2 = 3$ (from sets 1, 2 and 3), total $4 \bullet 3 = 12$. $C_4^4 = 1$ (from sets 4, 5, 6 and 7), $C_3^1 = 3$ (from sets 1, 2 and 3), total $1 \bullet 3 = 3$. In total we have $6 + 12 + 3 = 21$ intersecting empty sets for five subsets, i.e. all.

The case of six intersecting subsets. Calculate similarly. $C_4^3 = 4$ (from sets 4, 5, 6 and 7), $C_3^3 = 1$ (from sets 1, 2 and 3), total $4 \bullet 1 = 4$. $C_4^4 = 1$ (from sets 4, 5, 6 and 7), $C_3^2 = 3$ (from sets 1, 2 and 3), only $1 \bullet 3 = 3$. In total we have $4 + 3 = 7$ intersecting empty sets for six subsets, i.e. all.

The case of seven intersecting subsets. Calculate similarly. $C_4^4 = 1$ (from sets 4, 5, 6 and 7), $C_3^3 = 1$ (from sets 1, 2 and 3), a total of $1 \bullet 1 = 1$ intersecting empty sets for seven subsets, i.e. all.

Solution for L = 7.

$$\Theta^{L=2} = \Theta_1 + \Theta_2 - \Theta_{2//1}$$

$$\Theta^{L=2} = 2^2 + 2^1 - 2^0 = 5,$$

$$\Theta^{L=3} = \Theta^{L=2} \cdot 2^3 + \Theta_3 - \Theta_{3//1}$$

$$\Theta^{L=3} = 5 \cdot 2^3 + 2^3 - 2^2 = 44,$$

$$\Theta^{L=4} = \Theta^{L=3} \cdot 2^4 + \Theta_4 - (\Theta_{4//2} + \Theta_{4//1}) + \Theta_{4//2//1} - \Theta_{3//2} + \Theta_{3//2//1}$$

$$\Theta^{L=4} = 44 \cdot 2^4 + 2^6 - \left(2^4 + 2^5\right) + 2^3 - 2^5 + 2^4 = 712,$$

$$\Theta^{L=5} = \Theta^{L=4} \cdot 2^5 + \Theta_5 - (\Theta_{5//2} + \Theta_{5//1}) + \Theta_{5//2//1}$$

$$\Theta^{L=5} = 712 \cdot 2^5 + 2^{10} - \left(2^8 + 2^9\right) + 2^7 = 23\,168,$$

$$\Theta^{L=6} =$$

$$\Theta^{L=5} \cdot 2^6 + \Theta_6 - (\Theta_{6//3} + \Theta_{6//2} + \Theta_{6//1}) + \Theta_{6//3//2} + \Theta_{6//3//1} + \Theta_{6//2//1} - \Theta_{6//3//2//1}$$
$$- (\Theta_{4//3} + \Theta_{5//3}) + \Theta_{4//3//2} + \Theta_{4//3//1} + \Theta_{5//3//2} + \Theta_{5//3//1}$$

$$-(\Theta_{4//3//2//1}+\Theta_{5//3//2/1})$$

$$\Theta^{L=6}=$$

$$23\,168\cdot 2^{6}+2^{15}-\left(2^{12}+2^{13}+2^{14}\right)+2^{10}+2^{11}+2^{12}-2^{9}-\left(2^{14}+2^{13}\right)+2^{12}+2^{13}$$
$$+2^{11}+2^{12}-\left(2^{11}+2^{10}\right)=1\,484\,288,$$

$$\Theta^{L=7}=$$

$$\Theta^{L=6}\cdot 2^{7}+\Theta_{7}-(\Theta_{7//3}+\Theta_{7//2}+\Theta_{7//1})+\Theta_{7//3//2}+\Theta_{7//3//1}+\Theta_{7//2//1}-\Theta_{7//3//2//1}$$

$$\Theta^{L=7}=1\,484\,288\cdot 2^{7}+2^{21}-\left(2^{18}+2^{19}+2^{20}\right)+2^{16}+2^{17}+2^{18}-2^{15}=190\,676\,992.$$

Solution for L = 8.

$$\Theta^{L=8}=$$

$$\Theta^{L=7}\cdot 2^{8}+\Theta_{8}-(\Theta_{8//4}+\Theta_{8//3}+\Theta_{8//2}+\Theta_{8//1})+\Theta_{8//4//3}+\Theta_{8//4//2}+\Theta_{8//4//1}$$
$$+\Theta_{8//3//2}+\Theta_{8//3//1}+\Theta_{8//2//1}-(\Theta_{8//4//3//2}+\Theta_{8//4//3//1}+\Theta_{8//4//2//1}+\Theta_{8//3//2//1})$$
$$+\Theta_{8//4//3//2//1}-(\Theta_{5//4}+\Theta_{6//4}+\Theta_{7//4})+\Theta_{5//4//3}+\Theta_{5//4//2}+\Theta_{5//4//1}+\Theta_{6//4//3}$$
$$+\Theta_{6//4//2}+\Theta_{6//4//1}+\Theta_{7//4//3}+\Theta_{7//4//2}+\Theta_{7//4//1}-(\Theta_{5//4//3//2}+\Theta_{5//4//3//1}$$
$$+\Theta_{5//4//2//1}+\Theta_{6//4//3//2}+\Theta_{6//4//3/1}+\Theta_{6//4//2//1}+\Theta_{7//4//3//2}+\Theta_{7//4//3//1}$$
$$+\Theta_{7//4//2//1})+\Theta_{5//4//3//2//1}+\Theta_{6//4//3//2//1}+\Theta_{7//4//3//2//1}$$

$$\Theta^{L=8}=$$

$$190676992\cdot 2^{8}+2^{28}-\left(2^{24}+2^{25}+2^{26}+2^{27}\right)+2^{21}+2^{22}+2^{23}+2^{23}+2^{24}+2^{25}$$
$$-\left(2^{19}+2^{20}+2^{21}+2^{22}\right)+2^{18}-\left(2^{27}+2^{26}+2^{25}\right)+2^{24}+2^{25}+2^{26}+2^{23}+2^{24}+2^{25}$$
$$+2^{22}+2^{23}+2^{24}-\left(2^{22}+2^{23}+2^{24}+2^{21}+2^{22}+2^{23}+2^{20}+2^{21}+2^{22}\right)$$
$$+2^{21}+2^{22}+2^{19}$$
$$=48\,818\,814\,976.$$

To calculate the number of NLFSR, which satisfies Requirement 8, it is necessary to subtract from the total number of possible NLFSR (Λ_0) the calculated values of Θ, or, returning to the previously entered notation Λ^8:

For $L=2: 2^{n_L}-\Lambda^8=2^3-5=3$	$\frac{5}{2^3}=0.625$
For $L=3: 2^{n_L}-\Lambda^8=2^6-44=20$	$\frac{44}{2^6}=0.6875$
For $L=4: 2^{n_L}-\Lambda^8=2^{10}-712=312$	$\frac{712}{2^{10}}=0.6953125$
For $L=5: 2^{n_L}-\Lambda^8=2^{15}-23,168=9600$	$\frac{23,168}{2^{15}}=0.70703125$
For $L=6: 2^{n_L}-\Lambda^8=2^{21}-1,484,288=612,864$	$\frac{1,484,288}{2^{21}}=0.70776367185$
For $L=7: 2^{n_L}-\Lambda^8=2^{28}-190,676,992=77,758,464$	$\frac{1,190,676,992}{2^{28}}=0.7103271484375$
For $L=8: 2^{n_L}-\Lambda^8=2^{36}-48,818,814,976=$ 19,900,661,760	$\frac{48,818,814,976}{2^{36}}\approx 0.71040725708$

As you can see, the ratio of the number of NLFSR that do not meet the Requirement 8, to the total set of possible NLFSR, for a given size L, tends to a value close to 0.71. The above values for $L = 2, 3, 4, 5, 6, 7$ are fully confirmed by a computational experiment.

13.4 Requirement 9

13.4.1 Description Requirements 9

When considering the whole set of sequences that can generate NLFSR, there will be rings with periods less than the maximum possible. If we exclude all rings with a period less than $T_{\max} = 2^L - 1$, excluding the corresponding combinations a_{ij}, then there will be only a set of sequences (with the corresponding a_{ij}), which will be M-sequences. Thus, to find the set of combinations a_{ij} that generate the M-sequence, it is necessary and sufficient to determine and exclude the whole set of combinations a_{ij} that will generate sequences with periods T belonging to the interval $1 \leq T < T_{\max.}$

For certainty, we note that each sequence in the period will begin with 1 and, accordingly, end with 0. The next one after the final zero will be the beginning of the next period. This definition has one exception when $T = 1$, in which case the sequence in the period will consist of one 1 (and accordingly end also 1).

In the general case, using the introduced notation system (see Fig. 11.1), the feedback for the NLFSR, at time t, can be set as:

$$q_1(t+1) = \sum_{i=1}^{L} a_{ii} q_i(t) + \sum_{i=1}^{L-1} \sum_{j=i+1}^{L} a_{ij} q_i(t) q_j(t),$$

and the sequence that is generated:

$$Q = \{q_1(t), q_1(t+1), q_1(t+2), \ldots, q_1(t+i)\}.$$

It is obvious that the state of the register cell at time t corresponds to the value generated by the NLFSR at time $t - i$. Therefore, the states of their cells in the register are determined $q_1(t) = q_1(t + 1 - i)$.

The generated sequence with the period T is written in the form: $Q_T = q_{i+1}, q_{i+2}, q_{i+3}, \ldots, q_{i+T}, q_{i+1}, \ldots$ sequence with period T, can be written as:

$$\begin{gathered}\left[q_{i+1}, q_{i+2}, q_{i+3}, \ldots, q_{i+T}, q_{i+1}, \ldots, q_{i+L}\right] \text{ for } T < L; \\ \left[q_{i+1}, q_{i+2}, q_{i+3}, \ldots, q_L\right] \text{ for } T \geq L.\end{gathered}$$

Thus, the set of combinations of cell states in the register, or, equivalently, the generated sequences, can be divided into subsets of states that correspond to certain periods. As an example, write all possible sequences with $T < T_{max}$ for $L = 3\,(Tmax = 2^3 - 1 = 7)$, grouping them according to the periods:

$T = 1$	$T = 5$	$T = 6$
$Q^1_{T=1} = \underline{1}111111$	$Q^1_{T=5} = \underline{10000}10$	$Q^1_{T=6} = 1000001$
$T = 2$	$Q^2_{T=5} = \underline{11000}11$	$Q^2_{T=6} = 1100001$
$Q^1_{T=2} = \underline{10}10101$	$Q^3_{T=5} = \underline{10100}10$	$Q^3_{T=6} = 1010001$
$T = 3$	$Q^4_{T=5} = \underline{11100}11$	$Q^4_{T=6} = 1110001$
$Q^1_{T=3} = \underline{100}1001$	$Q^5_{T=5} = \underline{10010}10$	$Q^5_{T=6} = 1001001$
$Q^2_{T=3} = \underline{110}1101$	$Q^6_{T=5} = \underline{11010}11$	$Q^6_{T=6} = 1101001$
	$Q^7_{T=5} = \underline{10110}10$	$Q^7_{T=6} = 1011001$
$T = 4$	$Q^8_{T=5} = \underline{11110}11$	$Q^8_{T=6} = 1111001$
$Q^1_{T=4} = \underline{1000}100$		$Q^9_{T=6} = 1000101$
$Q^2_{T=4} = \underline{1100}110$		$Q^{10}_{T=6} = 1100101$
$Q^3_{T=4} = \underline{1010}101$		$Q^{11}_{T=6} = 1010101$
$T^4_{T=4} = \underline{1110}111$		$Q^{12}_{T=6} = 1110101$
		$Q^{13}_{T=6} = 1001101$
		$Q^{14}_{T=6} = 1101101$
		$Q^{15}_{T=6} = 1011101$
		$Q^{16}_{T=6} = 1111101$

The condition under which any of the periods will be repeated is the generation of the register value $q_i = q_i + T$ for all $i = 1, \ldots, T$. If the register, taken with arbitrary coefficients a_{ij}, satisfies the above condition, it is possible to state unequivocally, that such NLFSR generates a ring with period T.

By setting certain cell states, we check what value the register generates for the selected feedback coefficients. The test is performed using the appropriate *patterns*. Denote the pattern as, where k is the ordinal number of the pattern in the sequence for a given period T, which is checked.

Let's make templates of each of iterations in sequence of the above example.

For $T = 1$:

$$S^{1/1}_{T=1} = \begin{matrix} a_{11} = 1 & a_{12} = 1 & a_{13} = 1 & \\ & a_{22} = 1 & a_{23} = 1 & \Rightarrow 1 \\ & & a_{33} = 1 & \end{matrix}$$

Under $a_{ij} = 1$ we mean a significant coefficient a_{ij}, i.e. such a coefficient, the value of which affects the formation of the output bit. If $a_{ij} = 0$, then the specified coefficient is not significant, i.e. does not affect the formation of the original value. By $\Rightarrow 1$ we mean that the pattern must generate 1 or, equivalently, the sum of all significant coefficients a_{ij} must be an odd number. If $\Rightarrow 0$, then the template should generate 1, which is equivalent to the even sum of all significant coefficients a_{ij}.

Given the notation entered, here are the templates for $T = 2$:

$$S_{T=2}^{1/1} = \begin{array}{c} 1\,0\,1 \\ 0\,0 \\ 1 \end{array} \Rightarrow 0 \quad S_{T=2}^{1/2} = \begin{array}{c} 0\,0\,0 \\ 1\,0 \\ 0 \end{array} \Rightarrow 1$$

Thus, if the condition for patterns is fulfilled in NLFSR at the same time and (i.e. evenness of the sum of coefficients $a_{11} + a_{13} + a_{33}$ and oddness of coefficient a_{22}, and given that it is unique, therefore, $a_{22} = 1$), which is a necessary and sufficient condition for the formation of NLFSR length $L = 3$ sequence $Q = 10101010\ldots$.

Patterns for $T = 3$ have the form:

$$S_{T=3}^{1/1} = \begin{array}{c} 1\,0\,0 \\ 0\,0 \\ 0 \end{array} \Rightarrow 0 \quad S_{T=3}^{1/2} = \begin{array}{c} 0\,0\,0 \\ 1\,0 \\ 0 \end{array} \Rightarrow 0 \quad S_{T=3}^{1/3} = \begin{array}{c} 0\,0\,0 \\ 0\,0 \\ 1 \end{array} \Rightarrow 0$$

$$S_{T=3}^{2/1} = \begin{array}{c} 1\,1\,0 \\ 1\,0 \\ 0 \end{array} \Rightarrow 0 \quad S_{T=3}^{2/2} = \begin{array}{c} 0\,0\,0 \\ 1\,1 \\ 1 \end{array} \Rightarrow 1 \quad S_{T=3}^{2/3} = \begin{array}{c} 1\,0\,1 \\ 0\,0 \\ 1 \end{array} \Rightarrow 1$$

Patterns for T $= 4$ have the form:

$$S_{T=4}^{1/1} = \begin{array}{c} 1\,0\,0 \\ 0\,0 \\ 0 \end{array} \Rightarrow 0 \quad S_{T=4}^{1/2} = \begin{array}{c} 0\,0\,0 \\ 1\,0 \\ 0 \end{array} \Rightarrow 0 \quad S_{T=4}^{1/3} = \begin{array}{c} 0\,0\,0 \\ 0\,0 \\ 1 \end{array} \Rightarrow 0 \quad S_{T=4}^{1/4} = \begin{array}{c} 0\,0\,0 \\ 0\,0 \\ 0 \end{array} \Rightarrow 1$$

$$S_{T=4}^{2/1} = \begin{array}{c} 1\,1\,0 \\ 1\,0 \\ 0 \end{array} \Rightarrow 0 \quad S_{T=4}^{2/2} = \begin{array}{c} 0\,0\,0 \\ 1\,1 \\ 1 \end{array} \Rightarrow 0 \quad S_{T=4}^{2/3} = \begin{array}{c} 0\,0\,0 \\ 0\,0 \\ 1 \end{array} \Rightarrow 1 \quad S_{T=4}^{2/4} = \begin{array}{c} 1\,0\,0 \\ 0\,0 \\ 0 \end{array} \Rightarrow 1$$

$$S_{T=4}^{3/1} = \begin{array}{c} 1\,0\,1 \\ 0\,0 \\ 1 \end{array} \Rightarrow 0 \quad S_{T=4}^{3/2} = \begin{array}{c} 0\,0\,0 \\ 1\,0 \\ 0 \end{array} \Rightarrow 1 \quad S_{T=4}^{3/3} = \begin{array}{c} 1\,0\,1 \\ 0\,0 \\ 1 \end{array} \Rightarrow 0 \quad S_{T=4}^{3/4} = \begin{array}{c} 0\,0\,0 \\ 1\,0 \\ 0 \end{array} \Rightarrow 1$$

$$S_{T=4}^{4/1} = \begin{array}{c} 1\,1\,1 \\ 1\,1 \\ 1 \end{array} \Rightarrow 0 \quad S_{T=4}^{4/2} = \begin{array}{c} 0\,0\,0 \\ 1\,1 \\ 1 \end{array} \Rightarrow 1 \quad S_{T=4}^{4/3} = \begin{array}{c} 1\,0\,1 \\ 0\,0 \\ 1 \end{array} \Rightarrow 1 \quad S_{T=4}^{4/4} = \begin{array}{c} 1\,1\,0 \\ 1\,0 \\ 0 \end{array} \Rightarrow 1$$

Note that the templates $S_{T=4}^{1/k}$ and the corresponding sequence $Q = 10001000\ldots$ cannot be physically implemented according to the NLFSR design. Since filling all the cells of the register with zeros is a forbidden state and cannot form a unit at the output by any coefficients.

Also note that the sequence $Q_{T=4}^{3}$ and its corresponding patterns $S_{T=4}^{3/k}$ are actually the sum of the two subperiods $Q_{T=2}^{1}$ and correspond to the patterns $S_{T=2}^{1/1}$ and $S_{T=2}^{1/2}$.

Patterns for other periods are built similarly to the above. Other patterns also include patterns and their corresponding sequences, which either cannot be implemented due to the design features of the NLFSR, or they can be presented in the form of other periods.

You can use an alternative. Compile templates only for the sequences of the maximum period and check the NLFSR for compliance with these templates. However, this approach is equivalent to constructing a De Bruijn sequence and in practice for large values of L, as noted in Sect. 1, is difficult to calculate.

Summarizing the above example, we see that in order to weed out all possible combinations of feedback factors for $L = 3$, which will generate sequences with periods smaller than the maximum possible, it is necessary to analyze the type of NLFSR for compliance with nine patterns (*1* for $T = 1$; 1 for $T = 2$; 2 for $T = 3$; 2 for $T = 4$; 1 for $T = 5$; 2 for $T = 6$). Or compose and analyze two sets of templates for the maximum period.

For large values of L, starting at about $L = 15$, working with an array of templates for $T = T_{max}$ is a task that is difficult to implement for personal computers that use only this method. Moreover, the time spent on checking the patterns, significantly exceeds the time spent checking the period of the generated sequence.

13.4.2 Joint Application of Requirement 9 with Other Requirements

The test results of Requirement 9 for different values of T at $L = 7$ are shown in Table 13.9.

As can be seen from the table, the application of Requirement 1 excludes approximately half of the whole set under test—it is a sequence from No. 1 to No. 60.

Requirement 2 includes the complete set of combinations of Requirement 9 at $T = 1$, which corresponds to the sequence (1111 ...).

Requirement 3 significantly reduces (approximately 2^{L-1}) the set tested for compliance with Requirement 9.

Requirement 4 slightly reduces the set tested for compliance with Requirement 9. In Table 13.9 of the 36,866,264 combinations that meet Requirement 9 alone, the introduction of Requirement 4 reduced this set to 36,823,804 combinations, which is slightly more than 0.1%.

Requirement 5 also slightly reduces the set tested for compliance with Requirement 9: from 36,866,264 combinations that meet Requirement 9 only, Requirement 5 reduced this set to 36,546,060 combinations, which is approximately *0.9%*. In addition, the introduction of Requirement 5 completely excludes from the testing of Requirement 9 combinations of the form $(q_1 = 1, q_2 = 0, q_3 = 0, \ldots, q_L = 0)$ at $T = L$, because the rings of this type completely contradict the definition of Requirement 5.

Table 13.9 Test results of requirement 9 for different T at L = 7

No.	Test sequence	Period T	Number of combinations a_{ij} that do not meet the requirement 9		Number of combinations a_{ij} that do not meet the Requirement 9, taking into account the Requirements 1, 3, 5 та 7	Note
			Number	Relatively complete set (%)		
1	1	1	134,217,728	50.00	0	Requirement 2 is deleted
2	10	2	33,554,432	12.50	254,208	
3	100	3	12,582,912	469	95,264	
4	110	3	12,582,912	4.69	95,264	
5	1000	4	4,718,592	1.76	36,288	
6	1100	4	5,898,240	2.20	44,544	
7	1110	4	3,538,944	1.32	27,072	
8	10,000	5	2,027,520	0.76	15,456	
9	11,000	5	1,853,568	0.69	12,747	
10	10,100	5	1,790,208	0.67	11,308	
11	11,100	5	1,797,360	0.67	12,375	
12	10,010	5	0	0	0	Condition 3 requirement 9
13	11,010	5	1,740,912	0.65	11,046	
14	10,110	5	0	0	0	Condition 3 requirement 9
15	11,110	5	1,841,136	0.69	14,304	
16	100,000	6	859,840	0.32	6450	
17	110,000	6	798,319	0.30	5323	
18	101,000	6	841,509	0.31	6340	
19	111,000	6	571,336	0.21	4360	
20	110,100	6	732,709	0.27	4867	
21	101,100	6	720,309	0.27	4791	
22	111,100	6	787,727	0.29	5450	
23	100,010	6	0	0	0	Condition 3 requirement 9
24	110,010	6	0	0	0	Condition 3 requirement 9
25	111,010	6	743,002	0.28	6020	
26	100,110	6	0	0	0	Condition 3 requirement 9
27	101,110	6	0	0	0	Condition 3 requirement 9

(continued)

Table 13.9 (continued)

No.	Test sequence	Period T	Number of combinations a_{ij} that do not meet the requirement 9		Number of combinations a_{ij} that do not meet the Requirement 9, taking into account the Requirements 1, 3, 5 та 7	Note
			Number	Relatively complete set (%)		
28	111,110	6	763,376	0.28	5750	
29	1,000,000	7	375,528	0.14	0	Requiment 5 is deleted
30	1,100,000	7	357,532	0.13	2411	
31	1,010,000	7	362,174	0.13	2493	
32	1,110,000	7	353,565	0.13	2621	
33	1,001,000	7	422,293	0.16	2702	
34	1,101,000	7	0	0.00	0	Condition 5 requirement 9
35	1,011,000	7	0	0	0	Condition 5 requirement 9
36	1,111,000	7	346,761	0.13	2556	
37	1,000,100	7	0	0	0	Condition 3 requirement 9
38	1,100,100	7	421,657	0.16	2905	
39	1,010,100	7	444,925	0.17	2742	
40	1,110,100	7	675,028	0.25	4596	
41	1,001,100	7	0	0	0	Condition 3 requirement 9
42	1,101,100	7	410,955	0.15	2865	
43	1,011,100	7	657,040	0.24	4493	
44	1,111,100	7	325,793	0.12	2120	
45	1,000,010	7	0	0	0	Condition 3 requirement 9
46	1,100,010	7	0	0	0	Condition 3 requirement 9 or Condition 5 requirement 9
47	1,010,010	7	0	0	0	Condition 3 requirement 9
48	1,110,010	7	0	0	0	Condition 3 requirement 9

(continued)

Table 13.9 (continued)

No.	Test sequence	Period T	Number of combinations a_{ij} that do not meet the requirement 9		Number of combinations a_{ij} that do not meet the Requirement 9, taking into account the Requirements 1, 3, 5 та 7	Note
			Number	Relatively complete set (%)		
49	1,001,010	7	0	0	0	Condition 3 requirement 9
50	1,101,010	7	423,354	0.16	2664	
51	1,011,010	7	0	0	0	Condition 3 requirement 9
52	1,111,010	7	324,030	0.12	2103	
53	1,000,110	7	0	0	0	Condition 3 requirement 9
54	1,100,110	7	0	0	0	Condition 3 requirement 9
55	1,010,110	7	0	0	0	Condition 3 requirement 9
56	1,110,110	7	382,304	0.14	2368	
57	1,001,110	7	0	0	0	Condition 3 requirement 9
58	1,101,110	7	0	0	0	Condition 3 requirement 9
59	1,011,110	7	0	0	0	Condition 3 requirement 9
60	1,111,110	7	323,661	0.12	0	Condition 1 requirement 9

Requirement 6 is a special case of Requirement 9,·at $T = 2$ (sequence 1010101 …), so it contains the whole set of combinations that do not meet Requirement 6.

Requirement 7 also slightly intersects with the set of combinations a_{ij} that do not meet Requirement 9.

The full set of combinations of Requirement 8 (which includes Requirements 2 and 6) is included in Requirement 9.

13.4.3 Quantitative Assessment of the Application of the Requirements

In the Table 13.9 shows the results of calculations for NLFSR $L = 7$ to meet Requirement 9 for different variants of possible periods from 1 to L. The total number of possible combinations of a_{ij}, for this L, is 268,435,455. The total number of combinations of aij that did not pass Requirement 9 is 231,569,191 (86%). The remaining combinations (ie those that meet Requirement 9, with restrictions on the size of the test period) is 36,866,264 (13.7%). Additional application of Requirements 1, 3, 5 and 7 reduces the remaining set to 297,454 combinations, which corresponds to 0.11% of the total set.

Table 13.10 shows the results of calculations for NLFSR $L = 5$ for compliance only with Requirement 9 for different variants of possible periods from 1 to L (i.e.

Table 13.10 Test results of requirement 9 at $L = 5$

No.	Test sequence	Period T	Number of combinations a_{ij} that successfully passed the requirement 9 (taking into account the requirements 1 i 3)	
			Number	Relatively complete set (%)
Test results for individual T				
1	1	1	512	1.56
2	10	2	767	2.34
3	100	3	895	2.73
4	110	3	895	2.73
5	1000	4	959	2.93
6	1100	4	959	2.93
7	1110	4	959	2.93
8	10,000	5	960	2.93
9	11,000	5	992	3.03
10	10,100	5	992	3.03
11	11,100	5	992	3.03
12	11,010	5	992	3.03
Test results with simultaneous verification of all T of the same size				
3–4		3	783	2.39
5–7		4	846	2.58
8–12		5	860	2.62
Test results with simultaneous verification of all T from 1 to the specified value				
1–2		1–2	384	1.17
1–4		1–3	288	0.88
1–7		1–4	238	0.73
1–12		1–5	192	0.59

the depth of verification of compliance with Requirement 9). The total number of possible combinations of a_{ij}, for this L, is 32,768. The calculations were performed taking into account Requirements 1 and 3, which reduced the number of possible combinations of a_{ij} to 1024. The total number of RHLs that successfully passed compliance with Requirement 9 for the depth of verification to L, amounted to 192. Recall: for $L = 5$ the number of different M-NLFSR is 128.

As can be seen from Table 13.10, as the size of the test depth increases, the number of polynomials that do not pass Requirement 9 decreases significantly. As the depth of the test increases, the number of templates to be tested also increases proportionally, which increases the time to test. This allows you to justify the order of checks of the templates: from lower to higher value of T.

13.4.4 Conditions of the Zero Set Truncated by Requirement 9

There are situations when certain sequences (periods) do not cut off any sets of coefficients a_{ij}. Consider the following situations.

1. **Conditions for parity/oddness and Requirements 1–3**

The period $(q_1 = 1, q_2 = 1, q_3 = 1, \ldots, q_{L-1} = 1, q_L = 0)$ can generate a ring of the form

$$\begin{array}{c} 111\ldots10 \\ 011\ldots11 \\ 101\ldots11, \\ \ldots \\ 111\ldots01 \end{array}$$

a necessary condition for the existence of such a sequence is the parity of the coefficients a_{ij}:

$$\begin{array}{r} a_{11}a_{12}a_{13}\ldots a_{1(L-1)} \\ a_{22}a_{23}\ldots a_{2(L-1)} \\ \ldots \\ a_{(L-1)(L-1)} \end{array}.$$

Thus, only the coefficients of the last column, which are uniquely specified by Requirements 1 and 3, remain inactive in the pattern, and their sum is equal to one, i.e. an odd number. And the sum of even and odd numbers (in our case the whole set of coefficients a_{ij}) will always be an odd number, which contradicts Requirement 2 and, therefore, is already rejected. In Table 13.9, these are combinations under the ordinal No. 60.

2. **Condition of existence of subperiods**

The study period consists of subperiods that have already been tested and, accordingly, it will be a multiple of the expected value. As an example: for $L = 4$ we check the ring (1010), which can be represented as two subperiods from the sequence (10), respectively, the main period (i.e. the smallest period) will be equal to *2*, not the expected *4*; for $L = 6$ the ring (110,110) which can be given in the form of two subperiods (110) is checked, accordingly the basic period will be equal 3, instead of expected 6. In Table13.9 they are not given, as they were immediately excluded from consideration. What combinations are excluded by these conditions can be seen by analyzing the second column of this table.

3. **The condition of identical circles**

Of the possible initial states, there are sequences of periods that, at first glance, are different, but in detailed consideration—the same with precision to the shift. As an example, it is possible to consider NLFSR $L = 5$ and two periods, which are checked by Requirement 9: (10100) and (10010). They create two rings:

$$\begin{array}{ccc} 10100 & & 10010 \\ 01010 & & 01001 \\ 00101 & \text{and} & 10100 \\ 10010 & & 01010 \\ \underline{01001} & & \underline{00101} \\ 10100 & & 10010 \end{array}.$$

We see that these two rings have a period $T = 5$ and are identical to each other with some offset. Thus, one of the tested combinations will always have a zero set that is truncated, because all possible combinations of a_{ij} have already been tested during the previous test. In Table 13.9, these will be sequence tests under the ordinal No. 12, 14, 23, 24, 26, 27, 37, 41, 45, 46, 47, 48, 49, 51, 53, 54, 55, 57, 58, 59. A total of 20 sequences, which allowed to reduce by 1/3 the set of combinations tested. However, large values of T require a significant amount of time to exclude from consideration all combinations that meet this condition.

4. **Requirement Condition 5**

The combination No. 29 in Table 13.9, which corresponds to the period $(q_1 = 1, q_2 = 0, q_3 = 0, \ldots, q_L = 0)$, can be excluded from testing for compliance with Requirement 9 while testing to meet Requirement 5, because the Requirement 5 includes verification for such periods.

5. **The condition of parity**

There are cases when each of the templates for any period specifies the condition under which all without exception the coefficients a_{ij} will occur an odd number of times, which contradicts Requirement 2 or Requirement 9 with the period $L = 1$. As an example for $L = 7$ and $T = 6$ it there will be sequences (1101000) or (1011000).

6. **The condition of the ring**

This requirement can only be applied to circles with period $T > L$.

During the search of possible circles, forbidden combinations are observed, i.e. those that cannot be generated due to the design of the NLFSR itself. For $L = 5$ it will be, for example, a ring (1101000) with a period $T = 7$ of the form:

$$\begin{array}{c} 1001 \\ 0100 \\ 0010 \\ 0001 \\ 1000 \\ 0100 \\ \underline{0010} \\ 1001 \end{array}.$$

In the case of the identity of the third and seventh states of the NLFSR, it must generate different values, which is fundamentally impossible.

The introduction of the above conditions 1–5 in the calculation allowed to reduce the size of the array with templates and, as a consequence, the total number of sequences to be checked. Quantitative results of the introduction of conditions are given in Table 13.11. The values given with the previously performed optimizations in the calculation algorithm well demonstrate the possibilities of reducing the resources spent using the above conditions.

The test patterns of Requirement 9 need to be defined once for a given L, after which they can be applied during the tests for all combinations of coefficients a_{ij}.

13.4.5 Estimation of Spent Resources (RAM)

The number of possible periods for a given L (denote by i_L) consists of the sum of various sets of periods $T = 1, 2, \ldots, L$ (denote the number of such options for a single T by i_T). Accordingly, for each period, the number of templates (denoted as i_{TS}) increases with increasing L. Let's estimate the upper limit of this number.

By definition, the first value in the period should be 1 and the last 0. Therefore, these elements will be fixed, and all others can take any value. Thus we obtain that the maximum possible number for a given T will be determined by the relation:

Table 13.11 Total number of sequences from the requirements 9 patterns

L	3	4	5	6	7	8	9	10	11	12	13	14	15	16	17	18	19	20
Before the introduction of conditions	3	7	15	31	63	127	255	511	1023	2047	4095	8191	16,383	32,767	65,535	131,071	262,143	524,287
After the introduction of conditions	1	4	10	19	35	65	121	220	406	741	1371	2532	4712	8792	16,502	31,034	58,628	111,005

$$i_T \leq 2^{T-2}.$$

Note that the period $T = 1, 2$ is an exception. In both cases, only one period is possible—a sequence equal to $q_1 = 1$ (for $T = 1$) and $q_1 = 1,\ q_2 = 0$ (for $T = 0$). As a result, the number of possible periods for a given L can be calculated by the following ratio:

$$i_L = \sum_{k=1}^{L} i_{T=k}.$$

When checking for the ability of the NLFSR to generate any of the periods under test, you should check all possible combinations that will ensure the creation of a given ring. The number of templates corresponds to the value of the period being checked, i.e.:

$$i_{TS} = T.$$

Thus, the total number of patterns (denoted as S_i), which must be checked during the verification of the RHIF for compliance with Requirement 9, can be determined by the formula:

$$S_i = 1_{(k=1)} + 2_{(k=2)} + \sum_{k=3}^{L} \left(k \cdot 2^{k-2}\right).$$

During software implementation, when all patterns are placed in one array, the dimension of the array will be $2^{L-2} \cdot L \cdot (n_L + 1)$, where $n_L = L \cdot (L+1)/2$—the number of different coefficients a_{ij}. 1 is added to the value of n_L, because for each pattern you need to remember what number it should generate: 1 or 0.

As you can see, the dimension of this array increases with increasing L in degree, and already at $L = 20$ is measured in GB, which is a problem when implemented on personal computers due to limited memory.

Introducing various types of optimization into the algorithm allows you to significantly reduce the amount of memory spent to check each period, but not the general trend of increasing resources spent.

13.4.6 Estimation of Time Costs

Table 13.12 shows the time spent checking the whole set of combinations aij depending on the depth of verification of Requirement 9 and, accordingly, the number of patterns for different values of L. The result is given without checking the combinations themselves to generate M-sequence. In the Table 13.13 indicates the time

Table 13.12 Time spent on verification of requirement 9 (without checking the sequence period)

Depth of inspection, T	Time spent (s)					
	$L = 4$	$L = 5$	$L = 6$	$L = 7$	$L = 8$	$L = 9$[a]
0	<0.01	<0.01	<0.01	0.2813	38.7969	10,375
2	<0.01	<0.01	<0.01	0.3750	47.5313	12,750
3	<0.01	<0.01	<0.01	0.4063	59.4531	16,600
2–3	<0.01	<0.01	<0.01	0.4375	60.1719	16,559
4	<0.01	<0.01	<0.01	0.5000	68.4844	18,150
2–4	<0.01	<0.01	<0.01	0.5469	77.3125	21,949
5		<0.01	<0.01	0.7343	103.313	28,243
2–5		<0.01	<0.01	0.8438	112.375	31,942
6			<0.01	0.9219	137.063	38,658
2–6			0.0156	1.0781	156.906	44,493
7				1.4375	202.031	61,683
2–7				1.5469	219.219	62,335
8					361.078	102,751
2–8					296.984	95,796
9						186,261
2–9						143,883

[a]Estimated time obtained by forecasting

spent on similar tests, but with verification for the generation of M-sequences.

According to the results given in Tables 13.12 and 13.13, it is possible to draw the following conclusions:

1. The previously stated result is confirmed that with the increase of the register size the time for verification of Requirement 9 according to the power law increases.
2. Most combinations of coefficients a_{ij}, for a single period of the ring under test, also give rings, but with a smaller period. This explains the fact that the time spent on checking the templates only for $T = 8$ (*at* $L = 8$) is 1.2 times greater than the time spent checking the templates with $T = 2 - 8$, and indicates the priority of the sequence of checks, starting with smaller values of T.
3. Increasing the depth of verification of Requirement 9 does not reduce the time spent searching for M-NLFSR. During testing, for each L there is an optimal value of T. For $L = 7$, the optimal period is in the range of $T = 2–5$, for $L = 8$ within f($T = 2–6$ and for $L = 9$—at $T = 2–7$. As the number of checked periods increases, on the one hand, the number of combinations to be checked per M-sequence generation decreases. On the other hand, the number of patterns and, accordingly, the time to check them increases, which leads to an increase in the total time spent.

Table 13.13 Time spent on verification of requirement 9 (with sequence check)

Depth of inspection, T	$L = 4$		$L = 5$		$L = 6$		$L = 7$		$L = 8$		$L = 9$[a]
	Number of combinations	Time spent (s)	Number of combinations	Time spent (s)	Number of combinations	Time spent (s)	Number of combinations	Time spent (s)	Number of combinations	Time spent (s)	Time spent (s)
2–4	16	<0.01	208	<0.01	7216	0.047	463,680	3.438			
2–5			184	0.016	6039	0.047	386,444	3.250	49,915,956	701.84	343,301
2–6					984	0.047	337,093	3.484	43,618,366	675.47	322,810
2–7							297,454	3.609	38,011,514	694.98	309,319
2–8									34,169,520	763.52	320,032
2–9											345,337

[a]Estimated time obtained by forecasting

13.4.7 Optimization of Pattern Construction

Given Requirement 2 (the sum of all elements is an even number), it allows to consider two parts of it, chosen in any way, instead of the whole set of coefficients a_{ij}. If the sum of the first part is an even number, then the sum of the second is also even, and if the first is odd, then the second is also odd.

At $L = 20$, the amount of RAM in which the array of patterns is written exceeds 1 GB, and further increase in the size of the register makes it impossible to store the entire set of patterns.

To reduce the size of the array with patterns and increase the speed of the algorithm you need:

- In accordance with Requirements 1 and 3, do not write or check the L column (all elements in the column, except for the a_{LL} element, are zero), but calculate its parity. Therefore, make parity adjustments using the rule:

$$\begin{aligned}&(\text{even number}) + (\text{even number}) = (\text{even number});\\&(\text{even number}) + (\text{odd number}) = (\text{odd number});\\&(\text{odd number}) + (\text{odd number}) = (\text{even number}).\end{aligned}$$

  This optimization reduces the size of the array with templates in L times.

- Check the number of significant coefficients; if more than half of them, it is rational to check the rest. In this case, the above parity rule is also applied.

This optimization reduces the size of the array with templates in $L \bullet (L - 1) / 2$ times.

The introduction of this optimization into the algorithm slightly (≈15%) increases the speed of the algorithm and significantly reduces the size of the array with templates (in times, and the more L, the greater the saving of RAM).

13.4.8 Conclusions on the Application of the Requirement 9

The application of Requirement 9 in the analysis of the form and in the mutual arrangement of the feedback coefficients a_{ij} in the NLFSR allows, in theory, to exclude all registers that do not generate M-Sequence. The amount of excluded NLFSR is limited by the amount of available memory and time for inspection.

Sharing Requirement 9 with other Requirements can significantly reduce computational resources and increase the percentage of the set of NLFSR that do not generate a sequence. For For $L = 7$ verification of the type a_{ij} for compliance with Requirements 1, 2, 5, 7 and Requirement 9, at the depth of verification $T = 1, \ldots, L$, allowed to exclude from consideration 99.89% of NLFSR, which are guaranteed not to generate M-Sequence.

The practical significance of the presented technique is that it makes it possible to reduce the set of NLFSR under study, excluding registers that do not generate M-sequence, and thus significantly increase the search speed M-NLFSR.

13.5 Conclusions

1. Nine Requirements for feedback coefficients are formulated and substantiated. The essence of the Requirements is reduced to the analysis of the number, location in the register, the mutual location of the feedback coefficients in the NLFSR. Failure to comply with any of the nine Requirements indicates the inability of the NLFSR to form an M-sequence.
2. For Requirements 1 to 6, a strict estimate of the upper limit of the number of NLFSR, which do not meet Requirements 1–6 and clearly can not generate M-Sequence:

$$\begin{aligned}\Lambda^{1,2,3,4,5,6} &= 2^{n_L-1} + \frac{1}{2}\cdot 2^{n_L-1} \\ &+ \frac{1}{4}\cdot\Big\{2^{n_L-(L-1)}\cdot\left(2^{L-1}-1\right) \\ &+ \frac{1}{2^{L-1}}\cdot\Big(2^{\frac{L^2-k}{4}+L} + 2^{n_L-(L-1)} + 2^{n_L-2} \\ &- \Big[2^{\frac{L^2-k}{4}+L-\frac{L-k}{2}} + k\cdot 2\cdot 2^{\frac{L^2-k}{4}+L-2} + 2^{n_L-L-1} - k\cdot 2\cdot 2^{\frac{L^2-k}{4}-\frac{L-k}{2}+L-2}\Big]\Big)\Big\},\end{aligned}$$

where $k = 0$—for even L;
$k = 1$—for odd L.

3. It is shown that already at $L = 6$ more than 99% of NLFSR do not meet Requirements 1–4. When a polynomial of Requirements 1 to 6 is presented, more than 99.9% of the whole set of polynomials is cut off already at $L = 9$, and as the value of L increases, this percentage will increase in degree.
4. A quantitative assessment of the contribution of each of the Requirements from 1 to 6 to the rejection of NLFSR that do not generate M-Sequence is given. Recommendations for optimizing their application are given.
5. The formula for the complete set of combinations a_{ij}, which do not meet Requirement 8:

$$\Lambda^8 = 2^{n_L}$$

$$\cdot\left(\sum_{j=1}^{L}2^{-i}-\sum_{i=2}^{L}\sum_{j=1}^{(i-1)\geq\frac{L}{2}}2^{-i-j}+\sum_{i=3}^{L}\sum_{j=2}^{(i-1)\leq\frac{L}{2}}\sum_{k=1}^{(j-1)}2^{-i-j-k}\right.$$
$$\left.-\sum_{i=4}^{L}\sum_{j=3}^{(i-1)\leq\frac{L}{2}}\sum_{k=2}^{(j-1)}\sum_{t=1}^{(k-1)}2^{-i-j-k-t}+\cdots\right).$$

It is calculated that the ratio of the number of NLFSR, which do not meet the Requirement 8, to the total set of possible NLFSR, for a given value of L, tends to a number close to 0.71.

6. Based on the available results, the formula for estimating the number of M-NLFSR depending on L is given:

$$\Lambda_{\max}^{1}\leq 2^{\frac{L(L-1)}{2}-L+2}.$$

7. The application of Requirements 1–9 allowed the calculation of the exact number of M-NLFSR for values L from 1 to 9. As follows from the calculations, for $L=9$ the number of M-NLFSR is 519,239,794, which corresponds to 0.0076% of the total set of NLFSR and in 1,014,140 more than the whole set of NLFSR, or more in 10,817,496 than M-NLFSR.
8. The application of Requirement 9 makes it possible to completely exclude all NLFSR that do not generate M-sequence. However, for large L (20 and above), this task is more complex than the simple search of all NLFSR and verification of the sequence period that was generated. The quantitative results of the resources spent on testing Requirement 9—memory and time—are given.

Chapter 14
Method of Synthesis M-NLFSR. Hardware and Software Search M-NLFSR Second Order. List M-NLFSR

14.1 Development of Mathematical Apparatus Based on Solutions of the System of Linear Equations for Optimization of M-NLFSR Search

14.1.1 Restoration of NLFSR in a Known Sequence Using the SLAE Solution

In the general case, taking into account the introduced notation system (see Fig. 11.1), the feedback for the NLFSR can be given as follows:

$$q_1(t+1) = \sum_{i=1}^{L} a_{ii} q_i(t) + \sum_{i=1}^{L-1} \sum_{j=i+1}^{L} a_{ij} q_i(t) q_j(t) \tag{14.1}$$

As an example, we decompose expression (14.1) for NLFSR dimension $L = 3$ cells:

$$q_1(t+1) = a_{11} \cdot q_1(t) + a_{22} \cdot q_2(t) + a_{33} \cdot q_3(t) + a_{12} \cdot q_1(t) \cdot q_2(t) \\ + a_{13} \cdot q_1(t) \cdot q_3(t) + q_{23} \cdot q_2(t) \cdot q_3(t)$$

Or the same in matrix form:

$$\left(q_1(t)\ q_2(t)\ q_3(t)\ q_1(t) \cdot q_2(t)\ q_1(t) \cdot q_3(t)\ q_2(t) \cdot q_3(t) \right)$$

A. A. Kuznetsov et al., *Stream Ciphers in Modern Real-time IT Systems*, Studies in Systems, Decision and Control 375,
https://doi.org/10.1007/978-3-030-79770-6_14

$$\times \begin{pmatrix} a_{11} \\ a_{22} \\ a_{33} \\ a_{12} \\ a_{13} \\ a_{23} \end{pmatrix} = q_1(t+1)$$

Similar expressions can be written for the moment of time $(t+1)$, $(t+2)$ and so on. Thus, by linearizing the nonlinear part of Eq. (14.1) for our example, it is possible to make a system of six already linear equations.

$$\begin{pmatrix} q_1(t) & q_2(t) & q_3(t) & q_1(t)\cdot q_2(t) & q_1(t)\cdot q_3(t) & q_2(t)\cdot q_3(t) \\ q_1(t+1) & q_2(t+1) & q_3(t+1) & q_1(t+1)\cdot q_2(t+1) & q_1(t+1)\cdot q_3(t+1) & q_2(t+1)\cdot q_3(t+1) \\ q_1(t+2) & q_2(t+2) & q_3(t+2) & q_1(t+2)\cdot q_2(t+2) & q_1(t+2)\cdot q_3(t+2) & q_2(t+2)\cdot q_3(t+2) \\ q_1(t+3) & q_2(t+3) & q_3(t+3) & q_1(t+3)\cdot q_2(t+3) & q_1(t+3)\cdot q_3(t+3) & q_2(t+3)\cdot q_3(t+3) \\ q_1(t+4) & q_2(t+4) & q_3(t+4) & q_1(t+4)\cdot q_2(t+4) & q_1(t+4)\cdot q_3(t+4) & q_2(t+4)\cdot q_3(t+4) \\ q_1(t+5) & q_2(t+5) & q_3(t+5) & q_1(t+5)\cdot q_2(t+5) & q_1(t+5)\cdot q_3(t+5) & q_2(t+5)\cdot q_3(t+5) \end{pmatrix}$$

$$\times \begin{pmatrix} a_{11} \\ a_{22} \\ a_{33} \\ a_{12} \\ a_{13} \\ a_{23} \end{pmatrix} = \begin{pmatrix} q_1(t+1) \\ q_1(t+2) \\ q_1(t+3) \\ q_4(t+4) \\ q_1(t+5) \\ q_1(t+6) \end{pmatrix} \tag{14.2}$$

Assuming the feedback coefficients a_{ij} are unknown, it is possible to find them by solving the resulting system of linear algebraic equations (SLAE). To solve it, you need to know $\frac{L\cdot(L+1)}{2}+L$ the values of q_i. However, it is not necessary to take a continuous sequence of q_i values to solve SLAE, we can choose any, but belonging to one NLFSR, sequences with $L+1$, consecutive, q_i values, i.e. $\{q_0^k, q_1^k, q_2^k, \ldots, q_L^k\}$.

Taking into account the above, we can rewrite SLAE (14.2) as follows:

$$\begin{pmatrix} q_1^1 & q_2^1 & q_3^1 & q_1^1\cdot q_2^1 & q_1^1\cdot q_3^1 & q_2^1\cdot q_3^1 \\ q_1^2 & a_2^2 & q_3^2 & q_1^2\cdot q_2^2 & q_1^2\cdot q_3^2 & q_2^2\cdot q_3^2 \\ q_1^3 & q_2^3 & q_3^3 & q_1^3\cdot q_2^3 & q_1^3\cdot q_3^3 & q_2^3\cdot q_3^3 \\ q_1^4 & q_2^4 & q_3^4 & q_1^4\cdot q_2^4 & q_1^4\cdot q_3^4 & q_2^4\cdot q_3^4 \\ q_1^5 & q_2^5 & q_3^5 & q_1^5\cdot q_2^5 & q_1^5\cdot q_3^5 & q_2^5\cdot q_3^5 \\ q_1^6 & q_2^6 & q_3^6 & q_1^6\cdot q_2^6 & q_1^6\cdot q_3^6 & q_2^6\cdot q_3^6 \end{pmatrix} \times \begin{pmatrix} a_{11} \\ a_{22} \\ a_{33} \\ a_{12} \\ a_{13} \\ a_{23} \end{pmatrix} = \begin{pmatrix} q_0^1 \\ q_0^2 \\ q_0^3 \\ q_0^4 \\ q_0^5 \\ q_0^6 \end{pmatrix} \tag{14.3}$$

To simplify the notation, we denote the first matrix by the symbol Q, the matrix with the feedback coefficients a_{ij} by the symbol A, and their product by B. The introduced notations will allow us to write the system (14.3) in a more compact form:

$$Q \times A = B \tag{14.4}$$

The matrices Q, A and B can be similarly compiled for any value of L NLFSR. In what follows, we will mean the matrices Q, A, and B for an arbitrary value of L, which are formed as follows:

$$Q = \left(q_{i=\{1,\dots,L\}}^{k=\{1,\dots,L\cdot(L+1)/2\}};\ q_{i=\{1,\dots,L\}}^{k=\{1,\dots,L\cdot(L+1)/2\}} \cdot q_{j=\{i+1,\dots,L\}}^{k=\{1,\dots,L\cdot(L+1)/2\}}\right)$$
$$A = \left(a_{ij=\{1,\dots,L\}j=\{i,\dots,L\}}\right)$$
$$B = \left(q_0^{k=\{1,\dots,L\cdot(L+1)/2\}}\right)$$

SLAE (14.4) will have a solution only if the rows of the matrix Q are linearly independent, i.e. the matrix Q must not be degenerate. This requirement imposes restrictions on the choice of sequences, which will consist of (14.4), failure to comply with this requirement will lead to ambiguity of the decision.

The solution of SLAE (14.4) can be presented in the form:

$$A = Q^{-1} \times B, \tag{14.5}$$

where Q^{-1} is the matrix inverse to the matrix Q.

14.1.2 *Selection of Forming Sequences for Assembly of SLAE*

Finding the inverse matrix and expressing linearly dependent coefficients through the basis.

In Chap. 13, it was shown that for M-NLFSR all a_{iL} coefficients are uniquely defined by Requirements 1 and 3. This fact allows to reduce the dimension of the matrix of unknown coefficients A on L elements.

As shown in Sect. 13.4.3, the probability of rejection of Polynomials by Requirement 9 for different periods decreases with increasing period, i.e. the depth of verification of Requirement 9. Thus, it is advisable to start testing in Requirement 9 for the possibility of forming a ring with period $T = 1$, then the possibility formation of a ring with a period $T = 2$ and further in the direction of increasing the value of T.

We demonstrate the order of addition of SLAE (14.4) for NLFSR with $L = 4$.

Table 14.1 shows all possible rings with periods from 1 to $T = 4$, as well as the state of the cells in the register and the value that this state must generate to create the corresponding ring. The correspondence that is checked between the values of the state of the register and the value that this state can generate, will be called, as in the previous section,—the check pattern.

Taking into account the known a_{iL} ($a_{LL} = 1$ according to Requirement 1, and $a_{iL} = 1$ for all $i = \{1, \dots, L-1\}$ according to Requirement 3) we write the following relations (templates $S_{T=n}^{r/k}$, where k is the ordinal number of the template

Table 14.1 Possible rings with periods from 1 to T = 4 and the corresponding templates

Possible period in the ring values	Sequence in the ring period	Verifiable templates
$T = 1$	{1}	$(1\ 1\ 1\ 1) \to 1$
$T = 2$	$\{1\ 0\}$	$(1\ 0\ 1\ 0) \to 0$ $(0\ 1\ 0\ 1) \to 1$
$T = 3$	$\{1\ 0\ 0\}$	$(1\ 0\ 0\ 1) \to 0$ $(0\ 1\ 0\ 0) \to 0$ $(0\ 0\ 1\ 0) \to 1$
	$\{1\ 1\ 0\}$	$(1\ 1\ 0\ 1) \to 0$ $(0\ 1\ 1\ 0) \to 1$ $(1\ 0\ 1\ 1) \to 1$
$T = 4$	$\{1\ 0\ 0\ 0\}$	$(1\ 0\ 0\ 0) \to 0$ $(0\ 1\ 0\ 0) \to 0$ $(0\ 0\ 1\ 0) \to 0$ $(0\ 0\ 0\ 1) \to 1$
	$\{1\ 1\ 0\ 0\}$	$(1\ 1\ 0\ 0) \to 0$ $(0\ 1\ 1\ 0) \to 0$ $(0\ 0\ 1\ 1) \to 1$ $(1\ 0\ 0\ 1) \to 1$

in sequence for the period T) between the states of the first $L - 1$ cells of the NLFSR and the generated bit, in which, according to Requirement 9, there will be a ring with a period less than the maximum possible ($T_{\max}$).

$$
\begin{aligned}
\text{Ring}\{1\}: \quad & S_{T=1}^{1/2} : (1\ 1\ 1) \to q_0^1 = 0. \\
\text{Ring}\{1\ 0\}: \quad & S_{T=2}^{1/1} : (1\ 0\ 1) \to q_0^2 = 0 \quad S_{T=2}^{1/2} : (0\ 1\ 0) \to q_0^3 = 0. \\
\text{Ring}\{1\ 0\ 0\}: \quad & S_{T=3}^{1/1} : (1\ 0\ 0) \to q_0^4 = 1 \quad S_{T=3}^{1/2} : (0\ 1\ 0) \to q_0^3 = 0 \\
& S_{T=3}^{1/3} : (0\ 0\ 1) \to q_0^5 = 1 \\
\text{Ring}\{1\ 1\ 0\}: \quad & S_{T=3}^{2/1} : (1\ 1\ 0) \to q_0^6 = 1 \quad S_{T=3}^{2/2} : (0\ 1\ 1) \to q_0^7 = 1 \\
& S_{T=3}^{2/3} : (1\ 0\ 1) \to q_0^2 = 0 \\
\text{Ring}\{1\ 0\ 0\ 0\}: \quad & S_{T=4}^{1/1} : (1\ 0\ 0) \to q_0^4 = 0 \quad S_{T=4}^{1/2} : (0\ 1\ 0) \to q_0^3 = 0 \\
& S_{T=4}^{1/3} : (0\ 0\ 1) \to q_0^5 = 0 \quad S_{T=4}^{1/4} : (0\ 0\ 0) \to q_0^8 = 0 \\
\text{Ring}\{1\ 1\ 0\ 0\}: \quad & S_{T=4}^{2/1} : (1\ 1\ 0) \to q_0^6 = 0 \quad S_{T=4}^{2/2} : (0\ 1\ 1) \to q_0^7 = 0 \\
& S_{T=4}^{2/3} : (0\ 0\ 1) \to q_0^5 = 0 \quad S_{T=4}^{2/4} : (1\ 0\ 0) \to q_0^4 = 0
\end{aligned}
$$

Only when all patterns are executed at the same time, for each of the circles there will be a corresponding period in the sequence generated.

So we have eight different combinations. Since only six coefficients a_{ij} are unknown in the NLFSR, in our case, six equations will be enough for an unambiguous and unique solution of SLAE. Let's call these relations basic. The two relations can be expressed in terms of the base by their linear combination.

The first and obvious combination, using which we obtain a nondegenerate matrix, is the zero combination. The rest is determined after reducing the matrix Q to a single. In parallel, we compose a unit matrix and, performing the same actions with it as with the matrix Q, we obtain from it the inverse matrix Q^{-1}. Recall that arithmetic operations are performed in $GF(2)$.

Thus, after several trivial actions we obtain the basic matrix and the ratio for non-basic coefficients:

$$Q = \begin{pmatrix} 1\,1\,1\,1\,1\,1 \\ 1\,0\,1\,0\,1\,0 \\ 0\,1\,0\,0\,0\,0 \\ 1\,0\,0\,0\,0\,0 \\ 0\,0\,1\,0\,0\,0 \\ 1\,1\,0\,1\,0\,0 \end{pmatrix} \tag{14.6}$$

$$q_0^7 = q_0^6 + q_0^5 + q_0^4 + q_0^3 + q_0^2 + q_0^1$$
$$q_0^8 = 0$$

The next step is to reduce the diagonal matrix to a single one. Performing similar steps, we obtain the inverse matrix to the matrix Q:

$$Q^{-1} = \begin{pmatrix} 0\,0\,0\,1\,0\,0 \\ 0\,0\,1\,0\,0\,0 \\ 0\,0\,0\,0\,1\,0 \\ 0\,0\,1\,1\,0\,1 \\ 0\,1\,0\,1\,1\,0 \\ 1\,1\,0\,1\,0\,1 \end{pmatrix}$$

Let's check the correctness of finding the inverse matrix.

$$Q^{-1} \times Q = \begin{pmatrix} 0\,0\,0\,1\,0\,0 \\ 0\,0\,1\,0\,0\,0 \\ 0\,0\,0\,0\,1\,0 \\ 0\,0\,1\,1\,0\,1 \\ 0\,1\,0\,1\,1\,0 \\ 1\,1\,0\,1\,0\,1 \end{pmatrix} \times \begin{pmatrix} 1\,1\,1\,1\,1\,1 \\ 1\,0\,1\,0\,1\,0 \\ 0\,1\,0\,0\,0\,0 \\ 1\,0\,0\,0\,0\,0 \\ 0\,0\,1\,0\,0\,0 \\ 1\,1\,0\,1\,0\,0 \end{pmatrix} = \begin{pmatrix} 1\,0\,0\,0\,0\,0 \\ 0\,1\,0\,0\,0\,0 \\ 0\,0\,1\,0\,0\,0 \\ 0\,0\,0\,1\,0\,0 \\ 0\,0\,0\,0\,1\,0 \\ 0\,0\,0\,0\,0\,1 \end{pmatrix}$$

We have shown how it is desirable to make SLAE from the coefficients included in Requirement 9. The method of finding the inverse matrix and expressing linearly dependent coefficients through the basic coefficients is suitable for any value of L.

14.1.3 Search M-NLFSR Using SLAE

Given the above, to search for M-NLFSR it is enough to go through all possible solutions of SLAE, i.e. coefficients q_0^i, $i = \{1, \ldots, L \cdot (L-1)/2\}$. At the same time check whether Requirement 9 is met or not. If the coefficients are checked, then this combination will generate an M-sequence. Substituting the found combination in (14.5), i.e. by multiplying two matrices, we obtain a matrix of coefficients a_{ij} for M-NLFSR.

The advantage of this approach, compared to checking all patterns according to Requirement 9, is as follows:

- it is not necessary to store in memory all the patterns of Requirement 9, it is enough to know only their serial number and what value they should generate, which greatly reduces the amount of memory used. As an example, the amount of RAM consumed by the NLFSR testing program for compliance with the Requirements for $L = 30$ and $T = 15$ did not exceed 12 MB;
- there is no need to check the coefficients a_{ij} for compliance with the coefficients of the pattern, it is enough to check only their solution, which reduces the time spent on checking each pattern. The time spent on the test, for various parameters, is discussed in more detail below.

However, it should be noted that in order to test all the requirements of Requirements 9, it is necessary to know all the coefficients of resolution of the patterns q_0^i, and most of them are expressed in terms of basic q_0^i. The time spent calculating all q_0^i through the basic q_0^i, followed by checking for pattern compliance, becomes unreasonably large compared to simply checking the generated sequence itself. Therefore, it is necessary to find a compromise between the number of patterns to be checked and the subsequent check of the period of the sequence itself. Since the number of excluded combinations decreases with increasing depth of verification of Requirement 9, it will be sufficient to check the period from $T = 1$ to L to exclude more than 99% of combinations that do not generate M-sequence.

As another advantage of the method, it should be noted that the Requirement 9 pattern for $T = 1$ must always generate 1 to create a corresponding ring. Therefore, it can be accepted $q_0^1 = 1$ at first without considering the opposite meaning. And this requirement does not pass 50% of combinations from all set of possible combinations that allows to reduce twice the set tested.

Here is an example of searching M-NLFSR for $L = 4$, using the above matrices.

Total coefficients a_{ij} in NLFSR dimension $L = 4$:

$$n_L = \frac{L \cdot (L+1)}{2} = \frac{4 \cdot (4+1)}{2} = 10,$$

of which you can make $2^{n_L} = 2^{10} = 1024$ different combinations. By fixing the coefficients a_{iL}, we reduce the number of different a_{ij} to $\frac{L \cdot (L-1)}{2} = \frac{4 \cdot (4-1)}{2} = 6$ and the set of different combinations to $2^6 = 64$. By fixing $q_0^1 = 1$, we reduce the set of combinations to $2^5 = 32$ (from the initial 1024!). Let's limit ourselves to testing patterns up to L, that is, the maximum considered depth of verification for Requirement 9 will be $T = 4$. The coefficient q_0^7 is calculated from (14.6). The results of the calculations are shown in Table 14.2.

Thus, of all the possible set of different combinations (1024), 32 combinations (3.1%) were tested, 15 of which did not pass the test for compliance with Requirement 9 with a depth of verification from 1 to 4. Of the remaining 17 (1.7%) combinations only one combination did not generate an M-sequence.

Table 14.3 summarizes the above values, as well as similar quantitative indicators for other calculated sizes L. In all calculations, we limit the depth of verification of Requirement 9 from 1 to L.

14.2 Model with Simplified Formalized Description of Second-Order Nonlinear Feedback

For practical applications, it is not always necessary to find the entire set of NLFSR that generate the M-sequence, in addition, for large register sizes, the full set of M-NLFSR will not be physically possible to save. In most cases, it is sufficient to find a number of M-NLFSR for a given value of L. It is possible to reduce the number of elements included in (14.4), equating some of them to zero or one. For example, the feedback matrix, in the absence of feedback from the first cell (both linear and nonlinear) will look like this:

$$\begin{matrix} 0 & 0 & 0 & \dots & 0 \\ & a_{22} & a_{23} & \dots & a_{2L} \\ & & a_{33} & \dots & a_{3L} \\ & & & \dots & \dots \\ & & & & a_{LL} \end{matrix}$$

This simplification of the feedback model to k elements, from the initial n_L, will reduce the total set tested by $2^{n_L - k}$ times and, accordingly, reduce the computation time and the amount of memory used. In addition, it is possible to specify the mandatory presence or absence of feedback from some coefficients at the design stage of the NLFSR.

Table 14.2 Example of M-NLFSR search for L = 4 according to Requirement 9

№	$q_0^1\, q_0^2\, q_0^3\, q_0^4\, q_0^5\, q_0^6$	q_0^7	Patterns that the combination matches (i.e. there will be a corresponding ring)	M-NLFSR search from (14.5)
1	2	3	4	5
1.	1 0 0 0 0 0	1	$S_{T=2}^{1/1}, S_{T=2}^{1/2}$	–
2.	1 1 0 0 0 0	0	$S_{T=4}^{1/1}, S_{T=4}^{1/2}, S_{T=4}^{1/3}, S_{T=4}^{1/4}$	–
3.	1 0 1 0 0 0	0	$S_{T=4}^{2/1}, S_{T=4}^{2/2}, S_{T=4}^{2/3}, S_{T=4}^{2/4}$	–
4.	1 1 1 0 0 0	1	–	0 1 1 0 1 0 0 0 0 1
5.	1 0 0 1 0 0	0	$S_{T=2}^{1/1}, S_{T=2}^{1/2}$	–
6.	1 1 0 1 0 0	1	–	1 1 0 0 0 1 0 0 0 1
7.	1 0 1 1 0 0	1	–	1 0 1 0 1 0 0 0 0 1
8.	1 1 1 1 0 0	0	–	1 0 0 0 1 1 0 0 0 1
9.	1 0 0 0 1 0	0	$S_{T=2}^{1/1}, S_{T=2}^{1/2}$	–
10.	1 1 0 0 1 0	1		0 0 0 0 0 0 0 1 0 1

(continued)

Table 14.2 (continued)

№	$q_0^1\ q_0^2\ q_0^3\ q_0^4\ q_0^5\ q_0^6$	q_0^7	Patterns that the combination matches (i.e. there will be a corresponding ring)	M-NLFSR search from (14.5)
1	2	3	4	5
11.	1 0 1 0 1 0	1		0 1 1 0 1 1 0 1 0 1
12.	1 1 1 0 1 0	0		0 1 0 0 1 0 0 1 0 1
13.	1 0 0 1 1 0	1	$S_{T=2}^{1/1}, S_{T=2}^{1/2}$	–
14.	1 1 0 1 1 0	0	$S_{T=3}^{1/1}, S_{T=3}^{1/2}, S_{T=3}^{1/3}$	–
15.	1 0 1 1 1 0	0	–	–
16.	1 1 1 1 1 0	1	–	1 0 1 0 1 1 0 1 0 1
17.	1 0 0 0 0 1	0	$S_{T=2}^{1/1}, S_{T=2}^{1/2}$	–
18.	1 1 0 0 0 1	1	$S_{T=4}^{1/1}, S_{T=4}^{1/2}, S_{T=4}^{1/3}, S_{T=4}^{1/4}$	–
19.	1 0 1 0 0 1	1	$S_{T=3}^{2/1}, S_{T=3}^{2/2}, S_{T=3}^{2/3}$	–
20.	1 1 1 0 0 1	0	–	0 0 1 0 1 1 0 0 0 1
21.	1 0 0 1 0 1	1	$S_{T=2}^{1/1}, S_{T=2}^{1/2}$	–
22.	1 1 0 1 0 1	0	–	1 0 0 0 0 0 0 0 0 1

(continued)

Table 14.2 (continued)

№	$q_0^1\, q_0^2\, q_0^3\, q_0^4\, q_0^5\, q_0^6$	q_0^7	Patterns that the combination matches (i.e. there will be a corresponding ring)	M-NLFSR search from (14.5)
1	2	3	4	5
23.	1 0 1 1 0 1	0	–	1 1 1 0 1 1 0 0 0 1
24.	1 1 1 1 0 1	1	–	1 1 0 0 1 0 0 0 0 1
25.	1 0 0 0 1 1	1	$S_{T=2}^{1/1}, S_{T=2}^{1/2}$	–
26.	1 1 0 0 1 1	0	–	0 1 0 0 0 1 0 1 0 1
27.	1 0 1 0 1 1	0	–	0 0 1 0 1 0 0 1 0 1
28.	1 1 1 0 1 1	1	–	0 0 0 0 1 1 0 1 0 1
29.	1 0 0 1 1 1	0	$S_{T=2}^{1/1}, S_{T=2}^{1/2}$	–
30.	1 1 0 1 1 1	1	$S_{T=3}^{1/1}, S_{T=3}^{1/2}, S_{T=3}^{1/3}$	–
31.	1 0 1 1 1 1	1	$S_{T=3}^{2/1}, S_{T=3}^{2/2}, S_{T=3}^{2/3}$	–
32.	1 1 1 1 1 1	0	–	1 1 1 0 1 0 0 1 0 1

Table 14.3 Summary test results Requirements 9

L	Total possible combinations	Number of tested combinations	Passed the test on Requirement 9 ($T = 1 \ldots L$)	Found M-NLFSR	Number of tested conditions Requirements 9	Time spent searching (s)
4	1024	32	15	16	6	<0.001
5	32,768	512	192	128	12	<0.001
6	2,097,152	16,384	5711	1952	21	0.046
7	268,435,456	1,048,576	302,102	64,056	37	4.760
8	68,719,476,736	134,217,728	34,435,282	4,017,998	67	958.5
			($T = 1 \ldots 7$) 38,620,118			991.0
			($T = 1 \ldots 6$) 44,306,685			928.5
			($T = 1 \ldots 5$) 50,684,976			864.3
			($T = 1 \ldots 4$) 61,341,696			929.6

By introducing additional restrictions on the type of polynomial, we at the same time reduce the number of M-NLFSR that can be found, and with very large restrictions in the set under test, the desired combinations of feedback coefficients may not be found.

As an example, take the NLFSR dimension $L = 8$ and introduce a restriction—the lack of feedback from the first element, i.e. $a_{1j} = 0$ for all $j = 1, \ldots, L$. Then from the first and second elements ($a_{1j} = 0, a_{2j} = 0$ for all $j = i, \ldots, L$) and so on. The results of the calculations are shown in Table 14.4.

Table 14.4 Calculation of the set of NLFSR under the entered conditions $a_{ij} = 0$ for all $j = i, \ldots, L$ and some i

$L = 8$ $a_{ij} = 0$	Passed the test on Requirement 9 ($T = 1 \ldots L$)	Found M-NLFSR	Searching time (s)
–	34,435,282	4,017,998	958.5
for $i = 1$	270,769	31,567	4.65
for $i = 1, 2$	4334	576	0.07
for $i = 1 \ldots 3$	141	28	<0.001
for $i = 1 \ldots 4$	10	0	<0.001

14.2.1 Finding M-NLFSR for Large Registers

The above simplified feedback model can be applied to any coefficients, but we equate to zero the coefficients from the first cells, guided solely by the criterion of the speed of PRS generation, namely:

1. For software implementation, as will be shown in the next section, it is better to group non-zero coefficients from the cell, and the element $a_{LL} = 1$ and, therefore, it is forbidden to change it, so it is better to zero the initial coefficients.
2. For hardware implementation, the first coefficients are not involved, allow to make a kind of delay line, which will increase the generation rate, as the calculation of all feedback factors can be performed not for one cycle, but for two (or more) cycles.
3. It is possible to create several implementations of feedback functions, such as in the stream cipher Grain, which will increase the number of values generated per clock cycle.

As you can see from Table 14.4, as the number of zero lines increases, the set of combinations to be tested decreases, and the total time for the test decreases, but the set of found M- NLFSR decreases significantly.

Similarly, it is possible to further reduce the set under test and, accordingly, the time spent on the entire test, if you additionally equate to zero some nonlinear feedback factors.

As an example, consider the NLFSR dimension $L = 28$. The total number of possible combinations: $1.65 \cdot 10^{122}$. Introduced restrictions: all linear coefficients $a_{11} = a_{12} = 0$ and all nonlinear coefficients, except $a_{25\_26}, a_{25\_27}, a_{26\_27}$, are equal to zero. With such restrictions, the set of possible combinations is 536 870 912.

Table 14.5 shows the quantitative indicators of the set that meets Requirement 9 when applying different depth of verification. The time during which testing was performed under the imposed restrictions is also given.

14.2.2 Search Results for M-NLFSR—List of M-NLFSR Dimension from L = 19 to L = 25

Using the simplified model and the method outlined in this section, M-NLFSR with dimensions from $L = 19$ to $L = 25$ were found. The polynomials corresponding to the M-NLFSR found, as well as the time spent searching in each group, are given below.

$L = 19$:

$$x^{19} + x^{17} + x^{14} + x^{13} + x^{12} + x^{11} + x^{8} + x^{6} + x^{4} + x^{3} + x^{16} \cdot x^{17} + x^{17} \cdot x^{18}$$

$$x^{19} + x^{18} + x^{17} + x^{16} + x^{15} + x^{14} + x^{13} + x^{8} + x^{4} + x^{2} + x^{1} + x^{16} \cdot x^{17}$$

Table 14.5 Quantitative indicators of the set that meets the Requirement 9 for $L = 28$ with additional restrictions

Depth of verification, T	Number of combinations that meet the Requirement 9	Decrease of the required set in comparison with the previous value	Testing time
1-9	28,981,418		12 min
1-10	27,176,863	−1,804,555	17 min
1-11	25,266,290	−1,910,573	21 min
1-12	23,859,686	−1,406,604	33 min
1-13	22,364,226	−1,495,460	01 h 03 min
1-14	20,931,576	−1,432,650	01 h 44 min
1-15	19,786,133	−1,145,443	03 h 53 min
1-16	18,679,988	−1,106,145	10 h 38 min
1-17	17,619,849	−1,060,139	18 h 03 min
1-18	16,818,357	−801,492	36 h 34 min

$$+x^{16}\cdot x^{18}+x^{17}\cdot x^{18}$$

$$x^{19}+x^{13}+x^{12}+x^{11}+x^{10}+x^{8}+x^{3}+x^{2}+x^{16}\cdot x^{17}+x^{17}\cdot x^{18}$$

$$x^{19}+x^{15}+x^{14}+x^{10}+x^{9}+x^{7}+x^{6}+x^{5}+x^{2}+x^{16}\cdot x^{17}$$

$$x^{19}+x^{17}+x^{16}+x^{15}+x^{14}+x^{13}+x^{12}+x^{11}+x^{10}+x^{7}+x^{3}+x^{16}\cdot x^{17}$$
$$+x^{16}\cdot x^{18}+x^{17}\cdot x^{18}$$

$$x^{19}+x^{18}+x^{16}+x^{14}+x^{10}+x^{9}+x^{5}+x^{3}+x^{2}+x^{1}+x^{16}\cdot x^{17}+x^{16}\cdot x^{18}$$

$$x^{19}+x^{16}+x^{15}+x^{14}+x^{13}+x^{12}+x^{11}+x^{10}+x^{7}+x^{3}+x^{16}\cdot x^{17}+x^{17}\cdot x^{18}$$

$$x^{19}+x^{13}+x^{12}+x^{11}+x^{5}+x^{3}+x^{16}\cdot x^{17}+x^{16}\cdot x^{18}$$

$$x^{19}+x^{18}+x^{16}+x^{15}+x^{14}+x^{13}+x^{8}+x^{4}+x^{2}+x^{1}+x^{16}\cdot x^{17}+x^{17}\cdot x^{18}$$

$$x^{19}+x^{14}+x^{13}+x^{12}+x^{11}+x^{8}+x^{6}+x^{4}+x^{3}+x^{16}\cdot x^{17}$$
$$+x^{16}\cdot x^{18}+x^{17}\cdot x^{18}$$

$$x^{19}+x^{17}+x^{16}+x^{15}+x^{13}+x^{11}+x^{9}+x^{8}+x^{7}+x^{6}+x^{5}$$
$$+x^{3}+x^{16}\cdot x^{17}+x^{16}\cdot x^{18}$$

$$x^{19}+x^{18}+x^{17}+x^{16}+x^{12}+x^{10}+x^{9}+x^{7}+x^{6}+x^{3}+x^{16}\cdot x^{17}+x^{16}\cdot x^{18}$$

$$x^{19}+x^{17}+x^{13}+x^{12}+x^{11}+x^{10}+x^{8}+x^{3}+x^{2}+x^{16}\cdot x^{17}$$
$$+x^{16}\cdot x^{18}+x^{17}\cdot x^{18}$$

Search performed in 52 min.
$L = 20$:

$$x^{20}+x^{17}+x^{16}+x^{13}+x^{11}+x^{10}+x^{9}+x^{8}+x^{4}+x^{3}+x^{1}+x^{17}\cdot x^{18}$$

$+ x^{17} \cdot x^{19} + x^{18} \cdot x^{19}$

$x^{20} + x^{19} + x^{18} + x^{16} + x^{10} + x^{5} + x^{3} + x^{17} \cdot x^{18}$

$x^{20} + x^{18} + x^{17} + x^{16} + x^{15} + x^{14} + x^{12} + x^{10} + x^{8} + x^{5} + x^{3} + x^{2}$
$+ x^{1} + x^{17} \cdot x^{18} + x^{17} \cdot x^{19} + x^{18} \cdot x^{19}$

$x^{20} + x^{15} + x^{14} + x^{12} + x^{11} + x^{10} + x^{4} + x^{2} + x^{17} \cdot x^{18} + x^{17} \cdot x^{19}$

$x^{20} + x^{19} + x^{18} + x^{17} + x^{16} + x^{15} + x^{10} + x^{9} + x^{8} + x^{5} + x^{4} + x^{3}$
$+ x^{1} + x^{17} \cdot x^{18} + x^{17} \cdot x^{19} + x^{18} \cdot x^{19}$

$x^{20} + x^{17} + x^{14} + x^{13} + x^{8} + x^{5} + x^{4} + x^{17} \cdot x^{19}$

$x^{20} + x^{19} + x^{14} + x^{13} + x^{12} + x^{11} + x^{8} + x^{7} + x^{6} + x^{4} + x^{3} + x^{2}$
$+ x^{17} \cdot x^{18} + x^{18} \cdot x^{19}$

$x^{20} + x^{16} + x^{13} + x^{12} + x^{10} + x^{9} + x^{8} + x^{7} + x^{6} + x^{4} + x^{2} + x^{17} \cdot x^{18}$

$x^{20} + x^{18} + x^{15} + x^{8} + x^{7} + x^{2} + x^{17} \cdot x^{18} + x^{17} \cdot x^{19}$

$x^{20} + x^{19} + x^{18} + x^{14} + x^{13} + x^{12} + x^{11} + x^{8} + x^{7} + x^{6} + x^{4} + x^{3}$
$+ x^{2} + x^{17} \cdot x^{18} + x^{17} \cdot x^{19} + x^{18} \cdot x^{19}$

$x^{20} + x^{19} + x^{18} + x^{17} + x^{16} + x^{15} + x^{14} + x^{13} + x^{12} + x^{5} + x^{4} + x^{3} + x^{1} + x^{17} \cdot x^{19}$

$x^{20} + x^{18} + x^{14} + x^{13} + x^{9} + x^{8} + x^{5} + x^{1} + x^{17} \cdot x^{19} + x^{18} \cdot x^{19}$

$x^{20} + x^{17} + x^{16} + x^{15} + x^{14} + x^{12} + x^{10} + x^{8} + x^{5} + x^{3} + x^{2}$
$+ x^{1} + x^{17} \cdot x^{18} + x^{18} \cdot x^{19}$

$x^{20} + x^{19} + x^{17} + x^{16} + x^{15} + x^{10} + x^{9} + x^{8} + x^{5} + x^{4} + x^{3} + x^{1}$
$+ x^{17} \cdot x^{18} + x^{18} \cdot x^{19}$

$x^{20} + x^{19} + x^{17} + x^{16} + x^{15} + x^{8} + x^{7} + x^{6} + x^{5} + x^{3} + x^{17} \cdot x^{18} + x^{17} \cdot x^{19}$

$x^{20} + x^{18} + x^{17} + x^{16} + x^{13} + x^{11} + x^{10} + x^{9} + x^{8} + x^{4} + x^{3} + x^{1}$
$+ x^{17} \cdot x^{18} + x^{18} \cdot x^{19}$

Search conducted in 5 h 36 min.
$L = 21$:

$x^{21} + x^{19} + x^{18} + x^{15} + x^{11} + x^{10} + x^{9} + x^{8} + x^{6} + x^{4}$
$+ x^{18} \cdot x^{19} + x^{18} \cdot x^{20}$

$x^{21} + x^{20} + x^{19} + x^{16} + x^{14} + x^{13} + x^{11} + x^{9} + x^{8} + x^{7} + x^{6} + x^{4}$
$+ x^{2} + x^{1} + x^{18} \cdot x^{19} + x^{18} \cdot x^{20}$

$x^{21} + x^{19} + x^{16} + x^{15} + x^{14} + x^{13} + x^{11} + x^{10} + x^{9} + x^{8} + x^{4}$
$+ x^{1} + x^{18} \cdot x^{19} + x^{18} \cdot x^{20}$

$x^{21} + x^{18} + x^{17} + x^{16} + x^{12} + x^{8} + x^{7} + x^{6} + x^{3} + x^{1}$

$$+ x^{18} \cdot x^{19} + x^{18} \cdot x^{20}$$
$$x^{21} + x^{20} + x^{17} + x^{16} + x^{15} + x^{14} + x^{13} + x^{10} + x^{9} + x^{7} + x^{5}$$
$$+ x^{4} + x^{2} + x^{18} \cdot x^{20}$$
$$x^{21} + x^{20} + x^{17} + x^{16} + x^{15} + x^{13} + x^{12} + x^{10} + x^{7} + x^{6} + x^{4}$$
$$+ x^{2} + x^{1} + x^{18} \cdot x^{20}$$
$$x^{21} + x^{14} + x^{11} + x^{10} + x^{7} + x^{6} + x^{5} + x^{2} + x^{1} + x^{18} \cdot x^{20}$$
$$x^{21} + x^{20} + x^{17} + x^{14} + x^{7} + x^{4} + x^{3} + x^{2} + x^{1} + x^{18} \cdot x^{20}$$
$$x^{21} + x^{18} + x^{15} + x^{14} + x^{13} + x^{8} + x^{6} + x^{5} + x^{1} + x^{18} \cdot x^{20}$$

Search conducted in 8 h 4 min.
$L = 22$:

$$x^{22} + x^{20} + x^{19} + x^{18} + x^{17} + x^{15} + x^{14} + x^{8} + x^{7} + x^{5} + x^{4} + x^{19} \cdot x^{21}$$

Search conducted in 2 h 25 min.
L = 23:

$$x^{23} + x^{22} + x^{21} + x^{11} + x^{10} + x^{6} + x^{4} + x^{20} \cdot x^{21}$$

Search conducted in 24 h 53 min.
$L = 24$:

$$x^{24} + x^{22} + x^{21} + x^{19} + x^{16} + x^{15} + x^{14} + x^{13} + x^{12} + x^{10}$$
$$+ x^{7} + x^{6} + x^{4} + x^{21} \cdot x^{23}$$
$$x^{24} + x^{22} + x^{21} + x^{19} + x^{17} + x^{16} + x^{13} + x^{12} + x^{11} + x^{10}$$
$$+ x^{6} + x^{5} + x^{21} \cdot x^{22} + x^{21} \cdot x^{23}$$

Search took 36 h 47 min.
$L = 25$:

$$x^{25} + x^{23} + x^{22} + x^{21} + x^{20} + x^{19} + x^{14} + x^{12} + x^{9}$$
$$+ x^{4} + x^{22} \cdot x^{23} + x^{22} \cdot x^{24}$$
$$x^{25} + x^{22} + x^{20} + x^{18} + x^{14} + x^{12} + x^{11} + x^{9} + x^{6}$$
$$+ x^{4} + x^{22} \cdot x^{23} + x^{22} \cdot x^{24}$$

Search was conducted in 18 days 2 h 30 min.

It is worth noting that this method of M-NLFSR search can also be used to find M-LFSR (as a special case). For example, all M-LFSR were searched for $L = 21$ and 22. 84,672 and 120,032 polynomials were found, respectively, which completely coincides with the results given in [1]. Time was spent on the calculation of 7 and 17 h, respectively.

The calculations used the clipping of polynomials with $T = 1, \ldots, 13$ and $T = 1, \ldots, 6$. By increasing the depth of verification of Requirement 9, it is possible to significantly reduce the calculation time, such a task did not stand during the calculations and therefore was not performed.

14.3 Search for M-NLFSR with the Use of Hardware Implemented on FPGA

This section provides a method of searching M-NLFSR, which implements both software and hardware approach. This technique is based on the results of practical generalization of previously obtained results in order to implement an acceptable, in terms of hardware and time indicators, M-NLFSR search algorithm. As an example, the results of the implementation of the proposed method of searching M-NLFSR dimension up to $L = 30$ cells inclusive.

A non-standard hardware solution for the implementation of cryptographic algorithms is the use of programmable logic integrated circuits (FPGA—FieldProgrammable Gate Array) [2], which allow you to design digital devices using high-level hardware description languages, which reduces the complexity of development and allows IP reuse [3].

FPGAs, on the one hand, are universal and reliably repeatable, and on the other hand, parameterized, i.e. there is a possibility of parametric adjustment for a specific project. FPGAs have high reliability, which is ensured by thorough testing [4].

Programmable logic is attractive due to the ability to provide performance close to ASIC (Application Specific Integrated Circuit) technology, achieving high bandwidth, high design flexibility and maintaining low power consumption [5].

In addition, one of the important advantages of implementing cryptographic algorithms on FPGA is the ability to build a parallel and asynchronous architecture, which is superior in performance to solutions based on microprocessors, GPU or CPU [6].

The specifics of FPGA implementation makes it possible to solve most tasks both sequentially and in parallel, which significantly affects the required amount of resources for project implementation. In the general case, the decision on FPGA involves the implementation of repetitive operations, often in the pipeline, which allows you to significantly optimize the process of finding NLFSR.

14.3.1 M-NLFSR Search Method

The basis of the M-NLFSR search method is the software implementation of the Requirements for the type of polynomial and the subsequent hardware verification of the set of combinations for the generation of M-sequences.

Previously, the feedback coefficients aij are subject to restrictions that equate some of the coefficients to zero. The total number of variables (initially not equated to zero or one) of the coefficients a_{ij} is denoted by the symbol n_1. As shown earlier, the number of coefficients a_{ij} for NLFSR dimension L is calculated by the ratio $n_L = L \cdot (L+1)/2$, therefore, n_1 can take values in the range $0 \le n_1 \le n_L$.

By introducing this limitation, we reduce the total set of possible polynomials to a value acceptable in practice, but thereby reduce the number of potentially present polynomials corresponding to M-NLFSR. If it is necessary to find all M-NLFSR for a given L, then such restrictions are not imposed, but it is necessary to assess the possibilities of computer memory costs, which increase exponentially with increasing L and estimate the time spent searching so that the result does not exceed acceptable values.

The calculation can be divided into three stages, which can be performed both simultaneously and sequentially.

The *first stage (analytical)*: determines the structure of the future NLFSR, i.e. the feedback coefficients that must be tested in the following stages. Requirements 1 and 3 are applied to the NLFSR, some aij are fixed in order to reduce the total set of forming polynomials and the depth of verification of Requirement 9 is determined analytically.

The second stage (software): with the help of software implementation the set aij under test is reduced by rejecting NLFSR which do not form M-sequence. The total volume of rejected polynomials is selected in the range of 90% and more than 99.99% of their initial number. The volume of rejected polynomials is limited by available memory resources and is optimized over time.

It is desirable to balance the time costs between the second and third stage. The more time and memory is spent on the second stage, the more registers that are not M-NLFSR are discarded and, therefore, less time will be spent on the third stage. But as the number of rejected polynomials increases, after a certain value, the time spent on additional rejection of interest will be significantly greater than the time spent checking this number during the third stage.

The program stage consists of linearization of the problem, formation and solution of a system of linear algebraic equations for Requirement 9, checking the feedback coefficients for compliance with Requirements 2 and 9.

As a result of software calculations of the second stage, we obtain a set of polynomials acceptable (in terms of time) for verification in the third stage. Let us denote the number of such polynomials as k_0^{FPGA}.

The third stage (hardware): check on the hardware-software complex with the use of the hardware part, implemented on the FPGA, the period of the formed sequence.

The computational complex performs a direct check of the set of polynomials obtained after the second stage for the possibility of generating Msequences. The computer complex includes: hardware implemented on FPGA; software program running on a computer; switching equipment that provides interaction between hardware and software.

The advantage of FPGA-based hardware, in this case, is the simple creation of IP cores with a flexible configuration system for a specific implementation of NLFSR and the ability to simultaneously calculate a significant (hundreds or more, depending on the type of FPGA and L) number of different NLFSR. While on personal computers—from two to eight processes simultaneously (depending on the computer model and installed equipment).

As will be shown below, during the transfer of the computational process from the CPU to the FPGA, the performance (speed) of each of the NLFSR is preserved. This circumstance allows to increase the overall productivity of the M- NLFSR search complex by tens (up to hundreds) times in comparison with the use of a personal computer alone.

14.3.2 The Structure of the Hardware and Software

The complete structure of the computing tool used for the calculations is shown in Fig. 14.1.

The hardware consists of a board with a built-in chip from Altera family Cyclone IV E FPGA EP4CE10E22C8N and a built-in quartz generator at 50 MHz. FPGAs are made on an optimized 60 nm process and include 10,320 logic elements (Les—Logic Elements), as well as two blocks of phase-locked loop (PLL—phaselocked loop).

The communication equipment is a USBTTL converter that acts as a Universal Asynchronous ReceiverTransmitter (UART) between the FPGA and the computer. To increase the exchange rate, two USBTTL converters were used, operating in parallel at a data rate of 921,600 baud. Thus, up to 1,474,560 bits of useful information are transmitted per second. 4 packets of 8 bits were spent on the transmission of feedback

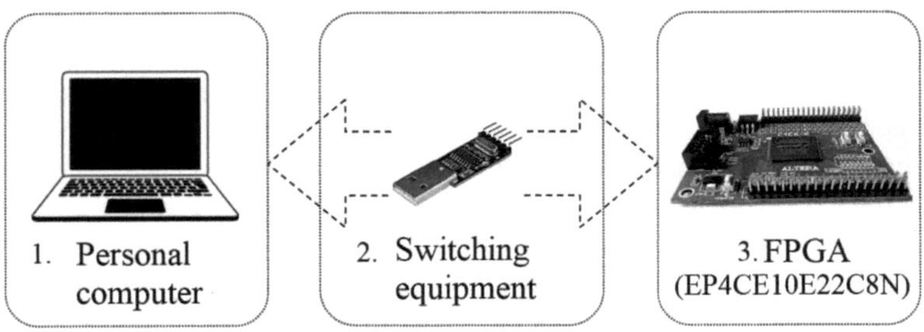

Fig. 14.1 The structure of the hardware and software complex from the search M-NLFSR

coefficients for one HVAC. Information for transmission on 150 such registers took (including software processing and forwarding) about 0.1 s.

The software part implemented the following main functions: reading data from the hard disk; formation of packets for forwarding to FPGA; synchronous (for two USBTTL converters) sending data to the specified COM port; reception via COM port of information from FPGA; decoding, visualization and storage on disk of the results of the inspection carried out in FPGA.

14.3.3 Implementation of NLFSR on FPGA

The FPGA project is written in Verilog HDL, and Altera Quartus Prime Version 16.0.0 CAD was used for compilation and modeling. An example of a block diagram implemented in the project is shown in Fig. 14.2 and consists of the following main modules:

NLFSR—a module in which the NLFSR is implemented. The number of LEs used in this module depends on the dimensionality of the NLFSR, as described in more detail below.

UART_USB1 (UART_USB2)—a module that performs the functions of decoding received packets on the UART channel, generating packets and sending information back to the UART channel (packets are sent back only through UART_USB1). The module uses 62 LEs.

AltPLL is a module for setting a clock frequency other than a quartz oscillator, uses a built-in PLL unit and does not use LEs.

Aij_modul is the main module that initializes the feedback coefficients for the NLFSR. After initialization of all NLFSR launches their work (NLFSR modules) and controls their formation of Msequence. At the end of the cycle, the module reset the system and informs the UART module to send the result: whether there was at least one M- NLFSR per cycle or not. The module contains 38 LEs of constant quantity to ensure the operation of the module, as well as a variable number of LEs used for storage and initialization of feedback coefficients, as well as NLFSR

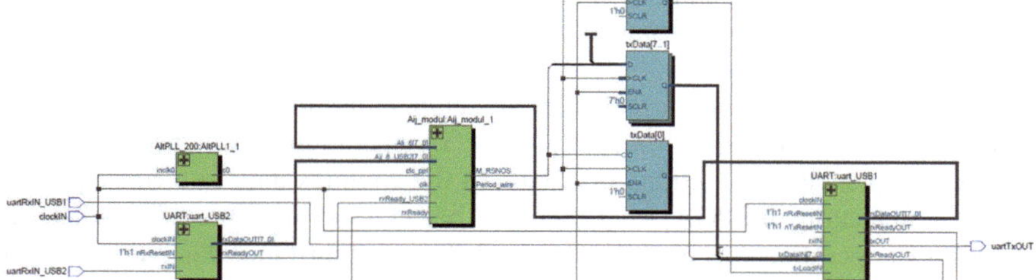

Fig. 14.2 Block diagram of the project

modules. The total number of LEs depends on the number of NLFSR modules and the variable feedback factors used in them.

14.3.4 FPGA Performance

Under the performance of the hardware we understand the number of tested NLFSR per unit time. It is easy to see that the performance of the hardware consists of the following components:

- Time spent on one cycle in the NLFSR module (denote as t_{cycle}^{FPGA}). Under the cycle we understand the implementation of cyclical changes in the state of NLFSR, during which all possible combinations are passed for M-NLFSR and the system returns to the initial state. Given that the registers change their state in one clock cycle of the reference frequency, such clocks should be 2^L. When you increase the size of the register by one cell, the number of cycles will double, which, other things being equal, will double the time to perform one cycle.
- The number of NLFSR modules, ie the number of NLFSR tested simultaneously for one cycle (denote this number as k_{NLFSR}^{FPGA}). The parameter depends on the number of LEs involved in the operation of one NLFSR and the total capacity of the used FPGA.
- Operating clock frequency. The higher the clock frequency, the less time is spent on each cycle. But when a certain maximum clock frequency ($f_{\max}^{FPGA}$) is reached, the FPGA elements do not have time to fully work out and the FPGA ceases to operate stably, which leads to errors in determining the M-NLFSR.
- The time spent on receiving from the software part of the packages with feedback coefficients for NLFSR, sending the result back and the work of the program part on the formation of new packages. However, for values of $L > 25$, the time spent on it is an order of magnitude less than the time spent on calculations directly in the FPGA, and with increasing L, its share will decrease.
- As it k_{NLFSR}^{FPGA} increases, the total number of LEs involved increases accordingly. As an example, the number of LEs involved in FPGA to search for M-NLFSR at $L = 29$ and $n_1 = 28$ depending on k_{NLFSR}^{FPGA} is shown in Fig. 14.3.

Due to the increase in the number of LEs used in FPGAs, their relative position and the increase in the physical distance between the interacting elements in the chip, there is a decrease $f_{\max}^{FPGA}$.

Also $f_{\max}^{FPGA}$ significantly affected by the temperature factor. Thus, when determining the maximum stable frequency of FPGA operation, it was noticed that at the same frequencies during long operation of the complex and, as a consequence, insignificant heating of FPGA, frequent errors began to occur at those frequencies at which stable operation was initially observed. This circumstance can partly explain the slight deviation observed in Figs. 14.4 and 14.5.

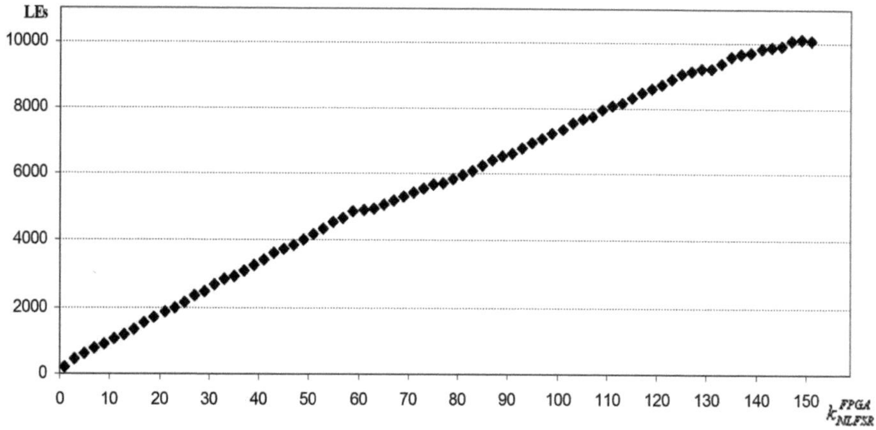

Fig. 14.3 Dependence of LEs involved in FPGA on k_{NLFSR}^{FPGA} for $L = 29$

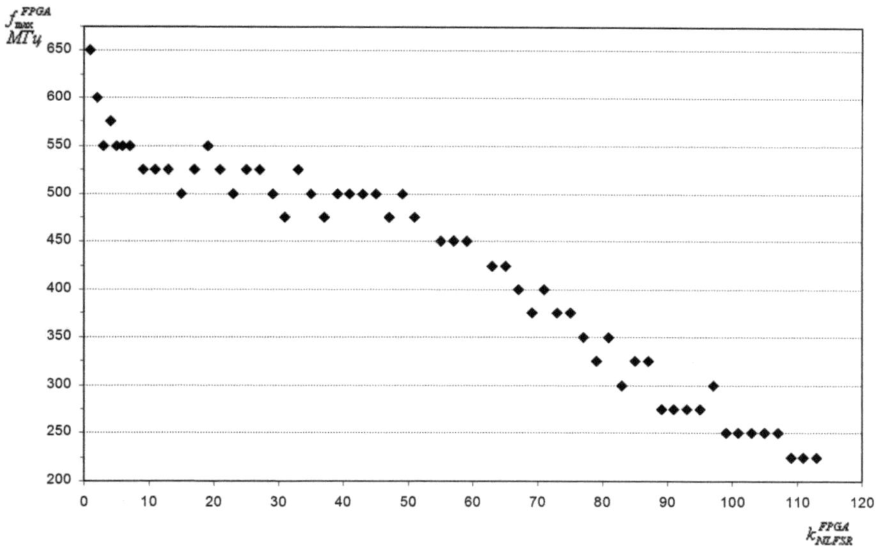

Fig. 14.4 Dependence $f_{\max}^{FPGA}$ on k_{NLFSR}^{FPGA} for $L = 28$

This fact should be taken into account when designing generators using FPGA. Thus, the use of an active cooling system for FPGA to search for M-NLFSR allowed to increase $f_{\max}^{FPGA}$ from 250 MHz for $L = 29$ to 300 MHz for $L = 30$.

Consider in more detail how the above characteristics change from the number of NLFSR tested simultaneously, and n1. Figure 14.4 shows the results of observations $f_{\max}^{FPGA}$ depending on the k_{NLFSR}^{FPGA} for NLFSR with $L = 29$ and $n_1 = 28$.

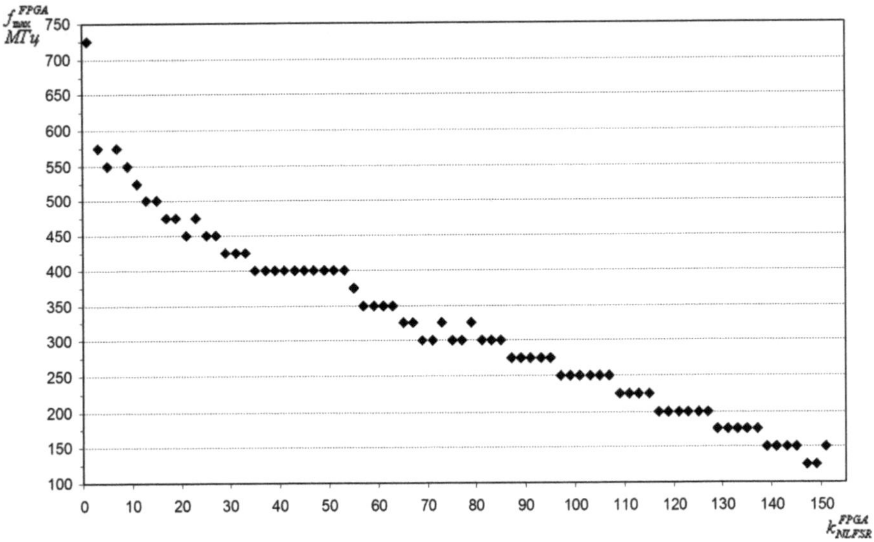

Fig. 14.5 Dependence $f_{\max}^{FPGA}$ on k_{NLFSR}^{FPGA} for $L = 29$

Taking into account the obtained results, the FPGA configuration was modified for $L = 29$, which allowed to increase the performance for = 1 to = 725 MHz, but required a significant increase in the number of LEs used in the *Aij_modul* module. Figure 14.5 shows the results depending on the NLFSR with $L = 29$ and $n_1 = 28$.

During the determination $f_{\max}^{FPGA}$, the clock change was performed discretely, in increments of 25 MHz. Knowing $f_{\max}^{FPGA}$, it is possible to determine t_{cycle}^{FPGA} for a different number of NLFSR, which are simultaneously calculated in FPGA. Figure 14.6 shows the following dependence for NLFSR with $L = 29$ and $n_1 = 28$.

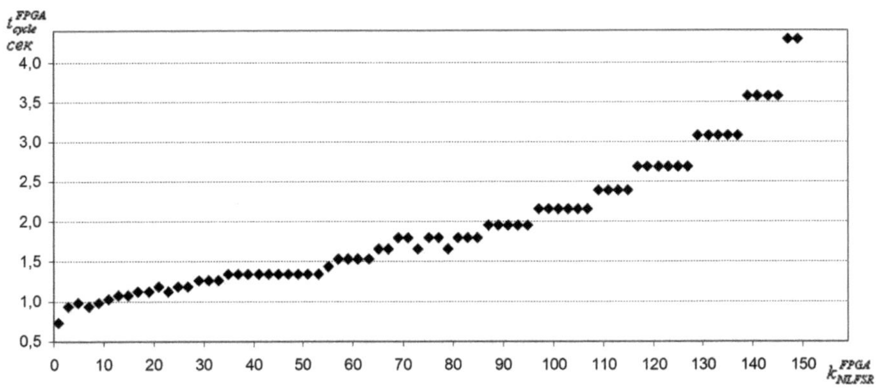

Fig. 14.6 Dependence t_{cycle}^{FPGA} (in seconds) on k_{NLFSR}^{FPGA} for $L = 29$

Based on the obtained data, we can determine the time t_0^{FPGA} required to search for M- NLFSR in the hardware, analyzing all k_0^{FPGA} polynomials. Time spent (in hours) will be calculated by the following ratio:

$$t_0^{FPGA} = \frac{1}{3600сек} \cdot \frac{k_0^{FPGA}}{k_{NLFSR}^{FPGA}} \cdot t_{cycle}^{FPGA}.$$

The estimated time spent searching for M-NLFSR for $L = 29$ under the imposed restrictions, under which $n_1 = 28$, depending on the number of NLFSR, simultaneously calculated in FPGA, is shown in Fig. 14.7.

As can be seen from the graph, the maximum performance of the system is not achieved with the maximum use of all FPGA elements due to the need to reduce the clock frequency. The optimal number of NLFSR, which is simultaneously checked, is in the range from 79 to 103, after which the time spent on the search begins to increase.

Figure 14.8 shows a graph of the depth of verification Requirements 9: time spent on testing in the software part (marker—●); time spent by FPGA for testing k_0^{FPGA} combinations, at the rate of 100 combinations per cycle, and $t_{cycle}^{FPGA} = 3.5$ s (marker—■, the value for T > 17 was determined by prediction); t_0^{FPGA} is calculated as the sum of the first two values (marker—▲). As can be seen, the optimal value for the reduction t_0^{FPGA} corresponds to $T = 1$–17.

The fourteenth section will show the generation time of 1 GB generator in the software implementation language Assembler, optimized for different NLFSR.

According to research, the time spent on the generation of 1 GB is in the range from 16 to 76 s, depending on the size and type of polynomial.

For comparison, Table 14.6 shows the time spent on generating 1 GB of NLFSR implemented in FPGA for different clock speeds f^{FPGA}.

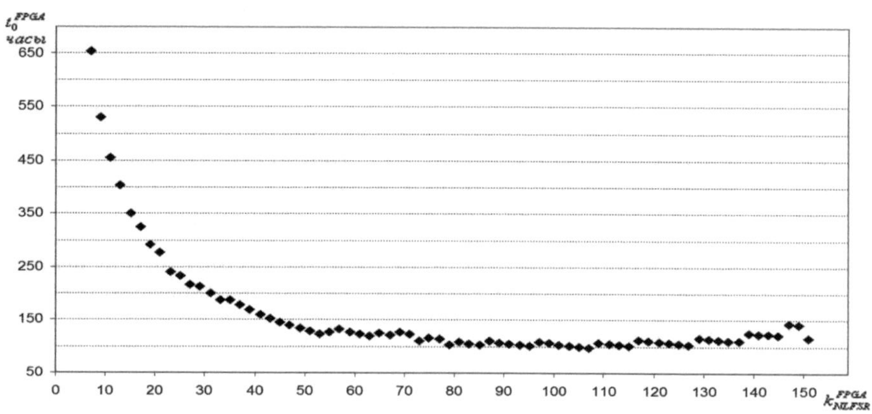

Fig. 14.7 Dependence t_0^{FPGA} (in hours) on k_{NLFSR}^{FPGA} for $L = 29$

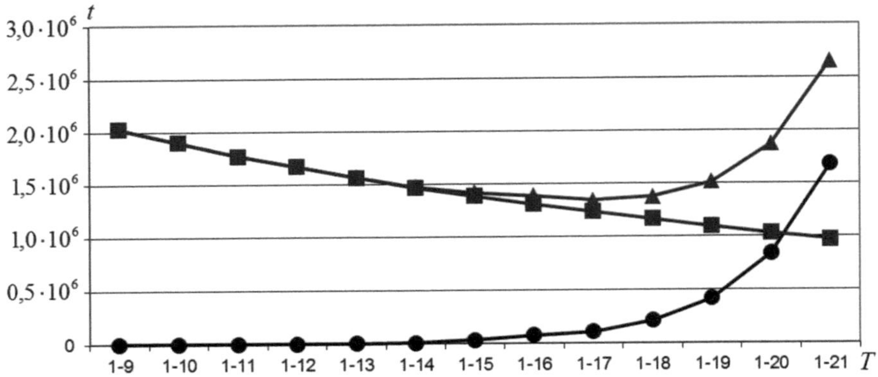

Fig. 14.8 Calculation of time *(t, seconds)* spent on search M-NLFSR, depending on the depth of verification Requirements 9 (T)

Table 14.6 Generation time of 1 GB using FPGA at different f^{FPGA}

f^{FPGA}	50 MHz	100 MHz	200 MHz	400 MHz	600 MHz	700 MHz	725 MHz
t	172 s	86 s	43 s	21.5 s	14.3 s	12.3 s	11.8 s

As can be seen from Table 14.6, the use of a FPGA-based hardware component to generate a PSP can double or more than the CPU.

It should be noted that the above results are valid for the experimental model of FPGA. Using another FPGA model, it is likely to get higher system efficiency. Thus, in [7] the results of the efficiency of hardware implementations of ciphers of the GRACES family for different FPGA families are presented. Unfortunately, Cyclone IV was not present in the studied families, but from the FPGAs described in [7], the Stratix IV and Stratix V families showed more than twice the efficiency of the FPGAs of the Cyclone II or Cyclone V family.

14.3.5 Search Results for M-NLFSR—List of M-NLFSR Dimension from L = 26 to L = 30

Using the method described in the work, a search for M-NLFSR with dimensions $L = 26$, 27, 28, 29 and 30. Polynomials, constraints and search parameters, as well as time costs are given below.

$L = 26$:

$$x^{26} + x^{24} + x^{23} + x^{20} + x^{18} + x^{14} + x^{12} + x^6 + x^5 + x^4 + x^{23} \cdot x^{25} + x^{24} \cdot x^{25}$$

$$x^{26} + x^{25} + x^{22} + x^{21} + x^{19} + x^{17} + x^{16} + x^{15} + x^{13} + x^{12}$$

$$+x^6+x^5+x^4+x^{23}\cdot x^{25}$$
$$x^{26}+x^{24}+x^{22}+x^{21}+x^{20}+x^{19}+x^{17}+x^{16}+x^{15}+x^{14}+x^{12}$$
$$+x^{11}+x^8+x^7+x^4+x^{23}\cdot x^{24}+x^{23}\cdot x^{25}+x^{24}\cdot x^{25}$$
$$x^{26}+x^{22}+x^{21}+x^{20}+x^{19}+x^{17}+x^{16}+x^{15}+x^{14}+x^{12}+x^{11}$$
$$+x^8+x^7+x^4+x^{23}\cdot x^{24}+x^{24}\cdot x^{25}$$

Search took 43 h 48 min.

The search was conducted in two stages: the first—10 h 23 min—the time spent on the previous program calculation (verification of compliance with Requirement 9 with the depth of verification $T = 1, \ldots, 18$; restrictions: linear coefficients $a_{1j} = a_{2j} = 0$; nonlinear coefficients are all zero, except $a_{23\_24}, a_{23\_25}, a_{24\_25}$. The introduced restrictions reduced the number of possible combinations from $4.6 \cdot 10^{105}$ to 67,108,864, of which only 4,205,758 met the Requirement 9); the second—33 h 25 min—time spent on hardware verification of other combinations for M-sequence generation (verification was performed in 140 combinations per cycle at a clock frequency of 50 MHz, data transmission to two parallel channels at a rate of 128,000 baud).

$L = 27$:

$$x^{27}+x^{26}+x^{25}+x^{24}+x^{19}+x^{16}+x^{15}+x^{11}+x^8+x^6+x^5+x^3$$
$$+x^{24}\cdot x^{25}+x^{24}\cdot x^{26}$$
$$x^{27}+x^{26}+x^{19}+x^{16}+x^{15}+x^{13}+x^{12}+x^{11}+x^{10}+x^9+x^8+x^7$$
$$+x^6+x^5+x^4+x^3+x^{24}\cdot x^{26}+x^{25}\cdot x^{26}$$
$$x^{27}+x^{24}+x^{22}+x^{20}+x^{19}+x^{17}+x^{16}+x^{13}+x^{10}+x^8$$
$$+x^7+x^6+x^4+x^{25}\cdot x^{26}$$
$$x^{27}+x^{24}+x^{23}+x^{17}+x^{14}+x^9+x^8+x^7+x^5+x^4+x^3+x^{25}\cdot x^{26}.$$

Search was conducted in 33 h 04 min.

The search was conducted in two stages: the first—14 h 52 min—time spent on the previous program calculation (verification of compliance with Requirement 9 of $T = 1, \ldots, 18$; restrictions: linear coefficients $a_{1j} = a_{2j} = 0$; nonlinear coefficients all are zero, except $a_{24\_25}, a_{24\_26}, a_{25\_26}$. The restrictions introduced have reduced the number of possible combinations from $6.2 \cdot 10^{113}$ to 134,217,728, of which only 8,409,664 met Requirement 9); the second—18 h 12 min—time spent on hardware verification of other combinations for M-sequence generation (verification was performed in 133 combinations per cycle at a clock frequency of 175 MHz, data transmission to two parallel channels at a speed of 921,600 baud).

$L = 28$:

$$x^{28}+x^{27}+x^{25}+x^{24}+x^{22}+x^{20}+x^{19}+x^{16}+x^{14}+x^{13}+x^{10}$$

$$+x^8+x^5+x^{25}\cdot x^{27}$$
$$x^{28}+x^{27}+x^{26}+x^{25}+x^{24}+x^{23}+x^{22}+x^{21}+x^{20}+x^{16}+x^{15}$$
$$+x^{14}+x^{13}+x^{12}+x^{11}+x^{10}+x^7+x^6+x^{25}\cdot x^{26}+x^{25}\cdot x^{27}$$
$$x^{28}+x^{25}+x^{20}+x^{19}+x^{17}+x^{16}+x^{13}+x^{11}+x^{10}+x^9+x^8$$
$$+x^7+x^6+x^3+x^{25}\cdot x^{27}+x^{26}\cdot x^{27}$$
$$x^{28}+x^{24}+x^{23}+x^{20}+x^{18}+x^{17}+x^{16}+x^{14}+x^{13}$$
$$+x^{12}+x^5+x^{25}\cdot x^{27}$$
$$x^{28}+x^{27}+x^{26}+x^{25}+x^{15}+x^{14}+x^{11}+x^{10}$$
$$+x^{25}\cdot x^{26}+x^{25}\cdot x^{27}$$
$$x^{28}+x^{25}+x^{24}+x^{21}+x^{20}+x^{18}+x^{17}+x^{16}+x^{15}+x^{14}$$
$$+x^{13}+x^{12}+x^{10}+x^9+x^5+x^3+x^{25}\cdot x^{27}+x^{26}\cdot x^{27}$$

Search was conducted for 3 days 01 h 25 min.

The search was conducted in two stages: the first—18 h 03 min—time spent on the previous program calculation (verification of compliance with Requirement 9 of $T = 1, \ldots, 17$; restrictions: linear coefficients $a_{1j} = a_{2j} = 0$; nonlinear coefficients all are zero except $a_{25\_26}, a_{25\_27}, a_{26\_27}$.The restrictions introduced reduced the number of possible combinations from $1.7 \cdot 10^{122}$ to 268,435,456, of which only 17,619,849 met Requirement 9); the second—2 days 07 h 22 min—time spent on hardware verification of other combinations for M-sequence generation (verification was performed in 127 combinations per cycle at a clock frequency of 200 MHz, data transmission in two parallel channels at a speed of 921,600 baud).

$L = 29$:

$$x^{29}+x^{28}+x^{27}+x^{26}+x^{25}+x^{24}+x^{23}+x^{20}+x^{16}+x^{12}$$
$$+x^{11}+x^6+x^4+x^{26}\cdot x^{27}$$
$$x^{29}+x^{28}+x^{27}+x^{26}+x^{25}+x^{21}+x^{19}+x^{15}+x^{10}+x^9$$
$$+x^8+x^6+x^4+x^{26}\cdot x^{28}$$

Search was conducted in 5 days 13 h 26 min.

The search was conducted in two stages: the first—14 h 54 min—the time spent on the previous program calculation (verification of compliance with Requirement 9 with $T = 1, \ldots, 17$; restrictions: linear coefficients $a_{1j} = a_{2j} = a_{3j} = 0$; nonlinear coefficients are all zero, except $a_{26\_27}, a_{26\_28}, a_{27\_28}$. The introduced restrictions reduced the number of possible combinations from $8.9 \cdot 10^{130}$ to 268,435,456, of which only 17,619,713 met the Requirement 9); the second—4 days 22 h 32 min—time spent on hardware verification of other combinations for M-sequence generation (verification was performed in 96 combinations per cycle at a clock frequency of 250 MHz, data transmission to two parallel channels at a speed of 921,600 baud).

$L = 30$:

$$x^{30} + x^{28} + x^{27} + x^{26} + x^{25} + x^{22} + x^{21} + x^{17} + x^{15} + x^{12}$$
$$+ x^{11} + x^{4} + x^{27} \cdot x^{28} + x^{27} \cdot x^{29}$$

Search was conducted in 16 days 6 h 37 min.

The search was conducted in two stages: the first—1 day 6 h 21 min—the time spent on the previous program calculation (verification of compliance with Requirement 9 with $T = 1, \ldots, 17$; restrictions: linear coefficients $a_{1j} = a_{2j} = a_{3j} = 0$, nonlinear coefficients are all zero, except a_{27_28}, a_{27_29}, a_{28_29}. The introduced restrictions have reduced the number of possible combinations from 9.5×10^{139} to 536,870,912, of which only 35,239,039 met the Requirement 9); second—15 days 0 h 16 min—time spent on hardware verification of the rest of the combination to generate M-sequence (verification was performed in 101 combinations per cycle at a clock frequency of 300 MHz, using active cooling FPGA, data transmission in two parallel channels at 921,600 baud).

It is worth noting that it took 15 days to fully test the entire set of combinations, with the first M-NLFSR was found 46 h after the start of the hardware.

The use of a hardware component has significantly reduced the search time for M-NLFSR. As an example, when searching by software at $L = 25$, 18 days were spent, and when using the hardware component for $L = 26$—less than two days. The estimated time for which a similar search would be performed for $L = 26$, using the software method, would be more than 3 months. Given the modifications introduced into the search algorithm for $L = 27$, 28, 29 and 30, it is possible to expect a threefold increase in performance for $L = 26$ when using the hardware component.

Thus, using this method, we get an overall increase in productivity for searching M-NLFSR more than 150 times compared to the method that uses only the CPU resources of a personal computer. And the use of more expensive and powerful FPGAs will further reduce the time to search for M-NLFSR.

14.3.6 Conclusions on the Use of FPGA in the Search for M-NLFSR

The method of M-NLFSR search by means of a hardware-software complex is resulted. Based on the methodology, as an example, the search result of nonlinear polynomials with powers 26, 27, 28, 29 and 30, by which M-sequences are generated.

The stated technique allows to carry out search of M-NLFSR with the set design characteristics. In all the above polynomials of degrees with $L = 22$ and up to 30 there is no feedback from the first few coefficients, which allows them to implement simultaneous generation of several bits of the original sequence by increasing the number of implementations of feedback functions (similar to the stream cipher Grain).

The use of hardware components based on FPGA can significantly increase the performance of the M-NLFSR search complex. In the calculations, the time spent searching for M-NLFSR was reduced by more than 150 times, compared to using only the computing power of a personal computer.

The possibilities of productivity of NLFSR realized on the basis of FPGA are given. Recommendations for optimizing the performance of the M-NLFSR search complex are given.

A comparison of the generator performance at the NLFSR, performed using software and hardware implementation. It is shown that the hardware implementation based on FPGA single NLFSR is not inferior in performance to its software counterpart, and in the case of using a generator with several parallel NLFSR can significantly exceed it.

Implemented stable FPGA operation based on NLFSR with a clock speed (lasing speed in NLFSR) equal to 725 MHz.

14.4 Search for MRZNZZ with the Application of Parallel Calculation at GPU

14.4.1 Use of Parallel Computing Technologies in GPUs

In the last few years, a promising area of increasing the computing power of computer systems has been the transfer of computing from the central processing unit (CPU—from the English processing unit) to graphics processors (GPU—from the English graphics processing units). This concept has been supported by major graphics accelerator manufacturers, including NVIDIA.

At this time, the processors used in general-purpose computers have reached the clock speed of about 3 GHz. Therefore, further productivity growth is carried out by parallelization [8]. Intel and AMD processor parallel execution tools include sets of instructions for performing vector operations, as well as the implementation of several processor cores on a single chip—real or virtual (using HyperThreading). However, the possibilities for parallelization of traditional processors are limited. This is hindered by their (starting with 8086) consistent architecture and a rather complex set of instructions, which requires, for quick implementation, quite complex elements of the processor—the pipeline, the system for predicting transitions, caches and more.

The NVIDIA graphics card architecture, which allows them to be used for non-graphical computing, is called CUDA (Common Unified Device Architecture), and the developed specification [9] facilitated the process of creating algorithms on the GPU. CUDA technology was first introduced by NVIDIA in 2007 [10]. CUDA can be used on GPUs manufactured by NVIDIA, such as GeForce, Quadro, Tesla. The latest line of devices (Tesla) has no video output at all and is designed exclusively for high-performance computing.

NVIDIA distributes a number of examples that demonstrate the solution of some problems in various fields of mathematics, physics and computer science. The increase in performance in these tasks during the transition from CPU to GPU is from 10 to 100 times.

Since the graphics card is a separate computing module with its own computing node and RAM, the program for CUDA code is divided into two parts: host—executed on the CPU; device—runs on the GPU.

Concurrency in CUDA is provided as follows: when running the code for execution, not one, but a group of threads is created—Fig. 14.9. The whole group of threads created during code execution is called a grid. The elements of the grid are blocks. They form a one-dimensional or two-dimensional structure. Each block, in turn, consists of a group of threads combined into a one-dimensional, two-dimensional, or three-dimensional array. The number of blocks in the grid and threads in the block is set at code start.

A parallel program in a messaging model is a set of regular sequential programs that are processed simultaneously. Typically, each of these serial programs runs on its own processor and has access to its own local memory. A clear advantage in such a computing organization is the ability to write and debug a program on a single-processor system.

To achieve high efficiency in massively parallel computing, you need to consider many factors: the architectural features of the GPU, the speed and order of memory access, the mechanisms of synchronization between computing threads. An important role is also played by the adaptability of the algorithm to run in parallel. The transfer of simple arithmetic operations to the GPU gives a performance advantage over the CPU when using even the simplest models of graphics adapters [11].

The CPU architecture allows the use of almost any parallel computing templates, while the CUDA architecture allows only one instruction to be executed on all its cores at any given time [12].

Therefore, the CUDA architecture can only be effectively applied to computations with high parallelism and intensive arithmetic. All functions implemented on the GPU do not support recursion and have some other limitations that are not present in the CPU architecture.

Modern GPUs provide a good value for money, performance and power consumption. Most supercomputer clusters have a high rating in the TOP500 due to the use of GPUs [13]. However, as specialized devices designed for streaming the same type of data, GPUs on many algorithms do not show a significant advantage over processors of traditional architecture [14]. This is due to the fact that modern GPUs use SIMD architecture [15] and specially designed memory controllers to work with it [16]. GPUs are "massively parallel" processors, ie computing devices that run a large (compared to CPU) number of computational threads.

The study [17] compares the efficiency of CPU and GPU implementations of some combinatorial algorithms used in cryptanalysis. It is shown that the use of special control flow transformation techniques can significantly compensate for performance losses that occur due to inefficient execution of conditional transitions on the SIMD

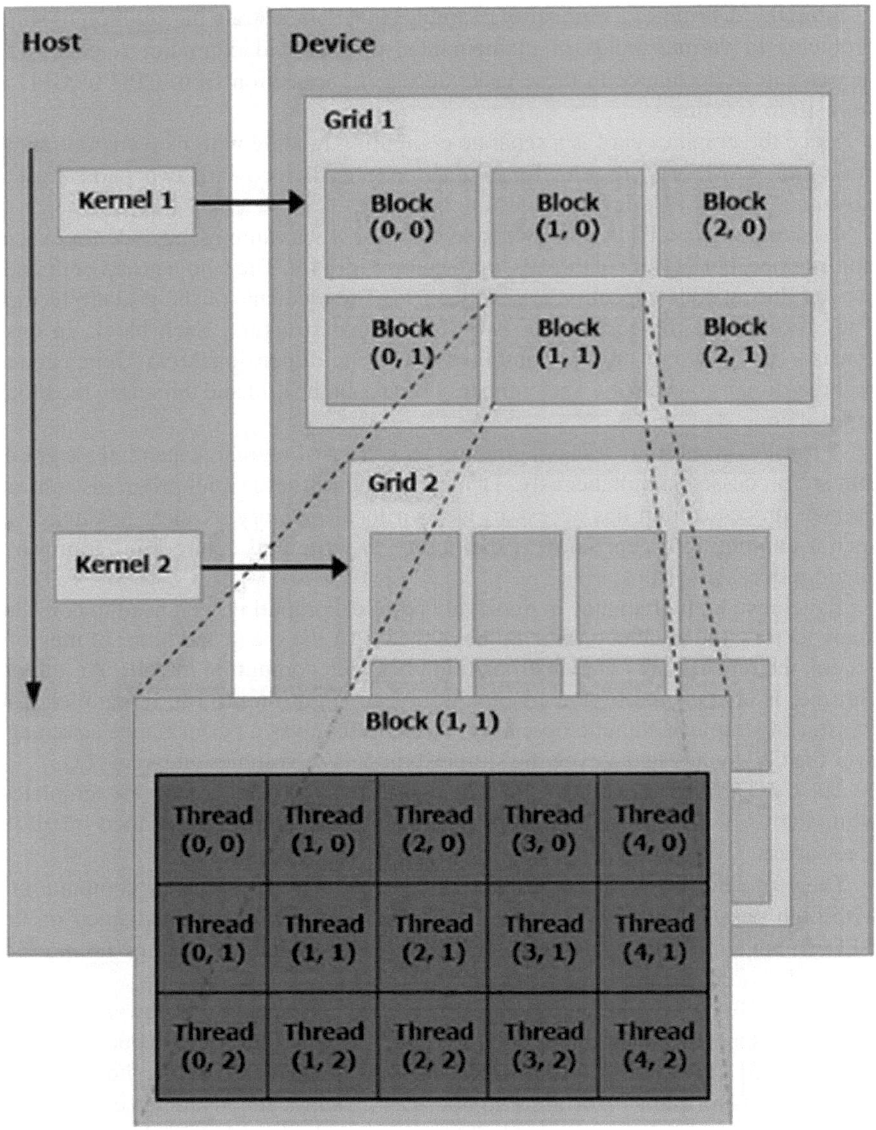

Fig. 14.9 The structure of the organization of threads on the GPU

device. However, the limitations imposed by the memory mechanisms used in modern GPUs for the considered algorithms are insurmountable.

14.4.2 GPU Application for Streaming Encryption Systems

Here is a study of the possibilities of transferring part of the calculations for the generation of PRS from CPU to GPU and comparing the performance of the obtained solutions. By performance we mean the number of bits generated per unit time.

To assess the capabilities of the GPU used M-NLFSR, taken from the algorithm that was presented in 2008 at the international competition eSTREAM Achterbahn128/80 version 1.1.

It should be noted that the Achterbahn stream cipher was primarily developed for hardware implementation. Some of its possibilities, such as the use of multiple implementation of the feedback function using the parallel implementation of the NLFSR, which allows to increase the number of generated bits per clock cycle, we have not considered.

The feedback functions for all 13 M-NLFSR (A0, …, A12) used in the stream cipher Achterbahn128/80 were performed according to the optimized implementation given in [18]. The feedback functions for each of the 13 registers have the same logical depth [19], which provides approximately the same generation rate for each register.

In addition to the M-NLFSR taken from the construction of the stream cipher Achterbahn128/80, the M-NLFSR found by us having the size $L = 30$ was also investigated (see Sect. 14.3.5). Let's denote its feedback function as A30.

Microsoft Visual Studio 2015 software and software platform for organizing parallel computing on NVIDIA CUDA 8.0 graphics processors were used to compile the software code. The calculations were performed on a personal computer running 64-bit Windows 7 SP 1, with an Intel Core i53210M CPU 2.5 GHz processor and NVIDIA GeForce GT 630 M video processor.

First, we present the results of measuring the rate of M-NLFSR generation on the CPU and GPU. Table 14.7 shows the time (t_{GB}), in seconds, spent on generating 1 GB (8,589,934,592 iterations) of data by each of the M-NLFSR A0, …, A12 both separately and for all together.

It is necessary to pay attention that here and further under generation we understand idle run of all registers necessary number of times without full transfer or record of the received values. During the actual generation, it is necessary to take into account the transfer of the generated values, which will also take some time.

As you can see, thanks to the optimized implementation, the execution time of each of the registers on both the CPU and the GPU is approximately the same (spread of no more than 14% for the CPU and 11% for the GPU). Thus, if we run on the GPU each of the NLFSR in parallel in a separate thread, we will get the output bit without significant downtime of any of the registers.

Table 14.8 shows the test results for the simultaneous calculation of iterations from A_0 to A_{12} register. Because the A_{12} used a 33-cell register, all registers used 64-bit values for versatility, with which the GPU runs slower than 32-bit values. The calculations were performed in one stream. Each register was calculated in its own

Table 14.7 Time spent on generating 1 GB for CPU (t_{GB}^{C}) and GPU (t_{GB}^{G})

M-PCHOC	A_0	A_1	A_2	A_3	A_4	A_5	A_6	A_7	A_8	A_9	A_{10}	A_{11}	A_{12}	$A_0 \ldots A_{12}$
t_{GB}^{C}(s)	107.4	112.5	118.1	115.6	117.7	114.2	115.3	118.3	128.4	130.5	140.9	106.8	133.8	133.8
t_{GB}^{G}(s)	2408	2457	2310	2719	2621	2736	2506	2834	2556	2589	2687	2556	2294	26,033

Table 14.8 Time costs for generation of 1 GB on GPU at simultaneous generation of several M-NLFSR

n_{trd}		13	26	39	52	65	78	91	104	117	130	143	156	169	182
t_{GB}^{G}(s)	One block	26,033	26,476	26,509	26,509	26,689	29,474	29,540	30,113	30,572	30,867	34,258	34,373	34,603	34,865
	Two blocks	26,476	26,492	26,492	26,492	26,705	29,720	29,605	30,244	30,523	30,900	34,717	34,504	34,799	34,881

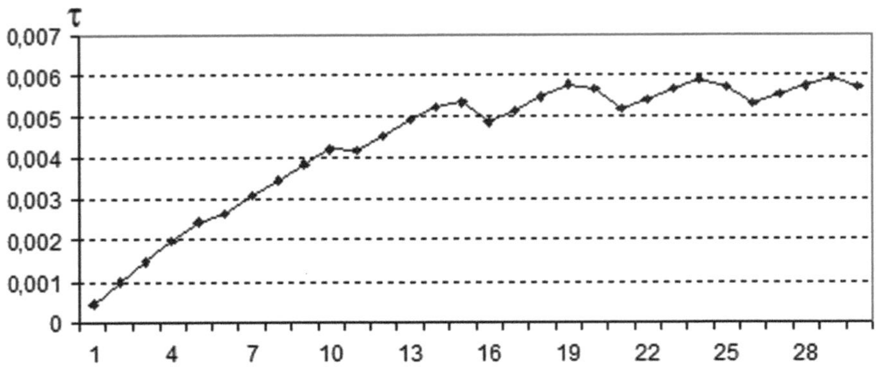

Fig. 14.10 Dependence of generation performance (τ) on GPU on the number of registers calculated simultaneously from A0 to A12

thread, i.e. a total of 13 threads were run to calculate one set of 13 registers. The total number of threads used is denoted by the symbol n_{trd}.

As can be seen from Table 14.8, as the number of simultaneously calculated registers increases, the time spent on calculations for each of the iterations increases according to the nonlinear law. Therefore, it is more appropriate to consider the generation performance (τ), which is defined as the number of values generated simultaneously by registers in all streams per unit time:

$$\tau = \frac{n_{trd}}{t_{GB}}\left[\frac{Gbyte}{second}\right]$$

The results obtained from Table 14.7 are more clearly shown in the graph in Fig. 14.10. As you can see, increasing the number of simultaneously calculated registers, the productivity reaches its limit value and a further increase in the number of registers that are calculated simultaneously, does not increase productivity.

Figures 14.11 and 14.12 show the performance results for the different number of blocks and threads involved in each block for the 32-bit and 64-bit variables used as registers.

As you can see, the GPU generation at the size of the variables used at 64 bits is much slower than at 32 bits.

The optimal value of performance corresponds to 16 blocks and the number of threads in the block, a multiple of 32. Figure 14.13 shows a graph of the performance of M-NLFSR A_{11} and A_{30} for 32-bit configuration at 16 blocks depending on the number of threads running in the block.

The maximum value for A_{11} is reached in 384 threads and corresponds to the generation at a rate of 0.328 GB/s (equivalent to the generation of 1 GB in 3 s). As we can see, the maximum values are formed when using the number of threads multiples of 32, which corresponds to the size of the warp (warp)—the number of physically simultaneously performed threads [20].

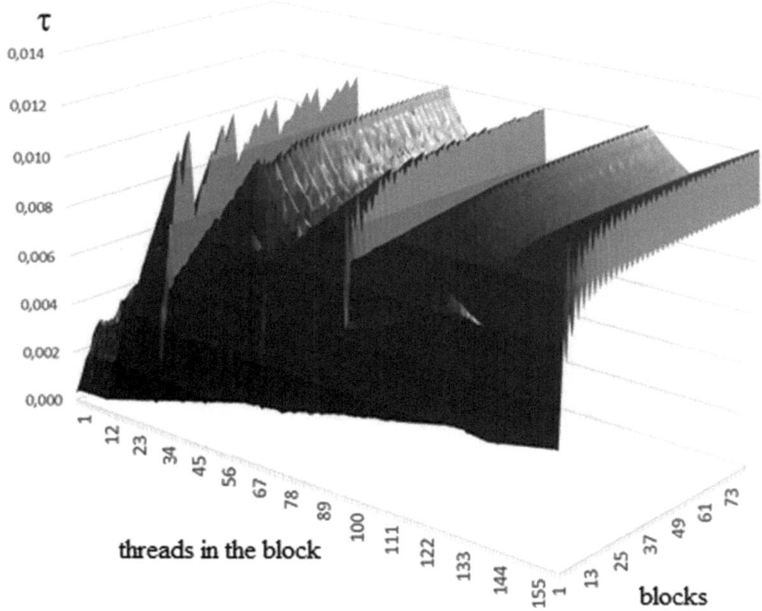

Fig. 14.11 Dependence of productivity (τ) of generation on GPU on number of blocks and threads in the block at dimension of registers in 64 bits

Using two or four streams instead of one does not increase the overall maximum performance.

When used for CPU generation only, generation was performed at a speed of 0.0119 GB/s (equivalent to 1 GB generation in 84 s). Thus, when using the GPU, performance increased 28 times.

It is worth noting that when using machine-oriented low-level language (Assembler) it is possible to increase the speed of generation of RZNZZ on the CPU. As will be shown in Sect. 14.9, it is possible to obtain a generation speed of 1 GB in a few tens of seconds, depending on the type of polynomial formed. Translating part of the code to the GPU in the PTX Assembler did not show a significant increase in performance.

In the case of replacing MLSNOS from the Achterbahn algorithm with a polynomial with second-order nonlinearity—A30 and equal other conditions, the average lasing speed on the CPU was 0.015 GB/s (equivalent to 1 GB lasing in 65.8 s). At GPU—0.626 GB/s (equivalent to 1 GB generation in 1.6 s).

Thus, when using GPU and M-NLFSR of the second order, which has a simpler, in terms of computational costs, the structure, there was an increase in performance by more than 41 times.

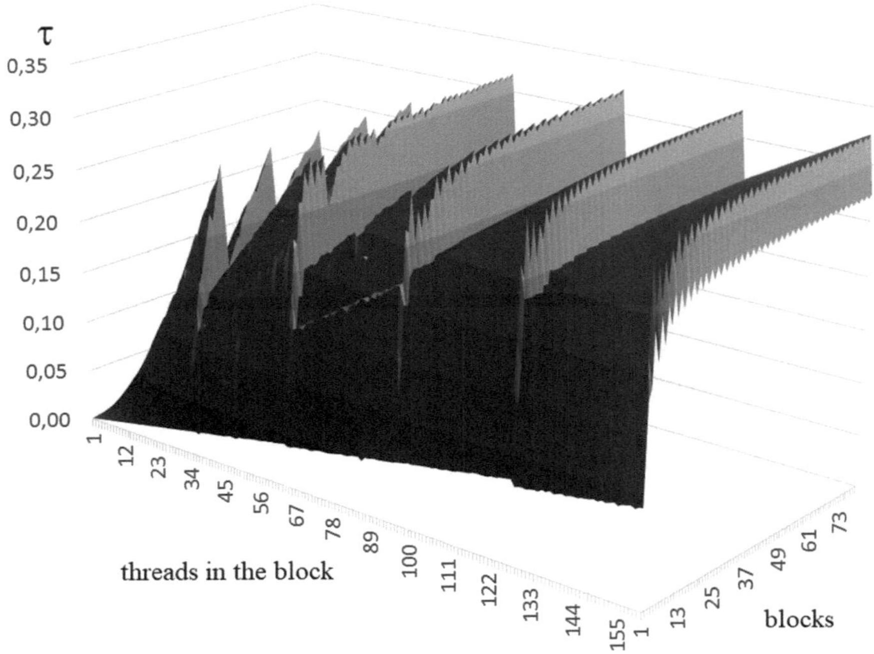

Fig. 14.12 Dependence of productivity (τ) of generation on GPU on number of blocks and threads in the block at dimension of registers in 32 bits

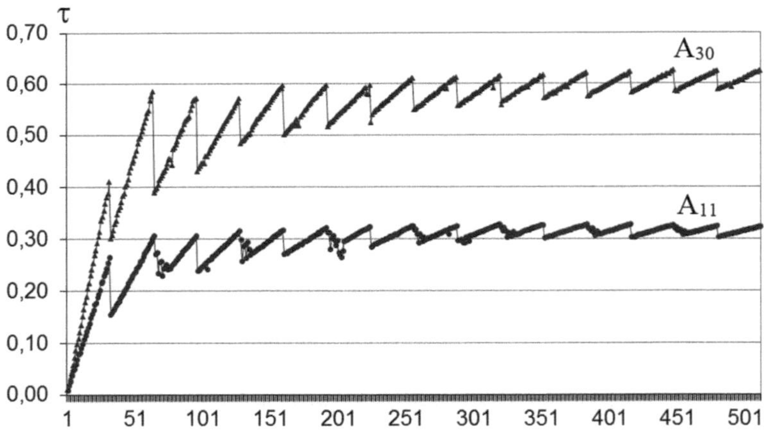

Fig. 14.13 Generation performance (τ) on the GPU depending on the number of threads in a block at 16 blocks

14.4.3 GPU Application in the Second Stage of M-NLFSR Search Obtained M-NLFSR Dimension $L = 31, 32$

However, the obtained result is significantly inferior to the productivity of NLFSR using FPGA calculations. However, the computing power of the GPU can be used at the stage of previous calculations, which are characterized by the use of a significant amount of RAM, which makes inefficient the use of FPGA-based technologies.

The calculations showed a reduction in the time spent in the second stage by more than five times, and as a result the following M-NLFSR were obtained:

$L = 31$:

$$x^{31} + x^{27} + x^{24} + x^{22} + x^{21} + x^{17} + x^{15} + x^{12} + x^{11} + x^{19} + \\ + x^{20} \cdot x^{22} + x^{20} \cdot x^{23} + x^{20} \cdot x^{25} + x^{20} \cdot x^{27} + x^{20} \cdot x^{29} + x^{20} \cdot x^{30}$$

$$x^{31} + x^{30} + x^{27} + x^{26} + x^{25} + x^{23} + x^{19} + x^{17} + x^{15} + x^{14} + x^{13} + x^{11} + x^{10} + \\ + x^{20} \cdot x^{21} + x^{20} \cdot x^{23} + x^{20} \cdot x^{27} + x^{20} \cdot x^{28} + x^{20} \cdot x^{29}$$

$$x^{31} + x^{29} + x^{28} + x^{26} + x^{24} + x^{23} + x^{21} + x^{18} + x^{17} + x^{16} + x^{15} + x^{13} + x^{10} + \\ + x^{20} \cdot x^{21} + x^{20} \cdot x^{22} + x^{20} \cdot x^{23} + x^{20} \cdot x^{25} + x^{20} \cdot x^{26} + x^{20} \cdot x^{27} + x^{20} \cdot x^{30}$$

Search was conducted in 16 days 22 h 16 min.

The search was conducted in three stages: the first—2 days 13 h—the time spent on the formation of Requirements 9 with $T = 1, \ldots, 20$; the second—1 day 9 h—time spent on direct verification of compliance with Requirement 9 (during the verification, parallel computing was used on the GPU, the program consumed 413,692 KB of RAM); the third—13 days 0 h 16 min—spent on hardware inspection.

Restrictions are introduced: linear coefficients $a_{ii} = 0$ for all $i = 1, \ldots, 8$; $a_{31\_31} = 1$; all nonlinear coefficients $a_{ij} = 0$, except $i = 20$ and $j = 21, \ldots, 30$. Thus, 22 linear and 10 nonlinear feedback coefficients (32 in total) could take the free value in testing. Of the initial $2^{496} \approx 2 \cdot 10^{149}$ possible combinations of a_{ij}, $2^{32} = 4\,294\,967\,296$ NLFSR were tested, of which 245,553,141 (5.7%) NLFSR correspond to Requirements 9 (with depth of check $T = 1, \ldots, 20$).

$L = 32$:

$$x^{32} + x^{30} + x^{29} + x^{20} + x^{18} + x^{16} + x^{14} + x^{13} + \\ + x^{19} \cdot x^{22} + x^{19} \cdot x^{24} + x^{19} \cdot x^{25} + x^{19} \cdot x^{27} + x^{19} \cdot x^{29} + x^{19} \cdot x^{31}$$

$$x^{32} + x^{30} + x^{28} + x^{27} + x^{20} + x^{19} + x^{16} + x^{15} + x^{14} + x^{13} + \\ + x^{19} \cdot x^{20} + x^{19} \cdot x^{22} + x^{19} \cdot x^{25} + x^{19} \cdot x^{27} + x^{19} \cdot x^{30} + x^{19} \cdot x^{31}$$

$$x^{32} + x^{29} + x^{26} + x^{25} + x^{24} + x^{23} + x^{22} + x^{20} + x^{18} + x^{17} + x^{16} + x^{14} + x^{13} + \\ + x^{19} \cdot x^{21} + x^{19} \cdot x^{22} + x^{19} \cdot x^{24} + x^{19} \cdot x^{26} + x^{19} \cdot x^{27} + x^{19} \cdot x^{28} + x^{19} \cdot x^{31}$$

Search was conducted in 5 days 19 h 42 min.

The search was conducted in three stages: the first—1 day 2 h 4 min—time spent on the formation of Requirements 9 with $T = 1, \ldots, 20$; the second—17 h 28 min—time spent on direct verification of compliance with Requirement 9 (parallel GPU computing was used during the verification); the third—4 days 0 h 10 min—spent on hardware inspection (checked 30% of the total volume in which the above MLRS was found, and the inspection was stopped).

Restrictions are introduced: linear coefficients $a_{ii} = 0$ for all $i = 1, \ldots, 12$; $a_{31\_31} = 1$; all nonlinear coefficients $a_{ij} = 0$, except $i = 19$ and $j = 20, \ldots, 31$. Thus, 19 linear and 12 nonlinear feedback coefficients could take the free value in testing (31 in total). From the initial $2^{528} \approx 8.8 \times 10^{158}$ possible combinations $\mathrm{a_{ij}}$ $2^{31} = 2147\,483\,648$ NLFSR were tested, of which Requirements 9 (with depth of check $\mathrm{T} = 1, \ldots, 20$) correspond to 122,835,043 (5.7%) NLFSR.

Six DE2 mock-ups with a built-in Altera Cyclone II EP2C35F672C6 chip were used as a hardware platform to search for LRSNOS with dimension $\mathrm{L} = 31, 32$ (the test was performed in 127 combinations per cycle at a clock frequency of 150 MHz; 4 output bits were generated per clock cycle). quadruple implementation of the feedback function, using active cooling of FPGA; data transmission to one channel at a speed of 921,600 Baud). The use of more powerful hardware increased productivity up to fifteen times in the third phase of M-NLFSR search.

As you can see, in all found M-NLFSR dimension L = 31 there is no feedback from the first 9 cells, and in M-NLFSR dimension L = 32—from the first 12 cells, so it is possible to implement simultaneous generation of up to 10 output bits in the first case and up to 13 in the second due to the repeated implementation of the feedback function, which will increase up to 10 and up to 13 times, respectively, the rate of PRS generation.

In addition, the inherent nonlinearity only from the 20th cell for L = 31 and from the 19th for L = 32 reduces the number of operations required to generate one bit. Thus, for the first of the above M-NLFSR dimension L = 31 to obtain the output bit you need to perform 15 XOR operations and 1 AND operation. For the first of the above M-NLFSR dimension L = 32—you need to perform 13 XOR operations and 1 AND operation. For which, in both cases, only 5 Les need to be used in the FPGA models used for testing.

14.5 Conclusions

1. A mathematical model based on the solution of a system of linear equations has been developed, thanks to which it was possible to significantly optimize the calculations performed during the verification of compliance with Requirement 9 and reduce the computational complexity of enumeration methods and procedures for synthesis of nonlinear shift registers $\left(0,95 \cdot 2^{\frac{L\cdot(L+1)}{2}-k}\right)$ times.

The application of the proposed mathematical model allowed to find M-NLFSR with dimension from L = 19 to L = 25.

2. Possibilities of hardware realization of NLFSR on the basis of FPGA are investigated. The productivity of various NLFSR realized on the basis of FPGA is given. Implemented stable FPGA operation for PRS generation based on NLFSR with a clock frequency (generation speed in NLFSR) equal to 725 MHz.
3. The hardware-software complex is developed and the method of search of M-NLFSR is offered. The method includes the implementation of verification (using the CPU and GPU) of the NLFSR for compliance with the Requirements with subsequent hardware (using FPGA) verification of the generation of M-Sequences.

   As an example of realization of possibilities of a hardware-software complex the result of search of nonlinear polynomials with degrees 26, 27, 28, 29, 30, 31 and 32 by means of which M-sequences are generated is resulted.
4. The proposed method of searching M-NLFSR is unique, the stated search approach is not found in any of the available sources. The stated method is an alternative to the existing one and its efficiency is not inferior to the existing methods of M-NLFSR search.
5. The found polynomials have the nonlinearity of the whole second order, and not the fourth—sixth, as in the Achterbahn flow encryption algorithm128 [19]. In this case, the average search time of one NLFSR, which generates a sequence of the maximum period of size L = 27, according to the information and the author's method given in [21], was 21 h. To search for M-NLFSR according to the proposed method, the average search time of the register of the same size was 8.25 h. It is worth noting that the equipment we use to search for NLFSR is much inferior to the complex used in [21].
6. The proposed method allows the use of parallel computing at almost all stages, thereby increasing the possibilities for its application.

   Using the method allows you to specify at the design stage the desired amount of feedback to be used in the register, as well as the maximum number and location of possible feedback factors, thereby searching for M-NLFSR under the specified specifications.

   As an example, in all found M-NLFSR dimension L = 32 there is no feedback from the first 12 cells, so it is possible to implement the simultaneous formation of up to 13 output bits due to multiple implementation of the feedback function, which will increase up to 13 times the generation rate PRS.
7. On the basis of the received results development of streaming ciphers of the increased cryptographic stability is possible. The use of the obtained nonlinear polynomials will allow to design stream ciphers, which are analogs of modern developments in this field, and the application of the described search technique will allow to find M-NLFSR of a higher degree.

References

1. Koopman, P.: Maximal Length LFSR Feedback Terms. [Electronic resourse]. Access mode: https://users.ece.cmu.edu/~koopman/lfsr/
2. Kulikova A.S., Lysenko I.V.: Realization of multiversion current cryptoconversion of data with use of keyless hashfunctions on programmable logic. Inf. Process. Syst. 7 (105). 22–26 (2012)
3. Kulanov V., Kharchenko V., Perepelitsyn, A.: Parameterized IP infrastructures for faulttolerant FPGAbased systems: development, assessment, casestudy [Text]. In: Proceedings of IEEE EastWest Design & Test Symposium (EWDTS 2019), 2009, pp. 322–325
4. Perepelitsyn A.E.: The use of parameterized IP infrastructures for the development of embedded fault-tolerant systems on FPGAs. Radio Electron. Comput. Syst. 2016, 5 (79), 104–112
5. Kolesnik I.N., Kulanov V.A., Perepelitsyn, A.E.: Analysis of the use of FPGA technology as part of the cloud infrastructure. Radio Electron. Comput. Syst. 2016, Issue 6 (80), pp. 130–135
6. Perepelitsyn, A.: FPGA technologies in medical equipment: electrical impedance tomography [Text]. In: Perepelitsyn, A., Shulga D. (eds.) Proceedings of IEEE EastWest Design & Test Symposium (EWDTS 2012). 2012, pp. 437–440
7. Klyucharyov P.G.: Productivity and Efficiency of Hardware Implementation of Stream Ciphers Based on Generalized Cellular Automata. Science and Education: Scientific Publishing House of MSTU. N.E. Bauman [Electronic resource]. 2013, No. 10. http://technomag.bmstu.ru/file/669391.html?__s=1 (access date: October 7, 2016)
8. Babenko, L.K., Ischukova, E.A., Sidorov, I.D.: Parallel algorithms for solving information security problems. 2nd edn, Stereotype. M .: Hot line—Telecom, 2014. 304 s
9. What is a CUDA? [Electronic resourse]. http://www.nvidia.ru/object/what_is_cuda_new_ru.html (date of death: 07.10.2016)
10. Sanders, D., Kendrot, E.: CUDA technology in examples: an introduction to the programming of GPUs. Per. from English Slinkina A.A., scientific editor A. Boreskov M: DMK Press, 2015. 232p.
11. Vereshchak, M.I., Nelasaya, A.V.: Peculiarities of the implementation of the arithmetic library of arbitrary accuracy on graphic accelerators for cryptographic applications. Inf. Process. Syst. 2011. VIP. 7(97), 55–59
12. Fayzullin R.T., Svench A.A., Khnykin I.G.: Application of a hybrid supercomputer system in cryptanalysis problems. Doklady TUSUR, 2010. No. 1 (21), part 1, pp. 61–63
13. TOP500 Supercomputer Site [Electronic resourse]. http://www.top500.org (access date: 07.10.2016).
14. Lee, V.W.: Debunking the 100X GPU versus CPU myth: an evaluation of throughput computing on CPU and GPU. ACM SIGARCH Comput. Arch. News. ACM **38**(3), 451–460 (2010). https://doi.org/10.1145/1816038.1816021
15. Flynn, M.: Some computer organizations and their effectiveness. IEEE Trans. Comput. **100**(9), 948–960 (1972). https://doi.org/10.1109/tc.1972.5009071
16. CUDA C Best Practices Guide—CUDA SDK v.6.0—NVIDIA corp. Electronic resourse]. 2014. http://docs.nvidia.com/cuda (access date: 07.10.2016)
17. Bulavintsev V.G.: Comparison of the efficiency of CPU and GPU implementations of some combinatorial algorithms on the problems of inverting cryptographic functions. Vestnik SUSU. Series "Computational Mathematics and Computer Science". 2015, **4**(3), 67–84. https://doi.org/10.14529/cmse150306
18. Achterbahn Stream Cipher Site [Electronic resource]. http://www.matpack.de/achterbahn/achterbahn12880.zip (date of death: 10/07/2016)
19. Gammel B., Gottfert R., Kniffler, O.: Achterbahn128/80: design and analysis. In SASC'2007: Workshop Record of The State of the Art of Stream Ciphers. 2007,pp. 152–165
20. Kazyonnov A.M.: Fundamentals of CUDA and OpenCL Technology M.: Textbook. 2013, 66p.
21. Lidl R., Niederreiter, H.: Introduction to Finite Fields and their Applications (Revisited Edition). Cambridge University Press, Cambridge, [Electronic resource]. 1994. http://math.boisestate.edu/~liljanab/MATH508/FiniteFields_and_Applications.pdf (access date: 07.10.2016)

Chapter 15
Research of Second-Order Properties of NLFSR. Comparative Analysis of M-NLFSR and M-LFSR

15.1 Assessment of Linear Complexity

The linear complexity (Li) of PRS is the shortest shift register, which generates a given periodic sequence, provided that the first L values of the sequence are the initial filling of the register [1–8].

Estimation of linear complexity is one of the main parameters of the system. Any sequence that can be generated automatically (linear or nonlinear) over a finite field has finite linear complexity. Thus, it is possible to build an algorithm that will determine the linear complexity of any sequence, regardless of the method of its generation. To calculate Li, we use the Burleckamp-Messi algorithm, the essence of which is described in detail in [9–11].

Linear complexity is a measure of the complexity of the generated sequence [12]. Considering a certain segment of the sequence, it is possible to build a sequence of identical source on the basis of the existing sequence and with the help of an equivalent system. Finding the linear complexity of PRS makes it possible to build a circuit that generates a sequence similar to the original, while knowledge of the structure of the circuit that generated the original sequence is unnecessary. Thus, the high linear complexity of PRS is a necessary but not sufficient condition for the practical stability of PRS generators. Li was calculated for the complete set of M-NLFSR for $L = 4,\ 5,\ 6,\ 7,\ 8,\ 9$. The results are shown in Figs. 15.1, 15.2, 15.3, 15.4, 15.5, and 15.6.

For a higher dimension of M-NLFSR ($L = 10,\ \ldots,\ 17$) the linear complexity was calculated only for individual registers.

Based on the results, the linear complexity of M-NLFSR in most cases (62.5% for $L = 4$; 79.7% for $L = 5$; 46.9% for $L = 6$; 86.5% for $L = 7$; 54.9% for $L = 8$; 68.6% for $L = 9$) is one less than the maximum possible period ($T_{\max} = 2^L - 1$), and with a high probability (77.4% for $L = 6$; 99% for $L = 7$, 98.8% for $L = 8$, 99.98% for $L = 9$) is in the range from $(T_{\max} - T_{\max} \cdot 5\%)$ to $T_{\max} - 1$.

A. A. Kuznetsov et al., *Stream Ciphers in Modern Real-time IT Systems*,
Studies in Systems, Decision and Control 375,
https://doi.org/10.1007/978-3-030-79770-6_15

Li	Amount of M-NLFSR
4	2
12	4
14	10

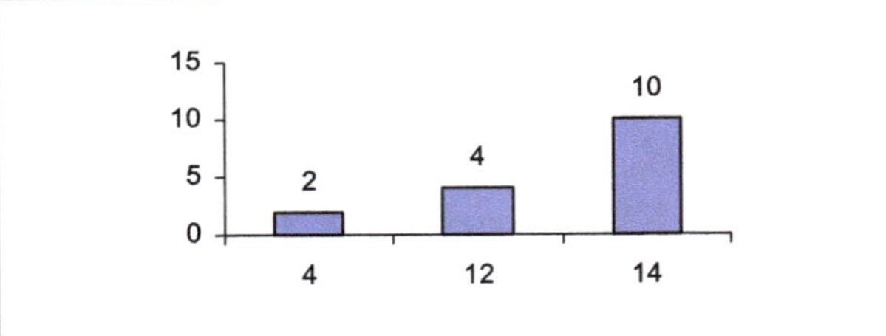

Fig. 15.1 Distribution of the number of M-NLFSR depending on their linear complexity (Li) for dimension L = 4

Li	Amount of M-NLFSR
5	6
25	20
30	102

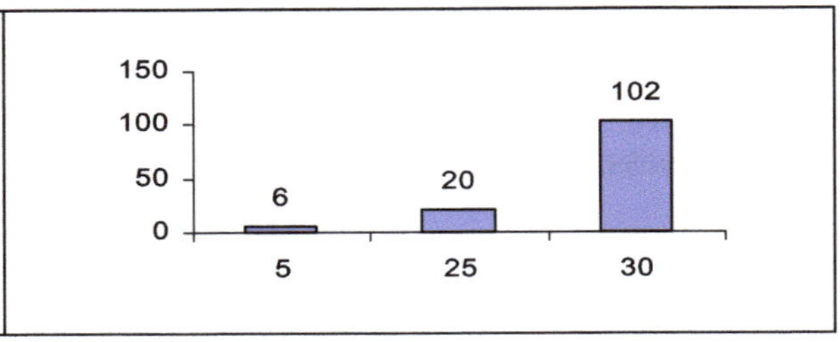

Fig. 15.2 The distribution of the number of M-NLFSR depending on their linear complexity (Li) for dimension $L = 5$

Li	Amount of M-NLFSR
6	6
45	4
48	4
50	8
51	20
53	48
54	68
56	158
57	126
59	256
60	338
62	916

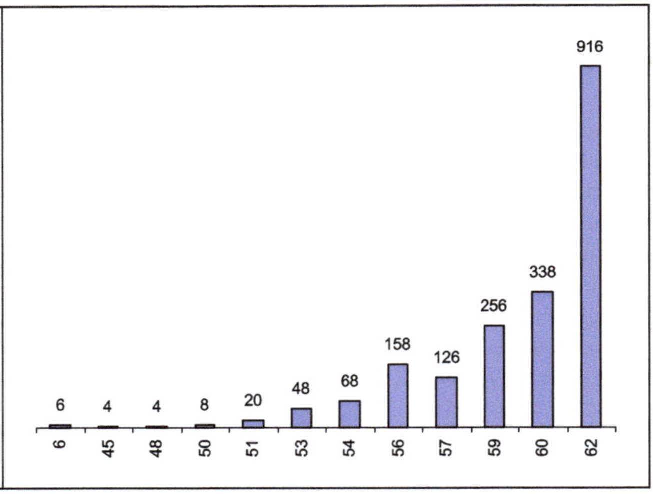

Fig. 15.3 The distribution of the number of M-NLFSR depending on their linear complexity (Li) for dimension $L = 6$

Li	Amount of M-NLFSR
7	18
105	22
112	594
119	8 044
126	55 378

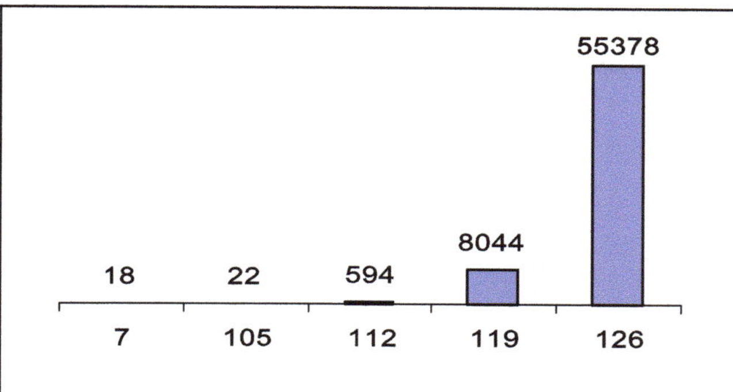

Fig. 15.4 The distribution of the number of M-NLFSR depending on their linear complexity (Li) for dimension $L = 7$

Li	Amount of M-NLFSR
8	16
208	2
214	2
218	2
220	6
222	18
224	26
226	96
228	204
230	726
232	1 020
234	2 914
236	5 954
238	18 168
240	17 578
242	52 538
244	96 554
246	289 222
248	147 584
250	440 762
252	738 236
254	2 206 370

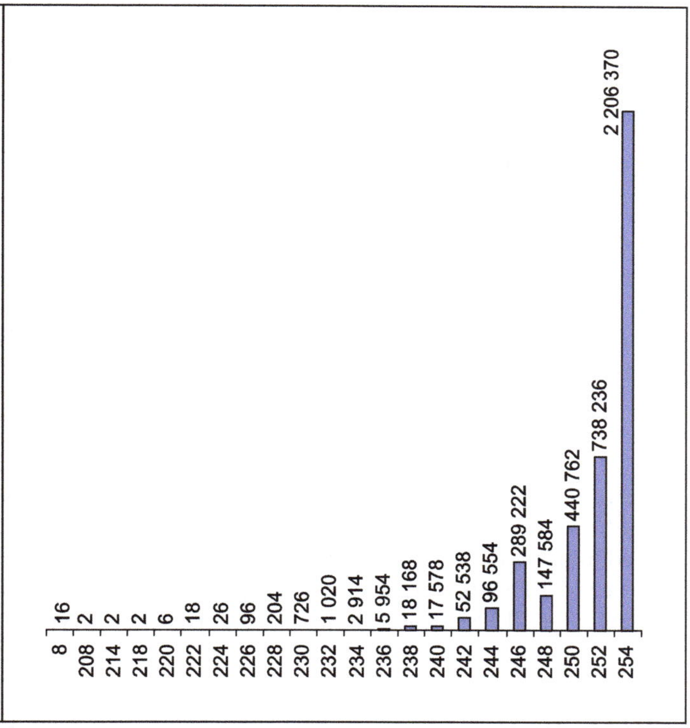

Fig. 15.5 The distribution of the number of M-NLFSR depending on their linear complexity (Li) for dimension $L = 8$

Li	**Amount of M-NLFSR**
9	48
456	2
459	2
462	6
465	34
468	30
471	504
474	1 866
477	1 386
480	21 112
483	74 458
486	42 542
489	599 644
492	2 097 832
495	795 706
498	11 163 082
501	39 051 092
504	7 268 192
507	101 824 464
510	356 297 792

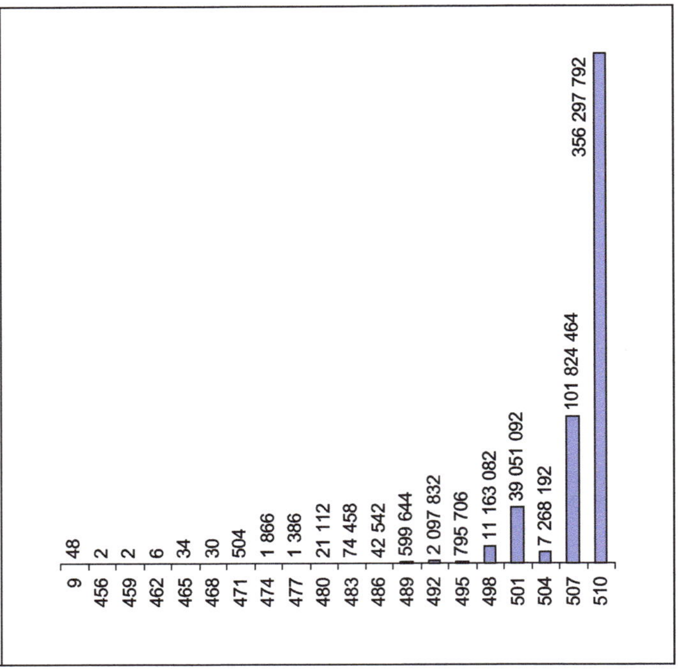

Fig. 15.6 The distribution of the number of M-NLFSR depending on their linear complexity (Li) for dimension $L = 9$

Thus, it can be argued that the linear complexity of an arbitrarily taken M-NLFSR large dimension *(L > 6)* with a high probability will be equal to the maximum value:

$$Li_{\max} \approx 2^L - 2.$$

The results, when $Li = L$, correspond to the special case when M-NLFSR is M-NLFSR.

In addition to the linear complexity itself, the profile of linear complexity for different M-NLFSR was studied. As an example, Fig. 15.7 shows a profile of linear complexity for NLFSR of length $L = 7$ for a nonlinear polynomial:

$$x^7 + x^6 + x^5 + x^4 + x^3 + x^2 + x^1 + x^1 \cdot x^2 + x^1 \cdot x^3 + x^1 \cdot x^4 + x^1 \cdot x^5 \\ + x^2 \cdot x^3 + x^2 \cdot x^5 + x^3 \cdot x^4 + x^3 \cdot x^6 + x^4 \cdot x^5 + x^4 \cdot x^6 + x^5 \cdot x^6 + 1.$$

As in the case of a truly random sequence, the linear complexity profile of M-NLFSR corresponds to the mathematical expectation with small fluctuations.

These results are also well confirmed by the results obtained by statistical tests of the NIST STS package.

Table 15.1 shows all M-NLFSR with dimension $L = 4$ and the corresponding equivalent NLFSR obtained using the Burleckamp–Messi algorithm. The

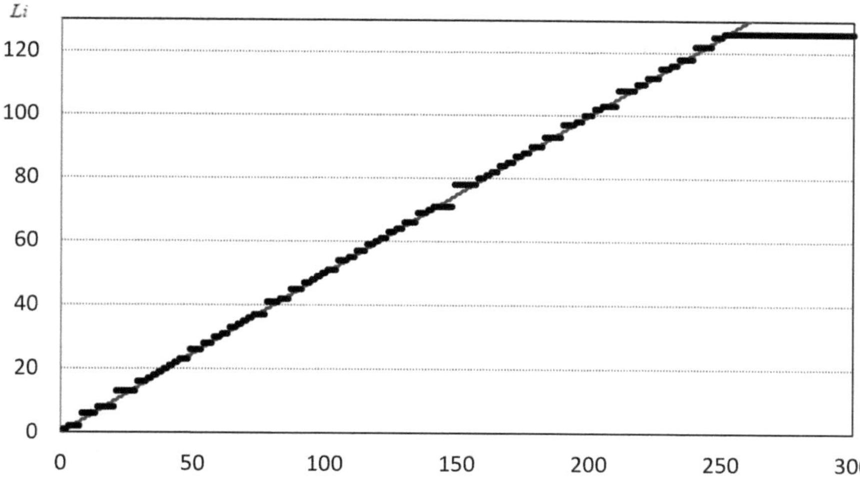

Fig. 15.7 Linear complexity profile (points) and mathematical expectation of linear complexity (solid line) for M-NLFSR of length $L = 7$

Table 15.1 M-NLFSR for $L = 4$ (except M-NLFSR) and equivalent NLFSR

№	M-NLFSR (symmetrical M-NLFSR)	*Li*	NLFSR equivalent
1	$x^4+x^2+x^2\cdot x^1+x^1+1$ $(x^4+x^3+x^3\cdot x^2+x^2+1)$	14	$x^{14}+x^{13}+x^{12}+x^{11}+x^{10}+x^9+x^8+$ $+x^7+x^6+x^5+x^4+x^3+x^2+x^1+1$
2	$x^4+x^2+x^3\cdot x^1+x^1+1$ $(x^4+x^3+x^3\cdot x^1+x^2+1)$	12	$x^{12}+x^9+x^6+x^3+1$
3	$x^4+x^2+x^3\cdot x^1+x^2\cdot x^1+1$ $(x^4+x^3\cdot x^1+x^3\cdot x^2+x^2+1)$	14	$x^{14}+x^{13}+x^{12}+x^{11}+x^{10}+x^9+x^8+$ $+x^7+x^6+x^5+x^4+x^3+x^2+x^1+1$
4	$x^4+x^3\cdot x^2+x^2\cdot x^1+x^1+1$ $(x^4+x^3+x^3\cdot x^2+x^2\cdot x^1+1)$	14	$x^{14}+x^{13}+x^{12}+x^{11}+x^{10}+x^9+x^8+$ $+x^7+x^6+x^5+x^4+x^3+x^2+x^1+1$
5	$x^4+x^3\cdot x^2+x^2+x^1+1$ $(x^4+x^3+x^2\cdot x^1+x^2+1)$	14	$x^{14}+x^{13}+x^{12}+x^{11}+x^{10}+x^9+x^8+$ $+x^7+x^6+x^5+x^4+x^3+x^2+x^1+1$
6	$x^4+x^3\cdot x^2+x^3\cdot x^1+x^2\cdot x^1+x^2+x^1+1$ $(x^4+x^3\cdot x^2+x^3\cdot x^1+x^3+x^2\cdot x^1+x^2+1)$	14	$x^{14}+x^{13}+x^{12}+x^{11}+x^{10}+x^9+x^8+$ $+x^7+x^6+x^5+x^4+x^3+x^2+x^1+1$
7	$x^4+x^3\cdot x^1+x^3+x^2\cdot x^1+x^2+x^1+1$ $(x^4+x^3+x^3\cdot x^1+x^3\cdot x^2+x^2+x^1+1)$	12	$x^{12}+x^9+x^6+x^3+1$

parentheses indicate symmetric M-NLFSR, which have the same linear complexity and equivalent NLFSR, respectively. The table did not consider M-NLFSR.

As we can see, for ordinal №№ 1, 3, 4, 5, 6 (i.e. for those cases where the linear complexity Li = 14) the equivalent NLFSR consists of all 1. Thus, it will not generate the M-sequence. The sum of all values, modulo 2, equivalent to NLFSR will correspond to the fifteenth value that remains in the M-sequence. The same NLFSR will generate different rings for all the given equivalent M-NLFSR. Moreover, it will have at least one more ring with a period $T = 15$ (in the case of initial filling with one zero and all other units).

15.2 Assessment of Quadratic Complexity

The quadratic complexity (Li^2) of PRS will be called the smallest length of the second-order NLFSR, with which it is possible to reproduce the original sequence.

The restoration of the second-order NLFSR is carried out by solving a system of linear equations according to the method described below.

The state of the first cell can be written using the values of each of the registers at time t:

$$q_1(t+1) = \sum_{i=1}^{L} a_{ii} q_i(t) + \sum_{i=1}^{L-1} \sum_{j=i+1}^{L} a_{ij} q_i(t) q_j(t).$$

We know the states of the cells at each moment of time, we denote them by the vector of states at a certain moment of time t: $\vec{Q}(t) = q_1(t), q_2(t), q_3(t), q_4(t)$. The vector, with the feedback coefficients a_{ij}, we denote by $\vec{A}$. Therefore, to find the unknown coefficients it is enough to solve a system of linear equations of the form:

$$\begin{gathered}\vec{Q}(t) \times \vec{A} = q_1(t+1) \\ \vec{Q}(t+1) \times \vec{A} = q_1(t+2) \\ \cdots \\ \vec{Q}(t+n_L-1) \times \vec{A} = q_1(t+n_L)\end{gathered}$$

where $n_L = \frac{L\cdot(L+1)}{2}$ is the number of coefficients a_{ij} for a given dimension L NLFSR.

Thus, to solve the system of linear equations, and accordingly find the quadratic complexity of Li^2, we need to know $n_L + L$ values of a continuous sequence at the output of the system under study.

If the system has no solutions, then it is not possible to generate the selected sequence using the selected dimension L NLFSR. We sort, gradually increasing all values of L and check values of the received coefficients a_{ij} on the following combinations generated with initial filling $\vec{Q}(t+n_L-L)$. In this case, if the generated and original sequences coincide, we conclude that the desired value of Li^2 is found, otherwise we continue the search further.

The quadratic complexity of the sequence generated by M-NLFSR is equal to the size of the register, i.e.:

$$Li^2 = L.$$

15.3 Estimation of the Amount of the Amount of Two or More Sequences Generated by M-NLFSR

The sequence period (T) is the number of bits generated by the register to the value when the generated sequence begins to repeat. Sequence period is one of the most important characteristics in studying the properties of the generated sequence [12]. If the period of the generated sequence is too short, the different parts of the plaintext will be transformed in an identical way, which is a serious weakness of the encryption scheme. Knowing a piece of plaintext, it is possible to recover the corresponding fragment of the encryption sequence.

The appearance of a fragment of the encryption sequence in many other parts of the encryption sequence, allows you to successfully open other encrypted texts. In addition, if there is only encrypted text converted by the same encryption sequence, it is possible, in pairs, to obtain a sequence that is equivalent to the sum of open messages and does not depend on the encryption sequence. Based on the available plaintext statistics, it is possible to recover both the plaintext of the message and the encryption sequence itself [1, 4–7, 13].

Since the NLFSR is a machine of finite state machines and can be in only one of the admissible internal states, the maximum possible period will be $T_{\max} = 2^L - 1$.

The sum of the NLFSR sequences (denoted by it as Q^{sum}) will mean the sequence obtained as a result of the operation of bitwise addition ($\oplus$) of two or more sequences from NLFSR. The generation of sequences in each of the NLFSR involved in the addition is carried out independently of each other. A generalized block diagram of adding sequences is shown in Fig. 15.8.

Consider what the *total period (T^{sum})* will have a sequence obtained by adding two or more sequences generated by M-NLFSR. By T^{sum} we mean the smallest period obtained. Denote by T^1 the period of the sequence from the first M-NLFSR with length L^1, through T^2—the period of sequence from the second M-NLFSR with length L^2, and so on.

Graphically, the summation of periods can be represented as shown in Fig. 15.9. Rhombuses indicate the relative length of the period for each sequence. If the periods of different sequences coincide (in the figure such cases are connected by lines), which corresponds to the return of the NLFSR to the initial state, the total sequence begins to repeat.

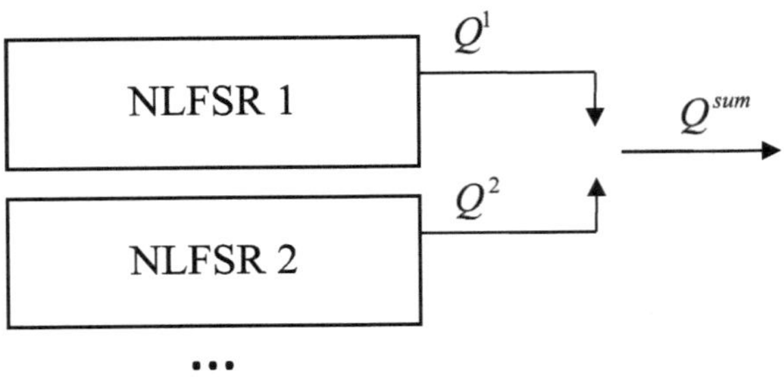

Fig. 15.8 Block diagram of adding sequences from several second-order NLFSR

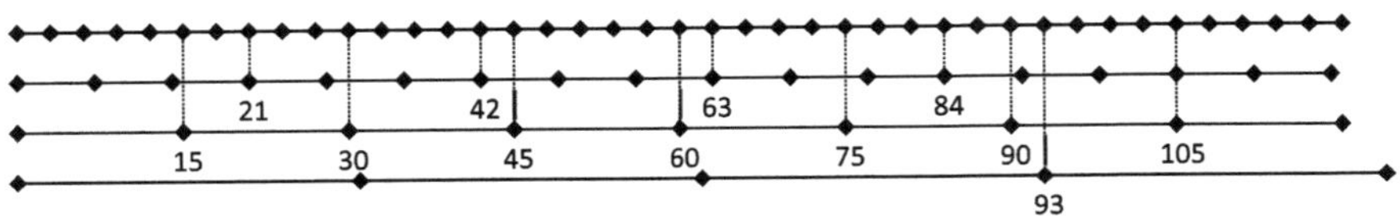

Fig. 15.9 Graphical representation of adding periods from 4 different NLFSR: $L^1 = 2, T^1 = 2^2 - 1 = 3$; $L^2 = 3, T^2 = 2^3 - 1 = 7$; $L^3 = 4, T^3 = 2^4 - 1 = 15$; $L^4 = 5, T^4 = 2^5 - 1 = 31$

First, we present the results obtained for individual cases.

Adding two identical NLFSR with different initial fillings will not change the period of the sequence received as a result of their addition, i.e. $T^{sum} = T^1 = T^2$.

The period of the sum of two sequences from different NLFSR, but of the same size (i.e. with the same L and respectively with $T^1 = T^2$), which are in the same initial state, will also be equal to the initial period, i.e. $T^{sum} = T^1 = T^2$.

Computational studies of T^{sum} sequences from the sum of two, three and four sequences of several different (including different L) M-NLFSR were performed.

The total period can be expressed using the least common multiple (LCM) or the largest common divisor (LCD). In generalized form, you can write the smallest period of the sequence, which is formed by adding several sequences from M- NLFSR, as follows:

$$T^{sum} = LCM\left(T^1, T^2, T^3, \ldots\right).$$

As we can see from the obtained results, $T^{sum} = T^1, T^2, T^3, \ldots$ only in cases when L^1, L^2, L^3 and so on are mutually prime numbers. In the given tables, in brackets, it is specified how many times the resulting period concerning product of the periods decreases. In cases where the values in parentheses are absent, correspond to the mutually simple dimension of NLFSR.

The software implementation of the calculation of the least common multiple is convenient to implement through the connection between the LCM and the LCD:

$$LCM(a, b) = \frac{a \cdot b}{LCD(a, b)}$$

LCD is determined using the well-known Euclidean algorithm. LCM of several numbers can be found through the successive finding of the LCM of two numbers, i.e.:

$$LCM(T^1, T^2, T^3, T^4, \ldots) = LCM(T^1, LCM(T^2, LCM(T^3, T^4, \ldots)))$$

It should be noted that the multiplicity of periods is uniquely determined by the multiplicity L of forming polynomials. Thus, the resulting product period of the two M-NLFSR can be recorded as:

$$T^{sum} = \frac{T^1 \cdot T^2}{2^{LCD(L^1, L^2)} - 1} = \frac{\left(2^{L^1} - 1\right) \cdot \left(2^{L^2} - 1\right)}{2^{LCD(L^1, L^2)} - 1}.$$

In the case of non-compliance with the requirement of mutually prime generators L, there may be situations where even an increase in the number of NLFSR, from which the addition of the sequence, does not lead to an increase in the resulting period. As an example, it is possible to add the sequence from M-NLFSR with $L = 4, 5, 6, 20$ for the following combinations:

$$Q(L = 6) \oplus Q(L = 20);$$
$$Q(L = 4) \oplus Q(L = 6) \oplus Q(L = 20);$$
$$Q(L = 5) \oplus Q(L = 6) \oplus Q(L = 20);$$
$$Q(L = 4) \oplus Q(L = 5) \oplus Q(L = 6) \oplus Q(L = 20).$$

As a result of all the above combinations, the period of the total sequence will be the same: $T^{sum} = 22020075$.

Based on the above result, we obtain that to generate a sequence of equivalent NLFSR with $L = 120$, we can use *4* NLFSR length 23, 29, 31, 37 and the simple addition function ($T^{sum} = T^1 \cdot T^2 \cdot T^3 \cdot T^4 \approx 2^{23} \cdot 2^{29} \cdot 2^{31} \cdot 2^{37} = 2^{120}$). Note that the period of this system will be $T = 2^{120} - 1 \approx 10^{36}$, which is sufficient for continuous operation of the generator without overlapping the range at a speed of 1 G/s for more than 10^{19} years. This approach significantly expands the possibilities of using already found M-NLFSR and building on their basis generators with specified characteristics.

It is worth noting that some types of analysis of such systems will be easier to perform, in terms of memory consumed, by analyzing the addition of several shorter polynomials than one large one. Thus, for the above example, analyzing the matrix of coefficients of polynomial size $L = 120$, it is necessary to analyze the matrix of coefficients of size $n_L = \frac{120 \cdot (120+1)}{2} = 7260$, while for the synthesis of 4 NLFSR in memory you need to keep matrices in the sum of $\frac{23 \cdot 24}{2} + \frac{29 \cdot 30}{2} + \frac{31 \cdot 32}{2} + \frac{37 \cdot 38}{2} = 276 + 435 + 496 + 703 = 1910$ cells, which is 3.8 times smaller.

As calculations have shown, the sum of sequences from several NLFSR is a commutative and associative value, ie the resulting sequence is not affected by the order of addition of sequences.

15.4 Assessment of Linear and Quadratic Complexity Sum of Sequences

The profile of linear complexity from the sum of the sequences of two NLFSR with lengths $L^1 = 4$ and $L^2 = 7$ is given in Fig. 15.10. The profile has the same appearance as one NLFSR, and is close to the mathematically expected. This statement is also well confirmed by statistical tests.

We show the total linear (Li_{sum}) and quadratic (Li^2_{sum}) complexity of the sequence obtained by adding two or more sequences generated by M-NLFSR.

Denote by the symbol Li_1 the linear and by the symbol Li_1^2 the quadratic complexity of the sequence from the first M-NLFSR dimension L^1, and the symbol Li_2 the linear and the symbol Li_2^2 the quadratic complexity of the sequence from the second M-NLFSR dimension L^2, and so on.

Adding sequences from two identical NLFSR with different initial fillings will not fundamentally change the value of the linear complexity of the resulting sequence obtained by adding the original sequences, i.e. $Li_{sum} \approx Li$. But at the same time Li^2_{sum}, as a rule, will increase, i.e. $Li^2_{sum} \geq Li^2$.

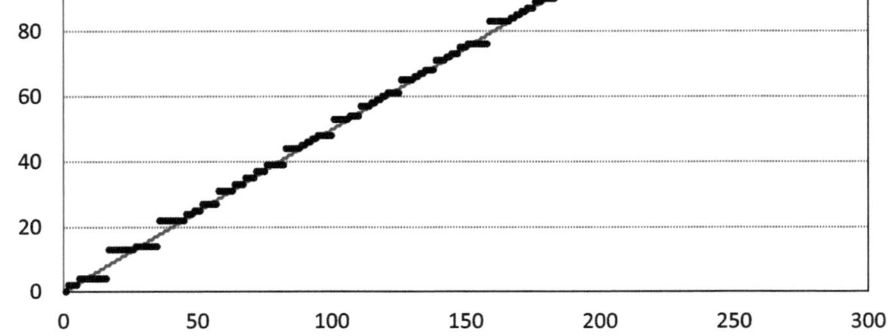

Fig. 15.10 Linear complexity profile (points) and mathematical expectation of linear complexity (solid line) from the sum of two M-NLFSR of lengths $L^1 = 4$ and $L^2 = 7$

Studies of the dependence of Li_{sum} and Li^2_{sum} the sequence on the sum of several different (including different lengths L) NLFSR. Summarizing the result, it is possible to draw the following conclusions:

$Li^{sum} = Li_1 + Li_2$ for L^1 and L^2 mutually prime values;
$\max(Li_1, Li_2) < Li^{sum} < Li_1 + Li_2$ for L^1 and L^2 not mutually prime values;
$Li^{sum} \approx Li_1$ if $L^1 = n \cdot L^2$ where $n = 1, 2, 3, \ldots$;
$Li^2_{sum} > Li^2_1 \cdot Li^2_2$ for L^1 and L^2 mutually prime values;
$Li^2_{sum} > \max\left(Li^2_1, Li^2_2\right)$ for L^1 and L^2 not mutually prime values.

It should be noted that the added quadratic complexity, as a rule, significantly exceeds the product Li^2_1 and Li^2_2, in the case of increasing L^1 and L^2, this excess also increases in degree.

If we add the sequences from NLFSR and LFSR (for example, NLFSR with $L^1 = 7$, $Li^1 = 126$ and LFSR with $L^2 = 6$, $Li^2 = 6$), then the added linear complexity, in the studied cases, also consisted (in the given example it will be $Li^{sum} = 132 = 126 + 6$).

15.5 Analysis of the Applicability of Decimation and Properties of Group Addition to M-NLFSR

Decimation of the M-sequence by the index $k(k = 1, 2, 3, \ldots)$ is a sample of k elements of this M-sequence. If the period *(T)* of the M-sequence and the decimation coefficients k are mutually prime values (i.e. LCD $(T, k) = 1$), then the decimation is considered proper or normal.

For LFSR, as shown in [11, 14], the result of any normal decimation is also M-sequence. If the decimation coefficients belong to the same cyclotomic class, they will correspond to the same sequence with an accuracy of a certain shift. If the decimation coefficients belong to different cyclotomic classes, then the decimation results in a sequence other than the derivative. Moreover, as stated in Ref. [11], if you go through all the decimation coefficients from 1 to $2^L - 1$ (then it makes no sense, because, due to its cyclicity, the sequences will be repeated), it is possible to obtain all possible M-sequences for LFSR for a given L, and using the Burleckamp–Messi algorithm, it is possible to reconstruct the polynomial-forming LFSR.

As an example, consider the polynomial LFSR of the fourth degree: $x^4 + x + 1$. The states of the shift registers for each iteration, as well as decimation for different coefficients are shown in Fig. 15.11.

As you can see, the decimation by coefficients belonging to the cyclotomic class $\{1, 2, 4, 8\}$ corresponds to the sequence generated by the polynomial $x^4 + x + 1$. Decimation by coefficients belonging to the cyclometric class $\{7, 11, 13, 14\}$, forms a symmetric sequence and corresponds to the symmetric forming polynomial $x^4 + x^3 + 1$.

LFSR					Decimation coefficients (k)															Group addition	
q_1	q_2	q_3	q_4		**1**	**2**	3	**4**	5	6	**7**	**8**	9	10	**11**	12	**13**	**14**	15	q_1+q_3	q_1+q_2
1	1	1	1		1	1	1	1	1	1	1	1	1	1	1	1	1	1	1	0	0
0	1	1	1		1	1	1	0	1	0	1	1	0	0	1	0	0	0	1	1	1
1	0	1	1		1	0	0	1	0	0	0	1	1	1	1	0	1	0	1	0	1
0	1	0	1		1	0	0	0	1	1	0	0	0	1	1	0	0	0	1	0	1
1	0	1	0		0	1	0	1	1	0	0	1	0	0	0	1	1	1	1	0	1
1	1	0	1		1	0	1	1	0	1	1	0	1	1	0	1	1	0	1	1	0
0	1	1	0		0	0	1	0	1	0	0	1	0	1	0	0	1	0	1	1	1
0	0	1	1		1	0	0	0	1	0	0	1	1	0	1	0	1	1	1	1	0
1	0	0	1		1	1	0	1	0	1	1	0	0	1	0	0	0	1	1	1	1
0	1	0	0		0	1	0	0	1	0	1	0	0	1	0	1	0	0	1	0	1
0	0	1	0		0	1	1	0	1	1	0	1	1	0	1	1	0	1	1	1	0
0	0	0	1		1	1	1	0	0	0	1	0	0	1	1	0	1	0	1	0	0
1	0	0	0		0	0	0	1	1	0	0	0	1	1	0	0	0	1	1	1	1
1	1	0	0		0	1	0	1	1	1	1	0	0	0	1	0	0	1	1	1	0
1	1	1	0		0	0	0	1	0	0	1	1	0	1	0	1	1	1	1	0	0
				$T =$	15	15	5	15	3	5	15	15	5	3	15	5	15	15	1	15	15

Fig. 15.11 The state of LFSR on the basis of the forming polynomial $x^4 + x + 1$, its decimation and group addition

Property of group addition [15]: if you add (modulo 2) some bits of the LFSR, then the result of such an operation will be M-sequence, the same as at the output of the LFSR, only with some offset. Thus, adding different bits of LFSR, it is possible to receive all $2^L - 1$ shift. An example of group addition is shown in Fig. 15.11. In the case of adding the values 1 and 3 of the LFSR cells, we obtain the original M sequence with a shift of 7 values, when adding the values of 1 and 2 LFSR cells—with an offset of 11 values.

Using the properties of group addition, knowing the forming polynomial and performing preliminary calculations [15], it is possible to obtain a sequence that will be generated by LFSR through a certain number of iterations, not running all register states to the desired iteration, which is impossible with large L values.

Applying the above rules of decimation and group addition to NLFSR in the form in which they apply to LFSR, we see that they are not fulfilled (see Fig. 15.12). Moreover, the obtained sequence as a result of decimation (for example, at $k = 2, 4, 11, 13$) cannot be generated in the shift register of length $L = 4$, because the sequence contains identical tuples of length 4, but has different subsequent values, which in principle cannot be implemented in NLFSR. It is also fundamentally impossible to implement in the NLFSR dimension $L = 4$ non-zero state after filling all the cells of the register with values equal to zero, which occur in the sequence for $k = 7, 8$.

NLFSR					Decimation coefficients (k)															Group addition	
q_1	q_2	q_3	q_4		1	2	3	4	5	6	7	8	9	10	11	12	13	14	15	q_1+q_3	q_1+q_2
1	1	1	1		1	1	1	1	1	1	1	1	1	1	1	1	1	1	1	0	0
0	1	1	1		1	1	1	0	0	1	0	1	1	0	1	0	0	0	1	1	1
0	0	1	1		1	0	1	1	0	0	0	1	1	0	0	1	1	0	1	1	0
1	0	0	1		1	1	1	0	1	1	1	1	0	1	1	1	1	0	1	1	1
0	1	0	0		0	1	0	1	0	1	0	1	1	0	0	1	0	1	1	0	1
1	0	1	0		0	0	1	0	0	1	0	0	1	0	0	1	0	0	1	0	1
1	1	0	1		1	0	1	1	1	1	0	1	1	1	1	0	1	1	1	1	0
0	1	1	0		0	0	1	0	0	0	0	1	1	0	1	1	1	1	1	1	1
1	0	1	1		1	1	1	1	0	1	1	0	0	0	0	1	0	0	1	0	1
0	1	0	1		1	1	0	1	1	1	1	0	1	1	1	1	0	1	1	0	1
0	0	1	0		0	0	1	0	0	1	0	0	1	0	0	1	0	0	1	1	0
0	0	0	1		1	0	1	0	0	1	1	0	1	0	1	0	1	0	1	0	0
1	0	0	0		0	1	1	1	1	0	1	1	1	1	0	1	1	1	1	1	1
1	1	0	0		0	1	1	0	0	1	1	0	0	0	1	1	0	1	1	1	0
1	1	1	0		0	0	0	1	0	1	1	0	1	0	0	1	1	1	1	0	0
				$T =$	15	-	5	-	3	5	-	-	5	3	-	5	-	15	1	-	

Fig. 15.12 State of NLFSR on the basis of forming polynomial $x^4 + x^2 \cdot x^1 + x^2 + x^1 + 1$, its decimation and group addition

For M-LFSR with forming polynomial $x^4 + x + 1$, which generates a sequence of bits *(Q)* for a fixed offset *(n)*, is determined by the following relationships between the initial filling of the register *(q_1, q_2, q_3, q_4):*

$$
\begin{array}{lll}
Q_{n=1} = q_1 + q_4 & Q_{n=6} = q_2 + q_4 & Q_{n=11} = q_1 + q_2 \\
Q_{n=2} = q_1 + q_3 + q_4 & Q_{n=7} = q_1 + q_3 & Q_{n=12} = q_4 \\
Q_{n=3} = q_1 + q_2 + q_3 + q_4 & Q_{n=8} = q_1 + q_2 + q_4 & Q_{n=13} = q_3 \\
Q_{n=4} = q_2 + q_3 + q_4 & Q_{n=9} = q_3 + q_4 & Q_{n=14} = q_2 \\
Q_{n=5} = q_1 + q_2 + q_3 & Q_{n=10} = q_2 + q_3 & Q_{n=15} = q_1
\end{array}
$$

If we take as a basis the principles of combinatorics, then the number of placements with repetitions of two elements {0,1} for L is defined as $\overline{A}_2^L = 2^L$ corresponding to the possible number of different linear combinations of register fills. But given that a combination of single zeros is not considered meaningless, the number of possible combinations will be defined as $2^L - 1$, which corresponds to the M-sequence sequence.

For M-NLFSR with forming nonlinear polynomial $x^4 + x^2 \cdot x^1 + x^2 + x^1 + 1$, which generates a sequence of bits for a fixed offset, is determined by the following relations between the initial filling of the register:

$$
\begin{aligned}
Q_{n=1} &= q_1 + q_2 + q_4 + q_1 \cdot q_2 \\
Q_{n=2} &= q_1 + q_2 + q_3 + q_4 + q_1 \cdot q_2 + q_1 \cdot q_4 \\
Q_{n=3} &= q_1 + q_3 + q_4 + q_1 \cdot q_2 + q_1 \cdot q_3 + q_1 \cdot q_4 + q_2 \cdot q_3 + q_3 \cdot q_4 + q_1 \cdot q_2 \cdot q_3 \\
Q_{n=4} &= q_2 + q_3 + q_4 + q_1 \cdot q_2 + q_1 \cdot q_3 + q_1 \cdot q_4 + q_2 \cdot q_4 + q_3 \cdot q_4 \\
&\quad + q_1 \cdot q_2 \cdot q_3 + q_1 \cdot q_2 \cdot q_4 \\
Q_{n=5} &= q_3 + q_1 \cdot q_2 + q_1 \cdot q_3 + q_1 \cdot q_4 + q_2 \cdot q_4 + q_3 \cdot q_4 + q_1 \cdot q_2 \cdot q_3 \\
&\quad + q_1 \cdot q_2 \cdot q_4 + q_1 \cdot q_3 \cdot q_4 \\
Q_{n=6} &= q_1 + q_1 \cdot q_2 + q_1 \cdot q_3 + q_1 \cdot q_4 + q_2 \cdot q_3 + q_2 \cdot q_4 + q_3 \cdot q_4 + q_1 \cdot q_2 \cdot q_3 \\
&\quad + q_1 \cdot q_2 \cdot q_4 + q_1 \cdot q_3 \cdot q_4 + q_2 \cdot q_3 \cdot q_4 \\
Q_{n=7} &= q_4 + q_1 \cdot q_2 + q_1 \cdot q_3 + q_1 \cdot q_4 + q_2 \cdot q_3 + q_2 \cdot q_4 + q_3 \cdot q_4 \\
&\quad + q_1 \cdot q_2 \cdot q_4 + q_1 \cdot q_3 \cdot q_4 + q_2 \cdot q_3 \cdot q_4 \\
Q_{n=8} &= q_1 + q_2 + q_3 + q_1 \cdot q_2 + q_1 \cdot q_3 + q_1 \cdot q_4 + q_2 \cdot q_3 + q_2 \cdot q_4 + q_3 \cdot q_4 \\
&\quad + q_1 \cdot q_3 \cdot q_4 + q_2 \cdot q_3 \cdot q_4 \\
Q_{n=9} &= q_1 + q_2 + q_4 + q_1 \cdot q_4 + q_2 \cdot q_3 + q_2 \cdot q_4 + q_3 \cdot q_4 + q_2 \cdot q_3 \cdot q_4 \\
Q_{n=10} &= q_2 + q_3 + q_4 + q_3 \cdot q_4 \\
Q_{n=11} &= q_1 + q_2 + q_3 + q_2 \cdot q_3 \\
Q_{n=12} &= q_4 \\
Q_{n=13} &= q_3 \\
Q_{n=14} &= q_2 \\
Q_{n=15} &= q_1
\end{aligned}
$$

Thus, from the above results we can draw the following conclusions:

1. Using the decimation of the NLFSR M-sequence, we do not obtain the M-sequence and, as a consequence, it is impossible to obtain all (or some, except symmetric at $q = T - 1$) NLFSR, restoring the forming polynomial sequence.
2. For NLFSR, as well as for LFSR, it is possible to calculate the ratio for the construction of a sequence equivalent to the original with some given offset.
3. In the case of NLFSR, the principle of group addition takes a more complex form, and is no longer just an "addition". Nonlinear terms from the product of two, three and more cells of registers also take part in the construction of a sequence based on this principle. That is, in the general case, only the second-order NLFSR is not enough for construction, the application of NLFSR of the third and higher order is necessary.
4. The complexity of constructing relationships to calculate sequences with a predetermined offset is more complex than for LFSR.
5. The number of combinations that can be included in the expression to calculate the M-sequence with a given offset is defined according to the principles of combinatorics as $2^{2^L-1} - 1$ (see Sect. 15.8), while for LFSR their number will be equal to $2^L - 1$. Thus, the probability that an arbitrary combination of feedback

coefficients for NLFSR will generate M-sequence is much less than the similar probability for LFSR.

These factors increase the cryptographic stability of systems built using NLFSR, from conducting analytical attacks, which advantageously increases their reliability in comparison with LFSR.

15.6 Study of Statistical Characteristics of prs, Which are Generated M-NLFSR

To study statistical security, we will use the NIST STS testing method and the method of statistical testing of cryptographic algorithms, which is based on estimating the mathematical expectation of the number of tests of the cryptoalgorithm under study (description of the method is given in Chapter 10). Applying this approach, an experimental study of the cryptographic properties of NLFSR was conducted.

15.6.1 Setting up a Computational Experiment

During testing, the NIST STS statistical package used the parameters recommended by the developer [13]. During each test, the sequence length was set to 10^6, 100 such sequences were used, and the binary data format in the input file was used (i.e., each byte contains 8 bits). Thus, the minimum sequence length should be $10^6 \bullet 10^2 = 10^8$ values, which corresponds to a file of approximately 12 MB (10^8/1024 KB/1024 bytes/1024 Kb).

From the above conditions, a file with data for analysis of more than 12 MB was selected. To generate a series of sequences longer than 10^8 values, we used a data stream obtained by adding sequences of two or more M-NLFSR with lengths L from 4 to 23, in various combinations. The following are the forming nonlinear polynomials by which the M-sequence was formed:

$$L = 4 \quad x^4 + x^2 + x^1 + x^1 \cdot x^2 + 1$$

$$L = 5 \quad x^5 + x^1 + x^1 \cdot x^3 + x^2 \cdot x^3 + 1$$

$$L = 6 \quad x^6 + x^2 + x^1 + x^1 \cdot x^2 + 1$$

$$L = 7 \quad x^7 + x^2 + x^1 + x^2 \cdot x^6 + 1$$

$$L = 8 \quad x^8 + x^2 + x^1 + x^1 \cdot x^2 + x^1 \cdot x^3 + x^1 \cdot x^5 + 1$$

$$L = 9 \qquad x^9 + x^2 + x^1 \cdot x^2 + x^1 \cdot x^2 + x^1 \cdot x^3 + x^1 \cdot x^7 + x^2 \cdot x^4 + 1$$

$$L = 10 \qquad x^{10} + x^2 + x^1 \cdot x^6 + x^1 \cdot x^7 + 1$$

$$L = 11 \qquad x^{11} + x^2 + x^1 + x^1 \cdot x^4 + x^1 \cdot x^5 + x^2 \cdot x^6 + 1$$

$$L = 12 \qquad x^{12} + x^2 + x^1 \cdot x^2 + x^1 \cdot x^8 + x^1 \cdot x^9 + x^1 \cdot x^{10} + 1$$

$$L = 13 \qquad x^{13} + x^2 + x^1 \cdot x^4 + x^1 \cdot x^5 + x^1 \cdot x^{12} + x^2 \cdot x^4 + 1$$

$$L = 14 \qquad x^{14} + x^2 + x^1 + x^1 \cdot x^2 + x^1 \cdot x^3 + x^1 \cdot x^4 + x^1 \cdot x^5 + x^1 \cdot x^6 + x^1 \cdot x^9 + x^1 \cdot x^{10} + 1$$

$$L = 15 \qquad x^{15} + x^2 + x^1 \cdot x^6 + x^1 \cdot x^{11} + x^1 \cdot x^{14} + x^2 \cdot x^6 + 1$$

$$L = 16 \qquad x^{16} + x^2 + x^1 \cdot x^2 + x^1 \cdot x^4 + x^1 \cdot x^6 + x^1 \cdot x^7 + x^1 \cdot x^8 + x^1 \cdot x^{10} + 1$$

$$L = 17 \qquad x^{17} + x^2 + x^1 \cdot x^5 + x^1 \cdot x^7 + x^1 \cdot x^8 + x^1 \cdot x^9 + x^1 \cdot x^{13} + x^1 \cdot x^{14} + 1$$

$$L = 18 \qquad x^{18} + x^2 + x^1 \cdot x^7 + x^1 \cdot x^9 + x^1 \cdot x^{11} + x^1 \cdot x^{17} + 1$$

$$L = 19 \qquad x^{19} + x^2 + x^1 \cdot x^3 + x^1 \cdot x^5 + x^1 \cdot x^{14} + x^1 \cdot x^{16} + 1$$

$$L = 20 \qquad x^{20} + x^2 + x^1 \cdot x^3 + x^1 \cdot x^4 + x^1 \cdot x^6 + x^1 \cdot x^7 + x^1 \cdot x^8 + x^1 \cdot x^{11} + x^1 \cdot x^{15} + x^1 \cdot x^{18} + 1$$

$$L = 21 \qquad x^{21} + x^2 + x^1 \cdot x^2 + x^1 \cdot x^3 + x^1 \cdot x^5 + x^1 \cdot x^7 + x^1 \cdot x^{10} + x^1 \cdot x^{11} + x^1 \cdot x^{13} + x^2 \cdot x^{16} + x^1 \cdot x^{19} + x^1 \cdot x^{20} + 1$$

$$L = 22 \qquad x^{22} + x^2 + x^1 \cdot x^2 + x^1 \cdot x^4 + x^1 \cdot x^6 + x^1 \cdot x^7 + x^1 \cdot x^9 + x^1 \cdot x^{10} + x^1 \cdot x^{11} + x^1 \cdot x^{14} + x^1 \cdot x^{15} + x^1 \cdot x^{16} + x^1 \cdot x^{18} + x^1 \cdot x^{19} + 1$$

$$L = 23 \qquad x^{23} + x^2 + x^1 \cdot x^4 + x^1 \cdot x^5 + x^1 \cdot x^6 + x^1 \cdot x^7 + x^1 \cdot x^8 + x^1 \cdot x^{10} + x^1 \cdot x^{13} + x^1 \cdot x^{16} + x^1 \cdot x^{18} + x^1 \cdot x^{19} + x^1 \cdot x^{20} + x^1 \cdot x^{22} + 1$$

The addition of M-NLFSR provided that the length of the tested values did not xceed the total period (T^{sum}) of the data stream.

A total of 3259 tests were performed with the NIST STS statistical package. About 377 GB were studied, which was generated from 37 different constructions based on M-NLFSR.

15.6.2 Analysis of Results

The test results of the NIST STS statistical package of pseudo-random sequences formed from some of the constructs listed in the previous section are shown in Table 15.2. The table does not show the results from sequences with a maximum number of tests less than 100.

The designs showed good results, without systemic or pronounced shortcomings (in the statistical sense). An example of a statistical portrait is shown in Fig. 15.13.

Of all the 3259 tests of the various designs that were tested, only two showed the worst result of all the tests for the individual $Min(Pj) < 0.9$. The addition of sequences from $L^1 = 5$ and $L^2 = 23$ showed $Min(Pj) = 0.84$, and the obvious systemic shortcomings showed the sequence from adding $L^1 = 14$ and $L^2 = 15$, in which $\text{Min}(Pj) = 0.01$. Two more sequences ($L^1 = 5$ and $L^2 = 22$, $L^1 = 6$ and $L^2 = 23$) showed $\text{Min}(Pj) = 0.9$. These four sequences have $T^{sum} \leq 64$ MB, other constructions that form a larger T^{sum} have a value of $\text{Min}(Pj) > 0.9$.

As can be seen from Table 15.2, the mathematical expectation of the number of passed tests tends to a value of 186.8 ± 1 (by the criterion of successful passing $Pj \geq 0.96$) and to a value of 132.4 ± 2 (for the criterion $Pj \to 0.99$), with confidence probability close to 1. In this case, the worst tests were tests in which the minimum number of tests passed $\text{Min}(X_i) = 179$, and the average worst of all tests for a single design $\text{Min}(Pj) = 0.92 - 0.93$.

In order to summarize the results and analyze for systemic deficiencies identified in each of the tests, averaging of $N = 500$ tests was performed for the total sequence obtained from the addition of three NLFSR ($L^1 = 21$, $L^2 = 22$ and $L^3 = 23$). Averaging was performed for each j test according to the ratio:

$$P_j^* = \frac{1}{N}\sum_{i=1}^{N} P_{ji}$$

The statistical portrait of the obtained results of averaged tests is shown in Fig. 15.14a. Figure 15.14b shows the average statistical portrait for a structure of two NLFSR ($L^1 = 21$, $L^2 = 22$) with $N = 100$. As we can see, the success rate for each of the 188 tests tends to 99%.

For comparison, Table 15.3 shows the results of tests by a similar method of pseudo-random sequences of modern stream ciphers, which were obtained from [16–18].

Table 15.2 The results of statistical testing of structures based on M-NLFSR

$L^1 \oplus L^2 \oplus L^3$	$P_j \geq 0.96$		$P_j \geq 0.99$		$\text{Min}(X_i)$	N	$\text{Min}(Pj)$
	$\tilde{m}[X]$	$P_o(\epsilon = 1)$	$\tilde{m}[X]$	$P_o(\epsilon = 2)$			
$11 \oplus 23$	186.65	1	132.70	0.9947	182	100	0.91
$12 \oplus 23$	186.74	1	133.25	0.9868	183	100	0.93
$13 \oplus 23$	186.79	1	133.29	0.9935	181	100	0.92
$14 \oplus 23$	186.69	1	132.80	0.9904	183	100	0.92
$15 \oplus 23$	186.91	1	132.16	0.9983	183	100	0.92
$16 \oplus 23$	186.86	1	131.78	0.9876	183	100	0.92
$17 \oplus 23$	186.89	1	132.92	0.9965	184	100	0.93
$18 \oplus 23$	186.47	1	132.10	0.8327	182	100	0.92
$19 \oplus 23$	186.82	1	132.94	0.9955	183	100	0.92
$20 \oplus 23$	186.80	1	132.58	0.9952	179	100	0.92
$21 \oplus 23$	186.73	1	132.34	0.9945	180	100	0.94
$22 \oplus 23$	186.71	1	133.08	0.9907	179	100	0.93
$13 \oplus 22$	186.79	1	131.70	0.9931	182	100	0.92
$14 \oplus 22$	186.70	1	131.69	0.9981	181	100	0.92
$15 \oplus 22$	186.95	1	131.92	0.9848	183	100	0.92
$16 \oplus 22$	186.83	1	131.88	0.9956	181	100	0.91
$17 \oplus 22$	187.05	1	132.93	0.9949	180	100	0.93
$18 \oplus 22$	186.93	1	131.93	0.9938	183	100	0.93
$19 \oplus 22$	186.61	1	132.12	0.9967	182	100	0.93
$20 \oplus 22$	186.92	1	131.95	0.9922	184	100	0.93
$21 \oplus 22$	186.65	1	132.44	0.9976	181	100	0.92
$20 \oplus 21$	186.94	1	132.46	0.9762	181	100	0.93
$4 \oplus 11 \oplus 23$	186.69	1	131.89	0.9945	182	100	0.92
$15 \oplus 19 \oplus 23$	186.93	1	132.72	0.9998	181	200	0.92
$21 \oplus 22 \oplus 23$	186.76	1	132.19	1	180	500	0.91

As you can see, the studied structure, built only with the help of NLFSR (without any additional complication algorithms) in its statistical characteristics is not inferior to the world-famous modern streaming encryption generators.

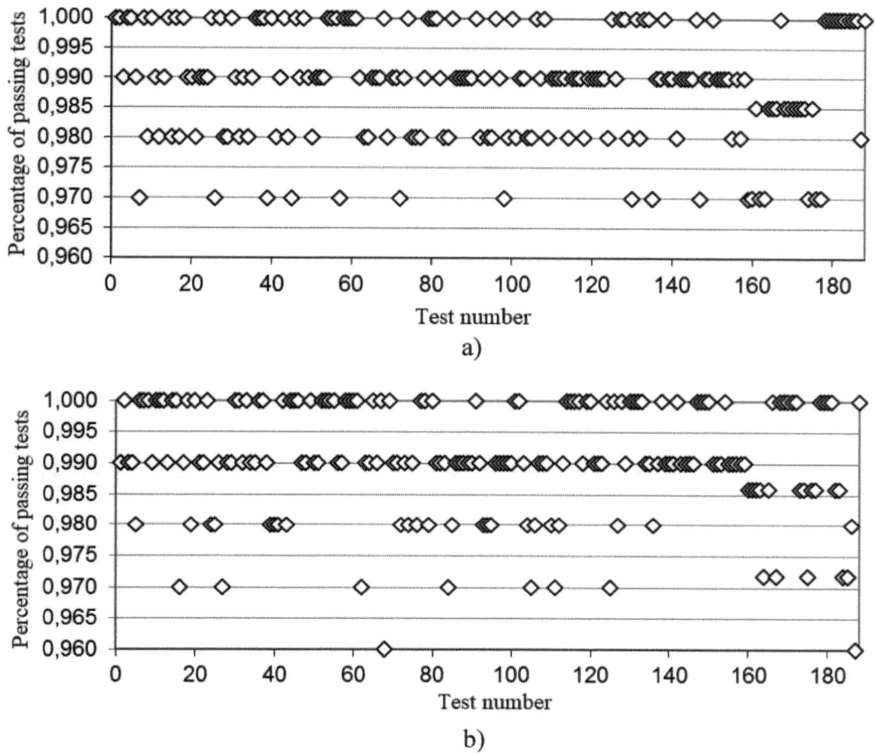

Fig. 15.13 Statistical portrait of the sequence sample from the addition of M-NLFSR: **a** $L^1 = 21$ and $L^2 = 20$; **b** $L^1 = 4$, $L^2 = 11$ and $L^3 = 23$

15.7 Comparison of the Volume of the M-LFSR and M-NLFSR Ensembles for GF(2) and in Field Extensions GF(2^2)

The term *ensemble volume* will mean the number of different M-NLFSR for a fixed value of L.

The comparison was performed on the following parameters:

T_{max} is the length of the maximum period that can be generated by a register with a given L (the table shows the values in bits, not in machine words);
n_L is the number of feedback coefficients a_{ij} used in the design;
k is the number of possible combinations of shift registers;
M_0 is the maximum possible number of different feedback combinations for a given L at which the register will generate the M-sequence. This value corresponds to the desired ensemble volume for each L established experimentally.

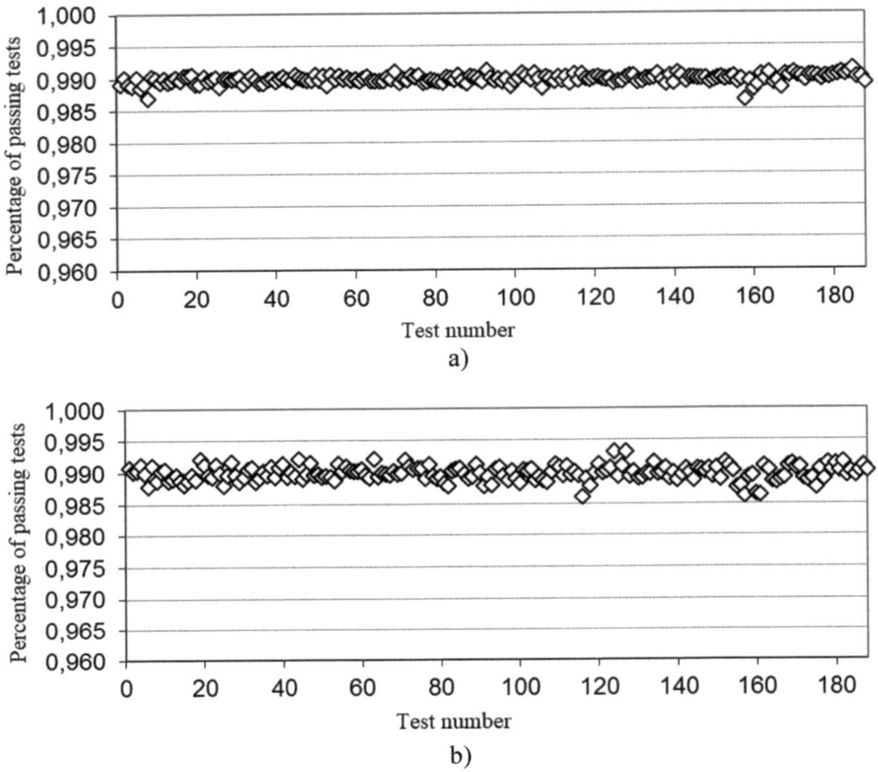

Fig. 15.14 The average statistical portrait of the total sequence from the addition of M- NLFSR: **a** $L^1 = 21$, $L^2 = 22$ and $L^3 = 23$ averaged over $N = 500$; **b** $L^1 = 21$ and $L^2 = 22$ averaged over $N = 100$

The results of the comparison of LFSR and NLFSR of the second order are given in Table 15.4 for *GF(2)* and in Table 15.5 for *GF(*2^2*)*. It should be noted that LFSR is a special case of NLFSR, in the case when all nonlinear feedback coefficients are equal to zero, i.e. $a_{ij} = 0$ for $i \neq j$.

Additionally, the number of M-NLFSR was counted as a special case during the search for M-LFSR. Starting with $L = 10$ and up to the length of registers $L = 22$, due to the significant time costs, only M-LFSR was searched and the volume of the found combinations with known values taken from sources was controlled [19, 20].

The values of the results given in Tables 15.4 and 15.5, except M_0, can be calculated by the formulas given in Table 15.6. It should be noted that for the extension of the field *GF(*2^3*)* and higher order there will be more than one generating polynomial $g(x)$ and, possibly, due to this, the ratio for calculating n_L and, consequently, the set k will change.

As can be seen from the results shown in Table 15.6, the number of different combinations that can be made for NLFSR, significantly exceeds the same number

Table 15.3 Results of statistical testing of streaming encryption algorithms

Algorithms	$Pj \geq 0.96$		$Pj \geq 0.99$		$\text{Min}(X_i)$	N
	$\tilde{m}[X]$	$P_o(\epsilon = 1)$	$\tilde{m}[X]$	$P_o(\epsilon = 2)$		
HC-256_iv_const	186.96	1	134.20	1	182	
Snow2.0_iv_const	186.79	1	133.59	1	182	
Grain_iv_const	186.91	1	133.05	1	182	
Hc-128_iv_const	186.97	1	132.84	1	183	
MICKEY_iv_const	186.80	1	131.62	1	183	
Rabbit_iv_const	186.72	1	131.67	1	179	
Salsa20_iv_const	186.97	1	133.64	1	183	
Sosemanuk_iv_const	186.76	1	132.98	1	180	
Strumok_iv_const	186.85	1	132.59	1	180	
Grain_key_const	186.78	1	131.63	1	182	
Hc-128_key_const	186.52	1	132.49	1	176	
MICKEY_key_const	186.78	1	133.17	1	181	
Rabbit_key_const	186.65	1	133.01	1	180	
Salsa20_key_const	186.83	1	132.61	1	180	
Sosemanuk_key_const	186.77	1	132.42	1	181	
HC-256_key_const	186.60	1	132.05	1	179	
Snow2.0_key_const	186.88	1	133.39	1	182	
Strumok_key_const	186.75	1	131.64	1	180	
HC-256	186.53	1	130.88	1	181	
Snow2.0	186.60	1	133.19	1	178	
Strumok	186.94	1	132.97	1	183	100
ГОСТ-28147–89	186.52	1	132.34	0.99	182	100
FIPS-197	186.89	1	133.96	0.99	183	100
Kalina	186.82	1	132.16	0.99	183	100
Camellia	186.76	1	132.76	0.99	182	100
TDEA	186.60	1	133.28	0.99	180	100
ATS-128_K_iv	186.63	1	127.07	1	185	
Enocoro_K_iv	187.17	1	132.92	1	185	
MUGI_K_iv	186.50	1	132.23	1	185	

for LFSR. Moreover, the difference increases with increasing L in degree, and if for $L = 5$ the number of different combinations for NLFSR exceeded the same number for NLFSR more than 1000 times, then for $L = 30$ such a difference was 10^{130}.

The tendency of a significant advantage in favor of NLFSR remains for a more important parameter—the volume of the ensemble. Already at $L = 9$ the number of M-NLFSR exceeds the number of M- LFSR more than 107 times, and the predicted difference for L = 30 is 10^{115}.

Table 15.4 Comparison of LFSR and NLFSR of the second order for GF(2)

L	$GF(2)$						
	T_{max}	LFSR			NLFSR		
		n_L	k	M_0	n_L	k	M_0
2	3	2	4	**1**	3	8	**1**
3	7	3	8	**2**	6	64	**2**
4	15	4	16	**2**	10	1024	**16**
5	31	5	32	**6**	15	32,768	**128**
6	63	6	64	**6**	21	2,097,152	**1952**
7	127	7	128	**18**	28	$2.6\cdot10^{8}$	**64,056**
8	255	8	256	**16**	36	$6.8\cdot10^{10}$	**4,017,998**
9	511	9	512	**48**	45	$3.5\cdot10^{13}$	**519,239,794**
10	1023	10	1024	**60**	55	$3.6\cdot10^{16}$	$1\cdot10^{11}$ (2)
11	2047	11	2048	**176**	66	$7.4\cdot10^{19}$	$7\cdot10^{13}$ (2)
12	4095	12	4096	**144**	78	$3.0\cdot10^{23}$	$7\cdot10^{16}$ (2)
13	8191	13	8192	**630**	91	$2.5\cdot10^{27}$	$1\cdot10^{20}$ (2)
14	16,383	14	16,384	**756**	105	$4.1\cdot10^{31}$	$6\cdot10^{23}$ (2)
15	32,767	15	32,768	**1800**	120	$1.3\cdot10^{36}$	$5\cdot10^{27}$ (2)
16	65,535	16	65,536	**2048**	136	$8.7\cdot10^{40}$	$8\cdot10^{31}$ (2)
17	131,071	17	131,072	**7710**	153	$1.1\cdot10^{46}$	$3\cdot10^{36}$ (2)
18	262,143	18	262,144	**7776**	171	$3.0\cdot10^{51}$	$2\cdot10^{41}$ (2)
19	524,287	19	524,288	**27,594**	190	$1.6\cdot10^{57}$	$2\cdot10^{46}$ (2)
20	1,048,575	20	1,048,576	**24,000**	210	$1.6\cdot10^{63}$	$6\cdot10^{51}$ (2)
21	2,097,151	21	2,097,152	**84,672**	231	$3.5\cdot10^{69}$	$3\cdot10^{57}$ (2)
22	4,194,303	22	4,194,304	**120,032**	253	$1.4\cdot10^{76}$	$3\cdot10^{63}$ (2)
23	$8.4\cdot10^{6}$	23	$8.4\cdot10^{6}$	356,960 (1)	276	$1.2\cdot10^{83}$	$7\cdot10^{69}$ (2)
24	$1.7\cdot10^{7}$	24	$1.7\cdot10^{7}$	276,480 (1)	300	$2.0\cdot10^{90}$	$3\cdot10^{76}$ (2)
25	$3.4\cdot10^{7}$	25	$3.4\cdot10^{7}$	1,296,000 (1)	325	$6.8\cdot10^{97}$	$2\cdot10^{83}$ (2)
26	$6.7\cdot10^{7}$	26	$6.7\cdot10^{7}$	1,719,900 (1)	351	$4.6\cdot10^{105}$	$4\cdot10^{90}$ (2)
27	$1.3\cdot10^{8}$	27	$1.3\cdot10^{8}$	4,202,496 (1)	378	$6.2\cdot10^{113}$	$1\cdot10^{98}$ (2)
28	$2.7\cdot10^{8}$	28	$2.7\cdot10^{8}$	4,741,632 (1)	406	$1.7\cdot10^{122}$	$9\cdot10^{105}$ (2)
29	$5.4\cdot10^{8}$	29	$5.4\cdot10^{8}$	18,407,808 (1)	435	$8.9\cdot10^{130}$	$1\cdot10^{114}$ (2)
30	$1.0\cdot10^{9}$	30	$1.0\cdot10^{9}$	17,820,000 (1)	465	$9.5\cdot10^{139}$	$3\cdot10^{122}$ (2)

(1) The value of M_0 for LFSR according to the values of $L = 23–30$ was taken from [19, 20]
(2) The value of M_0 for NLFSR according to the values of $L = 10–30$ is given as the upper estimate M_0, calculated by expression $M_0 \leq 2^{\frac{L(L-1)}{2} - L + 2}$.

Table 15.5 Comparison of LFSR and NLFSR of the second order for GF(2^2)

L	$FG(2^2)$						
	T_{max}	LFSR			NLFSR		
		n_L	k	M_0	n_L	k	M_0
2	15	4	16	**4**	6	64	**4**
3	63	6	64	**12**	12	4096	**12**
4	255	8	256	**32**	20	1048,576	**152**
5	1023	10	1024	**120**	30	1,073,741,824	**7896**
6	4095	12	4096	**288**	42	$4.4 \cdot 10^{12}$	
7	16,383	14	16,384	**1512**	56	$7.2 \cdot 10^{16}$	
8	65,535	16	65,536	**4096**	72	$4.7 \cdot 10^{21}$	
9	262,143	18	262,144	**15,552**	90	$1.2 \cdot 10^{27}$	

Table 15.6 Ratios for the main design characteristics

Parameter	$FG(2)$		$FG(2^{m=2})$	
	LFSR	NLFSR	LFSR	NLFSR
$T_{max} =$	$2^L - 1$		$2^{m \cdot L} - 1$	
$n_L =$	L	$\frac{L \cdot (L+1)}{2}$	$m \cdot L$	$m \cdot \frac{L \cdot (L+1)}{2}$
$k =$	2^{n_L}			

All of the above will also be true for registers constructed in Galois extended fields (see Table 15.6). Moreover, this advantage for GF (2^2) is higher than similar indicators for GF (2) and with increasing value of L, this advantage (predicted) will also increase by the power law.

Analyzing the above results, we can conclude: NLFSR give a significant advantage over LFSR in the number of possible different structures (the ability to implement different PRS generators), which will generate M-sequences at the same value of L. This statement increases the cryptographic stability of NLFSR compared to LFSR breaking the algorithm due to the exponential increase in the number of possible options.

NLFSR is better to use when there is a need to implement a large number of incompatible PRS generators with the possibility of using the same basic register (for example, in industrial production for different consumers). Also NLFSR will be expedient to use for fast replacement of algorithm of work of the generator, in case of its compromise, in principle without changing structure of the generator. By using the values of the coefficients a_{ij} as a parameter of the long-term key, it is possible to further increase the cryptographic stability of the generator, in terms of the probability of keystream overlap, by more frequent replacement of the feedback coefficients compared to LFSR.

15.8 Quantitative Calculation of the set of NLFSR Third and More High Order

In the previous section, LFSR and NLFSR of the second order, such a quantitative estimate of the number of different generating polynomials (n_L) are considered. Expressions for determining n_L can also be obtained based on the principles of combinatorics, considering the number of cells in the register L as a general population, and the product between different cells—as a sample of r disordered set. Then the number of combinations of L elements by $r\left(C_L^r\right)$ can be calculated by the formula:

$$C_L^r = \frac{L!}{r! \cdot (L-r)!}$$

where do we get for the corresponding LFSR at different sizes of the register:

$L = 2 \Rightarrow C_2^1 = \frac{2!}{1! \cdot (2-1)!} = 2$ corresponds to a set of coefficients $\{x_1; x_2\}$;
$L = 3 \Rightarrow C_3^1 = \frac{3!}{1! \cdot (3-1)!} = 3$ corresponds to a set of coefficients $\{x_1; x_2; x_3\}$;
$L = 4 \Rightarrow C_4^1 = \frac{4!}{1! \cdot (4-1)!} = 4$ corresponds to a set of coefficients $\{x_1; x_2; x_3; x_4\}$, etc.

Thus, for the linear case, the size of many different combinations will be determined by the expression:

$$C_L^1 = \frac{L!}{1! \cdot (L-1)!} = L.$$

The nonlinearity of the second order will correspond to $r = 2$ and the number of different nonlinear coefficients between two different cells of the register can be calculated as follows:

$L = 2 \Rightarrow C_2^2 = \frac{2!}{2! \cdot (2-2)!} = 1$ corresponds to a set of coefficients $\{x_1 \cdot x_2\}$;
$L = 3 \Rightarrow C_3^2 = \frac{3!}{2! \cdot (3-2)!} = 3$ corresponds to a set of coefficients $\{x_1 \cdot x_2; x_1 \cdot x_3; x_2 \cdot x_3\}$;
$L = 4 \Rightarrow C_4^2 = \frac{4!}{2! \cdot (4-2)!} = 6$ corresponds to a set of coefficients $\{x_1 \cdot x_2; x_1 \cdot x_3; x_1 \cdot x_4; x_2 \cdot x_3; x_2 \cdot x_4; x_3 \cdot x_4\}$, etc.

Thus, the number of nonlinear combinations of the second order will be determined by the ratio:

$$C_L^2 = \frac{L!}{2! \cdot (L-2)!} = \frac{L \cdot (L-1)}{2}.$$

The total number of combinations that can be composed of linear and nonlinear coefficients of the second order is given by the expression:

$$n_L = C_L^1 + C_L^2 = \frac{L!}{1! \cdot (L-1)!} + \frac{L!}{2! \cdot (L-2)!} = \frac{2 \cdot L! + L! \cdot (L-1)}{2! \cdot (L-1)!}$$
$$= \frac{L! \cdot (L+1)}{2! \cdot (L-1)!} = \frac{L \cdot (L+1)}{2},$$

which corresponds to the expression given in Table 15.6.

Consider now the nonlinearity of the third order, that is, when in the forming polynomial there are products that include up to three elements. The design of the third-order NLFSR for $L = 4$ is shown in Fig. 15.15.

Third-order nonlinearity, which is equivalent to the number of combinations for $r = 3$ between three different register cells, is calculated for different register sizes as:

$L = 3 \Rightarrow C_3^3 = \frac{3!}{3! \cdot (3-3)!} = 1$ corresponds to a set of coefficients $\{x_1 \cdot x_2 \cdot x_3\}$;
$L = 4 \Rightarrow C_4^3 = \frac{4!}{3! \cdot (4-3)!} = 4$ corresponds to a set of coefficients $\{\{x_1 \cdot x_2 \cdot x_3; x_1 \cdot x_2 \cdot x_4; x_1 \cdot x_3 \cdot x_4; x_2 \cdot x_3 \cdot x_4\}$, etc.

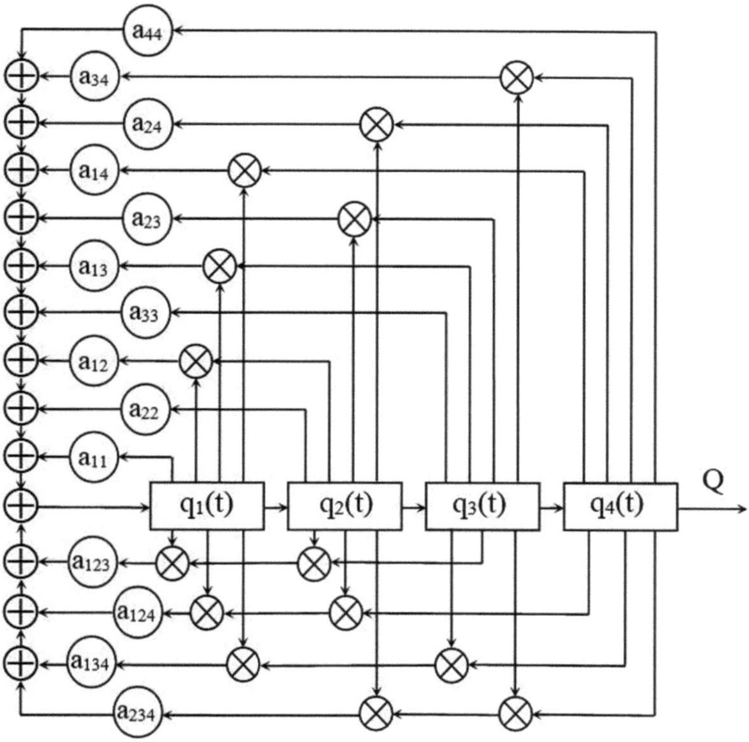

Fig. 15.15 The design of the third-order NLFSR for $L = 4$

Thus, the number of nonlinear combinations of the third order will be determined by the ratio:

$$C_L^3 = \frac{L!}{3! \cdot (L-3)!} = \frac{L \cdot (L-1) \cdot (L-2)}{6}.$$

The total number of combinations that can be composed of linear and nonlinear coefficients of the second and third order is given by the expression:

$$\begin{aligned} n_L^{r=3} &= C_L^1 + C_L^2 + C_L^3 \\ &= \frac{L!}{1! \cdot (L-1)!} + \frac{L!}{2! \cdot (L-2)!} + \frac{L!}{3! \cdot (L-3)!} \\ &= \frac{L \cdot \left(L^2 + 5\right)}{6}. \end{aligned}$$

Here is an example of calculating the number of combinations with nonlinearity of the fourth order, i.e. $r = 4$ for different L:

$L = 4 \Rightarrow C_4^4 = \frac{4!}{4! \cdot (4-4)!} = 1$ corresponds to a set of coefficients $\{x_1 \cdot x_2 \cdot x_3 \cdot x_4\}$;
$L = 5 \Rightarrow C_5^4 = \frac{5!}{4! \cdot (5-4)!} = 5$ corresponds to a set of coefficients $\{x_1 \cdot x_2 \cdot x_3 \cdot x_4;\ x_1 \cdot x_2 \cdot x_3 \cdot x_5;\ x_1 \cdot x_2 \cdot x_4 \cdot x_5;\ x_1 \cdot x_3 \cdot x_4 \cdot x_5;\ x_2 \cdot x_3 \cdot x_4 \cdot x_5\}$, etc.

The number of nonlinear combinations of the fourth order will be determined by the ratio:

$$C_L^4 = \frac{L!}{4! \cdot (L-4)!} = \frac{L \cdot (L-1) \cdot (L-2) \cdot (L-3)}{24}$$

The total number of combinations that can be composed of linear and nonlinear coefficients of the second, third and fourth order is given by the expression:

$$\begin{aligned} n_L^{r=4} &= C_L^1 + C_L^2 + C_L^3 + C_L^4 \\ &= \frac{L!}{1! \cdot (L-1)!} + \frac{L!}{2! \cdot (L-2)!} + \frac{L!}{3! \cdot (L-3)!} + \frac{L!}{4! \cdot (L-4)!} \\ &= L \cdot \left(1 + \frac{(L-1)}{2!} + \frac{(L-1) \cdot (L-2)}{3!} + \frac{(L-1) \cdot (L-2) \cdot (L-3)}{4!}\right) \end{aligned}$$

Carrying out similar calculations for the nonlinearity of the sixth, seventh and so on to the tenth order, we obtain the results shown in Table 15.7.

Summarizing the results for an arbitrary order of nonlinearity, we obtain a formula for determining the number of feedback coefficients used in the construction in the case of nonlinearity of the rth order for the register of dimension L:

Table 15.7 The number of possible combinations (C_L^r) of nonlinearity of the third order for different register sizes L

L	Order of nonlinearity, r									
	1	2	3	4	5	6	7	8	9	10
2	2	1	–	–	–	–	–	–	–	–
3	3	3	1	–	–	–	–	–	–	–
4	4	6	4	1	–	–	–	–	–	–
5	5	10	10	5	1	–	–	–	–	–
6	6	15	20	15	6	1	–	–	–	–
7	7	21	35	35	21	7	1	–	–	–
8	8	28	56	70	56	28	8	1	–	–
9	9	36	84	126	126	84	36	9	1	–
10	10	45	120	210	252	210	120	45	10	1

$$n_L^r = \sum_{i=1}^{r} C_L^i = L \cdot \left(1 + \sum_{i=1}^{r-1} \frac{\prod_{j=1}^{i} (L-j)}{(1+i)!} \right).$$

In the extreme case, when the nonlinearity of the maximum order is involved, for a given register size (i.e. $r = L$) $\mathrm{n_L}$ takes the form:

$$n_L^{r=L} = 2^L - 1,$$

which in terms of combinatorics corresponds to the number of placements with repetitions of two elements {0,1} for L and is denoted as: $\overline{A}_2^L = 2^L$. But, since the combination of only zeros is not considered by us as meaningless, then $n_L^{r=L} = \overline{A}_2^L - 1$.

The calculated values n_L^r for the nonlinearity of the r order are given in Table 15.8.

As an example, Table 15.9 shows the number of feedback coefficients in NLFSR, as well as the number of different combinations that can be made from them with maximum nonlinearity for $L = 32$ and $L = 64$. For comparison, we give similar values for LFSR and NLFSR of the second order.

We see that the number of different generating polynomials, by means of which it is possible to construct NLFSR of the maximum order, by many orders of magnitude exceeds the similar number for LFSR or NLFSR of the second order. In this case, if the trend in the number of M-NLFSR will also continue, as for the nonlinearity of the second order, then the nonlinearity of the third and higher order has every reason to believe that there is a huge number of different M-NLFSR even with small registers.

Table 15.8 The number of feedback coefficients () used in the NLFSR, with nonlinearity of the r order

L	Order of nonlinearity, r									
	1	2	3	4	5	6	7	8	9	10
2	2	3	–	–	–	–	–	–	–	–
3	3	6	7	–	–	–	–	–	–	–
4	4	10	14	15	–	–	–	–	–	–
5	5	15	25	30	31	–	–	–	–	–
6	6	21	41	56	62	63	–	–	–	–
7	7	28	63	98	119	126	127	–	–	–
8	8	36	92	162	218	246	254	255	–	–
9	9	45	129	255	381	465	501	510	511	–
10	10	55	175	385	637	847	967	1012	1022	1023

Table 15.9 Comparative table of the number of possible combinations for NLFSR with nonlinearity of different order

L	LFSR		NLFSR of second order		NLFSR of maximal order	
	n_L	k	n_L	k	$n_L^{r=L}$	k
32	32	$4.3 \cdot 10^9$	528	$8.8 \cdot 10^{158}$	4,294,967,295	$\sim 10^{1,292,820,256}$
64	64	$1.8 \cdot 10^{19}$	2080	$1.4 \cdot 10^{626}$	18,446,744,073,709,551,615	$\sim 10^{5,552,620,721,900,791,247}$

15.9 Evaluation of M-NLFSR and M-LFSR Productivity in the Software Implementation of the Algorithm

First, it is necessary to separately state the fact that the rate of generation (performance) of PRS of an algorithm is influenced by many factors, for example: the platform on which the calculation is performed; the approach and method by which the operations laid down in the algorithm are implemented at the program level; used compiler. With the best optimization of the program code and the appropriate choice of compiler, sometimes you can achieve a better result than from the implementation of a potentially faster "generator" in an inappropriate way. We do not claim that the approach used in the paper is the best. These results make it possible to quantitatively compare the algorithms of generators based on LFSR and NLFSR.

FASM (flat assembler version 1.71.51) was used to compile all examples. The calculations were performed on a personal computer (64-bit Windows 7 SP 1, Intel Core i53210M CPU 2.5 GHz).

From (11.1) and (11.2) it is seen that the calculation of the next bit can be divided into two blocks: in the first block of calculations the sum of the linear coefficients is calculated; in the second block—calculating the sum of nonlinear coefficients.

The second block is divided into sub-blocks for each q_i. In each sub-block the value of the register q_n is checked, where and when $q_n = 1$ the parity of the sum of the values of the registers q_j (for $j = \{\overline{n+1, L}\}$) is checked and only for those values of j for which $a_{nj} \neq 0$. If the sum of the values of the registers q_j is an odd number, then the value of the pre-calculated for the linear component changes, which after completion of checks of all sub-blocks is fed to the input of the register.

Applying the above approach to calculate the next state in the NLFSR, can be performed in addition from one to $L - 1$ blocks of operations, depending on the present nonlinear feedback coefficients.

The first block (for linear coefficients) contains 7 processor instructions, the second—8 instructions for each subblock. In addition, a parity flag (RFLAGS flag register PF bit) was used in each block to check the parity of the register. However, the parity flag was used only at values of $L \leq 8$. For registers of length $L > 8$, the continuous fragment of the register was divided into parts with a length of *8* bits and each of these parts was checked separately. Additional partitioning leads to an increase in the number of operations and, consequently, to an increase in the time spent on generation.

Table 15.10 shows the results of measuring the time spent on generating one GB of data according to the above algorithm. Only M-sequences were generated. In the case of non-sequence generation, series of only ones or zeros may occur, which will affect the result of the experiment.

As you can see, the generation based on NLFSR takes about 1.2–1.6 times more time than the generation of LFSR. It is worth paying special attention to three circumstances.

The first. In the method used, the number of operations for LFSR does not depend on the number of feedbacks (if there is such a relationship, it is so small that it was not monitored during testing). But, regardless of whether only one linear coefficient is not equal to zero or all, the speed of the entire cycle is not affected. An example is the forming polynomials №s 5 and 6 (Table 15.10). In the first case, the polynomial forming is dense, and in the second—discharged. Similarly, for NLFSR (polynomials №s 7 and 8 of Table 15.10), if all nonlinear coefficients have a common term, then all feedbacks can be calculated in one subblock.

Another. It is possible to select polynomials for which the non-zero feedback coefficients in the subblock will be at a distance of no more than 8 bits from each other. This will allow you to not split the calculation into blocks of 8 bits and calculate them simultaneously. An example is a polynomial under № 4, which is calculated for the time equal to the time spent on calculating the register three times smaller (polynomials (1, 2, 3).

Third. The time spent on one cycle in the NLFSR increases (by approximately 30%) with each new combination qn. For example, in polynomial № 9 all different nonlinear combinations are present, which leads to a twofold increase in computation time. Therefore, it makes sense to use polynomials in NLFSR, which will have only combinations of nonlinear coefficients included in one subblock (as an example of a polynomial №s 10, 11).

Table 15.10 Time to generate 1 GB of M-NLFSR, by software implementation

№	Polynomial representation	Generation time (s)
LFSR[(15.1)]		
	$x^8 + x^7 + x^2 + x^1 + 1$	21
2	$x^8 + x^7 + x^6 + x^5 + x^2 + x^1 + 1$	21
3	$x^8 + x^7 + x^6 + x^3 + x^2 + x^1 + 1$	21
4	$x^{24} + x^{23} + x^{22} + x^{17} + 1$	21
5	$x^{24} + x^{23} + x^{22} + x^{21} + x^{20} + x^{19} + x^{18} + x^{17} + x^{16} + x^{15} + x^{14} +$ $+ x^{13} + x^{12} + x^{11} + x^{10} + x^9 + x^8 + x^7 + x^5 + x^3 + x^2 + x^1 + 1$	34
6	$x^{24} + x^{10} + x^4 + x^3 + 1$	34
NLFSR		
	$x^8 + x^4 \cdot x^2 + x^3 \cdot x^2 + x^1 + 1$	34
8	$x^8 + x^7 + x^5 + x^4 + x^6 \cdot x^1 + x^4 \cdot x^1 + x^3 \cdot x^1 + x^2 \cdot x^1 + 1$	34
9	$x^8 + x^7 + x^7 \cdot x^6 + x^6 + x^7 \cdot x^5 + x^6 \cdot x^5 + x^5 + x^7 \cdot x^4 +$ $+ x^6 \cdot x^4 + x^5 \cdot x^4 + x^4 + x^7 \cdot x^3 + x^6 \cdot x^3 + x^5 \cdot x^3 + x^4 \cdot x^3 +$ $+ x^3 + x^7 \cdot x^2 + x^6 \cdot x^2 + x^5 \cdot x^2 + x^2 + x^7 \cdot x^1 + x^1 + 1$	76
10	$x^{24} + x^{23} \cdot x^{21} + x^{22} + x^{21} + x^{19} + x^{16} + x^{15} + x^{14} + x^{13} + x^{12} +$ $+ x^{10} + x^7 + x^6 + x^4 + 1$	43
11	$x^{24} + x^{23} \cdot x^{21} + x^{22} + x^{22} \cdot x^{21} + x^{21} + x^{19} + x^{17} + x^{16} + x^{13} +$ $+ x^{12} + x^{11} + x^{10} + x^6 + x^5 + 1$	43

[15.1)] Polynomials for M-LFSR are taken from work [19]

It is possible to increase the speed of the algorithm by optimizing the calculations for certain feedbacks. Thus, for $L = 8$ (polynomial № 1) under the same conditions it was possible to generate one GB in 16 s, and for the NLFSR corresponding to the polynomial № 9—in 41 s. However, the universality of the method is lost and the algorithm must be adapted to individual polynomials.

As shown in the fourth section, the hardware implementation managed to implement a similar generation of one GB in 11.8 s.

15.10 Distribution of Number of Nonlinears Feedback in M-NLFSR

As shown in the previous section, the selection of feedback coefficients directly affects the speed of the algorithm (in the software implementation). Therefore, we

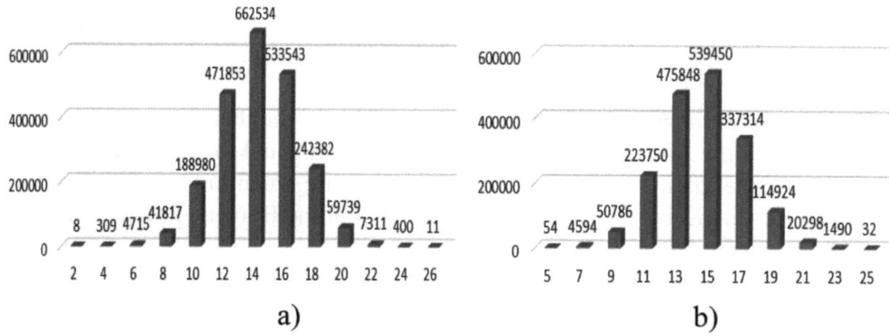

Fig. 15.16 Distribution of the number of nonlinear coefficients for M-NLFSR at linear: **a** x^9+x^1+1; **b** $x^9+x^8+x^7+x^6+x^5+x^4+x^3+x^2+x^1+1$

pay special attention to the distribution of the number of nonlinear feedbacks in M-NLFSR (for any linear feedback coefficients).

An analysis was performed on the distribution of the number of M-NLFSR by the numbers of non-zero nonlinear feedback coefficients for a fixed combination of linear coefficients for the whole set of M-NLFSR at $L \leq 9$.

As an example, Fig. 15.16 shows two histograms of the distribution of the number of nonlinear coefficients $a_{ij} \neq 0$ for two different coefficients a_{ii} at $L = 9$ (even number of linear coefficients, feedback aii, according to the linear part forming the polynomial $x^9 + x^1 + 1$ is presented in Fig. 15.16a; the odd number aij corresponds to the polynomial $x^9 + x^8 + x^7 + x^6 + x^5 + x^4 + x^3 + x^2 + x^1 + 1$—Fig. 15.16b. For other combinations of linear coefficients, the distribution is approximately the same.

Thus, if you look for (or design) NLFSR, it is optimal to search for non-zero nonlinear coefficients, calculating their number close to *L/2*. If we take only one or two (or almost all) nonlinear connections, then the number of possible M-NLFSR will be equal to a few units (regardless of which we take linear feedback).

15.11 Research of NLFSR as a Filter or Combining Functions

We introduce some definitions that we will use in the future [21].

A Boolean function is a Boolean mapping of a view $f : GF_2^L \to GF_2$. Boolean functions will be represented as polynomials (Zhegalkin polynomial or algebraic normal form—ANF) over the field GF (2):

$$f(x_1, x_2, \ldots, x_L)$$
$$= a_0 + \sum_{i=1}^{L} a_i x_i + \sum_{i=1}^{L-1} \sum_{j=i+1}^{L} a_{ij} x_i x_j + \sum_{i=1}^{L-2} \sum_{j=i+1}^{L-1} \sum_{m=j+1}^{L} a_{ijm} x_i x_j x_m + \ldots,$$

as well as a table. The table of the Boolean function $f(x)$ consists of rows of arguments of the function $x_1, x_2, \ldots, x_L$, arranged in lexicographic order, and values of the function, i.e. $x_1, x_2, \ldots, x_L, f(x_1, x_2, \ldots, x_L)$.

The algebraic power def (f) of the function f is the number of variables in the longest term of the ANF, at which the coefficient is not equal to zero. A function of degree not higher than 1 is an affine function. The case when in the affine function $a_0 = 0$, corresponds to *a linear function.* The set of affine Boolean functions from L variables is denoted as A_L.

The weight of the Boolean function or the weight of Hamming is called the number of units in the vector and is denoted as $wt(f)$ or $wt(x)$.

The Hamming distance between the Boolean functions f and g is the Hamming distance between the vectors of their values $dist(f, g) = wt(f \oplus g)$.

The statements and theorems given in this section, in order to reduce the amount of work, are given without proof. Proofs are publicly available, they can be found, for example, in [21–28]. All values given in this section were experimentally tested on the whole set of NLFSR (LFSR as a special case) for sizes from $L = 4$ to 9 cells inclusive.

15.11.1 Balance

M-NLFSR, as well as M-LFSR, generate a modified De Bruijn sequence, and if we add to the consideration the state of filling all cells with zero values, the resulting function will be balanced. With equal and independent choice of arguments of the Boolean function f, the probabilities of its values, respectively, are equal to $P(1) = wt(f)/2^L$ $P(0) = 1 - wt(f)/2^L$.

15.11.2 The Presence of Prohibitions

M-NLFSR are functions that do not have prohibitions. This follows from the fact that NLFSR form a De Bruijn sequence, which by definition has all possible combinations of sequences.

However, care should be taken because a well-balanced filtering function in one form or another transfers the properties of the input sequence to the properties of the generated sequence [29]. For example, in [30] a new criterion was established, which ideologically states the following: "the filtering function preserves prohibitions (in the appropriate sense) if and only when it is completely balanced." Accordingly, if the input of the function receives a "far" from the random sequence, then the output of its statistical properties will be bad.

15.11.3 Correlation Immunity

The presence of a correlation immune function of degree m means that the values of the function $Z = f(X)$ are statistically independent of any set of not more than m components of an arbitrary vector of the argument $X = GF(2)^L$. This is equivalent to the condition that the output of the transformation is not "leaked" information about the vectors coming to the input of the transformation and having a Hamming weight of not more than *m*.

A Boolean function *f* is called *a correlation immune order m*, $1 \leq m \leq L$, if for any set of numbers *m* of variables $1 \leq i_1 < i_2 < \ldots < i_m \leq L$ random variables $X = \left(x_{i_1}, x_{i_2}, \ldots, x_{i_m}\right)$ and $Y = f(x_1, x_2, \ldots, x_L)$ are independent.

We can prove that the correlation immune order m function from *L* variables is correlation immune of arbitrary lower order. Thus, the Boolean function *f* corresponds to some maximum order of its correlation immunity $m_{\max}$, which is denoted by $cor(f)$.

The case when $m = L$, occurs only when $f = const$. The maximum correlation immunity of degree $m = L - 1$ is reached only by affine functions, ie cryptographically weak. In addition, if f is balanced and $cor(f) = L - 2$, then the function *f* is also affine. Thus, it makes sense to consider the order of correlation immunity m only in the range $1 \leq m \leq L - 3$.

A balanced correlation-immune function of order *m* is called *m*-stable function. Formally, any balanced Boolean function can be considered as 0-stable and an arbitrary Boolean function as *(−1)* stable. By analogy with $cor(f)$, the notation for *the maximum order of stability* is introduced:

$$sut(f) = \begin{cases} -1, & \text{if } f \text{ is not balanced}, \\ \text{cor}(f) & \text{if } f \text{ is balanced}. \end{cases}$$

Siegenthaler inequality. If f is a correlation-immune order of m function on $GF(2)^L$, then:

1) $def(f) \leq L - m$;
2) if *f* is balanced and $sut(f) = m \leq L - 2$, then $def(f) + sut(f) \leq L - 1$.

Siegenthaler's inequality is one of the many contradictions in the cryptographic properties of functions to each other: a high order of correlated immune function leads to its low degree, and vice versa.

If the function *f* is balanced, $sut(f) = m \leq L - 2$ and $def(f) = L - m - 1$, then *f* is called optimal.

Whence we have the optimal *f* for LFSR $m = L - 1 - def(f) = L - 2$ and for NLFSR of the second order $m = L - 1 - def(f) = L - 3$. Values of the maximum order of stability for optimal functions, depending on register lengths are given in Table 15.11.

Thus, we determined the upper limit of values for m stable functions.

Table 15.11 Values of the maximum order of stability for optimal functions

		L					
		4	5	6	7	8	9
$sut(f)$	M-LFSR	2	3	4	5	6	7
	Second order MNLFSR	1	2	3	4	5	6

The correlation immunity of the whole set of M-LFSR and M-NLFSR of the second order with dimension L = 4, ..., 9 was investigated in the work. The results are summarized in Table 15.12.

As you can see from Table 15.12, M-NLFSR of the second order are achieved values for optimal functions (marked in the table as "m-opt") in all studied L. However, there is a very large proportion (approximately half of the whole set of M-NLFSR of the second order), which has no correlation immunity. The coefficients aij, which correspond to M-NLFSR and are the optimal functions, are given in Tables 15.13, 15.14, 15.15, 15.16, 15.17, 15.18.

In the Tables 15.13, 15.14, 15.15, 15.16 and 15.17, the coefficients of a polynomial are indicated in the left column, and the symmetric coefficients corresponding to this polynomial are indicated in the right column.

15.11.4 Nonlinearity

The nonlinearity of the function f is the distance from f to the class of affine functions. We denote the nonlinearity of the function f by N_f:

$$N_f = dist(f, A_L) = \min_{g \in A_L} dist(f, g).$$

In the case of an even L, the maximum possible value of nonlinearity is equal to $2^{L-1} - 2^{(L/2)^{-1}}$, functions that have such nonlinearity are allocated to a separate class—"bentfunctions". In the case of an odd L, the exact value of the maximum nonlinearity is not known and represents a complex combinatorial problem [25]. However, not all benthic functions are balanced (unlike M- NLFSR), which makes them vulnerable to statistical analysis.

The following statements show that the higher the order of correlated immune function, the lower the upper limit of its nonlinearity.

If f is balanced and m-stable, $m \le L - 2$. Then $N_f \le 2^{L-1} - 2^{m+1}$.

By analogy with the concept of optimal function, a special name is introduced for the m-stable function of the maximum possible nonlinearity.

If the function f with $GF(2)^L$ is balanced, $sut(f) = m \le L - 2$ and $N_f = 2^{L-1} - 2^{m+1}$, then f is called m-saturated.

Table 15.12 Distribution of the number of registers depending on the maximum resistance for M- LFSR and M- NLFSR of the second order

	Усього	$sut(f)$							
		$m = 0$	$m = 1$	$m = 2$	$m = 3$	$m = 4$	$m = 5$	$m = 6$	$m = 7$
$L = 4$									
MLFSR	2	0	2	0	–	–	–	–	–
M NLFSR of second order	14	4	10 *m-opt*	–	–	–	–	–	–
$L = 5$									
MLFSR	6	0	2	0	4 *m-opt*	–	–	–	–
M NLFSR of second order	122	64	52	6 *m-opt*	–	–	–	–	–
$L = 6$									
MLFSR	6	0	2	0	4	0	–	–	–
M NLFSR of second order	1946	788	1044	76	38 *m-opt*	–	–	–	–
$L = 7$									
MLFSR	18	0	4	0	10	0	4 *m-opt*	–	–
M NLFSR of second order	64,038	33,988	25,578	4090	378	4 *m-opt*	–	–	–
$L = 8$									
MLFSR	16	0	0	0	12	0	4	0	–
M NLFSR of second order	4,017,982	1,686,218	2,120,124	194,798	16,612	188	42 *m-opt*	–	–

(continued)

Table 15.12 (continued)

	Усього	$sut(f)$							
		$m = 0$	$m = 1$	$m = 2$	$m = 3$	$m = 4$	$m = 5$	$m = 6$	$m = 7$
$L = 9$									
MLFSR	48	0	2	0	16	0	28	0	2 *m-opt*
M NLFSR of second order	519,239,746	284,956,836	208,843,948	24,325,344	1,091,568	21,192	848	10 *m-opt*	–

Table 15.13 M-NLFSR second order dimension $L = 4$, with nonlinearity $N_f = 4$ and the maximum stability $sut(f) = 1$

№	$a_{11}a_{12}a_{13}a_{14}a_{22}a_{23}a_{24}a_{33}a_{34}a_{44}$	$a_{11}a_{12}a_{13}a_{14}a_{22}a_{23}a_{24}a_{33}a_{34}a_{44}$
1	1010 100 00 1	0010 100 10 1
2	0110 100 00 1	0010 110 00 1
3	1100 010 00 1	0100 010 10 1
4	1000 110 00 1	0100 100 10 1
5	1110 110 00 1	0110 110 10 1

Table 15.14 M-NLFSR second order dimension L = 6, with nonlinearity $N_f = 16$ and maximum stability $sut(f) = 3$

№	$a_{11} \ldots a_{16}a_{22} \ldots a_{26} \ldots a_{55}a_{56}a_{66}$	$a_{11} \ldots a_{16}a_{22} \ldots a_{26} \ldots a_{55}a_{56}a_{66}$
1	100010 10000 1000 100 00 1	000010 10000 1000 100 10 1
2	001010 10000 1000 100 00 1	000010 10000 1010 100 00 1
3	100000 01010 1000 100 00 1	000100 10000 1100 000 10 1
4	101000 10000 0010 100 00 1	001000 10000 0010 100 10 1
5	100000 10000 1010 100 00 1	001000 10000 1000 100 10 1
6	101010 10000 1010 100 00 1	001010 10000 1010 100 10 1
7	100000 00010 1010 100 00 1	001100 10000 1000 000 10 1
8	111000 00010 1010 100 00 1	001100 10000 1010 010 10 1
9	100000 00110 1110 100 00 1	001100 11100 1000 000 10 1
10	100100 10000 1000 010 00 1	010000 00010 1000 100 10 1
11	100000 10000 1100 010 00 1	010000 01000 1000 100 10 1
12	100000 00010 1000 110 00 1	010100 10000 1000 000 10 1
13	100000 10000 1110 110 00 1	011000 11000 1000 100 10 1
14	100100 10000 1000 000 10 1	100000 00010 1000 100 10 1
15	100000 01100 1000 000 10 1	100000 00100 1100 000 10 1
16	101000 10000 0100 000 10 1	100000 01000 0010 100 10 1
17	100000 11000 0100 000 10 1	100000 01000 0100 100 10 1
18	100000 11100 1100 000 10 1	100000 01100 1100 100 10 1
19	100000 10000 0110 000 10 1	101000 01000 0000 100 10 1

Table 15.15 M-NLFSR second order dimension $L = 5$, with nonlinearity $N_f = 8$ and the maximum stability $sut(f) = 2$

№	$a_{11} \ldots a_{15}a_{22} \ldots a_{25}a_{33} \ldots a_{35}a_{44}a_{45}a_{55}$	$a_{11} \ldots a_{15}a_{22} \ldots a_{25}a_{33} \ldots a_{35}a_{44}a_{45}a_{55}$
1	10000 1110 100 00 1	00100 1100 100 10 1
2	10110 1000 000 10 1	10010 0010 100 10 1
3	11010 0000 100 10 1	10010 1000 010 10 1

Table 15.16 M- NLFSR second order dimension $L = 7$, with nonlinearity $N_f = 32$ and maximum stability $sut(f) = 4$

№	$a_{11} \ldots a_{17}a_{22} \ldots a_{27} \ldots a_{66}a_{67}a_{77}$	$a_{11} \ldots a_{17}a_{22} \ldots a_{27} \ldots a_{66}a_{67}a_{77}$
1	1010000 100000 10100 1000 000 10 1	1000000 001000 10000 1010 100 10 1
2	1001100 100000 10000 0000 100 10 1	1000000 100010 00010 1000 100 10 1

Table 15.19 summarizes the calculated values according to the above formulas of the maximum possible nonlinearity of the balanced function depending on its stability.

However, the nonlinearity values in Table 15.19 are not necessarily achievable. Denote by $N_{f\,\max}(L, m)$ the maximum possible nonlinearity of the m-stable Boolean function given on $GF(2)^L$, and give the upper bound for the nonlinearity of m-stable functions.

From the above it follows that $N_{f\,\max}(L, -1) = 2^{L-1} - 2^{L/2-1}$, this value can be achieved only for even L. If f is a balanced function and L is an even value, then $N_{f\,\max}(L, m) = 2^{L-1} - 2^{L/2-1} - 2^{m+1}$ [23].

In Ref. [31] it is indicated that for odd L and $L \leq 7$, $N_{f\,\max}(L, -1) = 2^{L-1} - 2^{(L-1)/2}$, but for odd L and $L \geq 15$, the inequality $N_{f\,\max}(L, -1) > 2^{L-1} - 2^{(L-1)/2}$ is valid.

For $m \geq L - 2$, according to the Siegenthaler inequality $def(f) \leq 1$, where $N_{f\,\max}(L, m) = 0$. Also in Ref. [31] refer to the proved inequality $N_{f\,\max}(L, L-3) = 2^{L-2}$ and the hypothesis that $N_{f\,\max}(L, L-4) = 2^{L-1} - 2^{L-3}$. In addition, some exact values of $N_{f\,\max}(L, m)$ for small L and m:

$N_{f\,\max}(4, 0) = 4;$
$N_{f\,\max}(5, -1) = N_{f\,\max}(5, 0) = N_{f\,\max}(5, 1) = 12;$
$N_{f\,\max}(6, 0) = 26; N_{f\,\max}(6, 1) = N_{f\,\max}(6, 2) = 24;$
$N_{f\,\max}(7, -1) = N_{f\,\max}(7, 0) = N_{f\,\max}(7, 1) = 56.$

These results do not contradict the results obtained in this paper and given below.

Table 15.20 summarizes the obtained distribution of the nonlinearity of the whole set of M-NLFSR dimension $L = 4, \ldots, 9$ depending on the size of M-NLFSR and correlation immunity.

As we can see from the above results, the second-order M-NLFSR simultaneously achieve the maximum possible stability and maximum nonlinearity. Moreover, all optimal functions are also m-saturated (in Table 15.20 marked—"m-sat"). In addition, most M-NLFSR, which are not m-saturated functions by definition, achieve the maximum possible result for $N_{f\,\max}(L, m)$ above.

Second-order M-NLFSR corresponding to m-saturated functions are given above in Tables 15.13, 15.14, 15.15, 15.16, 15.17 and 15.18.

Analyzing the obtained results, we see that symmetric M-NLFSR have the same indicators $sut(f)$ and N_f. All studied M-NLFSR have $N_f \geq 2^{L-2}$. M-saturated functions are not dense polynomials.

Table 15.17 M-NLFSR second order dimension $L = 8$, with nonlinearity $N_f = 64$ and maximum stability $sut(f) = 5$

№	$a_{11} \dots a_{18} a_{22} \dots a_{28} \dots a_{77} a_{78} a_{88}$	$a_{11} \dots a_{18} a_{22} \dots a_{28} \dots a_{77} a_{78} a_{88}$
1	11000100 1000010 100000 10000 1000 010 00 1	01000100 0000010 100000 10000 1000 110 10 1
2	10000000 1000000 100000 10010 1000 010 00 1	01010000 0000000 100000 10000 1000 100 10 1
3	10000000 1000000 100010 10000 0010 100 00 1	00101000 1000000 000000 10000 1000 100 10 1
4	10001100 1000000 100000 10000 0010 110 00 1	01100000 1000010 000010 10000 1000 100 10 1
5	10000000 1000010 000010 10000 1000 100 00 1	00001100 1000000 100000 10000 0000 100 10 1
6	10000000 1000000 010010 10000 1000 100 00 1	00001000 1000000 100000 11000 0000 100 10 1
7	10001010 1000000 001010 10000 1000 100 00 1	00001010 1000000 101010 10000 0000 100 10 1
8	10000000 1100100 000010 10000 1000 110 00 1	01001000 1000100 100000 10000 0100 100 10 1
9	10000000 0010010 100000 10100 1000 110 00 1	01000100 1010000 100000 10100 1000 000 10 1
10	10000000 1000010 100000 10000 1000 100 00 1	00000100 1000000 100000 10000 1000 100 10 1
11	10000000 1000000 100010 10000 1000 100 00 1	00001000 1000000 100000 10000 1000 100 10 1
12	10000000 1000000 100000 11000 0100 000 10 1	10000000 0100000 010000 10000 1000 100 10 1
13	10000000 1000000 100000 10000 0100 010 10 1	11000000 0100000 000000 10000 1000 100 10 1
14	10000000 1000000 100000 01100 1010 010 10 1	11100000 0010000 110000 00000 1000 100 10 1
15	11100000 0000100 100100 10000 1000 000 10 1	10000000 0001100 100000 10000 1010 010 10 1
16	10000000 1000000 100000 11100 1100 000 10 1	10000000 0110000 110000 10000 1000 100 10 1
17	10000000 1000000 100000 01010 0000 100 10 1	10010000 1000000 010000 00000 1000 100 10 1
18	10000000 1000000 100000 01010 0100 110 10 1	11010000 1100000 010000 00000 1000 100 10 1
19	10000000 1000000 011000 10000 0000 100 10 1	10000000 1000000 001000 11000 0000 100 10 1
20	10000000 1000000 100000 10000 0010 100 10 1	10100000 1000000 000000 10000 1000 100 10 1

(continued)

Table 15.17 (continued)

№	$a_{11} \ldots a_{18} a_{22} \ldots a_{28} \ldots a_{77} a_{78} a_{88}$	$a_{11} \ldots a_{18} a_{22} \ldots a_{28} \ldots a_{77} a_{78} a_{88}$
21	10000000 1000000 100000 10000 0110 110 10 1	11100000 1100000 000000 10000 1000 100 10 1

Table 15.18 M-NLFSR second order dimension $L = 9$, with nonlinearity $N_f = 128$ and maximum stability $sut(f) = 6$

№	$a_{11} a_{12} a_{13} a_{14} a_{15} a_{16} a_{17} a_{18} a_{19} a_{22} \ldots a_{29} \ldots a_{88} a_{89} a_{99}$
1	100000000 10010010 1000000 100000 10000 1000 100 00 1
2	000000100 10000000 1000000 100100 10000 1000 100 10 1
3	100000000 10000000 1000000 101010 10000 0000 100 10 1
4	100010000 10000000 0010000 100000 10000 1000 100 10 1
5	100000000 10000000 1011000 100000 00000 1000 100 10 1
6	100000000 10000000 1001000 001000 10000 1000 100 10 1
7	100001000 10000000 1000000 100000 01000 1000 100 10 1
8	100000000 10000000 1100010 000000 10000 1000 100 10 1
9	100000000 10000100 1000000 100000 00100 1000 100 10 1
10	100000000 10100100 1000000 000000 10000 1000 100 10 1

Table 15.19 The value of the nonlinearity of m-saturated functions depends from their maximum stability

		$sut(f)$					
		1	2	3	4	5	6
N_f	$L = 4$	4	0	–	–	–	–
	$L = 5$	12	8	0	–	–	–
	$L = 6$	28	24	16	0	–	–
	$L = 7$	60	56	48	32	0	–
	$L = 8$	124	120	112	96	64	0
	$L = 9$	252	248	240	224	192	128

15.12 Cryptographic Resistance Assessment Model as a Basic Element of Current Symmetric Encryption Schemes. Comparisons by Developed Model of NLFSR and LFSR Assessments

We formulate a model for estimating the cryptographic stability of the NLFSR, which takes into account the indicators introduced in the second and studied in the fifth section. The model assesses the ability of NLFSR to withstand some common

Table 15.20 Distribution of the number of registers depending on the nonlinearity and maximum resistance for M-NLFSR of the second order

N_f	$sut(f)$						
	$m = 0$	$m = 1$	$m = 2$	$m = 3$	$m = 4$	$m = 5$	$m = 6$
$L = 4$							
4	$4^{(1)}$	m-sat 10	–	–	–	–	–
0	0	0	0	–	–	–	–
$L = 5$							
12	$56^{(1)}$	0	–	–	–	–	–
8	8	52	m-sat 6	–	–	–	–
0	0	0	0	0	–	–	–
$L = 6$							
24	740	$856^{(1)}$	0	–	–	–	–
16	48	188	76	m-sat 38	–	–	–
0	0	0	0	0	0	–	–
$L = 7$							
56	$26{,}324^{(1)}$	0	0	–	–	–	–
48	7624	24,862	3596	0	–	–	–
32	40	716	494	378	m-sat 4	–	–
0	0	0	0	0	0	0	–
$L = 8$							
112	1,620,992	1,737,690	0	0	–	–	–
96	65 078	380,856	192,572	14,270	0	–	–
64	148	1 578	2 226	2 342	188	m-sat 42	–
0	0	0	0	0	0	0	0
$L = 9$							
240	216,743,896	0	0	0	–	–	–
224	67,714,544	203,967,024	19,364,756	0	0	–	–
192	498,196	4,872,526	4,953,980	1,079,370	18,642	0	–
128	200	4398	6 608	12 198	2 550	848	m-sat 10
0	0	0	0	0	0	0	0

(1) The N_f values are maximal for the data m, L and correspond to the $N_{f\,\max}(L, m)$ specified in [23]

vulnerability-based attacks that include any of the indicators included in the proposed model.

Let us denote by W_{6i}^{P3H33} unconditional, by W_{yi}^{P3H33} conditional evaluation criteria for NLFSR, and by $W_{\phi i}^{P3H33}$ the criteria considered within the filtering or combining functions, where i meet the selected criterion. Values W_{6i}^{P3H33}, W_{yi}^{P3H33} and $W_{\phi i}^{P3H33}$ are calculated in the range [0,2]. We will consider: $W_i^{P3H33} = 0$, if ith indicator has a worse value than the corresponding indicator for LFSR; $W_i^{P3H33} = 0$, if the indicators for NLFSR and for LFSR are the same; $W_i^{P3H33} = 2$, if its indicator is better than the corresponding indicator for LFSR.

The total normalized criterion for evaluating the NLFSR (W_6^{P3H33}—for unconditional, W_y^{P3H33}—for conditional criteria and W_ϕ^{P3H33}—for the filtering or combining function) is defined as:

$$W_6^{P3H33} = \frac{\sum_i W_i^{P3H33}}{n}, \; W_y^{P3H33} = \frac{\sum_i W_{yi}^{P3H33}}{n} \text{ and } W_\phi^{P3H33} = \frac{\sum_i W_{\phi i}^{P3H33}}{n},$$

where n is the number of indicators that are included in the unconditional (in our case $n = 6$), conditional ($n = 7$) criteria and criteria considered for filtering or combining functions ($n = 4$). The value of W^{P3H33} makes it possible to quantitatively compare the design of registers with linear and nonlinear feedback and, as a consequence, to make a reasonable conclusion about cryptographic stability.

The summary results of the obtained stability indicators and the values of the evaluation criteria are shown in Table 15.21 (for unconditional), in Table 15.22 (for conditional criteria) and in Table 15.23 (for filtering or combining functions).

From the above summary data we obtain the total normalized unconditional evaluation criterion $W_6^{P3H33} = 1, 5$, the total normalized conditional evaluation criterion $W_y^{P3H33} = 1, 43$ and the total normalized evaluation criterion for filtering or combining functions $W_\phi^{P3H33} = 1, 25$.

According to the introduced evaluation system, NLFSR of the second order have a clear and significant advantage in comparison with LFSR on the number of possible constructions, including the constructions forming M Sequence; by linear complexity—both Msequences from a single M-NLFSR, and the sum of several M-sequences; M-sequences generated by NLFSR successfully pass statistical testing, in contrast to M-sequences generated by NLFSR; it is impossible (in the classical sense) to apply decimation and properties of group addition to M-NLFSR; M- NLFSR have

Table 15.21 Summary indicators of the resilience of NLFSR for unconditional criteria

Indicator	M-LFSR	M-NLFSR	W_{6i}^{P3H33}
Linear complexity (Li)	$Li = L$	$Li \approx 2^L - 2$	2
Linear complexity of the total sequence (Li_{sum})	$Li_{sum} \leq L_1 + L_2 + \ldots$	$Li_{sum} \leq \left(2^{L_1} - 2\right) + +\left(2^{L_2} - 2\right) + \ldots$	2
Quadratic complexity (Li^2)	$Li^2 = L$		1
Quadratic complexity of total sequence (Li_{sum}^2)	$Li_{sum}^2 > \max\left(Li_1^2, Li_2^2\right)$		1
Period of the total sequence (T^{sum})	$T^{sum} = HCK\left(T^1, T^2, T^3, \ldots\right)$		1
Statistical properties	no	yes	2

Table 15.22 Summary indicators of the resilience of NLFSR for conditional criteria

Indicator	LFSR	NLFSR	W_{yi}^{P3H33}
Volume of different shift registers with a fixed size L (M)	$M = 2^L$	$M = 2^{\frac{L(L+1)}{2}}$	2
Volume of the ensemble of M registers (M_0)	$M_0 = \frac{phi(2^L-1)}{L}$	$M_0 \leq 2^{\frac{L(L-1)}{2}-L+2}$	2
Indicator	M-LFSR	M-NLFSR	
Correspondence of the profile of linear complexity to mathematical expectation	yes	yes	1
Correspondence of the profile of linear complexity of the total sequence to the mathematical expectation	yes	yes	1
Possibility of application of decimation	yes	no	2
Ability to apply the properties of group addition	yes	yes (increased complexity)	1
High generation speed	yes	yes	1

Table 15.23 Summary indicators of the resilience of the NLFSR for the criteria considered in the framework of filtering or combining functions

Indicator	M-LFSR	M-NLFSR	$W_{\phi i}^{P3H33}$
Balance	yes	yes	1
The presence of prohibitions	no	no	1
Correlation immunity $(cor(f))$	The maximum value is reached only at odd L	$cor(f) \leq L - 1 - def(f)$, the maximum value is reached	1
Nonlinearity (N_f)	0	$N_f \geq 2^{L-2}$, the maximum value is reached	2

high indicators of nonlinearity and reach the maximum value. Other indicators have the same values.

This model is not exhaustive and is a development of existing models for assessing the cryptographic stability of PRS, generated by NLFSR second order for their practical application in symmetric streaming encryption systems.

The proposed system of criteria and indicators allows to build an effective mechanism for assessing the cryptographic stability of the second-order NLFSR as a basic element of streaming encryption schemes and reduces the uncertainty of the expert to evaluate the components of the streaming encryption algorithm.

The parameters included in the evaluation model sufficiently characterize the NLFSR as a basic element of the streaming encryption system. The introduced efficiency evaluation criteria allowed to make a substantiated quantitative assessment of the advantage of using NLFSR instead of LFSR in stream encryption algorithms.

It is useful to use the described model to study the efficiency of the components of both known streaming encryption algorithms and the developed algorithms, as well as for a scientifically sound decision on the feasibility of using the selected design of streaming encryption. The model is a step towards creating a comprehensive methodology for estimating streaming encryption systems based on registers with nonlinear feedback.

15.13 Conclusions on Chap. 15

1. The linear complexity of an arbitrarily taken high-dimensional M-NLFSR with a high degree of probability corresponds to the maximum period that can be generated by this register, and is defined as:

$$Li \approx 2^L - 2.$$

The linear complexity profile of M-NLFSR corresponds to the mathematical expectation, as well as for a truly random sequence.

2. The period of the sequence formed by adding the sequence from different M-NLFSR is equal to the product of the periods of the original sequences (i.e. $T^{sum} = T^1 \cdot T^2 \cdot T^3 \cdot \ldots$) under the condition of mutually simple values of L forming M-NLFSR.
3. The Period of the Sequence from the Addition of Sequences of Different M-NLFSR, Provided not Mutually Simple Values of L, is Determined by the Ratio:

$$T^{sum} = LCM\left(T^1, T^2, T^3, \ldots\right)$$

4. The Linear and Quadratic Complexity of the Sum of Sequences from M-NLFSR is Determined by the Following Relations:
   - $Li^{sum} = Li_1 + Li_2$ for L^1 and L^2 mutually prime values;
   - $\max(Li_1, Li_2) < Li^{sum} < Li_1 + Li_2$ for L^1 and L^2 not mutually prime values;
   - $Li^{sum} \approx Li_1$ if $L^1 = n \cdot L^2$ where $n = 1, 2, 3, \ldots$;
   - $Li^2_{sum} > Li^2_1 \cdot Li^2_2$ for L^1 and L^2 mutually prime values;
   - $Li^2_{sum} > \max\left(Li^2_1, Li^2_2\right)$ for L^1 and L^2 not mutually prime values.
5. Using the decimation of the NLFSR M-sequence, we do not obtain the M-sequence and, as a consequence, it is not possible to obtain all (or some,

except for symmetric polynomials) NLFSR, restoring the forming polynomial by the obtained sequence by its decimation.

6. For NLFSR as Well as for LFSR, It is Possible to Calculate the Ratio for Construction of the Sequence Equivalent to Initial with Some Set Shift.
7. In the case of NLFSR, the principle of group addition takes a more complex form and is no longer just an "addition". Nonlinear terms from the product of two, three and more cells of registers are also involved in the construction of a sequence based on this principle. That is, in the general case, for the construction will not be enough only NLFSR second order, it is necessary to use NLFSR third and higher order.
8. The procedure for constructing relationships to calculate sequences with a predetermined offset for NLFSR is more complex than for LFSR.
9. The probability that an arbitrary combination of feedback factors for NLFSR will generate M sequence is much less than the similar probability for LFSR.
10. It is shown that NLFSR have a clear and significant advantage over LFSR in the number of possible different structures that will generate M-sequences at the same value of L. The values for the complete set of M-NLFSR length up to $L = 9$ for $GF(2)$ and up to $L = 5$ for $GF(2^2)$.
11. The number of different generating polynomials, with which you can construct NLFSR r-order, is determined by the ratio $k = 2^{n_L^r}$, the number of feedback coefficients is calculated by the formula:

$$n_L^r = L \cdot \left(1 + \sum_{i=1}^{r-1} \frac{\prod_{j=1}^{i} (L - j)}{(1 + i)!} \right)$$

As a last resort, when $r = L$, which corresponds to the maximum order NLFSR, n_L^r will be defined as:

$$n_L^{r=L} = 2^L - 1.$$

The number of different NLFSR r-order significantly exceeds the same number for LFSR or NLFSR second order and potentially may also be many times greater than the number of registers that generate M-sequence.

12. Designs on the basis of NLFSR allow to form the ciphering scale which on the statistical properties does not concede to the sequences formed by some random processes. That is, the studied constructs of stream generation of pseudo-random sequences have a high statistical safety index. The initial sequence is statistically indistinct compared to the implementation of a random process, which indicates a potentially high cryptographic property of the keystream generator, which contains NLFSR.

A total of 3259 tests were performed with the NIST STS statistical package. About 377 GB were studied, which was generated from 37 different constructions based on M-NLFSR.

13. It is shown that in the software implementation the time spent on the generation of NLFSR, compared with the time spent on the generation of LFSR, increases by approximately 1.2–1.6 times.
14. Boolean Functions Corresponding to M-NLFSR Are Balanced and Have no Prohibitions.
15. About half of the Boolean functions corresponding to M- NLFSR are correlated immune functions, some of which reach the maximum possible stability and are m-optimal functions.
16. All Boolean functions corresponding to M-NLFSR, with the exception of M-LFSR, have nonlinearity $N_f \geq 2^{L-2}$ and some reach the maximum value, i.e. are m-saturated functions.

All m-optimal and m-saturated functions corresponding to the M-NLFSR of the second order are given for the dimension $L = 4, ..., 9$.

17. A model for estimating the cryptographic stability of NLFSR as a basic element of streaming symmetric encryption schemes is proposed. According to the developed model, comparisons of NLFSR and LFSR are made, the quantitative estimation of cryptographic stability of NLFSR is resulted.

These factors increase the crypto-resistance of systems based on NLFSR and thus increase their reliability compared to systems that use LFSR.

References

1. Mao, W.: Modern cryptography. Theory and practice /Per. with English—M.: Izd, p. 768. Williams House (2005)
2. Yesin, V.I., Kuznetsov, O.O., Soroka, L.S.: Security of Information Systems and Technologies, p. 632. H.: KhNU. V.N. Karazina (2013)
3. Zadiraka, V., Oleksyk, O.: Computer Cryptology. Textbook, p. 502 (2002)
4. Anin, B.Yu.: Protection of Computer Information, c. 384. SPb, BKhV St. Petersburg (2000)
5. Alferov, A.P., Zubov A.Yu., Kuzmin A.S., Cheremushkin A.V.: Fundamentals of Cryptography, p. 480. Helios ARV (2005)
6. Gorbenko, I.D., Gorbenko, Y.I.: Applied Cryptology: Monograph, 1st and 2nd edn, p. 868. KNURE, Fort, Kharkiv (2012)
7. Gorbenko, I.D., Gorbenko, Y.I.: Applied Cryptology: Textbook, 1st and 2nd edn., p. 878. Kharkiv, KNURE, Fort (2012)
8. Orlova, S.Yu.: Methodology for evaluating the effectiveness of stream ciphers. In: Legal, Regulatory and Metrological Security Systems and Information Protection in Ukraine, vol. 9, pp. 141–152. Science and Technology Zbirnik (2004)
9. Berlekamp, E.R.: Algebraic Coding Theory, p. 447. McGrawHill, NY (1968)
10. McWilliams, F.J.: The Theory of ErrorCorrecting Codes, p. 762. F.J. McWilliams, N.J. Sloane, NorthHolland (1978)

11. Zakharov, I.D., Zhiganov, A.A.: The use of generating polynomials of m-sequences in the construction of pseudo-random code scales. Izv. Univ. Instrum. Making **54**(6), 49–55 (2011)
12. Karazin, V.N., Kuznetsov, O.O., Malakhov, S.V.: Research of flow symmetric ciphers and flow modes of block symmetric ciphers: report on GDR (intermediate), code "Stream". Analysis and comparative studies of modern algorithms for streaming cryptocurrency, p. 233. V.N. Karazina, KhNU (2015)
13. NIST Special Publication 80022. A Statistical Test Suite for Random and Pseudorandom Number Generators for Cryptographic Applications. [Electronic resourse]. Access mode: http://csrc.nist.gov/publications/nistpubs/80022rev1a/SP80022rev1a.pdf
14. Morozov, V.V., Murashko, I.A.: Calculation of phase shifts of the M sequence. [Electronic resource] (2011). http://libeldoc.bsuir.by/bitstream/123456789/3338/1/CalculationofphaseshiftoftheMsequence.PDF. Date of beast 10 Jul 2016
15. Borisenko, N.P., Gusarov, A.V., Krivonos, A.P.: On the possibility of generating primitive polynomials of a given degree and quick calculation of the shift of the output RSLOS sequence by a given number of ticks. In: Proceedings of the XII Internal and Scientific Conference on Informatization and Information Security of Law Enforcement Systems, pp. 334–339 (2003).
16. Kuznetsov, A.A., Mordvinov, R.I., Kolovanova, E.P., Samoilova, A.V.: Methods of statistical testing of cryptographic algorithms. In: Special Telecommunication Systems and Information Security, vol. 1, no. 25, pp. 54–61. Kyiv (2014)
17. Karazin, V.N., Kuznetsov, O.O., Malakhov, S.V.: Research of flow symmetric ciphers and flow modes of block symmetric ciphers: GDR report (final), stream code "Stream", vol. 2. Development of proposals for the project of the algorithm of stream symmetric encryption and substantiation of its properties, p. 73. V.N. Karazina, KhNU (2015)
18. Kuznetsov, O.O., Lutsenko, M.S., Andrushkevich, A.V., Melkozerova, O.M., Novikova, D.V., Loban, A.V.: Statistical research of modern streaming ciphers. In: Applied Radio Electronics, vol. 15, no. 3, pp. 167–178. KhNURE, Kharkiv (2016)
19. Koopman, P.: Maximal Length LFSR Feedback Terms. [Electronic resource]. Access mode: https://users.ece.cmu.edu/~koopman/lfsr/
20. Pseudo Random Number Generation Using Linear Feedback Shift Registers. [Electronic resource] (2010). https://www.maximintegrated.com/en/appnotes/index.mvp/id/4400. Access date 07 Oct 2016
21. Mukhachev, V.A., Khoroshko, V.A.: Methods of Practical Cryptography, p. 215. Polygraph Consulting LLC (2005)
22. Gorodilova, A.A.: From cryptoanalysis of a cipher to a cryptographic property of a Boolean function. Appl. Discrete Math. **3**(33), 16–44 (2016)
23. Pankratova, I.A.: Boolean Functions in Cryptography: Textbook Allowance, p. 88. Publishing House of Tomsk State University, Tomsk (2014)
24. Tokareva, N.N.: Generalizations of bent functions. Work review. Discrete Anal. Oper. Res. **17**(1), 33–62 (2010)
25. Tokareva, N.N.: Nonlinear Boolean Functions: Bent Functions and Their Generalizations, 180 s. LAP LAMBERT Academic Publishing, Saarbrucken, Germany (2011). ISBN: 9783843309042
26. Agafonova, I.V.: Cryptographic properties of nonlinear Boolean functions. In: Discrete Seminar, Harmonious, Analysis and Geometer Modeling, pp. 1–24. DHA & CAGD, St. Petersburg (2007)
27. Shevelev, Yu.P.: Discrete Math. Part 1: Set Theory. In: Boolean Algebra (Automated Learning Technology "Symbol"): Textbook, p. 118. Tomsk state CNT of Control Systems and Radio Electronics (2003)
28. Moldovyan, A.A.: Cryptography. Speed Ciphers, p. 496. BHV Peterburg (2002)
29. Logachev, O.A., Salnikov, A.A., Smyshlyaev, S.V., Yashchenko, V.V.: Boolean Functions in Coding Theory and Cryptology, 2nd edn, p. 584. MCCNMO (2012)
30. Smyshlyaev, S.V.: On cryptographic weaknesses of some classes of binary sequence transformations. Appl. Discrete Math. **1**, 5–15 (2010)
31. Tarannikov, Yu.V.: On correlation-immune and stable Boolean functions. Math. Problems Cybernet. Fizmatlit 11, 91–148 (2002)

Chapter 16
Stream Symmetric Cipher "Strumok"

The new algorithm is based on SSC "Strumok" with 512 and 1024 bit keys, which was previously studied in [1–7] and in Chap. 10.

Due to the improvement of the structure and basic transformations it was possible to increase the speed of formation of the key flow with the possibility of parallelization of certain operations.

The "Strumok" is built according to the SNOW 2.0-like scheme of the summing generator and is based on increasing (compared to SNOW) the length of the secret key and the initialization vector, which allows you to reliably apply the stream cipher even with quantum methods of cryptographic analysis. According to the results of research, "Strumok" provides a high rate of key stream formation (over 17 Gbps), which exceeds most of the known algorithms, and is suitable for use in the post-quantum environment [1–3]. To date, this algorithm has already been adopted as the national standard of SSC of Ukraine National Standard DSTU 8845 [8].

16.1 Justification of the Mathematical Structure of the Cipher

Consider the main results obtained from the development, research and improvement of the mathematical structure of the new PSS "Strumok".

The "Strumok" algorithm is a symmetric stream cipher, the first version of which was developed in 2015 and has been repeatedly improved [1–3]. SSC "Strumok" has a simple scheme and is focused on a 64-bit computing platform, which allowed it to become extremely fast. The cipher is based on a secret internal state consisting of eighteen 64-bit blocks: 16 cells of the shift register register with linear feedback and two registers of a finite state machine. A secret key of size $K = 256/512$ bits and an initialization vector of size $IV = 256$ bits are used as input data [1–3]. In this chapter

A. A. Kuznetsov et al., *Stream Ciphers in Modern Real-time IT Systems*,
Studies in Systems, Decision and Control 375,
https://doi.org/10.1007/978-3-030-79770-6_16

we substantiate the main cryptographic elements of the cipher, the key system and the structure of streaming cryptoconversion.

16.1.1 Justification of General Requirements for the Key System Perspective Symmetric Stream Cipher for Post-Quantum Application

In recent years, there has been a significant body of research on quantum computers, machines that use quantum mechanical phenomena to solve mathematical problems that are complex and unsolvable for ordinary computers. If large-scale quantum computers are ever built, they will be able to compromise many of the cryptosystems currently in use [9–17]. This may jeopardize the confidentiality and integrity of digital communication on the Internet, in other areas of life. The purpose of post-quantum cryptography (also called quantum-resistant cryptography) is to develop cryptographic systems that will be protected from quantum and classical computers and can interact with existing protocols and communication networks.

Over the last twenty years, the theory of quantum computing has become much more developed. Quantum algorithms that reach exponential acceleration have identified several problems with modeling in physics, number theory, and topology. However, the list of problems involving the acceleration of quantum computations exponentially remains relatively small. In contrast, more modest accelerations have been developed for a wide range of tasks related to finding, colliding, and evaluating Boolean functions.

In particular, Grover's algorithm [9] improves quadratic acceleration on unstructured search problems. The problem to be solved by Grover's method can be formulated as follows. Suppose there is an unordered database (list) of N elements, and let there be one element in it that has some property (which is easy to check). You need to find this item. Thus, we will formulate the problem of searching in terms of an exponentially disordered database with $N = 2n$ elements, among which one element is numbered in a special way. The problem is that you need to find this element. Elementary probability theory shows that if we look at k elements, we have the probability k/N of finding the element we need. Therefore, it is necessary to create $O\ (N)$ queries to the database to find the required element with any constant (independent of N) probability [9, 18]. Thus, using Grover's algorithm, you can find the secret key of symmetric encryption for time $\sqrt{K}$, where K is the size of the key space. There are many classic algorithms in which the procedure is repeated many times to achieve the best result. Repeating the quantum procedure may improve the result for a while, but after a sufficient number of repetitions, the result becomes worse again. A quantum procedure is a unitary transformation that rotates in a complex space. Therefore, while re-applying the quantum transformation can bring the current state closer and closer to the state we need for some time, further application of the quantum transformation may bypass the desired state and delay the correct decision. Therefore, in

Table 16.1 Resources for the quantum solution of the symmetric cryptosystem key search problem

Key lengths, k	Number of qubits, k	Quantum time, $(\pi/4)\sqrt{2k}$	Average time for a classic computer
57	57	$2.8 \times 10^8 < 2^{29}$	$2^{57} \approx 6.39 \times 10^{16}$
82	82	$1.93 \times 10^{12} < 2^{41}$	$2^{82} \approx 3.03 \times 10^{24}$
113	113	$1.06 \times 10^{17} < 2^{57}$	$2^{113} \approx 9.20 \times 10^{33}$
129	129	$2.76 \times 10^{19} < 2^{65}$	$2^{129} \approx 6.03 \times 10^{38}$
258	258	$5.03 \times 10^{38} < 2^{129}$	$2^{258} \approx 2.05 \times 10^{77}$

order to get a good result with repeated quantum transformations, it is very important to know when to stop [9–14].

Table 16.1 shows the corresponding key lengths k of symmetric cryptosystems required to solve the problem using the Grover algorithm, the number of qubits, as well as the number of quantum steps [9]. It is assumed that the vector of Boolean functions of the corresponding symmetric cryptocurrency is calculated in one quantum step, which in the General case does not correspond to reality. This assumption corresponds to the simplification of the task of finding the key and leads to quantum complexity, which can be considered as the lower (not always achievable) limit of stability. The classic complexity given in the table is called average because it corresponds to the average key search time on a classic computer.

If a quantum step can ever be compared in time with a classical one, and if we assume the presence of a quantum register of length k qubits, where k is the length of a symmetric key, it will suffice to double the key length for a symmetric cryptographic algorithm to obtain equivalent complexity. At the same time, the cipher, which is absolutely stable according to Shannon, remains absolutely stable, because its hacking is not solved by the method of search and on a classic computer. In the case of symmetric ciphers, it is also possible to use shorter quantum registers to solve overriding problems that arise in the process of cryptanalysis due to, for example, decomposition of the original problem, if the attacked cryptoalgorithm allows. However, in principle it does not change anything.

Thus, from the point of view of ensuring the established levels of stability at possible post-quantum application, the general requirements to the key system of the perspective stream symmetric cipher are the following:

- to ensure a high level of stability (which requires at least 2^{128} quantum steps), the length of the key must be at least 256 bits;
- to ensure an extremely high level of stability (which requires at least 2^{256} quantum steps), the key length must be at least 512 bits.

The following generator "Strumok" is built on SNOW 2.0 similar scheme of the summing generator (generator SNOW2.0 is defined in the National Standard DSTU ISO/IEC 180334: 2015), it uses a 256-bit initialization vector and a 256-bit or 512-bit secret key.

Depending on the length of the secret key, two modes are used:

- mode with a 256-bit key, denoted as "Strumok-256";
- mode with a 512-bit key, denoted as "Strumok-512".

The generator provides high and ultra-high levels of stability and is designed to protect important restricted information owned by the state, in particular to ensure the confidentiality of information during its processing taking into account current trends in the development of cryptographic science and the possible application of quantum cryptographic analysis.

16.1.2 General Cipher Parameters

The "Strumok" algorithm is based on the classical scheme of the summing generator [19, 20], similar to the SNOW2.0 generator, which is defined in ISO/IEC 180,334: 2011 [21].

The "Strumok" algorithm uses a 256-bit IV initialization vector and a 256-bit or 512-bit secret key K and provides a high and ultra-high level of stability, taking into account the possible application of quantum cryptographic analysis. The crypto algorithm is focused on 64-bit computing systems, so the word size is defined as 64 bits.

The main structural components of the generator are the linear feedback shift register (LFSR) and the finitestate machine (FSM), in which the nonlinear transformation is performed. The input data is used to initialize the state variable $S_i (i \geq 0)$, which consists of eighteen 64-bit blocks, which include two components:

- 16 variables $s^{(i)}$—cells of the shift register with linear feedback: $s^{(i)} = (s_{15}^{(i)}, s_{14}^{(i)}, \ldots, s_0^{(i)})$;
- two registers of a finite automaton $r^{(i)}$: $r^{(i)} = (r_2^{(i)}, r_1^{(i)})$.

At the output we get a key stream (cipher gamut), which is formed from 64-bit words Z_i. Schematic representation of the operation of the generator of key flows "Strumok" at any time is shown in Fig. 16.1. The time dependence variable i is not given.

Feedbacks in the LFSR are constructed by a primitive polynomial over the field GF (2^{64})

$$f(x) = x^{16} + x^{13} + \alpha^{-1} x^{11} + \alpha,$$

where α is the root of the primitive polynomial over the field GF (2^8)

$$g(z) = z^8 + g_7 z^7 + \ldots + g_1 z + g_0$$

In turn, the field GF (2^8) is constructed by a primitive polynomial over the field GF(2)

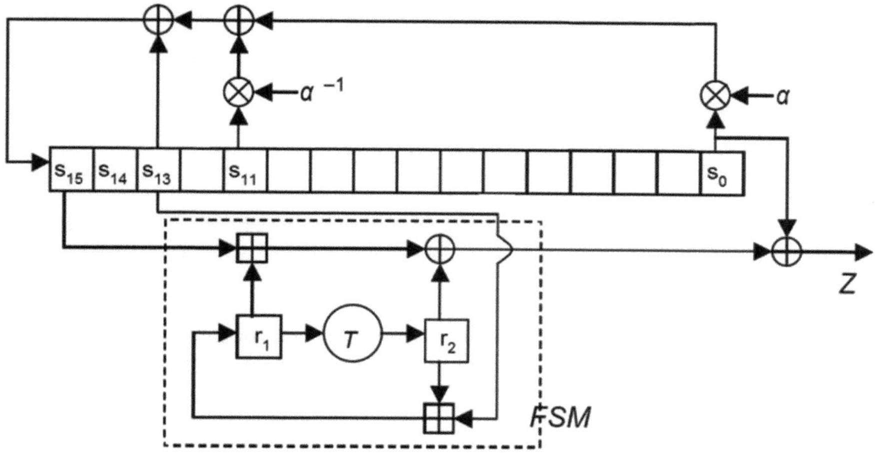

Fig. 16.1 Schematic representation of the generator of key streams "Strumok" in the mode of generating a range of ciphers (key stream)

$$p(y) = y^8 + y^4 + y^3 + y^2 + 1,$$

and the coefficients of the polynomial $g(z)$ are given through the degree of the primitive element β of the field GF (2^8), i.e. β is the root of the polynomial $p(y)$.

So we have a tower of fields:

$$GF(2) \subset GF(2^8) \subset GF(2^{64}) \subset GF(2^{1024}),$$

where:

- the field GF (2^{1024}) is given by the feedback taps LFSR as a factor ring $GF(2^{64})[x]/(f(x))$;
- the field *GF* (2^{64}) is given as a factor ring $GF(2^8)[z]/(g(z))$;
- the field *GF* (2^8) is given as a factor ring $GF(2)[y]/(p(y))$.

Therefore, the period of the original sequence of the LFSR is maximum and equal to $2^{1024} - 1$.

Structurally in the algorithm "Strumok" can be divided into three main functions:

(1) the initialization function *Init*, which takes as input the key *K* (256 bits or 512 bits) and the initialization vector IV (256 bits), and produces the initial value of the state variable

$$S_0 = (s^{(0)}, r^{(0)});$$

(2) the function of the next state Next, which takes the input state variable

$$S_i = (s^{(i)}, r^{(i)})$$

and produces the following value of the state variable

$$S_{i+1} = (s^{(i+1)}, r^{(i+1)}).$$

The *Next* function can be executed in two modes, depending on the method of iteration—as part of the implementation or as part of the normal mode of generating the original data;

(3) the function of the key stream *Strm*, which receives at the input the state variable $S_i = (s^{(i)}, r^{(i)})$ and produces a 64-bit key stream Z_i at the output.

16.1.3 *Substantiation of the Mathematical Structure of the Stream Symmetric Cipher “Strumok”*

The main elements of the streaming cipher (see Fig. 16.1) are LFSR and CA, the mathematical structure of which is decisive in assessing the properties of cryptographic transformation.

16.1.3.1 Substantiation of the Mathematical Structure of LFSR

To determine the structure of the first basic element—LFSR, it is necessary to properly form and investigate the properties of the polynomial $g(z) = z^8 + g_7 z^7 + \vdots + g_1 z + g_0$, which directly specifies a specific type of linear recurrent transformation.

During the study using the computer algebra system MAGMA [22], four variants of the polynomial g (z) were formed:

(1) $g(z) = z^8 + \beta^{170} z^7 + \beta^{166} z^6 + \beta^2 z^5 + \beta^{224} z^4 + \beta^{70} z^3 + \beta^2,$

or at hexadecimal representation of coefficients:

$$g(z) = z^8 + D7z^7 + 3Fz^6 + 04z^5 + 12z^4 + 5Ez^3 + 04;$$

(2) $g(z) = x^8 + \beta^{153} z^7 + \beta^{63} z^6 + \beta^{172} z^5 + \beta^{186} z^4 + \beta^{123} z^3 + \beta^{184} z^2 + \beta^{242},$

or at hexadecimal representation of coefficients:

$$g(z) = z^8 + 92z^7 + A1z^6 + 7Bz^5 + 6Ez^4 + C5z^3 + 95z^2 + B0;$$

(3) $g(z) = z^8 + \beta^{228} z^7 + \beta^{237} z^6 + \beta^{200} z^5 + \beta^{37} z^4 + \beta^{64} z^3 + \beta^{64} z^2 + \beta^{149},$

Table 16.2 The number of monomials in LFSR feedback polynomials in different ways of their representation

No	The number of monomials when represented in the field				
	$GF(2)$	$GF(2^2)$	$GF(2^4)$	$GF(2^8)$	$GF(2^{16})$
1	457	277	122	62	28
2	455	291	143	71	28
3	457	294	146	71	28
4	457	294	141	71	28

or at hexadecimal representation of coefficients:

$$g(z) = z^8 + 3Dz^7 + 8Bz^6 + 1Cz^5 + 4Az^4 + 5Fz^3 + 5Fz^2 + A4;$$

(4) $g(z) = z^8 + \beta^{14}z^7 + \beta^{151}z^6 + \beta^{158}z^5 + \beta^{117}z^4 + \beta^{95}z^3 + \beta^8 z^2 + \beta^{112}$,

or at hexadecimal representation of coefficients:

$$g(z) = z^8 + 13z^7 + AAz^6 + B7z^5 + EDz^4 + E2z^3 + 1Dz^2 + 70,$$

where $\beta = y$ is a primitive element of the field GF (2^8), the root of the binary polynomial $p(y) = y^8 + y^4 + y^3 + y^2 + 1$ (in the hexadecimal representation $\beta = 02$).

These polynomials are formed in such a way that the sequences at the output of the LFSR have certain cryptographic properties. In particular, Table 16.2 shows the data on the number of monomials in the feedback polynomials of the LFSR for different ways of their representation and for different variants of the polynomial *g(z)*.

In particular, when choosing the first option, i.e.

$$g(z) = z^8 + \beta^{170}z^7 + \beta^{166}z^6 + \beta^2 z^5 + \beta^{224}z^4 + \beta^{70}z^3 + \beta^2.$$

The operation of the LFSR can be set in several equivalent ways, for example (see Annex A):

- binary polynomial $a(x)$ of degree 1024 with 457 monomials;
- polynomial $b(x)$ of degree 512 with 277 monomials, the coefficients of which are given as the power of the primitive element $\gamma \in GF(2^2)$;
- polynomial $c(x)$ of degree 256 with 122 monomials, the coefficients of which are given as the degree of a primitive element $\eta \in GF(2^4)$;
- polynomial $d(x)$ of degree 128 with 62 monomials, the coefficients of which are given as the degree of a primitive element $\mu \in GF(2^8)$;
- polynomial e(x) of degree 64 with 28 monomials, the coefficients of which are given as the degree of the primitive element $\lambda \in GF(2^{16})$.

Thus, in any representation of the LFSR, about half of its cells are involved in the formation of feedback, which provides resistance against a certain class of cryptographic attacks.

16.1.3.2 Substantiation of the Mathematical Structure of CA

The construction of a finite automaton and its main properties are determined by the nonlinear transformation T. For compatibility of implementations of cryptographic algorithms "Strumok" and "Kalina" $\pi_0, \pi_1, \pi_2, \pi_3$ (see Appendix B).

16.1.4 Init Internal State Initialization Function

The internal state initialization function Init is described as follows.

Input: 256 or 512-bit key K, 256-bit initialization vector IV.

Output: initial value of the state variable $S_0 = (s^{(0)}, r^{(0)})$.

The key for the version of the streaming cipher "Strumok-256" can be given in the form of four 64-bit words:

$$K = (K_3, K_2, K_1, K_0)$$

and for a 512-bit key, in the form of eight 64-bit words:

$$K = (K_7, K_6, K_5, K_4, K_3, K_2, K_1, K_0),$$

where K_3 and K_7, respectively for 256 and 512 bits—the most significant words, and K_0—the least significant.

The initialization vector can be represented as four 64-bit words:

$$IV = (IV_3, IV_2, IV_1, IV_0),$$

where IV_3 is the most significant word and IV_0 is the least significant.

1. The Key Value is Entered in 16 LFSR Cells.

For the version with a 256-bit key, the following operations are performed:

$$\begin{aligned} s_{15}^{(-33)} &= \neg K_0, s_{14}^{(-33)} = K_1, s_{13}^{(-33)} = \neg K_2, s_{12}^{(-33)} = K_3, s_{11}^{(-33)} \\ &= K_0, s_{10}^{(-33)} = \neg K_1, \\ s_9^{(-33)} &= K_2, s_8^{(-33)} = K_3, s_7^{(-33)} = \neg K_0, s_6^{(-33)} = \neg K_1, s_5^{(-33)} \\ &= K_2 \oplus IV_3, s_4^{(-33)} = K_3, \end{aligned}$$

$$s_3^{(-33)} = K_0 \oplus IV_2, s_2^{(-33)} = K_1 \oplus IV_1, s_1^{(-33)} = K_2, s_0^{(-33)} = K_3 \oplus IV_0.$$

For the version with the 512-bit key K, the following operations are performed:

$$s_{15}^{(-33)} = K_0, s_{14}^{(-33)} = \neg K_1, s_{13}^{(-33)} = K_2, s_{12}^{(-33)} = K_3, s_{11}^{(-33)} = \neg K_7, s_{10}^{(-33)} = K_5, s_9^{(-33)} = \neg K_6,$$

$$s_8^{(-33)} = K_4 \oplus IV_3, s_7^{(-33)} = \neg K_0, s_6^{(-33)} = K_1, \; s_5^{(-33)} = K_2 \oplus IV_2, s_4^{(-33)} = K_3, s_3^{(-33)} = K_4 \oplus IV_1,$$

$$s_2^{(-33)} = K_5, s_1^{(-33)} = K_6, s_0^{(-33)} = K_7 \oplus IV_0.$$

2. Perform 32 Initiating Clocks in Without Generating a Key Stream, I.E. 4 Complete Cycles. Formally, It is Presented as Follows:

$$S_{-1} = Next^{32}(S_{-33}, INIT),$$

which means 32 iterations to perform the *Next* function in *INIT* initialization mode, $S_{-33} = (s^{(-33)}, r^{(-33)})$—the values of the state variable calculated in the previous step.

3. The initial value of the state variable $S_0 = (s^{(0)}, r^{(0)})$ is calculated by the rule: $S_0 = Next(S_{-1})$, i.e. by performing the function *Next* in normal mode.
4. The initial value $S_0 = (s^{(0)}, r^{(0)})$ is displayed.

16.1.5 Next State Function Next

The function of the next state *Next* is described as follows.

Input: State variable $S_i = (s^{(i)}, r^{(i)})$, selected mode (normal, or initialization mode).

Output: The next value of the state variable $S_{i+1} = (s^{(i+1)}, r^{(i+1)})$.

1. A nonlinear substitution is performed to update the value of the finite state register $r_2^{(i+1)}$. To do this, calculate the value of the function T: $r_2^{(i+1)} = T(r_1^{(i)})$.
2. The value of the finite state register $r_1^{(i+1)}$ is updated. To do this, calculate the value:

$$r_1^{(i+1)} = r_2^{(i)} +_{64} s_{13}^{(i)}$$

where $+_{64}$ denotes the operation of adding integers modulo 2^{64} (in Fig. 16.2 this operation is denoted as $\boxplus$.

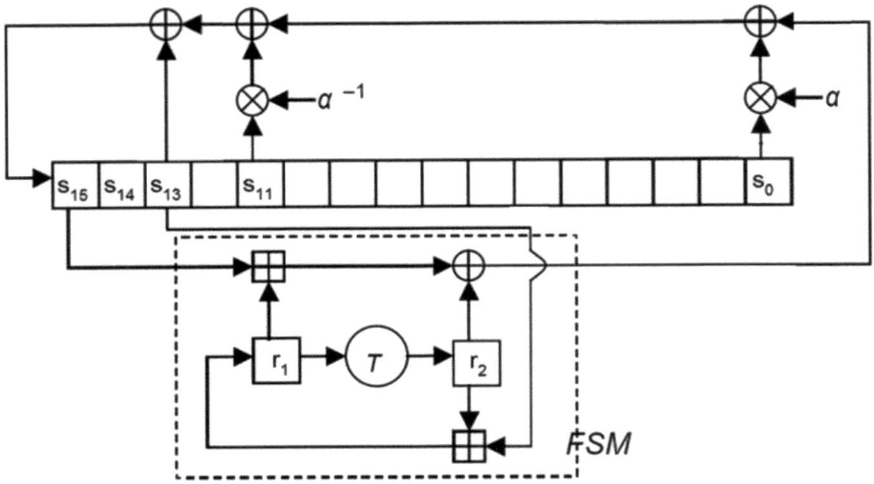

Fig. 16.2 Schematic representation of the stream generator "Strumok" in the mode of initialization of the *Next* function

3. Updates the Value of 15 *LFSR* Cells

$$s_j^{(i+1)} = s_{j+1}^{(i)}$$

for all j = 0, 1,..., 14.

4. The value of the 16th cell of the *LFSR* is updated. If the normal mode of the *Next* function is set, the value of this cell is calculated according to the rule:

$$s_{15}^{(i+1)} = (s_0^{(i)} \otimes \alpha) \oplus (s_{11}^{(i)} \otimes \alpha^{-1}) \oplus s_{13}^{(i)}.$$

If the *INIT* initialization mode of the *Next* function is set, the value is calculated according to the rule:

$$s_{15}^{(i+1)} = FSM(s_{15}^{(i)}, r_1^{(i)}, r_2^{(i)}) \oplus (s_0^{(i)} \otimes \alpha) \oplus (s_{11}^{(i)} \otimes \alpha^{-1}) \oplus s_{13}^{(i)}.$$

The operations of multiplication $\otimes$ by α and by α^{-1}, as well as the essence of the FSM function are explained below.

5. The Value of the State Variable $S_i = (s^{(i)}, r^{(i)})$ is Calculated and Displayed.

Schematic representation of the generator of key streams "Strumok" when performing the function *Next* in the initialization mode is given in Fig. 16.2.

16.1.6 *Key Stream Function* Strm

The key stream function *Strm* is described as follows.

Input: State variable $S_i = (s^{(i)}, r^{(i)})$.

Output: 64-bit key stream Z_i.

1. The value is calculated

$$Z_i = FSM(s_{15}^{(i)}, r_1^{(i)}, r_2^{(i)}) \oplus s_0^{(i)}.$$

2. The Initial Value of Z_i is Displayed.

16.1.7 *Function of Finite Automaton* FSM

The finite state function is denoted as *FSM (x, y, z)* and is described as follows.

Input: three 64-bit lines *x, y, z*.

Output: 64-bit string *q*.

1. Calculate the value of $q = (x +_{64} y) \oplus z$
2. The initial value of q is displayed.

16.1.8 *Nonlinear Substitution Function* T

The nonlinear substitution function T implements the permutation of the elements of the finite field GF(2^{64}) using the components of the national standard of block symmetric cryptocurrency DSTU 7624: 2014 [23].

Input: 64-bit string *w*.

Output: 64-bit string $T = T(w)$.

1. The input 64-bit string w is divided into sub-blocks w_j by 8 bits:

$$w = (w_7, w_6, w_5, w_4, w_3, w_2, w_1, w_0),$$

2. For each sub-block w_j the substitution from the algorithm DSTU 7624: 2014 is performed using four tabular transformations $\pi_0, \pi_1, \pi_2, \pi_3$ (see appendix B). Execution of the function T with the help of these transformations is schematically (on the example of subblocks w_j, in the hexadecimal representation) shown in Fig. 16.3.

The result is the original $r = (r_7, r_6, r_5, r_4, r_3, r_2, r_1, r_0)$:

$$r_j = \pi_{j \bmod 4}[w_j],$$

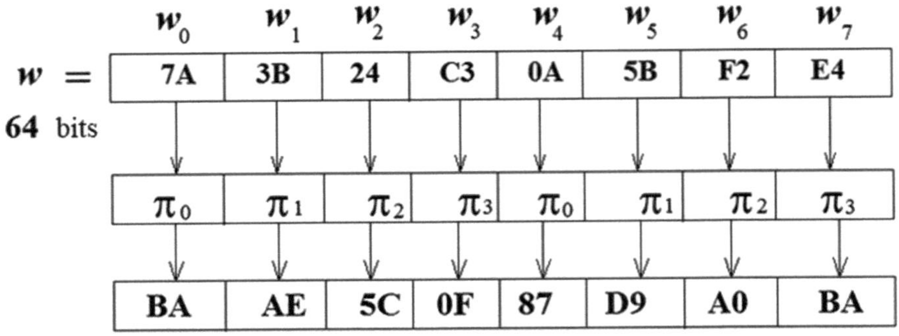

Fig. 16.3 Schematic representation of the substitution procedure $T = T(w)$

where $j = 0, 1, \ldots, 7$.

3. The vector is calculated

$$q = (q_7, q_6, q_5, q_4, q_3, q_2, q_1, q_0)$$

as a rule:

$$\begin{pmatrix} q_0 \\ q_1 \\ q_2 \\ q_3 \\ q_4 \\ q_5 \\ q_6 \\ q_7 \end{pmatrix} = \begin{pmatrix} 01\ 01\ 05\ 01\ 08\ 06\ 07\ 04 \\ 04\ 01\ 01\ 05\ 01\ 08\ 06\ 07 \\ 07\ 04\ 01\ 01\ 05\ 01\ 08\ 06 \\ 06\ 07\ 04\ 01\ 01\ 05\ 01\ 08 \\ 08\ 06\ 07\ 04\ 01\ 01\ 05\ 01 \\ 01\ 08\ 06\ 07\ 04\ 01\ 01\ 05 \\ 05\ 01\ 08\ 06\ 07\ 04\ 01\ 01 \\ 01\ 05\ 01\ 08\ 06\ 07\ 04\ 01 \end{pmatrix} \cdot \begin{pmatrix} r_0 \\ r_1 \\ r_2 \\ r_3 \\ r_4 \\ r_5 \\ r_6 \\ r_7 \end{pmatrix},$$

where the elements of the matrix (in the hexadecimal representation) and the vectors r and q are interpreted as elements of the finite field GF(2^8), given as a factor ring $GF(2)[y]/(p(y))$.

This operation can be written in abbreviated form (as in DSTU 7624: 2014):

$$q_i = (\upsilon \ggg i)r^T,$$

where $\upsilon = (01, 01, 05, 01, 08, 06, 07, 04)$, $\ggg i$ is operation of cyclic shift on i digits to the right, $i = 0, 1, \ldots, 7$,

$$r^T = (r_0, r_1, r_2, r_3, r_4, r_5, r_6, r_7)^T.$$

4. The original value q is displayed, which is interpreted as a 64-bit string.

16.2 Substantiation of the Main Cryptographic Elements of the Stream Symmetric Cipher "STRUMOK"

Stream symmetric cipher "Strumok" is focused on efficient software implementation on 64-bit computing platforms. In order to optimize the complexity of certain mathematical transformations, this section develops recommendations for the implementation of nonlinear substitution and multiplication in finite fields. This allows you to significantly reduce the amount of basic operations and significantly increase the speed of key stream generation on 64-bit computing platforms.

16.2.1 Rationale for Fast Execution of Nonlinear Substitution Operation

Precalculation tables are usually used to quickly perform a nonlinear substitution operation. For example, the implementation of nonlinear substitution focused on 32-bit computing systems stream symmetric cipher SNOW 2.0 involves the formation of 4 tables of precalculations [21]. Indeed, because the SNOW 2.0 algorithm uses nonlinear substitution tables from the AES block cipher (Advanced Encryption Standard) [24, 25], certain operations can be grouped and performed on 32-bit systems by applying precalculated values.

The various steps of the AES algorithm's round conversion can be combined into a single set of sample tables that allow fast transformations to be performed on processors with a word length of 32 or higher. To do this, denote one column of the output of the round as e, and one column of the input of the round as a. Denote byte $a_{i,j}$ byte in row i and column j, denote j by column of internal state a. To add keys and convert MixColumn we have [25]:

$$\begin{bmatrix} e_{0,j} \\ e_{1,j} \\ e_{2,j} \\ e_{3,j} \end{bmatrix} = \begin{bmatrix} d_{0,j} \\ d_{1,j} \\ d_{2,j} \\ d_{3,j} \end{bmatrix} \oplus \begin{bmatrix} k_{0,j} \\ k_{1,j} \\ k_{2,j} \\ k_{3,j} \end{bmatrix} \text{ and } \begin{bmatrix} d_{0,j} \\ d_{1,j} \\ d_{2,j} \\ d_{3,j} \end{bmatrix} = \begin{bmatrix} 02 & 03 & 01 & 01 \\ 01 & 02 & 03 & 01 \\ 01 & 01 & 02 & 03 \\ 03 & 01 & 01 & 02 \end{bmatrix} \begin{bmatrix} c_{0,j} \\ c_{1,j} \\ c_{2,j} \\ c_{3,j} \end{bmatrix}.$$

For ShiftRow and ByteSub conversions of the AES cipher we have [25]:

$$\begin{bmatrix} c_{0,j} \\ c_{1,j} \\ c_{2,j} \\ c_{3,j} \end{bmatrix} = \begin{bmatrix} b_{0,j} \\ b_{1,j-C1} \\ b_{2,j-C2} \\ b_{3,j-C3} \end{bmatrix} \text{ and } b_{i,j} = \mathsf{S}\left[a_{i,j}\right].$$

In this expression, the column indices should be taken modulo the number of Nb columns [25].

The expressions above can be combined as follows:

$$\begin{bmatrix} e_{0,j} \\ e_{1,j} \\ e_{2,j} \\ e_{3,j} \end{bmatrix} = \begin{bmatrix} 02\ 03\ 01\ 01 \\ 01\ 02\ 03\ 01 \\ 01\ 01\ 02\ 03 \\ 03\ 01\ 01\ 02 \end{bmatrix} \begin{bmatrix} \mathsf{S}[a_{0,j}] \\ \mathsf{S}[a_{1,j-C1}] \\ \mathsf{S}[a_{2,j-C2}] \\ \mathsf{S}[a_{3,j-C3}] \end{bmatrix} \oplus \begin{bmatrix} k_{0,j} \\ k_{1,j} \\ k_{2,j} \\ k_{3,j} \end{bmatrix}.$$

Matrix multiplication can be expressed through a linear combination of vectors:

$$\begin{bmatrix} e_{0,j} \\ e_{1,j} \\ e_{2,j} \\ e_{3,j} \end{bmatrix} = \mathrm{S}[a_{0,j}] \begin{bmatrix} 02 \\ 01 \\ 01 \\ 03 \end{bmatrix} \oplus \mathrm{S}[a_{1,j-C1}] \begin{bmatrix} 03 \\ 02 \\ 01 \\ 01 \end{bmatrix} \oplus \mathrm{S}[a_{2,j-C2}] \begin{bmatrix} 01 \\ 03 \\ 02 \\ 01 \end{bmatrix}$$
$$\oplus \mathrm{S}[a_{3,j-C3}] \begin{bmatrix} 01 \\ 01 \\ 03 \\ 02 \end{bmatrix} \oplus \begin{bmatrix} k_{0,j} \\ k_{1,j} \\ k_{2,j} \\ k_{3,j} \end{bmatrix}.$$

Define the tables of precalculations from T_0 to T_3:

$$\mathsf{T}_0[a] = \begin{bmatrix} \mathsf{S}[a] \cdot 02 \\ \mathsf{S}[a] \\ \mathsf{S}[a] \\ \mathsf{S}[a] \cdot 03 \end{bmatrix}, \quad \mathsf{T}_1[a] = \begin{bmatrix} \mathsf{S}[a] \cdot 03 \\ \mathsf{S}[a] \cdot 02 \\ \mathsf{S}[a] \\ \mathsf{S}[a] \end{bmatrix},$$
$$\mathsf{T}_2[a] = \begin{bmatrix} \mathsf{S}[a] \\ \mathsf{S}[a] \cdot 03 \\ \mathsf{S}[a] \cdot 02 \\ \mathsf{S}[a] \end{bmatrix}, \quad \mathsf{T}_3[a] = \begin{bmatrix} \mathsf{S}[a] \\ \mathsf{S}[a] \\ \mathsf{S}[a] \cdot 03 \\ \mathsf{S}[a] \cdot 02 \end{bmatrix}.$$

These are four tables with 256 inputs of 4-byte words, occupying up to 4 Kbytes of memory. Using these tables, the round conversion of the AES cipher can be expressed as [25]:

$$e_j = \mathsf{T}_0[a_{0,j}] \oplus \mathsf{T}_1[a_{1,j-C1}] \oplus \mathsf{T}_2[a_{2,j-C2}] \oplus \mathsf{T}_3[a_{3,j-C3}] \oplus k_j.$$

Using the above approach, we group the appropriate mathematical transformations for nonlinear replacement of the cipher "Strumok".

The result is a method of fast calculation of the vector

$$(q_0, q_1, q_2, q_3, q_4, q_5, q_6, q_7) = Q$$

through a linear combination of sample 8-byte values from 8 precalculated tables, each of which is indexed by 256 values:

$$Q^T = T_0[w_0] \oplus T_1[w_1] \oplus T_2[w_2] \oplus T_3[w_3] \oplus T_4[w_4] \oplus T_5[w_5] \oplus T_6[w_6] \oplus T_7[w_7],$$

where:

$$T_0[a] = \begin{pmatrix} 01 \\ 04 \\ 07 \\ 06 \\ 08 \\ 01 \\ 05 \\ 01 \end{pmatrix} \cdot \pi_0[a], \quad T_1[a] = \begin{pmatrix} 01 \\ 01 \\ 04 \\ 07 \\ 06 \\ 08 \\ 01 \\ 05 \end{pmatrix} \cdot \pi_1[a], \quad T_2[a] = \begin{pmatrix} 05 \\ 01 \\ 01 \\ 04 \\ 07 \\ 06 \\ 08 \\ 01 \end{pmatrix} \cdot \pi_2[a]$$

$$T_3[a] = \begin{pmatrix} 01 \\ 05 \\ 01 \\ 01 \\ 04 \\ 07 \\ 06 \\ 08 \end{pmatrix} \cdot \pi_3[a], \quad T_4[a] = \begin{pmatrix} 08 \\ 01 \\ 05 \\ 01 \\ 01 \\ 04 \\ 07 \\ 06 \end{pmatrix} \cdot \pi_0[a], \quad T_5[a] = \begin{pmatrix} 06 \\ 08 \\ 01 \\ 05 \\ 01 \\ 01 \\ 04 \\ 07 \end{pmatrix} \cdot \pi_1[a]$$

$$T_6[a] = \begin{pmatrix} 07 \\ 06 \\ 08 \\ 01 \\ 05 \\ 01 \\ 01 \\ 04 \end{pmatrix} \cdot \pi_2[a], \quad T_7[a] = \begin{pmatrix} 04 \\ 07 \\ 06 \\ 08 \\ 01 \\ 05 \\ 01 \\ 01 \end{pmatrix} \cdot \pi_3[a].$$

Applying four tabular transformations $\pi_0, \pi_1, \pi_2, \pi_3$ from the algorithm DSTU 7624: 2014, we can directly calculate the eigenvalues of the tables $T_i[a]$, $i = 0, 1, ..., 7$, which are given in Appendix C.

Practical use of tables of constants $T_i[a]$, $i = 0, 1, ..., 7$ allows to reduce considerably quantity of operations, in particular the function of nonlinear substitution is calculated on seven XOR operations on 64-bit lines.

16.2.2 Substantiation of Fast Execution of Multiplication Operation on α^{-1} in Arithmetic of Field GF (2^{64})

To quickly perform the multiplication operation on α in the arithmetic of the field GF (2^{64}), you can also use precalculation tables Mul_α with 256 rows of 64 bits each.

The polynomial defining the feedback LFSR has the form $f(x) = x^{16} + x^{13} + \alpha^{-1}x^{11} + \alpha$, where α and α^{-1} belong to the field GF (2^{64}), and $\alpha = z$ is the root of the primitive over the field GF(2^8) the polynomial g(z). Thus, each cell LFSR stores a 64-bit sequence w, which is represented as 8 sub-blocks w_j of 8 bits each:

$$w = (w_7, w_6, w_5, w_4, w_3, w_2, w_1, w_0),$$

which are interpreted as polynomial coefficients

$$w(z) \in GF(2^8)[z]/(g(z)).$$

If $\alpha = z$ is the root of the primitive polynomial $g(z) = z^8 + g_7z^7 + ... + g_1z + g_0$ over GF(2^8), then we have:

$$\begin{aligned} w(z) \cdot \alpha = (w_7z^8 + w_6z^7 + \cdots + w_1z^2 + w_0z) mod g(z) \equiv \\ \equiv (w_6 + w_7g_7)z^7 + (w_5 + w_7g_6)z^6 + \cdots + (w_0 + w_7g_1)z + (w_7g_0)z^0 = \\ = w_{\ll 8}(z) + w_{\gg 56}(z) \cdot g'(z), \end{aligned}$$

where the polynomials $w_{\ll 8}(z)$, $w_{\gg 56}(z)$ and $g'(z)$ have the form:

$$w_{\ll 8}(z) = w_6z^7 + w_5z^6 + \cdots + w_0z,$$

$$w_{\gg 56}(z) = w_7,$$

$$g'(z) = g_7z^7 + g_6z^6 + \cdots + g_1z + g_0.$$

Taking into account the obtained expressions, the fast execution of the multiplication operation on α in the arithmetic of the field GF (2^{64}) can be realized using the following algorithm.

16.2.3 Algorithm for Fast Execution of Multiplication Operation by α^{-1} in Arithmetic of the GF Field (2^{64})

Input: 64-bit string w representing the field element GF (2^{64}).

Output: 64-bit string $w' = w \otimes \alpha$, representing the field element GF (2^{64}).

1. The Value is Calculated

$$w' = (w \ll 8) \oplus Mul_{\alpha}[w \gg 56], \tag{16.15.1}$$

where:

$w \ll 8$ is the result of the shift to the left (towards the higher bits) of the 64-bit line w by 8 bits with the filling of the lower bits with zero values;

$w \gg 56$ is the result of the shift to the right (towards the lower bits) of the 64-bit string w by 56 bits with the filling of the higher bits with zero values. The eight lower bits of the vector $w \gg 56$ are interpreted as an element of the field GF(2^8) for indexing the precalculation table Mul_{α};

Mul_{α} is a table-constant of 256 rows of 64 bits in each (precalculation table),

$Mul_{\alpha}[c]$ is a 64-bit value of the precalculation table in a row with index c, where c is an element of the field GF (2^8), $Mul_{\alpha}[c]$ is an element of the field GF(2^{64}).

2. The Initial Value w' is Displayed.

The binary sequences $w \ll 8$ and $w \gg 56$ in formula (16.15.1) correspond to the binary representation of the coefficients of the polynomials $w_{\ll 8}(z)$ and $w_{\gg 56}(z)$. Therefore, the calculation of $w(z) \cdot \alpha$ in the arithmetic of the field GF(2^{64}) corresponds to formula (16.15.1), where 256 values of the table $Mul_{\alpha}[w_7]$ are calculated as 64-bit sequences with binary representation of the coefficients $(w_7 g_7, w_7 g_6, \ldots, w_7 g_1, w_7 g_0)$ polynomial

$$w_{\gg 56}(z) \cdot g'(z) = w_7(g_7 z^7 + g_6 z^6 + \cdots + g_1 z + g_0)$$

for each of the 256 possible values of $w_7 \in GF(2^8)$.

The eigenvalues of the table $Mul_{\alpha}[w_7]$ are given in Appendix D.

16.2.4 Substantiation of Fast Execution of Multiplication Operation by α^{-1} in Arithmetic of the GF Field (2^{64})

Using the above approach, it is possible to implement a fast execution of the multiplication operation by α^{-1} in the arithmetic of the field GF(2^{64}). Indeed, if $\alpha = z$ is the root of a primitive polynomial over GF(2^8)

$$g(z) = z^8 + g_7 z^7 + \cdots + g_1 z + g_0,$$

then we have:

$$\alpha^8 = g_7\alpha^7 + \cdots + g_1\alpha + g_0\alpha^0,$$

or

$$\alpha^7 = g_7\alpha^6 + \cdots + g_1\alpha^0 + g_0\alpha^{-1},$$

so,

$$g_0^{-1}\alpha^7 + g_0^{-1}g_7\alpha^6 + \cdots + g_0^{-1}g_1\alpha^0 = \alpha^{-1} = z^{-1}.$$

Then

$$\begin{aligned} w(z)\alpha^{-1} &= w_7z^6 + w_6z^5 + \ldots + w_1z^0 + w_0z^{-1} \\ & w_7z^6 + w_6z^5 + \cdots + w_1z^0 \\ &+ w_0(g_0^{-1}z^7 + g_0^{-1}g_7z^6 + \cdots + g_0^{-1}g_1z^0) \\ &= (w_0g_0^{-1})z^7 + (w_7 + w_0g_0^{-1}g_7)z^6 \\ &+ \cdots + (w_1 + w_0g_0^{-1}g_1)z^0 \\ &= w_{>>8}(z) + w_0(z) \cdot g''(z), \end{aligned}$$

where polinomials $w_{\gg 8}(z)$, $w_0(z)$ and $g''(z)$ have the form:

$$w_{\gg 8}(z) = w_7z^6 + w_6z^5 + \cdots + w_2z + w_1,$$

$$w_0(z) = w_0,$$

$$g''(z) = g_0^{-1}z^7 + g_0^{-1}g_7z^6 + \cdots + g_0^{-1}g_2z + g_0^{-1}g_1.$$

Multiplication by α^{-1} in the arithmetic of the field GF(2^{64}) is realized using the precalculation table $Mul_{\alpha^{-1}}$ of 256 rows of 64 bits each.

16.2.5 Algorithm for Fast Execution of Multiplication Operation by α^{-1} in Arithmetic of Field GF(2^{64})

Input: 64-bit string w representing the field element GF(2^{64}).

Output: 64-bit string $w' = w \otimes \alpha^{-1}$, representing the field element GF (2^{64}).

1. The Value is Calculated

$$w' = (w \gg 8) \oplus Mul_{\alpha^{-1}}[w\&\gamma], \qquad (16.15.2)$$

where:

$w \gg 8$ is the result of the shift to the right (towards the lower bits) of the 64-bit line w by 8 bits with the filling of the higher bits with zero values;

$w\&\gamma$ is the result of the bit conjunction of a 64-bit string w and a 64-bit string γ, which in hexadecimal represents is γ = 00000000000000FF. The eight lower bits of the vector w & γ are interpreted as an element of the field GF(2^8) for indexing the precalculation table $Mul_{\alpha^{-1}}$;

$Mul_{\alpha^{-1}}$—table of constants of 256 rows of 64 bits in each (table of precalculations);

$Mul_{\alpha^{-1}}[c]$—64-bit value of the table is precalculated in the line with index c, where c is an element of the field GF(2^8), $Mul_{\alpha}[c]$ is an element of the field GF(2^{64}).

2. The Initial Value w' is Displayed.

The binary sequences $w \gg 8$ and γ in (15.2) correspond to the binary representation of the coefficients of the polynomials $w_{\gg 8}(z)$ and $w_0(z)$. Thus, the calculation of $w(z) \cdot \alpha^{-1}$ in the arithmetic of the field GF(2^{64}) corresponds to formula (16.15.2), where 256 values of the table $Mul_{\alpha^{-1}}[w_0]$ are calculated as 64-bit sequences at binary representation of coefficients ($w_0 g_0^{-1}$, $w_0 g_0^{-1} g_7$, ..., $w_0 g_0^{-1} g_2$, $w_0 g_0^{-1} g_1$) polynomial

$$w_0(z) \cdot g''(z) = w_0(g_0^{-1} z^7 + g_0^{-1} g_7 z^6 + \cdots + g_0^{-1} g_2 z + g_0^{-1} g_1)$$

for each of the 256 possible values of.

The eigenvalues of the table are given in Appendix D.

To verify the correctness of all justified transformations in this section, test vectors have been developed, which are given in Appendix E. In particular, for various values of initialization vectors and key data, the appendix provides intermediate values of calculations and initial values of key flow. The practical application of test vectors greatly simplifies the correct implementation of cryptocurrencies and allows you to ensure their correctness and compatibility with the implementations of different developers.

16.3 Evaluation of «STRUMOK» Cipher Speed

Modern cryptographic algorithm must meet high security and reliability requirements, and must provide high performance.

The speed of the key stream generator depends on the speed of initialization of cryptoconversions (key deployment, calculation of precalculation tables, etc.) and directly on the speed of formation of the encryption range (bandwidth of the generator). The minimum amount of memory required to implement the developed cipher is also important for performance. This parameter determines the limit on the minimum amount of memory for fast cryptocurrency conversion, including first-level cache memory of modern processors.

In this section, studies of the rate of formation of the encryption range and the rate of initial initialization of the developed SSC "Strumok". The obtained estimates are compared with other streaming cryptoconversions, which are standardized at the international and/or national levels. In particular, the results of comparative studies of the rate of formation of the key flow using software implementations of algorithms MICKEY, Decim, Trivium, SOSEMANUK, SNOW 2.0, Salsa20, Rabbit, HC, AES.

16.3.1 Substantiation of the Method of Research of the Speed of Streaming Cryptoconversion

In streaming algorithms for cryptographic transformation, information is presented and processed in the form of sequences, where the symbols of the plaintext are encrypted using the symbols of the key stream (or keystream encryption).

According to the methodology developed, applied and researched during the international project eSTREAM to identify new current ciphers suitable for widespread use (organized by the European Union), the assessment of the speed of streaming cryptoconversion is performed on several partial indicators [26]. In particular, testing of streaming ciphers according to the eSTREAM method is carried out on three main groups of indicators, namely [26]:

(1) Compliance API;
(2) Correctness of implementation;
(3) Productivity (Performance) implementation.

Compliance API (API (application programming interface)—a set of ready-made classes, procedures, functions, structures and constants provided by the program (library, service) or operating system for use in external software products) **testing** is to verify the following components:

- necessary interfaces, correctness of source code compilation, etc.;
- correct settings of key lengths, internal states, etc.;
- the correctness of the fact that calls to the same functions with the same parameters give the same results, regardless of how they alternate;
- the correctness of the fact that the encryption reproduces the same encrypted text when submitting the same plaintext; correct conversion of the same ciphertext into the source text.

Correctness testing of source code implementation on different platforms is checked by generating and comparing test vectors.

Performance testing of streaming ciphers is performed according to various partial criteria. In particular, stream ciphers can be used in various application situations and applications, each of which makes special demands on the efficiency of the cryptographic primitive. The current version of the eSTREAM test structure [26] uses four performance indicators, namely:

- ***Encryption speed for long streams***. It is on this indicator that stream ciphers have the most potential advantage over block ciphers, so this estimate is the most important criterion in many applications. The test method measures the speed of cryptoconversion by encrypting a long stream, such as 1 GB in length. The encryption rate is measured in cycles/bytes and is calculated by measuring the number of bytes encrypted over a specified time. Note that the installation time of the initialization key or vector is not considered in this test;
- ***Speed of encryption of short packets***. While block cipher is probably the best choice for encrypting very short packets, it's still interesting to determine at what length the streaming cipher starts to gain in performance. Moreover, streaming ciphers, whose encryption speed is not greatly degraded for small packets, may have a clear advantage in applications that use a wide range of packet sizes. Packet encryption speed is measured separately for packets of different lengths. Each measurement involves setting a certain value of the initialization vector. Packet lengths (40, 576 and 1500 bytes) were chosen to most objectively reflect possible cases of telecommunication traffic on the Internet;
- ***Dexterity***. When a cryptographic application needs to encrypt many threads in parallel on one processor, its performance will depend not only on the encryption speed, but also on the time spent switching from one session to another. This service data is usually determined by the number of bytes that must be saved or restored during each context switch. To create a picture of the flexibility of different applications, the test environment performs the following test: first it initiates a large number of sessions (filling 16 MB of RAM), and then encrypts plaintext streams in short blocks around 256 bytes, each time moving from one session to another;
- ***Speed of setting the key and the initialization vector***. This is the latest test in the test, which separately measures the efficiency of setting the key and the initialization vector. This figure is probably the least critical, given that the efficiency of setting the initialization vectors is already reflected in the encryption speed of the packets and that the time to install the key will usually be insignificant compared to the work required to create and exchange keys.

Thus, the method of testing streaming cryptographic transformations Estream, which was developed, applied and researched during an international project to identify new current ciphers suitable for widespread use in the European Union, is the most detailed and reasonable. It can be used not only to determine the compliance of interfaces, the correctness of software implementation and calls to individual functions, but also to investigate the speed of encryption on various indicators that demonstrate the advantages and disadvantages of streaming cryptoconversion in different application situations. Therefore, during further comparative studies we will use some indicators from this methodology.

16.3.2 Comparative Studies of the Speed of Streaming Cryptoconversion

According to the chosen method of research of the speed of streaming cryptoconversion, the following partial indicators are used in this work:

(1) encryption speed of long streams (1 GB) without taking into account the time to install the key and the initialization vector;
(2) encryption speed of short data packets (40, 576 and 1500 bytes), taking into account the setting of a certain value of the initialization vector and the secret key. In particular, the following settings were selected during the research:

 - encryption of 350 packets of 40 bytes (used 10 keys, encrypted 35 packets on each key);
 - encryption of 120 packets of 576 bytes (10 keys are used, 12 packets on each key are encrypted);
 - encryption of 50 packets of 1500 bytes (1 key is used, 50 packets are encrypted on this key);

(3) initialization rate of key parameters (key and initialization vector).

Testing was performed on various computing platforms, in particular the following equipment was used:

(1) with Intel Core i34005U 1.7 GHz processor (cache memory 2 × 32 KB of the first level and 512 KB of the second level), 6 GB of DDR3 798 MHz RAM and Windows 10 operating system;
(2) with Intel Core i57200U 2.5 GHz processor (cache memory 4 × 32 KB of the first level and 512 KB of the second level), 8 GB of DDR4 2133 MHz RAM and Windows 10 operating system;
(3) with Intel Core i77700 3.6 GHz processor (64 KB first level cache and 1024 KB second level memory), 32 GB DDR3 2133 MHz RAM and Windows 10 operating system;
(4) with Intel Core i97980XE 2.60 GHz processor (cache memory 18 × 32 KB first level and 1024 KB second level), 64 GB DDR4 2133 MHz RAM and Windows 10 operating system;
(5) with AMD FX6300 Six Core 3.5 GHz processor (cache memory 94 × 32 KB first level and 6 MB second level), 8 GB DDR3 688 MHz RAM and Windows 10 operating system:

The following world-renowned and nationally and/or internationally standardized cryptographic algorithms were included in the comparative studies (they were discussed in detail in Chaps. 6, 8, 9).

Stream Symmetric Cipher MICKEY, an improved version 2.0 of which was introduced in 2005 by Steve Babbage and Matthew Dodd [27] (stands for Mutual Irregular Clocking KEYstream generator). It is intended for hardware platforms with limited resources, ie the MICKEY stream cipher was developed on the second profile as a hardware-oriented cipher. An 80-bit key and an initialization vector of up to 80 bits are used to initialize the initial state. The maximum possible length of the key stream is 2^{40} bits on one key, but using different initialization vectors of the same length. The MICKEY encryption algorithm has a simple hardware implementation, but at the same time provides a high level of security. Due to the use of irregular movement of shift registers, as well as new methods, high resistance to certain cryptanalytic attacks is provided.

Streaming symmetric cipher Decim—a hardware-specific algorithm developed by Com Berbain, Oliver Billet, Anne Cantou, Nicolas Courtois, Blondin Debray, Henry Gilbert, Louis Gubin, Aline Guzier, Louis Granbulan, Cederic Thorne Lardin and Mardu, Marder [21]. This is a hardware-oriented streaming cipher with an 80-bit key and a 64-bit initialization vector (IV), which was submitted to the eSTREAM streaming encryption project (did not pass the third stage of the competition). The Decim design is based on a non-linear shift register filter with linear feedback and an irregular pseudorandom sequence rarefaction mechanism—ABSG. As a result, Decim has low hardware complexity. After identifying certain shortcomings in [28], this algorithm was improved. The new version of Decim, called Decim^{v2}, looks more secure and has less hardware implementation than the previous version of Decim.

Streaming symmetric cipher Trivium is a symmetric hardware-oriented parallel synchronous streaming cipher. The authors of the cipher are Christoph De Cannier and Bart Prenel. The cipher was presented in 2008 in the eSTREAM project with the profile "hardware-oriented ciphers" [29]. *Trivium* is the simplest cipher of the eSTREAM project (second profile), which demonstrates excellent cryptographic results. According to the specification, the Trivium algorithm is a parallel stream cipher designed to generate 2^{64} bits of the key stream from 80 bits of the secret key and 80 bits of the initialization vector. The cipher is bit-oriented. It is also standardized as a cipher of light cryptography [30].

The stream symmetric cipher *Sosemanuk* is a synchronous software-oriented stream cipher that corresponds to the first profile of the eECRYPT competition [31]. Developed in 2004 by a group of French scientists. The key length can be selected between 128 and 256 bits. The cipher works with a 128-bit initial value, and, according to the developers of the algorithm, any key length reaches 128-bit protection. The Sosemanuk algorithm uses some basic principles of the SNOW 2.0 streaming cipher and some transformations derived from the SERPENT block cipher.

Stream symmetric cipher SNOW 2.0 is a key stream generator [21], which uses as input 128 or 256-bit secret key K and 128-bit initialization vector IV. The cipher is word-oriented. The authors of the algorithm are Thomas Johansson and Patrick Ekdal. The algorithm was standardized in ISO/IEC 180,334. The key size can be 128 and 256 bits, the initialization vector is 128 bits. For SNOW 2.0, the maximum recommended number of bits of the key stream produced on one pair (K, IV) is $23 \bullet 2^{50}$ bits. This limitation is justified in terms of ensuring the stability of the algorithm against cryptanalytic attacks.

Streaming symmetric cipher *Salsa20*. This is a system of streaming transformation, which was developed by Daniel Bernstein [32]. The algorithm won the eSTREAM competition in the first profile (software-oriented algorithms, ie stream ciphers for software use with high bandwidth). A 256-bit key, a 64-bit nonce, and a 64-bit key stream block position are used to initialize the internal state. The maximum length of the pseudo-random key sequence is 2^{70} bits.

Stream symmetric cipher *Rabbit*, the developers of the algorithm are Martin Boesgaard, Mette Westerager, Thomas Pedersen, Jesper Christiansen and Ove Scavinus. This is a high-speed streaming cipher, which was first presented in 2003 at the 10th FSE Symposium. In May 2005, this cipher was presented at the eStream competition in the first profile "software-oriented algorithms" [33]. The algorithm uses a 128-bit key and a 64-bit initialization vector. Up to 2^{67} bits of the key stream can be generated on one key/vector pair. Standardized in ISO/IEC 180,334 [21].

The stream symmetric HC cipher was developed in 2004 by cryptographer Hongjun Wu and first published in 2004 [34]. HC128 was presented at the eSTREAM competition, which aimed to create European standards for streaming encryption systems. HC256 is a simple, secure, software-oriented cipher with efficient implementation and can be used freely. A simplified version of HC128 was presented on eSTREAM in the first profile. A 256-bit key and a 256-bit initialization vector are used for initialization. The recommended maximum length of the key sequence is 2^{128}.

Block symmetric cipher AES, standardized in the United States as FIPS197 [24]. Internationally standardized in ISO/IEC 180,333 [35]. Uses a key length of 128, 192 or 256 bits. Depending on the length of the key, there are 10, 12 or 14 rounds of encryption. AES is based on the principle known as the "permutation replacement network" and thus has a fast hardware and software implementation. In output feedback mode, this cipher can be used as a stream.

DSTU 7624: 2014 (Kalina) is a symmetric block cipher with a block and key length of 128/256/512 bits and 10 to 18 rounds (according to the selected block and key lengths). In 2014, the Kalina algorithm was declared the national standard for block encryption in Ukraine. During research in [36–40] it was found that the performance of BSS "Kalina" is compared with the properties of the American national encryption standard AES (FIPS197) [24, 25, 41–44]. Therefore, the tables below show data only for AES (which are almost identical to the results for BSC "Kalina").

16.3.2.1 Comparison with Inter Core i34005U 1.70 GHz

Tables 16.3, 16.4, 16.5, 16.6 and 16.7 show the results of testing the performance of streaming cryptographic transformations, which were obtained on a computer system with an Inter Core i34005U 1.70 GHz processor. In particular, Table 16.3 shows the results of testing the encryption speed of a long data stream, Tables 16.4, 16.5 and 16.6 show the data for encrypting data packets, and Table 16.7 shows the results of the rate of setting keys and initialization vectors.

According to the results of measuring the encryption speed of a long packet (Table 16.3), the algorithm "Strumok" should be noted. Even on a rather slow computing platform, this cipher provides extremely high key stream generation rates (6.8 Gbit/s). The closest prototype—the algorithm SNOW 2.0 provides a speed 1.5 times less (4.2

Table 16.3 The obtained results of the encryption speed of the long stream (length 1 GB)

Algorithm	Speed	
	Cycles/byte	Mbit/s
Mickey	381	36
Decim	1859	7
Strumok-256	2	6801
Strumok-512	2	6843
Trivium	7	2075
Sosemanuk	5	2522
Snow2.0–128	3	4213
Snow2.0–256	3	4228
Salsa20	10	1360
Rabbit	7	1833
HC-128	2	6032
HC-256	6	2195
AES-128	11	1266
AES-256	15	908

Table 16.4 The results of encryption speed of 350 data packets of 40 bytes each

Algorithm	Speed	
	Cycles/byte	Mbit/s
Mickey	883	15
Decim	2615	5
Strumok-256	23	589
Strumok-512	20	675
Trivium	32	429
Sosemanuk	20	663
Snow2.0–128	17	794
Snow2.0–256	16	848
Salsa20	14	957
Rabbit	22	622
HC-128	373	36
HC-256	2367	6
AES-128	15	911
AES-256	21	647

Gbps). The HC algorithm with a 128-bit key provides a high lasing speed (6.0 Gbps), but its enhanced version with a 256-bit key is three times slower (2.2 Gbit/s).

Thus, according to the criterion of complexity/stability, the undisputed leader in terms of encryption speed of a long data stream is the developed cryptographic

Table 16.5 The results of encryption speed of 120 data packets of 576 bytes each

Algorithm	Speed	
	Cycles/byte	Mbit/s
Mickey	405	33
Decim	1905	7
Strumok-256	3	4221
Strumok-512	3	4127
Trivium	8	1636
Sosemanuk	5	2537
Snow2.0–128	4	3435
Snow2.0–256	4	3413
Salsa20	10	1369
Rabbit	9	1519
HC-128	28	485
HC-256	169	80
AES-128	11	1265
AES-256	15	895

Table 16.6 The results of the encryption speed of 50 data packets of 1500 bytes each

Algorithm	Speed	
	Cycles/byte	Mbit/s
Mickey	384	35
Decim	1867	7
Strumok-256	2	5556
Strumok-512	2	5556
Trivium	7	1893
Sosemanuk	4	3046
Snow2.0-128	4	3822
Snow2.0-256	4	3822
Salsa20	10	1345
Rabbit	8	1739
HC-128	12	1101
HC-256	69	197
AES-128	11	1277
AES-256	15	906

algorithm "Strumok" (with 256-bit and 512-bit key—"Strumok-256" and "Strumok-512", respectively).

According to the results of research on the speed of encryption of data packets, the following should be noted. In the case of short data packets, block symmetric

Table 16.7 The results of the speed of setting keys and initialization vectors are obtained

Algorithm	Speed	
	Cycles/setting	Settings/s
Mickey	20,171	84,076
Decim	39,850	42,564
Strumok-256	475	3,571,429
Strumok-512	472	3,597,122
Trivium	979	1,730,104
Sosemanuk	492	3,424,658
Snow2.0-128	402	4,237,288
Snow2.0-256	403	4,201,681
Salsa20	2	50,000,000
Rabbit	451	3,759,398
HC-128	14,718	115,234
HC-256	94,474	17,953
AES-128	0.26	50,000,000
AES-256	0.27	48,148,148

ciphers have the advantage, as expected. Only Salsa20 and SNOW 2.0 ciphers can compete with block ciphers according to this criterion. Stream cipher "Stream" at a comparable length of the key loses by 20–30% in the speed of keystream formation. However, if the packet length is increased to 576 bytes and higher, it again becomes the undisputed leader, its speed is 20–30% higher than other alternatives.

According to the results of measuring the speed of setting keys and initialization vectors, block ciphers are preferred (in this case, the key deployment time was measured). Among streaming ciphers, the Salsa20 cipher with a very simple scheme of setting keys and initialization vectors is preferred. Next are the ciphers SNOW 2.0, Rabbit, "Strumok" etc.

Thus, according to the last criterion, the developed streaming cipher "Strumok" shows the average results of the rate of installation of keys and initialization vectors.

According to the integrated performance estimation, the Stream stream cipher should be preferred, which provides a huge speed of encryption of long streams and data packets.

16.3.2.2 Comparison with Intel Core i57200U 2.50 GHz

The results of testing the speed of streaming cryptographic transformations, which were obtained on a computer system with an Intel Core i57200U 2.50 GHz processor, are shown in Tables 16.8, 16.9, 16.10, 16.11 and 16.12.

The obtained results confirm the conclusion about high speed indicators of the developed flow symmetric cipher. For example, when encrypting a long stream of

Table 16.8 The obtained results of the encryption speed of the long stream (length 1 GB)

Algorithm	Speed	
	Cycles/byte	Mbit/s
Mickey	316	68
Decim	1545	14
Strumok-256	1	12,840
Strumok-512	1	12,861
Trivium	6	3785
Sosemanuk	4	4865
Snow2.0–128	3	8037
Snow2.0–256	3	8025
Salsa20	8	2690
Rabbit	6	3743
HC-128	2	10,857
HC-256	6	3861
AES-128	9	2470
AES-256	13	1231

Table 16.9 The results of encryption speed of 350 data packets of 40 bytes each

Algorithm	Speed	
	Cycles/byte	Mbit/s
Mickey	753	29
Decim	2302	9
Strumok-256	17	1273
Strumok-512	17	1244
Trivium	27	794
Sosemanuk	16	1318
Snow2.0-128	13	1647
Snow2.0-256	13	1698
Salsa20	12	1867
Rabbit	19	1155
HC-128	302	72
HC-256	2407	9
AES-128	13	1697
AES-256	18	1231

data (see Table 16.8), the algorithm "Strumok" provides the rate of formation of the key stream over 12 Gbps. The closest analogue—the SNOW 2.0 algorithm provides a speed of about 8.0 Gbps, and the HC algorithm with a 128-bit key—a little more

Table 16.10 The results of encryption speed of 120 data packets of 576 bytes each

Algorithm	Speed	
	Cycles/byte	Mbit/s
Mickey	349	62
Decim	1623	13
Strumok-256	3	8253
Strumok-512	3	8132
Trivium	7	2989
Sosemanuk	4	4851
Snow2.0-128	3	6662
Snow2.0-256	3	6662
Salsa20	8	2711
Rabbit	6	3351
HC-128	23	957
HC-256	196	111
AES-128	9	2436
AES-256	13	1728

Table 16.11 The results of the encryption speed of 50 data packets of 1500 bytes each

Algorithm	Speed	
	Cycles/byte	Mbit/s
Mickey	330	66
Decim	1598	14
Strumok-256	2	10,714
Strumok-512	2	10,909
Trivium	6	3448
Sosemanuk	4	5769
Snow2.0-128	3	7317
Snow2.0-256	3	7407
Salsa20	8	2643
Rabbit	6	3571
HC-128	10	2198
HC-256	83	263
AES-128	9	2449
AES-256	13	1724

than 10 Gbps. However, the enhanced version of HC256 is three times slower and provides speeds of less than 4 Gbps.

Table 16.12 The results of the speed of setting keys and initialization vectors are obtained

Algorithm	Speed	
	Cycles/setting	Settings/s
Mickey	17,102	158,579
Decim	35,157	77,148
Strumok-256	411	6,578,947
Strumok-512	401	6,140,350
Trivium	847	3,505,128
Sosemanuk	430	6,250,000
Snow2.0-128	324	8,333,333
Snow2.0-256	327	8,333,333
Salsa20	1	500,000,000
Rabbit	341	7,936,507
HC-128	11,867	228,519
HC-256	94,981	28,552
AES-128	0.24	50,000,000
AES-256	0.28	48,148,148

In terms of data packet encryption rate, the above trend persists: in the case of short data packets, block symmetric ciphers have an advantage; when increasing the packet length to 576 bytes and above, the "Strumok" algorithm is preferred.

The results of measuring the speed of setting keys and initialization vectors on the Intel Core i57200U 2.50 GHz platform show the same ratios that were obtained earlier. In particular, among streaming ciphers, the Salsa20 cipher has an advantage, followed by SNOW 2.0, Rabbit, Stream and other ciphers.

Thus, according to the integrated assessment, the cipher "Strumok" has an advantage, which provides high rates of data encryption.

16.3.2.3 Comparison with Intel Core I77700 3.6 GHz

Tables 16.13, 16.14, 16.15, 16.16 and 16.17 show the results of testing the speed of streaming cryptocurrency using a computer system with Intel Core i77700 3.6 GHz processor (cache memory 64 KB first level and 1024 KB second level), 32 GB DDR3 2133 MHz RAM and Windows 10 operating system.

The obtained results of the long-stream encryption rate demonstrate ultra-high performance of the cryptoalgorithm "Strumok". In particular, at different key lengths (256 and 512 bits) the cipher provides keystream formation with speeds of 17.4 and 17.6 Gbps. In fact, this means that even with a single personal computer, the software implementation of the Stream algorithm allows cryptographic conversion of huge amounts of data streams, for example, simultaneously encrypting about 8500 digital E1 streams with a bandwidth of 2048 bits each.

Table 16.13 The obtained results of the encryption speed of the long stream (length 1 GB)

Algorithm	Speed	
	Cycles/byte	Mbit/s
Mickey	310	93
Decim	1519	19
Strumok-256	2	17,406
Strumok-512	2	17,631
Trivium	6	5069
Sosemanuk	4	6476
Snow2.0-128	3	10,663
Snow2.0-256	3	10,580
Salsa20	8	3624
Rabbit	6	4943
HC-128	2	15,044
HC-256	5	6152
AES-128	9	3290
AES-256	12	2321

Table 16.14 The results of encryption speed of 350 data packets of 40 bytes each

Algorithm	Speed	
	Cycles/byte	Mbit/s
Mickey	737	39
Decim	2294	13
Strumok-256	17	1723
Strumok-512	17	1697
Trivium	27	1057
Sosemanuk	16	1750
Snow2.0-128	13	2196
Snow2.0-256	13	2154
Salsa20	13	2286
Rabbit	17	1672
HC-128	299	97
HC-256	1832	16
AES-128	13	2240
AES-256	18	1600

According to the indicators of data packet encryption speed and setting of initialization vectors and keys, the above trends are preserved.

Table 16.15 The results of encryption speed of 120 data packets of 576 bytes each

Algorithm	Speed	
	Cycles/byte	Mbit/s
Mickey	342	84
Decim	1600	18
Strumok-256	3	9216
Strumok-512	3	10,842
Trivium	8	3814
Sosemanuk	4	6430
Snow2.0-128	3	8919
Snow2.0-256	3	8777
Salsa20	8	3477
Rabbit	7	4424
HC-128	22	1280
HC-256	130	221
AES-128	9	3291
AES-256	13	2160

Table 16.16 The results of the encryption speed of 50 data packets of 1500 bytes each

Algorithm	Speed	
	Cycles/byte	Mbit/s
Mickey	323	89
Decim	1569	18
Strumok-256	2	11,538
Strumok-512	2	14,634
Trivium	6	4478
Sosemanuk	4	7595
Snow2.0-128	3	9836
Snow2.0-256	3	9836
Salsa20	8	3429
Rabbit	6	3953
HC-128	10	2941
HC-256	53	545
AES-128	9	3297
AES-256	13	2247

16.3.2.4 Comparison with Intel Core i97980XE 2.60 GHz

Tables 16.18, 16.19, 16.20, 16.21 and 16.22 show the results of testing the performance of streaming cryptographic transformations, which were obtained on a

Table 16.17 The results of the speed of setting keys and initialization vectors are obtained

Algorithm	Speed	
	Cycles/setting	Settings/s
Mickey	16,751	214,961
Decim	34,393	104,668
Strumok-256	407	8,771,930
Strumok-512	394	9,090,909
Trivium	834	4,310,345
Sosemanuk	422	8,474,576
Snow2.0-128	320	11,363,636
Snow2.0-256	319	11,363,636
Salsa20	1	5,000,000,000
Rabbit	343	10,638,298
HC-128	11,831	304,321
HC-256	72,435	49,697
AES-128	0.27	50,000,000
AES-256	0.21	64,285,714

Table 16.18 The obtained results of the encryption speed of the long stream (length 1 GB)

Algorithm	Speed	
	Cycles/byte	Mbit/s
Mickey	222	93
Decim	1085	19
Strumok-256	1	17,197
Strumok-512	1	17,221
Trivium	4	5131
Sosemanuk	3	6564
Snow2.0-128	2	10,643
Snow2.0-256	2	10,836
Salsa20	6	3626
Rabbit	4	4802
HC-128	1	15,182
HC-256	3	6515
AES-128	6	3341
AES-256	9	2374

computer system with Intel Core i97980XE 2.60 GHz processor (cache memory 18 · 32 KB first level and 1024 KB second level), RAM 64 GB DDR4 2133 MHz and Windows 10. This is the most powerful computing system that has been used in experimental studies. At the same time, the results show that the use of the Intel

Table 16.19 The results of encryption speed of 350 data packets of 40 bytes each

Algorithm	Speed	
	Cycles/byte	Mbit/s
Mickey	530	39
Decim	1541	13
Strumok-256	12	1723
Strumok-512	12	1723
Trivium	19	1077
Sosemanuk	12	1778
Snow2.0-128	9	2286
Snow2.0-256	9	2286
Salsa20	8	2489
Rabbit	12	1672
HC-128	213	97
HC-256	1273	16
AES-128	9	2240
AES-256	12	1671

Table 16.20 The results of encryption speed of 120 data packets of 576 bytes each

Algorithm	Speed	
	Cycles/byte	Mbit/s
Mickey	246	84
Decim	1114	19
Strumok-256	2	11,059
Strumok-512	2	11,059
Trivium	5	4036
Sosemanuk	3	6505
Snow2.0-128	2	8919
Snow2.0-256	2	8919
Salsa20	6	3638
Rabbit	5	4286
HC-128	16	1295
HC-256	91	227
AES-128	6	3272
AES-256	9	2333

Core i97980XE 2.60 GHz processor does not give a significant advantage compared to, for example, the Intel Core i77700 3.6 GHz processor. The encryption speed is mainly affected by the clock speed and the amount of first-level cache memory. The number of processor cores does not have a significant effect.

Table 16.21 The results of the encryption speed of 50 data packets of 1500 bytes each

Algorithm	Speed	
	Cycles/byte	Mbit/s
Mickey	232	89
Decim	1096	19
Strumok-256	1	14,634
Strumok-512	1	14,285
Trivium	4	4615
Sosemanuk	3	7692
Snow2.0-128	2	10,000
Snow2.0-256	2	9836
Salsa20	6	3550
Rabbit	5	4580
HC-128	7	2970
HC-256	37	560
AES-128	6	3315
AES-256	9	2362

Table 16.22 The results of the speed of setting keys and initialization vectors are obtained

Algorithm	Speed	
	Cycles/setting	Settings/s
Mickey	12,043	215,239
Decim	24,191	107,135
Strumok-256	285	9,090,909
Strumok-512	283	9,090,909
Trivium	599	4,310,345
Sosemanuk	303	8,620,690
Snow2.0-128	226	11,363,636
Snow2.0-256	226	11,363,636
Salsa20	1	5,000,000,000
Rabbit	265	9,803,922
HC-128	8379	309,597
HC-256	50,812	51,010
AES-128	0.14	50,000,000
AES-256	0.15	46,666,667

The obtained research results, given in Tables 16.18, 16.19, 16.20, 16.21 and 16.22, coincide with the previous results. When encrypting a long stream of data, high performance rates (over 17 Gbps) are provided. Unfortunately, the use of an 18-core Intel Core i97980XE 2.60 GHz processor does not significantly increase the encryption speed, a significant factor is the processor clock speed and the amount of first-level cache memory.

16.3.3 *Comparison with AMD FX6300 SixCoreProcessor 3.50 GHz*

The results of comparative studies of the speed of streaming cryptocurrencies on a computer system with an AMD FX6300 SixCoreProcessor 3.50 GHz processor are shown in Tables 16.23, 16.24, 16.25, 16.26 and16.27.

The obtained results generally coincide with the previous ones both in terms of the rate of transformations and in terms of the ratio of speeds in different testing conditions. However, it should be noted that the advantage of the "Strumok" algorithm has decreased over other algorithms. For example, "Strumok" wins only when encrypting a long stream. The SNOW 2.0 algorithm is more suitable for encrypting short packets in terms of speed.

Given the possible post-quantum application, only the Stream algorithm has increased key lengths and initialization vector. Therefore, in terms of stability/complexity, the cipher "Strumok" should still be preferred.

Table 16.23 The obtained results of the encryption speed of the long stream (length 1 GB)

Algorithm	Speed	
	Cycles/byte	Mbit/s
Mickey	518	54
Decim	2828	10
Strumok-256	5	5872
Strumok-512	4	6326
Trivium	8	3347
Sosemanuk	8	3637
Snow2.0-128	5	5630
Snow2.0-256	5	5822
Salsa20	17	1688
Rabbit	11	2490
HC-128	3	9417
HC-256	8	3622
AES-128	16	1804
AES-256	21	1329

Table 16.24 The results of encryption speed of 350 data packets of 40 bytes each

Algorithm	Speed	
	Cycles/byte	Mbit/s
Mickey	1300	22
Decim	4377	6
Strumok-256	40	687
Strumok-512	40	709
Trivium	40	700
Sosemanuk	26	1067
Snow2.0-128	25	1120
Snow2.0-256	24	1155
Salsa20	23	1204
Rabbit	32	875
HC-128	534	53
HC-256	3804	7
AES-128	22	1258
AES-256	28	982

Table 16.25 The results of encryption speed of 120 data packets of 576 bytes each

Algorithm	Speed	
	Cycles/byte	Mbit/s
Mickey	577	49
Decim	2932	10
Strumok-256	7	3950
Strumok-512	7	4001
Trivium	11	2608
Sosemanuk	8	3393
Snow2.0-128	6	4767
Snow2.0-256	6	4808
Salsa20	17	1603
Rabbit	12	2353
HC-128	39	715
HC-256	271	104
AES-128	15	1813
AES-256	21	1310

Table 16.26 The results of the encryption speed of 50 data packets of 1500 bytes each

Algorithm	Speed	
	Cycles/byte	Mbit/s
Mickey	547	51
Decim	2874	9
Strumok-256	6	5000
Strumok-512	5	5172
Trivium	9	3045
Sosemanuk	7	3774
Snow2.0-128	5	5217
Snow2.0-256	5	5309
Salsa20	17	1653
Rabbit	11	2564
HC-128	17	1662
HC-256	109	258
AES-128	15	1881
AES-256	21	1354

Table 16.27 The results of the speed of setting keys and initialization vectors are obtained

Algorithm	Speed	
	Cycles/setting	Settings/s
Mickey	29,277	119,933
Decim	61,819	56,811
Strumok-256	930	3,759,398
Strumok-512	869	4,065,041
Trivium	1181	2,976,190
Sosemanuk	588	5,952,380
Snow2.0-128	549	6,410,256
Snow2.0-256	539	6,493,506
Salsa20	2	5,000,000,000
Rabbit	613	5,747,126
HC-128	20,914	167,954
HC-256	151,362	23,201
AES-128	0.65	50,000,000
AES-256	0.62	52,419,355

16.4 Substantiation of Flow Characteristics Symmetric Encryption by Stability/Complexity Ratio

This section estimates the number of operations and the minimum amount of memory required for software implementation of the developed algorithm "Strumok". The obtained results are compared with the nearest prototype—SNOW 2.0 streaming encryption algorithm, using both the absolute number of operations and the relative rate of specific complexity of the cryptoalgorithm.

16.4.1 Estimate the Number of Operations and the Minimum Amount of Memory, Required for the Implementation of the Cipher "Strumok"

Analysis of the structure and software implementation of the streaming cryptoconversion algorithm "Strumok" and the main functions used in this SSC, shows that during the implementation of the generation of the key stream (encryption range) the following computational operations and transformations are used.

(1) To implement the *Next* function, you must perform the following steps:

- in the first step it is necessary to perform a nonlinear substitution. In fast computation, this procedure requires 7 exclusive disjunction operations (XOR operations or addition modulo two) over 8 64-bit vectors, which are read from the precalculation tables,. To index these tables it is necessary to perform 7 shift operations and 8 XOR operations;
- in the second step it is necessary to update the value of the register of the finite automaton by performing one operation of adding integers modulo 2^{64};
- in the third step, the values of 15 cells of the shift register are updated by shifting 15 64-bit words;
- in the fourth step it is necessary to perform 2 XOR operations: one multiplication operation by α and one multiplication operation by. In turn, multiplication by α requires 2 shift operations and one XOR operation with one value read from the precalculation table. Multiplication by requires 1 shift operation, 1 XOR operation, and 1 bit conjunction operation.

(2) To implement the Strm function, it is necessary to calculate the value of the finite state function and perform one XOR operation. To calculate the function of a finite automaton, in turn, it is necessary to perform 1 operation of adding integers modulo 2^{64} and 1 operation XOR.

Thus, to generate 64 bits of the key stream according to the specification of the algorithm "Strumok" it is necessary to perform the following operations (over 64-bit words):

- 21 operations of exclusive disjunction;
- 10 shift operations;
- 2 operations of adding integers modulo 2^{64};
- 1 bit conjunction operation;
- 10 readings from pre-calculation tables;

– update 15 words by shifting them.

To store precalculation data, the implementation of the "Strumok" algorithm requires 10 sample tables, which are permanently stored in the memory of the computer system (mainly in the cache memory of the first level):

– 8 tables are required for storage of nonlinear replacement nodes;
– 1 table is required for fast implementation of multiplication by a finite field element;
– 1 table is needed for fast implementation of multiplication by the inverse element of a finite field.

In addition to pre-calculation tables, the implementation of the algorithm also involves storing in memory the internal state of the generator and state variables.

In general, to store precalculation tables, the implementation of the algorithm requires the allocation of 20 KB = 20,480 bytes of memory. 128 bytes of memory are required to store the internal state of the generator, and another 16 bytes require storage of state variables. Thus, the total amount of memory required to implement the developed algorithm "Strumok" is 20624 bytes.

Similar calculations of the amount of memory required to implement the nearest prototype, the SNOW 2.0 algorithm, show the following.

(1) To implement the *Next* function, you must perform the following steps:

   – in the first step it is necessary to perform a nonlinear substitution. In fast computation, this procedure requires 3 operations of exclusive disjunction (XOR operations or addition modulo two) over 4 32-bit vectors, which are read from the tables of precalculations $T_i[a]$, $i = 0, 1, ..., 3$. To index these tables, 3 shift operations and 4 XOR operations must be performed;
   – in the second step it is necessary to update the value of the register $r_1^{(i+1)}$ of the finite automaton by performing one operation of adding integers modulo 2^{32};
   – in the third step, the values of 15 cells of the shift register are updated by shifting 15 32-bit words;
   – in the fourth step it is necessary to perform 2 XOR operations: one multiplication operation by α and one multiplication operation by α^{-1}. In turn, multiplication by α requires 2 shift operations and one XOR operation with one value read from the precalculation table. Multiplication by requires 1 shift operation, 1 XOR operation, and 1 bit conjunction operation.

(2) To implement the *Strm* function, it is necessary to calculate the value of the finite state function and perform one XOR operation. To calculate the function of a finite automaton, in turn, it is necessary to perform 1 operation of adding integers modulo 2^{32} and 1 operation XOR.

Thus, to generate 32 bits of the key stream according to the specification of the SNOW 2.0 algorithm, it is necessary to perform the following operations (over 32-bit words):

- 11 operations of exclusive disjunction;
- 6 shift operations;
- 2 operations of adding integers modulo 2^{32};
- 1 bit conjunction operation;
- 6 readings from pre-calculation tables;
- update 15 words by shifting them.

To store precalculation data, the implementation of the Snow 2.0 algorithm requires 6 sample tables, which are permanently stored in the memory of the computer system (mainly in the cache memory of the first level):

- 4 tables are required for storage of nonlinear replacement nodes;
- 1 table is required for fast implementation of multiplication by a finite field element;
- 1 table is needed for fast implementation of multiplication by the inverse element of a finite field.

In addition to pre-calculation tables, the implementation of the algorithm also involves storing in memory the internal state of the generator and state variables.

In general, to store precalculation tables, the implementation of the algorithm requires the allocation of 6 Kbytes = 6144 bytes of memory. The internal state of the generator requires 64 bytes of memory and another 8 bytes require the storage of state variables. Thus, the total amount of memory required to implement the developed SNOW 2.0 algorithm is 6216 bytes.

The results of the comparative analysis of the complexity of the implementation of the algorithm SSC SNOW 2.0 and the developed algorithm SSC "Strumok" are given in Table 16.28.

Table 16.28 The results of a comparative analysis of the complexity of the implementation of streaming encryption algorithms SNOW 2.0 and "Strumok" by the number of operations performed and the amount of memory

Computing resource or operation	Number of operations or amount of memory	
	SNOW 2.0	«Струмок»
Exclusive disjunction operation (XOR)	11	21
Shear operation	6	10
Addition operation by module 2^{32}	2	–
Addition operation by module 2^{64}	–	2
Bitwise conjunction operation	1	1
Read operation from precalculation tables	6	10
The amount of memory required to implement, bytes	6144	20,624
The amount of key stream that generates the generator, bits	32	64

The last row of Table 16.28 shows the amount of key flow that the generator generates in one cycle of its operation. That is, the data in the table on the cost of certain operations, conversions and memory should be normalized to this value.

16.4.2 Substantiation of the Characteristics of the Cipher "Strumok" in Terms of Stability/Complexity

To substantiate the characteristics of the project of stream symmetric encryption "Strumok" in terms of stability/complexity, we will use estimates of the specific complexity of the implementation of the cryptoalgorithm in comparison with the nearest prototype—the algorithm SNOW 2.0.

The main purpose of developing the "Strumok" algorithm was to increase the speed (compared to SNOW 2.0) without reducing the cryptographic properties. Since the developed algorithm is performed according to the same general scheme as the Snow 2.0 prototype, and taking into account the results of studies of cryptographic properties of the formed key sequences, we will assume that the stability of the algorithm "Strumok" is not less than the corresponding characteristics of SNOW 2.0. Therefore, the substantiation of the characteristics of the SSC project in terms of stability/complexity is reduced to determining the advantage over the prototype algorithm in terms of reducing the specific complexity of the practical implementation.

The corresponding estimates of the specific complexity of the implementation of SSC Snow 2.0 and SSC "Strumok" are given in Table 16.29. This data is calculated taking into account the data of Table 16.28 and the size of the word SSC, ie the number of bits of the key stream that generates the generator for one conversion cycle.

Table 16.29 The results of the comparative analysis of the specific complexity of the implementation of the algorithms SSC SNOW 2.0 and "Strumok"

Computing resource or operation	Number of operations or amount of memory	
	Snow 2.0	«Струмок»
Exclusive disjunction operation (XOR)	11/32 ≈ 0,34	21/64 ≈ 0,33
Shear operation	6/32 ≈ 0,19	10/64 ≈ 0,16
Addition operation by module 2^{32}	2/32 ≈ 0,06	–
Addition operation by module 2^{64}	–	2/64 ≈ 0,03
Bitwise conjunction operation	1/32 ≈ 0,03	1/64 ≈ 0,02
Read operation from precalculation tables	6/32 ≈ 0,19	10/64 ≈ 0,16
The amount of memory required to implement, bytes	6144/32 ≈ 192	20,624/64 ≈ 322,25

Taking into account the data of Table 16.29, it can be stated that the relative time complexity of implementation (as a specific number of operations to the number of generated bits) for SSC "Strumok" is comparable or less than similar indicators for SSC SNOW 2.0. The relative capacitive complexity of the implementation of SSC "Strumok" is one and a half times higher than SSC SNOW 2.0. However, the latter figure, as shown below, is insignificant, as the main condition for the rapid implementation of cryptographic algorithms is to limit the amount of cache memory of the first level of the processor of the applied computer system.

The processor cache (level 1—L_1, level 2—L_2 and level 3—L_3)—memory, access time to which is several clock cycles, and which has a size of tens of kilobytes. This is the fastest (after zero) level of the memory hierarchy, because this memory is essentially an integral part of the processor, as it is located on the same chip and is part of the functional blocks. In modern processors, L_1 is usually divided into two caches—a command cache (instructions) and a data cache (so-called Harvard architecture). Most processors cannot function without L_1. The L_1 runs at CPU frequency and, in general, reverse every measure can be performed to it. It is often possible to perform several read/write operations at the same time.

Thus, cache memory is used to perform the following, including cryptographic functions [45, 46]:

- to increase computer performance;
- for temporary storage of the contents of RAM cells;
- to speed up the exchange between the processor and the non-volatile storage device.

The results of a brief comparative analysis of the properties of processors from different manufacturers are given in Tables 16.30 and 16.31. Table 16.30 shows the characteristics of some Intel Core i7 series processors, and Table 16.31 shows the characteristics of AMD processors.

Symbols used in Tables 16.30 and 16.31: TDP—Thermal Design Power.

The analysis shows that most modern processors manufactured by the world's leading manufacturers (Intel and AMD) differ in the size of the cache of the second and third levels, as well as the clock frequency and the corresponding heat output. However, in terms of the level of memory of the first level cache, almost all the considered processors have the same parameters—the limit of the amount of memory of the level 1 cache is 64 KB.

Thus, the general requirements for modern cryptographic algorithms regarding the amount of memory are limited to 64 KB, i.e. the total amount of preprocessing tables for fast implementation of ciphers should not exceed the amount of memory of the first level cache. As research has shown that SSC "Strumok" requires 20,624 bytes for its implementation (which is less than the level 1 memory cache limit), an increase in memory compared to SSC SNOW 2.0 will not lead to a decrease. the speed of operation of the generator and, in particular, to reduce the speed of formation of the key stream (encryption range).

This conclusion is confirmed by the data of the experimental evaluation of the speed of streaming cryptocurrencies, which are given in subsection 16.3. Indeed, the

Table 16.30 Characteristics of "Core i7" processors manufactured by Intel

«Bloomfield» (45 nm)					
Model	Clock frequency	Cores	L2	L3	TDP
Core i7920	2.67 GHz	4	4 × 256 KB	8 MB	130 W
Core i7930	2.8 GHz	4	4 × 256 KB	8 MB	130 W
Core i7940	2.93 GHz	4	4 × 256 KB	8 MB	130 W
Core i7950	3.07 GHz	4	4 × 256 KB	8 MB	130 W
Core i7960	3.2 GHz	4	4 × 256 KB	8 MB	130 W
Core i7965 Extreme Edition	3.2 GHz	4	4 × 256 KB	8 MB	130 W
Core i7975 Extreme Edition	3.33 GHz	4	4 × 256 KB	8 MB	130 W
«Gulftown» (32 nm)					
Model	Clock frequency	Cores	L2	L3	**TDP**
Core i7970	3.2 GHz	6	6 × 256 KB	12 MB	130 W
Core i7980	3.33 GHz	6	6 × 256 KB	12 MB	130 W
Core i7980X	3.33 GHz	6	6 × 256 KB	12 MB	130 W
Core i7990X	3.47 GHz	6	6 × 256 KB	12 MB	130 W

rate of formation of the encryption range (key stream) using the proposed algorithm SSC "Strumok" in most cases exceeds 1.5 times the corresponding results for the nearest prototype SSC SNOW 2.0. This result is confirmed by coincidence with the theoretical calculations, the results of which are given in table 16.29.

Thus, a set of individual factors, namely:

- the rate of formation of PRS;
- initial initialization rate;
- the minimum amount of memory required for the implementation of PSS;
- provided cryptographic properties of the encryption range (key stream),

The most reasonable technical solution in terms of stability/complexity is developed by SSC "Strumok".

16.5 Conclusions in this Chapter

In recent years, the requirements for modern streaming encryption algorithms have increased significantly: on the one hand, it is necessary to ensure high speed cryptographic transformation (over 10 Gbps), on the other—to effectively resist the latest methods of cryptographic analysis, including quantum computational methods. From the point of view of ensuring the established levels of stability in the possible post-quantum application, the general requirements for the key system of advanced flow symmetric cipher are as follows: to ensure a high level of stability (requiring at least 2128 quantum steps), the key length must be at least 256 bits; to ensure an extremely

Table 16.31 Characteristics of AMD processors

AMD Phenom™ II X6 Processor					
Model	Clock frequency (GHz)	L2 (MB)	L3 (MB)	TDP	CMOS Technology
1100 T*	3.3	3 MB	6 MB	125 W	45 nm SOI
1090 T*	3.2	3 MB	6 MB	125 W	45 nm SOI
1075 T	3.0	3 MB	6 MB	125 W	45 nm SOI
1065 T	2.9	3 MB	6 MB	95 W	45 nm SOI
1055 T	2.8	3 MB	6 MB	125 W	45 nm SOI
1045 T	2.7	3 MB	6 MB	95 W	45 nm SOI
AMD Phenom™ II X4 Processor					
980*	3.7 GHz	2 MB	6 MB	125 W	45 nm SOI
975*	3.6 GHz	2 MB	6 MB	125 W	45 nm SOI
970*	3.5 GHz	2 MB	6 MB	125 W	45 nm SOI
965*	3.4 GHz	2 MB	6 MB	125 W	45 nm SOI
955*	3.2 GHz	2 MB	6 MB	125 W	45 nm SOI
AMD FX Processors					
FX 9590	4.7/5.0 GHz	8 MB	8 MB	220 W	32 nm SOI
FX 9370	4.4/4.7 GHz	8 MB	8 MB	220 W	32 nm SOI
FX 8350	4.0/4.2 GHz	8 MB	8 MB	125 W	32 nm SOI
FX 8320	3.5/4.0 GHz	8 MB	8 MB	125 W	32 nm SOI
FX 6350	3.9/4.2 GHz	6 MB	8 MB	125 W	32 nm SOI
FX 6300	3.5/4.1 GHz	6 MB	8 MB	95 W	32 nm SOI
FX 6200	3.8/4.1 GHz	6 MB	8 MB	125 W	32 nm SOI
FX 6100	3.3/3.9 GHz	6 MB	8 MB	95 W	32 nm SOI
FX 4350	4.2/4.3 GHz	4 MB	8 MB	125 W	32 nm SOI
FX 4300	3.8/4.0 GHz	4 MB	4 MB	95 W	32 nm SOI
FX 4130	3.8/3.9 GHz	4 MB	4 MB	125 W	32 nm SOI
FX 4100	3.6/3.8 GHz	4 MB	8 MB	95 W	32 nm SOI

high level of stability (which requires at least 2256 quantum steps), the key length must be at least 512 bits.

The SSC "Strumok" generator is built according to the SNOW 2.0-like scheme of the summing generator (the SNOW 2.0 generator is defined in DSTU ISO/IEC 180,334: 2015), it uses a 256-bit initialization vector and a 256-bit or 512-bit secret key. The generator provides high and ultra-high levels of stability and is designed to protect important information with limited access owned by the state, in particular to ensure the confidentiality of information during its processing, taking into account current trends in cryptographic science and possible application of quantum cryptographic analysis.

The symmetric streaming cryptographic transformation algorithm uses the "Strumok" key stream generator. Depending on the length of the secret key, two modes are used:

mode with a 256-bit key, denoted as "Strumok-256";
mode with a 512-bit key, denoted as "Strumok-512".

LFSR and CA, which are built on the principles of the world-famous SNOW2.0 algorithm, were chosen as the two main elements of the "Strumok" generator. The main differences of the proposed generator are as follows:

- feedback taps in LFSR are built on the primitive over the field GF (2^{64}) polynomial *f(x)* of degree 16. The field GF(2^{64}) is constructed as an extension of the field GF(2^8), which, in turn, is built on the primitive over the field GF(2) by the polynomial p(y). Thus, we have a tower of fields. Therefore, the period of the original sequence of LFSR is the maximum and is equal to 2^{1024}—1 (in the algorithm SNOW 2.0 the period is equal to 2^{512}—1);
- for the compatibility of implementations of cryptographic algorithms as a transformation T (the main component of CA, which determines its cryptographic properties), the use of nonlinear transformation SSC "Kalina" (defined by the national standard BSS of Ukraine DSTU 7624: 2014) is proposed.

Thus, the generator of the key stream "Strumok" is built on SNOW 2.0 similar scheme of the summing generator; the increased lengths of the secret key and the initialization vector allow to reliably apply the stream cipher even taking into account quantum methods of cryptographic analysis.

The optimization of cryptoconversions is based on the idea of using pre-calculated data. Precalculation tables do not require multiplication calculations during encryption and nonlinear substitution, which will significantly reduce the number of operations to process the input data block. Given the proposed optimization of calculations, the developed algorithm "Strumok" is able to provide high performance and operate efficiently on different computing platforms. It does not require large amounts of memory, to ensure the normal operation of the cipher enough cache memory level 1 of modern processors. This provides a high speed of calculations, and, consequently, a high speed of key stream generation.

According to the mathematical structure of the stream generator "Strumok" to produce one word of the key sequence, it is necessary to perform two complex multiplication operations in the Galois field. This multiplication can be implemented using integer multiplication, but it can reduce the speed of the algorithm. For reasons of ensuring the speed of the cipher, to perform these operations, 2 tables of precalculations are formed. The preprocessing algorithm is part of the cryptosystem, so it also contributes to the problem of general secrecy (stability) of the cipher. This part is not as critical as the direct encryption algorithm itself. In addition, significant computational resources can be used to implement precalculations, which simplifies the task of constructing the necessary precalculation procedures. The idea of high-speed encryption is based on the construction of precalculation tables, having which will not need to use the transformation for each byte of the word, which will significantly reduce the number of operations to process the block of input data.

Thus, given the proposed optimization of calculations by using pre-formed tables of precalculations, the algorithm "Strumok" is able to provide high performance and operate efficiently on different computing platforms. In particular, the results of research show that in comparison with the best world analogues (streaming cryptoalgorithms MICKEY, Decim, Trivium, SOSEMANUK, SNOW 2.0, Salsa20, Rabbit, HC and streaming mode of BSC AES) developed SSC "Strumok" provides the highest speed., the bandwidth of this generator exceeds 17 Gbps. Such high estimates of speed are primarily the result of a simple and computationally efficient structure of the algorithm.

According to the results of comparative studies of the initial initialization rate of generators, it was found that among the considered algorithms the lowest initialization time of cryptoconversions have SSC Salsa20 and SNOW 2.0, which are somewhat inferior to the developed algorithm "Strumok". This advantage of the considered algorithms is a consequence of a simple and computationally efficient scheme of deployment of encryption keys. It should be noted that the simplification of the key deployment scheme and the lack of pre-calculation tables in some cryptographic algorithms may be a disadvantage in terms of security and/or speed of key stream formation. For example, the key scheme of the Salsa20 algorithm consists of filling the key sequence in a certain way with key bits, which can be considered too simple a scheme and cause the emergence of effective cryptanalytic attacks on the key deployment scheme in the future.

In the case of encrypting short data packets, block symmetric ciphers have the advantage in performance, as expected. Only Salsa20 and SNOW 2.0 ciphers can compete with block ciphers according to this criterion. Stream cipher "Strumok" at a comparable length of the key loses by 20–30% in the speed of keystream formation. However, when the packet length is increased to 576 bytes and above, it usually has an advantage again, its speed is 20–30% higher than other alternatives.

Thus, according to the criterion of complexity/stability in most practically significant cases of application, the developed streaming cryptoalgorithm "Stream" (with 256-bit and 512-bit key—"Strumok-256" and "Strumok-512", respectively) has an advantage.

The obtained results of the comparative analysis show that the relative time complexity of implementation (as a specific number of operations to the number of generated bits) for SSC "Strumok" is comparable or less than similar indicators for SSC SNOW 2.0. The relative capacitive complexity of the implementation of SSC "Strumok" is one and a half times higher than SSC SNOW 2.0. However, the latter indicator is insignificant, as the main condition for the rapid implementation of cryptocurrency algorithms is the requirement to "fit" into the limit on the amount of cache memory of the first level.

The analysis shows that most modern processors manufactured by the world's leading manufacturers differ in the size of the cache of the second and third levels, as well as the clock frequency and the corresponding heat output. However, in terms of the level of memory of the first level cache, almost all the considered processors have the same parameters—the limit of the amount of memory of the level 1 cache is 64 KB. Therefore, the general requirements for modern cryptographic algorithms regarding the amount of memory are limited to 64 KB, i.e. the total amount of pre-calculation tables for fast implementation of ciphers should not exceed the amount of memory of the 1st level cache. As research has shown that SSC “Strumok” requires 20,624 bytes for its implementation (which is less than the level 1 memory cache limit), the increase in memory compared to SSC SNOW 2.0 will not lead to a decrease the speed of operation of the generator and, in particular, to reduce the speed of formation of the key stream (encryption range).

Thus, the set of individual factors: the rate of formation of PRS; initial initialization rate; the minimum amount of memory required for the implementation of SSC; provided with cryptographic properties of the encryption range (key stream),—the most reasonable technical solution in terms of stability/complexity is developed by SSC “Strumok”.

References

1. Kuznetsov, O., Lutsenko, M., Ivanenko, D.: Strumok stream cipher: Specification and basic properties. In: 2016 Third International Scientific-Practical Conference Problems of Infocommunications Science and Technology (PIC S&T), pp. 59–62. IEEE, Kharkiv, Ukraine (2016). https://doi.org/10.1109/INFOCOMMST.2016.7905335
2. Gorbenko, I., Kuznetsov, A., Gorbenko, Y., Vdovenko, S., Tymchenko, V., Lutsenko, M.: Studies on statistical analysis and performance evaluation for some stream ciphers. Int. J. Comput. **18**, 82–88 (2019)
3. Gorbenko, I., Kuznetsov, A., Tymchenko, V., Gorbenko, Y., Kachko, O.: Experimental studies of the modern symmetric stream ciphers. In: 2018 International Scientific-Practical Conference Problems of Infocommunications. Science and Technology (PIC S T), pp. 125–128 (2018). https://doi.org/10.1109/INFOCOMMST.2018.8632058
4. Gorbenko, I., Kuznetsov, O., Gorbenko, Y., Alekseychuk, A., Tymchenko, V.: Strumok keystream generator. In: 2018 IEEE 9th International Conference on Dependable Systems, Services and Technologies (DESSERT), pp. 294–299. IEEE, Kyiv, Ukraine (2018). https://doi.org/10.1109/DESSERT.2018.8409147
5. Kuznetsov, A., Frolenko, V., Eremin, E., Zavgorodnia, O.: Research of cross-platform stream symmetric ciphers implementation. In: 2018 IEEE 9th International Conference on Dependable Systems, Services and Technologies (DESSERT), pp. 300–305 (2018). https://doi.org/10.1109/DESSERT.2018.8409148
6. Kuznetsov, A.A., Potii, A.V., Poluyanenko, N.A., Stelnik, I.V.: Nonlinear functions of complication for symmetric stream ciphers. TRE **78** (2019). https://doi.org/10.1615/TelecomRadEng.v78.i9.10
7. Kuznetsov, O., Potii, O., Perepelitsyn, A., Ivanenko, D., Poluyanenko, N.: Lightweight stream ciphers for Green IT Engineering. In: Kharchenko, V., Kondratenko, Y., Kacprzyk, J. (eds.) Green IT Engineering: Social, Business and Industrial Applications, pp. 113–137. Springer International Publishing, Cham (2019). https://doi.org/10.1007/978-3-030-00253-4_6
8. DSTU 8845: 2019. Information technology. Cryptrographic information protection. Algorithm of symmetric streaming re-creation. Kiev; Subsidiary “UkrNDNTS” (2019), 50 p.
9. Grover, L.K.: A fast quantum mechanics algorithm for database search [Text]. Grover, L.K.: Proceeding of the 28th ACM Symposium on Theory of Computation. New York: ACM Press. (1996), pp. 212–219

10. Deutsch, D.: Quantum Theory, the Church Turing Principle and the Universal Quantum Computer. Proc. Roy. Soc. Lond. **V.A400**(1818), 97–117 (1985)
11. Nielsen, M., Chang, I.: Quantum calculations and quantum information. In: Sluggish, M.N., Ostrovsky, P.M. (eds.) M.: Мир (2006), 824 p. (Per. with English)
12. Feynman, R.: Quantum mechanical computers. Found. Phys. **16**(6), 507–531 (1986)
13. Von Neumann, J.: Mathematische Grudlagen der Quantenmechanik. Springer, Berlin (1932)
14. Baumester, D., Eckert, A., Zeilinger, A.: Physics of quantum information. M.: Postmarket (2002), 376 p
15. Raussendorf, R., Browne, D.E., Briegel, H.J.: Measurementbased quantum computation on cluster states. Phys. Rev. A. **68**, 022312 (2003)
16. Cleve, R., Ekert, A., Macchiavello, C., Mosca, M.: Quantum algorithms revisited. Phil. Trans. Royal Soc. Lond. **A454**. 339–354 (1998)
17. Proos, J., Zalka, C.: Shor's Discrete Logarithm Quantum Algorithm for Elliptic Curves. arXiv.quantph/0301141 v2 (2004)
18. Gorbenko, I.D., Kuznetsov, O.O., Potiy, O.V., Gorbenko, Y.I., Ganzya, R.S., Ponomar, V.A.: Postquantum cryptography and mechanisms of its implementation. Radiotechnics: All-Ukrainian. Interscience. Scientific Technology Sat, Issue 186 (2016). KhNURE, Kharkiv, pp. 32–52
19. Gorbenko, Y.I., Potiy, A.V., Izbenko, Y.A., Orlova, S.Y.: Analysis of current encryption schemes submitted to the European competition NESSIE. Legal, regulatory and metrological support of the information protection system in Ukraine: scientific and technical collection, Vip. 5 (2002), pp. 92–110
20. Babbage S.: A Space/Time TradeOff in Exhaustive Search Attacks on Stream Ciphers. Vodafone Ltd, Newbury, UK. 09.04.1996
21. ISO/IEC 180334:2011: Information technology—Security techniques—Encryption algorithms—Part 4: Stream ciphers. [Electronic resource]. Access mode: http://www.iso.org/iso/home/store/catalogue_ics/catalogue_detail_ics.htm?csnumber=54532
22. Magma Computational Algebra System. [Electronic resourse]. Access mode: http://magma.maths.usyd.edu.au/magma/
23. DSTU 7624: 2014. Information Technology. Cryptographic information protection. Symmetric block transformation algorithm. [Text]. Enter. 01–07–2015. Ministry of Economic Development of Ukraine, Kyiv (2015)
24. FIPS197: Advanced Encryption Standard (AES). National Institute of Standards and Technology (2001). [Electronic resource]. Access mode: http://csrc.nist.gov/publications/fips/fips197/fips197.pdf
25. Daemen, J., Rijmen, V.: The Rijdael Block Cipher, AES Proposal, 9.04.2003. 45 p
26. eSTREAM Optimized Code HOWTO [Electronic resourse]. Access mode: http://www.ecrypt.eu.org/stream/perf/
27. The eSTREAM Project—eSTREAM Phase 3: MICKEY (Portfolio Profile 2). [Electronic resource]. Access mode: http://www.ecrypt.eu.org/stream/mickeypf.html
28. Wu, H., Preneel, B.: Cryptanalysis of Stream Cipher Decim. Available at http://www.ecrypt.eu.org/stream/
29. The eSTREAM Project—eSTREAM Phase 3: Trivium (Portfolio Profile 2). [Electronic resource]. Access mode: http://www.ecrypt.eu.org/stream/triviumpf.html
30. ISO/IEC 291923:2012: Information technology—Security techniques—Lightweight cryptography—Part 3: Stream ciphers. [Electronic resource]. Access mode: http://www.iso.org/iso/home/store/catalogue_tc/catalogue_detail.htm?csnumber=56426
31. The eSTREAM Project—eSTREAM Phase 3. SOSEMANUK (Portfolio Profile 1). [Electronic resource]. Access mode: http://www.ecrypt.eu.org/stream/sosemanukpf.html
32. The eSTREAM Project—eSTREAM Phase 3. Salsa20 (Portfolio Profile 1). [Electronic resource]. Access mode: http://www.ecrypt.eu.org/stream/salsa20pf.html
33. The eSTREAM Project—eSTREAM Phase 3. Rabbit (Portfolio Profile 1). [Electronic resource]. Access mode: http://www.ecrypt.eu.org/stream/rabbitpf.html

34. The eSTREAM Project eSTREAM Phase 3. HC (Portfolio Profile 1). [Electronic resource]. Access mode: http://www.ecrypt.eu.org/stream/hcpf.html
35. «ISO/IEC 180333:2010. Information technology—Security techniques—Encryption algorithms—Part 3: Block ciphers». [Online]. Internet: https://www.iso.org/standard/54531.html
36. Gorbenko, I.D., Dolgov, V.I., Oleynikov, R.V., Ruzhentsev, V.I., Mikhailenko, M.S., Gorbenko, Y.I., Totsky, A.S., Kazmina, S.V.: Perspective block cipher "Kalina"—main provisions and specification. Appl. Radio Electron. (2) (2007)
37. Oliynykov, R., Gorbenko, I., Kazymyrov, O. Ruzhentsev, V., Kuznetsov, O., Gorbenko, Y., Dyrda, O., Dolgov, V., Pushkaryov, A., Mordvinov, R., Kaidalov, D.: A New Encryption Standard of Ukraine: The Kalina Block Cipher. Cryptology ePrint Archive, Report 2015/650 (2015). [Electronic resourse]. Access mode: https://eprint.iacr.org/2015/650
38. Gorbenko, I.D., Dolgov, V.I., Oleynikov, R.V., Ruzhentsev, V.I., Mikhailenko, M.S., Gorbenko, Y.I., Neyvanov, A.V.: Principles of construction and properties of block symmetric cipher "Kalina". Appl. Radio Electron. (2) (2007)
39. Gorbenko, I.D.: Perspective block symmetric code "Kalina"—the main position and specificity. In: Gorbenko, I.Д., Dolgov, V.I., Oleynikov, R.V., Ruzhentsev, V.I., Mikhailenko, M.S., Gorbenko, Y.I., Totsky, O.S., Kazmina, S.V. (eds.) Applied Radio Electronics, vol. 6, issue 2 (2007), pp. 195–208
40. Gorbenko, I.D.: The principles of stimulation and power of the block symmetric cipher "Kalina". In: I.D. Gorbenko, V.I. Dolgov, R.V. Oliynykov and others (eds.) Applied Radioelectronics. KhTURE, Kharkov, vol. 6, Issue 2 (2007), pp. 147–157
41. Daemen, J: Annex to AES Proposal Rijndael. http://www.nist.gov/aes
42. National Institute of Standards and Technology, FIPS 197: «Advanced Encryption Standard» (Nov. 2001). http://www.nist.gov/aes
43. Daemen, J., Rijmen, V.: AES submission, Document on Rijndael, Version 2, (September 1999), pp 1–45
44. Daemen, J., Rijmen, V.: The Design of Rijndael: AES—The Advanced Encryption Standard, Information Security and Cryptography. Springer, Berlin (2002)
45. Menezis, A., Van Arshot, P., Watson, S.: Guide to applied cryptography. CRC Press, electronic copy, 662 p (1997)
46. GitHub: Powerful collaboration, code review, and code management for open source and private projects. Public projects are always free. [Electronic resourse]. Access mode: https://github.com/floodyberry/supercop/tree/master/crypto_stream/snow20

Chapter 17
Conclusions

1. Models, methods and computational algorithms for cryptographic information protection are widely used at present to provide various information security services, in particular they are a critical component of the national confidential communication system, play an important role in developing new measures and innovative technologies for national security of Ukraine. in the information sphere. Thus, the development, research, implementation and operation of cryptographic information security is an extremely important and urgent problem of national importance for the development of national information infrastructure and the creation of advanced information technologies.
2. Symmetrical cryptoconversions, in particular SSC and streaming modes of BSC application, in which information is fed and processed in the form of an infinite stream, i.e. a sequence that can hypothetically be of infinite length, have become the most widely used for information protection in modern ITS. Their main advantage is the establishment of a certain relationship between the individual symbols of the data stream, which provides additional protection against the imposition of false information or erroneous modes of operation of security equipment or terminal equipment of telecommunications systems and networks. Accordingly, streaming cryptographic transformation is more trusted because secure data streams cannot be distorted by the intentional or unintentional actions of users and attackers or by any random natural factors or factors. Among the advantages of streaming cryptoalgorithms should also be noted the increased indicators of cryptographic stability and speed with the possibility of parallel computing.
3. Development, research and substantiation of conditions for the use of modern streaming cryptographic algorithms is an extremely complex and time-consuming task. The standardized SSC must provide a high level of stability, have the required level of performance and operate efficiently on different

A. A. Kuznetsov et al., *Stream Ciphers in Modern Real-time IT Systems*,
Studies in Systems, Decision and Control 375,
https://doi.org/10.1007/978-3-030-79770-6_17

computing platforms, which is ensured by the efficiency of various components: the scheme of deployment of the secret key (key schedule); the selected basic structure of the algorithm; linear and nonlinear transformations, etc.

4. Analysis of indicators and criteria of efficiency of SSC and flow regimes of BSC allowed to allocate two main groups: unconditional criteria and indicators characterizing critical properties of system which non-fulfillment or reduction below admissible limits will lead to non-fulfillment of basic functions of system (indicators and criteria of stability), characterizing the vulnerability to the impact of various attacks); conditional indicators and criteria that characterize the desired properties, the implementation of which will improve certain performance indicators, and failure or lowering the allowable limits will not lead to dysfunction of the system (software and hardware implementation criteria and indicators, design and technological and others). Taking into account the specific conditions of functioning of the system of cryptographic protection of information and information and telecommunication system in general, the considered indicators and criteria of efficiency allow to carry out a complex estimation of efficiency of PN and flow modes of BSC.
5. For the construction of PN schemes and flow regimes SSC the most widely used today are: theoretical information approach, which uses general theoretical positions, mathematical models and assumptions from the theory of information and the theory of secret systems; system-theoretical approach that uses the methods of systems theory and systems analysis, in particular to determine the criteria and indicators of the effectiveness of SSC; a theoretically complex approach based on the use of complexity theory methods, in particular, the problem of decryption by an attacker is associated with the solution of some extremely complex mathematical problem; theoretical-stochastic approach aimed at constructing so-called randomized streaming cryptoconversions.
6. The most common mathematical apparatus for describing the operation of SSC or flow modes BSC is the use of methods of theory of automata. In the general case, streaming cryptoconversion is represented by equations that use a formalized representation of the next state function and the exit function. The output function is usually implemented by adding (for example, modulo 2) the result of the next state function (encryption range) to the information sequence. To ensure stable encryption, the keystream must be random (pseudo-random), because in fact it is an extended key in PRS, and it is on its randomness, irreversibility and unpredictability that the stability of streaming cryptocurrency is based.
7. Known methods of streaming cryptoconversion are divided into two general groups: ciphers that are self-synchronized (asynchronous ciphers) and have the ability to continue the correct decryption in the case when the cipher sequences fall out of synchronization with the range of the transmitting encoder; synchronous streaming cryptoconversins, for which the encryption keystream is formed independently of the plaintext stream and the ciphertext stream. The most common are synchronous cryptoconversions, which have

the ability to use a PRS generator to pre-generate an encryption keystream, and cryptoconversion of plaintext to perform a simple output function with the possibility of parallelization, which will significantly increase the speed of cryptoconversion. Thus, the development of an effective synchronous streaming cryptoconversion is reduced to the construction of a generator with the specified cryptographic properties of the source sequence (key stream).

8. The analysis showed that the known methods of construction of synchronous SSC are divided into two large groups: with uniform movement of shift registers and PRS generators with non-uniform movement (shift) of registers. In the first case, all LFSR used in the PRS generator are shifted by the same clock signals, i.e. the movement of the registers is uniform (same). This greatly simplifies the implementation of generators and usually increases their speed. In comparison with this approach, in generators with non-uniform motion, different LFSR can be shifted by different clock signals. This is realized either through the use of the so-called control register (one of the used in the circuit of the generator LFSR performs the function of controlling the clock frequency of other LFSR), or by constructing generators in which the clock frequency of individual LFSR is determined by the logic of feedback or certain functional transformations between the outputs or contents of LFSR.
   In the circuit of the generator LFSR performs the function of controlling the clock frequency of other LFSR), or by constructing generators in which the clock frequency of individual LFSR is determined by the logic of feedback or certain functional transformations between the outputs or contents of LFSR.
9. Synchronous streaming cryptoconversions with uniform movement of shift registers have become the most widespread in modern information and telecommunication systems. This direction is attractive for its simplicity and convenience in the synthesis and analysis of various cryptographic properties of the generator (periodic and statistical properties, speed, linear complexity, etc.). It is synchronous streaming cryptoconversions with uniform movement of shift registers that have gained the greatest development and standardization at the national and international levels in recent years.
10. The most trusted and widespread are the world-famous streaming ciphers, which are standardized at the international or national level:
   ISO/IEC 180334 «Information technology—Security techniques—Encryption algorithms—Part 4: Stream ciphers»; ISO/IEC 10116 «Information technology—Security techniques—Modes of operation for an nbit block cipher»; NIST Special Publication 80038 «Block Cipher Modes»; DSTU 7624: 2014 "Information technology. Cryptographic information protection. Symmetric block transformation algorithm"; NIST Special Publication 80090A «Recommendation for Random Number Generation Using Deterministic Random Bit Generators»; DSTU GOST 28147: 2009 "Information processing systems. Cryptographic protection. Cryptographic transformation algorithm"; GOST R ISO/IEC 1011693 "Information technology. Operating modes for n-bit block encryption algorithm"; eSTREAM—an international project to detect

new stream ciphers; ISO/IEC 291923 «Information technology—Security techniques—Lightweight cryptography».

11. Streaming should include the following BSS modes: counter (CTR), cipher feedback (Cipher Feedback—CFB), cipher block (Cipher Block Chaining—CBC), casing with cipher feedback. (Output Feedback—OFB). The specifications of these modes are known, they are given in many national and international standards, in particular in NIST Special Publication 80038 "Block Cipher Modes" and DSTU GOST 28147: 2009 "Information processing systems. Cryptographic protection. Cryptographic transformation algorithm". However, the national standards of the USA and Ukraine define the largest list of modes of application of BSC, i.e. these normative documents provide the most complete field of use of cryptographic protection of information by means of BSC. The analysis of BSC streaming modes showed that their application allows to provide a high level of cryptographic stability and speed to ensure the confidentiality of information in certain applications.
12. The obtained results of experimental researches of statistical safety of SSC "Kalina" testify to high characteristics of the investigated cryptographic primitive. The formed pseudo-random sequences do not differ in the statistical sense from the random sequence, i.e. the hypothesis that the studied cryptoalgorithm is a generator, the original sequences of which are the implementation of a random process, is not refuted. The results of studies of the e-codebook mode in each round separately showed that the studied BSC (its basic transformation) has increased statistical security. Already after the first round of encryption, the original sequences have the appearance of pseudo-random, with a high probability we can say that the generated sequence is statistically secure. That is, BSC "Kalina" well approximates the generators of "random" sequences, the hypothesis that it is a generator, the original sequences of which are the implementation of a random process, in most cases is not refuted. The results of statistical studies of different modes of streaming encryption (CTR, CFB, CBC, OFB) on the full version of the algorithm show that BSC "Kalina" in any set mode generates pseudo-random sequences, i.e. with high probability and accuracy we can say that the studied algorithm cryptocurrency conversion will be statistically secure in any practical application.
13. The obtained results of experimental studies of statistical properties of pseudo-random sequences, which are formed by block symmetric ciphers AES (FIPS197) and GOST 2814789 (DSTU GOST 28147: 2009), indicate high characteristics of the studied cryptographic primitives. Virtually every cryptoalgorithm can be considered as a generator of pseudo-random sequences, the work of which is well approximated by a random process. Regarding the specific values of statistical security, it should be noted a slight advantage of BSC "Kalina", because the minimum value of the number of passed statistical tests in BSC "Kalina" is the highest among the studied cryptoalgorithms.
14. Based on the results of comparative studies, the recommendations on the modes of application of BSC "Kalina" in various cryptographic applications are substantiated. It is recommended that you limit the total length of messages that

are protected using a single key depending on the size of the base conversion block. The obtained results of statistical studies show that each of the studied modes forms sequences that are similar to random ones, which indicates the high statistical security of BSC in different cryptographic applications.

15. The analysis and comparative studies of standardized algorithms for streaming cryptographic transformation, the specification of which is given in ISO/IEC 180334, showed that the considered stream ciphers (algorithms MUGI, SNOW 2.0, Rabbit, DEcim, KCipher2) are built on the scheme of key stream generators. are built using both the technique of linear shift registers and cryptographic elements of block ciphers. In particular, the generators use nonlinear complication functions, which are built using nonlinear replacement units BSC AES (FIPS197). This, on the one hand, greatly simplifies the analysis of the strength of the cipher, using a proven mathematical apparatus and known statements about certain cryptographic properties; on the other hand, it allows the use of existing crypto-primitives and, possibly, their software implementations, which significantly accelerates the implementation of algorithms on different computer systems.
16. The US standard NIST Special Publication 80090A defines functional models of deterministic random bit generators, conceptual mechanisms and basic requirements for such generators, mechanisms of individual functions and specifications of certain algorithms, including generators based on the use of hashing functions, BSC and numerical problems. Studies have shown that the greatest speed in the formation of PRS is provided by generators built using hashing functions and BSC. This is due to the simplicity of mathematical operations and the ability to optimize them for implementation on modern computer systems. However, from the point of view of statistical safety the best indicators are shown by PRS generators which are built on application of theoretical and numerical problems. In particular, the generator Dual Elliptic Curve Deterministic RBG—Dual_EC_DRBG is built using the mathematical apparatus of cryptography on elliptic curves and the problem of cryptanalysis of such a generator is to solve the theoretical problem of discrete logarithm in a group of points of an elliptic curve. Studies have shown that the statistical portrait of the generator on elliptic curves is the best, all statistical tests with a high probability show that the formed PRS is not statistically different from the implementation of a random process. In addition, such generators comply with the model of evidence-based security and trust in their performance is higher. It should be noted that the Dual Elliptic Curve Deterministic RBG generator was contained in a previous version of the US national standard NIST Special Publication 80090A [1], which provides recommendations for the construction of PRS generators using various mathematical methods, covering the method using transformations in a group of elliptical points. curve. And although in the updated version of the standard [2] this generator was excluded due to certain shortcomings, its properties should also be taken into account and studied in order to study different areas of construction of reliable and safe pseudo-random sequence generators.

17. The analysis of streaming cryptocurrencies, selected as the winners of the international project eSTREAM, showed that according to their specification, the corresponding algorithms are divided into two functional groups. The first group of algorithms—software-oriented stream ciphers, which include: HC128; Rabbit; Salsa20/12; SOSEMANUK. The second group of hardware-oriented ciphers includes Grain algorithms; MICKEY and Trivium. Analysis of the structure and main applied functions of software-oriented stream cipher conversion (HC128, Rabbit, Salsa 20/12, SOSEMANUK) showed that the corresponding algorithms use mainly simple and computationally efficient bitwise addition operations modulo 2, AND, cyclic shifts, other logical functions, with the ability to apply all the benefits of parallelization of the processes of formation of the key flow. The state of the generator is presented, as a rule, either in the form of words or in a tabular way, which greatly simplifies the implementation of calculations in software implementation. An exception is the SOSEMANUK algorithm, which uses a shift register of length 10 to represent the state of the generator, but the small length of this register does not significantly reduce the performance even in software implementation. It should also be noted that almost all algorithms use elements of BSS constructions, in particular, the SOSEMANUK algorithm uses nonlinear replacement nodes constructed using the AES BSC block (FIPS197). Analysis of hardware-oriented ciphers, which include Grain, MICKEY and Trivium algorithms, showed that their structure usually uses long shift registers. This allows the hardware implementation for one time cycle of the device to change the state of the key stream generator, which allows to achieve extremely high rates of PRS formation rate. To ensure cryptographic stability, a nonlinear function is also added to the generator structure, which is implemented either by nonlinear feedback in the shift registers or, for example, through the use of a cryptographic Boolean function with high nonlinearities. The complexity of cryptanalysis also increases due to the use of uneven movement of registers, which complicates the original PRS.
18. International standard ISO/IEC 291923 "Information technology—Security techniques—Lightweight cryptography" is devoted to the presentation of streaming algorithms for lightweight cryptography, the implementation of which is focused on cheap and unpretentious to the speed and amount of memory devices. Among the specified algorithms, the Trivium stream cipher is also considered as the winner of the international eSTREAM project, which studied its properties in detail and noted all the advantages and disadvantages of the algorithm. The second Enocoro streaming cipher is very similar to the MUGI cipher specified in the international standard ISO/IEC 180334, and is in fact a simplified version for simple systems with low computing resources. The application in Enocoro of a nonlinear complication function, built taking into account the BSC AES block (FIPS-197), is also aimed at providing the required stability indicators with high requirements for the rate of formation of the original PRS.

19. The statistical research methodology considered in this monograph and the results obtained using it can be considered as a primary analysis of the cryptographic properties of generators, as such statistical tests do not take into account the generator's own structure. The results of testing various streaming cryptocurrencies confirm their high cryptographic performance. All studied ciphers showed a large number of successfully passed tests: 130–134 according to the criterion $P_j \geq 0.99$ and 186–187 according to the criterion $P_j \geq 0.96$. It should be noted the high statistical security of the algorithm SSC "Strumok", which revealed certain properties of the random bit generator. In particular, the properties of the formed sequences are not inferior to PRS, which are formed by world-famous streaming cryptographic algorithms, in particular ciphers HC-256 and SNOW 2.0. In addition, for SSC "Strumok" the minimum values of the number of passed statistical tests by the criterion $P_j \geq 0.96$ are higher than in these algorithms, which indicates a slight advantage of the statistical security of the algorithm "Strumok".
20. The results of statistical testing by the DIEHARD package for symmetric ciphers AES, Enocoro, Grain, HC, MICKEY, MUGI, Salsa20, Sosemanuk, Trivium and SSC "Strumok" show that for most indicators all streaming cryptographic algorithms have comparable statistical security indicators. DIEHARD probability distribution tests were additionally performed for SSC "Strumok", the results of which show that both for the version of "Strumok" with a key length of 512 bits and for the version with a length of 1024 bits, the probability distribution corresponds to a uniform law with small fluctuations (in these studies used a previous version of the algorithm "Strumok", which was later improved). This further supports the conclusion about the high statistical security of the algorithm "Strumok".
21. The monograph highlights the problem of development and theoretical substantiation of the method of synthesis of shift registers with nonlinear feedback of a given length and established design characteristics for their use in symmetric streaming encryption schemes. The analysis of methods of stream symmetric encryption, research of directions of their development due to synthesis of registers of shift of the set length with nonlinear feedbacks is carried out.

    It is shown that one of the priorities of international and national projects is the development and research of modern flow symmetric ciphers, one of the promising approaches is the use of design based on shift registers with nonlinear feedback.

    A comparative analysis of methods for the synthesis of shift registers with nonlinear feedback, resulting in a lack of well-developed mathematical apparatus for analysis and synthesis of streaming encryption algorithms based on shift registers with nonlinear feedback, which significantly inhibits their potential development. The synthesis of shift registers with nonlinear feedback with a guaranteed period length remains an open problem. A systematic algorithm for the synthesis of shift registers with nonlinear feedback has not yet been discovered.

22. Unconditional and conditional stability indicators are defined, as well as indicators considered in the framework of filtering or combining functions for shift registers with nonlinear feedback, which are considered as one of the basic elements of streaming encryption schemes.
    Nine requirements for feedback coefficients are formulated and substantiated, which should be met by shift registers with nonlinear feedback, which generate a sequence of the maximum period. For Requirements 1 to 6, a strict estimate of the upper limit of the number of shift registers with nonlinear feedback that does not meet Requirements 1–6 is given. A similar expression is obtained for Requirement 8.
    It is shown that already at length L = 6 more than 99% of shift registers with nonlinear feedback of the second order do not meet Requirements from 1 to 4. With representation to a polynomial of Requirements from 1 to 6, already at the size of the register at L = 9 more than 99.9% of the whole set of polynomials is truncated, and as the value of L increases, this percentage will increase in degree.
    The application of Requirements 1 to 9 allowed the calculation of the exact number of shift registers with nonlinear feedback of the second order, forming a sequence of the maximum period for values of L from 1 to 9. For L = 9 the number of such registers is 519239794, which corresponds to 0.0076% of the whole set of shift registers with nonlinear feedback, and 1,014,140 more than the whole set of shift registers with linear feedback or more than 10,817,496 than the shift registers with linear feedback, forming a sequence of the maximum period.
23. The obtained results in general allowed to improve the method of synthesis of nonlinear shift registers with the maximum period of sequences to reduce the computational complexity of the search of registers with given design characteristics. The developed hardware and software tool for the synthesis of shift registers with nonlinear feedback, which form a sequence of the maximum period and meet the specified design characteristics. The tool combines the computing power of the CPU, GPU and FPGA. Due to the multithreading of the proposed method and studies of the application of GPU and FPGA, appropriate proposals for the use of the method on the hardware of the synthesis of shift registers with nonlinear feedback of the second order, forming a sequence of the maximum period.
    As a realization of the possibilities of the method the result of search of shift registers with nonlinear feedback of the second order which form a sequence of the maximum period and correspond to the set design characteristics, dimension L = 25, 26, 27, 28, 29, 30, 31 and 32 is given. and empirical evaluation of the stability indicators of the generated shift registers with nonlinear feedback, their experimental verification and implementation during the development of stream symmetric encryption schemes.
24. The main characteristics and properties of shift registers with nonlinear feedback of the second order are studied, in particular: linear complexity and profile of linear complexity; the period of the sequence formed as a result of adding

a sequence from different shift registers with nonlinear feedback, forming the sequence of the maximum period, as well as their linear and quadratic complexity; the decimation and the principle of group addition of Msequence of shift registers with nonlinear feedbacks are investigated; found a complete set of shift registers with nonlinear feedback of the second order, forming a sequence of the maximum period with dimension up to L = 9 for GF(2) and up to L = 5 for GF(2^2); an analytical estimate of the number of shift registers with nonlinear feedback forming the sequence of the maximum period depending on L is given; the statistical security indicator is analyzed; calculated indicators of nonlinearity and correlation immunity; the time costs for the generation of shift registers with nonlinear feedback M-sequences of software and hardware are studied.

25. The results of substantiation of the structure of the modern symmetric stream cipher, key system and basic cryptographic elements are given. In particular, the mathematical structure of the main cryptographic transformations of the new stream cipher “Strumok” is substantiated.
The SSC “Strumok” generator is built according to the SNOW 2.0-like scheme of the summing generator (the SNOW 2.0 generator is defined in DSTU ISO/IEC 180334: 2015), it uses a 256-bit initialization vector and a 256-bit or 512-bit secret key. The generator provides high and ultra-high levels of stability and is designed to protect important information with limited access owned by the state, in particular to ensure the confidentiality of information during its processing, taking into account current trends in cryptographic science and possible application of quantum cryptographic analysis.
26. In the algorithm of symmetric streaming cryptographic transformation “Strumok”, depending on the length of the secret key, two modes are used:

- mode with a 256-bit key, denoted as “Strumok-256”;
- mode with a 512-bit key, denoted as “Strumok-512”.
LFSR and CA, built on the principles of the world-famous SNOW 2.0 algorithm, were chosen as the two main elements of the “Strumok” generator. The main differences of the proposed generator are as follows:
- feedback taps in LFSR are built on the primitive over the field GF(2^{64}) polynomial $f(x)$ of degree 16. The field GF(2^{64}) is constructed as an extension of the field GF(2^8), which, in turn, is built on the primitive over the field GF(2) by the polynomial p(y). Thus, we have a tower of fields: $GF(2) \subset GF(2^8) \subset GF(2^{64}) \subset GF(2^{1024})$. Therefore, the period of the original sequence of LFSR is the maximum and is equal to 2^{1024}—1 (in the algorithm SNOW 2.0 the period is equal to 2^{512}—1);
- for the compatibility of implementations of cryptographic algorithms as a transformation T (the main component of CA, which determines its cryptographic properties), the use of nonlinear transformation BSC “Kalina” (defined by the national standard BSS of Ukraine DSTU 7624: 2014) is proposed.

27. The basis of optimization of cryptocurrencies of the cipher “Strumok” is the idea of using pre-calculated data. Precalculation tables do not require multiplication calculations during encryption and nonlinear substitution, which will significantly reduce the number of operations to process the input data block. Given the proposed optimization of calculations, the developed algorithm “Strumok” is able to provide high performance and operate efficiently on different computing platforms. It does not require large amounts of memory, to ensure the normal operation of the cipher enough cache memory level 1 of modern processors. This provides a high speed of calculations, and, consequently, a high speed of key stream generation.
28. Studies of the speed of the developed algorithm and the minimum amount of memory required for the implementation of the algorithm have shown that SSC “Strumok” is able to provide high computational efficiency. In particular, in comparison with the best world analogues (streaming cryptoalgorithms MICKEY, Decim, Trivium, SOSEMANUK, SNOW 2.0, Salsa20, Rabbit, HC and streaming modes BSC AES) developed SSC “Strumok” provides the highest rates of PRS formation Gbit/s. Such high estimates of speed are primarily the result of a simple and computationally efficient structure of the algorithm.

    Among the considered algorithms the shortest time of initialization of cryptocurrencies has SSC Salsa20 and SNOW 2.0 to which the developed algorithm “Strumok” is a little inferior. This advantage is a consequence of a simple and computationally efficient encryption key deployment scheme. But the simplification of the key deployment scheme and the lack of pre-calculation tables in some cryptographic algorithms may be a disadvantage in terms of security and/or speed of key stream formation.
29. The obtained results of the comparative analysis show that the relative time complexity of implementation (as a specific number of operations to the number of generated bits) for SSC “Strumok” is comparable or less than similar indicators for SSC SNOW 2.0. The relative capacitive complexity of the implementation of SSC “Strumok” is one and a half times higher than SSC SNOW 2.0. However, the latter indicator is insignificant, as the main condition for the rapid implementation of cryptocurrency algorithms is the requirement to “fit” into the limit on the amount of cache memory of the first level. The general requirements for modern cryptographic algorithms regarding the amount of memory are limited to 64 Kbytes, i.e. the total amount of pre-calculation tables for fast implementation of ciphers should not exceed the amount of memory of the first level cache. Since the results of the research showed that SSC “Strumok” requires 20,624 bytes for its implementation (this is less than the limit of the amount of memory of the 1st level cache), the increase in memory compared to SSC Snow 2.0 will not lead to a decrease the speed of operation of the generator and, in particular, to reduce the speed of formation of the key stream (encryption range).
30. By a set of individual factors: the rate of formation of PRS; initial initialization rate; the minimum amount of memory required to implement SSC; provided

cryptographic properties of the encryption range (key stream) the most reasonable technical solution in terms of stability/complexity is developed by SSC "Strumok". It provides high performance and is able to operate efficiently on various computing platforms.

References

1. NIST Special Publication 80090A: Recommendation for Random Number Generation Using Deterministic Random Bit Generators. [Electronic resourse]. Access mode http://csrc.nist.gov/publications/nistpubs/800-90A/SP800-90A.pdf.
2. Barker, E., Kelsey, J.: Recommendation for random number generation using deterministic random bit generators. National Institute of Standards and Technology (2015, June), 101 p. [Online]. Internet https://doi.org/10.6028/NIST.SP.80090Ar1

Appendix A

LFSR feedback polynomials with different ways of presenting them (for $g(z) = z^8 + \beta^{170}z^7 + \beta^{166}z^6 + \beta^2z^5 + \beta^{224}z^4 + \beta^{70}z^3 + \beta^2$) the cipher "Strumok".

A. A. Kuznetsov et al., *Stream Ciphers in Modern Real-time IT Systems*, Studies in Systems, Decision and Control 375,
https://doi.org/10.1007/978-3-030-79770-6

$$
\begin{aligned}
a(x) = {} & x^{1024} + x^{992} + x^{989} + x^{984} + x^{980} + x^{979} + x^{974} + x^{970} + x^{967} + x^{962} + x^{960} + x^{959} \\
& + x^{957} + x^{953} + x^{952} + x^{951} + x^{950} + x^{947} + x^{946} + x^{945} + x^{944} + x^{943} + x^{942} + x^{939} + x^{937} \\
& + x^{935} + x^{932} + x^{928} + x^{925} + x^{924} + x^{922} + x^{916} + x^{915} + x^{914} + x^{913} + x^{912} + x^{911} + x^{907} \\
& + x^{905} + x^{902} + x^{899} + x^{898} + x^{897} + x^{895} + x^{893} + x^{892} + x^{891} + x^{889} + x^{886} + x^{880} + x^{877} \\
& + x^{876} + x^{875} + x^{871} + x^{867} + x^{864} + x^{863} + x^{862} + x^{860} + x^{859} + x^{855} + x^{852} + x^{849} + x^{848} \\
& + x^{844} + x^{842} + x^{841} + x^{839} + x^{838} + x^{837} + x^{835} + x^{833} + x^{832} + x^{831} + x^{830} + x^{829} + x^{827} \\
& + x^{826} + x^{825} + x^{824} + x^{822} + x^{814} + x^{813} + x^{812} + x^{811} + x^{806} + x^{805} + x^{804} + x^{803} + x^{802} \\
& + x^{799} + x^{798} + x^{797} + x^{794} + x^{792} + x^{790} + x^{786} + x^{785} + x^{783} + x^{782} + x^{779} + x^{776} + x^{775} \\
& + x^{774} + x^{773} + x^{771} + x^{770} + x^{769} + x^{766} + x^{765} + x^{764} + x^{763} + x^{762} + x^{759} + x^{756} + x^{755} \\
& + x^{751} + x^{750} + x^{749} + x^{748} + x^{746} + x^{745} + x^{742} + x^{740} + x^{737} + x^{736} + x^{734} + x^{731} + x^{727} \\
& + x^{726} + x^{723} + x^{722} + x^{721} + x^{720} + x^{718} + x^{717} + x^{716} + x^{715} + x^{708} + x^{707} + x^{702} + x^{701} \\
& + x^{700} + x^{697} + x^{696} + x^{695} + x^{693} + x^{690} + x^{688} + x^{686} + x^{684} + x^{681} + x^{680} + x^{678} + x^{675} \\
& + x^{673} + x^{672} + x^{670} + x^{669} + x^{668} + x^{666} + x^{664} + x^{661} + x^{659} + x^{658} + x^{654} + x^{652} + x^{651} \\
& + x^{648} + x^{646} + x^{644} + x^{643} + x^{642} + x^{640} + x^{637} + x^{633} + x^{630} + x^{627} + x^{624} + x^{622} + x^{621} \\
& + x^{619} + x^{618} + x^{617} + x^{615} + x^{614} + x^{613} + x^{612} + x^{608} + x^{603} + x^{598} + x^{596} + x^{594} + x^{593} \\
& + x^{592} + x^{590} + x^{586} + x^{585} + x^{584} + x^{580} + x^{578} + x^{576} + x^{573} + x^{570} + x^{569} + x^{564} + x^{561} \\
& + x^{559} + x^{558} + x^{556} + x^{555} + x^{554} + x^{552} + x^{551} + x^{550} + x^{548} + x^{546} + x^{545} + x^{544} + x^{543} \\
& + x^{541} + x^{540} + x^{538} + x^{537} + x^{536} + x^{535} + x^{533} + x^{530} + x^{529} + x^{528} + x^{526} + x^{521} + x^{519} \\
& + x^{518} + x^{516} + x^{515} + x^{514} + x^{513} + x^{509} + x^{507} + x^{505} + x^{502} + x^{501} + x^{498} + x^{493} + x^{492} \\
& + x^{490} + x^{488} + x^{486} + x^{484} + x^{483} + x^{480} + x^{476} + x^{475} + x^{473} + x^{471} + x^{469} + x^{465} + x^{464} \\
& + x^{463} + x^{461} + x^{454} + x^{453} + x^{452} + x^{450} + x^{449} + x^{448} + x^{440} + x^{439} + x^{438} + x^{437} + x^{432} \\
& + x^{431} + x^{430} + x^{426} + x^{424} + x^{423} + x^{421} + x^{420} + x^{418} + x^{417} + x^{415} + x^{410} + x^{405} + x^{400} \\
& + x^{399} + x^{397} + x^{396} + x^{394} + x^{393} + x^{392} + x^{391} + x^{390} + x^{388} + x^{387} + x^{385} + x^{384} + x^{383} \\
& + x^{382} + x^{380} + x^{379} + x^{377} + x^{374} + x^{371} + x^{369} + x^{368} + x^{367} + x^{364} + x^{362} + x^{359} + x^{356} \\
& + x^{355} + x^{352} + x^{349} + x^{347} + x^{339} + x^{337} + x^{335} + x^{334} + x^{333} + x^{332} + x^{328} + x^{324} + x^{322} \\
& + x^{321} + x^{319} + x^{317} + x^{314} + x^{313} + x^{312} + x^{308} + x^{307} + x^{306} + x^{305} + x^{304} + x^{303} + x^{302} \\
& + x^{301} + x^{300} + x^{299} + x^{298} + x^{297} + x^{296} + x^{294} + x^{292} + x^{291} + x^{289} + x^{286} + x^{285} + x^{284} \\
& + x^{283} + x^{281} + x^{280} + x^{278} + x^{275} + x^{273} + x^{270} + x^{269} + x^{268} + x^{265} + x^{263} + x^{259} + x^{256} \\
& + x^{255} + x^{254} + x^{251} + x^{250} + x^{249} + x^{245} + x^{242} + x^{240} + x^{239} + x^{238} + x^{237} + x^{234} + x^{233} \\
& + x^{232} + x^{231} + x^{230} + x^{229} + x^{228} + x^{226} + x^{225} + x^{221} + x^{220} + x^{217} + x^{216} + x^{213} + x^{212} \\
& + x^{211} + x^{205} + x^{203} + x^{201} + x^{198} + x^{197} + x^{196} + x^{193} + x^{192} + x^{191} + x^{190} + x^{188} + x^{186} \\
& + x^{183} + x^{182} + x^{179} + x^{178} + x^{177} + x^{175} + x^{174} + x^{173} + x^{172} + x^{171} + x^{170} + x^{169} + x^{167} \\
& + x^{165} + x^{164} + x^{158} + x^{156} + x^{155} + x^{154} + x^{153} + x^{152} + x^{149} + x^{148} + x^{145} + x^{144} + x^{141} \\
& + x^{140} + x^{136} + x^{133} + x^{130} + x^{128} + x^{127} + x^{126} + x^{115} + x^{114} + x^{111} + x^{108} + x^{106} + x^{105} \\
& + x^{101} + x^{100} + x^{98} + x^{97} + x^{94} + x^{93} + x^{92} + x^{87} + x^{79} + x^{77} + x^{76} + x^{60} + x^{57} + x^{49} + {} \\
& x^{46} + 1
\end{aligned}
$$

$b(x) = x^{512} + \gamma x^{480} + \gamma^2 x^{477} + x^{474} + \gamma x^{472} + \gamma x^{468} + \gamma x^{467} + \gamma x^{462} + \gamma x^{458} + x^{456} +$
$\gamma x^{455} + \gamma x^{450} + \gamma^2 x^{447} + x^{445} + x^{444} + \gamma^2 x^{441} + \gamma^2 x^{440} + x^{439} + \gamma^2 x^{438} + \gamma x^{437} +$
$\gamma x^{436} + \gamma^2 x^{435} + x^{434} + x^{433} + \gamma^2 x^{432} + \gamma x^{431} + x^{430} + \gamma x^{429} + x^{428} + x^{427} + x^{426} +$
$\gamma x^{425} + \gamma^2 x^{424} + x^{423} + \gamma x^{422} + \gamma x^{421} + \gamma x^{420} + x^{419} + \gamma x^{417} + x^{413} + \gamma^2 x^{412} + \gamma x^{411}$
$+ \gamma x^{409} + \gamma^2 x^{408} + \gamma^2 x^{405} + x^{402} + \gamma x^{401} + \gamma x^{400} + \gamma x^{398} + \gamma x^{397} + x^{396} + \gamma x^{395} +$
$\gamma x^{392} + \gamma^2 x^{391} + x^{390} + \gamma x^{389} + \gamma^2 x^{388} + x^{387} + \gamma x^{385} + \gamma x^{382} + \gamma^2 x^{380} + \gamma x^{379} +$
$\gamma x^{378} + x^{376} + \gamma x^{375} + x^{374} + x^{373} + \gamma x^{372} + \gamma^2 x^{371} + x^{370} + x^{369} + \gamma^2 x^{368} + \gamma x^{366} +$
$\gamma x^{365} + \gamma^2 x^{364} + x^{363} + \gamma^2 x^{362} + \gamma x^{361} + x^{360} + x^{359} + \gamma^2 x^{357} + \gamma x^{356} + x^{355} +$
$\gamma^2 x^{354} + \gamma^2 x^{352} + x^{351} + \gamma x^{350} + \gamma x^{348} + \gamma^2 x^{347} + \gamma x^{346} + \gamma^2 x^{345} + x^{344} + \gamma^2 x^{343} +$
$\gamma x^{340} + x^{339} + \gamma^2 x^{338} + x^{337} + x^{336} + x^{334} + \gamma x^{333} + x^{331} + \gamma^2 x^{330} + \gamma x^{326} + \gamma x^{325} +$
$\gamma x^{324} + \gamma^2 x^{323} + \gamma x^{322} + \gamma x^{321} + \gamma^2 x^{320} + \gamma^2 x^{318} + \gamma^2 x^{317} + x^{316} + \gamma x^{315} + x^{314} +$
$x^{313} + \gamma^2 x^{312} + \gamma x^{309} + \gamma^2 x^{308} + x^{307} + \gamma^2 x^{306} + x^{305} + \gamma^2 x^{304} + \gamma^2 x^{303} + x^{300} +$
$\gamma x^{299} + \gamma^2 x^{297} + \gamma x^{296} + \gamma^2 x^{295} + \gamma x^{293} + \gamma x^{292} + x^{291} + \gamma x^{289} + \gamma^2 x^{287} + \gamma x^{285} +$
$x^{284} + \gamma^2 x^{283} + \gamma^2 x^{282} + x^{281} + \gamma x^{279} + x^{278} + \gamma x^{277} + \gamma x^{276} + \gamma^2 x^{275} + x^{274} + \gamma x^{273}$
$+ \gamma^2 x^{272} + \gamma x^{271} + \gamma x^{270} + \gamma x^{269} + \gamma^2 x^{266} + \gamma x^{265} + x^{263} + \gamma^2 x^{262} + \gamma x^{261} + \gamma^2 x^{259}$
$+ \gamma x^{258} + \gamma x^{257} + x^{256} + \gamma x^{255} + \gamma x^{252} + x^{251} + \gamma x^{250} + \gamma x^{249} + x^{248} + \gamma^2 x^{247} +$
$\gamma x^{246} + x^{244} + \gamma^2 x^{242} + x^{241} + \gamma x^{240} + \gamma^2 x^{239} + x^{238} + \gamma^2 x^{237} + x^{236} + \gamma x^{235} +$
$\gamma^2 x^{234} + x^{233} + x^{231} + x^{230} + x^{229} + \gamma^2 x^{228} + \gamma^2 x^{227} + x^{224} + \gamma^2 x^{223} + \gamma x^{222} + \gamma^2 x^{219}$
$+ x^{218} + x^{216} + x^{212} + \gamma^2 x^{210} + x^{209} + \gamma x^{207} + \gamma^2 x^{206} + x^{205} + x^{204} + \gamma x^{203} + \gamma x^{200} +$
$\gamma x^{199} + x^{197} + \gamma^2 x^{195} + \gamma^2 x^{194} + \gamma^2 x^{192} + x^{191} + \gamma x^{190} + x^{189} + x^{188} + x^{187} + x^{186} +$
$\gamma x^{184} + \gamma x^{183} + \gamma^2 x^{182} + x^{181} + \gamma^2 x^{180} + \gamma^2 x^{179} + x^{178} + \gamma x^{177} + \gamma^2 x^{176} + \gamma x^{174} +$
$\gamma^2 x^{173} + \gamma^2 x^{172} + \gamma x^{170} + \gamma x^{169} + x^{168} + x^{166} + x^{165} + \gamma x^{164} + \gamma^2 x^{163} + \gamma^2 x^{160} +$
$\gamma^2 x^{159} + x^{158} + \gamma^2 x^{156} + \gamma^2 x^{155} + \gamma x^{154} + \gamma x^{153} + \gamma^2 x^{150} + \gamma^2 x^{149} + \gamma x^{144} + \gamma x^{142}$
$+ \gamma x^{141} + x^{140} + x^{138} + \gamma x^{137} + \gamma^2 x^{136} + \gamma x^{134} + x^{133} + x^{130} + \gamma^2 x^{128} + x^{127} + \gamma^2 x^{123}$
$+ x^{120} + x^{115} + \gamma x^{114} + \gamma^2 x^{111} + \gamma^2 x^{109} + \gamma^2 x^{108} + x^{106} + \gamma^2 x^{105} + \gamma x^{104} + \gamma^2 x^{103} +$
$x^{101} + x^{100} + \gamma x^{99} + \gamma x^{98} + x^{97} + x^{94} + x^{93} + \gamma x^{92} + \gamma x^{90} + \gamma^2 x^{87} + x^{79} + \gamma^2 x^{77} +$
$x^{76} + \gamma^2 x^{60} + \gamma^2 x^{57} + x^{49} + x^{46} + \gamma$

$c(x) = x^{256} + x^{236} + \eta^{13}x^{224} + \eta^{3}x^{221} + \eta^{2}x^{218} + \eta^{11}x^{216} + \eta^{13}x^{212} + \eta^{13}x^{211} + \eta^{10}x^{208}$
$+ \eta^{9}x^{206} + \eta^{5}x^{205} + \eta^{6}x^{204} + \eta^{13}x^{202} + \eta^{3}x^{201} + \eta^{4}x^{200} + \eta^{13}x^{199} + \eta^{2}x^{198} + \eta^{3}x^{196} +$
$\eta^{9}x^{194} + \eta^{5}x^{193} + x^{192} + \eta^{2}x^{191} + \eta x^{189} + \eta^{8}x^{188} + \eta^{5}x^{187} + \eta^{2}x^{186} + \eta^{12}x^{185} + \eta^{9}x^{184}$
$+ \eta^{9}x^{183} + \eta^{2}x^{182} + \eta^{10}x^{180} + \eta^{4}x^{179} + \eta x^{178} + \eta x^{177} + \eta^{13}x^{176} + \eta^{13}x^{175} + \eta^{2}x^{173} +$
$\eta x^{171} + \eta^{12}x^{170} + x^{169} + \eta^{10}x^{168} + \eta^{3}x^{167} + \eta^{8}x^{166} + \eta x^{165} + \eta^{14}x^{164} + \eta^{6}x^{163} + \eta x^{162}$
$+ \eta^{10}x^{161} + \eta^{13}x^{160} + \eta^{10}x^{159} + \eta^{14}x^{158} + \eta^{9}x^{157} + \eta^{8}x^{156} + \eta^{4}x^{155} + \eta^{3}x^{154} + \eta^{5}x^{153}$
$+ \eta^{12}x^{152} + \eta^{5}x^{151} + \eta^{12}x^{150} + \eta^{2}x^{149} + \eta^{7}x^{147} + \eta^{4}x^{146} + \eta^{11}x^{145} + \eta^{14}x^{144} + \eta^{8}x^{143}$
$+ \eta^{9}x^{142} + \eta^{11}x^{141} + \eta^{3}x^{140} + \eta^{7}x^{139} + \eta^{6}x^{138} + \eta^{8}x^{137} + \eta^{6}x^{136} + \eta x^{134} + \eta^{5}x^{132} +$
$\eta^{7}x^{131} + \eta^{6}x^{130} + \eta^{7}x^{129} + \eta^{12}x^{126} + \eta^{14}x^{125} + \eta^{14}x^{124} + \eta^{8}x^{123} + \eta^{5}x^{122} + \eta x^{121} +$
$\eta^{7}x^{120} + \eta^{9}x^{119} + \eta^{6}x^{118} + \eta^{9}x^{116} + \eta^{4}x^{115} + x^{114} + \eta^{4}x^{113} + \eta^{8}x^{112} + \eta^{14}x^{111} +$
$\eta^{13}x^{110} + \eta^{11}x^{109} + \eta^{14}x^{108} + \eta^{12}x^{106} + \eta^{14}x^{105} + \eta^{14}x^{104} + \eta^{8}x^{103} + \eta^{8}x^{101} + \eta^{13}x^{100}$
$+ \eta^{5}x^{99} + \eta^{4}x^{98} + x^{97} + x^{94} + \eta x^{93} + \eta^{7}x^{90} + \eta^{5}x^{88} + \eta^{14}x^{87} + \eta x^{82} + x^{79} + \eta^{7}x^{77} +$
$\eta^{12}x^{76} + \eta^{6}x^{70} + \eta x^{64} + \eta^{14}x^{60} + \eta^{14}x^{57} + \eta x^{52} + \eta^{4}x^{49} + \eta^{4}x^{46} + \eta x^{44} + \eta$
$d(x) = x^{128} + \mu^{68}x^{113} + \mu^{170}x^{112} + \mu^{68}x^{110} + \mu^{170}x^{109} + \mu^{222}x^{108} + \mu^{170}x^{106} + x^{104} +$
$\mu^{85}x^{103} + \mu^{68}x^{101} + \mu^{85}x^{100} + \mu^{64}x^{98} + \mu^{204}x^{97} + \mu^{42}x^{96} + \mu^{85}x^{94} + \mu^{168}x^{93} +$
$\mu^{45}x^{92} + \mu^{216}x^{90} + \mu^{198}x^{89} + \mu^{227}x^{88} + \mu^{170}x^{87} + \mu^{198}x^{86} + \mu^{238}x^{85} + \mu^{166}x^{84} +$
$\mu^{198}x^{83} + \mu^{174}x^{82} + \mu^{240}x^{81} + \mu^{2}x^{80} + \mu^{15}x^{79} + \mu^{183}x^{78} + \mu^{162}x^{77} + \mu^{57}x^{76} +$
$\mu^{234}x^{75} + \mu^{234}x^{72} + \mu^{201}x^{71} + \mu^{133}x^{70} + \mu^{77}x^{68} + \mu^{75}x^{66} + \mu^{19}x^{65} + \mu^{224}x^{64} +$
$\mu^{245}x^{60} + \mu^{19}x^{59} + \mu^{245}x^{57} + \mu^{2}x^{56} + \mu^{2}x^{55} + \mu^{166}x^{54} + \mu^{2}x^{53} + \mu^{224}x^{52} + \mu^{2}x^{50} +$
$\mu^{32}x^{49} + \mu^{66}x^{48} + \mu^{32}x^{46} + \mu^{70}x^{45} + \mu^{191}x^{44} + \mu^{70}x^{42} + \mu^{70}x^{39} + \mu^{2}x^{38} + \mu^{2}x^{35} +$
$\mu^{138}x^{33} + \mu^{70}x^{27} + \mu^{70}x^{24} + \mu^{2}$
$e(x) = x^{64} + \lambda^{35723}x^{59} + \lambda^{35723}x^{56} + \lambda^{31114}x^{54} + \lambda^{35723}x^{53} + x^{52} + \lambda^{35723}x^{50} + \lambda^{38725}x^{49}$
$+ \lambda^{9749}x^{48} + \lambda^{38725}x^{46} + \lambda^{39327}x^{45} + \lambda^{64933}x^{44} + \lambda^{9515}x^{43} + \lambda^{39327}x^{42} + \lambda^{39327}x^{39} +$
$\lambda^{45840}x^{38} + \lambda^{9515}x^{37} + \lambda^{45840}x^{35} + \lambda^{12517}x^{33} + \lambda^{31716}x^{32} + \lambda^{61827}x^{27} + \lambda^{31716}x^{26} +$
$\lambda^{61827}x^{24} + \lambda^{62830}x^{22} + \lambda^{36325}x^{16} + \lambda^{36325}x^{13} + \lambda^{6513}x^{11} + \lambda^{602}$

Appendix B

Nonlinear tables of substitutions of the cipher "Strumok" (hexadecimal representation).

B.1 Substitution π_0

A8	43	5F	06	6B	75	6C	59	71	DF	87	95	17	F0	D8	09
6D	F3	1D	CB	C9	4D	2C	AF	79	E0	97	FD	6F	4B	45	39
3E	DD	A3	4F	B4	B6	9A	0E	1F	BF	15	E1	49	D2	93	C6
92	72	9E	61	D1	63	FA	EE	F4	19	D5	AD	58	A4	BB	A1
DC	F2	83	37	42	E4	7A	32	9C	CC	AB	4A	8F	6E	04	27
2E	E7	E2	5A	96	16	23	2B	C2	65	66	0F	BC	A9	47	41
34	48	FC	B7	6A	88	A5	53	86	F9	5B	DB	38	7B	C3	1E
22	33	24	28	36	C7	B2	3B	8E	77	BA	F5	14	9F	08	55
9B	4C	FE	60	5C	DA	18	46	CD	7D	21	B0	3F	1B	89	FF
EB	84	69	3A	9D	D7	D3	70	67	40	B5	DE	5D	30	91	B1
78	11	01	E5	00	68	98	A0	C5	02	A6	74	2D	0B	A2	76
B3	BE	CE	BD	AE	E9	8A	31	1C	EC	F1	99	94	AA	F6	26
2F	EF	E8	8C	35	03	D4	7F	FB	05	C1	5E	90	20	3D	82
F7	EA	0A	0D	7E	F8	50	1A	C4	07	57	B8	3C	62	E3	C8
AC	52	64	10	D0	D9	13	0C	12	29	51	B9	CF	D6	73	8D
81	54	C0	ED	4E	44	A7	2A	85	25	E6	CA	7C	8B	56	80

.

A. A. Kuznetsov et al., *Stream Ciphers in Modern Real-time IT Systems*, Studies in Systems, Decision and Control 375,
https://doi.org/10.1007/978-3-030-79770-6

B.2 Substitution π_1

CE	BB	EB	92	EA	CB	13	C1	E9	3A	D6	B2	D2	90	17	F8
42	15	56	B4	65	1C	88	43	C5	5C	36	BA	F5	57	67	8D
31	F6	64	58	9E	F4	22	AA	75	0F	02	B1	DF	6D	73	4D
7C	26	2E	F7	08	5D	44	3E	9F	14	C8	AE	54	10	D8	BC
1A	6B	69	F3	BD	33	AB	FA	D1	9B	68	4E	16	95	91	EE
4C	63	8E	5B	CC	3C	19	A1	81	49	7B	D9	6F	37	60	CA
E7	2B	48	FD	96	45	FC	41	12	0D	79	E5	89	8C	E3	20
30	DC	B7	6C	4A	B5	3F	97	D4	62	2D	06	A4	A5	83	5F
2A	DA	C9	00	7E	A2	55	BF	11	D5	9C	CF	0E	0A	3D	51
7D	93	1B	FE	C4	47	09	86	0B	8F	9D	6A	07	B9	B0	98
18	32	71	4B	EF	3B	70	A0	E4	40	FF	C3	A9	E6	78	F9
8B	46	80	1E	38	E1	B8	A8	E0	0C	23	76	1D	25	24	05
F1	6E	94	28	9A	84	E8	A3	4F	77	D3	85	E2	52	F2	82
50	7A	2F	74	53	B3	61	AF	39	35	DE	CD	1F	99	AC	AD
72	2C	DD	D0	87	BE	5E	A6	EC	04	C6	03	34	FB	DB	59
B6	C2	01	F0	5A	ED	A7	66	21	7F	8A	27	C7	C0	29	D7

.

B.3 Substitution π_2

93	D9	9A	B5	98	22	45	FC	BA	6A	DF	02	9F	DC	51	59
4A	17	2B	C2	94	F4	BB	A3	62	E4	71	D4	CD	70	16	E1
49	3C	C0	D8	5C	9B	AD	85	53	A1	7A	C8	2D	E0	D1	72
A6	2C	C4	E3	76	78	B7	B4	09	3B	0E	41	4C	DE	B2	90
25	A5	D7	03	11	00	C3	2E	92	EF	4E	12	9D	7D	CB	35
10	D5	4F	9E	4D	A9	55	C6	D0	7B	18	97	D3	36	E6	48
56	81	8F	77	CC	9C	B9	E2	AC	B8	2F	15	A4	7C	DA	38
1E	0B	05	D6	14	6E	6C	7E	66	FD	B1	E5	60	AF	5E	33
87	C9	F0	5D	6D	3F	88	8D	C7	F7	1D	E9	EC	ED	80	29
27	CF	99	A8	50	0F	37	24	28	30	95	D2	3E	5B	40	83
B3	69	57	1F	07	1C	8A	BC	20	EB	CE	8E	AB	EE	31	A2
73	F9	CA	3A	1A	FB	0D	C1	FE	FA	F2	6F	BD	96	DD	43
52	B6	08	F3	AE	BE	19	89	32	26	B0	EA	4B	64	84	82
6B	F5	79	BF	01	5F	75	63	1B	23	3D	68	2A	65	E8	91
F6	FF	13	58	F1	47	0A	7F	C5	A7	E7	61	5A	06	46	44
42	04	A0	DB	39	86	54	AA	8C	34	21	8B	F8	0C	74	67

.

B.4 Substitution π_3

68	8D	CA	4D	73	4B	4E	2A	D4	52	26	B3	54	1E	19	1F
22	03	46	3D	2D	4A	53	83	13	8A	B7	D5	25	79	F5	BD
58	2F	0D	02	ED	51	9E	11	F2	3E	55	5E	D1	16	3C	66
70	5D	F3	45	40	CC	E8	94	56	08	CE	1A	3A	D2	E1	DF
B5	38	6E	0E	E5	F4	F9	86	E9	4F	D6	85	23	CF	32	99
31	14	AE	EE	C8	48	D3	30	A1	92	41	B1	18	C4	2C	71
72	44	15	FD	37	BE	5F	AA	9B	88	D8	AB	89	9C	FA	60
EA	BC	62	0C	24	A6	A8	EC	67	20	DB	7C	28	DD	AC	5B
34	7E	10	F1	7B	8F	63	A0	05	9A	43	77	21	BF	27	09
C3	9F	B6	D7	29	C2	EB	C0	A4	8B	8C	1D	FB	FF	C1	B2
97	2E	F8	65	F6	75	07	04	49	33	E4	D9	B9	D0	42	C7
6C	90	00	8E	6F	50	01	C5	DA	47	3F	CD	69	A2	E2	7A
A7	C6	93	0F	0A	06	E6	2B	96	A3	1C	AF	6A	12	84	39
E7	B0	82	F7	FE	9D	87	5C	81	35	DE	B4	A5	FC	80	EF
CB	BB	6B	76	BA	5A	7D	78	0B	95	E3	AD	74	98	3B	36
64	6D	DC	F0	59	A9	4C	17	7F	91	B8	C9	57	1B	E0	61

Appendix C

Tables constants $T_i[a]$, $i = 0, 1, \ldots, 7$ cipher "Strumok" (hexadecimal representation).

C.1 Table T_0

A832A829D77F9AA8	4352432297D41143	5F3E5FC2DF80615F	061E063014121806
6BDA6B7F670CB16B	75BC758F2356C975	6CC16C477519AD6C	592059F2CB927959
71A871AF3B4AD971	DF84DFB6F8275BDF	87A1874C35B22687	95FB95DC59CC6E95
174B17B872655C17	F017F0D31AEAE7F0	D89FD88EEA3247D8	092D0948363F2409
6DC46D4F731EA96D	F318F3CB10E3EBF3	1D691DE84E53741D	CBC0CB16804B0BCB
C9CAC9068C4503C9	4D644D52B3FE294D	2C9C2C7DE8C4B02C	AF29AF11C56A86AF
798079EF0B72F979	E047E0537A9AA7E0	97F197CC55C26697	FD2EFDBB34C9D3FD
6FCE6F5F7F10A16F	4B7A4B62A7EC314B	454C451283C60945	39DD39D596AFE439
3EC63EED84BAF83E	DD8EDDA6F42953DD	A315A371ED4EB6A3	4F6E4F42BFF0214F
B45EB4C99F2BEAB4	B654B6D99325E2B6	9AC89AA47BE1529A	0E360E70242A380E
1F631FF8425D7C1F	BF79BF91A51AC6BF	154115A87E6B5415	E142E15B7C9DA3E1
49704972ABE23949	D2BDD2DED6046FD2	93E593EC4DDE7693	C6F9C67EAE683FC6
92E092E44BD97292	72A772B73143D572	9EDC9E8463FD429E	61F8612F5B3A9961
D1B2D1C6DC0D63D1	63F2633F57349163	FA35FA8326DCCFFA	EE71EE235EB09FEE
F403F4F302F6F7F4	197D19C8564F6419	D5A6D5E6C41173D5	AD23AD01C9648EAD
582558FACD957D58	A40EA449FF5BAAA4	BB6DBBB1BD06D6BB	A11FA161E140BEA1
DC8BDCAEF22E57DC	F21DF2C316E4EFF2	83B5836C2DAE3683	37EB37A5B285DC37
4257422A91D31542	E453E4736286B7E4	7A8F7AF7017BF57A	32FA328DAC9EC832
9CD69C946FF34A9C	CCDBCC2E925E17CC	AB3DAB31DD7696AB	4A7F4A6AA1EB354A
8F898F0C058A068F	6ECB6E577917A56E	04140420181C1004	27BB2725D2F59C27
2E962E6DE4CAB82E	E75CE76B688FBBE7	E24DE2437694AFE2	5A2F5AEAC19B755A
96F496C453C56296	164E16B074625816	23AF2305CAE98C23	2B872B45FAD1AC2B

A. A. Kuznetsov et al., *Stream Ciphers in Modern Real-time IT Systems*, Studies in Systems, Decision and Control 375,
https://doi.org/10.1007/978-3-030-79770-6

C2EDC25EB6742FC2	65EC650F43268965	66E36617492F8566	0F330F78222D3C0F
BC76BC89AF13CABC	A937A921D1789EA9	474647028FC80147	415841329BDA1941
34E434BDB88CD034	4875487AADE53D48	FC2BFCB332CED7FC	B751B7D19522E6B7
6ADF6A77610BB56A	88928834179F1A88	A50BA541F95CAEA5	530253A2F7A45153
86A4864433B52286	F93AF99B2CD5C3F9	5B2A5BE2C79C715B	DB90DB96E03B4BDB
38D838DD90A8E038	7B8A7BFF077CF17B	C3E8C356B0732BC3	1E661EF0445A781E
22AA220DCCEE8822	33FF3385AA99CC33	24B4243DD8FC9024	2888285DF0D8A028
36EE36ADB482D836	C7FCC776A86F3BC7	B240B2F98B39F2B2	3BD73BC59AA1EC3B
8E8C8E04038D028E	77B6779F2F58C177	BA68BAB9BB01D2BA	F506F5FB04F1F3F5
144414A0786C5014	9FD99F8C65FA469F	0828084030382008	551C5592E3B64955
9BCD9BAC7DE6569B	4C614C5AB5F92D4C	FE21FEA33EC0DFFE	60FD60275D3D9D60
5C315CDAD5896D5C	DA95DA9EE63C4FDA	187818C050486018	4643460A89CF0546
CDDECD26945913CD	7D947DCF136EE97D	21A52115C6E78421	B04AB0E98737FAB0
3FC33FE582BDFC3F	1B771BD85A416C1B	8997893C11981E89	FF24FFAB38C7DBFF
EB60EB0B40AB8BEB	84AE84543FBB2A84	69D0696F6B02B969	3AD23ACD9CA6E83A
9DD39D9C69F44E9D	D7ACD7F6C81F7BD7	D3B8D3D6D0036BD3	70AD70A73D4DDD70
67E6671F4F288167	405D403A9DDD1D40	B55BB5C1992CEEB5	DE81DEBEFE205FDE
5D345DD2D38E695D	30F0309DA090C030	91EF91FC41D07E91	B14FB1E18130FEB1
788578E70D75FD78	1155118866774411	0105010806070401	E556E57B6481B3E5
0000000000000000	68D568676D05BD68	98C298B477EF5A98	A01AA069E747BAA0
C5F6C566A46133C5	020A02100C0E0802	A604A659F355A2A6	74B974872551CD74
2D992D75EEC3B42D	0B270B583A312C0B	A210A279EB49B2A2	76B37697295FC576
B345B3F18D3EF6B3	BE7CBE99A31DC2BE	CED1CE3E9E501FCE	BD73BD81A914CEBD
AE2CAE19C36D82AE	E96AE91B4CA583E9	8A988A241B91128A	31F53195A697C431
1C6C1CE04854701C	EC7BEC3352BE97EC	F112F1DB1CEDE3F1	99C799BC71E85E99
94FE94D45FCB6A94	AA38AA39DB7192AA	F609F6E30EF8FFF6	26BE262DD4F29826
2F932F65E2CDBC2F	EF74EF2B58B79BEF	E86FE8134AA287E8	8C868C140F830A8C
35E135B5BE8BD435	030F03180A090C03	D4A3D4EEC21677D4	7F9E7FDF1F60E17F
FB30FB8B20DBCBFB	051105281E1B1405	C1E2C146BC7D23C1	5E3B5ECAD987655E
90EA90F447D77A90	20A0201DC0E08020	3DC93DF58EB3F43D	82B082642BA93282
F70CF7EB08FFFBF7	EA65EA0346AC8FEA	0A220A503C36280A	0D390D682E23340D
7E9B7ED71967E57E	F83FF8932AD2C7F8	500D50BAFDAD5D50	1A721AD05C46681A
C4F3C46EA26637C4	071B073812151C07	57165782EFB84157	B862B8A9B70FDAB8
3CCC3CFD88B4F03C	62F7623751339562	E348E34B7093ABE3	C8CFC80E8A4207C8
AC26AC09CF638AAC	520752AAF1A35552	64E9640745218D64	1050108060704010
D0B7D0CEDA0A67D0	D99AD986EC3543D9	135F13986A794C13	0C3C0C602824300C
125A12906C7E4812	298D2955F6DFA429	510851B2FBAA5951	B967B9A1B108DEB9
CFD4CF3698571BCF	D6A9D6FECE187FD6	73A273BF3744D173	8D838D1C09840E8D
81BF817C21A03E81	5419549AE5B14D54	C0E7C04EBA7A27C0	ED7EED3B54B993ED
4E6B4E4AB9F7254E	4449441A85C10D44	A701A751F552A6A7	2A822A4DFCD6A82A
85AB855C39BC2E85	25B12535DEFB9425	E659E6636E88BFE6	CAC5CA1E864C0FCA
7C917CC71569ED7C	8B9D8B2C1D96168B	5613568AE9BF4556	80BA807427A73A80

C.2 Table T_1

D1CE3E9E501FCECE	6DBBB1BD06D6BBBB	60EB0B40AB8BEBEB	E092E44BD9729292
65EA0346AC8FEAEA	C0CB16804B0BCBCB	5F13986A794C1313	E2C146BC7D23C1C1
6AE91B4CA583E9E9	D23ACD9CA6E83A3A	A9D6FECE187FD6D6	40B2F98B39F2B2B2
BDD2DED6046FD2D2	EA90F447D77A9090	4B17B872655C1717	3FF8932AD2C7F8F8
57422A91D3154242	4115A87E6B541515	13568AE9BF455656	5EB4C99F2BEAB4B4
EC650F4326896565	6C1CE04854701C1C	928834179F1A8888	52432297D4114343
F6C566A46133C5C5	315CDAD5896D5C5C	EE36ADB482D83636	68BAB9BB01D2BABA
06F5FB04F1F3F5F5	165782EFB8415757	E6671F4F28816767	838D1C09840E8D8D
F53195A697C43131	09F6E30EF8FFF6F6	E9640745218D6464	2558FACD957D5858
DC9E8463FD429E9E	03F4F302F6F7F4F4	AA220DCCEE882222	38AA39DB7192AAAA
BC758F2356C97575	330F78222D3C0F0F	0A02100C0E080202	4FB1E18130FEB1B1
84DFB6F8275BDFDF	C46D4F731EA96D6D	A273BF3744D17373	644D52B3FE294D4D
917CC71569ED7C7C	BE262DD4F2982626	962E6DE4CAB82E2E	0CF7EB08FFFBF7F7
2808403038200808	345DD2D38E695D5D	49441A85C10D4444	C63EED84BAF83E3E
D99F8C65FA469F9F	4414A0786C501414	CFC80E8A4207C8C8	2CAE19C36D82AEAE
19549AE5B14D5454	5010806070401010	9FD88EEA3247D8D8	76BC89AF13CABCBC
721AD05C46681A1A	DA6B7F670CB16B6B	D0696F6B02B96969	18F3CB10E3EBF3F3
73BD81A914CEBDBD	FF3385AA99CC3333	3DAB31DD7696ABAB	35FA8326DCCFFAFA
B2D1C6DC0D63D1D1	CD9BAC7DE6569B9B	D568676D05BD6868	6B4E4AB9F7254E4E
4E16B07462581616	FB95DC59CC6E9595	EF91FC41D07E9191	71EE235EB09FEEEE
614C5AB5F92D4C4C	F2633F5734916363	8C8E04038D028E8E	2A5BE2C79C715B5B
DBCC2E925E17CCCC	CC3CFD88B4F03C3C	7D19C8564F641919	1FA161E140BEA1A1
BF817C21A03E8181	704972ABE2394949	8A7BFF077CF17B7B	9AD986EC3543D9D9
CE6F5F7F10A16F6F	EB37A5B285DC3737	FD60275D3D9D6060	C5CA1E864C0FCACA
5CE76B688FBBE7E7	872B45FAD1AC2B2B	75487AADE53D4848	2EFDBB34C9D3FDFD
F496C453C5629696	4C451283C6094545	2BFCB332CED7FCFC	5841329BDA194141
5A12906C7E481212	390D682E23340D0D	8079EF0B72F97979	56E57B6481B3E5E5
97893C11981E8989	868C140F830A8C8C	48E34B7093ABE3E3	A0201DC0E0802020
F0309DA090C03030	8BDCAEF22E57DCDC	51B7D19522E6B7B7	C16C477519AD6C6C
7F4A6AA1EB354A4A	5BB5C1992CEEB5B5	C33FE582BDFC3F3F	F197CC55C2669797
A3D4EEC21677D4D4	F762375133956262	992D75EEC3B42D2D	1E06301412180606
0EA449FF5BAAA4A4	0BA541F95CAEA5A5	B5836C2DAE368383	3E5FC2DF80615F5F
822A4DFCD6A82A2A	95DA9EE63C4FDADA	CAC9068C4503C9C9	0000000000000000
9B7ED71967E57E7E	10A279EB49B2A2A2	1C5592E3B6495555	79BF91A51AC6BFBF
5511886677441111	A6D5E6C41173D5D5	D69C946FF34A9C9C	D4CF3698571BCFCF
360E70242A380E0E	220A503C36280A0A	C93DF58EB3F43D3D	0851B2FBAA595151
947DCF136EE97D7D	E593EC4DDE769393	771BD85A416C1B1B	21FEA33EC0DFFEFE
F3C46EA26637C4C4	4647028FC8014747	2D0948363F240909	A4864433B5228686
270B583A312C0B0B	898F0C058A068F8F	D39D9C69F44E9D9D	DF6A77610BB56A6A
1B073812151C0707	67B9A1B108DEB9B9	4AB0E98737FAB0B0	C298B477EF5A9898

7818C05048601818	FA328DAC9EC83232	A871AF3B4AD97171	7A4B62A7EC314B4B
74EF2B58B79BEFEF	D73BC59AA1EC3B3B	AD70A73D4DDD7070	1AA069E747BAA0A0
53E4736286B7E4E4	5D403A9DDD1D4040	24FFAB38C7DBFFFF	E8C356B0732BC3C3
37A921D1789EA9A9	59E6636E88BFE6E6	8578E70D75FD7878	3AF99B2CD5C3F9F9
9D8B2C1D96168B8B	43460A89CF054646	BA807427A73A8080	661EF0445A781E1E
D838DD90A8E03838	42E15B7C9DA3E1E1	62B8A9B70FDAB8B8	32A829D77F9AA8A8
47E0537A9AA7E0E0	3C0C602824300C0C	AF2305CAE98C2323	B37697295FC57676
691DE84E53741D1D	B12535DEFB942525	B4243DD8FC902424	1105281E1B140505
12F1DB1CEDE3F1F1	CB6E577917A56E6E	FE94D45FCB6A9494	88285DF0D8A02828
C89AA47BE1529A9A	AE84543FBB2A8484	6FE8134AA287E8E8	15A371ED4EB6A3A3
6E4F42BFF0214F4F	B6779F2F58C17777	B8D3D6D0036BD3D3	AB855C39BC2E8585
4DE2437694AFE2E2	0752AAF1A3555252	1DF2C316E4EFF2F2	B082642BA9328282
0D50BAFDAD5D5050	8F7AF7017BF57A7A	932F65E2CDBC2F2F	B974872551CD7474
0253A2F7A4515353	45B3F18D3EF6B3B3	F8612F5B3A996161	29AF11C56A86AFAF
DD39D596AFE43939	E135B5BE8BD43535	81DEBEFE205FDEDE	DECD26945913CDCD
631FF8425D7C1F1F	C799BC71E85E9999	26AC09CF638AACAC	23AD01C9648EADAD
A772B73143D57272	9C2C7DE8C4B02C2C	8EDDA6F42953DDDD	B7D0CEDA0A67D0D0
A1874C35B2268787	7CBE99A31DC2BEBE	3B5ECAD987655E5E	04A659F355A2A6A6
7BEC3352BE97ECEC	140420181C100404	F9C67EAE683FC6C6	0F03180A090C0303
E434BDB88CD03434	30FB8B20DBCBFBFB	90DB96E03B4BDBDB	2059F2CB92795959
54B6D99325E2B6B6	EDC25EB6742FC2C2	0501080607040101	17F0D31AEAE7F0F0
2F5AEAC19B755A5A	7EED3B54B993EDED	01A751F552A6A7A7	E36617492F856666
A52115C6E7842121	9E7FDF1F60E17F7F	988A241B91128A8A	BB2725D2F59C2727
FCC776A86F3BC7C7	E7C04EBA7A27C0C0	8D2955F6DFA42929	ACD7F6C81F7BD7D7

C.3 Table T_2

93EC4DDE769393E5	D986EC3543D9D99A	9AA47BE1529A9AC8	B5C1992CEEB5B55B
98B477EF5A9898C2	220DCCEE882222AA	451283C60945454C	FCB332CED7FCFC2B
BAB9BB01D2BABA68	6A77610BB56A6ADF	DFB6F8275BDFDF84	02100C0E0802020A
9F8C65FA469F9FD9	DCAEF22E57DCDC8B	51B2FBAA59515108	59F2CB9279595920
4A6AA1EB354A4A7F	17B872655C17174B	2B45FAD1AC2B2B87	C25EB6742FC2C2ED
94D45FCB6A9494FE	F4F302F6F7F4F403	BBB1BD06D6BBBB6D	A371ED4EB6A3A315
623751339562 62F7	E4736286B7E4E453	71AF3B4AD97171A8	D4EEC21677D4D4A3
CD26945913CDCDDE	70A73D4DDD7070AD	16B074625816164E	E15B7C9DA3E1E142
4972ABE239494970	3CFD88B4F03C3CCC	C04EBA7A27C0C0E7	D88EEA3247D8D89F
5CDAD5896D5C5C31	9BAC7DE6569B9BCD	AD01C9648EADAD23	855C39BC2E8585AB
53A2F7A451535302	A161E140BEA1A11F	7AF7017BF57A7A8F	C80E8A4207C8C8CF

2D75EEC3B42D2D99	E0537A9AA7E0E047	D1C6DC0D63D1D1B2	72B73143D57272A7
A659F355A2A6A604	2C7DE8C4B02C2C9C	C46EA26637C4C4F3	E34B7093ABE3E348
7697295FC57676B3	78E70D75FD787885	B7D19522E6B7B751	B4C99F2BEAB4B45E
0948363F2409092D	3BC59AA1EC3B3BD7	0E70242A380E0E36	41329BDA19414158
4C5AB5F92D4C4C61	DEBEFE205FDEDE81	B2F98B39F2B2B240	90F447D77A9090EA
2535DEFB942525B1	A541F95CAEA5A50B	D7F6C81F7BD7D7AC	03180A090C03030F
1188667744111155	0000000000000000	C356B0732BC3C3E8	2E6DE4CAB82E2E96
92E44BD9729292E0	EF2B58B79BEFEF74	4E4AB9F7254E4E6B	12906C7E4812125A
9D9C69F44E9D9DD3	7DCF136EE97D7D94	CB16804B0BCBCBC0	35B5BE8BD43535E1
1080607040101050	D5E6C41173D5D5A6	4F42BFF0214F4F6E	9E8463FD429E9EDC
4D52B3FE294D4D64	A921D1789EA9A937	5592E3B64955551C	C67EAE683FC6C6F9
D0CEDA0A67D0D0B7sssss	7BFF077CF17B7B8A	18C0504860181878	97CC55C2669797F1
D3D6D0036BD3D3B8	36ADB482D83636EE	E6636E88BFE6E659	487AADE53D484875
568AE9BF45565613	817C21A03E8181BF	8F0C058A068F8F89	779F2F58C17777B6
CC2E925E17CCCCDB	9C946FF34A9C9CD6	B9A1B108DEB9B967	E2437694AFE2E24D
AC09CF638AACAC26	B8A9B70FDAB8B862	2F65E2CDBC2F2F93	15A87E6B54151541
A449FF5BAAA4A40E	7CC71569ED7C7C91	DA9EE63C4FDADA95	38DD90A8E03838D8
1EF0445A781E1E66	0B583A312C0B0B27	05281E1B14050511	D6FECE187FD6D6A9
14A0786C50141444	6E577917A56E6ECB	6C477519AD6C6CC1	7ED71967E57E7E9B
6617492F856666E3	FDBB34C9D3FDFD2E	B1E18130FEB1B14F	E57B6481B3E5E556
60275D3D9D6060FD	AF11C56A86AFAF29	5ECAD987655E5E3B	3385AA99CC3333FF
874C35B2268787A1	C9068C4503C9C9CA	F0D31AEAE7F0F017	5DD2D38E695D5D34
6D4F731EA96D6DC4	3FE582BDFC3F3FC3	8834179F1A888892	8D1C09840E8D8D83
C776A86F3BC7C7FC	F7EB08FFFBF7F70C	1DE84E53741D1D69	E91B4CA583E9E96A
EC3352BE97ECEC7B	ED3B54B993EDED7E	807427A73A8080BA	2955F6DFA429298D
2725D2F59C2727BB	CF3698571BCFCFD4	99BC71E85E9999C7	A829D77F9AA8A832
50BAFDAD5D50500D	0F78222D3C0F0F33	37A5B285DC3737EB	243DD8FC902424B4
285DF0D8A0282888	309DA090C03030F0	95DC59CC6E9595FB	D2DED6046FD2D2BD
3EED84BAF83E3EC6	5BE2C79C715B5B2A	403A9DDD1D40405D	836C2DAE368383B5
B3F18D3EF6B3B345	696F6B02B96969D0	5782EFB841575716	1FF8425D7C1F1F63
073812151C07071B	1CE04854701C1C6C	8A241B91128A8A98	BC89AF13CABCBC76
201DC0E0802020A0	EB0B40AB8BEBEB60	CE3E9E501FCECED1	8E04038D028E8E8C
AB31DD7696ABAB3D	EE235EB09FEEEE71	3195A697C43131F5	A279EB49B2A2A210
73BF3744D17373A2	F99B2CD5C3F9F93A	CA1E864C0FCACAC5	3ACD9CA6E83A3AD2
1AD05C46681A1A72	FB8B20DBCBFBFB30	0D682E23340D0D39	C146BC7D23C1C1E2
FEA33EC0DFFEFE21	FA8326DCCFFAFA35	F2C316E4EFF2F21D	6F5F7F10A16F6FCE
BD81A914CEBDBD73	96C453C5629696F4	DDA6F42953DDDD8E	432297D411434352
52AAF1A355525207	B6D99325E2B6B654	0840303820080828	F3CB10E3EBF3F318
AE19C36D82AEAE2C	BE99A31DC2BEBE7C	19C8564F6419197D	893C11981E898997
328DAC9EC83232FA	262DD4F2982626BE	B0E98737FAB0B04A	EA0346AC8FEAEA65
4B62A7EC314B4B7A	640745218D6464E9	84543FBB2A8484AE	82642BA9328282B0
6B7F670CB16B6BDA	F5FB04F1F3F5F506	79EF0B72F9797980	BF91A51AC6BFBF79

0108060704010105	5FC2DF80615F5F3E	758F2356C97575BC	633F5734916363F2
1BD85A416C1B1B77	2305CAE98C2323AF	3DF58EB3F43D3DC9	68676D05BD6868D5
2A4DFCD6A82A2A82	650F4326896565EC	E8134AA287E8E86F	91FC41D07E9191EF
F6E30EF8FFF6F609	FFAB38C7DBFFFF24	13986A794C13135F	58FACD957D585825
F1DB1CEDE3F1F112	47028FC801474746	0A503C36280A0A22	7FDF1F60E17F7F9E
C566A46133C5C5F6	A751F552A6A7A701	E76B688FBBE7E75C	612F5B3A996161F8
5AEAC19B755A5A2F	063014121806061E	460A89CF05464643	441A85C10D444449
422A91D315424257	0420181C10040414	A069E747BAA0A01A	DB96E03B4BDBDB90
39D596AFE43939DD	864433B5228686A4	549AE5B14D545419	AA39DB7192AAAA38
8C140F830A8C8C86	34BDB88CD03434E4	2115C6E7842121A5	8B2C1D96168B8B9D
F8932AD2C7F8F83F	0C602824300C0C3C	74872551CD7474B9	671F4F28816767E6

C.4 Table T_3

676D05BD6868D568	1C09840E8D8D838D	1E864C0FCACAC5CA	52B3FE294D4D644D
BF3744D17373A273	62A7EC314B4B7A4B	4AB9F7254E4E6B4E	4DFCD6A82A2A822A
EEC21677D4D4A3D4	AAF1A35552520752	2DD4F2982626BE26	F18D3EF6B3B345B3
9AE5B14D54541954	F0445A781E1E661E	C8564F6419197D19	F8425D7C1F1F631F
0DCCEE882222AA22	180A090C03030F03	0A89CF0546464346	F58EB3F43D3DC93D
75EEC3B42D2D992D	6AA1EB354A4A7F4A	A2F7A45153530253	6C2DAE368383B583
986A794C13135F13	241B91128A8A988A	D19522E6B7B751B7	E6C41173D5D5A6D5
35DEFB942525B125	EF0B72F979798079	FB04F1F3F5F506F5	81A914CEBDBD73BD
FACD957D58582558	65E2CDBC2F2F932F	682E23340D0D390D	100C0E0802020A02
3B54B993EDED7EED	B2FBAA5951510851	8463FD429E9EDC9E	8866774411115511
C316E4EFF2F21DF2	ED84BAF83E3EC63E	92E3B64955551C55	CAD987655E5E3B5E
C6DC0D63D1D1B2D1	B074625816164E16	FD88B4F03C3CCC3C	17492F856666E366
A73D4DDD7070AD70	D2D38E695D5D345D	CB10E3EBF3F318F3	1283C60945454C45
3A9DDD1D40405D40	2E925E17CCCCDBCC	134AA287E8E86FE8	D45FCB6A9494FE94
8AE9BF4556561356	4030382008082808	3E9E501FCECED1CE	D05C46681A1A721A
CD9CA6E83A3AD23A	DED6046FD2D2BDD2	5B7C9DA3E1E142E1	B6F8275BDFDF84DF
C1992CEEB5B55BB5	DD90A8E03838D838	577917A56E6ECB6E	70242A380E0E360E
7B6481B3E5E556E5	F302F6F7F4F403F4	9B2CD5C3F9F93AF9	4433B5228686A486
1B4CA583E9E96AE9	42BFF0214F4F6E4F	FECE187FD6D6A9D6	5C39BC2E8585AB85
05CAE98C2323AF23	3698571BCFCFD4CF	8DAC9EC83232FA32	BC71E85E9999C799
95A697C43131F531	A0786C5014144414	19C36D82AEAE2CAE	235EB09FEEEE71EE
0E8A4207C8C8CFC8	7AADE53D48487548	D6D0036BD3D3B8D3	9DA090C03030F030
61E140BEA1A11FA1	E44BD9729292E092	329BDA1941415841	E18130FEB1B14FB1
C050486018187818	6EA26637C4C4F3C4	7DE8C4B02C2C9C2C	AF3B4AD97171A871

B73143D57272A772	1A85C10D44444944	A87E6B5415154115	BB34C9D3FDFD2EFD
A5B285DC3737EB37	99A31DC2BEBE7CBE	C2DF80615F5F3E5F	39DB7192AAAA38AA
AC7DE6569B9BCD9B	34179F1A88889288	8EEA3247D8D89FD8	31DD7696ABAB3DAB
3C11981E89899789	946FF34A9C9CD69C	8326DCCFFAFA35FA	275D3D9D6060FD60
0346AC8FEAEA65EA	89AF13CABCBC76BC	37513395626 2F762	602824300C0C3C0C
3DD8FC902424B424	59F355A2A6A604A6	29D77F9AA8A832A8	3352BE97ECEC7BEC
1F4F28816767E667	1DC0E0802020A020	96E03B4BDBDB90DB	C71569ED7C7C917C
5DF0D8A028288828	A6F42953DDDD8EDD	09CF638AACAC26AC	E2C79C715B5B2A5B
BDB88CD03434E434	D71967E57E7E9B7E	8060704010105010	DB1CEDE3F1F112F1
FF077CF17B7B8A7B	0C058A068F8F898F	3F5734916363F263	69E747BAA0A01AA0
281E1B1405051105	A47BE1529A9AC89A	2297D41143435243	9F2F58C17777B677
15C6E7842121A521	91A51AC6BFBF79BF	25D2F59C2727BB27	48363F2409092D09
56B0732BC3C3E8C3	8C65FA469F9FD99F	D99325E2B6B654B6	F6C81F7BD7D7ACD7
55F6DFA429298D29	5EB6742FC2C2EDC2	0B40AB8BEBEB60EB	4EBA7A27C0C0E7C0
49FF5BAAA4A40EA4	2C1D96168B8B9D8B	140F830A8C8C868C	E84E53741D1D691D
8B20DBCBFBFB30FB	AB38C7DBFFFF24FF	46BC7D23C1C1E2C1	F98B39F2B2B240B2
CC55C2669797F197	6DE4CAB82E2E962E	932AD2C7F8F83FF8	0F4326896565EC65
E30EF8FFF6F609F6	8F2356C97575BC75	3812151C07071B07	20181C1004041404
72ABE23949497049	85AA99CC3333FF33	736286B7E4E453E4	86EC3543D9D99AD9
A1B108DEB9B967B9	CEDA0A67D0D0B7D0	2A91D31542425742	76A86F3BC7C7FCC7
477519AD6C6CC16C	F447D77A9090EA90	0000000000000000	04038D028E8E8C8E
5F7F10A16F6FCE6F	BAFDAD5D50500D50	0806070401010501	66A46133C5C5F6C5
9EE63C4FDADA95DA	028FC80147474647	E582BDFC3F3FC33F	26945913CDCDDECD
6F6B02B96969D069	79EB49B2A2A210A2	437694AFE2E24DE2	F7017BF57A7A8F7A
51F552A6A7A701A7	7EAE683FC6C6F9C6	EC4DDE769393E593	78222D3C0F0F330F
503C36280A0A220A	3014121806061E06	636E88BFE6E659E6	45FAD1AC2B2B872B
C453C5629696F496	71ED4EB6A3A315A3	E04854701C1C6C1C	11C56A86AFAF29AF
77610BB56A6ADF6A	906C7E4812125A12	543FBB2A8484AE84	D596AFE43939DD39
6B688FBBE7E75CE7	E98737FAB0B04AB0	642BA9328282B082	EB08FFFBF7F70CF7
A33EC0DFFEFE21FE	9C69F44E9D9DD39D	4C35B2268787A187	DAD5896D5C5C315C
7C21A03E8181BF81	B5BE8BD43535E135	BEFE205FDEDE81DE	C99F2BEAB4B45EB4
41F95CAEA5A50BA5	B332CED7FCFC2BFC	7427A73A8080BA80	2B58B79BEFEF74EF
16804B0BCBCBC0CB	B1BD06D6BBBB6DBB	7F670CB16B6BDA6B	97295FC57676B376
B9BB01D2BABA68BA	EAC19B755A5A2F5A	CF136EE97D7D947D	E70D75FD78788578
583A312C0B0B270B	DC59CC6E9595FB95	4B7093ABE3E348E3	01C9648EADAD23AD
872551CD7474B974	B477EF5A9898C298	C59AA1EC3B3BD73B	ADB482D83636EE36
0745218D6464E964	4F731EA96D6DC46D	AEF22E57DCDC8BDC	D31AEAE7F0F017F0
F2CB927959592059	21D1789EA9A937A9	5AB5F92D4C4C614C	B872655C17174B17
DF1F60E17F7F9E7F	FC41D07E9191EF91	A9B70FDAB8B862B8	068C4503C9C9CAC9
82EFB84157571657	D85A416C1B1B771B	537A9AA7E0E047E0	2F5B3A996161F861

C.5 Table T_4

D77F9AA8A832A829	97D4114343524322	DF80615F5F3E5FC2	14121806061E0630
670CB16B6BDA6B7F	2356C97575BC758F	7519AD6C6CC16C47	CB927959592059F2
3B4AD97171A871AF	F8275BDFDF84DFB6	35B2268787A1874C	59CC6E9595FB95DC
72655C17174B17B8	1AEAE7F0F017F0D3	EA3247D8D89FD88E	363F2409092D0948
731EA96D6DC46D4F	10E3EBF3F318F3CB	4E53741D1D691DE8	804B0BCBCBC0CB16
8C4503C9C9CAC906	B3FE294D4D644D52	E8C4B02C2C9C2C7D	C56A86AFAF29AF11
0B72F979798079EF	7A9AA7E0E047E053	55C2669797F197CC	34C9D3FDFD2EFDBB
7F10A16F6FCE6F5F	A7EC314B4B7A4B62	83C60945454C4512	96AFE43939DD39D5
84BAF83E3EC63EED	F42953DDDD8EDDA6	ED4EB6A3A315A371	BFF0214F4F6E4F42
9F2BEAB4B45EB4C9	9325E2B6B654B6D9	7BE1529A9AC89AA4	242A380E0E360E70
425D7C1F1F631FF8	A51AC6BFBF79BF91	7E6B5415154115A8	7C9DA3E1E142E15B
ABE2394949704972	D6046FD2D2BDD2DE	4DDE769393E593EC	AE683FC6C6F9C67E
4BD9729292E092E4	3143D57272A772B7	63FD429E9EDC9E84	5B3A996161F8612F
DC0D63D1D1B2D1C6	5734916363F2633F	26DCCFFAFA35FA83	5EB09FEEEE71EE23
02F6F7F4F403F4F3	564F6419197D19C8	C41173D5D5A6D5E6	C9648EADAD23AD01
CD957D58582558FA	FF5BAAA4A40EA449	BD06D6BBBB6DBBB1	E140BEA1A11FA161
F22E57DCDC8BDCAE	16E4EFF2F21DF2C3	2DAE368383B5836C	B285DC3737EB37A5
91D315424257422A	6286B7E4E453E473	017BF57A7A8F7AF7	AC9EC83232FA328D
6FF34A9C9CD69C94	925E17CCCCDBCC2E	DD7696ABAB3DAB31	A1EB354A4A7F4A6A
058A068F8F898F0C	7917A56E6ECB6E57	181C100404140420	D2F59C2727BB2725
E4CAB82E2E962E6D	688FBBE7E75CE76B	7694AFE2E24DE243	C19B755A5A2F5AEA
53C5629696F496C4	74625816164E16B0	CAE98C2323AF2305	FAD1AC2B2B872B45
B6742FC2C2EDC25E	4326896565EC650F	492F856666E36617	222D3C0F0F330F78
AF13CABCBC76BC89	D1789EA9A937A921	8FC8014747464702	9BDA194141584132
B88CD03434E434BD	ADE53D484875487A	32CED7FCFC2BFCB3	9522E6B7B751B7D1
610BB56A6ADF6A77	179F1A8888928834	F95CAEA5A50BA541	F7A45153530253A2
33B5228686A48644	2CD5C3F9F93AF99B	C79C715B5B2A5BE2	E03B4BDBDB90DB96
90A8E03838D838DD	077CF17B7B8A7BFF	B0732BC3C3E8C356	445A781E1E661EF0
CCEE882222AA220D	AA99CC3333FF3385	D8FC902424B4243D	F0D8A0282888285D
B482D83636EE36AD	A86F3BC7C7FCC776	8B39F2B2B240B2F9	9AA1EC3B3BD73BC5
038D028E8E8C8E04	2F58C17777B6779F	BB01D2BABA68BAB9	04F1F3F5F506F5FB
786C5014144414A0	65FA469F9FD99F8C	3038200808280840	E3B64955551C5592
7DE6569B9BCD9BAC	B5F92D4C4C614C5A	3EC0DFFEFE21FEA3	5D3D9D6060FD6027
D5896D5C5C315CDA	E63C4FDADA95DA9E	50486018187818C0	89CF05464643460A
945913CDCDDECD26	136EE97D7D947DCF	C6E7842121A52115	8737FAB0B04AB0E9
82BDFC3F3FC33FE5	5A416C1B1B771BD8	11981E898997893C	38C7DBFFFF24FFAB
40AB8BEBEB60EB0B	3FBB2A8484AE8454	6B02B96969D0696F	9CA6E83A3AD23ACD
69F44E9D9DD39D9C	C81F7BD7D7ACD7F6	D0036BD3D3B8D3D6	3D4DDD7070AD70A7
4F28816767E6671F	9DDD1D40405D403A	992CEEB5B55BB5C1	FE205FDEDE81DEBE
D38E695D5D345DD2	A090C03030F0309D	41D07E9191EF91FC	8130FEB1B14FB1E1

0D75FD78788578E7	6677441111551188	0607040101050108	6481B3E5E556E57B
0000000000000000	6D05BD6868D56867	77EF5A9898C298B4	E747BAA0A01AA069
A46133C5C5F6C566	0C0E0802020A0210	F355A2A6A604A659	2551CD7474B97487
EEC3B42D2D992D75	3A312C0B0B270B58	EB49B2A2A210A279	295FC57676B37697
8D3EF6B3B345B3F1	A31DC2BEBE7CBE99	9E501FCECED1CE3E	A914CEBDBD73BD81
C36D82AEAE2CAE19	4CA583E9E96AE91B	1B91128A8A988A24	A697C43131F53195
4854701C1C6C1CE0	52BE97ECEC7BEC33	1CEDE3F1F112F1DB	71E85E9999C799BC
5FCB6A9494FE94D4	DB7192AAAA38AA39	0EF8FFF6F609F6E3	D4F2982626BE262D
E2CDBC2F2F932F65	58B79BEFEF74EF2B	4AA287E8E86FE813	0F830A8C8C868C14
BE8BD43535E135B5	0A090C03030F0318	C21677D4D4A3D4EE	1F60E17F7F9E7FDF
20DBCBFBFB30FB8B	1E1B140505110528	BC7D23C1C1E2C146	D987655E5E3B5ECA
47D77A9090EA90F4	C0E0802020A0201D	8EB3F43D3DC93DF5	2BA9328282B08264
08FFFBF7F70CF7EB	46AC8FEAEA65EA03	3C36280A0A220A50	2E23340D0D390D68
1967E57E7E9B7ED7	2AD2C7F8F83FF893	FDAD5D50500D50BA	5C46681A1A721AD0
A26637C4C4F3C46E	12151C07071B0738	EFB8415757165782	B70FDAB8B862B8A9
88B4F03C3CCC3CFD	5133956262F76237	7093ABE3E348E34B	8A4207C8C8CFC80E
CF638AACAC26AC09	F1A35552520752AA	45218D6464E96407	6070401010501080
DA0A67D0D0B7D0CE	EC3543D9D99AD986	6A794C13135F1398	2824300C0C3C0C60
6C7E4812125A1290	F6DFA429298D2955	FBAA5951510851B2	B108DEB9B967B9A1
98571BCFCFD4CF36	CE187FD6D6A9D6FE	3744D17373A273BF	09840E8D8D838D1C
21A03E8181BF817C	E5B14D545419549A	BA7A27C0C0E7C04E	54B993EDED7EED3B
B9F7254E4E6B4E4A	85C10D444449441A	F552A6A7A701A751	FCD6A82A2A822A4D
39BC2E8585AB855C	DEFB942525B12535	6E88BFE6E659E663	864C0FCACAC5CA1E
1569ED7C7C917CC7	1D96168B8B9D8B2C	E9BF45565613568A	27A73A8080BA8074

C.6 Table T_5

501FCECED1CE3E9E	06D6BBBB6DBBB1BD	AB8BEBEB60EB0B40	D9729292E092E44B
AC8FEAEA65EA0346	4B0BCBCBC0CB1680	794C13135F13986A	7D23C1C1E2C146BC
A583E9E96AE91B4C	A6E83A3AD23ACD9C	187FD6D6A9D6FECE	39F2B2B240B2F98B
046FD2D2BDD2DED6	D77A9090EA90F447	655C17174B17B872	D2C7F8F83FF8932A
D315424257422A91	6B5415154115A87E	BF45565613568AE9	2BEAB4B45EB4C99F
26896565EC650F43	54701C1C6C1CE048	9F1A888892883417	D411434352432297
6133C5C5F6C566A4	896D5C5C315CDAD5	82D83636EE36ADB4	01D2BABA68BAB9BB
F1F3F5F506F5FB04	B8415757165782EF	28816767E6671F4F	840E8D8D838D1C09
97C43131F53195A6	F8FFF6F609F6E30E	218D6464E9640745	957D58582558FACD
FD429E9EDC9E8463	F6F7F4F403F4F302	EE882222AA220DCC	7192AAAA38AA39DB
56C97575BC758F23	2D3C0F0F330F7822	0E0802020A02100C	30FEB1B14FB1E181

275BDFDF84DFB6F8	1EA96D6DC46D4F73	44D17373A273BF37	FE294D4D644D52B3
69ED7C7C917CC715	F2982626BE262DD4	CAB82E2E962E6DE4	FFFBF7F70CF7EB08
3820080828084030	8E695D5D345DD2D3	C10D444449441A85	BAF83E3EC63EED84
FA469F9FD99F8C65	6C5014144414A078	4207C8C8CFC80E8A	6D82AEAE2CAE19C3
B14D545419549AE5	7040101050108060	3247D8D89FD88EEA	13CABCBC76BC89AF
46681A1A721AD05C	0CB16B6BDA6B7F67	02B96969D0696F6B	E3EBF3F318F3CB10
14CEBDBD73BD81A9	99CC3333FF3385AA	7696ABAB3DAB31DD	DCCFFAFA35FA8326
0D63D1D1B2D1C6DC	E6569B9BCD9BAC7D	05BD6868D568676D	F7254E4E6B4E4AB9
625816164E16B074	CC6E9595FB95DC59	D07E9191EF91FC41	B09FEEEE71EE235E
F92D4C4C614C5AB5	34916363F2633F57	8D028E8E8C8E0403	9C715B5B2A5BE2C7
5E17CCCCDBCC2E92	B4F03C3CCC3CFD88	4F6419197D19C856	40BEA1A11FA161E1
A03E8181BF817C21	E2394949704972AB	7CF17B7B8A7BFF07	3543D9D99AD986EC
10A16F6FCE6F5F7F	85DC3737EB37A5B2	3D9D6060FD60275D	4C0FCACAC5CA1E86
8FBBE7E75CE76B68	D1AC2B2B872B45FA	E53D484875487AAD	C9D3FDFD2EFDBB34
C5629696F496C453	C60945454C451283	CED7FCFC2BFCB332	DA1941415841329B
7E4812125A12906C	23340D0D390D682E	72F979798079EF0B	81B3E5E556E57B64
981E898997893C11	830A8C8C868C140F	93ABE3E348E34B70	E0802020A0201DC0
90C03030F0309DA0	2E57DCDC8BDCAEF2	22E6B7B751B7D195	19AD6C6CC16C4775
EB354A4A7F4A6AA1	2CEEB5B55BB5C199	BDFC3F3FC33FE582	C2669797F197CC55
1677D4D4A3D4EEC2	33956262F7623751	C3B42D2D992D75EE	121806061E063014
5BAAA4A40EA449FF	5CAEA5A50BA541F9	AE368383B5836C2D	80615F5F3E5FC2DF
D6A82A2A822A4DFC	3C4FDADA95DA9EE6	4503C9C9CAC9068C	0000000000000000
67E57E7E9B7ED719	49B2A2A210A279EB	B64955551C5592E3	1AC6BFBF79BF91A5
7744111155118866	1173D5D5A6D5E6C4	F34A9C9CD69C946F	571BCFCFD4CF3698
2A380E0E360E7024	36280A0A220A503C	B3F43D3DC93DF58E	AA5951510851B2FB
6EE97D7D947DCF13	DE769393E593EC4D	416C1B1B771BD85A	C0DFFEFE21FEA33E
6637C4C4F3C46EA2	C80147474647028F	3F2409092D094836	B5228686A4864433
312C0B0B270B583A	8A068F8F898F0C05	F44E9D9DD39D9C69	0BB56A6ADF6A7761
151C07071B073812	08DEB9B967B9A1B1	37FAB0B04AB0E987	EF5A9898C298B477
486018187818C050	9EC83232FA328DAC	4AD97171A871AF3B	EC314B4B7A4B62A7
B79BEFEF74EF2B58	A1EC3B3BD73BC59A	4DDD7070AD70A73D	47BAA0A01AA069E7
86B7E4E453E47362	DD1D40405D403A9D	C7DBFFFF24FFAB38	732BC3C3E8C356B0
789EA9A937A921D1	88BFE6E659E6636E	75FD78788578E70D	D5C3F9F93AF99B2C
96168B8B9D8B2C1D	CF05464643460A89	A73A8080BA807427	5A781E1E661EF044
A8E03838D838DD90	9DA3E1E142E15B7C	0FDAB8B862B8A9B7	7F9AA8A832A829D7
9AA7E0E047E0537A	24300C0C3C0C6028	E98C2323AF2305CA	5FC57676B3769729
53741D1D691DE84E	FB942525B12535DE	FC902424B4243DD8	1B1405051105281E
EDE3F1F112F1DB1C	17A56E6ECB6E5779	CB6A9494FE94D45F	D8A0282888285DF0
E1529A9AC89AA47B	BB2A8484AE84543F	A287E8E86FE8134A	4EB6A3A315A371ED
F0214F4F6E4F42BF	58C17777B6779F2F	036BD3D3B8D3D6D0	BC2E8585AB855C39
94AFE2E24DE24376	A35552520752AAF1	E4EFF2F21DF2C316	A9328282B082642B
AD5D50500D50BAFD	7BF57A7A8F7AF701	CDBC2F2F932F65E2	51CD7474B9748725

A45153530253A2F7	3EF6B3B345B3F18D	3A996161F8612F5B	6A86AFAF29AF11C5
AFE43939DD39D596	8BD43535E135B5BE	205FDEDE81DEBEFE	5913CDCDDECD2694
5D7C1F1F631FF842	E85E9999C799BC71	638AACAC26AC09CF	648EADAD23AD01C9
43D57272A772B731	C4B02C2C9C2C7DE8	2953DDDD8EDDA6F4	0A67D0D0B7D0CEDA
B2268787A1874C35	1DC2BEBE7CBE99A3	87655E5E3B5ECAD9	55A2A6A604A659F3
BE97ECEC7BEC3352	1C10040414042018	683FC6C6F9C67EAE	090C03030F03180A
8CD03434E434BDB8	DBCBFBFB30FB8B20	3B4BDBDB90DB96E0	927959592059F2CB
25E2B6B654B6D993	742FC2C2EDC25EB6	0704010105010806	EAE7F0F017F0D31A
9B755A5A2F5AEAC1	B993EDED7EED3B54	52A6A7A701A751F5	2F856666E3661749
E7842121A52115C6	60E17F7F9E7FDF1F	91128A8A988A241B	F59C2727BB2725D2
6F3BC7C7FCC776A8	7A27C0C0E7C04EBA	DFA429298D2955F6	1F7BD7D7ACD7F6C8

C.7 Table T_6

769393E593EC4DDE	43D9D99AD986EC35	529A9AC89AA47BE1	EEB5B55BB5C1992C
5A9898C298B477EF	882222AA220DCCEE	0945454C451283C6	D7FCFC2BFCB332CE
D2BABA68BAB9BB01	B56A6ADF6A77610B	5BDFDF84DFB6F827	0802020A02100C0E
469F9FD99F8C65FA	57DCDC8BDCAEF22E	5951510851B2FBAA	7959592059F2CB92
354A4A7F4A6AA1EB	5C17174B17B87265	AC2B2B872B45FAD1	2FC2C2EDC25EB674
6A9494FE94D45FCB	F7F4F403F4F302F6	D6BBBB6DBBB1BD06	B6A3A315A371ED4E
956262F762375133	B7E4E453E4736286	D97171A871AF3B4A	77D4D4A3D4EEC216
13CDCDDECD269459	DD7070AD70A73D4D	5816164E16B07462	A3E1E142E15B7C9D
394949704972ABE2	F03C3CCC3CFD88B4	27C0C0E7C04EBA7A	47D8D89FD88EEA32
6D5C5C315CDAD589	569B9BCD9BAC7DE6	8EADAD23AD01C964	2E8585AB855C39BC
5153530253A2F7A4	BEA1A11FA161E140	F57A7A8F7AF7017B	07C8C8CFC80E8A42
B42D2D992D75EEC3	A7E0E047E0537A9A	63D1D1B2D1C6DC0D	D57272A772B73143
A2A6A604A659F355	B02C2C9C2C7DE8C4	37C4C4F3C46EA266	ABE3E348E34B7093
C57676B37697295F	FD78788578E70D75	E6B7B751B7D19522	EAB4B45EB4C99F2B
2409092D0948363F	EC3B3BD73BC59AA1	380E0E360E70242A	1941415841329BDA
2D4C4C614C5AB5F9	5FDEDE81DEBEFE20	F2B2B240B2F98B39	7A9090EA90F447D7
942525B12535DEFB	AEA5A50BA541F95C	7BD7D7ACD7F6C81F	0C03030F03180A09
4411115511886677	0000000000000000	2BC3C3E8C356B073	B82E2E962E6DE4CA
729292E092E44BD9	9BEFEF74EF2B58B7	254E4E6B4E4AB9F7	4812125A12906C7E
4E9D9DD39D9C69F4	E97D7D947DCF136E	0BCBCBC0CB16804B	D43535E135B5BE8B
4010105010806070	73D5D5A6D5E6C411	214F4F6E4F42BFF0	429E9EDC9E8463FD
294D4D644D52B3FE	9EA9A937A921D178	4955551C5592E3B6	3FC6C6F9C67EAE68
67D0D0B7D0CEDA0A	F17B7B8A7BFF077C	6018187818C05048	669797F197CC55C2
6BD3D3B8D3D6D003	D83636EE36ADB482	BFE6E659E6636E88	3D484875487AADE5

45565613568AE9BF	3E8181BF817C21A0	068F8F898F0C058A	C17777B6779F2F58
17CCCCDBCC2E925E	4A9C9CD69C946FF3	DEB9B967B9A1B108	AFE2E24DE2437694
8AACAC26AC09CF63	DAB8B862B8A9B70F	BC2F2F932F65E2CD	5415154115A87E6B
AAA4A40EA449FF5B	ED7C7C917CC71569	4FDADA95DA9EE63C	E03838D838DD90A8
781E1E661EF0445A	2C0B0B270B583A31	1405051105281E1B	7FD6D6A9D6FECE18
5014144414A0786C	A56E6ECB6E577917	AD6C6CC16C477519	E57E7E9B7ED71967
856666E36617492F	D3FDFD2EFDBB34C9	FEB1B14FB1E18130	B3E5E556E57B6481
9D6060FD60275D3D	86AFAF29AF11C56A	655E5E3B5ECAD987	CC3333FF3385AA99
268787A1874C35B2	03C9C9CAC9068C45	E7F0F017F0D31AEA	695D5D345DD2D38E
A96D6DC46D4F731E	FC3F3FC33FE582BD	1A8888928834179F	0E8D8D838D1C0984
3BC7C7FCC776A86F	FBF7F70CF7EB08FF	741D1D691DE84E53	83E9E96AE91B4CA5
97ECEC7BEC3352BE	93EDED7EED3B54B9	3A8080BA807427A7	A429298D2955F6DF
9C2727BB2725D2F5	1BCFCFD4CF369857	5E9999C799BC71E8	9AA8A832A829D77F
5D50500D50BAFDAD	3C0F0F330F78222D	DC3737EB37A5B285	902424B4243DD8FC
A0282888285DF0D8	C03030F0309DA090	6E9595FB95DC59CC	6FD2D2BDD2DED604
F83E3EC63EED84BA	715B5B2A5BE2C79C	1D40405D403A9DDD	368383B5836C2DAE
F6B3B345B3F18D3E	B96969D0696F6B02	415757165782EFB8	7C1F1F631FF8425D
1C07071B07381215	701C1C6C1CE04854	128A8A988A241B91	CABCBC76BC89AF13
802020A0201DC0E0	8BEBEB60EB0B40AB	1FCECED1CE3E9E50	028E8E8C8E04038D
96ABAB3DAB31DD76	9FEEEE71EE235EB0	C43131F53195A697	B2A2A210A279EB49
D17373A273BF3744	C3F9F93AF99B2CD5	0FCACAC5CA1E864C	E83A3AD23ACD9CA6
681A1A721AD05C46	CBFBFB30FB8B20DB	340D0D390D682E23	23C1C1E2C146BC7D
DFFEFE21FEA33EC0	CFFAFA35FA8326DC	EFF2F21DF2C316E4	A16F6FCE6F5F7F10
CEBDBD73BD81A914	629696F496C453C5	53DDDD8EDDA6F429	11434352432297D4
5552520752AAF1A3	E2B6B654B6D99325	2008082808403038	EBF3F318F3CB10E3
82AEAE2CAE19C36D	C2BEBE7CBE99A31D	6419197D19C8564F	1E898997893C1198
C83232FA328DAC9E	982626BE262DD4F2	FAB0B04AB0E98737	8FEAEA65EA0346AC
314B4B7A4B62A7EC	8D6464E964074521	2A8484AE84543FBB	328282B082642BA9
B16B6BDA6B7F670C	F3F5F506F5FB04F1	F979798079EF0B72	C6BFBF79BF91A51A
0401010501080607	615F5F3E5FC2DF80	C97575BC758F2356	916363F2633F5734
6C1B1B771BD85A41	8C2323AF2305CAE9	F43D3DC93DF58EB3	BD6868D568676D05
A82A2A822A4DFCD6	896565EC650F4326	87E8E86FE8134AA2	7E9191EF91FC41D0
FFF6F609F6E30EF8	DBFFFF24FFAB38C7	4C13135F13986A79	7D58582558FACD95
E3F1F112F1DB1CED	0147474647028FC8	280A0A220A503C36	E17F7F9E7FDF1F60
33C5C5F6C566A461	A6A7A701A751F552	BBE7E75CE76B688F	996161F8612F5B3A
755A5A2F5AEAC19B	1806061E06301412	05464643460A89CF	0D444449441A85C1
15424257422A91D3	100404140420181C	BAA0A01AA069E747	4BDBDB90DB96E03B
E43939DD39D596AF	228686A4864433B5	4D545419549AE5B1	92AAAA38AA39DB71
0A8C8C868C140F83	D03434E434BDB88C	842121A52115C6E7	168B8B9D8B2C1D96
C7F8F83FF8932AD2	300C0C3C0C602824	CD7474B974872551	816767E6671F4F28

C.8 Table T_7

6868D568676D05BD	8D8D838D1C09840E	CACAC5CA1E864C0F	4D4D644D52B3FE29
7373A273BF3744D1	4B4B7A4B62A7EC31	4E4E6B4E4AB9F725	2A2A822A4DFCD6A8
D4D4A3D4EEC21677	52520752AAF1A355	2626BE262DD4F298	B3B345B3F18D3EF6
545419549AE5B14D	1E1E661EF0445A78	19197D19C8564F64	1F1F631FF8425D7C
2222AA220DCCEE88	03030F03180A090C	464643460A89CF05	3D3DC93DF58EB3F4
2D2D992D75EEC3B4	4A4A7F4A6AA1EB35	53530253A2F7A451	8383B5836C2DAE36
13135F13986A794C	8A8A988A241B9112	B7B751B7D19522E6	D5D5A6D5E6C41173
2525B12535DEFB94	79798079EF0B72F9	F5F506F5FB04F1F3	BDBD73BD81A914CE
58582558FACD957D	2F2F932F65E2CDBC	0D0D390D682E2334	02020A02100C0E08
EDED7EED3B54B993	51510851B2FBAA59	9E9EDC9E8463FD42	1111551188667744
F2F21DF2C316E4EF	3E3EC63EED84BAF8	55551C5592E3B649	5E5E3B5ECAD98765
D1D1B2D1C6DC0D63	16164E16B0746258	3C3CCC3CFD88B4F0	6666E36617492F85
7070AD70A73D4DDD	5D5D345DD2D38E69	F3F318F3CB10E3EB	45454C451283C609
40405D403A9DDD1D	CCCCDBCC2E925E17	E8E86FE8134AA287	9494FE94D45FCB6A
565613568AE9BF45	0808280840303820	CECED1CE3E9E501F	1A1A721AD05C4668
3A3AD23ACD9CA6E8	D2D2BDD2DED6046F	E1E142E15B7C9DA3	DFDF84DFB6F8275B
B5B55BB5C1992CEE	3838D838DD90A8E0	6E6ECB6E577917A5	0E0E360E70242A38
E5E556E57B6481B3	F4F403F4F302F6F7	F9F93AF99B2CD5C3	8686A4864433B522
E9E96AE91B4CA583	4F4F6E4F42BFF021	D6D6A9D6FECE187F	8585AB855C39BC2E
2323AF2305CAE98C	CFCFD4CF3698571B	3232FA328DAC9EC8	9999C799BC71E85E
3131F53195A697C4	14144414A0786C50	AEAE2CAE19C36D82	EEEE71EE235EB09F
C8C8CFC80E8A4207	484875487AADE53D	D3D3B8D3D6D0036B	3030F0309DA090C0
A1A11FA161E140BE	9292E092E44BD972	41415841329BDA19	B1B14FB1E18130FE
18187818C0504860	C4C4F3C46EA26637	2C2C9C2C7DE8C4B0	7171A871AF3B4AD9
7272A772B73143D5	444449441A85C10D	15154115A87E6B54	FDFD2EFDBB34C9D3
3737EB37A5B285DC	BEBE7CBE99A31DC2	5F5F3E5FC2DF8061	AAAA38AA39DB7192
9B9BCD9BAC7DE656	8888928834179F1A	D8D89FD88EEA3247	ABAB3DAB31DD7696
898997893C11981E	9C9CD69C946FF34A	FAFA35FA8326DCCF	6060FD60275D3D9D
EAEA65EA0346AC8F	BCBC76BC89AF13CA	6262F76237513395	0C0C3C0C60282430
2424B4243DD8FC90	A6A604A659F355A2	A8A832A829D77F9A	ECEC7BEC3352BE97
6767E6671F4F2881	2020A0201DC0E080	DBDB90DB96E03B4B	7C7C917CC71569ED
282888285DF0D8A0	DDDD8EDDA6F42953	ACAC26AC09CF638A	5B5B2A5BE2C79C71
3434E434BDB88CD0	7E7E9B7ED71967E5	1010501080607040	F1F112F1DB1CEDE3
7B7B8A7BFF077CF1	8F8F898F0C058A06	6363F2633F573491	A0A01AA069E747BA
05051105281E1B14	9A9AC89AA47BE152	434352432297D411	7777B6779F2F58C1
2121A52115C6E784	BFBF79BF91A51AC6	2727BB2725D2F59C	09092D0948363F24
C3C3E8C356B0732B	9F9FD99F8C65FA46	B6B654B6D99325E2	D7D7ACD7F6C81F7B
29298D2955F6DFA4	C2C2EDC25EB6742F	EBEB60EB0B40AB8B	C0C0E7C04EBA7A27
A4A40EA449FF5BAA	8B8B9D8B2C1D9616	8C8C868C140F830A	1D1D691DE84E5374
FBFB30FB8B20DBCB	FFFF24FFAB38C7DB	C1C1E2C146BC7D23	B2B240B2F98B39F2

9797F197CC55C266	2E2E962E6DE4CAB8	F8F83FF8932AD2C7	6565EC650F432689
F6F609F6E30EF8FF	7575BC758F2356C9	07071B073812151C	0404140420181C10
4949704972ABE239	3333FF3385AA99CC	E4E453E4736286B7	D9D99AD986EC3543
B9B967B9A1B108DE	D0D0B7D0CEDA0A67	424257422A91D315	C7C7FCC776A86F3B
6C6CC16C477519AD	9090EA90F447D77A	0000000000000000	8E8E8C8E04038D02
6F6FCE6F5F7F10A1	50500D50BAFDAD5D	0101050108060704	C5C5F6C566A46133
DADA95DA9EE63C4F	47474647028FC801	3F3FC33FE582BDFC	CDCDDECD26945913
6969D0696F6B02B9	A2A210A279EB49B2	E2E24DE2437694AF	7A7A8F7AF7017BF5
A7A701A751F552A6	C6C6F9C67EAE683F	9393E593EC4DDE76	0F0F330F78222D3C
0A0A220A503C3628	06061E0630141218	E6E659E6636E88BF	2B2B872B45FAD1AC
9696F496C453C562	A3A315A371ED4EB6	1C1C6C1CE0485470	AFAF29AF11C56A86
6A6ADF6A77610BB5	12125A12906C7E48	8484AE84543FBB2A	3939DD39D596AFE4
E7E75CE76B688FBB	B0B04AB0E98737FA	8282B082642BA932	F7F70CF7EB08FFFB
FEFE21FEA33EC0DF	9D9DD39D9C69F44E	8787A1874C35B226	5C5C315CDAD5896D
8181BF817C21A03E	3535E135B5BE8BD4	DEDE81DEBEFE205F	B4B45EB4C99F2BEA
A5A50BA541F95CAE	FCFC2BFCB332CED7	8080BA807427A73A	EFEF74EF2B58B79B
CBCBC0CB16804B0B	BBBB6DBBB1BD06D6	6B6BDA6B7F670CB1	7676B37697295FC5
BABA68BAB9BB01D2	5A5A2F5AEAC19B75	7D7D947DCF136EE9	78788578E70D75FD
0B0B270B583A312C	9595FB95DC59CC6E	E3E348E34B7093AB	ADAD23AD01C9648E
7474B974872551CD	9898C298B477EF5A	3B3BD73BC59AA1EC	3636EE36ADB482D8
6464E9640745218D	6D6DC46D4F731EA9	DCDC8BDCAEF22E57	F0F017F0D31AEAE7
59592059F2CB9279	A9A937A921D1789E	4C4C614C5AB5F92D	17174B17B872655C
7F7F9E7FDF1F60E1	9191EF91FC41D07E	B8B862B8A9B70FDA	C9C9CAC9068C4503
5757165782EFB841	1B1B771BD85A416C	E0E047E0537A9AA7	6161F8612F5B3A99

Appendix D

Tables constants $Mul_{\alpha}[a] Mul_{\alpha^{-1}}[a]$ cipher “Strumok”.
(hexadecimal representation).

D.1 Table $Mul_{\alpha}[a]$

0000000000000000	D73F04125E000004	B37E0824BC000008	64410C36E200000C
7BFC104865000010	ACC3145A3B000014	C882186CD9000018	1FBD1C7E8700001C
F6E52090CA000020	21DA248294000024	459B28B476000028	92A42CA62800002C
8D1930D8AF000030	5A2634CAF1000034	3E6738FC13000038	E9583CEE4D00003C
F1D7403D89000040	26E8442FD7000044	42A9481935000048	95964C0B6B00004C
8A2B5075EC000050	5D145467B2000054	3955585150000058	EE6A5C430E00005C
073260AD43000060	D00D64BF1D000064	B44C6889FF000068	63736C9BA100006C
7CCE70E526000070	ABF174F778000074	CFB078C19A000078	188F7CD3C400007C
FFB3807A0F000080	288C846851000084	4CCD885EB3000088	9BF28C4CED00008C
844F90326A000090	5370942034000094	37319816D6000098	E00E9C048800009C
0956A0EAC50000A0	DE69A4F89B0000A4	BA28A8CE790000A8	6D17ACDC270000AC
72AAB0A2A00000B0	A595B4B0FE0000B4	C1D4B8861C0000B8	16EBBC94420000BC
0E64C047860000C0	D95BC455D80000C4	BD1AC8633A0000C8	6A25CC71640000CC
7598D00FE30000D0	A2A7D41DBD0000D4	C6E6D82B5F0000D8	11D9DC39010000DC
F881E0D74C0000E0	2FBEE4C5120000E4	4BFFE8F3F00000E8	9CC0ECE1AE0000EC
837DF09F290000F0	5442F48D770000F4	3003F8BB950000F8	E73CFCA9CB0000FC
E37B1DF41E00001D	344419E640000019	500515D0A2000015	873A11C2FC000011
98870DBC7B00000D	4FB809AE25000009	2BF90598C7000005	FCC6018A99000001
159E3D64D400003D	C2A139768A000039	A6E0354068000035	71DF315236000031
6E622D2CB100002D	B95D293EEF000029	DD1C25080D000025	0A23211A53000021
12AC5DC99700005D	C59359DBC9000059	A1D255ED2B000055	76ED51FF75000051
69504D81F200004D	BE6F4993AC000049	DA2E45A54E000045	0D1141B710000041

A. A. Kuznetsov et al., *Stream Ciphers in Modern Real-time IT Systems*,
Studies in Systems, Decision and Control 375,
https://doi.org/10.1007/978-3-030-79770-6

E4497D595D00007D	3376794B03000079	5737757DE1000075	8008716FBF000071
9FB56D113800006D	488A690366000069	2CCB653584000065	FBF46127DA000061
1CC89D8E1100009D	CBF7999C4F000099	AFB695AAAD000095	788991B8F3000091
67348DC67400008D	B00B89D42A000089	D44A85E2C8000085	037581F096000081
EA2DBD1EDB0000BD	3D12B90C850000B9	5953B53A670000B5	8E6CB128390000B1
91D1AD56BE0000AD	46EEA944E00000A9	22AFA572020000A5	F590A1605C0000A1
ED1FDDB3980000DD	3A20D9A1C60000D9	5E61D597240000D5	895ED1857A0000D1
96E3CDFBFD0000CD	41DCC9E9A30000C9	259DC5DF410000C5	F2A2C1CD1F0000C1
1BFAFD23520000FD	CCC5F9310C0000F9	A884F507EE0000F5	7FBBF115B00000F1
6006ED6B370000ED	B739E979690000E9	D378E54F8B0000E5	0447E15DD50000E1
DBF63AF53C00003A	0CC93EE76200003E	688832D180000032	BFB736C3DE000036
A00A2ABD5900002A	77352EAF0700002E	13742299E5000022	C44B268BBB000026
2D131A65F600001A	FA2C1E77A800001E	9E6D12414A000012	4952165314000016
56EF0A2D9300000A	81D00E3FCD00000E	E59102092F000002	32AE061B71000006
2A217AC8B500007A	FD1E7EDAEB00007E	995F72EC09000072	4E6076FE57000076
51DD6A80D000006A	86E26E928E00006E	E2A362A46C000062	359C66B632000066
DCC45A587F00005A	0BFB5E4A2100005E	6FBA527CC3000052	B885566E9D000056
A7384A101A00004A	70074E024400004E	14464234A6000042	C3794626F8000046
2445BA8F330000BA	F37ABE9D6D0000BE	973BB2AB8F0000B2	4004B6B9D10000B6
5FB9AAC7560000AA	8886AED5080000AE	ECC7A2E3EA0000A2	3BF8A6F1B40000A6
D2A09A1FF900009A	059F9E0DA700009E	61DE923B45000092	B6E196291B000096
A95C8A579C00008A	7E638E45C200008E	1A22827320000082	CD1D86617E000086
D592FAB2BA0000FA	02ADFEA0E40000FE	66ECF296060000F2	B1D3F684580000F6
AE6EEAFADF0000EA	7951EEE8810000EE	1D10E2DE630000E2	CA2FE6CC3D0000E6
2377DA22700000DA	F448DE302E0000DE	9009D206CC0000D2	4736D614920000D6
588BCA6A150000CA	8FB4CE784B0000CE	EBF5C24EA90000C2	3CCAC65CF70000C6
388D270122000027	EFB223137C000023	8BF32F259E00002F	5CCC2B37C000002B
4371374947000037	944E335B19000033	F00F3F6DFB00003F	27303B7FA500003B
CE680791E8000007	19570383B6000003	7D160FB55400000F	AA290BA70A00000B
B59417D98D000017	62AB13CBD3000013	06EA1FFD3100001F	D1D51BEF6F00001B
C95A673CAB000067	1E65632EF5000063	7A246F181700006F	AD1B6B0A4900006B
B2A67774CE000077	6599736690000073	01D87F507200007F	D6E77B422C00007B
3FBF47AC61000047	E88043BE3F000043	8CC14F88DD00004F	5BFE4B9A8300004B
444357E404000057	937C53F65A000053	F73D5FC0B800005F	20025BD2E600005B
C73EA77B2D0000A7	1001A369730000A3	7440AF5F910000AF	A37FAB4DCF0000AB
BCC2B733480000B7	6BFDB321160000B3	0FBCBF17F40000BF	D883BB05AA0000BB
31DB87EBE7000087	E6E483F9B9000083	82A58FCF5B00008F	559A8BDD0500008B
4A2797A382000097	9D1893B1DC000093	F9599F873E00009F	2E669B956000009B
36E9E746A40000E7	E1D6E354FA0000E3	8597EF62180000EF	52A8EB70460000EB
4D15F70EC10000F7	9A2AF31C9F0000F3	FE6BFF2A7D0000FF	2954FB38230000FB
C00CC7D66E0000C7	1733C3C4300000C3	7372CFF2D20000CF	A44DCBE08C0000CB
BBF0D79E0B0000D7	6CCFD38C550000D3	088EDFBAB70000DF	DFB1DBA8E90000DB

D.2 *Table* $Mul_{\alpha^{-1}}[a]$[1]

0000000000000000	47FCC6018A990000	8EE59102092F0000	C919570383B60000
01D73F04125E0000	462BF90598C70000	8F32AE061B710000	C8CE680791E80000
02B37E0824BC0000	454FB809AE250000	8C56EF0A2D930000	CBAA290BA70A0000
0364410C36E20000	4498870DBC7B0000	8D81D00E3FCD0000	CA7D160FB5540000
047BFC1048650000	43873A11C2FC0000	8A9E6D12414A0000	CD62AB13CBD30000
05ACC3145A3B0000	42500515D0A20000	8B49521653140000	CCB59417D98D0000
06C882186CD90000	41344419E6400000	882D131A65F60000	CFD1D51BEF6F0000
071FBD1C7E870000	40E37B1DF41E0000	89FA2C1E77A80000	CE06EA1FFD310000
08F6E52090CA0000	4F0A23211A530000	8613742299E50000	C1EFB223137C0000
0921DA2482940000	4EDD1C25080D0000	87C44B268BBB0000	C0388D2701220000
0A459B28B4760000	4DB95D293EEF0000	84A00A2ABD590000	C35CCC2B37C00000
0B92A42CA6280000	4C6E622D2CB10000	8577352EAF070000	C28BF32F259E0000
0C8D1930D8AF0000	4B71DF3152360000	82688832D1800000	C5944E335B190000
0D5A2634CAF10000	4AA6E03540680000	83BFB736C3DE0000	C443713749470000
0E3E6738FC130000	49C2A139768A0000	80DBF63AF53C0000	C727303B7FA50000
0FE9583CEE4D0000	48159E3D64D40000	810CC93EE7620000	C6F00F3F6DFB0000
10F1D7403D890000	570D1141B7100000	9E14464234A60000	D9E88043BE3F0000
1126E8442FD70000	56DA2E45A54E0000	9FC3794626F80000	D83FBF47AC610000
1242A94819350000	55BE6F4993AC0000	9CA7384A101A0000	DB5BFE4B9A830000
1395964C0B6B0000	5469504D81F20000	9D70074E02440000	DA8CC14F88DD0000
148A2B5075EC0000	5376ED51FF750000	9A6FBA527CC30000	DD937C53F65A0000
155D145467B20000	52A1D255ED2B0000	9BB885566E9D0000	DC444357E4040000
1639555851500000	51C59359DBC90000	98DCC45A587F0000	DF20025BD2E60000
17EE6A5C430E0000	5012AC5DC9970000	990BFB5E4A210000	DEF73D5FC0B80000
18073260AD430000	5FFBF46127DA0000	96E2A362A46C0000	D11E65632EF50000
19D00D64BF1D0000	5E2CCB6535840000	97359C66B6320000	D0C95A673CAB0000
1AB44C6889FF0000	5D488A6903660000	9451DD6A80D00000	D3AD1B6B0A490000
1B63736C9BA10000	5C9FB56D11380000	9586E26E928E0000	D27A246F18170000
1C7CCE70E5260000	5B8008716FBF0000	92995F72EC090000	D565997366900000
1DABF174F7780000	5A5737757DE10000	934E6076FE570000	D4B2A67774CE0000
1ECFB078C19A0000	593376794B030000	902A217AC8B50000	D7D6E77B422C0000
1F188F7CD3C40000	58E4497D595D0000	91FD1E7EDAEB0000	D601D87F50720000
20FFB3807A0F0000	67037581F0960000	AE1A228273200000	E9E6E483F9B90000
21288C8468510000	66D44A85E2C80000	AFCD1D86617E0000	E831DB87EBE70000
224CCD885EB30000	65B00B89D42A0000	ACA95C8A579C0000	EB559A8BDD050000
239BF28C4CED0000	6467348DC6740000	AD7E638E45C20000	EA82A58FCF5B0000
24844F90326A0000	63788991B8F30000	AA61DE923B450000	ED9D1893B1DC0000
2553709420340000	62AFB695AAAD0000	ABB6E196291B0000	EC4A2797A3820000
2637319816D60000	61CBF7999C4F0000	A8D2A09A1FF90000	EF2E669B95600000
27E00E9C04880000	601CC89D8E110000	A9059F9E0DA70000	EEF9599F873E0000

280956A0EAC50000	6FF590A1605C0000	A6ECC7A2E3EA0000	E11001A369730000
29DE69A4F89B0000	6E22AFA572020000	A73BF8A6F1B40000	E0C73EA77B2D0000
2ABA28A8CE790000	6D46EEA944E00000	A45FB9AAC7560000	E3A37FAB4DCF0000
2B6D17ACDC270000	6C91D1AD56BE0000	A58886AED5080000	E27440AF5F910000
2C72AAB0A2A00000	6B8E6CB128390000	A2973BB2AB8F0000	E56BFDB321160000
2DA595B4B0FE0000	6A5953B53A670000	A34004B6B9D10000	E4BCC2B733480000
2EC1D4B8861C0000	693D12B90C850000	A02445BA8F330000	E7D883BB05AA0000
2F16EBBC94420000	68EA2DBD1EDB0000	A1F37ABE9D6D0000	E60FBCBF17F40000
300E64C047860000	77F2A2C1CD1F0000	BEEBF5C24EA90000	F91733C3C4300000
31D95BC455D80000	76259DC5DF410000	BF3CCAC65CF70000	F8C00CC7D66E0000
32BD1AC8633A0000	7541DCC9E9A30000	BC588BCA6A150000	FBA44DCBE08C0000
336A25CC71640000	7496E3CDFBFD0000	BD8FB4CE784B0000	FA7372CFF2D20000
347598D00FE30000	73895ED1857A0000	BA9009D206CC0000	FD6CCFD38C550000
35A2A7D41DBD0000	725E61D597240000	BB4736D614920000	FCBBF0D79E0B0000
36C6E6D82B5F0000	713A20D9A1C60000	B82377DA22700000	FFDFB1DBA8E90000
3711D9DC39010000	70ED1FDDB3980000	B9F448DE302E0000	FE088EDFBAB70000
38F881E0D74C0000	7F0447E15DD50000	B61D10E2DE630000	F1E1D6E354FA0000
392FBEE4C5120000	7ED378E54F8B0000	B7CA2FE6CC3D0000	F036E9E746A40000
3A4BFFE8F3F00000	7DB739E979690000	B4AE6EEAFADF0000	F352A8EB70460000
3B9CC0ECE1AE0000	7C6006ED6B370000	B57951EEE8810000	F28597EF62180000
3C837DF09F290000	7B7FBBF115B00000	B266ECF296060000	F59A2AF31C9F0000
3D5442F48D770000	7AA884F507EE0000	B3B1D3F684580000	F44D15F70EC10000
3E3003F8BB950000	79CCC5F9310C0000	B0D592FAB2BA0000	F72954FB38230000
3FE73CFCA9CB0000	781BFAFD23520000	B102ADFEA0E40000	F6FE6BFF2A7D0000

Appendix E

Examples for checking the code "Stream" (hexadecimal representation).

E.1 256-Bit Key

E.1.1 Key Initialization Vector and Key Sequence

(1) Initialization vector $IV = (IV_3, IV_2, IV_1, IV_0)$:

IV_3:0000000000000000 IV_2:0000000000000000
IV_1:0000000000000000 IV_0:0000000000000000

Key $K = (K_3, K_2, K_1, K_0)$:

K_3:8000000000000000 K_2:0000000000000000
K_1:0000000000000000 K_0:0000000000000000

Key stream output Z:

```
Z0:E442D15345DC66CA   Z1:F47D700ECC66408A
Z2:B4CB284B5477E641   Z3:A2AFC9092E4124B0
Z4:728E5FA26B11A7D9   Z5:E6A7B9288C68F972
Z6:70EB3606DE8BA44C   Z7:ACED7956BD3E3DE7
```


A. A. Kuznetsov et al., *Stream Ciphers in Modern Real-time IT Systems*,
Studies in Systems, Decision and Control 375,
https://doi.org/10.1007/978-3-030-79770-6

(2) Initialization vector $IV = (IV_3, IV_2, IV_1, IV_0)$:

```
IV3:0000000000000000   IV2:0000000000000000
IV1:0000000000000000   IV0:0000000000000000
```

Key $K = (K_3, K_2, K_1, K_0)$:

```
K3:AAAAAAAAAAAAAAAA   K2:AAAAAAAAAAAAAAAA
K2:AAAAAAAAAAAAAAAA   K0:AAAAAAAAAAAAAAAA
```

Key stream output Z:

```
Z0:A7510B38C7A95D1D   Z1:CD5EA28A15B8654F
Z2:C5E2E2771D0373B2   Z3:98AE829686D5FCEE
Z4:45BDDF65C523DBB8   Z5:32A93FCDD950001F
Z6:752A7FB588AF8C51   Z7:9DE92736664212D4
```

(3) Initialization vector $IV = (IV_3, IV_2, IV_1, IV_0)$:

```
IV3:0000000000000004   IV2:0000000000000003
IV1:0000000000000002   IV0:0000000000000001
```

Key $K = (K_3, K_2, K_1, K_0)$:

```
K3:8000000000000000   K2:0000000000000000
K1:0000000000000000   K0:0000000000000000
```

Key stream output Z:

```
Z0:FE44A2508B5A2ACD   Z1:AF355B4ED21D2742
Z2:DCD7FDD6A57A9E71   Z3:5D267BD2739FB5EB
Z4:B22EEE96B2832072   Z5:C7DE6A4CDAA9A847
Z6:72D5DA93812680F2   Z7:4A0ACB7E93DA2CE0
```

(4) Initialization vector $IV = (IV_3, IV_2, IV_1, IV_0)$:

```
IV3:0000000000000004   IV2:0000000000000003
IV1:0000000000000002   IV0:0000000000000001
```

Key $K = (K_3, K_2, K_1, K_0)$:

```
K3:AAAAAAAAAAAAAAAA   K2:AAAAAAAAAAAAAAAA
K2:AAAAAAAAAAAAAAAA   K0:AAAAAAAAAAAAAAAA
```

Key stream output Z:

```
Z0:E6D0EFD9CEA5ABCD   Z1:1E78BA1A9B0E401E
Z2:BCFBEA2C02BA0781   Z3:1BD375588AE08794
Z4:5493CF21E114C209   Z5:66CD5D7CC7D0E69A
Z6:A5CDB9F3380D07FA   Z7:2940D61A4D4E9CE4
```

E.1.2 Sample of Internal States

Initialization vector $IV = (IV_3, IV_2, IV_1, IV_0)$:

```
IV3:0000000000000004   IV2:0000000000000003
IV1:0000000000000002   IV0:0000000000000001
```

Key $K = (K_3, K_2, K_1, K_0)$:

```
K3:8000000000000000   K2:0000000000000000
K1:0000000000000000   K0:0000000000000000
```

Internal state in time -34.

```
15:FFFFFFFFFFFFFFFF   14:0000000000000000
13:FFFFFFFFFFFFFFFF   12:8000000000000000
11:0000000000000000   10:FFFFFFFFFFFFFFFF
09:0000000000000000   08:8000000000000000
07:FFFFFFFFFFFFFFFF   06:FFFFFFFFFFFFFFFF
05:0000000000000000   04:8000000000000000
03:0000000000000000   02:0000000000000000
01:0000000000000000   00:8000000000000000
```

Internal state in time -33.

```
15:FFFFFFFFFFFFFFFF   14:0000000000000000
13:FFFFFFFFFFFFFFFF   12:8000000000000000
```

```
11:0000000000000000   10:FFFFFFFFFFFFFFFF
09:0000000000000000   08:8000000000000000
07:FFFFFFFFFFFFFFFF   06:FFFFFFFFFFFFFFFF
05:0000000000000004   04:8000000000000000
03:0000000000000003   02:0000000000000002
01:0000000000000000   00:8000000000000001
r1:0000000000000000   r2:0000000000000000
```

Internal state in time −32.

```
15:DBF63AF53C00013A   14:FFFFFFFFFFFFFFFF
13:0000000000000000   12:FFFFFFFFFFFFFFFF
11:8000000000000000   10:0000000000000000
09:FFFFFFFFFFFFFFFF   08:0000000000000000
07:8000000000000000   06:FFFFFFFFFFFFFFFF
05:FFFFFFFFFFFFFFFF   04:0000000000000004
03:8000000000000000   02:0000000000000003
01:0000000000000002   00:0000000000000000
r1:FFFFFFFFFFFFFFFF   r2:14E6CC3F14E6CC3F
```

Internal state in time −31.

```
15:CF90F6CA28E6CD06   14:DBF63AF53C00013A
13:FFFFFFFFFFFFFFFF   12:0000000000000000
11:FFFFFFFFFFFFFFFF   10:8000000000000000
09:0000000000000000   08:FFFFFFFFFFFFFFFF
07:0000000000000000   06:8000000000000000
05:FFFFFFFFFFFFFFFF   04:FFFFFFFFFFFFFFFF
03:0000000000000004   02:8000000000000000
01:0000000000000003   00:0000000000000002
r1:14E6CC3F14E6CC3F   r2:BCF371DDBCF371DD
```

Internal state in time −30.

```
15:517AD92BAB43EA98   14:CF90F6CA28E6CD06
13:DBF63AF53C00013A   12:FFFFFFFFFFFFFFFF
11:0000000000000000   10:FFFFFFFFFFFFFFFF
09:8000000000000000   08:0000000000000000
07:FFFFFFFFFFFFFFFF   06:0000000000000000
```

```
05:8000000000000000   04:FFFFFFFFFFFFFFFF
03:FFFFFFFFFFFFFFFF   02:0000000000000004
01:8000000000000000   00:0000000000000003
r1:BCF371DDBCF371DC   r2:E38BD2D9E38BD2D9
```

Internal state in time −29.

```
15:3613A325B7BC8C97   14:517AD92BAB43EA98
13:CF90F6CA28E6CD06   12:DBF63AF53C00013A
11:FFFFFFFFFFFFFFFF   10:0000000000000000
09:FFFFFFFFFFFFFFFF   08:8000000000000000
07:0000000000000000   06:FFFFFFFFFFFFFFFF
05:0000000000000000   04:8000000000000000
03:FFFFFFFFFFFFFFFF   02:FFFFFFFFFFFFFFFF
01:0000000000000004   00:8000000000000000
r1:BF820DCF1F8BD413   r2:5E3B32CAD987095E
```

Internal state in time −28.

```
15:49C9DA01CFAB5B37   14:3613A325B7BC8C97
13:517AD92BAB43EA98   12:CF90F6CA28E6CD06
11:DBF63AF53C00013A   10:FFFFFFFFFFFFFFFF
09:0000000000000000   08:FFFFFFFFFFFFFFFF
07:8000000000000000   06:0000000000000000
05:FFFFFFFFFFFFFFFF   04:0000000000000000
03:8000000000000000   02:FFFFFFFFFFFFFFFF
01:FFFFFFFFFFFFFFFF   00:0000000000000004
r1:2DCC2995026DD664   r2:C078ECF575BCD573
```

Internal state in time −27.

```
15:669436480CE60A71   14:49C9DA01CFAB5B37
13:3613A325B7BC8C97   12:517AD92BAB43EA98
11:CF90F6CA28E6CD06   10:DBF63AF53C00013A
09:FFFFFFFFFFFFFFFF   08:0000000000000000
07:FFFFFFFFFFFFFFFF   06:8000000000000000
05:0000000000000000   04:FFFFFFFFFFFFFFFF
03:0000000000000000   02:8000000000000000
01:FFFFFFFFFFFFFFFF   00:FFFFFFFFFFFFFFFF
r1:11F3C6212100C00B   r2:32816DFE63910730
```

Internal state in time −26.

```
15:D3A628153E6D58CD  14:669436480CE60A71
13:49C9DA01CFAB5B37  12:3613A325B7BC8C97
11:517AD92BAB43EA98  10:CF90F6CA28E6CD06
09:DBF63AF53C00013A  08:FFFFFFFFFFFFFFFF
07:0000000000000000  06:FFFFFFFFFFFFFFFF
05:8000000000000000  04:0000000000000000
03:FFFFFFFFFFFFFFFF  02:0000000000000000
01:8000000000000000  00:FFFFFFFFFFFFFFFF
r1:689511241B4D93C7  r2:6F72A1049C69B875
```

Internal state in time −25.

```
15:1CA82D2A21FAB3E7  14:D3A628153E6D58CD
13:669436480CE60A71  12:49C9DA01CFAB5B37
11:3613A325B7BC8C97  10:517AD92BAB43EA98
09:CF90F6CA28E6CD06  08:DBF63AF53C00013A
07:FFFFFFFFFFFFFFFF  06:0000000000000000
05:FFFFFFFFFFFFFFFF  04:8000000000000000
03:0000000000000000  02:FFFFFFFFFFFFFFFF
01:0000000000000000  00:8000000000000000
r1:B93C7B066C1513AC  r2:9680E3F5B5591C0C
```

Internal state in time −24.

```
15:127A734C8D856D58  14:1CA82D2A21FAB3E7
13:D3A628153E6D58CD  12:669436480CE60A71
11:49C9DA01CFAB5B37  10:3613A325B7BC8C97
09:517AD92BAB43EA98  08:CF90F6CA28E6CD06
07:DBF63AF53C00013A  06:FFFFFFFFFFFFFFFF
05:0000000000000000  04:FFFFFFFFFFFFFFFF
03:8000000000000000  02:0000000000000000
01:FFFFFFFFFFFFFFFF  00:0000000000000000
r1:FD151A3DC23F267D  r2:B6C5DF4E874642B8
```

Internal state in time −23.

```
15:AEE6C23CBE6722FB  14:127A734C8D856D58
13:1CA82D2A21FAB3E7  12:D3A628153E6D58CD
```

```
11:669436480CE60A71   10:49C9DA01CFAB5B37
09:3613A325B7BC8C97   08:517AD92BAB43EA98
07:CF90F6CA28E6CD06   06:DBF63AF53C00013A
05:FFFFFFFFFFFFFFFF   04:0000000000000000
03:FFFFFFFFFFFFFFFF   02:8000000000000000
01:0000000000000000   00:FFFFFFFFFFFFFFFF
r1:8A6C0763C5B39B85   r2:D82E146B04E05BA3
```

Internal state in time −22.

```
15:867C48F1904C4F15   14:AEE6C23CBE6722FB
13:127A734C8D856D58   12:1CA82D2A21FAB3E7
11:D3A628153E6D58CD   10:669436480CE60A71
09:49C9DA01CFAB5B37   08:3613A325B7BC8C97
07:517AD92BAB43EA98   06:CF90F6CA28E6CD06
05:DBF63AF53C00013A   04:FFFFFFFFFFFFFFFF
03:0000000000000000   02:FFFFFFFFFFFFFFFF
01:8000000000000000   00:0000000000000000
r1:F4D6419526DB0F8A   r2:0EC0F27AD0FBDB43
```

Internal state in time −21.

```
15:13AD4E55049A85DC   14:867C48F1904C4F15
13:AEE6C23CBE6722FB   12:127A734C8D856D58
11:1CA82D2A21FAB3E7   10:D3A628153E6D58CD
09:669436480CE60A71   08:49C9DA01CFAB5B37
07:3613A325B7BC8C97   06:517AD92BAB43EA98
05:CF90F6CA28E6CD06   04:DBF63AF53C00013A
03:FFFFFFFFFFFFFFFF   02:0000000000000000
01:FFFFFFFFFFFFFFFF   00:8000000000000000
r1:213B65C75E81489B   r2:A6DB6721BD55D12C
```

Internal state in time −20.

```
15:17096A3E30ACC729   14:13AD4E55049A85DC
13:867C48F1904C4F15   12:AEE6C23CBE6722FB
11:127A734C8D856D58   10:1CA82D2A21FAB3E7
09:D3A628153E6D58CD   08:669436480CE60A71
07:49C9DA01CFAB5B37   06:3613A325B7BC8C97
```

```
05:517AD92BAB43EA98   04:CF90F6CA28E6CD06
03:DBF63AF53C00013A   02:FFFFFFFFFFFFFFFF
01:0000000000000000   00:FFFFFFFFFFFFFFFF
r1:55C2295E7BBCF427   r2:37B926DACBADE0D3
```

Internal state in time −19.

```
15:EB6BF6CBFCAA6E20   14:17096A3E30ACC729
13:13AD4E55049A85DC   12:867C48F1904C4F15
11:AEE6C23CBE6722FB   10:127A734C8D856D58
09:1CA82D2A21FAB3E7   08:D3A628153E6D58CD
07:669436480CE60A71   06:49C9DA01CFAB5B37
05:3613A325B7BC8C97   04:517AD92BAB43EA98
03:CF90F6CA28E6CD06   02:DBF63AF53C00013A
01:FFFFFFFFFFFFFFFF   00:0000000000000000
r1:BE356FCC5BFA2FE8   r2:914C57600B3AA47E
```

Internal state in time −18.

```
15:DCC7CD945399D888   14:EB6BF6CBFCAA6E20
13:17096A3E30ACC729   12:13AD4E55049A85DC
11:867C48F1904C4F15   10:AEE6C23CBE6722FB
09:127A734C8D856D58   08:1CA82D2A21FAB3E7
07:D3A628153E6D58CD   06:669436480CE60A71
05:49C9DA01CFAB5B37   04:3613A325B7BC8C97
03:517AD92BAB43EA98   02:CF90F6CA28E6CD06
01:DBF63AF53C00013A   00:FFFFFFFFFFFFFFFF
r1:A4F9A5B50FD52A5A   r2:0E4ED9EDAC9C7FA5
```

Internal state in time −17.

```
15:FA1E9D90C89209FA   14:DCC7CD945399D888
13:EB6BF6CBFCAA6E20   12:17096A3E30ACC729
11:13AD4E55049A85DC   10:867C48F1904C4F15
09:AEE6C23CBE6722FB   08:127A734C8D856D58
07:1CA82D2A21FAB3E7   06:D3A628153E6D58CD
05:669436480CE60A71   04:49C9DA01CFAB5B37
03:3613A325B7BC8C97   02:517AD92BAB43EA98
01:CF90F6CA28E6CD06   00:DBF63AF53C00013A
r1:2558442BDD4946CE   r2:00EA7127D085D7B7
```

Internal state in time −16.

15:6E31AC6466F04991	14:FA1E9D90C89209FA
13:DCC7CD945399D888	12:EB6BF6CBFCAA6E20
11:17096A3E30ACC729	10:13AD4E55049A85DC
09:867C48F1904C4F15	08:AEE6C23CBE6722FB
07:127A734C8D856D58	06:1CA82D2A21FAB3E7
05:D3A628153E6D58CD	04:669436480CE60A71
03:49C9DA01CFAB5B37	02:3613A325B7BC8C97
01:517AD92BAB43EA98	00:CF90F6CA28E6CD06
r1:EC5667F3CD3045D7	r2:02A95CF26FFC8BBF

Internal state in time −15.

15:886B00BA81577683	14:6E31AC6466F04991
13:FA1E9D90C89209FA	12:DCC7CD945399D888
11:EB6BF6CBFCAA6E20	10:17096A3E30ACC729
09:13AD4E55049A85DC	08:867C48F1904C4F15
07:AEE6C23CBE6722FB	06:127A734C8D856D58
05:1CA82D2A21FAB3E7	04:D3A628153E6D58CD
03:669436480CE60A71	02:49C9DA01CFAB5B37
01:3613A325B7BC8C97	00:517AD92BAB43EA98
r1:DF712A86C3966447	r2:60F52CA312B6CAE4

Internal state in time −14.

15:4A6066D44F152BE3	14:886B00BA81577683
13:6E31AC6466F04991	12:FA1E9D90C89209FA
11:DCC7CD945399D888	10:EB6BF6CBFCAA6E20
09:17096A3E30ACC729	08:13AD4E55049A85DC
07:867C48F1904C4F15	06:AEE6C23CBE6722FB
05:127A734C8D856D58	04:1CA82D2A21FAB3E7
03:D3A628153E6D58CD	02:669436480CE60A71
01:49C9DA01CFAB5B37	00:3613A325B7BC8C97
r1:5B13CA33DB48D4DE	r2:144F74B588BEDF26

Internal state in time −13.

15:28DF1E00ED7C9876	14:4A6066D44F152BE3
13:886B00BA81577683	12:6E31AC6466F04991

```
11:FA1E9D90C89209FA   10:DCC7CD945399D888
09:EB6BF6CBFCAA6E20   08:17096A3E30ACC729
07:13AD4E55049A85DC   06:867C48F1904C4F15
05:AEE6C23CBE6722FB   04:127A734C8D856D58
03:1CA82D2A21FAB3E7   02:D3A628153E6D58CD
01:669436480CE60A71   00:49C9DA01CFAB5B37
r1:82812119EFAF28B7   r2:7C42E661E8100591
```

Internal state in time −12.

```
15:E41D6D1FB745170F   14:28DF1E00ED7C9876
13:4A6066D44F152BE3   12:886B00BA81577683
11:6E31AC6466F04991   10:FA1E9D90C89209FA
09:DCC7CD945399D888   08:EB6BF6CBFCAA6E20
07:17096A3E30ACC729   06:13AD4E55049A85DC
05:867C48F1904C4F15   04:AEE6C23CBE6722FB
03:127A734C8D856D58   02:1CA82D2A21FAB3E7
01:D3A628153E6D58CD   00:669436480CE60A71
r1:04ADE71C69677C14   r2:13F11FEBD41D023A
```

Internal state in time −11.

```
15:923058D0493B3B36   14:E41D6D1FB745170F
13:28DF1E00ED7C9876   12:4A6066D44F152BE3
11:886B00BA81577683   10:6E31AC6466F04991
09:FA1E9D90C89209FA   08:DCC7CD945399D888
07:EB6BF6CBFCAA6E20   06:17096A3E30ACC729
05:13AD4E55049A85DC   04:867C48F1904C4F15
03:AEE6C23CBE6722FB   02:127A734C8D856D58
01:1CA82D2A21FAB3E7   00:D3A628153E6D58CD
r1:5E5186C023322E1D   r2:D71522BCCAF9FB65
```

Internal state in time −10.

```
15:ED16129B2C88905D   14:923058D0493B3B36
13:E41D6D1FB745170F   12:28DF1E00ED7C9876
11:4A6066D44F152BE3   10:886B00BA81577683
09:6E31AC6466F04991   08:FA1E9D90C89209FA
07:DCC7CD945399D888   06:EB6BF6CBFCAA6E20
```

```
05:17096A3E30ACC729   04:13AD4E55049A85DC
03:867C48F1904C4F15   02:AEE6C23CBE6722FB
01:127A734C8D856D58   00:1CA82D2A21FAB3E7
r1:FFF440BDB87693DB   r2:3616F8CB4D3331EE
```

Internal state in time −9.

```
15:1A492ACD428FF082   14:ED16129B2C88905D
13:923058D0493B3B36   12:E41D6D1FB745170F
11:28DF1E00ED7C9876   10:4A6066D44F152BE3
09:886B00BA81577683   08:6E31AC6466F04991
07:FA1E9D90C89209FA   06:DCC7CD945399D888
05:EB6BF6CBFCAA6E20   04:17096A3E30ACC729
03:13AD4E55049A85DC   02:867C48F1904C4F15
01:AEE6C23CBE6722FB   00:127A734C8D856D58
r1:1A3465EB047848FD   r2:737C1BCDCECA0B0E
```

Internal state in time −8.

```
15:7E8D68598E2E2D97   14:1A492ACD428FF082
13:ED16129B2C88905D   12:923058D0493B3B36
11:E41D6D1FB745170F   10:28DF1E00ED7C9876
09:4A6066D44F152BE3   08:886B00BA81577683
07:6E31AC6466F04991   06:FA1E9D90C89209FA
05:DCC7CD945399D888   04:EB6BF6CBFCAA6E20
03:17096A3E30ACC729   02:13AD4E55049A85DC
01:867C48F1904C4F15   00:AEE6C23CBE6722FB
r1:05AC749E18054644   r2:E5ACFCF6B4C0A351
```

Internal state in time −7.

```
15:BAFA8735D3BAFE42   14:7E8D68598E2E2D97
13:1A492ACD428FF082   12:ED16129B2C88905D
11:923058D0493B3B36   10:E41D6D1FB745170F
09:28DF1E00ED7C9876   08:4A6066D44F152BE3
07:886B00BA81577683   06:6E31AC6466F04991
05:FA1E9D90C89209FA   04:DCC7CD945399D888
03:EB6BF6CBFCAA6E20   02:17096A3E30ACC729
01:13AD4E55049A85DC   00:867C48F1904C4F15
r1:D2C30F91E14933AE   r2:B29E23F85B47094A
```

Internal state in time −6.

15:C97BCB951614E621	14:BAFA8735D3BAFE42
13:7E8D68598E2E2D97	12:1A492ACD428FF082
11:ED16129B2C88905D	10:923058D0493B3B36
09:E41D6D1FB745170F	08:28DF1E00ED7C9876
07:4A6066D44F152BE3	06:886B00BA81577683
05:6E31AC6466F04991	04:FA1E9D90C89209FA
03:DCC7CD945399D888	02:EB6BF6CBFCAA6E20
01:17096A3E30ACC729	00:13AD4E55049A85DC
r1:CCE74EC59DD6F9CC	r2:747C86F618506935

Internal state in time −5.

15:F4B557B586ABCF93	14:C97BCB951614E621
13:BAFA8735D3BAFE42	12:7E8D68598E2E2D97
11:1A492ACD428FF082	10:ED16129B2C88905D
09:923058D0493B3B36	08:E41D6D1FB745170F
07:28DF1E00ED7C9876	06:4A6066D44F152BE3
05:886B00BA81577683	04:6E31AC6466F04991
03:FA1E9D90C89209FA	02:DCC7CD945399D888
01:EB6BF6CBFCAA6E20	00:17096A3E30ACC729
r1:F309EF4FA67E96CC	r2:81409600F59CBF20

Internal state in time −4.

15:95055FEB17A98191	14:F4B557B586ABCF93
13:C97BCB951614E621	12:BAFA8735D3BAFE42
11:7E8D68598E2E2D97	10:1A492ACD428FF082
09:ED16129B2C88905D	08:923058D0493B3B36
07:E41D6D1FB745170F	06:28DF1E00ED7C9876
05:4A6066D44F152BE3	04:886B00BA81577683
03:6E31AC6466F04991	02:FA1E9D90C89209FA
01:DCC7CD945399D888	00:EB6BF6CBFCAA6E20
r1:3C3B1D36C957BD62	r2:DE2DAC08A42568A7

Internal state in time −3.

15:144EF0620652BED3	14:95055FEB17A98191
13:F4B557B586ABCF93	12:C97BCB951614E621

```
11:BAFA8735D3BAFE42   10:7E8D68598E2E2D97
09:1A492ACD428FF082   08:ED16129B2C88905D
07:923058D0493B3B36   06:E41D6D1FB745170F
05:28DF1E00ED7C9876   04:4A6066D44F152BE3
03:886B00BA81577683   02:6E31AC6466F04991
01:FA1E9D90C89209FA   00:DCC7CD945399D888
r1:A7A9779DBA3A4EC8   r2:477B227C35F364DD
```

Internal state in time −2.

```
15:15166D44EF78947C   14:144EF0620652BED3
13:95055FEB17A98191   12:F4B557B586ABCF93
11:C97BCB951614E621   10:BAFA8735D3BAFE42
09:7E8D68598E2E2D97   08:1A492ACD428FF082
07:ED16129B2C88905D   06:923058D0493B3B36
05:E41D6D1FB745170F   04:28DF1E00ED7C9876
03:4A6066D44F152BE3   02:886B00BA81577683
01:6E31AC6466F04991   00:FA1E9D90C89209FA
r1:3C307A31BC9F3470   r2:124A36DFFDE03710
```

Internal state in time −1.

```
15:F425899289129044   14:15166D44EF78947C
13:144EF0620652BED3   12:95055FEB17A98191
11:F4B557B586ABCF93   10:C97BCB951614E621
09:BAFA8735D3BAFE42   08:7E8D68598E2E2D97
07:1A492ACD428FF082   06:ED16129B2C88905D
05:923058D0493B3B36   04:E41D6D1FB745170F
03:28DF1E00ED7C9876   02:4A6066D44F152BE3
01:886B00BA81577683   00:6E31AC6466F04991
r1:A74F96CB1589B8A1   r2:8CCEC14B2B6B8112
```

Internal state in time 0.

```
15:EA249CB2F04184B9   14:F425899289129044
13:15166D44EF78947C   12:144EF0620652BED3
11:95055FEB17A98191   10:F4B557B586ABCF93
09:C97BCB951614E621   08:BAFA8735D3BAFE42
07:7E8D68598E2E2D97   06:1A492ACD428FF082
```

```
05:ED16129B2C88905D   04:923058D0493B3B36
03:E41D6D1FB745170F   02:28DF1E00ED7C9876
01:4A6066D44F152BE3   00:886B00BA81577683
r1:A11DB1AD31BE3FE5   r2:FD6DEC8A2BF298D0
Z0:FE44A2508B5A2ACD
```

Internal state in time 1.

```
15:30E8416E1DEABEE7   14:EA249CB2F04184B9
13:F425899289129044   12:15166D44EF78947C
11:144EF0620652BED3   10:95055FEB17A98191
09:F4B557B586ABCF93   08:C97BCB951614E621
07:BAFA8735D3BAFE42   06:7E8D68598E2E2D97
05:1A492ACD428FF082   04:ED16129B2C88905D
03:923058D0493B3B36   02:E41D6D1FB745170F
01:28DF1E00ED7C9876   00:4A6066D44F152BE3
r1:128459CF1B6B2D4C   r2:A639A6A7A45DE092
Z1:Af355b4ed21d2742
```

Internal state in time 2.

```
15:CFDBE9BE1A6A21CF   14:30E8416E1DEABEE7
13:EA249CB2F04184B9   12:F425899289129044
11:15166D44EF78947C   10:144EF0620652BED3
09:95055FEB17A98191   08:F4B557B586ABCF93
07:C97BCB951614E621   06:BAFA8735D3BAFE42
05:7E8D68598E2E2D97   04:1A492ACD428FF082
03:ED16129B2C88905D   02:923058D0493B3B36
01:E41D6D1FB745170F   00:28DF1E00ED7C9876
r1:9A5F303A2D7070D6   r2:9E33FA2E0FDC94A2
Z2:DCD7FDD6A57A9E71
```

Internal state in time 3.

```
15:2361A5A4DEF28A8D   14:CFDBE9BE1A6A21CF
13:30E8416E1DEABEE7   12:EA249CB2F04184B9
11:F425899289129044   10:15166D44EF78947C
09:144EF0620652BED3   08:95055FEB17A98191
07:F4B557B586ABCF93   06:C97BCB951614E621
```

```
05:BAFA8735D3BAFE42   04:7E8D68598E2E2D97
03:1A492ACD428FF082   02:ED16129B2C88905D
01:923058D0493B3B36   00:E41D6D1FB745170F
r1:885896E1001E195B   r2:12812A481BCA010C
Z3:5D267BD2739FB5EB
```

Internal state in time 4.

```
15:80952427ADA3A3C0   14:2361A5A4DEF28A8D
13:CFDBE9BE1A6A21CF   12:30E8416E1DEABEE7
11:EA249CB2F04184B9   10:F425899289129044
09:15166D44EF78947C   08:144EF0620652BED3
07:95055FEB17A98191   06:F4B557B586ABCF93
05:C97BCB951614E621   04:BAFA8735D3BAFE42
03:7E8D68598E2E2D97   02:1A492ACD428FF082
01:ED16129B2C88905D   00:923058D0493B3B36
r1:43696BB639B4BFF3   r2:E3E0399B1CE078F7
Z4:B22EEE96B2832072
```

Internal state in time 5.

```
15:0F0B7D3E96245639   14:80952427ADA3A3C0
13:2361A5A4DEF28A8D   12:CFDBE9BE1A6A21CF
11:30E8416E1DEABEE7   10:EA249CB2F04184B9
09:F425899289129044   08:15166D44EF78947C
07:144EF0620652BED3   06:95055FEB17A98191
05:F4B557B586ABCF93   04:C97BCB951614E621
03:BAFA8735D3BAFE42   02:7E8D68598E2E2D97
01:1A492ACD428FF082   00:ED16129B2C88905D
r1:B3BC2359374A9AC6   r2:E80FD8403B4FC8E5
Z5:C7DE6A4CDAA9A847
```

Internal state in time 6.

```
15:586DAC9FA2DB3DA0   14:0F0B7D3E96245639
13:80952427ADA3A3C0   12:2361A5A4DEF28A8D
11:CFDBE9BE1A6A21CF   10:30E8416E1DEABEE7
09:EA249CB2F04184B9   08:F425899289129044
07:15166D44EF78947C   06:144EF0620652BED3
```

```
05:95055FEB17A98191   04:F4B557B586ABCF93
03:C97BCB951614E621   02:BAFA8735D3BAFE42
01:7E8D68598E2E2D97   00:1A492ACD428FF082
R1:0B717DE51A425372   r2:0B43DADA7EB4E162
Z6:72D5DA93812680F2
```

Internal state in time 7.

```
15:874F28CA919B4B89   14:586DAC9FA2DB3DA0
13:0F0B7D3E96245639   12:80952427ADA3A3C0
11:2361A5A4DEF28A8D   10:CFDBE9BE1A6A21CF
09:30E8416E1DEABEE7   08:EA249CB2F04184B9
07:F425899289129044   06:15166D44EF78947C
05:144EF0620652BED3   04:95055FEB17A98191
03:F4B557B586ABCF93   02:C97BCB951614E621
01:BAFA8735D3BAFE42   00:7E8D68598E2E2D97
r1:8bd8ff022c588522   r2:27af84eba007d1dc
Z7:4A0ACB7E93DA2CE0
```

E.2 512-Bit Key

E.2.1 Key Initialization Vector and Key Sequence

(1) Initialization vector $IV = (IV_3, IV_2, IV_1, IV_0)$:

```
IV3:0000000000000000   IV2:0000000000000000
IV1:0000000000000000   IV0:0000000000000000
```

Key $K = (K_7, K_6, K_5, K_4, K_3, K_2, K_1, K_0)$:

```
K7:8000000000000000   K6:0000000000000000
K5:0000000000000000   K4:0000000000000000
K3:0000000000000000   K2:0000000000000000
K1:0000000000000000   K0:0000000000000000
```

Key stream output Z:

```
Z0:F5B9AB51100F8317   Z1:898EF2086A4AF395
Z2:59571FECB5158D0B   Z3:B7C45B6744C71FBB
Z4:FF2EFCF05D8D8DB9   Z5:7A585871E5C419C0
Z6:6B5C4691B9125E71   Z7:A55BE7D2B358EC6E
```

(2) Initialization vector $IV = (IV_3, IV_2, IV_1, IV_0)$:

```
IV3:0000000000000000   IV2:0000000000000000
IV1:0000000000000000   IV0:0000000000000000
```

Key $K = (K_7, K_6, K_5, K_4, K_3, K_2, K_1, K_0)$:

```
K7:AAAAAAAAAAAAAAAA   K6:AAAAAAAAAAAAAAAA
K5:AAAAAAAAAAAAAAAA   K4:AAAAAAAAAAAAAAAA
K3:AAAAAAAAAAAAAAAA   K2:AAAAAAAAAAAAAAAA
K1:AAAAAAAAAAAAAAAA   K0:AAAAAAAAAAAAAAAA
```

Key stream output Z:

```
Z0:D2A6103C50BD4E04   Z1:DC6A21AF5EB13B73
Z2:DF4CA6CB07797265   Z3:F453C253D8D01876
Z4:039A64DC7A01800C   Z5:688CE327DCCB7E84
Z6:41E0250B5E526403   Z7:9936E478AA200F22
```

(3) Initialization vector $IV = (IV_3, IV_2, IV_1, IV_0)$:

```
IV3:0000000000000004   IV2:0000000000000003
IV1:0000000000000002   IV0:0000000000000001
```

Key $K = (K_7, K_6, K_5, K_4, K_3, K_2, K_1, K_0)$:

```
K7:8000000000000000   K6:0000000000000000
K5:0000000000000000   K4:0000000000000000
K3:0000000000000000   K2:0000000000000000
K1:0000000000000000   K0:0000000000000000
```

Key stream output Z:

```
Z0:cca12eae8133aaaa   Z1:528d85507ce8501d
Z2:da83c7fe3e1823f1   Z3:21416ebf63b71a42
Z4:26d76d2bf1a625eb   Z5:eec66ee0cd0b1efc
Z6:02dd68f338a345a8   Z7:47538790a5411adb
```

(4) Initialization vector $IV = (IV_3, IV_2, IV_1, IV_0)$:

```
IV3:0000000000000004   IV2:0000000000000003
IV1:0000000000000002   IV0:0000000000000001
```

Key $K = (K_7, K_6, K_5, K_4, K_3, K_2, K_1, K_0)$:

```
K7:AAAAAAAAAAAAAAAA   K6:AAAAAAAAAAAAAAAA
K5:AAAAAAAAAAAAAAAA   K4:AAAAAAAAAAAAAAAA
K3:AAAAAAAAAAAAAAAA   K2:AAAAAAAAAAAAAAAA
K1:AAAAAAAAAAAAAAAA   K0:AAAAAAAAAAAAAAAA
```

Key stream output Z:

```
Z0:965648E775C717D5   Z1:A63C2A7376E92DF3
Z2:0B0EB0BBD47CA267   Z3:EA593D979AE5BD39
Z4:D773B5E5193CAFE1   Z5:B0A26671D259422B
Z6:85B2AA326B280156   Z7:511ACE6451435F0C
```

E.2.2 Sample of Internal States

Initialization vector $IV = (IV_3, IV_2, IV_1, IV_0)$:

```
IV3:0000000000000004   IV2:0000000000000003
IV1:0000000000000002   IV0:0000000000000001
```

Key $K = (K_7, K_6, K_5, K_4, K_3, K_2, K_1, K_0)$:

```
K7:8000000000000000   K6:0000000000000000
K5:0000000000000000   K4:0000000000000000
K3:0000000000000000   K2:0000000000000000
K1:0000000000000000   K0:0000000000000000
```

Internal state in time −34.

15:0000000000000000	14:FFFFFFFFFFFFFFFF
13:0000000000000000	12:0000000000000000
11:7FFFFFFFFFFFFFFF	10:0000000000000000
09:FFFFFFFFFFFFFFFF	08:0000000000000000
07:FFFFFFFFFFFFFFFF	06:0000000000000000
05:0000000000000000	04:0000000000000000
03:0000000000000000	02:0000000000000000
01:0000000000000000	00:8000000000000000

Internal state in time −33.

15:0000000000000000	14:FFFFFFFFFFFFFFFF
13:0000000000000000	12:0000000000000000
11:7FFFFFFFFFFFFFFF	10:0000000000000000
09:FFFFFFFFFFFFFFFF	08:0000000000000004
07:FFFFFFFFFFFFFFFF	06:0000000000000000
05:0000000000000003	04:0000000000000000
03:0000000000000002	02:0000000000000000
01:0000000000000000	00:8000000000000001
r1:0000000000000000	r2:0000000000000000

Internal state in time −32.

15:2D77AEF5E982FEC5	14:0000000000000000
13:FFFFFFFFFFFFFFFF	12:0000000000000000
11:0000000000000000	10:7FFFFFFFFFFFFFFF
09:0000000000000000	08:FFFFFFFFFFFFFFFF
07:0000000000000004	06:FFFFFFFFFFFFFFFF
05:0000000000000000	04:0000000000000003
03:0000000000000000	02:0000000000000002
01:0000000000000000	00:0000000000000000
r1:0000000000000000	r2:14E6CC3F14E6CC3F

Internal state in time −31.

15:C66E9D35029BCD05	14:2D77AEF5E982FEC5
13:0000000000000000	12:FFFFFFFFFFFFFFFF
11:0000000000000000	10:0000000000000000

```
09:7FFFFFFFFFFFFFFF   08:0000000000000000
07:FFFFFFFFFFFFFFFF   06:0000000000000004
05:FFFFFFFFFFFFFFFF   04:0000000000000000
03:0000000000000003   02:0000000000000000
01:0000000000000002   00:0000000000000000
r1:14E6CC3F14E6CC3E   r2:14E6CC3F14E6CC3F
```

Internal state in time −30.

```
15:CFB3A54B0364557C   14:C66E9D35029BCD05
13:2D77AEF5E982FEC5   12:0000000000000000
11:FFFFFFFFFFFFFFFF   10:0000000000000000
09:0000000000000000   08:7FFFFFFFFFFFFFFF
07:0000000000000000   06:FFFFFFFFFFFFFFFF
05:0000000000000004   04:FFFFFFFFFFFFFFFF
03:0000000000000000   02:0000000000000003
01:0000000000000000   00:0000000000000002
r1:14E6CC3F14E6CC3F   r2:F9F9C809BFCDBAC3
```

Internal state in time −29.

```
15:C61583769B869842   14:CFB3A54B0364557C
13:C66E9D35029BCD05   12:2D77AEF5E982FEC5
11:0000000000000000   10:FFFFFFFFFFFFFFFF
09:0000000000000000   08:0000000000000000
07:7FFFFFFFFFFFFFFF   06:0000000000000000
05:FFFFFFFFFFFFFFFF   04:0000000000000004
03:FFFFFFFFFFFFFFFF   02:0000000000000000
01:0000000000000003   00:0000000000000000
r1:277176FFA950B988   r2:E38BD2D9E38BD2D9
```

Internal state in time −28.

```
15:C863B59AA5C74E16   14:C61583769B869842
13:CFB3A54B0364557C   12:C66E9D35029BCD05
11:2D77AEF5E982FEC5   10:0000000000000000
09:FFFFFFFFFFFFFFFF   08:0000000000000000
07:0000000000000000   06:7FFFFFFFFFFFFFFF
05:0000000000000000   04:FFFFFFFFFFFFFFFF
```

```
03:0000000000000004   02:FFFFFFFFFFFFFFFF
01:0000000000000000   00:0000000000000003
r1:A9FA700EE6279FDE   r2:C3B90F3B03F1D821
```

Internal state in time −27.

```
15:085C65B2A1D3E157   14:C863B59AA5C74E16
13:C61583769B869842   12:CFB3A54B0364557C
11:C66E9D35029BCD05   10:2D77AEF5E982FEC5
09:0000000000000000   08:FFFFFFFFFFFFFFFF
07:0000000000000000   06:0000000000000000
05:7FFFFFFFFFFFFFFF   04:0000000000000000
03:FFFFFFFFFFFFFFFF   02:0000000000000004
01:FFFFFFFFFFFFFFFF   00:0000000000000000
r1:936CB48607562D9D   r2:A67FA917402138A8
```

Internal state in time −26.

```
15:BD4EA7C1DF4835D3   14:085C65B2A1D3E157
13:C863B59AA5C74E16   12:C61583769B869842
11:CFB3A54B0364557C   10:C66E9D35029BCD05
09:2D77AEF5E982FEC5   08:0000000000000000
07:FFFFFFFFFFFFFFFF   06:0000000000000000
05:0000000000000000   04:7FFFFFFFFFFFFFFF
03:0000000000000000   02:FFFFFFFFFFFFFFFF
01:0000000000000004   00:FFFFFFFFFFFFFFFF
r1:6C952C8DDBA7D0EA   r2:69281A094F875F1B
```

Internal state in time −25.

```
15:B7316352DE888C3E   14:BD4EA7C1DF4835D3
13:085C65B2A1D3E157   12:C863B59AA5C74E16
11:C61583769B869842   10:CFB3A54B0364557C
09:C66E9D35029BCD05   08:2D77AEF5E982FEC5
07:0000000000000000   06:FFFFFFFFFFFFFFFF
05:0000000000000000   04:0000000000000000
03:7FFFFFFFFFFFFFFF   02:0000000000000000
01:FFFFFFFFFFFFFFFF   00:0000000000000004
r1:318BCFA3F54EAD31   r2:E1D244C8843EFCD0
```

Internal state in time −24.

15:9FE1404DB407A670	14:B7316352DE888C3E
13:BD4EA7C1DF4835D3	12:085C65B2A1D3E157
11:C863B59AA5C74E16	10:C61583769B869842
09:CFB3A54B0364557C	08:C66E9D35029BCD05
07:2D77AEF5E982FEC5	06:0000000000000000
05:FFFFFFFFFFFFFFFF	04:0000000000000000
03:0000000000000000	02:7FFFFFFFFFFFFFFF
01:0000000000000000	00:FFFFFFFFFFFFFFFF
r1:EA2EAA7B2612DE27	r2:D281055C75828C61

Internal state in time −23.

15:4E0F5DA1AF9E05B0	14:9FE1404DB407A670
13:B7316352DE888C3E	12:BD4EA7C1DF4835D3
11:085C65B2A1D3E157	10:C863B59AA5C74E16
09:C61583769B869842	08:CFB3A54B0364557C
07:C66E9D35029BCD05	06:2D77AEF5E982FEC5
05:0000000000000000	04:FFFFFFFFFFFFFFFF
03:0000000000000000	02:0000000000000000
01:7FFFFFFFFFFFFFFF	00:0000000000000000
r1:8FCFAD1E54CAC234	r2:C4A495773E5A72DE

Internal state in time −22.

15:7206E3D7B21FEAE5	14:4E0F5DA1AF9E05B0
13:9FE1404DB407A670	12:B7316352DE888C3E
11:BD4EA7C1DF4835D3	10:085C65B2A1D3E157
09:C863B59AA5C74E16	08:C61583769B869842
07:CFB3A54B0364557C	06:C66E9D35029BCD05
05:2D77AEF5E982FEC5	04:0000000000000000
03:FFFFFFFFFFFFFFFF	02:0000000000000000
01:0000000000000000	00:7FFFFFFFFFFFFFFF
r1:7BD5F8CA1CE2FF1C	r2:A1B780B3F3E7BA36

Internal state in time −21.

15:D5E38389EF974193	14:7206E3D7B21FEAE5
13:4E0F5DA1AF9E05B0	12:9FE1404DB407A670

11:B7316352DE888C3E	10:BD4EA7C1DF4835D3
09:085C65B2A1D3E157	08:C863B59AA5C74E16
07:C61583769B869842	06:CFB3A54B0364557C
05:C66E9D35029BCD05	04:2D77AEF5E982FEC5
03:0000000000000000	02:FFFFFFFFFFFFFFFF
01:0000000000000000	00:0000000000000000
r1:4198C101A7EF60A6	r2:A385C4D8F549B62F

Internal state in time −20.

15:7B4D25AF78ED992A	14:D5E38389EF974193
13:7206E3D7B21FEAE5	12:4E0F5DA1AF9E05B0
11:9FE1404DB407A670	10:B7316352DE888C3E
09:BD4EA7C1DF4835D3	08:085C65B2A1D3E157
07:C863B59AA5C74E16	06:C61583769B869842
05:CFB3A54B0364557C	04:C66E9D35029BCD05
03:2D77AEF5E982FEC5	02:0000000000000000
01:FFFFFFFFFFFFFFFF	00:0000000000000000
r1:F195227AA4E7BBDF	r2:B846E5908CE5A151

Internal state in time −19.

15:BA41615D8BBD191B	14:7B4D25AF78ED992A
13:D5E38389EF974193	12:7206E3D7B21FEAE5
11:4E0F5DA1AF9E05B0	10:9FE1404DB407A670
09:B7316352DE888C3E	08:BD4EA7C1DF4835D3
07:085C65B2A1D3E157	06:C863B59AA5C74E16
05:C61583769B869842	04:CFB3A54B0364557C
03:C66E9D35029BCD05	02:2D77AEF5E982FEC5
01:0000000000000000	00:FFFFFFFFFFFFFFFF
r1:2A4DC9683F058C36	r2:2C86C538D10986AE

Internal state in time −18.

15:1198EDCEE1AC03B2	14:BA41615D8BBD191B
13:7B4D25AF78ED992A	12:D5E38389EF974193
11:7206E3D7B21FEAE5	10:4E0F5DA1AF9E05B0
09:9FE1404DB407A670	08:B7316352DE888C3E
07:BD4EA7C1DF4835D3	06:085C65B2A1D3E157

```
05:C863B59AA5C74E16   04:C61583769B869842
03:CFB3A54B0364557C   02:C66E9D35029BCD05
01:2D77AEF5E982FEC5   00:0000000000000000
r1:026A48C2C0A0C841   r2:8F92552284E6DBCD
```

Internal state in time −17.

```
15:9E7D381AC67E96FE   14:1198EDCEE1AC03B2
13:BA41615D8BBD191B   12:7B4D25AF78ED992A
11:D5E38389EF974193   10:7206E3D7B21FEAE5
09:4E0F5DA1AF9E05B0   08:9FE1404DB407A670
07:B7316352DE888C3E   06:BD4EA7C1DF4835D3
05:085C65B2A1D3E157   04:C863B59AA5C74E16
03:C61583769B869842   02:CFB3A54B0364557C
01:C66E9D35029BCD05   00:2D77AEF5E982FEC5
r1:0ADF7AD1FDD474F7   r2:4F72D384DC8C9252
```

Internal state in time −16.

```
15:631CBA7CD7AFD249   14:9E7D381AC67E96FE
13:1198EDCEE1AC03B2   12:BA41615D8BBD191B
11:7B4D25AF78ED992A   10:D5E38389EF974193
09:7206E3D7B21FEAE5   08:4E0F5DA1AF9E05B0
07:9FE1404DB407A670   06:B7316352DE888C3E
05:BD4EA7C1DF4835D3   04:085C65B2A1D3E157
03:C863B59AA5C74E16   02:C61583769B869842
01:CFB3A54B0364557C   00:C66E9D35029BCD05
r1:09B434E26849AB6D   r2:51358A89C799B8DB
```

Internal state in time −15.

```
15:3634C5786B202E79   14:631CBA7CD7AFD249
13:9E7D381AC67E96FE   12:1198EDCEE1AC03B2
11:BA41615D8BBD191B   10:7B4D25AF78ED992A
09:D5E38389EF974193   08:7206E3D7B21FEAE5
07:4E0F5DA1AF9E05B0   06:9FE1404DB407A670
05:B7316352DE888C3E   04:BD4EA7C1DF4835D3
03:085C65B2A1D3E157   02:C863B59AA5C74E16
01:C61583769B869842   00:CFB3A54B0364557C
r1:62CE7858A945BC8D   r2:97560BE288E3CB73
```

Internal state in time −14.

```
15:3D33CABFE3497789   14:3634C5786B202E79
13:631CBA7CD7AFD249   12:9E7D381AC67E96FE
11:1198EDCEE1AC03B2   10:BA41615D8BBD191B
09:7B4D25AF78ED992A   08:D5E38389EF974193
07:7206E3D7B21FEAE5   06:4E0F5DA1AF9E05B0
05:9FE1404DB407A670   04:B7316352DE888C3E
03:BD4EA7C1DF4835D3   02:085C65B2A1D3E157
01:C863B59AA5C74E16   00:C61583769B869842
r1:35D343FD4F626271   r2:685254F77715F646
```

Internal state in time −13.

```
15:3F430A9F8AE713C9   14:3D33CABFE3497789
13:3634C5786B202E79   12:631CBA7CD7AFD249
11:9E7D381AC67E96FE   10:1198EDCEE1AC03B2
09:BA41615D8BBD191B   08:7B4D25AF78ED992A
07:D5E38389EF974193   06:7206E3D7B21FEAE5
05:4E0F5DA1AF9E05B0   04:9FE1404DB407A670
03:B7316352DE888C3E   02:BD4EA7C1DF4835D3
01:085C65B2A1D3E157   00:C863B59AA5C74E16
r1:CB6F0F744EC5C88F   r2:81C42531D2DF775A
```

Internal state in time −12.

```
15:A103B7A8F53FEDEA   14:3F430A9F8AE713C9
13:3D33CABFE3497789   12:3634C5786B202E79
11:631CBA7CD7AFD249   10:9E7D381AC67E96FE
09:1198EDCEE1AC03B2   08:BA41615D8BBD191B
07:7B4D25AF78ED992A   06:D5E38389EF974193
05:7206E3D7B21FEAE5   04:4E0F5DA1AF9E05B0
03:9FE1404DB407A670   02:B7316352DE888C3E
01:BD4EA7C1DF4835D3   00:085C65B2A1D3E157
r1:B7F8EAAA3DFFA5D3   r2:8B8CB17818648F57
```

Internal state in time −11.

```
15:111E38563E889391   14:A103B7A8F53FEDEA
13:3F430A9F8AE713C9   12:3D33CABFE3497789
```

```
11:3634C5786B202E79   10:631CBA7CD7AFD249
09:9E7D381AC67E96FE   08:1198EDCEE1AC03B2
07:BA41615D8BBD191B   06:7B4D25AF78ED992A
05:D5E38389EF974193   04:7206E3D7B21FEAE5
03:4E0F5DA1AF9E05B0   02:9FE1404DB407A670
01:B7316352DE888C3E   00:BD4EA7C1DF4835D3
r1:C8C07C37FBAE06E0   r2:37E89D9669086345
```

Internal state in time −10.

```
15:49636E9CE984191D   14:111E38563E889391
13:A103B7A8F53FEDEA   12:3F430A9F8AE713C9
11:3D33CABFE3497789   10:3634C5786B202E79
09:631CBA7CD7AFD249   08:9E7D381AC67E96FE
07:1198EDCEE1AC03B2   06:BA41615D8BBD191B
05:7B4D25AF78ED992A   04:D5E38389EF974193
03:7206E3D7B21FEAE5   02:4E0F5DA1AF9E05B0
01:9FE1404DB407A670   00:B7316352DE888C3E
r1:772BA835F3EF770E   r2:02F24A2CCDFBF19E
```

Internal state in time −9.

```
15:FDBF67073BF2FBCE   14:49636E9CE984191D
13:111E38563E889391   12:A103B7A8F53FEDEA
11:3F430A9F8AE713C9   10:3D33CABFE3497789
09:3634C5786B202E79   08:631CBA7CD7AFD249
07:9E7D381AC67E96FE   06:1198EDCEE1AC03B2
05:BA41615D8BBD191B   04:7B4D25AF78ED992A
03:D5E38389EF974193   02:7206E3D7B21FEAE5
01:4E0F5DA1AF9E05B0   00:9FE1404DB407A670
r1:A3F601D5C33BDF88   r2:84AFBD591EAAA77E
```

Internal state in time −8.

```
15:63437982568378EC   14:FDBF67073BF2FBCE
13:49636E9CE984191D   12:111E38563E889391
11:A103B7A8F53FEDEA   10:3F430A9F8AE713C9
09:3D33CABFE3497789   08:3634C5786B202E79
07:631CBA7CD7AFD249   06:9E7D381AC67E96FE
```

05:1198EDCEE1AC03B2	04:BA41615D8BBD191B
03:7B4D25AF78ED992A	02:D5E38389EF974193
01:7206E3D7B21FEAE5	00:4E0F5DA1AF9E05B0
r1:95CDF5AF5D333B0F	r2:D356BE3628C44621

Internal state in time −7.

15:056A5661B3D9630F	14:63437982568378EC
13:FDBF67073BF2FBCE	12:49636E9CE984191D
11:111E38563E889391	10:A103B7A8F53FEDEA
09:3F430A9F8AE713C9	08:3D33CABFE3497789
07:3634C5786B202E79	06:631CBA7CD7AFD249
05:9E7D381AC67E96FE	04:1198EDCEE1AC03B2
03:BA41615D8BBD191B	02:7B4D25AF78ED992A
01:D5E38389EF974193	00:7206E3D7B21FEAE5
r1:1CBA2CD312485F3E	r2:7700A79F0675E6E6

Internal state in time −6.

15:9370D6202E81B223	14:056A5661B3D9630F
13:63437982568378EC	12:FDBF67073BF2FBCE
11:49636E9CE984191D	10:111E38563E889391
09:A103B7A8F53FEDEA	08:3F430A9F8AE713C9
07:3D33CABFE3497789	06:3634C5786B202E79
05:631CBA7CD7AFD249	04:9E7D381AC67E96FE
03:1198EDCEE1AC03B2	02:BA41615D8BBD191B
01:7B4D25AF78ED992A	00:D5E38389EF974193
r1:74C00EA64268E2B4	r2:256589018B269B37

Internal state in time −5.

15:88A6F6BFC2F96066	14:9370D6202E81B223
13:056A5661B3D9630F	12:63437982568378EC
11:FDBF67073BF2FBCE	10:49636E9CE984191D
09:111E38563E889391	08:A103B7A8F53FEDEA
07:3F430A9F8AE713C9	06:3D33CABFE3497789
05:3634C5786B202E79	04:631CBA7CD7AFD249
03:9E7D381AC67E96FE	02:1198EDCEE1AC03B2
01:BA41615D8BBD191B	00:7B4D25AF78ED992A
r1:88A90283E1AA1423	r2:B8447B87B25B15CE

Internal state in time −4.

15:238D816187C8DA42	14:88A6F6BFC2F96066
13:9370D6202E81B223	12:056A5661B3D9630F
11:63437982568378EC	10:FDBF67073BF2FBCE
09:49636E9CE984191D	08:111E38563E889391
07:A103B7A8F53FEDEA	06:3F430A9F8AE713C9
05:3D33CABFE3497789	04:3634C5786B202E79
03:631CBA7CD7AFD249	02:9E7D381AC67E96FE
01:1198EDCEE1AC03B2	00:BA41615D8BBD191B
r1:BDAED1E9663478DD	r2:A4D05469FE0FD798

Internal state in time −3.

15:3C0BDD1B2F92AE0E	14:238D816187C8DA42
13:88A6F6BFC2F96066	12:9370D6202E81B223
11:056A5661B3D9630F	10:63437982568378EC
09:FDBF67073BF2FBCE	08:49636E9CE984191D
07:111E38563E889391	06:A103B7A8F53FEDEA
05:3F430A9F8AE713C9	04:3D33CABFE3497789
03:3634C5786B202E79	02:631CBA7CD7AFD249
01:9E7D381AC67E96FE	00:1198EDCEE1AC03B2
r1:38412A8A2C9189BB	r2:9ED2DACF33E4382B

Internal state in time −2.

15:1644DD4202DD04A3	14:3C0BDD1B2F92AE0E
13:238D816187C8DA42	12:88A6F6BFC2F96066
11:9370D6202E81B223	10:056A5661B3D9630F
09:63437982568378EC	08:FDBF67073BF2FBCE
07:49636E9CE984191D	06:111E38563E889391
05:A103B7A8F53FEDEA	04:3F430A9F8AE713C9
03:3D33CABFE3497789	02:3634C5786B202E79
01:631CBA7CD7AFD249	00:9E7D381AC67E96FE
r1:2779D18EF6DD9891	r2:06E4DA32AE8A8149

Internal state in time −1.

15:B0D56F843B3CB9CF	14:1644DD4202DD04A3
13:3C0BDD1B2F92AE0E	12:238D816187C8DA42

11:88A6F6BFC2F96066	10:9370D6202E81B223
09:056A5661B3D9630F	08:63437982568378EC
07:FDBF67073BF2FBCE	06:49636E9CE984191D
05:111E38563E889391	04:A103B7A8F53FEDEA
03:3F430A9F8AE713C9	02:3D33CABFE3497789
01:3634C5786B202E79	00:631CBA7CD7AFD249
r1:2A725B9436535B8B	r2:349EAAA9B88E189F

Internal state in time 0.

15:CF850AE47AB01EFF	14:B0D56F843B3CB9CF
13:1644DD4202DD04A3	12:3C0BDD1B2F92AE0E
11:238D816187C8DA42	10:88A6F6BFC2F96066
09:9370D6202E81B223	08:056A5661B3D9630F
07:63437982568378EC	06:FDBF67073BF2FBCE
05:49636E9CE984191D	04:111E38563E889391
03:A103B7A8F53FEDEA	02:3F430A9F8AE713C9
01:3D33CABFE3497789	00:3634C5786B202E79
r1:70AA87C4E820C6AD	r2:BABA797F88C3617F
Z0:CCA12EAE8133AAAA	

Internal state in time 1.

15:7A50B6C128D2B5A1	14:CF850AE47AB01EFF
13:B0D56F843B3CB9CF	12:1644DD4202DD04A3
11:3C0BDD1B2F92AE0E	10:238D816187C8DA42
09:88A6F6BFC2F96066	08:9370D6202E81B223
07:056A5661B3D9630F	06:63437982568378EC
05:FDBF67073BF2FBCE	04:49636E9CE984191D
03:111E38563E889391	02:A103B7A8F53FEDEA
01:3F430A9F8AE713C9	00:3D33CABFE3497789
r1:D0FF56C18BA06622	r2:24EE426D2BD23C57
Z1:528D85507CE8501D	

Internal state in time 2.

15:5AE0FF3921A9A295	14:7A50B6C128D2B5A1
13:CF850AE47AB01EFF	12:B0D56F843B3CB9CF
11:1644DD4202DD04A3	10:3C0BDD1B2F92AE0E

```
09:238D816187C8DA42   08:88A6F6BFC2F96066
07:9370D6202E81B223   06:056A5661B3D9630F
05:63437982568378EC   04:FDBF67073BF2FBCE
03:49636E9CE984191D   02:111E38563E889391
01:A103B7A8F53FEDEA   00:3F430A9F8AE713C9
r1:D5C3B1F1670EF626   r2:D5647C4B3C47A883
Z2:DA83C7FE3E1823F1
```

Internal state in time 3.

```
15:8AB52CB97DD20A07   14:5AE0FF3921A9A295
13:7A50B6C128D2B5A1   12:CF850AE47AB01EFF
11:B0D56F843B3CB9CF   10:1644DD4202DD04A3
09:3C0BDD1B2F92AE0E   08:238D816187C8DA42
07:88A6F6BFC2F96066   06:9370D6202E81B223
05:056A5661B3D9630F   04:63437982568378EC
03:FDBF67073BF2FBCE   02:49636E9CE984191D
01:111E38563E889391   00:A103B7A8F53FEDEA
r1:A4E9872FB6F7C782   r2:AFDC6AFEA2412621
Z3:21416EBF63B71A42
```

Internal state in time 4.

```
15:705E07090CD663A6   14:8AB52CB97DD20A07
13:5AE0FF3921A9A295   12:7A50B6C128D2B5A1
11:CF850AE47AB01EFF   10:B0D56F843B3CB9CF
09:1644DD4202DD04A3   08:3C0BDD1B2F92AE0E
07:238D816187C8DA42   06:88A6F6BFC2F96066
05:9370D6202E81B223   04:056A5661B3D9630F
03:63437982568378EC   02:FDBF67073BF2FBCE
01:49636E9CE984191D   00:111E38563E889391
r1:2A2D21BFCB13DBC2   r2:AD427DB518C48912
Z4:26D76D2BF1A625EB
```

Internal state in time 5.

```
15:940103DDB03D83CF   14:705E07090CD663A6
13:8AB52CB97DD20A07   12:5AE0FF3921A9A295
11:7A50B6C128D2B5A1   10:CF850AE47AB01EFF
```

```
09:B0D56F843B3CB9CF   08:1644DD4202DD04A3
07:3C0BDD1B2F92AE0E   06:238D816187C8DA42
05:88A6F6BFC2F96066   04:9370D6202E81B223
03:056A5661B3D9630F   02:63437982568378EC
01:FDBF67073BF2FBCE   00:49636E9CE984191D
r1:08237CEE3A6E2BA7   r2:3B8180B7CE24A897
Z5:EEC66EE0CD0B1EFC
```

Internal state in time 6.

```
15:44F54931D2BFC58B   14:940103DDB03D83CF
13:705E07090CD663A6   12:8AB52CB97DD20A07
11:5AE0FF3921A9A295   10:7A50B6C128D2B5A1
09:CF850AE47AB01EFF   08:B0D56F843B3CB9CF
07:1644DD4202DD04A3   06:3C0BDD1B2F92AE0E
05:238D816187C8DA42   04:88A6F6BFC2F96066
03:9370D6202E81B223   02:056A5661B3D9630F
01:63437982568378EC   00:FDBF67073BF2FBCE
r1:C636AD714BF6B29E   r2:F449F9571DE7C64F
Z6:02DD68F338A345A8
```

Internal state in time 7.

```
15:C10385D438A104D7   14:44F54931D2BFC58B
13:940103DDB03D83CF   12:705E07090CD663A6
11:8AB52CB97DD20A07   10:5AE0FF3921A9A295
09:7A50B6C128D2B5A1   08:CF850AE47AB01EFF
07:B0D56F843B3CB9CF   06:1644DD4202DD04A3
05:3C0BDD1B2F92AE0E   04:238D816187C8DA42
03:88A6F6BFC2F96066   02:9370D6202E81B223
01:056A5661B3D9630F   00:63437982568378EC
r1:64A800602ABE29F5   r2:01BB7826909D4CFB
Z7:47538790A5411ADB
```

FSC
www.fsc.org

MIX
Papier aus verantwortungsvollen Quellen
Paper from responsible sources
FSC® C105338

If you have any concerns about our products,
you can contact us on
ProductSafety@springernature.com

In case Publisher is established outside the EU,
the EU authorized representative is:
Springer Nature Customer Service Center GmbH
Europaplatz 3, 69115 Heidelberg, Germany

Printed by Libri Plureos GmbH
in Hamburg, Germany